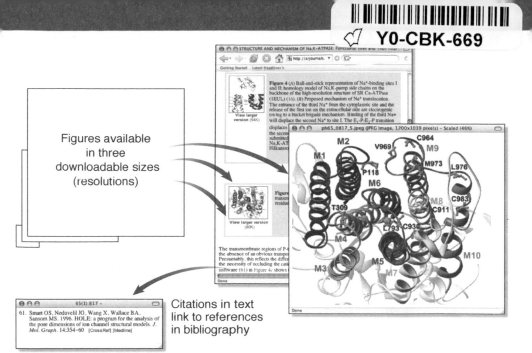
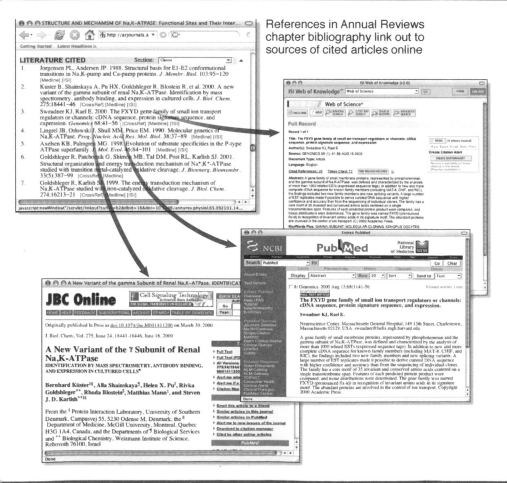

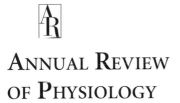

ANNUAL REVIEW
OF PHYSIOLOGY

ANNUAL REVIEW
OF PHYSIOLOGY

VOLUME 68, 2006

DAVID L. GARBERS, *Editor*
University of Texas, Southwestern Medical Center

DAVID JULIUS, *Associate Editor*
University of California, San Francisco

www.annualreviews.org   science@annualreviews.org   650-493-4400

ANNUAL REVIEWS
4139 El Camino Way • P.O. Box 10139 • Palo Alto, California 94303-0139

A̶R

ANNUAL REVIEWS
Palo Alto, California, USA

International Standard Serial Number: 0066-4278
International Standard Book Number: 0-8243-0368-7
Library of Congress Catalog Card Number: 39-15404

TYPESET BY TECHBOOKS, FALLS CHURCH, VIRGINIA
PRINTED AND BOUND BY THE SHERIDAN PRESS, HANOVER, PENNSYLVANIA

PREFACE

Joe Hoffman's service to the *Annual Review of Physiology* began in 1983 and spanned more than two decades. Joe first was the Associate Editor (under Editor Robert Berne) until 1988, when he became Editor. His editorship ended last volume; beginning this volume I am Editor, and David Julius the new Associate Editor. Under Joe's direction the *Annual Review of Physiology* stood as one of the leading sources of information for all of us in physiology as well as those in other disciplines. To follow in Joe's footsteps will be a demanding challenge for both of us, and yet because Joe has mapped a clear path as to the direction of this series, the task will be relatively easy for us if we stay the course. During his tenure as Editor he successfully kept the *Annual Review of Physiology* current with the most recent developments in physiology while maintaining a historical perspective. He did this in the face of alterations in physiology departmental names to include biophysics, cellular and molecular physiology, cell biology, integrative biology, and possibly others. New descriptors such as functional genomics, systems biology, networking, and translational physiology entered our vocabulary, and yet these modern "catchphrases" were in reality seemingly always at the heart of physiology. The new technologies, and in particular the genome projects, allowed us as physiologists to approach age-old questions in new ways. Joe Hoffman brought this new age to this Annual Reviews series, while yet insisting we maintain our perspective on the importance of the classical, evolutionary, and comparative elements. This approach will continue for the *Annual Review of Physiology*. All of us owe Joe Hoffman a great deal of gratitude for his accomplishments as Editor, and our Board will strive to live up to the expectations he fostered over his 17 years as Editor.

We are pleased that this year's prefatory chapter is written by Watt W. Webb on the "The Pleasures of Solving Impossible Problems of Experimental Physiology." Dr. Webb, Professor of Applied and Engineering Physics at Cornell University, has been a leader in the challenges of applying effective noninvasive methods to cells, which have pushed the physical limits of resolution in time, space, and sensitivity. We are also pleased that David Clapham organized a Special Topics Section on Trp channels, a particularly active area of research.

New members of the Editorial Board include David Julius as the new Associate Editor and Neurophysiology Section Editor, David Clapham as the Cell Physiology Section Editor, Holly Ingraham as the Endocrinology Section Editor, and James Anderson as the Gastrointestinal Physiology Section Editor. Jeffrey Robbins has accepted a second term as the Cardiovascular Physiology Section Editor. I would like to thank Bert O'Malley, Richard Aldrich, and John Williams for their past service as Section Editors and also thank Kate Horwitz for her one year of service as the Endocrinology Section Editor. Ms. Shirley Park now joins us as Production Editor.

All of us remain interested in comments and suggestions, and we encourage you to contact us at http://www.annualreviews.org. The current and previous chapters are available on our Web site.

David L. Garbers
Editor

Annual Review of Physiology
Volume 68, 2006

CONTENTS

INDEXES

ERRATA

An online log of corrections to *Annual Review of Physiology* chapters
may be found at http://physiol.annualreviews.org/errata.shtml

OTHER REVIEWS OF INTEREST TO PHYSIOLOGISTS

From the *Annual Review of Biochemistry*, Volume 74 (2005)

Structure and Physiologic Function of the Low-Density Lipoprotein Receptor,
Hyesung Jeon and Stephen C. Blacklow

From the *Annual Review of Biomedical Engineering*, Volume 7 (2005)

Structure and Mechanics of Healing Myocardial Infarcts, Jeffrey W. Holmes,
Thomas K. Borg, and James W. Covell

From the *Annual Review of Cell and Developmental Biology*, Volume 21 (2005)

Cajal Bodies: A Long History of Discovery, Mario Cioce
and Angus I. Lamond

Molecular Mechanisms of Steroid Hormone Signaling in Plants, Grégory Vert,
Jennifer L. Nemhauser, Niko Geldner, Fangxin Hong, and Joanne Chory

From the *Annual Review of Pharmacology and Toxicology*, Volume 46 (2006)

Accessory Proteins for G Proteins: Partners in Signaling, Motohiko Sato,
Joe B. Blumer, Violaine Simon, and Stephen M. Lanier

*Function of Retinoid Nuclear Receptors: Lessons from Genetic
and Pharmacological Dissections of the Retinoic Acid Signaling Pathway
During Mouse Embryogenesis*, Manuel Mark, Norbert B. Ghyselinck,
and Pierre Chambon

Regulation of Phospholipase C Isozymes by Ras Superfamily GTPases,
T. Kendall Harden and John Sondek

*Molecular Mechanism of 7TM Receptor Activation—A Global Toggle
Switch Model*, Thue W. Schwartz, Thomas M. Frimurer, Birgitte Holst,
Mette M. Rosenkilde, and Christian E. Elling

Watt W. Webb

Annu. Rev. Physiol. 2006. 68:1–28
doi: 10.1146/annurev.physiol.68.040504.151957
First published online as a Review in Advance on November 8, 2005

Commentary on the Pleasures of Solving Impossible Problems of Experimental Physiology

Watt W. Webb

*School of Applied and Engineering Physics, Cornell University, Ithaca, New York 14853;
email: www2@cornell.edu*

Key Words   fluorescence, multiphoton, microscopy, correlation, transduction

■ **Abstract**   This commentary presents a series of examples of "impossible experimental problems" that we have encountered over the years in addressing various challenging questions in physiology. We aim to show how stimulating the challenges of physiology can be and demonstrate how our naive invocation of methods from disparate fields of science and engineering has led to delightful resolutions of physiological challenges that were utterly new to this intrepid interdisciplinary researcher.

SEDUCTION BY SOME IMPOSSIBLE PROBLEMS OF PHYSIOLOGY

Chemical Kinetics in Physiology by Fluorescence Correlation Spectroscopy

This innocent materials scientist was dragged into molecular physiology in the late 1960s, when I was presented with the challenge of understanding how the genomes of the recently defined DNA double helix could be separated into individual DNA strands for transcription. Physical biochemist Elliot Elson addressed this question when he joined the Cornell faculty in 1968, before the enabling transcription enzymes were characterized and understood. My relevant perspective then was derived from the excitement of measuring (for the first time) the fascinating phase-phase interface fluctuations near continuous phase transitions of molecular mixtures as well as intraphase fluctuations in ^{3}He-^{4}He isotopic mixtures at temperatures down to $-272.5°$C and in quantum superconductors to test the forthcoming theories of their fluctuations and statistical thermodynamics of their phase transitions (cf. 1–6). By comparison, the biophysical challenges of cellular physiology looked feasible even at sparse biomolecular concentrations (thus displaying my ignorance of the complexity of biochemistry).

That naive innocence led us to develop Fluorescence Correlation Spectroscopy (FCS) (7, 8), which provided some experimental insight into the chemical kinetics

0066-4278/06/0315-0001$20.00

1

of the DNA double helix and which has subsequently delivered many other physiological insights. My realization that my favorite dynamical tool, quasi-elastic light scattering, was a poor measure of the kinetics of chemical reactions forced us to recognize the general effectiveness of molecular fluorescence as markers for biomolecular research and to generate the techniques for measuring the kinetics of sparse molecules that have now proven very useful in molecular cell biology.

But our invention was temporarily confounded by its instrumentation difficulties and so, after approximately 10 publications, we abandoned FCS for a decade. The faithful had to wait nearly 20 years before adequate infrastructure in computer capability, laser stability, and optical components was developed. It is ironic that our use of other indicators for correlation spectroscopies continued unabated. Now the use of FCS is almost a pleasure, yielding hundreds of research papers per year, which we finally recognized when the annual citations of our own FCS papers exceeded 360 during 2004. We now find FCS to be functional in cells as well as on cell membranes and broadly applicable in a variety of geometries (cf. 9–17).

FCS did lead us to recognize early on that the sensitivity of fluorescent markers, which can reach single-molecule sensitivity, provided a universal key to molecular signals in molecular cell physiology. We turned to measurement of molecular mobility on the lipid plus protein membranes of living cell surfaces, a challenge not then understood in the light of Singer and Nicolson's (18) fluid mosaic model and still controversial in the age of the mysterious lipid rafts, but improving, as we shall see. Recently our research has led us back to the statistical thermodynamics of the continuous phase transitions we had previously been studying, leading us to discover them in cell membranes too.

Cell Surface Molecular Mobilities

We circumvented the infrastructure obstacles of early FCS for measurement of local diffusion in membranes by fluorescently labeling cell surface molecules, photobleaching the fluorescence in a diffraction-limited spot, and monitoring microscopically the Fluorescence Photobleaching Recovery (FPR or FRAP) of the spot fluorescence as fresh fluorophore diffused in from the surrounding surface. Although at least three other groups had already invented this technique, we seem to have worked hardest to grind out results. To facilitate the grind, postdoc Dan Axelrod in 1976 wrote a clear and succinct description of the method that is still used and referenced, now approaching nearly 1000 citations (19). In contrast, our original FCS paper of 1972 (7) contained such formal statistical thermodynamics that it is hardly ever referenced, despite current widespread use of FCS.

The next relevant part of this story is my observation that our science does reflect the saying "what goes around comes around." This was true in particular in our continuing attempts to understand the heterogeneity of the molecular mobilities and distributions on cell surface membranes. Our first rude shock was the finding that membrane proteins on cell surfaces diffuse slower than lipids by factors of 10 to 100 (20). This contradicted the initially surprising theoretical result of Saffman & Delbruck (21) that the diffusion coefficients of transmembrane molecules on two-dimensional membranes should decrease only as the reciprocal logarithm of the molecular radius rather than as the reciprocal radius itself, as occurs in three-dimensional fluids.

This contradiction, and the broad spread of diffusion coefficients measured on the same cell, led us to track the trajectories of individual protein molecules as they diffuse and flow on living cell surfaces. This was first accomplished in response to a question raised by the famous medical duo Brown and Goldstein in their important medical physiology research on cholesterol control. A mutant of the low-density lipoprotein receptor (LDL-R), called the JD mutant, does not enter cells through coated pits to enable removal of LDL, leading to deadly cholesterol deposits in the affected patients. The question was whether the JD-LDL-R is unusually immobile. In labeling the LDL-R, we found that LDL itself could absorb up to 30 copies of a carbocyanine fluorophore and become so brightly labeled that we could track its trajectories on the living cell surface for many minutes (22). The LDL-R did show remarkably slow diffusion on living cell surfaces, where it stayed with little internalization. Another mutant, with a cytoplasmic tail severely truncated to leave only three amino acids, diffused just as slowly and stayed on the cell surface, eliminating cytoskeletal binding as the cause in this case. Soon, however, other researchers recognized that the JD mutant had a lesion in the cytoplasmic tail that defeated necessary binding to coated pits and thereby inhibited internalization. Nevertheless, this question led us into illuminating research on protein mobility on membranes. As molecular tracking on living cell surfaces advanced with time and effort, we achieved 15-nm resolution over 5-minute trajectories in 1988 (23) and reported it in excessive detail in 1994 (24). We can now reliably reach approximately 5 nm or better resolution (25) under ideal circumstances. Reliability and validity of interpretation are the important words here, with many misinterpretations of nanoscopic cell surface tracking in the literature of cell physiology.

Tracking of the LDR-R and later, the IgE receptor, which is famous for its (so-far) indefeatable function in activation of allergic responses, has shown clearly the heterogeneity of living cell surfaces. Simultaneous analysis over at least three decades of numerous individual trajectories of cell surface proteins with nanometer resolution clearly shows a broad diversity in molecular mobility behavior, even after separation of effects of membrane lipid flow that can often be seen as concerted quasi-linear drifts over large cell membrane areas. In 1994, we concocted a schematic picture of this cell surface heterogeneity on the basis of the observed diffusion heterogeneity, with variably viscous regions with no barriers between them (Figure 1). There is no evidence for corral fence–like barriers in any of our measurements, contrary to various erroneous interpretations of others' data in the literature. There is, however, one clear corral structure occurring in the red blood cell membrane, on which some integral proteins see a well-defined cytoskeletal cortex in a square array of fences that bind certain cell surface proteins (26).

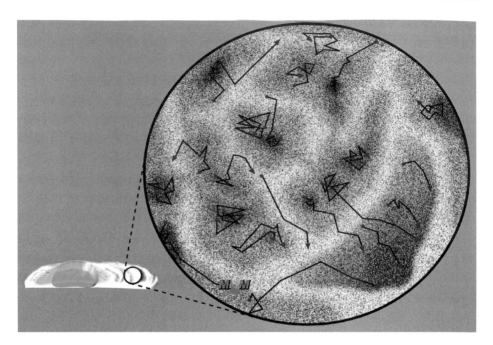

Figure 1   Tracking of anomalous molecular mobility on cell membranes with inhomogeneous fluctuations of potential energy barriers to mobility varying temporally and spatially
in distributed inhomogeneous fields (24, 25a, 27).

Our 1994 data show anomalous statistical physics in the molecular trajectories.
The usual "Brownian motion" random walk theory on "a level playing field" has
the mean square displacements growing linearly in time. But we inevitably find that
proteins on cell membranes and sometimes in viscous cytoplasm display what is
called anomalous subdiffusion (24, 27). The statistical physics of this phenomenon
can be understood by a simple model in which the two-dimensional membrane
is not "a level playing field" but is instead representable by molecular potential
energy fluctuations that vary in space and time, changing over the complete range
of length and times represented by the trajectory data (28). The heterogeneity of
the cell surface that we had seen in the trajectories (Figure 1) is reflected in the
values of the algebraic exponent α of time that fit the probability distributions of the
spreading of various trajectories, as in $\langle (\Delta r)^2 \rangle = \Gamma t^a$. Their distribution of values
for individual molecules varies from near 1.0 (conventional diffusion) smoothly
down to zero, with zero corresponding to a molecule immobile on the timescale
of the measurements. Feder et al. (27) showed that the previous representation
of anomalous diffusion phenomenon by a so-called immobile fraction in FPR (or
FRAP) experiments was generally wrong. However, this error of interpretation was
likely to be overlooked until the anomalous subdiffusion concept could be tested

by trajectory recording. This is so because the shapes of the FPR recovery curves fitted with a single classical diffusion coefficient and an "immobile fraction" were indistinguishable from those of anomalous subdiffusion unless valid FPR data were to be available over more than five decades in time, which still remains impossible. Unfortunately, FCS measurements of diffusion on membranes are also not quite sufficient to define anomalous diffusion. Instead, the phenomenon represented by the immobile fraction was shown by molecular tracking measurements to be due to anomalous subdiffusion, with a distribution among the molecular population of the values of the time exponent α (27).

THE IMPOSSIBLE CHALLENGE OF UNDERSTANDING INTERNALLY CONSISTENT PHYSICAL CHEMISTRY AND STATISTICAL PHYSICS OF "MEMBRANE RAFTS" PHYSIOLOGY

We thought five years ago that, in some ways, the heterogeneity of cell surface protein mobility was approaching full understanding, so we terminated our membrane research, which had generated approximately 75 publications from our laboratory over the years (cf. 29–33). But this "impossible problem" in cellular physiology had not been resolved, as I learned later in the numerous reports on the concept of so-called lipid rafts. The evening workshops on this subject at Biophysical Society meetings a few years ago infuriated me, as every speaker presented ideas that violated every basic principle of statistical thermodynamics that I had ever learned, used, or taught. Even further from the fundamentals, these reports blamed protein distribution heterogeneity on the lipids alone and ignored the reciprocal thermodynamic effects of membrane proteins and nonlinear events pumped by the active cell regulatory process. Tim Ryan, now a professor at Cornell Medical, had shown in a 1988 *Science* cover article that the proteins on a typical (mucosal mast) cell occupied approximately one third of the cell surface area and followed the statistical physics of the Langmuir adsorption isotherm, which implied the involvement of available site entropy on the lipid "continuum" (33).

This confusion led us to enter a Cornell–Cornell Medical–Keck Foundation collaborative project on cell membrane physiology and to contemplate the thermodynamics following from the lipid phase diagram research of Professor Jerry Feigenson at Cornell University (34). He had found that certain compositions of ternary lipid mixtures, such as cholesterol, sphingomyelin, and dioleoylphosphatidylcholine (DOPC), made reasonable models for the complex lipid mixtures in cell membranes, and we were able to measure the properties of these mixtures, using giant unilamellar vesicles (GUVs) as a model membrane system (35, 36).

Then Dr. Tobias Baumgart discovered fluorescent markers that distinguish the coexisting liquid phases in membranes (molecules of interest are still under study)

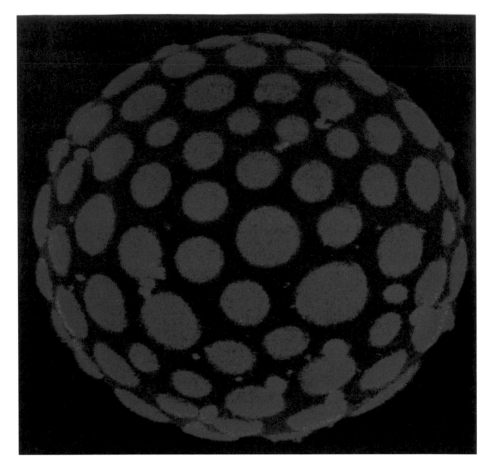

Figure 2   Imaging coexisting fluid domains in biomembrane models called giant unilamellar vesicles coupling curvature and line tension (37, 39).

and applied them to analyze the thermodynamics of their interactions in this region of coexistence, as described in our *Nature* cover article in 2003 (37). Notice in Figure 2 that the red phase (pseudocolor) representations are good circles placed in an orderly hexagonal pattern in the blue-labeled matrix. This tidy "polka-dot" order persists near and somewhat above room temperature, even if the average composition determined by position along the tie lines is shifted so that the blue matrix assumes the circular form and orderly pattern in a red matrix.

These tidy patterns persist for several days to approximately 30°C, but above that temperature, the phase separation displays differing order, and as the temperature increases to approximately 45°C, two phases of approximately equal areas approach each other in composition with differences shrinking approximately proportional to $[T_c\text{-}T]/T_c]^{1/8}$, where T_c is the temperature at which the phase separation vanishes in a continuous phase transition, leaving a single optically homogeneous lipid bilayer phase. For a range below T_c, this phase separation displays various untidy but interesting serpentine forms, suggesting the influence of critical fluctuations and decrease of the interphase energy per unit length, i.e., the tension of the interphase boundary. The fundamental details of behavior of the GUVs in this continuous phase transition region in which fluctuations dominate require delicate experiments now underway.

At a temperature near 23°C, when the GUVs are composed such that approximately equal areas of red- and blue-labeled phases occur, some of the vesicles eventually assume hourglass shapes with only one interphase boundary near the neck of the hourglass. The theory of the energetics of these shapes has been worked out over the past decade or so by Professor Reinhard Lipowsky and coworkers (38). Our experiments confirm their theory experimentally for the first time and detect, also for the first time, an earlier theoretical contribution called the Gaussian curvature energy, which appears only in an inhomogeneous membrane. The relevant physiological energy contributions also include the curvature energies of each phase, interphase tension, and pressure difference across the membrane, all integrated over the entire GUV (37, 39). It is ironic that the early curvature energy measurements on pure lipid membranes by our superb student and subsequent postdoc Marilyn Schneider (40) were occasionally attacked by theorists who had not realized that the Gaussian curvature energy appears only in heterogeneous membranes. It does remain to be determined whether Gaussian curvature energy can induce significant membrane heterogeneity.

There remain the questions of the origin of the long-range ordering of the pattern of arrangement of the coexisting phases in the polka-dot balloon pattern and more generally of the nature of the short-range order of ordered liquid phases. Professor David R. Nelson of Harvard University and colleagues (as derived by M. Gopalakrishnan, F. Juelicher, & D.R. Nelson, unpublished theory) have proposed an elegant ordered phase in this two-dimensional system associated with the concept of hexatic structures (41–44). Disclinations in them are thought to generate long-range interactions with an image repulsion in the surrounding disordered phase. This membrane elasticity effect is akin to the electrostatic image potentials across a dielectric surface. The question of a possible sharp behavioral transition— from coexisting hexatic plus disordered phases to the classical two-dimensional critical behavior—recalls Nelson's collaborative research with us more than thirty years ago (41, 45–47) on critical phenomena around phase transitions at ~0.6°K in isotopic mixtures of ^{3}He and ^{4}He. At ~0.7°K, tricritical compositional separation into two phases appears, with one phase showing dynamical order due to the quantum hydrodynamic coupling of atomic and thermal diffusion. Nelson et al. propose that in the lipid mixtures, molecular order in a hexatic structure appears up to a tricritical point, above which the classical two-dimensional critical phenomena appear. We have also considered another lipid-ordered liquid model assuming that strong curvature of the polka dots and the associated reversed curvatures of the matrix phase may lead to membrane stress accommodated first by the curved polka dots and much later relaxed by molecular redistribution. If this sounds confused upon reading, you are right; it still is.

Electrophysiology

Our first electrophysiology research attempted to understand the fluctuations of conductance as a few ion channels switched in small patches of membrane, like those Peter Lauger of the University of Konstanz, Germany, was measuring in the late 1970s, when I was most focused on fluctuations in many fields. Then, in 1980, Sakmann, Patlak, and Neher developed patch clamping, which provided gigaohm isolation of a micron squared of membrane to read the switching of the pico-ampere ion currents through a single channel. It is fun to recall that this generated shocking controversy about even the virtual idea of the existence of individual ion channels, and even the famous acetylcholine (ACh) receptor–associated channel was subject to dispute about its distribution and need for cofactors. Many scientists claimed that channels were loose ensembles of proteins or just defects in the lipid membrane, and the defect model appeared to be confirmed by the first attempts to reconstitute purified channels into membranes.

Our first significant single channel recording experiments were enabled by the development of membrane molecule purification and reconstitution methods by Cornell's great biochemist Ephraim Racker, with whom Rick Huganir had done ACh channel reconstitution research (47a). Dave Tank learned patch clamping on the chloride channel in my lab with visitor Chris Miller from Brandeis (48). We worked out a way to bulge multilayer lipid membrane aggregates called Banga-somes by osmotic stress to expose an area of lipid single bilayer to patch clamp them. The results showed that the ACh channel does consist of the hypothesized assembly of transmembrane subunits alone and that they display the same set of open and closed conductance states and switching kinetics that had been seen in the synapse membranes of cells (49). This experiment appeared to quiet the controversy.

When we first started patch clamping ion channels, I was shocked to find that the head of neurobiology here at Cornell had never heard of it when I wanted to recruit one of its heroes. Now, patch clamping is routine in our laboratory as elsewhere, and our recent studies have focused on functional neurological systems. We suspect that some electrophysiologists may regard our recent developments of membrane potential imaging and our generations of Ca^{2+} imaging in neurons, both of which are now ubiquitous, as insidious attempts to replace patch clamping (cf. 50, 51).

Some channel mysteries persisted for decades; for example, the basis for the gigaohm seal between membrane and glass pipette was not understood until 1994, when my student Lorinda Opsahl measured the adhesion force and the resistance as a function of contact area. We calculated that the thickness of the intermediate hydration layer between glass and membrane generated by the London dispersion polarization forces just accounted for the resistance (52). Lorinda also analyzed the membrane tension dependence of the only stress-dependent ion channel, Alamethicin, that has ever been understood physically (53). I was also interested in the fluctuations of protein structures and asked grad student Dan Mak to look at the conductance fluctuations of the largest ion channel that we knew, again Alamethicin, which, to our surprise, showed white noise out to 50 kHz, our measurement noise limit (54). He persists in elegant measurements of the most elusive channels of which I know, those in the nuclear membrane.

MEASURING AUDITORY TRANSDUCTION
MECHANISMS TO UNDERSTAND THE PHYSIOLOGY

Crickets

Here is a demanding physiology that requires subpicometer spatial resolution and cross-correlation over wide frequency ranges! During the mid-1970s Bob Capranica, Professor of Neurobiology and Behavior at Cornell, led us innocent new biophysicists into the delicate challenges of auditory physiology in order to test models of how two identical-appearing species of field crickets could distinguish each other for breeding by recognizing the signatory tonal difference of their identical chirp patterns. One hypothesis alleged differences in the mechanical resonances of their eardrums because the cricket's nervous system was judged insufficiently intelligent. So we developed an apparatus that sang monotone with an acoustic oscillator to the living, anchored crickets while we were shining a monochromatic laser on their eardrums, which are conveniently located on their "elbows." The scattered laser light was coherently collected and aligned to interfere with the incident laser light via a Michelson-Morley interferometer to measure the patterns of the nanoscopic motions of the eardrums. Because the living cricket's head motion and background scattering could disturb the measurements, the light was modulated and correlated at three different frequencies to compensate for this interference through three tuned lock-in detectors assembled by enterprising graduate student Paul Dragsten (55, 56). Unfortunately, we found no differences between the eardrum responses of the two species, voiding the conjectural mechanisms. But we do hear that other groups have later implicated organ pipe–like modes of the crickets' "arms," thus confirming Bob Capranica's idea of mechanical resolution.

Nevertheless, our sensitivity to the eardrum motion had allowed us to map the eardrum mode amplitudes to the remarkable precision of approximately 3 picometers (55, 56). The photophysical extension of this 1974 direction of research anticipated by some 30 years the physics underlying development of the modern technique of Optical Coherence Tomography, which utilizes the time-discriminated light scattered by tissue for deep imaging (57). I was pleased to summarize our own Michelson-Morley interferometry experiments on auditory transduction a few years ago when I was chosen for the Michelson-Morley Award for Optical Physics by Case Western Reserve University, where Michelson's interferometer was developed.

Vertebrates

Approximately 15 years after the cricket research, my great graduate student Winfried Denk chose auditory transduction mechanisms in vertebrates as the target of his PhD thesis. He used laser-illuminated Differential Interference Contrast (DIC) imaging microscopy with polarization-selected diode recording to obtain similar three-picometer resolution of the spectrum of the much more elusive motion of the sound-sensing microscopic hair bundles on the living sensory hair cells of the frog sacculus, to which we were introduced by Jim Hudspeth (58, 59). By simultaneously measuring the cellular membrane potential modulation and cross-correlating it with the random Brownian thermal motion, we were able to show that, at the low thermal fluctuation amplitudes of a few picometers, the physiological transduction process is characterized by linear mechano-electrical transduction (59, 60). These results led to understanding of the threshold sensitivity of hearing as limited by the thermal noise, unlike those of the other senses—vision, taste, and odor—which are quantum limited (61).

We culminated this research with an experiment that enabled us to prove that the mechanical transduction process in auditory transduction is a strain-sensitive ion channel near the tips of the hair-bundle processes. Our approach utilized ex vitro application of an early physiological side effect of one of the first antibiotics, an aminoglycoside that blocks hearing and that, unfortunately, was found to produce deafness by destroying patients' hair bundles, the hearing detectors. We took advantage of the known aminoglycoside-binding constant and kinetic coefficients to apply a 50% binding concentration and measure its new kinetic contribution to the frequency response correlation functions of the mechano-electrical transduction. It was possible to show unequivocally that the ion channel blockage that occurred by binding the drug also blocked the membrane potential modulation that triggers auditory transduction, thus demonstrating that this function of auditory physiology does depend on strain-sensitive ion channels, just as had long been expected (62).

Much later, in collaboration with a group of auditory physiology collaborators from three European countries, we were able, with colleague Professor Warren Zipfel, to extend methods and to carry out hair-bundle motion imaging on both inner and outer hair cells in the living mammalian cochlea, but sadly, our collaborators never returned to take advantage of our arduous methods development; thus, an opportunity awaits in research on mammalian auditory transduction physiology.... This development was based on multiphoton microscopy (MPM), which comprises the next section of this chapter. And I have to admit here that most of the preceding research topics have now come to depend on MPM and are utilized worldwide.

MULTIPHOTON MICROSCOPY: AN ULTIMATE TOOL FOR PHYSIOLOGY

Laser Microscopy

Laser microscopy has illuminated our biophysical experiments from their beginning, with our FCS measurements in 1970–72 of the chemical kinetics of the reaction of DNA with the intercalating drug ethidium. Later came the development of Fluorescence Photobleaching Recovery (FPR, or FRAP) (discussed above), which relies on fast switching of a focused, high-intensity laser beam to provide a perturbation whose recovery kinetics provides convenient molecular mobility measurements in cell membranes, a standard issue in membrane physiology. Then the first commercial apparatus for confocal laser scanning microscopy (LCSM), based on the first practical design by John White and Brad Amos in 1987, came from Bio-Rad Laboratories (63). It quickly became our principal fluorescence imaging tool for cellular physiology. The confocal aperture improves axial resolution as it is reduced in diameter to exclude out-of-focus fluorescence excited along the double conical laser illumination beam, but unfortunately, the improved three-dimensional resolution has severe costs in fluorescence image sensitivity. The limiting problem, however, comes from photodamage and photobleaching all along the illumination beam within the specimen. Another limitation quickly becomes apparent in trying to image physiological structures more than a few tens of microns into tissue because the quasi-random scattering of out-of-focus fluorescence can generate significant fluorescence background that enters the confocal aperture. Our invention and development of MPM, originally carried out as two-photon microscopy, was driven by the needs of physiological microscopy to avoid these obstacles in order to image deeper in tissue with minimum photobleaching and photodamage.

Was two-photon molecular excitation a reasonable expectation in the mid-1980s? Yes; in fact, Maria Goeppert-Mayer had carried out the quantum theoretical analysis in 1931 (64). But it was not until thirty years later, after the development of the laser, that sufficiently intense illumination became available so that two-photon molecular excitation became feasible and was demonstrated. Graduate student Jim Strickler took on the chore of demonstrating that we could do it somehow at Cornell with locally available lasers, and he did finally succeed. And in 1989 we achieved our first successful two-photon laser scanning images and our first two-photon caged reagent activation. (Thus, it required approximately 60 years for the science to advance from basic theory to practical application!)

Our development of MPM addressed three objectives. The first was to realize my vision of submicron 3-d-resolved UV-photoactivated pharmacology by utilizing "caged" neurotransmitter molecules on which an attached inactivating group could be released by UV photochemical illumination. By using simultaneous absorption of two infrared photons from a focused laser, we could achieve 3-d-resolved submicron submicroscopic release. In our first publication on this endeavor in 1990 (65), we did report two-photon uncaging of caged ATP, but the first significant physiological application was independently achieved later by co-inventor Winfried Denk to map the distribution of ACh receptors (66). Jim Strickler also used two-photon activation of photoresist activator to develop three-dimensional recording memory conceived as a 30-layer-thick compact disc (67, 68).

We found in 1997 that the powerful neuromodulator serotonin, which one photon absorbs at approximately 235 nm, could be effectively excited and imaged by simultaneous absorption of three photons at 700 nm (69). Later we were more surprised to discover that serotonin could be photodimerized by simultaneous absorption of four 700-nm photons corresponding to absorption of UV photon energy at ~170 nm and that the dimer then could be two-photon excited at 700 nm, corresponding to ~350-nm photon energy to emit green fluorescence (70).

But it was the second objective, development of MPM as the multiphoton fluorescence imaging technique for 3-d-resolved fluorescence imaging in physiological specimens, that has been the dominant application. It was this objective that motivated Winfried Denk to stay on in our laboratory, after finishing his PhD earlier in 1990, in order to assemble the necessary components, with Strickler's assistance, and to make MPM imaging work for the first time. And so it did work for the first time with the use of a new colliding pulse mode-locked laser at 630 nm that had been recently assembled by my colleague, Professor Frank Wise, and his students. We scanned its beam with a modified Bio-Rad 500 confocal microscope operated without benefit of the confocal aperture. Our first physiological images used the Hoechst 33,258 DNA stain, which can absorb one photon at approximately 350 nm, to image with MPM the chromosomes within live cultured pig kidney cells of type LLC-PK-1. Thirteen seconds of laser scanning produced bright images, and testing for photobleaching first showed deleterious effects only at over two minutes continuous imaging with no cellular degradation. In contrast, brief exposure to UV illumination to image the Hoechst stain by conventional widefield ultraviolet imaging killed the cells rapidly.

This comprised the beginning of MPM, the use of which has expanded rapidly, with hundreds of new papers per year citing the technique in their title, abstract, or text since the early 1990s. As a measure of MPM's application rate, our original publication (65) has now been cited more than 1400 times, and now it is cited at a rate well above 200 per year. Thus, it is now difficult to summarize the subsequent massive progress. I mention only some of our crucial developments and some of my favorite recent discoveries in our group from our >65 publications based on MPM; for more, I must refer the reader to our publication lists on our website at http://www.drbio.cornell.edu. DRBIO stands for the Developmental Resource for Biophysical Opto-Electronics, our NIH-NIBIB (National Institutes of Science-National Institute of Biomedical Imaging and Bioengineering) Resource, which has supported much of our research with the mission of developing biophysical methods and making them available through collaborations and disseminated through all available means. I must note that reviewers for both the National Institutes of Health and the National Science Foundation declined to support a series of our proposals to develop MPM and that, furthermore, most of our real innovations have been bootlegged, as was FCS, as I was supported for materials science research at that time.

Fluorophores for MPM Fluorescent Marker Molecules for Multiphoton Excitation

As we began MPM imaging, we realized the need to characterize appropriate fluorophores for MPM. The quantum mechanics of two-photon excitation associates it with an even parity selection rule for excitation, whereas one- and three-photon excitation are associated with odd parity transitions. This means that the wavelengths of the maximum cross sections for two-photon molecular excitation may be less than twice the optimum wavelengths for one-photon excitation of fluorescence. Fortunately, most fluorophore molecules are sufficiently complex that the first odd parity, 1S, excited state often is overlapped by the first even parity, 2S, excited state. However, in some cases, for example, rhodamine, the first strong two-photon excitation occurs at wavelengths far below the wavelength corresponding to twice the 1S wavelength. Then-graduate student Chris Xu, now Professor Xu at Cornell, carried out the key initial research yielding our first basic set of two-photon fluorophore excitation data and a general understanding of the problem, thus enabling efficient physiological applications (71–75). Now, apparatus developed by Professor Warren Zipfel allows us to accumulate excitation spectra efficiently, yielding a large collection of data, most of which are yet to be conveniently made generally available (we write guiltily). Joe Perry and Seth Marder (then of the University of Arizona), in a collaboration with our group, developed a series of fluorophores optimized for enormous two-photon excitation cross sections (76), but unfortunately their chemistry has inhibited useful physiological applications.

Our second major approach to two-photon excitation of fluorophores addressed the Green Fluorescent Proteins (GFPs) and their many mutants, which I think provide the most powerful new optical tools for physiology research. It had been observed that the GFP molecules blink repeatedly if excited at low intensity (77). We took up this problem with one of the first enhanced GFP (EGFP) mutants and found by FCS that it flickers on and off at a fast rate linearly proportional to bright one-photon illumination intensity and to the square of the two-photon excitation intensity. But the FCS measurements of the proportion of on and off time fractions due to the flicker are independent of intensity (10). We found in a long series of studies, led by Dr. Ahmed Heikal, of many GFP and of Red Fluorescent Protein (RFP) mutants that this flicker is ubiquitous and is due to transitions with low quantum efficiency ($\sim 10^{-3}$) back and forth only through the excited state manifold, to partition between the stable ground state and one or more other "ground" states. The secondary molecular ground states generally have significantly different excitation wavelength bands that make the partitioning dependent on excitation wavelength. Haupts et al. (10) also studied H^+ binding in the chromophore, which extinguishes the fluorescence; we used FCS to measure kinetics versus pH and temperature to determine the proton-binding free energy, enthalpy and entropy, and the on-and-off kinetic coefficients and activation energies. Heikal et al. (78) recently culminated our studies of the GFP mutant photophysics, added several more mutant studies, and summarized the photophysical processes. Figure 3 displays our photophysical transition model. This research is not in itself physiology, but the results have made possible more effective photophysics methods for important physiological

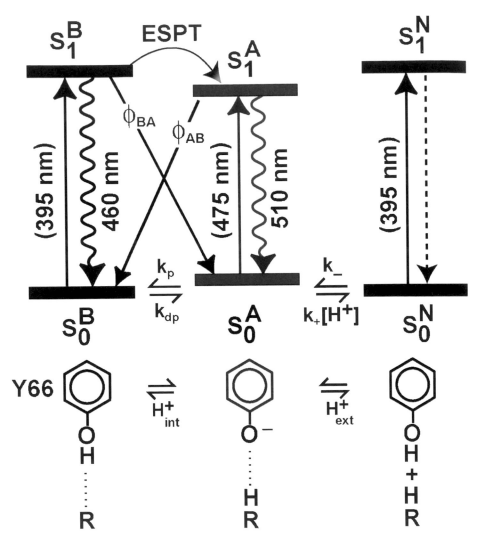

Figure 3 Schematic kinetic model demonstrates light- and pH-dependent interconversion between the various electronic states of GFP mutants (78). The direct transitions between the ground states of the center and left systems are extremely slow in most mutants and can occur only by internal transfer of a bound proton. On the other hand, external protonation (*right*) quenches fluorescence.

research. Now, new fluorescent proteins are being discovered and developed with such productivity that it is hard to keep up with the literature.

There has been recurrent interest in quantum dots (qdots), consisting of CdSe semiconducting crystals with ZnS monolayers to ~4 or 5 nm diameter that are then coated with multiple amphiphilic protective layers to ~30 nm diameter for use in the biological fluids implicit in physiological research. Qdots are resistant to photobleaching, can be excited effectively with very-low-intensity illumination, and have narrow emission bands facilitating multiple distinguishable fluorescence colors, but unfortunately each dot either blinks repeatedly or is entirely dark. We have studied qdots' very efficient two-photon excitation. One physiological application to monitoring through the skin of capillary blood flow velocity under the skin yielded the flow velocities and their modulation by breathing and the 600 heartbeats sec^{-1} of our live mouse subjects. We have recently succeeded in monitoring their quantum efficiency, blinking time distributions, and probabilities in physiological environments (78a). We have helped to develop other extraordinarily bright fluorescence markers, called Cornell Dots, that do *not* suffer from the blinking problem. These have approximately 20 times brighter fluorescence than rhodamine and entirely avoid photobleaching for at least two or three days in biological fluids. Their only problem is that, like water-stable qdots, their diameter is approximately 30 nm (79); nonetheless, they have been used successfully to label IgE receptors and observe their aggregation on living cells in culture.

What about the basic instrumentation for MPM research on physiological applications? We still use primarily our ancient, repeatedly repaired Bio-Rad 500 and 600 laser scanners because they are all we can afford and because their open access makes them easy to adapt to new challenges and to repair them. But we do also like the new Zeiss 510M MPM-integrated instrument, which has elegant software and is now sold under license from Cornell. However, it is interesting that many publications revealing the most elegant research with MPM were and are based on home-built assemblies of equipment for MPM. Femtosecond modelocked lasers suitable for MPM are now available in single-box, diode-pumped, broadly tunable but expensive designs from Spectra Physics and Coherent Optics. The trend to use even longer wavelengths has led to the use of the new, but still a bit rough, fiber-optic femtosecond lasers above one micron in the infrared. Our recent discovery of the broad wavelength suitability of solid-core air-sheathed optical fibers for high-power transmission of femtosecond pulses (80) should make future instrumental assemblies more amenable to demanding locations such as medical surgery environments and open the way to more flexible instrument design.

My recent favorite MPM-based publications from our laboratory include our "polka-dot balloon" (Figure 2) cover article in *Nature* by Dr. Tobias Baumgart on multiphase lipid membrane phase separations. This discussion omitted mention of its use of convenient stacks of MPM images to yield three-dimensional image reconstructions. The sequels of this research, of which there are more to come, are using the same methods to pursue the elusive understanding of cell membrane physiology (37, 39).

Another highly pleasing success is our report of MPM observations led by Dr. Karl Kasischke, MD, with crucial help from Dr. Harshad Vishwasrao, of the metabolic coupling between neurons and astrocytes in brain, using MPM imaging of NADH fluorescence as the sensitive metabolic indicator (81). Our results, recently published in *Science*, showed that electrical stimulation of the neurons rapidly exhausted their mitochondrial NADH by oxidative metabolism, causing their NADH fluorescence to crash. But the astrocyte NADH, mostly in their cytoplasm, did not decrease. Rather, it slowly increased by astrocytic glycolysis and overshot as it replenished the metabolic state of the nuclei. To our innocent surprise, this physiological result attracted more vigorous comments, both for and against, than I had ever encountered for our previous publications. A key to the success of this research was our development of the capability to resolve the astrocytes from the neurons over time as the soft brain-slice preparation slightly distorted. A second discovery in this area of research by Vishwasrao, then a graduate student, derives from an analysis of the conformational partition dependence of the fluorescence of intracellular NADH molecules as the metabolic state changes. These conformational states affect the fluorescence efficiency of NADH and therefore require calibrations of the NADH fluorescence versus NADH chemical potential to image accurate measures of the chemical potential of NADH as a quantitative measure of metabolic state. It was necessary to measure the distribution of NADH fluorescence decay times as functions of metabolic state, etc., to demonstrate the corrections as illustrated here and in Vishwasrao's *Journal of Biological Chemistry* "paper-of-the-week" cover article (82) (Figure 4).

Second Harmonic Generation

MPM has led us back to practical applications of nonlinear imaging with second harmonic generation (SHG) by noncentrosymmetric, polar, optically nonlinear biological structures. This approach was discovered some twenty years ago by Wilson & Sheppard (83) and more recently analyzed and applied in a series of effective research papers led by our former postdoc, now-professor, Jerome Mertz, currently at Boston University and previously at ESPCI in Paris. Dr. Rebecca Williams (84) has applied SHG imaging very effectively to physiological imaging and particularly to the analysis of natural structural assemblies of collagen fibrils. And again, following on from concepts developed by Professor Mertz, Dr. Dan Dombeck, then a graduate student and now a postdoctoral associate in my laboratory, began to use SHG imaging for fast membrane potential measurements. Along the way, Dombeck (85) also discovered that the SHG generated by microtubules in parallel alignment in neuronal axons provides good images of neuronal bundle distributions (Figure 5). Figure 5 shows a live hippocampal brain slice from rat with the axonal bundles called mossy fibers, shown by SHG (85) in pseudocolor red against a green MPM image of the free-zinc distribution recognized by a Zn^{2+}-sensitive fluorescent indicator. His work on SHG imaging of membrane potentials has recently revealed time resolution fast enough to resolve action potentials. SHG measurement with membrane potential–sensing dyes has

Figure 4 Metabolic dynamics in the brain are imaged using the fluorescence of endogenous NADH. Fluorescence measurements, however, are complicated by the dependence of the quantum efficiency of NADH on its free/bound states. Time-resolved fluorescence anisotropy discriminates between free/bound NADH and shows a preferential increase in free NADH during the normoxic (*blue curve*) to hypoxic (*red curve*) metabolic transition (82). Copyright 2005 by the American Society for Biochemistry and Molecular Biology. Reproduced with permission of the American Society for Biochemistry and Molecular Biology in the Other Book format, via Copyright Clearance Center.

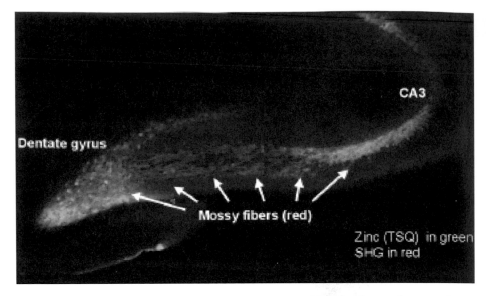

Figure 5   Pseudocolor images of brain slices showing "red" second harmonic generation by parallel microtubules in axons of the "mossy fiber" neurons and "green" fluorescent indicators of zinc ion in the hypoxic rat hippocampus (85).

the advantage over MPM fluorescence measurement of suppressing background fluorescence, although weak signals of even the best dyes still require averaging over approximately 50 events. To illustrate, a static SHG image of the membrane potential in Aplysia neurons that comprised a cover figure for the *Journal of Neuroscience* for one of Dan Dombeck's articles (51) appears as Figure 6.

But perhaps the culmination of our development of MPM will arise from recent success on in vivo imaging of cancer in transgenic mice, led by my colleagues Professor Zipfel and Dr. Williams in collaboration with Professor Alex Nikitin of Cornell's Molecular Medicine Department in the College of Veterinary Medicine. Our first venture in this direction several years ago, in collaboration with Dr. Brad Hyman at Harvard Medical and Massachusetts General Hospital, was in vivo imaging of the appearance and growth of the beta amyloid aggregates or plaques of Alzheimer's disease in a transgenic mouse model. This imaging showed that the plaques that appeared at their final size in fewer than three days could remain in situ for at least six months thereafter (86, 87). The key summary papers emphasizing Dr. Williams's cancer imaging appeared for the first time in *Proceedings of the National Academy of Sciences* and *Nature Biotechnology* (88, 89). Brain applications are illustrated in Figure 7, and an in vivo cancer image compared with conventional pathologists' absorption stains in fixed images in Figure 8. The transgenic mouse images suggest that MPM and SHG now are ready for clinical application.

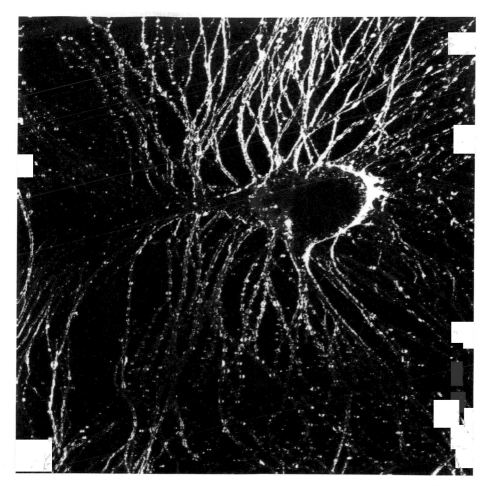

Figure 6   Second harmonic generation microscopy image of a primary cultured Aplysia neuron stained with the membrane electric field–sensitive dye DHPESBP (51). Copyright 2004 by the Society for Neuroscience.

CONCLUSIONS AND APPLICATIONS IN PHYSIOLOGY OF OUR SOLUTIONS TO IMPOSSIBLE PROBLEMS

Prospective

Where do we go from here? I think we go in two associated directions: utilizing collaborations with medical schools to apply our MPM methods in clinical medicine and heading back into heavy-duty engineering development (ugh) to engineer the endoscopic and surgery-compatible apparati for MPM. We believe that our in vivo imaging has demonstrated useful capability for diagnostic applications

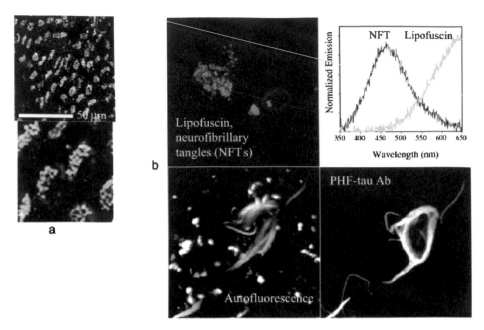

Figure 7 Multiphoton imaging of intrinsic fluorescence for neurophysiology. (*a*) Neuro-modulator fluorescence imaging of serotonin (5HT) (*red*) and NADH (*green*) in pineal gland. (*b*) Inappropriately cross-linked molecules (often associated with disease states). Intrinsic fluorescence of tau protein aggregates in neurofibrillary tangles and lipofusion droplets aggregating in brain. Upper right shows intrinsic emission spectra of NFT and lipofusion (87).

in clinical medicine, particularly in early cancer recognition in vivo and ex vivo by MPM imaging of intrinsic fluorescence and by SHG imaging of collagen and other nonlinear anisotropies. We think collagen signals may also be useful in orthopedic surgery. Are there obstacles remaining? Yes, indeed; it is necessary to validate our results to satisfy the professional medical pathologists, but that is already beginning to happen! We think the best approach is to find applications in which our images can be directly compared with those of the pathologists. In fact, two applications of MPM in urological surgery have been spontaneously and independently proposed by two surgeons at Cornell Weill Medical that provide this opportunity and that appear to have demonstrable applicability. We shall see. . . .

But supposing that we do achieve validation of our approaches, what engineering is needed for effective utilization in medicine? For biopsy samples, none is required because present standard MPM instrumentation is fully applicable. But the surgeons want to utilize MPM and SHG imaging internally during surgery and even in laparoscopic surgery. Surprisingly, we find that even SHG imaging has immediate surgical applications of existing knowledge in surgery. Yes, we think we know how solutions, in principle, exist for all of the obstacles. What are the problems then? We see the engineering of compact, flexible, durable, partially

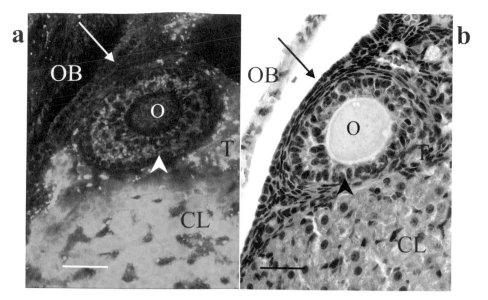

Figure 8   Mouse ovary fluorescence and histostain. (*a*) MPM intrinsic fluorescence (*green*) and SHG (*red*) of mouse ovary. (*b*) Hematoxylinyeosin-stained fixed preparation in pathologist's protocol. The ovarian epithelium (*arrow*), oocyte (O), granulosa cells (*arrowhead*), thecal cells (T), the corpus luteum (CL), and ovarian bursa (OB) are all clearly resolvable and resemble the histological image in *b*. Scale bars, 50 μm. See References 88 and 89 for more information.

sterilizable microscopies, sometimes internalizable endoscopic geometries, as the major challenges (Figure 9). As an experienced industrial "Research, Development, and Engineering" manager, I estimate any engineering stage development as costing at least 10 times the human effort and dollars of the underlying laboratory research. We think these culminating "impossible" problems in medical physiology are eventually solvable too. . . .

Innumerable "Impossible Problems" Persist to Challenge Present and Future Research in Physiology

Conspicuous among these "impossible problems" are innumerable questions about our nervous system: (*a*) What are the molecular, cellular, and system bases of long-term (lifetime) memory? (*b*) What is the actual integral connectivity of the nervous system of the mammalian brain at fundamental levels that enables its elegant functionality? (*c*) What are the initial deleterious events in the initiation of each of the neurodegenerative diseases? As we are only now recognizing the ubiquitous amyloids and aggregates as probable detritus of earlier initiation of damage, it is challenging that even the initial molecular events in nucleating these aggregates eludes us.

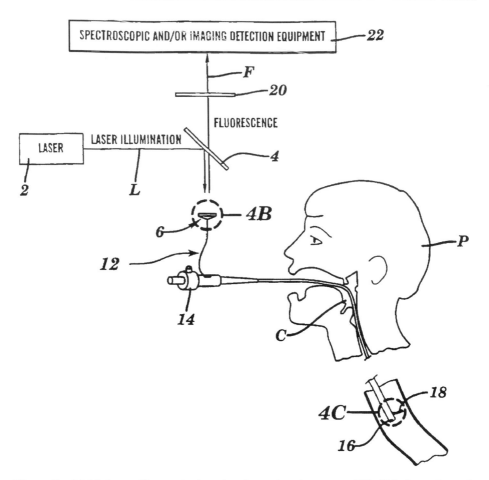

Figure 9 Multiphoton fiber optic–based endoscopic microscope (90). This is a schematic of a potential endoscopic MPM structure for imaging within the esophagus.

Cellular physiology still faces the challenge of understanding the molecular basis of membrane heterogeneity. The ubiquitous term "membrane rafts" in the plasma membrane hides the abysmal ignorance about the numerous underlying factors generating and governing cell membrane systematics. It even remains to introduce effectively the effects of the membrane proteins on this heterogeneity (except of course in a few special structures like synapses and coated pits). The anomalously slow protein mobility in plasma membranes remains puzzling in spite of considerable study. And the considerable effects on the plasma membrane of the flux of lipids and proteins between the endoplasmic reticulum/Golgi complex and the plasma membrane are gradually emerging as the results of excellent current research. Nevertheless, the effects of these dynamical fluxes on the plasma membrane behavior remain challenging.

Sensory system mechanisms and system structures remain challenging in spite of substantial exciting progress. The learning of odors, sounds of communication, and speech in the nervous system assembly appear to be fruitful current fields of physiological research with major promises of progress. But despite all of that promise, there are some simple but elusive features, such as the identity and sequence of the stress-sensitive ion channels on the hair cells underlying vertebrate hearing (58–62).

Undoubtedly, physiology continues to offer a multitude of "impossible problems" to challenge research.

Reflections

On reflection, the overriding pleasure of my entire career of "solving impossible problems" has been, and still is, my honor and pleasure in associations with a wonderful long series of interdisciplinary students, postdocs, collaborators, and colleagues in science and engineering. It is these brilliant, thoughtful, multidisciplinary associates who have made and are making the journey so interesting and rewarding.

And I think about the outstanding achievements of our alumni, who include two Vice Presidents of corporate giants, Lorinda Opsahl of General Electric and Dan Wack of KLA-Tencor, plus more than 60 professorial alumni. And recently Rob Engel, who left superconductivity physics research in my laboratory with an MS in order to switch to economics, was awarded the Nobel Prize in Economics (of all things!). I wonder whether this was partially my fault for telling him about my Business and Engineering Administration BS degree from MIT's Sloan School, an educational component most useful to me in dealing with the burdens of the modern academic researcher.

I particularly recall David Tank, who displayed a remarkable diversity and wide-ranging curiosity in our research. He later educated further a number of our subsequent students as postdocs in his department at Bell Labs, alias Lucent Technologies, and has continued to lead outstanding research. I have already mentioned Elliot Elson, who pulled me into biophysics after I had made the conscious decision to avoid it because of my terrible memory for names. Winfried Denk, whose innovative approaches to research and sensible approaches to experimental design I have always admired, drove so hard to get MPM working before he had to move on from Cornell. He still continues to innovate most importantly in methods for physiology. Watch out, neurophysiologists: He is converting from low-energy photons to high-energy electrons for imaging neural systems.

Here I have to admit that this sentiment extends far back to long before my intrepid venture into biophysics, alias physiology. My education really advanced at MIT, especially with my BS thesis supervisor Howard Taylor and my ScD supervisor Carl Wagner when I returned to MIT five years later. Several MIT professors also nucleated my interest in philosophical thought, which I culminated by leading Great Books Foundation classes in night school for three years with several of my industrial bosses in the classes. My respect goes back even to the New Mexico cowboy who showed me how to do my first job, at the age of 10, as a horse wrangler on the continental divide in New Mexico as I recovered enough health to be able to start school for the first time after prolonged childhood illnesses. My skipper, Leigh Brite, on MIT's sailing team steered us delightfully to innumerable intercollegiate national championships in the 1940s (MIT's heyday in intercollegiate sailing), which addicted me to an enduring preoccupation with yacht racing! Dr. Bill Forgeng, my colleague at Union Carbide and Carbon Research laboratories, led me to appreciate the powers and limitations of microscopy as we recorded the latest metallographic images on his 10×12 high-resolution photographic plates at 1000-times magnification. And my delight persists in my devotion to the lovely lady I met sailing on the Charles River, Page Chapman Webb, descendent of a Quaker Salem witch, who continues to mesmerize me, win sailboat races with me, and ask insightful questions after quietly listening to scientific discourses by our visiting scientists.

The *Annual Review of Physiology* is online at
http://physiol.annualreviews.org

LITERATURE CITED

1. Gilmer GH, Gilmore W, Huang J, Webb WW. 1965. Diffuse interface in a critical fluid mixture. *Phys. Rev. Lett.* 14:491–94

2. Huang JS, Webb WW. 1969. Diffuse interface in a critical fluid mixture. *J. Chem. Phys.* 50:3677–93

3. Huang JS, Webb WW. 1969. Viscous damping of thermal excitations on interface of critical fluid mixtures. *Phys. Rev. Lett.* 23:160–63

4. Lukens JE, Warburton RJ, Webb WW. 1970. Onset of quantized thermal fluctuations in one-dimensional superconductors. *Phys. Rev. Lett.* 25:1180–84

5. Henkels WH, Webb WW. 1971. Intrinsic fluctuations in the driven Josephson oscillator. *Phys. Rev. Lett.* 26:1164–67

6. Watts DR, Goldburg WI, Jackel LD, Webb WW. 1972. Preliminary observations of light scattering from the ^{3}He-^{4}He mixture near its consolute critical point. *J. Physiol. Paris* 33:C1–155

7. Magde D, Elson E, Webb WW. 1972. Thermodynamic fluctuations in a reacting system: Measurement by fluorescence correlation spectroscopy. *Phys. Rev. Lett.* 29:705–8

8. Elson EL, Webb WW. 1975. Concentration correlation spectroscopy: New biophysical probe based on occupation number fluctuations. *Annu. Rev. Biophys. Bioeng.* 4:311–34

9. Mertz J, Xu C, Webb WW. 1995. Single-molecule detection by two-photon-excited fluorescence. *Optics Letters* 20:2532–34

10. Haupts U, Maiti S, Schwille P, Webb WW. 1998. Dynamics of fluorescence fluctuations in green fluorescent protein observed by fluorescence correlation spectroscopy. *Proc. Natl. Acad. Sci. USA* 95:13573–78

11. Zipfel WR, Webb WW. 2001. In vivo diffusion measurements using multiphoton excited fluorescence photobleaching recovery (MPFPR) and fluorescence correlation spectroscopy (MPFCS). In *Methods in Cellular Imaging*, ed. A Periasamy, pp. 216–35. Oxford, England: Oxford Univ. Press

12. Kohler RH, Schwille P, Webb WW,

Hanson MR. 2000. Active protein transport through plastid tubules: Velocity quantified by fluorescence correlation spectroscopy. *J. Cell Sci.* 113:3921–30

13. Webb WW. 2001. Fluorescence correlation spectroscopy: Genesis, evolution, maturation and prognosis. In *Fluorescence Correlation Spectroscopy Theory and Applications*, ed. R Rigler, ES Elson, pp. 305–30. Berlin: Springer-Verlag

14. Webb WW. 2001. Fluorescence correlation spectroscopy: Inception, biophysical experimentations and prospectus. *Appl. Opt.* 40:3969–83

15. Hess ST, Huang S, Heikal AA, Webb WW. 2002. Biological and chemical applications of fluorescence correlation spectroscopy: A review. *Biochemistry* 41:697–705

16. Larson D, Ma YM, Vogt VM, Webb WW. 2003. Direct measurement of Gag–Gag interaction during retrovirus assembly with FRET and fluorescence correlation spectroscopy. *J. Cell Biol.* 162:1233–44

17. Foquet M, Korlach J, Zipfel WR, Webb WW, Craighead HG. 2004. Focal volume confinement by submicrometer-sized fluidic channels. *Anal. Chem.* 76:1618–26

18. Singer SJ, Nicolson GL. 1972. Fluid mosaic model of structure of cell membranes. *Science* 175:720–31

19. Axelrod D, Koppel DE, Schlessinger J, Elson E, Webb WW. 1976. Mobility measurement by analysis of fluorescence photobleaching recovery kinetics. *Biophys. J.* 16:1055–69

20. Elson EL, Schlessinger J, Koppel DE, Axelrod D, Webb WW. 1976. Measurement of lateral transport on cell surfaces. In *Measurement of Lateral Transport on Cell Surfaces*, ed. VT Marchesi, pp. 137–40. New York: Alan R. Liss, Inc.

21. Saffman PG, Delbruck M. 1975. Brownian motion in biological membranes. *Proc. Natl. Acad. Sci. USA* 72:3111–13

22. Barak LS, Webb WW. 1981. Fluorescent low-density lipoprotein for observation of dynamics of individual receptor complexes on cultured human fibroblasts. *J. Cell Biol.* 90:595–604

23. Ghosh RN, Webb WW. 1988. *Results of automated tracking of LDL receptors on cell surfaces.* Presented at Annu. Meet. Am. Biophys. Soc., 41st, New Orleans

24. Ghosh RN, Webb WW. 1994. Automated detection and tracking of individual and clustered cell-surface low-density-lipoprotein receptor molecules. *Biophys. J.* 66:1301–18

25. Thompson RE, Larson DR, Webb WW. 2002. Precise nanometer localization analysis for individual fluorescent probes. *Biophys. J.* 82:2775–83

25a. Schwille P, Haupts U, Maiti S, Webb WW. 1999. Molecular dynamics in living cells observed by fluorescence correlation spectroscopy with one- and two-photon excitation. *Biophys. J.* 77:2251–65

26. Sheetz MP. 1983. Membrane skeletal dynamics: Role in modulation of red-cell deformability, mobility of transmembrane proteins, and shape. *Semin. Hematol.* 20:175–88

27. Feder TJ, Brust-Mascher I, Slattery JP, Baird B, Webb WW. 1996. Constrained diffusion or immobile fraction on cell surfaces: A new interpretation. *Biophys. J.* 70:2767–73

28. Bouchaud JP, Georges A. 1990. Anomalous diffusion in disordered media: Statistical mechanisms, models and physical applications. *Phys. Rep.* 195:127–293

29. Thomas JL, Holowka D, Baird B, Webb WW. 1994. Large-scale coaggregation of fluorescent lipid probes with cell-surface proteins. *J. Cell Biol.* 125:795–802

30. Schlessinger J, Barak LS, Hammes GG, Yamada KM, Pastan I, et al. 1977. Mobility and distribution of a cell-surface glycoprotein and its interaction with other membrane components. *Proc. Natl. Acad. Sci. USA* 74:2909–13

31. Tank DW, Wu ES, Meers PR, Webb WW. 1982. Lateral diffusion of gramicidin-C in phospholipid multibilayers: Effects of

cholesterol and high gramicidin concentration. *Biophys. J.* 40:129–35

32. Bloom JA, Webb WW. 1983. Lipid diffusibility in the intact erythrocyte membrane. *Biophys. J.* 42:295–305

33. Ryan TA, Myers J, Holowka D, Baird B, Webb WW. 1988. Molecular crowding on the cell surface. *Science* 239:61–64

34. Smith AK, Buboltz J, Spink CH, Feigenson GW. 2003. Ternary phase diagram of the lipid mixture sphingomyelin/DOPC/cholesterol. *Biophys. J.* 84:372A

35. Schwille P, Korlach J, Webb WW. 1999. Fluorescence correlation spectroscopy with single-molecule sensitivity on cell and model membranes. *Cytometry* 36: 176–82

36. Korlach J, Schwille P, Webb WW, Feigenson GW. 1999. Characterization of lipid bilayer phases by confocal microscopy and fluorescence correlation spectroscopy. *Proc. Natl. Acad. Sci. USA* 96:8461–66

37. Baumgart T, Hess ST, Webb WW. 2003. Imaging coexisting fluid domains in biomembrane models coupling curvature and line tension. *Nature* 425:821–24

38. Lipowsky R, Dimova R. 2003. Domains in membranes and vesicles. *J. Phys. Condens. Matter* 15:S31–S45

39. Baumgart T, Das S, Webb WW, Jenkins JT. 2005. Membrane elasticity in giant vesicles with fluid phase coexistence. *Biophys. J.* 89:1067–80

40. Schneider MB, Jenkins JT, Webb WW. 1984. Thermal fluctuations of large quasispherical bimolecular phospholipid vesicles. *J. Phys.* 45:1457–72

41. Nelson DR. 1977. Recent developments in phase-transitions and critical phenomena. *Nature* 269:379–83

42. Nelson DR, Peliti L. 1987. Fluctuations in membranes with crystalline and hexatic order. *J. Phys.* 48:1085–92

43. Selinger JV, Nelson DR. 1988. Theory of hexatic-to-hexatic transitions. *Phys. Rev. Lett.* 61:416–19

44. Selinger JV, Nelson DR. 1989. Theory of transitions among tilted hexatic phases in liquid crystals. *Phys. Rev. A* 39:3135–47

45. Leiderer P, Watts DR, Webb WW. 1974. Light-scattering by He-3-He-4 mixtures near tricritical point. *Phys. Rev. Lett.* 33: 483–85

46. Leiderer P, Nelson DR, Watts DR, Webb WW. 1975. Tricritical slowing down of superfluid dynamics in He-3-He-4 mixtures. *Phys. Rev. Lett.* 34:1080–83

47. Leiderer P, Nelson DR, Watts DR, Webb WW. 1975. *Tricritical slowing down of superfluid dynamics in He-3-He-4 mixtures.* Presented at Proc. Intl. Cong. Low Temp. Phys., 14th, Amsterdam

47a. Huganir RL, Racker E. 1982. Properties of proteoliposomes reconstituted with acetylcholine receptor from *Torpedo californica. J. Biol. Chem.* 25:9372–78

48. Tank DW, Miller C, Webb WW. 1982. Isolated-patch recording from liposomes containing functionally reconstituted chloride channels from torpedo electroplax. *Proc. Natl. Acad. Sci. USA* 79:7749–53

49. Tank DW, Huganir RL, Greengard P, Webb WW. 1983. Patch-recorded single-channel currents of the purified and reconstituted torpedo acetylcholine receptor. *Proc. Natl. Acad. Sci. USA* 80:5129–33

50. Kloppenburg P, Zipfel WR, Webb WW, Harris-Warrick RM. 2000. Highly localized Ca^{2+} accumulation revealed by multiphoton microscopy in an identified motoneuron and its modulation by dopamine. *J. Neurosci.* 20:2523–33

51. Dombeck DA, Blanchard-Desce M, Webb WW. 2004. Optical recording of action potentials with second-harmonic generation microscopy. *J. Neurosci.* 24:999–1003

52. Opsahl LR, Webb WW. 1994. Lipid-glass adhesion in giga-sealed patch-clamped membranes. *Biophys. J.* 66:75–79

53. Opsahl LR, Webb WW. 1994. Transduction of membrane tension by the

ion-channel alamethicin. *Biophys. J.* 66: 71–74

54. Mak DOD, Webb WW. 1995. Molecular dynamics of alamethicin transmembrane channels from open-channel current noise analysis. *Biophys. J.* 69:2337–49

55. Dragsten PR, Webb WW, Paton JA, Capranic RR. 1974. Auditory membrane vibrations: Measurements at sub-angstrom levels by optical heterodyne spectroscopy. *Science* 185:55–57

56. Dragsten PR, Webb WW, Paton JA, Capranica RR. 1976. Light-scattering heterodyne interferometer for vibration measurements in auditory organs. *J. Acoust. Soc. Am.* 60:663–71

57. Fujimoto JG, Pitris C, Boppart SA, Brezinski ME. 2000. Optical coherence tomography: An emerging technology for biomedical imaging and optical biopsy. *Neoplasia* 2:9–25

58. Denk W, Webb WW. 1990. Optical measurement of picometer displacements of transparent microscopic objects. *Appl. Opt.* 29:2382–91

59. Denk W, Webb WW, Hudspeth AJ. 1989. Mechanical properties of sensory hair bundles are reflected in their Brownian motion measured with a laser differential interferometer. *Proc. Natl. Acad. Sci. USA* 86:5371–75

60. Denk W, Webb WW. 1992. Forward and reverse transduction at the limit of sensitivity studied by correlating electrical and mechanical fluctuations in frog saccular hair cells. *Hear. Res.* 60:89–102

61. Denk W, Webb WW. 1989. Thermal-noise-limited transduction observed in mechanosensory receptors of the inner ear. *Phys. Rev. Lett.* 63:207–10

62. Denk W, Keolian RM, Webb WW. 1992. Mechanical response of frog saccular hair bundles to the aminoglycoside block of mechanoelectrical transduction. *J. Neurophysiol.* 68:927–32

63. White JG, Amos WB, Fordham M. 1987. An evaluation of confocal versus conventional imaging of biological structures by fluorescence light microscopy. *J. Cell Biol.* 105:41–48

64. Goeppert-Mayer M. 1931. Elementary file with two quantum fissures. *Ann. Phys.* 9:273–94

65. Denk W, Strickler JH, Webb WW. 1990. Two-photon laser scanning fluorescence microscopy. *Science* 248:73–76

66. Denk W. 1994. Two-photon scanning photochemical microscopy: Mapping ligand-gated ion-channel distributions. *Proc. Natl. Acad. Sci. USA* 91:6629–33

67. Strickler JH, Webb WW. 1991. Three-dimensional optical data storage in refractive media by two-photon point excitation. *Opt. Lett.* 16:1780–82

68. Wu ES, Webb WW, Strickler JH, Harrell WR. 1992. Two-photon lithography for microelectronic application. *Proc. SPIE* 1674:776–82

69. Maiti S, Shear JB, Williams RM, Zipfel WR, Webb WW. 1997. Measuring serotonin distribution in live cells with three-photon excitation. *Science* 275:530–32

70. Shear JB, Xu C, Webb WW. 1997. Multiphoton-excited visible emission by serotonin solutions. *Photochem. Photobiol.* 65:931–36

71. Xu C, Guild J, Webb WW, Denk W. 1995. Determination of absolute two-photon excitation cross-sections by in-situ second-order autocorrelation. *Opt. Lett.* 20:2372–74

72. Xu C, Webb WW. 1996. Measurement of two-photon excitation cross sections of molecular fluorophores with data from 690 to 1050 nm. *J. Opt. Soc. B* 13:481–91

73. Xu C, Zipfel W, Shear JB, Williams RM, Webb WW. 1996. Multiphoton fluorescence excitation: New spectral windows for biological nonlinear microscopy. *Proc. Natl. Acad. Sci. USA* 93:10763–68

74. Xu C, Webb WW. 1997. Multiphoton excitation of molecular fluorophores and nonlinear laser microscopy. In *Topics in Fluorescence Spectroscopy: Volume 5: Nonlinear and Two-Photon-Induced*

Fluorescence, ed. J Lakowicz, pp. 471–540. New York: Plenum

75. Albota MA, Xu C, Webb WW. 1998. Two-photon fluorescence excitation cross sections of biomolecular probes from 690 to 960 nm. *Appl. Opt.* 37:7352–56

76. Albota M, Beljonne D, Bredas JL, Ehrlich JE, Fu JY, et al. 1998. Design of organic molecules with large two-photon absorption cross sections. *Science* 281:1653–56

77. Dickson RM, Cubitt AB, Tsien RY, Moerner WE. 1997. On/off blinking and switching behaviour of single green fluorescent protein molecules. *Nature* 388:355–58

78. Hess ST, Heikal AA, Webb WW. 2004. Fluorescence photoconversion kinetics in novel green fluorescent protein pH sensors (pHluorins). *J. Phys. Chem. B* 108: 10138–48

78a. Yao J, Larson DR, Vishwasrao HD, Zipfel WR, Webb WW. 2005. Blinking and non-radiant dark fraction of water-soluble quantum dots in aqueous solution. *Proc. Natl. Acad. Sci. USA* 102:14284–89

79. Ow H, Larson DR, Srivastava M, Baird BA, Webb WW, Wiesner U. 2005. Bright and stable core-shell fluorescent silica nanoparticles. *Nano Lett.* 5:113–17

80. Ouzounov DG, Moll KD, Foster MA, Zipfel WR, Webb WW, Gaeta AL. 2002. Delivery of nanojoule femtosecond pulses through large-core microstructured fibers. *Opt. Lett.* 27:1513–15

81. Kasischke KA, Vishwasrao HD, Fisher PJ, Zipfel WR, Webb WW. 2004. Neural activity triggers neuronal oxidative metabolism followed by astrocytic glycolysis. *Science* 305:99–103

82. Vishwasrao HD, Heikal AA, Kasischke KA, Webb WW. 2005. Conformational dependence of intracellular NADH on metabolic state revealed by associated fluorescence anisotropy. *J. Biol. Chem.* 280:25119–26

83. Wilson T, Sheppard CJR. 1984. *Theory and Practice of Scanning Optical Microscopy.* London: Academic. 213 pp.

84. Williams RM, Zipfel WR, Webb WW. 2005. Interpreting second harmonic generation images of collagen I fibrils. *Biophys. J.* 88:1377–86

85. Dombeck DA, Kasischke KA, Vishwasrao HD, Ingelsson M, Hyman BT, Webb WW. 2003. Uniform polarity microtubule assemblies imaged in native brain tissue by second-harmonic generation microscopy. *Proc. Natl. Acad. Sci. USA* 100: 7081–86

86. Christie RH, Bacskai BJ, Zipfel WR, Williams RM, Kajdasz ST, et al. 2001. Growth arrest of individual senile plaques in a model of Alzheimer's disease observed by in vivo multiphoton microscopy. *J. Neurosci.* 21:858–64

87. Bacskai BJ, Kajdasz ST, Christie RH, Zipfel WR, Williams RM, et al. 2001. Chronic imaging of amyloid plaques in the live mouse brain using multiphoton microscopy. In *Multiphoton Microscopy in the Biomedical Sciences*, ed. A Periasamy, PTC So, *Proc. SPIE* 4262:125–33. Bellingham, WA: Int. Soc. Opt. Eng.

88. Zipfel WR, Williams RM, Christie RH, Nikitin AY, Hyman BT, Webb WW. 2003. Live tissue intrinsic emission microscopy using multiphoton excited intrinsic fluorescence and second harmonic generation. *Proc. Natl. Acad. Sci. USA* 100:7075–80

89. Zipfel WR, Williams RM, Webb WW. 2003. Nonlinear magic: Multiphoton microscopy in the biosciences. *Nat. Biotechnol.* 21:1369–77

90. Webb WW. 2003. *U.S. Patent No. 6,839,586*

Annu. Rev. Physiol. 2006. 68:29–49
doi: 10.1146/annurev.physiol.68.040104.124530
Copyright © 2006 by Annual Reviews. All rights reserved
First published online as a Review in Advance on October 11, 2005

CARDIAC REGENERATION: Repopulating the Heart

Michael Rubart and Loren J. Field

*Herman B Wells Center for Pediatric Research and Krannert Institute of Cardiology,
Indiana University School of Medicine, Indianapolis, Indiana 46202-5225;
email: mrubartv@iupui.edu, ljfield@iupui.edu*

Key Words   cell transplantation, cardiomyocyte proliferation, stem cells

■ **Abstract**   Many forms of pediatric and adult heart disease result from a deficiency in cardiomyocyte number. Through repopulation of the heart with new cardiomyocytes (that is, induction of regenerative cardiac growth), cardiac disease potentially can be reversed, provided that the newly formed myocytes structurally and functionally integrate in the preexisting myocardium. A number of approaches have been utilized to effect regenerative growth of the myocardium in experimental animals. These include interventions aimed at enhancing the ability of cardiomyocytes to proliferate in response to cardiac injury, as well as transplantation of cardiomyocytes or myogenic stem cells into diseased hearts. Here we review efforts to induce myocardial regeneration. We also provide a critical review of techniques currently used to assess cardiac regeneration and functional integration of de novo cardiomyocytes.

INTRODUCTION

Many forms of pediatric and adult heart disease result from a deficiency in cardiomyocyte number. Therapeutic interventions aimed at reducing the extent of acute or chronic cardiomyocyte loss have proven to be effective at reducing morbidity and mortality in patients with heart disease (1). It has also been proposed that repopulating the hearts with new cardiomyocytes may have a beneficial effect, particularly if the new cells become structurally integrated within the heart and contribute to cardiac function. Repopulation of the heart with new cardiomyocytes (that is, induction of regenerative cardiac growth) potentially may lead to reversal of cardiac disease and therefore may represent a marked advance over current therapies that are aimed largely at salvaging at-risk myocardium.

A number of approaches have been utilized to effect regenerative growth of the myocardium in experimental animals. Initial efforts focused on enhancing the ability of cardiomyocytes to proliferate in response to cardiac injury. This entailed first determining the intrinsic capacity of cardiomyocytes to proliferate in normal and injured adult hearts (2), followed by efforts to enhance this capacity via genetic and/or cytokine based interventions (3, 4). More recently, the notion of delivering donor cardiomyocytes or myogenic stem cells into diseased hearts has garnered considerable enthusiasm (5). Encouraging results in experimental animals with skeletal myoblast or endothelial precursor cells transplantation has prompted several clinical trials (6). The recent identification of putative resident cardiomyogenic stem cells in the adult heart (7) offers yet another potential target population with which to induce regenerative growth of the myocardium. Here we review efforts to induce myocardial regeneration vial cell cycle activation and cell transplantation. We also provide a critical review of the strengths and weaknesses of techniques currently used to assess cardiac regeneration. It is hoped that this review will stimulate interest in the field.

CELL CYCLE ACTIVATION APPROACHES

A priori, it seems that induction of cell proliferation would provide the most direct way to repopulate the heart with cardiomyocytes. In this section, we review the approaches used to monitor cardiomyocyte cell cycle activity as well as various efforts to manipulate cardiomyocyte cell cycle activity in vivo.

Intrinsic Rate of Cardiomyocyte Cell Cycle Activity In Vivo

A thorough perusal of the literature reveals an abysmally wide range of values for cardiomyocyte cell cycle activity in normal and injured adult hearts. Undoubtedly, some of this is attributable to intrinsic differences between species as well as differences between individuals within a species. In addition, because many of the assays measure different events associated with cell cycle progression (i.e., expression of proteins involved with various aspects of cell cycle activity, DNA synthesis, presence of mitotic figures, increases in cell number, etc.), some variation in read out is to be expected. Despite these caveats, it is extremely difficult to reconcile the reported range of 0.0005% to 3% for cardiomyocyte cell cycle activity in normal adult hearts and the even greater range reported for injured hearts (2, 8, 9). Thus, for the uninitiated these data are quite confusing.

Much of the variation in the reported values for cardiomyocyte cell cycle activity can be traced to flaws in the identification of cardiomyocyte nuclei in histologic sections. Although cardiomyocyte cell bodies are readily identified using conventional histochemical stains or immune reactivity for myocyte-restricted proteins, identification of cardiomyocyte nuclei is more difficult. Because cardiomyocytes represent only 20% of the cells present in a given histologic section of the adult ventricle, the vast preponderance of nuclei reside in nonmyocytes. Moreover, the volume of nonmyocyte cell bodies are markedly smaller than that of cardiomyocytes. Nonmyocyte nuclei can thus easily be assigned to cardiomyocytes in sections containing both myocytes and nonmyocytes in the z plane.

Given the comparatively high levels of nonmyocyte proliferation and low levels of cardiomyocyte proliferation, misidentification of cardiomyocyte nuclei can dramatically increase the apparent rate of cell cycle activity. This is particularly true in injured hearts, in which noncardiomyocyte cell cycle activity is markedly increased. Although the use of confocal microscopy can circumvent these problems, many investigators present their data as stacked images, thereby negating the superb three-dimensional resolution of this system. Scanning confocal microscopy is also exceedingly time consuming, making it problematic to use for monitoring rare events.

Genetically modified animals with a cardiomyocyte-restricted, nuclear-localized reporter (10) have proven to be quite useful in monitoring cardiomyocyte nuclear tritiated thymidine incorporation in histologic sections. This approach reveals that 0.0005% of the ventricular cardiomyocytes exhibit DNA synthesis in uninjured adult hearts from mice maintained in a DBA/2J genetic background (11). In hearts with permanent coronary artery occlusion, DNA synthesis is present in only 0.004% of the infarct border zone cardiomyocytes at seven days postinjury (12). Thus, the intrinsic proliferative capacity of adult cardiomyocytes is quite low.

Manipulation of the Cardiomyocyte Cell Cycle In Vivo

The genetic modification of cells in vivo has provided a wealth of information regarding cardiomyocyte cell cycle regulation. Initial studies demonstrated that expression of the SV40 Large T-Antigen oncoprotein is sufficient to induce sustained cell cycle activity in atrial (13, 14) and ventricular (15) cardiomyocytes. Subsequent transgenic and gene targeting experiments have demonstrated that many genes influence cardiomyocyte cell cycle activity during development (3, 16).

Only a limited number of these genetic modifications have been shown to induce sustained ventricular cardiomyocyte cell cycle activity in adult cardiomyocytes. These include overexpression of SV40 Large T-Antigen (15), the D-type cyclins (12, 17), CDK-2 (18), dominant interfering TSC2 (19), dominant interfering p193 (20), dominant interfering p53 (20), cyclin A2 (21), IGF-1 (22), bcl-2 (23), and dominant interfering p38 MAP kinase (24). Similarly, combinatorial deletion of retinoblastoma family members can give rise to cardiomyocyte cell cycle activity in the adult heart (25). In all of these studies, altered gene expression occurred prior to terminal differentiation; as such, it is possible that similar modifications in genetically naive adult cardiomyocyte may not give rise to cell cycle activation. In this regard, it is of interest to note that adenoviral codelivery of CDK4 and a modified cyclin D molecule is sufficient to induce cell cycle activity in terminally differentiated adult cardiomyocytes in vivo (26). Similarly, conditional expression of a c-myc transgene in adult cardiomyocytes results in cardiac hypertrophy accompanied by DNA synthesis and formation of multinucleated cardiomyocytes (27). Thus, genetically naive adult cardiomyocytes remain responsive to at least a subset of the pathways identified by transgenic and gene targeting studies, with constitutive changes in gene expression.

Additional experiments have been performed to determine if cardiomyocyte cell cycle activation has a positive impact on heart structure and/or function following experimental injury in transgenic animals. In some cases, as exemplified by the IGF-1 (28, 29) and Bcl-2 (23) studies, transgene expression impacts both cell cycle activity and survival. As such, it was impossible to determine if the observed beneficial effect on cardiac structure and function results from transgene-induced cardiomyocyte proliferation or, alternatively, from reduced levels of cardiomyocyte apoptosis. In other cases, as exemplified by dominant interfering p193 expression, cardiomyocyte cell cycle activity is accompanied with favorable postinfarction ventricular remodeling (20) and a concomitant (albeit modest) improvement in cardiac function (30). However, given the absence of time-course studies, it was not clear if transgene-induced cell cycle activity reverses myocardial damage or simply lessens the degree of damage. Paradoxically, although cardiomyocyte cell cycle activity was induced in adult mice overexpressing CDK-2, these animals exhibited an aberrant hypertrophic response to surgically induced pressure overload (18).

One study has demonstrated a progressive, cell cycle–induced restoration of cardiac structure and function postinjury, consistent with regenerative growth of the myocardium. Adult transgenic mice expressing cyclin D2 in the myocardium have increased levels of baseline cardiomyocyte cell cycle activity (12). Myocardial infarction resulted in similar infarct size and reduced cardiac function in the cyclin D2 mice and their nontransgenic siblings. In the cyclin D2 animals, cell cycle activity results in a progressive reduction of infarct size and improved cardiac function (12; R. Hassink, K.B. Pasumarthi, H. Nakajima, M. Rubart, M. Soonpaa, et al., submitted manuscript). Indeed, by 180 days postinfarction, there is on average a 50% reduction in infarct size, and cardiac function is not statistically different from that in age-matched sham-operated transgenic animals. By contrast, no improvement is detected in cardiac architecture or function in the nontransgenic siblings at 180 days postinfarction.

Collectively, these studies demonstrate that cardiomyocyte cell cycle activity can readily be enhanced in genetically modified animals. Moreover, several studies suggest that cell cycle activity is able to reverse structural and/or functional defects following experimental injury to the myocardium. These emerging proof-of-concept studies strongly support the potential therapeutic value of cardiomyocyte cell cycle induction. However, successful transfer to clinical practice will likely require the development of small molecules that are able to mimic key aspects of the genetic manipulations described above.

CELL TRANSPLANTATION APPROACHES

Transplantation of donor myocytes or myogenic stem cells offers an alternative approach to repopulate the heart with cardiomyocytes. Below, we describe studies wherein transplantation of skeletal myoblasts, cardiomyocytes, and undifferentiated stem cells has been used in an effort to regenerate cardiac tissue. Efforts to promote regenerative growth via mobilization and/or homing of stem cells are also discussed.

Transplantation of Skeletal Myoblasts

That skeletal myoblasts form viable, long-term skeletal myotube grafts following transplantation into adult hearts first was reported in 1993 (32). Subsequent studies described the structural and functional consequences of skeletal myoblast transplantation into normal and injured hearts in a number of species (5). Efforts in this area were accelerated by the suggestion that myoblasts can generate functional cardiac-like muscle following their transplantation into rabbit hearts with cryoinjury (33). However, earlier studies (34) had clearly demonstrated that increasing voltage during stimulation of cardiac wound strips containing engrafted myotubes results in stepwise increases in tension development, implying that the graft myotubes are electrically insulated from one another. These data were confirmed via intracellular recording of myotube transmembrane action potentials in vibratome sections prepared from infarcted rat hearts following transplantation of genetically labeled skeletal myoblasts (35). Other studies have shown that engrafted myoblasts are able to form heterokaryons at extremely low frequency with host cardiomyocytes (36). A portion of these heterokaryons remains functionally coupled with the host myocardium and exhibits heterogeneity in intracellular calcium signaling (37). Although it remains possible that engrafted skeletal myotubes may undergo stretch-induced contraction, such activity (if it exists at all) most likely occurs during diastolic relaxation of the heart.

Given the absence of significant electromechanical coupling between donor myoblast–derived myotubes and the host myocardium, it is somewhat surprising that numerous studies reported improvement in left ventricular function following myoblast transplantation into experimentally injured hearts (5). It is likely that the beneficial impact on cardiac function is due to indirect activities of the donor cells (as are, for example, donor cell–induced angiogenesis, alteration of postinfarction remodeling, changes in the elastic properties of the scar tissue, etc.). These observations prompted a number of Phase I clinical trials designed to establish the safety and feasibility of autologous skeletal myoblast transplantation in patients in whom myocardial infarction resulted in severe left ventricular dysfunction (38–44). These trials revealed that the transplanted myoblasts form clusters of differentiated skeletal myotubes that are well aligned with host cardiomyocytes or embedded within scarred regions for as long as 18 months posttransplantation. At present, no conclusions can be made regarding the functional consequences of these procedures. However, given that the volume of donor-derived myocytes was quite small, any functional improvement would by necessity be due to indirect effects of the donor cells.

Transplantation of Cardiomyocytes

Cardiomyocytes appear to constitute the ideal donor cell for transplantation into adult hearts, as they possess the necessary structural and physiologic attributes to integrate functionally with resident myocardial cells. Initial studies demonstrated that fetal cardiomyocytes can be successfully engrafted into normal (45) or infarcted (46) myocardium. Immune histology as well as molecular and ultrastructural analyses revealed that donor fetal cardiomyocytes undergo terminal differentiation and form mature adult-like cells following transplantation. Moreover, intercalated disks (comprised of fascia adherens, desmosomes, and gap junctional complexes) were observed to couple donor and host cardiomyocytes (45, 47). Despite these circumstantial data, an additional eight years elapsed before definitive proof was obtained demonstrating that donor cardiomyocytes form a functional syncytium with the host myocardium. These experiments revealed that all donor cardiomyocytes that are in physical contact with host cardiomyocytes are also functionally coupled (48).

More than 22 studies reported improved cardiac function following transplantation of fetal or neonatal cardiomyocytes into experimentally injured hearts (5). In all cases, it was very clear that the volume and/or the anatomical position of the engrafted cells could not account for the observed functional improvement. It was, however, observed that the presence of cells appeared to lead to favorable postinjury ventricular remodeling (49). Although the functional improvement following cardiomyocyte transplantation is likely due to indirect effects of these donor cells, additional efforts to develop their potential clinical application are warranted, as they are the only transplanted donor cell type that has thus far been shown to participate in a functional syncytium with the host myocardium.

Three formidable obstacles must be overcome for cardiomyocyte transplantation to be clinically useful. First, strategies must be developed to enhance the volume of donor-derived grafts following transplantation. This in theory could be accomplished by enhancing donor cardiomyocyte survival and/or proliferation following transplantation. In the latter case, many of the manipulations described above for enhancing cardiomyocyte proliferation in vivo would likely be directly applicable to donor-derived cells. Second, sufficient numbers of donor cells must be readily available for transplantation. Toward that end, it has previously been shown that embryonic stem cell–derived cardiomyocytes form stable grafts following transplantation into immune-suppressed animals (50). Moreover, the generation of mouse embryonic stem cell–derived cardiomyocytes can be easily modified to produce large numbers of cells in bioreactor-like cultures (51–53). Laugwitz et al. (54) recently observed that cells that are isolated from neonatal hearts and that express the Islet-1 transcription factor can be amplified in vitro and differentiated into bona fide cardiomyocytes; these cells provide an alternative source of donor cells. Given these observations and the likelihood that additional sources of embryonic or adult stem cell–derived cardiomyocytes will be identified in the future, generation of sufficient numbers of donor cells will likely not be problematic. Finally, adequate immune suppression (55) or tolerance-inducing (56) protocols will need to be developed to safely and efficiently block rejection of allogenic stem cell–derived donor cardiomyocytes. If all three of the above criteria can be met, cardiomyocyte transplantation likely will be quite efficacious.

Transplantation of Stem Cells

Studies in the late 1990s suggested that adult-derived stem cells may possess a greater capacity to differentiate into noncanonical cell types in vitro and in vivo. This prompted the direct injection of a variety of multipotent adult-derived stem cells, including hematopoietic stem cells (HSCs), mesenchymal stem cells (MSCs), and endothelial precursor cells (EPCs), into normal and injured hearts. Interest in this approach peaked in 2001, when it was reported that transplantation of HSCs expressing high levels of the c-kit receptor led to rapid regeneration of more than 60% of the infarcted left ventricle in a mouse permanent coronary occlusion model (57). However, enthusiasm for this approach was tempered when other groups, using a variety of cardiomyogenic read outs, were unable to replicate the results in other experimental animals (58–60). Moreover, the marginal level of functional improvement following delivery of a variety of bone marrow–derived cells in patients with acute infarcts (61–64) or chronic disease (65–67) has thus far not been consistent with results showing marked regeneration of the ventricle in mice (57).

Initial enthusiasm regarding the potential of MSC transplantation to reconstitute myocardial tissue (68–70) has also been tempered by the recent observation that much of the reconstitutive effect is due to cardioprotective paracrine activity from the donor cells (71). Similarly, angiogenic paracrine activity has recently been suggested as the underlying mechanism of improved cardiac function following bone marrow–derived cell transplantation into injured hearts (72). The potential of EPC differentiation into cardiomyocytes has been demonstrated in vitro (73), but not yet in vivo. Nonetheless, the potential of exogenous EPCs to contribute to angiogenesis is well established.

Mobilization and/or Homing of Stem Cells

Bone marrow reconstitution studies using either unfractionated (74) or highly purified HSCs (75) from mice with ubiquitously expressed genetic tags suggested that extracardiac cells may contribute to the formation of cardiomyocytes. These observations were supported by the detection of y chromosome–containing cardiomyocytes in male transplant patients receiving a female donor heart, although the frequency at which this occurred varied greatly in different studies, with some groups failing to observe the phenomenon (76–79). This caveat notwithstanding, these data provided a potential mechanistic explanation for the apparent robust regeneration of infarcted myocardium following treatment with cytokines known to mobilize HSCs from the marrow (80). Unfortunately, other groups failed to recapitulate this finding (60, 81). At present, the effect of cytokine stimulation in animals with experimental cardiac injury is not clear, with one study suggesting a cardioprotective effect via apoptosis inhibition (82) and another suggesting abnormal wound healing with a concomitant reduction in cardiac function (83). Similar conflicting results regarding the benefit on cardiac function have been reported in initial clinical studies (84, 85). Nonetheless, it is clear that this type of intervention does not give rise to marked regenerative growth of the infarcted myocardium.

Several recent studies have identified stem cells with apparent cardiomyogenic activity within the adult heart itself. These include, among others, cells that express the c-kit receptor (86, 87) or sca-1 (88–90) or that exhibit the so-called side-population phenotype (91, 92). The exact nature of the cardiomyogenic stem cells in these different studies presently is not clear, as the markers used for their isolation were in some cases mutually exclusive. Moreover, the phenotypic characteristics and relative propensity of the stem cell-derived myocytes were markedly different in the different studies. For example, direct transplantation of c-kit-positive stem cells into hearts with permanent coronary artery ligation results in the formation of numerous small, poorly differentiated cells lacking obvious myofiber structure (86). Homing of sca-1-positive stem cells to hearts with ischemia-reperfusion injury results in the formation of well-differentiated cardiomyocytes predominantly at the infarct border zone, which arises at least in part via fusion events (90). In spite of the current level of confusion, the potential utility of adult heart–derived stem cells is great provided that their cardiomyogenic capacity is further validated.

ASSESSING REGENERATIVE GROWTH OF THE MYOCARDIUM

The ultimate success of cardiac repair strategies will depend on the number of de novo cardiomyocytes generated and on their ability to form a functional syncytium with the preexisting myocardium. Accordingly, quantitative assays to determine both the number of de novo cardiomyocytes and their functional status are needed.

Structural Analyses

MONITORING REGENERATIVE GROWTH   No single quantitative approach is currently available to determine directly the number of newly formed cardiac myocytes resulting from cardiomyocyte cell cycle reactivation and/or progenitor cell differentiation in the myocardium of experimental animals. However, combined use of complementary histological techniques was recently used to provide quantitative estimates of the extent of regenerative growth in a transgenic mouse model of infarct regression. Transgenic mice expressing cyclin D2 in the myocardium and their nontransgenic siblings were euthanized at increasing time intervals following coronary artery ligation (12). Serial coronal sections were sampled at fixed, regular intervals from the apex to the base, and the total number of cardiomyocytes was determined for each section. It was found that cell cycle activation markedly increases cardiomyocyte content in transgenic mice compared to infarcted wild-type mice hearts at 150 days postinfarction.

To exclude the possibility that differential cardiomyocyte hypertrophy confounded the assay (as cardiomyocyte volume increases, the number of cardiomyocytes per unit area of tissue diminishes, and vice versa), minimal cardiomyocyte diameter measurements were determined to obtain an index of cardiomyocyte hypertrophy at the cellular level (93). No differences in average minimal cardiomyocyte diameter were found between wild-type and cyclin D2 transgenic mice, compatible with the presence of de novo myocardium in the transgenic animals. Moreover, the relative area of viable cardiac tissue in each section was equivalent to that obtained by multiplying the total cardiomyocyte number per tissue section by their average cross-sectional area (as determined by minimal cardiomyocyte diameter measurements).

Thus, a variety of histological techniques can be employed in a complementary fashion to quantitatively assess regenerative cardiac growth, provided that rigorous morphological criteria are used. Owing to their relatively small sizes, mouse and rat hearts are ideally suited for studies designed to examine the effect of experimental interventions on regenerative cardiac growth, as these organs can readily be subjected to histological analyses in their entirety. Employing such an approach minimizes sampling errors, which may occur if only representative areas are selected for morphometric analysis. A major disadvantage of histological analyses is the need to euthanize the animals at fixed time intervals, raising the need for serial noninvasive imaging techniques to assess regenerative growth in vivo. Gadolinium-enhanced magnetic resonance imaging, which can measure acute and chronic infarct size with high precision (94, 95), would greatly complement the structural analyses described above.

MONITORING THE FATE OF CARDIOMYOGENIC CELLS   For cell transplantation–based interventions, the ability to distinguish donor-derived myocytes from pre-existing host myocytes is critical. One of two general approaches is typically used to accomplish this. The first relies on the expression of lineage-restricted reporter transgenes, whereas the second employs lineage-independent tags (either genetic or nongenetic) in combination with histologic analyses to monitor donor cell survival and cardiomyogenic differentiation.

Lineage-restricted reporter transgenes have been used to monitor the fate of fetal cardiomyocytes (45), HSCs (59), and adult heart–derived stem cells (96) following transplantation into normal or injured hearts. An example of this approach is shown in Figure 1. Fetal cardiomyocytes were isolated from transgenic mice expressing a nuclear localized beta-galactosidase reporter under a cardiomyocyte-restricted promoter. The cells were transplanted into a normal, nontransgenic heart, and the animals were allowed to recover. The recipient heart was subsequently harvested, sectioned, and stained with a chromogenic dye that produces robust blue signal in the presence of beta-galactosidase. Donor cells are readily distinguished from host cells by virtue of the blue signal in histologic sections. The signal is also detectable in ultrastructural analyses, in which it appears as a perinuclear, electron-dense precipitate. Because the reporter transgene is expressed only in cardiomyocytes, the presence of blue signal is indicative of cardiomyogenic differentiation. This type of approach is ideally suited to monitor the cardiomyogenic potential of donor cells as well as to investigate the effect of interventions aimed at enhancing donor cell survival and/or cardiomyogenic potential following intracardiac delivery.

A number of lineage-independent tags have been used to follow the fate of donor cells. These include the use of donor cells from transgenic mice with ubiquitously expressed reporter transgenes (57–60), use of donor cells labeled with gap junction–impermeable fluorescent dyes (90, 97, 98), and the monitoring of the presence of y chromosomes when male donor cells and female recipients are used (99). Although these tagging techniques readily allow one to determine if transplanted and/or homed stem cells are present, they do not indicate that cardiomyogenic

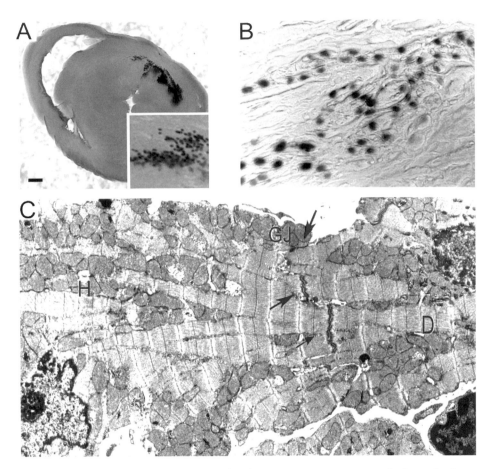

Figure 1   Use of a cardiomyocyte-restricted reporter transgene to track donor fetal cardiomyocytes following intracardiac transplantation into a nontransgenic recipient heart. The
donor cardiomyocytes express a nuclear-localized beta-galactosidase reporter transgene.
(*A*) Low-power view of a thick heart section (300 μm) from a heart with a fetal cardiomyocyte
transplant. The section was stained with the chromogenic dye X-GAL, which generates a
blue signal in the presence of beta-galactosidase. Donor cells are identified by virtue of their
blue nuclei. Individual cells can be readily visualized and counted under higher magnification
(*see inset*). (*B*) High-power photomicrograph of a histologic section from a heart with a fetal
cardiomyocyte transplant. The section was stained with X-GAL. (*C*) Ultrastructural view of
the junction between a host cardiomyocyte (H, *left*) and a donor cardiomyocyte (D, *right*).
The donor cardiomyocyte was identified by the presence of crystalloid X-GAL reaction products in the perinuclear region. Note the presence of intercalated disks (*red arrows*) and gap
junctions (GJ) at the donor-host junction.

differentiation has occurred. Additional analyses are necessary to confirm the cardiac phenotype of donor cells. For example, the expression of cardiomyocyte-specific structural proteins or transcription factors can be monitored, complemented with high-resolution imaging to assure that the signal originates from within a grafted or homed cell. Alternatively, electron microscopy can be used to determine if labeled cells exhibit specific ultrastructural attributes compatible with a cardiac phenotype (45, 47).

TRANSDIFFERENTIATION VERSUS CELL FUSION EVENTS   Analyses of cardiomyogenic differentiation are further complicated by the observation that some donor cell types fuse with host cardiomyocytes, thereby giving rise to apparent cardiomyogenic differentiation. Adequate techniques to detect donor-host cell fusion events must therefore be employed in cell transplantation/homing studies to distinguish these events from true differentiation events. One approach to accomplish this is based on site-specific recombination assays. For example, when a donor cell expressing cre recombinase fuses with a host cardiomyocyte carrying a conditionally active reporter transgene (e.g., *LacZ* or EGFP behind a lox-flanked stop signal), the cre recombinase catalyzes the excision of the stop signal, thereby resulting in the accumulation of reporter protein. Via this cre/lox approach, it was demonstrated that bone marrow–derived cells fuse with cardiac muscle in mice following marrow reconstitution (100). The same approach was employed to demonstrate fusion events between host cardiomyocytes and donor skeletal myoblasts (36) or adult heart–derived progenitor cells (90).

Alternatively, double-reporter systems wherein donor and recipient cells express different reporter transgenes can be used to detect fusion events. If the cell in question expresses both reporter transgenes, then it arose via cell fusion. This approach has been used to demonstrate fusion events between host cardiomyocytes and donor skeletal myoblasts (37) or marrow-derived stem cells (60). An example demonstrating the detection of myoblast-cardiomyocyte fusion events is shown in Figure 2. A potential pitfall of the recombination-based and double-reporter

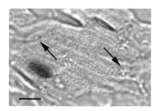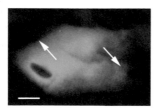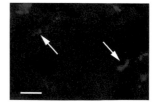

Figure 2   Use of a double-reporter transgene system to detect cell fusion events. EGFP-expressing skeletal myoblasts were transplanted into the heart of a mouse with cardiomyocyte-restricted expression of a nuclear-localized beta-galactosidase reporter transgene. A cell arising from fusion between a donor myoblast and a resident cardiomyocyte is identified by the presence of blue nuclear staining (following X-GAL staining; *left panel*) and green cytoplasmic EGFP fluorescence (*middle panel*). Connexin 43 (*red signal*; *right panel*) is present at the junctions between the fused cell and the host cardiomyocytes (*arrows*).

transgene-based assays is that negative results do not exclude the possibility that fusion has occurred, as reporter genes in the resulting heterokaryons may be inactivated or eliminated over time (100), thereby giving rise to apparent transdifferentiation events.

Functional Analyses

A number of analyses have been employed to determine if cell cycle activation and/or cell transplantation has a positive impact on cardiac function in injured hearts. In most instances the assays employed examined function at the whole organ level, e.g., via echocardiography, left ventricular pressure-volume measurements, magnetic resonance imaging, and positron emission tomography. Other studies relied on fractional shortening measurements of limited regions of the ventricular wall through the use of sono-micrometry analyses. In many of these analyses, cardiac function improved as a consequence of a genetic modification aimed at enhancing cardiomyocyte proliferation or, alternatively, as a consequence of cell transplantation or homing.

Although these analyses can quite accurately provide an index of cardiac function, they do not indicate the functional status of individual cardiomyocytes derived from the intervention (i.e., newly replicated cells, transplanted myocytes, stem cell–derived myocytes, etc.). Consequently, these assays are unable to identify the underlying mechanism of functional improvement. Specifically, they cannot distinguish between direct contraction of the donor cells vs. indirect, yet beneficial, effects imparted upon the surviving host myocardium. Although such information is not needed to determine if a given intervention has a benefit on cardiac function, the information is absolutely required if one wishes to devise strategies to enhance the efficacy of the intervention. The ability to monitor cardiomyocyte function at the individual cell level in intact hearts would thus be of considerable utility in assessing the mechanistic basis of improved cardiac function following cell cycle– and cell transplantation–based interventions.

Toward that end, two-photon laser scanning microscopy (TPLSM)–based fluorescence imaging systems have been used to monitor intracellular calcium ($[Ca^{2+}]_i$) transients in intact hearts (101). This approach has the advantage of providing thin

---

Figure 3   Imaging of $[Ca^{2+}]_i$ transients in a nontransgenic heart carrying donor cardiomyocytes expressing EGFP. (*A*) Full-frame image during continuous electrical stimulation at 2 Hz. Electrical stimulation results in uniform increases in rhod-2 fluorescence in both EGFP$^+$ donor and EGFP$^-$ host myocytes. Green scale bar: 20 μm. (*B*) Line-scan image of rhod-2 transients in seven juxtaposed myocytes during sinus rhythm and during stimulation at 2 and 4 Hz. The position of the line scan is denoted by the white line in panel A. Horizontal green scale bar: 20 μm; vertical green scale bar: 1000 ms. (*C*) Time course of changes in rhod-2 fluorescence in a host, i.e., EGFP$^-$ cardiomyocyte and a neighboring EGFP$^+$ donor cell. (*D*) Superimposition of normalized $[Ca^{2+}]_i$ transients in a donor and host myocyte during 1 and 2 Hz stimulation.

optical sections from deeper within biological specimens than is available with conventional single-photon confocal imaging systems (102). Intact hearts are perfused on a Langendorff apparatus with a calcium-sensitive fluorescent indicator (i.e., rhod-2) and with a compound to effect excitation-contraction uncoupling (i.e., cytochalasin D). The hearts are then transferred to the microscope stage and imaged during spontaneous depolarization or during stimulation at remote sites.

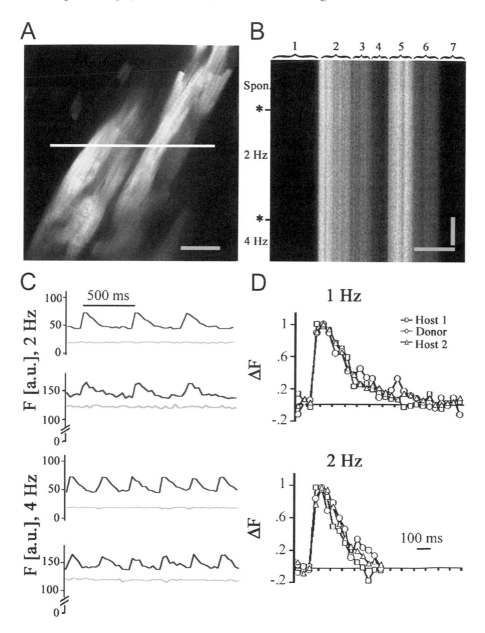

Action potential–induced increases in rhod-2 fluorescence, corresponding to increases in cytosolic calcium concentration, can readily be recorded in line-scan mode (101). This assay reveals that the kinetics of action potential–induced $[Ca^{2+}]_i$ transients in juxtaposed cells are almost indistinguishable from each other, which is indicative of their functional coupling.

An example demonstrating functional coupling between donor and host cells following the transplantation of fetal cardiomyocytes is depicted in Figure 3. Fetal cardiomyocytes isolated from transgenic mice expressing EGFP under the control of a cardiomyocyte-restricted promoter were injected into a nontransgenic recipient heart and subjected to TPLSM imaging 35 days later (48). Donor cardiomyocytes (which appear yellow owing to the overlay of green EGFP and red rhod-2 fluorescence) are well aligned with and morphologically indistinguishable from EGFP-negative host cardiomyocytes (*panel A*). Cyclic variations in rhod-2 fluorescence, due to electrically evoked $[Ca^{2+}]_i$ transients, appear simultaneously in all cells along the scan line (including those with EGFP fluorescence).

The line-scan image (*panel B*) produced by stacking successive line scans along the white line in panel A demonstrates that rhod-2 fluorescence transients occur simultaneously in EGFP-expressing donor cardiomyocytes and host cardiomyocytes at stimulation rates of 2 and 4 Hz. Time plots of spatially averaged traces for the red and green fluorescence present in cells 1 and 2 were generated from the line-scan data (*panel C*). The frequency dependence of the kinetics of $[Ca^{2+}]_i$ decline in donor cardiomyocytes was indistinguishable from that in neighboring host cardiomyocytes (*panel D*), indicating that removal of calcium ions from the cytosol is highly coordinated between host and donor cardiomyocytes. These data constitute direct proof that donor cardiomyocytes form a functional syncytium with the host myocardium. Furthermore, donor cardiomyocytes possess functional attributes that are necessary to develop force in synchrony with their immediate neighbors. This analysis has also been employed to monitor functional coupling between host cardiomyocytes and myoblast-derived cells (37) as well as to monitor functional activity in newly replicated cardiomyocytes (R. Hassink, K.B. Pasumarthi, H. Nakajima, M. Rubart, M. Soonpaa, et al., submitted manuscript).

TPLSM-based imaging of $[Ca^{2+}]_i$ dynamics is suitable to determine the functional fate of transplanted donor cells of various origins, provided they can be identified following engraftment. Combined with methods to assess cardiac contractile function, for example cine-MRI (104) or left ventricular pressure-volume measurements (105, 106), this approach should answer conclusively the important question of whether the presence of functionally competent de novo cardiomyocytes correlates with increased cardiac function.

SUMMARY

The prospect of inducing bona fide cardiac regeneration via cell cycle activation, cell transplantation, or stem cell mobilization is greater now than at any previous time. Numerous genetic pathways and cell types appear to be able to affect repair in experimental animals. This enthusiasm, however, must be tempered by the realization that the heart is the archetypal nonregenerative organ. Thus, results obtained with newly identified genetic pathways and stem cells with apparent cardio-regenerative activity must be reconciled with the large body of data failing to demonstrate a substantive regenerative reserve in the adult heart (2). The paucity of primary cardiomyocyte tumors in humans (107) further underscores the reticence of this organ with regard to undergoing regenerative growth. It is thus imperative that stringent criteria be used for assessment of cell cycle activity, cardiomyogenic activity, and functional activity following regenerative interventions. The goal is to advocate the best approaches to both rapidly and definitively establish the potential utility of a given intervention.

ACKNOWLEDGMENTS

We thank the National Heart, Lung, and Blood Institute for support. We also thank our many colleagues working in the field and apologize in advance for any relevant views/studies that were not included in the present review owing to space limitations.

The *Annual Review of Physiology* is online at
http://physiol.annualreviews.org

LITERATURE CITED

1. Jessup M, Brozena S. 2003. Heart failure. *N. Engl. J. Med.* 348:2007–18
2. Rumiantsev PP. 1991. *Growth and Hyperplasia of Cardiac Muscle Cells.* London/ New York: Harwood Acad. Publ. 376 pp.
3. Pasumarthi KB, Field LJ. 2002. Cardiomyocyte cell cycle regulation. *Circ. Res.* 90:1044–54
4. Dowell JD, Field LJ, Pasumarthi KB. 2003. Cell cycle regulation to repair the infarcted myocardium. *Heart Fail. Rev.* 8:293–303
5. Dowell JD, Rubart M, Pasumarthi KB, Soonpaa MH, Field LJ. 2003. Myocyte and myogenic stem cell transplantation in the heart. *Cardiovasc. Res.* 58:336–50
6. Murry CE, Field LJ, Menasche P. 2005. Cell-based cardiac repair: Reflections at the 10-year point. *Circulation* 112:3174–83
7. Dimmeler S, Zeiher AM, Schneider MD. 2005. Unchain my heart: the scientific foundations of cardiac repair. *J. Clin. Invest.* 115:572–83
8. Soonpaa MH, Field LJ. 1998. Survey of studies examining mammalian cardiomyocyte DNA synthesis. *Circ. Res.* 83:15–26
9. Anversa P, Kajstura J. 1998. Ventricular myocytes are not terminally differentiated in the adult mammalian heart. *Circ. Res.* 83:1–14
10. Soonpaa MH, Field LJ. 1994. Assessment of cardiomyocyte DNA synthesis during hypertrophy in adult mice. *Am. J. Physiol.* 266:H1439–45
11. Soonpaa MH, Field LJ. 1997. Assessment of cardiomyocyte DNA synthesis in normal and injured adult mouse hearts. *Am. J. Physiol.* 272:H220–26
12. Pasumarthi KB, Nakajima H, Nakajima HO, Soonpaa MH, Field LJ. 2005. Targeted expression of cyclin D2 results in cardiomyocyte DNA synthesis and infarct

regression in transgenic mice. *Circ. Res.* 96:110–18

13. Field LJ. 1988. Atrial natriuretic factor-SV40 T antigen transgenes produce tumors and cardiac arrhythmias in mice. *Science* 239:1029–33

14. Behringer RR, Peschon JJ, Messing A, Gartside CL, Hauschka SD, et al. 1988. Heart and bone tumors in transgenic mice. *Proc. Natl. Acad. Sci. USA* 85:2648–52

15. Katz EB, Steinhelper ME, Delcarpio JB, Daud AI, Claycomb WC, Field LJ. 1992. Cardiomyocyte proliferation in mice expressing alpha-cardiac myosin heavy chain-SV40 T-antigen transgenes. *Am. J. Physiol.* 262:H1867–76

16. Field LJ. 2004. Modulation of the cardiomyocyte cell cycle in genetically altered animals. *Ann. NY Acad. Sci.* 1015:160–70

17. Soonpaa MH, Koh GY, Pajak L, Jing S, Wang H, et al. 1997. Cyclin D1 overexpression promotes cardiomyocyte DNA synthesis and multinucleation in transgenic mice. *J. Clin. Invest.* 99:2644–54

18. Liao HS, Kang PM, Nagashima H, Yamasaki N, Usheva A, et al. 2001. Cardiac-specific overexpression of cyclin-dependent kinase 2 increases smaller mononuclear cardiomyocytes. *Circ. Res.* 88:443–50

19. Pasumarthi KB, Nakajima H, Nakajima HO, Jing S, Field LJ. 2000. Enhanced cardiomyocyte DNA synthesis during myocardial hypertrophy in mice expressing a modified TSC2 transgene. *Circ. Res.* 86:1069–77

20. Nakajima H, Nakajima HO, Tsai SC, Field LJ. 2004. Expression of mutant p193 and p53 permits cardiomyocyte cell cycle reentry after myocardial infarction in transgenic mice. *Circ. Res.* 94:1606–14

21. Chaudhry HW, Dashoush NH, Tang H, Zhang L, Wang X, et al. 2004. Cyclin A2 mediates cardiomyocyte mitosis in the postmitotic myocardium. *J. Biol. Chem.* 279:35858–66

22. Reiss K, Cheng W, Ferber A, Kajstura

J, Li P, et al. 1996. Overexpression of insulin-like growth factor-1 in the heart is coupled with myocyte proliferation in transgenic mice. *Proc. Natl. Acad. Sci. USA* 93:8630–35

23. Limana F, Urbanek K, Chimenti S, Quaini F, Leri A, et al. 2002. bcl-2 overexpression promotes myocyte proliferation. *Proc. Natl. Acad. Sci. USA* 99:6257–62

24. Engel FB, Schebesta M, Duong MT, Lu G, Ren S, et al. 2005. p38 MAP kinase inhibition enables proliferation of adult mammalian cardiomyocytes. *Genes Dev.* 19:1175–87

25. MacLellan WR, Garcia A, Oh H, Frenkel P, Jordan MC, et al. 2005. Overlapping roles of pocket proteins in the myocardium are unmasked by germ line deletion of p130 plus heart-specific deletion of Rb. *Mol. Cell. Biol.* 25:2486–97

26. Tamamori-Adachi M, Ito H, Sumrejkanchanakij P, Adachi S, Hiroe M, et al. 2003. Critical role of cyclin D1 nuclear import in cardiomyocyte proliferation. *Circ. Res.* 92:e12–19

27. Xiao G, Mao S, Baumgarten G, Serrano J, Jordan MC, et al. 2001. Inducible activation of c-Myc in adult myocardium in vivo provokes cardiac myocyte hypertrophy and reactivation of DNA synthesis. *Circ. Res.* 89:1122–29

28. Li Q, Li B, Wang X, Leri A, Jana KP, et al. 1997. Overexpression of insulin-like growth factor-1 in mice protects from myocyte death after infarction, attenuating ventricular dilation, wall stress, and cardiac hypertrophy. *J. Clin. Invest.* 100:1991–99

29. Li B, Setoguchi M, Wang X, Andreoli AM, Leri A, et al. 1999. Insulin-like growth factor-1 attenuates the detrimental impact of nonocclusive coronary artery constriction on the heart. *Circ. Res.* 84:1007–19

30. Hassink RJ, Nakajima H, Nakajima HO, Brutel de la Riviere A, Doevendans PA, Field LJ. 2002. Antagonization of p193 and p53 pro-apoptotic pathways has

positive effects on cardiac function after myocardial infarction. *Circulation* 106:II 130 (Abstr.)

31. Deleted in proof

32. Koh GY, Klug MG, Soonpaa MH, Field LJ. 1993. Differentiation and long-term survival of C2C12 myoblast grafts in heart. *J. Clin. Invest.* 92:1548–54

33. Taylor DA, Atkins BZ, Hungspreugs P, Jones TR, Reedy MC, et al. 1998. Regenerating functional myocardium: improved performance after skeletal myoblast transplantation. *Nat. Med.* 4:929–33

34. Murry CE, Wiseman RW, Schwartz SM, Hauschka SD. 1996. Skeletal myoblast transplantation for repair of myocardial necrosis. *J. Clin. Invest.* 98:2512–23

35. Leobon B, Garcin I, Menasche P, Vilquin JT, Audinat E, Charpak S. 2003. Myoblasts transplanted into rat infarcted myocardium are functionally isolated from their host. *Proc. Natl. Acad. Sci. USA* 100:7808–11

36. Reinecke H, Minami E, Poppa V, Murry CE. 2004. Evidence for fusion between cardiac and skeletal muscle cells. *Circ. Res.* 94:e56–60

37. Rubart M, Soonpaa MH, Nakajima H, Field LJ. 2004. Spontaneous and evoked intracellular calcium transients in donor-derived myocytes following intracardiac myoblast transplantation. *J. Clin. Invest.* 114:775–83

38. Menasche P, Hagege AA, Vilquin JT, Desnos M, Abergel E, et al. 2003. Autologous skeletal myoblast transplantation for severe postinfarction left ventricular dysfunction. *J. Am. Coll. Cardiol.* 41:1078–83

39. Herreros J, Prosper F, Perez A, Gavira JJ, Garcia-Velloso MJ, et al. 2003. Autologous intramyocardial injection of cultured skeletal muscle-derived stem cells in patients with non-acute myocardial infarction. *Eur. Heart. J.* 24:2012–20

40. Pagani FD, DerSimonian H, Zawadzka A, Wetzel K, Edge AS, et al. 2003. Autologous skeletal myoblasts transplanted to ischemia-damaged myocardium in humans. Histological analysis of cell survival and differentiation. *J. Am. Coll. Cardiol.* 41:879–88

41. Smits PC, van Geuns RJ, Poldermans D, Bountioukos M, Onderwater EE, et al. 2003. Catheter-based intramyocardial injection of autologous skeletal myoblasts as a primary treatment of ischemic heart failure: Clinical experience with six-month follow-up. *J. Am. Coll. Cardiol.* 42:2063–69

42. Siminiak T, Fiszer D, Jerzykowska O, Grygielska B, Rozwadowska N, et al. 2005. Percutaneous trans-coronary-venous transplantation of autologous skeletal myoblasts in the treatment of post-infarction myocardial contractility impairment: The POZNAN trial. *Eur. Heart. J.* 26:1188–95

43. Siminiak T, Kalawski R, Fiszer D, Jerzykowska O, Rzezniczak J, et al. 2004. Autologous skeletal myoblast transplantation for the treatment of postinfarction myocardial injury: Phase I clinical study with 12 months of follow-up. *Am. Heart J.* 148:531–37

44. Hagege AA, Carrion C, Menasche P, Vilquin JT, Duboc D, et al. 2003. Viability and differentiation of autologous skeletal myoblast grafts in ischaemic cardiomyopathy. *Lancet* 361:491–92

45. Soonpaa MH, Koh GY, Klug MG, Field LJ. 1994. Formation of nascent intercalated disks between grafted fetal cardiomyocytes and host myocardium. *Science* 264:98–101

46. Leor J, Patterson M, Quinones MJ, Kedes LH, Kloner RA. 1996. Transplantation of fetal myocardial tissue into the infarcted myocardium of rat. A potential method for repair of infarcted myocardium? *Circulation* 94:II332–36

47. Koh GY, Soonpaa MH, Klug MG, Pride HP, Cooper BJ, et al. 1995. Stable fetal cardiomyocyte grafts in the hearts of dystrophic mice and dogs. *J. Clin. Invest.* 96:2034–42

48. Rubart M, Pasumarthi KB, Nakajima H, Soonpaa MH, Nakajima HO, Field LJ. 2003. Physiological coupling of donor and host cardiomyocytes after cellular transplantation. *Circ. Res.* 92:1217–24

49. Li RK, Jia ZQ, Weisel RD, Mickle DA, Zhang J, et al. 1996. Cardiomyocyte transplantation improves heart function. *Ann. Thorac. Surg.* 62:654–61

50. Klug MG, Soonpaa MH, Koh GY, Field LJ. 1996. Genetically selected cardiomyocytes from differentiating embronic stem cells form stable intracardiac grafts. *J. Clin. Invest.* 98:216–24

51. Zweigerdt R, Burg M, Willbold E, Abts H, Ruediger M. 2003. Generation of confluent cardiomyocyte monolayers derived from embryonic stem cells in suspension: a cell source for new therapies and screening strategies. *Cytotherapy* 5:399–413

52. Zandstra PW, Bauwens C, Yin T, Liu Q, Schiller H, et al. 2003. Scalable production of embryonic stem cell-derived cardiomyocytes. *Tissue Eng.* 9:767–78

53. Schroeder M, Niebruegge S, Werner A, Willbold E, Burg M, et al. 2005. Embryonic stem cell differentiation and lineage selection in a stirred bench scale bioreactor with automated process control. *Biotech. Bioeng.* 92:920–33

54. Laugwitz KL, Moretti A, Lam J, Gruber P, Chen Y, et al. 2005. Postnatal isl[1+] cardioblasts enter fully differentiated cardiomyocyte lineages. *Nature* 433:647–53

55. Shapiro AM, Lakey JR, Ryan EA, Korbutt GS, Toth E, et al. 2000. Islet transplantation in seven patients with type 1 diabetes mellitus using a glucocorticoid-free immunosuppressive regimen. *N. Engl. J. Med.* 343:230–38

56. Strom TB, Field LJ, Ruediger M. 2004. Allogeneic stem cell-derived "repair unit" therapy and the barriers to clinical deployment. *J. Am. Soc. Nephrol.* 15:1133–39

57. Orlic D, Kajstura J, Chimenti S, Jakoniuk I, Anderson SM, et al. 2001. Bone marrow cells regenerate infarcted myocardium. *Nature* 410:701–5

58. Balsam LB, Wagers AJ, Christensen JL, Kofidis T, Weissman IL, Robbins RC. 2004. Haematopoietic stem cells adopt mature haematopoietic fates in ischaemic myocardium. *Nature* 428:668–73

59. Murry CE, Soonpaa MH, Reinecke H, Nakajima H, Nakajima HO, et al. 2004. Haematopoietic stem cells do not transdifferentiate into cardiac myocytes in myocardial infarcts. *Nature* 428:664–68

60. Nygren JM, Jovinge S, Breitbach M, Sawen P, Roll W, et al. 2004. Bone marrow-derived hematopoietic cells generate cardiomyocytes at a low frequency through cell fusion, but not transdifferentiation. *Nat. Med.* 10:494–501

61. Wollert KC, Meyer GP, Lotz J, Ringes-Lichtenberg S, Lippolt P, et al. 2004. Intracoronary autologous bone-marrow cell transfer after myocardial infarction: the BOOST randomised controlled clinical trial. *Lancet* 364:141–48

62. Assmus B, Schachinger V, Teupe C, Britten M, Lehmann R, et al. 2002. Transplantation of progenitor cells and regeneration enhancement in acute myocardial infarction (TOPCARE-AMI). *Circulation* 106:3009–17

63. Strauer BE, Brehm M, Zeus T, Kostering M, Hernandez A, et al. 2002. Repair of infarcted myocardium by autologous intracoronary mononuclear bone marrow cell transplantation in humans. *Circulation* 106:1913–18

64. Chen SL, Fang WW, Ye F, Liu YH, Qian J, et al. 2004. Effect on left ventricular function of intracoronary transplantation of autologous bone marrow mesenchymal stem cell in patients with acute myocardial infarction. *Am. J. Cardiol.* 94:92–95

65. Perin EC, Dohmann HF, Borojevic R, Silva SA, Sousa AL, et al. 2003. Transendocardial, autologous bone marrow cell transplantation for severe, chronic ischemic heart failure. *Circulation* 107:2294–302

66. Tse HF, Kwong YL, Chan JK, Lo G, Ho CL, Lau CP. 2003. Angiogenesis in

ischaemic myocardium by intramyocardial autologous bone marrow mononuclear cell implantation. *Lancet* 361:47–49

67. Fuchs S, Satler LF, Kornowski R, Okubagzi P, Weisz G, et al. 2003. Catheter-based autologous bone marrow myocardial injection in no-option patients with advanced coronary artery disease: a feasibility study. *J. Am. Coll. Cardiol.* 41:1721–24

68. Mangi AA, Noiseux N, Kong D, He H, Rezvani M, et al. 2003. Mesenchymal stem cells modified with Akt prevent remodeling and restore performance of infarcted hearts. *Nat. Med.* 9:1195–201

69. Toma C, Pittenger MF, Cahill KS, Byrne BJ, Kessler PD. 2002. Human mesenchymal stem cells differentiate to a cardiomyocyte phenotype in the adult murine heart. *Circulation* 105:93–98

70. Shake JG, Gruber PJ, Baumgartner WA, Senechal G, Meyers J, et al. 2002. Mesenchymal stem cell implantation in a swine myocardial infarct model: engraftment and functional effects. *Ann. Thorac. Surg.* 73:1919–26

71. Gnecchi M, He H, Liang OD, Melo LG, Morello F, et al. 2005. Paracrine action accounts for marked protection of ischemic heart by Akt-modified mesenchymal stem cells. *Nat. Med.* 11:367–68

72. Kinnaird T, Stabile E, Burnett MS, Shou M, Lee CW, et al. 2004. Local delivery of marrow-derived stromal cells augments collateral perfusion through paracrine mechanisms. *Circulation* 109:1543–49

73. Badorff C, Brandes RP, Popp R, Rupp S, Urbich C, et al. 2003. Transdifferentiation of blood-derived human adult endothelial progenitor cells into functionally active cardiomyocytes. *Circulation* 107:1024–32

74. Bittner RE, Schofer C, Weipoltshammer K, Ivanova S, Streubel B, et al. 1999. Recruitment of bone-marrow-derived cells by skeletal and cardiac muscle in adult dystrophic mdx mice. *Anat. Embryol.* 199:391–96

75. Jackson KA, Majka SM, Wang H, Pocius J, Hartley CJ, et al. 2001. Regeneration of ischemic cardiac muscle and vascular endothelium by adult stem cells. *J. Clin. Invest.* 107:1395–402

76. Quaini F, Urbanek K, Beltrami AP, Finato N, Beltrami CA, et al. 2002. Chimerism of the transplanted heart. *N. Engl. J. Med.* 346:5–15

77. Laflamme MA, Myerson D, Saffitz JE, Murry CE. 2002. Evidence for cardiomyocyte repopulation by extracardiac progenitors in transplanted human hearts. *Circ. Res.* 90:634–40

78. Hruban RH, Long PP, Perlman EJ, Hutchins GM, Baumgartner WA, et al. 1993. Fluorescence in situ hybridization for the Y-chromosome can be used to detect cells of recipient origin in allografted hearts following cardiac transplantation. *Am. J. Pathol.* 142:975–80

79. Glaser R, Lu MM, Narula N, Epstein JA. 2002. Smooth muscle cells, but not myocytes, of host origin in transplanted human hearts. *Circulation* 106:17–19

80. Orlic D, Kajstura J, Chimenti S, Limana F, Jakoniuk I, et al. 2001. Mobilized bone marrow cells repair the infarcted heart, improving function and survival. *Proc. Natl. Acad. Sci. USA* 98:10344–49

81. Deten A, Volz HC, Clamors S, Leiblein S, Briest W, et al. 2005. Hematopoietic stem cells do not repair the infarcted mouse heart. *Cardiovasc. Res.* 65:52–63

82. Harada M, Qin Y, Takano H, Minamino T, Zou Y, et al. 2005. G-CSF prevents cardiac remodeling after myocardial infarction by activating the Jak-Stat pathway in cardiomyocytes. *Nat. Med.* 11:305–11

83. Maekawa Y, Anzai T, Yoshikawa T, Sugano Y, Mahara K, et al. 2004. Effect of granulocyte-macrophage colony-stimulating factor inducer on left ventricular remodeling after acute myocardial infarction. *J. Am. Coll. Cardiol.* 44:1510–20

84. Valgimigli M, Rigolin GM, Cittanti C, Malagutti P, Curello S, et al. 2005. Use

of granulocyte-colony stimulating factor during acute myocardial infarction to enhance bone marrow stem cell mobilization in humans: clinical and angiographic safety profile. *Eur. Heart. J.* In press

85. Wang Y, Tagil K, Ripa RS, Nilsson JC, Carstensen S, et al. 2005. Effect of mobilization of bone marrow stem cells by granulocyte colony stimulating factor on clinical symptoms, left ventricular perfusion and function in patients with severe chronic ischemic heart disease. *Int. J. Cardiol.* 100:477–83

86. Beltrami AP, Barlucchi L, Torella D, Baker M, Limana F, et al. 2003. Adult cardiac stem cells are multipotent and support myocardial regeneration. *Cell* 114:763–76

87. Dawn B, Stein AB, Urbanek K, Rota M, Whang B, et al. 2005. Cardiac stem cells delivered intravascularly traverse the vessel barrier, regenerate infarcted myocardium, and improve cardiac function. *Proc. Natl. Acad. Sci. USA* 102:3766–71

88. Oh H, Chi X, Bradfute SB, Mishina Y, Pocius J, et al. 2004. Cardiac muscle plasticity in adult and embryo by heart-derived progenitor cells. *Ann. NY Acad. Sci.* 1015:182–89

89. Matsuura K, Nagai T, Nishigaki N, Oyama T, Nishi J, et al. 2004. Adult cardiac SCA-1-positive cells differentiate into beating cardiomyocytes. *J. Biol. Chem.* 279:11384–91

90. Oh H, Bradfute SB, Gallardo TD, Nakamura T, Gaussin V, et al. 2003. Cardiac progenitor cells from adult myocardium: homing, differentiation, and fusion after infarction. *Proc. Natl. Acad. Sci. USA* 100:12313–18

91. Hierlihy AM, Seale P, Lobe CG, Rudnicki MA, Megeney LA. 2002. The post-natal heart contains a myocardial stem cell population. *FEBS Lett.* 530:239–43

92. Martin CM, Meeson AP, Robertson SM, Hawke TJ, Richardson JA, et al. 2004. Persistent expression of the ATP-binding cassette transporter, Abcg2, identifies cardiac SP cells in the developing and adult heart. *Dev. Biol.* 265:262–75

93. Dubowitz V, Sewry CA, Fitzsimons RB. 1985. *Muscle Biopsy: A Practical Approach.* London/Philadelphia: Baillière Tindall. 720 pp.

94. Kim RJ, Fieno DS, Parrish TB, Harris K, Chen EL, et al. 1999. Relationship of MRI delayed contrast enhancement to irreversible injury, infarct age, and contractile function. *Circulation* 100:1992–2002

95. Fieno DS, Kim RJ, Chen EL, Lomasney JW, Klocke FJ, Judd RM. 2000. Contrast-enhanced magnetic resonance imaging of myocardium at risk: distinction between reversible and irreversible injury throughout infarct healing. *J. Am. Coll. Cardiol.* 36:1985–91

96. Messina E, De Angelis L, Frati G, Morrone S, Chimenti S, et al. 2004. Isolation and expansion of adult cardiac stem cells from human and murine heart. *Circ. Res.* 95:911–21

97. Muller-Ehmsen J, Peterson KL, Kedes L, Whittaker P, Dow JS, et al. 2002. Rebuilding a damaged heart: long-term survival of transplanted neonatal rat cardiomyocytes after myocardial infarction and effect on cardiac function. *Circulation* 105:1720–26

98. Kehat I, Khimovich L, Caspi O, Gepstein A, Shofti R, et al. 2004. Electromechanical integration of cardiomyocytes derived from human embryonic stem cells. *Nat. Biotechnol.* 22:1282–89

99. Barbash IM, Chouraqui P, Baron J, Feinberg MS, Etzion S, et al. 2003. Systemic delivery of bone marrow-derived mesenchymal stem cells to the infarcted myocardium: feasibility, cell migration, and body distribution. *Circulation* 108:863–68

100. Alvarez-Dolado M, Pardal R, Garcia-Verdugo JM, Fike JR, Lee HO, et al. 2003. Fusion of bone-marrow-derived cells with Purkinje neurons, cardiomyocytes and hepatocytes. *Nature* 425:968–73

101. Rubart M, Wang E, Dunn KW, Field LJ. 2003. Two-photon molecular excitation imaging of Ca^{2+} transients in Langendorff-perfused mouse hearts. *Am. J. Physiol. Cell Physiol.* 284:C1654–68

102. Centonze VE, White JG. 1998. Multiphoton excitation provides optical sections from deeper within scattering specimens than confocal imaging. *Biophys. J.* 75:2015–24

103. Deleted in proof

104. Wiesmann F, Ruff J, Engelhardt S, Hein L, Dienesch C, et al. 2001. Dobutamine-stress magnetic resonance microimaging in mice: acute changes of cardiac geometry and function in normal and failing murine hearts. *Circ. Res.* 88:563–69

105. Feldman MD, Mao Y, Valvano JW, Pearce JA, Freeman GL. 2000. Development of a multifrequency conductance catheter-based system to determine LV function in mice. *Am. J. Physiol. Heart Circ. Physiol.* 279:H1411–20

106. Georgakopoulos D, Kass DA. 2000. Estimation of parallel conductance by dual-frequency conductance catheter in mice. *Am. J. Physiol. Heart Circ. Physiol.* 279:H443–50

107. Butany J, Nair V, Naseemuddin A, Nair GM, Catton C, Yau T. 2005. Cardiac tumours: diagnosis and management. *Lancet Oncol.* 6:219–28

Annu. Rev. Physiol. 2006. 68:51–66
doi: 10.1146/annurev.physiol.68.040104.124629
First published online as a Review in Advance on September 28, 2005

ENDOTHELIAL-CARDIOMYOCYTE INTERACTIONS IN CARDIAC DEVELOPMENT AND REPAIR

Patrick C.H. Hsieh, Michael E. Davis, Laura K. Lisowski, and Richard T. Lee

Cardiovascular Division, Department of Medicine, Brigham and Women's Hospital, Harvard Medical School, Boston, Massachusetts 02115; email: phsieh@rics.bwh.harvard.edu, mdavis@rics.bwh.harvard.edu, lkl7@cornell.edu, rlee@rics.bwh.harvard.edu

Key Words   cell-cell interaction, regeneration, endothelial cell, heart failure

■ **Abstract**   Communication between endothelial cells and cardiomyocytes regulates not only early cardiac development but also adult cardiomyocyte function, including the contractile state. In the normal mammalian myocardium, each cardiomyocyte is surrounded by an intricate network of capillaries and is next to endothelial cells. Cardiomyocytes depend on endothelial cells not only for oxygenated blood supply but also for local protective signals that promote cardiomyocyte organization and survival. While endothelial cells direct cardiomyocytes, cardiomyocytes reciprocally secrete factors that impact endothelial cell function. Understanding how endothelial cells communicate with cardiomyocytes will be critical for cardiac regeneration, in which the ultimate goal is not simply to improve systolic function transiently but to establish new myocardium that is both structurally and functionally normal in the long term.

INTRODUCTION

Modern vascular biology has dispensed with the notion that endothelial cells provide a mere "passive lining" supportive role. We now recognize endothelial cells as dynamic regulators of vascular tone and growth of nearby cells. In fact, endothelial cells can participate in the development of many tissues, including the liver and pancreas (1). The myocardium is yet another example of a tissue in which many functions and responses have a surprising dependence on endothelial cells.

In the normal adult mammalian myocardium, there is at least one capillary next to every cardiomyocyte (Figure 1). The mass of cardiomyocytes in a mammalian heart is approximately 25 times that of cardiac endothelial cells, although the smaller endothelial cells outnumber cardiomyocytes by ~3:1 (2). Because diffusion of a given factor is inversely proportional to the square of distance the factor must travel, a minimal endothelial-cardiomyocyte cell-to-cell distance is crucial for cardiomyocytes to obtain oxygen and nutrients. Cardiomyocytes are typically

0066-4278/06/0315-0051$20.00
51

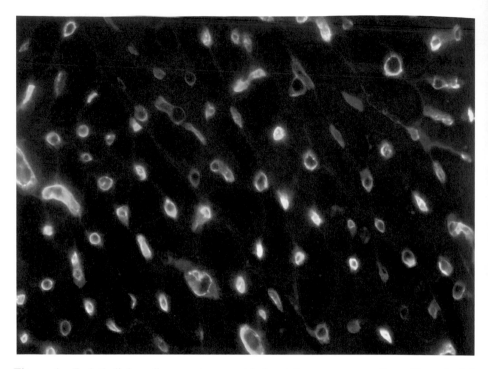

Figure 1   Endothelial-cardiomyocyte assembly in adult mouse myocardium. Normal adult mouse myocardium is stained with intravital perfusion techniques to demonstrate cardiomyocyte (outlined in *red*) and capillary (*green*; stained with isolectin-fluorescein) assembly. Nuclei are blue (Hoechst). Original magnification: 600X.

~10–100 μm in size, and the intercapillary distance in mammalian myocardium is ~15–50 μm (2, 3). This intricate anatomical arrangement of cardiomyocytes within the capillary network not only allows for physiological transport but also for local communication between endothelial cells and cardiomyocytes.

The ongoing molecular conversation between endothelial cells and cardiomyocytes is highly relevant to the recent excitement in promoting cardiac regeneration. The ultimate goal of myocardial regeneration is to rebuild a functional tissue that closely resembles mature myocardium, not just to improve systolic function transiently. Thus, regenerating myocardium will require rebuilding the vascular network along with the cardiomyocyte architecture. Here we review evidence demonstrating crucial molecular interactions between endothelial cells and cardiomyocytes. We first discuss endothelial-cardiomyocyte interactions during embryonic cardiogenesis, followed with morphological and functional characteristics of endothelial-cardiomyocyte interactions in mature myocardium. Finally, we consider strategies exploiting endothelial-cardiomyocyte interplay for cardiac regeneration.

ENDOTHELIAL-CARDIOMYOCYTE INTERACTIONS IN CARDIAC DEVELOPMENT

In mice, endoderm-derived growth factors from the primitive cardiac crescent (~E7.5) govern the formation of cardiac mesoderm, primary myocardium, and endocardium (4). Signals among endocardial cells, myocardial cells, and neural crest–derived cells control the endocardial endothelial cells undergoing the epithelial-to-mesenchymal transformation to form cardiac cushions and subsequent cardiac septa, valves, and the outflow tract (~E10.5–12.5) (5,6). Endocardial and myocardial cells may arise from the same cardiac mesodermal precursors and may thus share some of their specific lineage markers in early development (7). With formation of the primitive cardiac tube (~E8.5), endocardial and myocardial cells become separated by a dense layer of extracellular matrix, the cardiac jelly (5, 6). However, these two distinct cardiac cell layers continuously and reciprocally communicate through paracrine signaling during further development. The interdependence of endothelial and myocardial cells for cardiac development has been demonstrated by cell-restricted gene inactivation experiments (Figure 2).

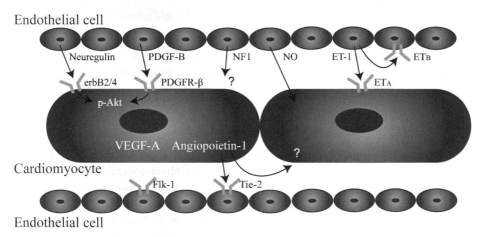

Figure 2 Endothelial-cardiomyocyte interactions through autocrine and paracrine signaling. Endothelial cells may secret signaling mediators modulating cardiomyocyte development (neuregulin, PDGF-B, and NF1), survival (neuregulin and PDGF-B), and contraction (NO and ET-1). Reciprocally, cardiomyocytes may also promote endothelial cell survival and assembly through VEGF-A and angiopoietin-1. Even more complicated is combined autocrine and paracrine signaling between endothelial cells and cardiomyocytes through ET-1 and angiopoietin-1. PDGF-B, platelet-derived growth factor-B; PDGFR-β, PDGF receptor-β; NF1, neurofibromatosis type 1; NO, nitric oxide; ET-1, endothelin-1; VEGF-A, vascular endothelial growth factor-A.

Signaling from Endothelial Cells to Cardiomyocytes in Development

NEUREGULIN-erbB   One well-characterized example of endothelial-cardiomyocyte interactions during early cardiogenesis is the neuregulin-erbB signaling pathway. Neuregulins are members of the epidermal growth factor family that are ligands for receptor tyrosine kinases of the erbB family. There are four neuregulins: neuregulin-1, -2, -3, and -4. During early cardiogenesis, endocardial endothelial cells produce neuregulin-1, and primary cardiomyocytes express the receptors erbB2 and erbB4. Paracrine signaling of the neuregulin-erbB2/4 system from endocardium to myocardium is essential for the formation of myocardial trabeculae and cardiac cushions (8). Mice lacking neuregulin or its receptors, erbB2 or erbB4, die in mid-embryogenesis because they lack myocardial trabeculae (9–11). Although the process of trabeculation is not well understood at the molecular level, cardiomyocyte proliferation and maturation are required. Neuregulin-1 can promote the proliferation, survival, and hypertrophic growth of cultured neonatal cardiomyocytes (12), and this may explain why this growth factor is essential for normal trabeculation.

NEUROFIBROMATOSIS TYPE 1   Another factor that can mediate endothelial-cardiomyocyte interactions during early cardiac development is neurofibromatosis type 1 (NF1). *NF1* is a tumor-suppressor gene that downregulates Ras activity, and mutations in *NF1* in humans may cause cardiovascular defects (13). In mice, conditional inactivation of *NF1* in endothelial cells using the Cre-lox system causes cardiac developmental defects in the myocardium and endocardial cushions (14). In addition to causing myocardial thinning, endothelial-specific inactivation of NF1 leads to enlarged endocardial cushions, ventricular septal defects, and double-outlet right ventricle. Similar defects are present in *Nf1$^{-/-}$* mice, but not in mice with cardiomyocyte-specific *NF1* inactivation (14), demonstrating that NF1 signaling from endothelial cells causes defects in myocardial development.

PLATELET-DERIVED GROWTH FACTOR-B   Platelet-derived growth factor (PDGF)-B and PDGF receptor-β are essential for mammalian cardiovascular development (15). During vasculogenesis, endothelial-derived PDGF-B recruits nearby pericytes to the nascent vascular structure (15); similarly, endothelial-derived PDGF-B may provide key signals to cardiomyocytes during development. Complete deletion of PDGF-B is lethal, whereas endothelial-restricted deletion of PDGF-B causes defects in the developing myocardium as well as vascular and glomerular abnormalities (16). Such cardiac defects include thinned myocardium, chamber dilation, hypertrabeculation, and septal abnormalities. Interestingly, thinning of the myocardium is no longer apparent at one month postnatally with endothelial-restricted deletion of PDGF-B. The Cre-lox endothelial-specific ablation method is only 60–90% efficient, which suggests either that postnatal PDGF-B from the minority of cells that did not delete the PDGF-B gene is sufficient to allow cardiomyocyte growth to compensate postnatally or that other growth factors drive this compensation (16).

Signaling from Cardiomyocytes to Endothelial Cells in Development

The examples of neuregulin-1, NF1, and PDGF-B demonstrate that signals from endothelial cells regulate the formation of primary myocardium. Similarly, signaling from myocardial cells to endothelial cells is also required for cardiac development. Two examples of myocardial-to-endothelial signaling are vascular endothelial growth factor (VEGF)-A and angiopoietin-1.

VASCULAR ENDOTHELIAL GROWTH FACTOR-A   VEGF-A is a key regulator of angiogenesis during embryogenesis. In mice, a mutation in VEGF-A causes endocardial detachment from an underdeveloped myocardium (17). A mutation in VEGF receptor-2 (or Flk-1) also results in failure of the endocardium and myocardium to develop (18). Furthermore, cardiomyocyte-specific deletion of VEGF-A results in defects in vasculogenesis/angiogenesis and a thinned ventricular wall (19), further confirming reciprocal signaling from the myocardial cell to the endothelial cell during cardiac development. Interestingly, this cardiomyocyte-selective VEGF-A-deletion mouse has underdeveloped myocardial microvasculature but preserved coronary artery structure, implying a different signaling mechanism for vasculogenesis/angiogenesis in the myocardium and in the epicardial coronary arteries.

Cardiomyocyte-derived VEGF-A also inhibits cardiac endocardial-to-mesenchymal transformation. This process is essential in the formation of the cardiac cushions and requires delicate control of VEGF-A concentration (20–22). A minimal amount of VEGF initiates endocardial-to-mesenchymal transformation, whereas higher doses of VEGF-A terminate this transformation (23). Interestingly, this cardiomyocyte-derived VEGF-A signaling for endocardial-to-mesenchymal transformation may be controlled by an endothelial-derived feedback mechanism through the calcineurin/NFAT pathway (24), demonstrating the importance of endothelial-cardiomyocyte interactions for cardiac morphogenesis.

ANGIOPOIETIN-1   Another mechanism of cardiomyocyte control of endothelial cells during cardiac development is the angiopoietin-Tie-2 system. Both angiopoietin-1 and angiopoietin-2 may bind to Tie-2 receptors in a competitive manner, but with opposite effects: Angiopoietin-1 activates the Tie-2 receptor and prevents vascular edema, whereas angiopoietin-2 blocks Tie-2 phosphorylation and increases vascular permeability. During angiogenesis/vasculogenesis, angiopoietin-1 is produced primarily by pericytes, and Tie-2 receptors are expressed on endothelial cells. Angiopoietin-1 regulates the stabilization and maturation of neovasculature; genetic deletion of angiopoietin-1 or Tie-2 causes a defect in early vasculogenesis/angiogenesis and is lethal (25, 26). Because in the early embryo the cardiac endocardium is one of the earliest vascular components (along with the dorsal aorta and yolk sac vessels) and the adult heart can be regarded as a fully vascularized organ, angiopoietin-Tie-2 signaling may also be required for early cardiac development. Indeed, mice with mutations in Tie-2 have underdeveloped endocardium and myocardium (27). These Tie-2 knockout mice display defects in the endocardium but have normal vascular morphology at E10.5, suggesting that the endocardial defect is the fundamental cause of death. In addition, a recent study showed that overexpression, and not deletion, of angiopoietin-1 from cardiomyocytes caused embryonic death between E12.5–15.5 due to cardiac hemorrhage (28). The mice had defects in the endocardium and myocardium and lack of coronary arteries, suggesting that, as with VEGF-A, a delicate control of angiopoietin-1 concentration is critical for early heart development.

ENDOTHELIAL-CARDIOMYOCYTE INTERACTIONS IN NORMAL CARDIAC FUNCTION

Cardiac Endothelial Cells Regulate Cardiomyocyte Contraction

The vascular endothelium senses the shear stress of flowing blood and regulates vascular smooth muscle contraction. It is therefore not surprising that cardiac endothelial cells—the endocardial endothelial cells as well as the endothelial cells of intramyocardial capillaries—regulate the contractile state of cardiomyocytes. Autocrine and paracrine signaling molecules released or activated by cardiac endothelial cells are responsible for this contractile response (Figure 2).

NITRIC OXIDE   Three different nitric oxide synthase isoenzymes synthesize nitric oxide (NO) from L-arginine. The neuronal and endothelial NO synthases (nNOS and eNOS, respectively) are expressed in normal physiological conditions, whereas the inducible NO synthase is induced by stress or cytokines. Like NO in the vessel, which causes relaxation of vascular smooth muscle, NO in the heart affects the onset of ventricular relaxation, which allows for a precise optimization of pump function beat by beat. Although NO is principally a paracrine effector secreted by cardiac endothelial cells, cardiomyocytes also express both nNOS and eNOS. Endothelial expression of eNOS exceeds that in cardiomyocytes by greater than 4:1 (29). Cardiomyocyte autocrine eNOS signaling can regulate β-adrenergic and muscarinic control of contractile state (30).

Barouch et al. (31) demonstrated that cardiomyocyte nNOS and eNOS may have opposing effects on cardiac structure and function. Using mice with nNOS or eNOS deficiency, they found that nNOS and eNOS have not only different localization in cardiomyocytes but also opposite effects on cardiomyocyte contractility; eNOS localizes to caveolae and inhibits L-type Ca^{2+} channels, leading to negative inotropy, whereas nNOS is targeted to the sarcoplasmic reticulum and facilitates Ca^{2+} release and thus positive inotropy (31). These results demonstrate that spatial confinement of different NO synthase isoforms contribute independently to the maintenance of cardiomyocyte structure and phenotype.

As indicated above, mutation of neuregulin or either of two of its cognate receptors, erbB2 and erbB4, causes embryonic death during mid-embryogenesis due to aborted development of myocardial trabeculation (9–11). Neuregulin also appears to play a role in fully developed myocardium. In adult mice, cardiomyocyte-specific deletion of erbB2 leads to dilated cardiomyopathy (32, 33). Neuregulin from endothelial cells may induce a negative inotropic effect in isolated rabbit papillary muscles (34). This suggests that, along with NO, the neuregulin signaling pathway acts as an endothelial-derived regulator of cardiac inotropism. In fact, the negative inotropic effect of neuregulin may require NO synthase because L-NMMA, an inhibitor of NO synthase, significantly attenuates the negative inotropy of neuregulin (34).

ENDOTHELIN-1   Endothelin (ET)-1 is a 21-amino acid peptide produced and released by the endothelial cells with a direct vasoconstrictive effect. In the myocardium, endothelin-1 may act in both autocrine and paracrine manners; it binds to ET_B receptors on cardiac endothelial cells and ET_A receptors on cardiomyocytes. When endothelin-1 binds to ET_B receptors, it results in the release of other signaling molecules (NO and prostaglandin I_2) rather than vasoconstriction. By contrast, when binding to the ET_A receptors on cardiomyocytes, endothelin-1 causes cardiomyocyte constriction, as observed when vascular smooth muscle cells are treated with endothelin-1 (35). These results imply that there may be a feedback mechanism between cardiac endothelial cells and cardiomyocytes for control of cardiomyocyte constriction through the endothelin-1 system.

There is now considerable evidence showing a role of the endothelin-1 system in the pathogenesis and progression of heart failure. Patients with heart failure exhibit increased plasma endothelin-1 levels and expression of myocardial endothelin receptors, and the extent of the increases correlate with disease severity (36). Indeed, it has been hypothesized that endothelin receptor antagonists may be a therapeutic approach for heart failure. In a double-blinded study, darusentan, an ET_A-receptor antagonist, improved cardiac index in patients with heart failure (37). However, in other clinical trials endothelin antagonists resulted in no benefit to or worsening of patient condition (38, 39).

ENDOTHELIAL-CARDIOMYOCYTE INTERACTIONS AND REGENERATION

Regeneration Occurs in Myocardium But Is Inadequate for Functional Improvement

In some nonmammalian higher organisms like amphibians and zebrafish, regeneration of the myocardium occurs after experimental myocardial injury (40, 41). In humans, however, cardiomyocytes that die following ischemia are not adequately replaced, resulting in loss of ventricular function (42). Recent evidence indicates that the adult heart is capable of some limited regeneration (43) and that adult mouse cardiomyocytes may be able to proliferate under specific conditions (44). The origins of regenerating cardiomyocytes are controversial. Studies have shown the presence of Y chromosome–positive bone marrow–derived cardiomyocytes in adult female hearts (45), indicating the possibility that bone marrow–derived precursors differentiate into cardiomyocytes or fuse with existing cardiomyocytes. Other studies suggest that resident cardiac stem cells are the source of regenerating myocardium (43, 46–48).

In part owing to the lack of understanding of how myocardium regenerates, many different cell types have been injected or infused into injured myocardium in an attempt to stimulate myocardial regeneration. Surprisingly, several different cell types have demonstrated the potential for benefit, including skeletal myoblasts, adult and neonatal cardiomyocytes, bone marrow stem cells, and embryonic stem cells (49). In some circumstances, there is poor survival of injected cells or incomplete differentiation of the implanted cells, and in others, lack of integration into the host myocardium (50). Bone marrow–derived progenitor cells have been shown to assist in vascularization of the myocardium following infarction and to improve cardiomyocyte survival (51, 52), although their capacity to transdifferentiate into cardiomyocytes is probably very limited (53).

Why does experimental cell therapy appear to benefit left ventricular function, even though most cells do not survive and even fewer appear to integrate functionally into the myocardium? There are many potential explanations (Figure 3), but one plausible theory is that injected cells stimulate angiogenesis, either as living cells or as they die. As discussed further below, endothelial cells can promote the survival of cardiomyocytes. Many types of injected cells, including endothelial progenitor cells, are rich sources of angiogenic factors (54). The potential for cell therapy to be an expensive method of introducing angiogenic factors suggests that similar benefits may be derived from injection of individual angiogenic factors alone, although the success of this approach to date has been limited (55).

Endothelial Cells Promote Cardiomyocyte Survival

The theory that angiogenic stimulation can improve systolic function is supported by recent studies demonstrating that endothelial cells can promote cardiomyocyte survival. Kuramochi et al. (56) tested the hypothesis that reactive oxygen species regulate neuregulin-erbB signaling. They found that neuregulin is a prosurvival factor for cardiomyocytes via the phosphatidylinositol-3-kinase-Akt pathway. Cardiac microvascular endothelial cells express neuregulin, and recombinant neuregulin-1β protects cardiomyocytes against anthracycline-, β-adrenergic receptor–, and H_2O_2-induced cardiomyocyte death (56–58). H_2O_2 induces neuregulin-1β release from endothelial cells in a concentration-dependent

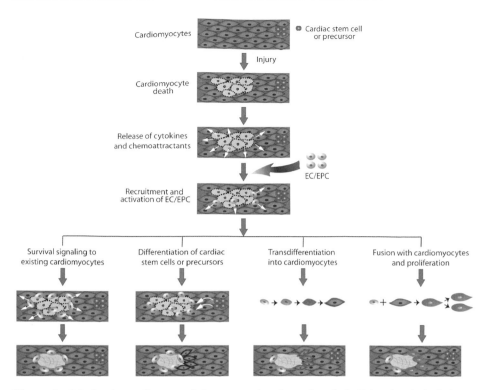

Figure 3 Mechanisms of myocardial regeneration through endothelial and endothelial progenitor cell therapies. After injury to a site, cardiomyocytes undergo apoptosis and necrosis and release cytokines and chemoattractants to recruit endothelial cells (EC) and endothelial progenitor cells (EPC) to the site. EC/EPC are activated and may promote myocardial regeneration through different mechanisms, including (*a*) releasing cardiomyocyte survival factors to protect fragile cardiomyocytes, (*b*) promoting differentiation of resident cardiac stem cells or progenitors for population into the injured area, (*c*) undergoing transdifferentiation into cardiomyocytes to replace dead cells, and (*d*) fusing with cardiomyocytes to facilitate cell proliferation for repair.

manner and conditioned medium from the cells activates erbB4 signaling in cardiomyocytes via paracrine mechanisms (56). Cocultured cardiac endothelial cells protect H_2O_2-induced cardiomyocyte apoptosis through neuregulin-erbB4 signaling.

Cardiomyocytes are also protected from apoptosis by endothelial cells in three-dimensional culture (59). In this setting, there is no blood flow to consider, and the effects of endothelial cells are thus due to secreted factors or to cell-cell contact. Preliminary data from our laboratory suggest that endothelial cells, when in contact with cardiomyocytes, secrete PDGF-B, which protects cardiomyocytes through phosphatidylinositol-3-kinase-Akt signaling (P.C.H. Hsieh, unpublished data). Interestingly, PDGF mediates cardiac microvascular endothelial cell hemostatic and angiogenic activity (60), and injection of PDGF may decrease the extent of myocardial infarction after coronary occlusion in young, but not in old, rat hearts (61, 62).

In addition to neuregulin and PDGF-B, another candidate endothelial protein mediating cardiomyocyte survival is angiopoietin-1. As described above, mice with genetic deletion of angiopoietin-1 die during development because of cardiac defects (26). These defects are thought to be due to impairment in endothelial functions such as vascular maturation and endocardial formation. Angiopoietin-1 not only promotes endothelial cell survival but can also improve cell survival in skeletal myocytes and cardiomyocytes (63). Furthermore, adenoviral overexpression of angiopoetin-1 reduces infarct size in experimental myocardial infarction (64).

Endothelial Cells Guide Cardiomyocyte Organization

During vasculogenesis, endothelial cells recruit mural cells to the outside of the growing tube, where they adhere and differentiate to form the media of the new vessel. Endothelial cells can also recruit cardiomyocytes in a similar fashion. When cultured in a three-dimensional scaffold, endothelial cells form typical vascular networks with capillary-like tubes, whereas primary cardiomyocytes form small islands of cells that die. When endothelial cells and cardiomyocytes are cultured together, however, the endothelial cells form tube-like structures and the cardiomyocytes position themselves on the outside of these tubes (59). These data suggest that an endothelial factor directs the assembly of cardiomyocytes on the capillary-like endothelial tube. Furthermore, endothelial cells promote the synthesis of Connexin43, a principal gap junction protein of cardiomyocytes. In the presence of endothelial cells, cardiomyocyte contraction is more coordinated, suggesting that the physiological coupling of cardiomyocytes is, in part, endothelial dependent. These data raise the intriguing hypothesis that achieving proper capillary-cardiomyocyte architecture in cardiac repair may provide electrical stability in addition to mechanical functional improvement.

Endothelial Cells Transdifferentiatate into Cardiomyocytes

Whereas endothelial cells may direct differentiation, survival, and organization of cardiomyocytes, some endothelial progenitor cells may have the capacity to transdifferentiate into cardiomyocytes. Condorelli et al. (65) reported that various types of endothelial cells are capable of expressing putative cardiomyocyte markers when cultured with cardiomyocytes. Surprisingly, this transdifferentiation may occur in both adult committed endothelial cells from different tissues as well as in putative endothelial progenitor cells (65–67). Transdifferentiation may require direct cell-cell contact between the endothelial progenitor cell and the cardiomyocyte (66, 68). Because in vivo studies suggest that transdifferentiation into cardiomyocytes is very rare (53), many experimental studies may show transdifferentiation because of cell fusion, which can explain the apparent phenotypic change. Indeed, injection of bone marrow–derived cells into the myocardium can cause fusion with cardiomyocytes without transdifferentiation into cardiomyocytes (69). Interestingly, fused cardiomyocytes express cardiac troponin-I and intercalate with neighboring cardiomyocytes with mature gap junction formation, implying a true potential mechanism for cardiac regeneration. Moreover, a recent study showed that endothelial and endothelial progenitor cells can fuse with cardiomyocytes in vitro and in vivo (70). The endothelial-cardiomyocyte fused cells express predominantly cardiomyocyte markers and, surprisingly, also express Ki67, phosphohistone H3, and cyclinB1 and reenter the G_2/M cell cycle. Although direct cardiomyocyte division was not observed after fusion, this study suggests that augmented cell fusion with endothelial cells may rescue injured cardiomyocytes.

Clinical Approaches Using Endothelial Cells for Cardiac Regeneration

To date, most studies using endothelial cells have aimed to enhance the endogenous postinfarct angiogenic response with growth factors or to improve angiogenesis with endothelial progenitor cell therapy. It is worth noting that injured myocardium is itself angiogenic initially and that native cardiomyocytes respond to hypoxia and other stress signals with the release of angiogenic factors, including VEGF (71). In humans, angiogenic factors are increased in the plasma and the myocardium after acute myocardial ischemia and infarction (72, 73). Thus, with therapeutic angiogenesis it generally is assumed that increasing angiogenic stimuli beyond the normal response to injury is beneficial. This is not always a good assumption, however, because excessive delivery of angiogenic factors can lead to aberrant vasculature and hemangiomas (74).

Ex vivo–expanded endothelial progenitor cells have shown promise in animal studies, improving cardiac function following myocardial infarction (51, 75, 76). This has led to several clinical trials including the TOPCARE-AMI (77) and MAGIC (78) trials. These studies used either endothelial progenitors or peripheral blood mononuclear cells for repair of the damaged myocardium (TOPCARE-AMI) or the progenitor-mobilizing factor granulocyte colony–stimulating factor (in MAGIC). The use of granulocyte colony–stimulating factor was associated with early restenosis within stented coronary arteries, suggesting that systemic mobilization of endothelial precursors may have adverse effects. It is far too early to draw conclusions from myocardial angiogenesis and endothelial cell therapy trials. It is possible that early attempts to overdrive angiogenesis will fail, whereas carefully timed delivery of angiogenic factors or endothelial precursors can successfully protect cardiomyocytes or even promote regeneration and functional integration of new cardiomyocytes. Future studies that define precisely how endothelial cells protect and stimulate cardiomyocytes may allow simple and practical ways to prevent heart failure.

CONCLUSIONS

Studies to date indicate that cardiac regeneration in mammals may be feasible, but the response is inadequate to preserve myocardial function after a substantial injury. Thus, understanding how normal myocardial structure can be regenerated in adult hearts is essential. It is clear that endothelial cells play a role in cardiac morphogenesis and most likely also in survival and function of mature cardiomyocytes. Initial attempts to promote angiogenesis in myocardium were based on the premise that persistent ischemia could be alleviated. However, it is also possible that endothelial-cardiomyocyte interactions are essential in normal cardiomyocyte function and for protection from injury. Understanding the molecular and cellular mechanisms controlling these cell-cell interactions will not only enhance our understanding of the establishment of vascular network in the heart but also allow the development of new targeted therapies for cardiac regeneration by improving cardiomyocyte survival and maturation.

The *Annual Review of Physiology* is online at
http://physiol.annualreviews.org

LITERATURE CITED

1. Cleaver O, Melton DA. 2003. Endothelial signaling during development. *Nat. Med.* 9: 661–68
2. Brutsaert DL. 2003. Cardiac endothelial-myocardial signaling: its role in cardiac growth, contractile performance, and rhythmicity. *Physiol. Rev.* 83:59–115
3. Korecky B, Hai CM, Rakusan K. 1982. Functional capillary density in normal and transplanted rat hearts. *Can. J. Physiol. Pharmacol.* 60:23–32
4. Lough J, Sugi Y. 2000. Endoderm and heart development. *Dev. Dyn.* 217:327–42
5. Harvey RP. 2002. Patterning the vertebrate heart. *Nat. Rev. Genet.* 3:544–56
6. Moorman AF, Christoffels VM. 2003. Cardiac chamber formation: development, genes, and evolution. *Physiol. Rev.* 83: 1223–67
7. Linask KK, Lash JW. 1993. Early heart development: dynamics of endocardial cell sorting suggests a common origin with cardiomyocytes. *Dev. Dyn.* 196:62–69
8. Marchionni MA. 1995. Cell-cell signalling: neu tack on neuregulin. *Nature* 378:334–35
9. Meyer D, Birchmeier C. 1995. Multiple essential functions of neuregulin in development. *Nature* 378:386–90
10. Gassmann M, Casagranda F, Orioli D, Simon H, Lai C, et al. 1995. Aberrant neural and cardiac development in mice lacking the ErbB4 neuregulin receptor. *Nature* 378:390–94
11. Lee KF, Simon H, Chen H, Bates B, Hung MC, et al. 1995. Requirement for neuregulin receptor erbB2 in neural and cardiac development. *Nature* 378:394–98
12. Zhao YY, Sawyer DR, Baliga RR, Opel DJ, Han X, et al. 1998. Neuregulins promote survival and growth of cardiac myocytes. Persistence of ErbB2 and ErbB4 expression in neonatal and adult ventricular myocytes. *J. Biol. Chem.* 273:10261–69
13. Lin AE, Birch PH, Korf BR, Tenconi R, Niimura M, et al. 2000. Cardiovascular malformations and other cardiovascular abnormalities in neurofibromatosis 1. *Am. J. Med. Genet.* 95:108–17

14. Gitler AD, Zhu Y, Ismat FA, Lu MM, Yamauchi Y, et al. 2003. Nf1 has an essential role in endothelial cells. *Nat. Genet.* 33:75–79

15. Hoch RV, Soriano P. 2003. Roles of PDGF in animal development. *Development* 130:4769–84

16. Bjarnegard M, Enge M, Norlin J, Gustafsdottir S, Fredriksson S, et al. 2004. Endothelium-specific ablation of PDGFB leads to pericyte loss and glomerular, cardiac and placental abnormalities. *Development* 131:1847–57

17. Haigh JJ, Gerber HP, Ferrara N, Wagner EF. 2000. Conditional inactivation of VEGF-A in areas of collagen2a1 expression results in embryonic lethality in the heterozygous state. *Development* 127:1445–53

18. Shalaby F, Rossant J, Yamaguchi TP, Gertsenstein M, Wu XF, et al. 1995. Failure of blood-island formation and vasculogenesis in Flk-1-deficient mice. *Nature* 376:62–66

19. Giordano FJ, Gerber HP, Williams SP, VanBruggen N, Bunting S, et al. 2001. A cardiac myocyte vascular endothelial growth factor paracrine pathway is required to maintain cardiac function. *Proc. Natl. Acad. Sci. USA* 98:5780–85

20. Miquerol L, Langille BL, Nagy A. 2000. Embryonic development is disrupted by modest increases in vascular endothelial growth factor gene expression. *Development* 127:3941–46

21. Dor Y, Camenisch TD, Itin A, Fishman GI, McDonald JA, et al. 2001. A novel role for VEGF in endocardial cushion formation and its potential contribution to congenital heart defects. *Development* 128:1531–38

22. Dor Y, Klewer SE, McDonald JA, Keshet E, Camenisch TD. 2003. VEGF modulates early heart valve formation. *Anat. Rec.* 271:202–8

23. Lambrechts D, Carmeliet P. 2004. Sculpting heart valves with NFATc and VEGF. *Cell* 118:532–34

24. Chang CP, Neilson JR, Bayle JH, Gestwicki JE, Kuo A, et al. 2004. A field of myocardial-endocardial NFAT signaling underlies heart valve morphogenesis. *Cell* 118:649–63

25. Sato TN, Tozawa Y, Deutsch U, Wolburg-Buchholz K, Fujiwara Y, et al. 1995. Distinct roles of the receptor tyrosine kinases Tie-1 and Tie-2 in blood vessel formation. *Nature* 376:70–74

26. Suri C, Jones PF, Patan S, Bartunkova S, Maisonpierre PC, et al. 1996. Requisite role of angiopoietin-1, a ligand for the TIE2 receptor, during embryonic angiogenesis. *Cell* 87:1171–80

27. Puri MC, Partanen J, Rossant J, Bernstein A. 1999. Interaction of the TEK and TIE receptor tyrosine kinases during cardiovascular development. *Development* 126:4569–80

28. Ward NL, Van Slyke P, Sturk C, Cruz M, Dumont DJ. 2004. Angiopoietin 1 expression levels in the myocardium direct coronary vessel development. *Dev. Dyn.* 229:500–9

29. Godecke A, Heinicke T, Kamkin A, Kiseleva I, Strasser RH, et al. 2001. Inotropic response to β-adrenergic receptor stimulation and anti-adrenergic effect of ACh in endothelial NO synthase-deficient mouse hearts. *J. Physiol.* 532:195–204

30. Champion HC, Georgakopoulos D, Takimoto E, Isoda T, Wang Y, et al. 2004. Modulation of in vivo cardiac function by myocyte-specific nitric oxide synthase-3. *Circ. Res.* 94:657–63

31. Barouch LA, Harrison RW, Skaf MW, Rosas GO, Cappola TP, et al. 2002. Nitric oxide regulates the heart by spatial confinement of nitric oxide synthase isoforms. *Nature* 416:337–39

32. Ozcelik C, Erdmann B, Pilz B, Wettschureck N, Britsch S, et al. 2002. Conditional mutation of the ErbB2 (HER2) receptor in cardiomyocytes leads to dilated cardiomyopathy. *Proc. Natl. Acad. Sci. USA* 99:8880–85

33. Crone SA, Zhao YY, Fan L, Gu Y, Minamisawa S, et al. 2002. ErbB2 is essential in the

prevention of dilated cardiomyopathy. *Nat. Med.* 8:459–65

34. Lemmens K, Fransen P, Sys SU, Brutsaert DL, DeKeulenaer GW. 2004. Neuregulin-1 induces a negative inotropic effect in cardiac muscle: role of nitric oxide synthase. *Circulation* 109:324–26

35. Rich S, McLaughlin VV. 2003. Endothelin receptor blockers in cardiovascular disease. *Circulation* 108:2184–90

36. Zolk O, Quattek J, Sitzler G, Schrader T, Nickenig G, et al. 1999. Expression of endothelin-1, endothelin-converting enzyme, and endothelin receptors in chronic heart failure. *Circulation* 99:2118–23

37. Luscher TF, Enseleit F, Pacher R, Mitrovic V, Schulze MR, et al. 2002. Hemodynamic and neurohumoral effects of selective endothelin A (ET(A)) receptor blockade in chronic heart failure: the Heart Failure ET(A) Receptor Blockade Trial (HEAT). *Circulation* 106:2666–72

38. O'Connor CM, Gattis WA, Adams KF Jr, Hasselblad V, Chandler B, et al. 2003. Tezosentan in patients with acute heart failure and acute coronary syndromes: results of the Randomized Intravenous TeZosentan Study (RITZ-4). *J. Am. Coll. Cardiol.* 41:1452–55

39. Packer M, McMurray J, Massie BM, Caspi A, Charlon V, et al. 2005. Clinical effects of endothelin receptor antagonism with bosentan in patients with severe chronic heart failure: results of a pilot study. *J. Card. Fail.* 11:12–20

40. Becker RO, Chapin S, Sherry R. 1974. Regeneration of the ventricular myocardium in amphibians. *Nature* 248:145–47

41. Poss KD, Wilson LG, Keating MT. 2002. Heart regeneration in zebrafish. *Science* 298:2188–90

42. Anversa P. 2000. Myocyte death in the pathological heart. *Circ. Res.* 86:121–24

43. Beltrami AP, Barlucchi L, Torella D, Baker M, Limana F, et al. 2003. Adult cardiac stem cells are multipotent and support myocardial regeneration. *Cell* 114:763–76

44. Engel FB, Schebesta M, Duong MT, Lu

G, Ren SX, et al. 2005. p38 MAP kinase inhibition enables proliferation of adult mammalian cardiomyocytes. *Genes Dev.* 19:1175–87

45. Deb A, Wang S, Skelding KA, Miller D, Simper D, et al. 2003. Bone marrow-derived cardiomyocytes are present in adult human heart: A study of gender-mismatched bone marrow transplantation patients. *Circulation* 107:1247–49

46. Oh H, Bradfute SB, Gallardo TD, Nakamura T, Gaussin V, et al. 2003. Cardiac progenitor cells from adult myocardium: homing, differentiation, and fusion after infarction. *Proc. Natl. Acad. Sci. USA* 100:12313–18

47. Laugwitz KL, Moretti A, Lam J, Gruber P, Chen YH, et al. 2005. Postnatal isl1[+] cardioblasts enter fully differentiated cardiomyocyte lineages. *Nature* 433:647–53

48. Rosenblatt-Velin N, Lepore MG, Cartoni C, Beermann F, Pedrazzini T. 2005. FGF-2 controls the differentiation of resident cardiac precursors into functional cardiomyocytes. *J. Clin. Invest.* 115:1724–33

49. Reffelmann T, Kloner RA. 2003. Cellular cardiomyoplasty: cardiomyocytes, skeletal myoblasts, or stem cells for regenerating myocardium and treatment of heart failure? *Cardiovasc. Res.* 58:358–68

50. Zimmermann WH, Schneiderbanger K, Schubert P, Didie M, Munzel F, et al. 2002. Tissue engineering of a differentiated cardiac muscle construct. *Circ. Res.* 90:223–30

51. Kocher AA, Schuster MD, Szabolcs MJ, Takuma S, Burkhoff D, et al. 2001. Neovascularization of ischemic myocardium by human bone-marrow-derived angioblasts prevents cardiomyocyte apoptosis, reduces remodeling and improves cardiac function. *Nat. Med.* 7:430–36

52. Itescu S, Kocher AA, Schuster MD. 2003. Myocardial neovascularization by adult bone marrow-derived angioblasts: strategies for improvement of cardiomyocyte function. *Heart Fail. Rev.* 8:253–58

53. Murry CE, Soonpaa MH, Reinecke H,

Nakajima H, Nakajima HO, et al. 2004. Haematopoietic stem cells do not trans-differentiate into cardiac myocytes in myocardial infarcts. *Nature* 428:664–68

54. Rehman J, Li JL, Orschell CM, March KL. 2003. Peripheral blood "endothelial progenitor cells" are derived from monocyte/macrophages and secrete angiogenic growth factors. *Circulation* 107:1164–69

55. Annex BH, Simons M. 2005. Growth factor-induced therapeutic angiogenesis in the heart: protein therapy. *Cardiovasc. Res.* 65:649–55

56. Kuramochi Y, Cote GM, Guo X, Lebrasseur NK, Cui L, et al. 2004. Cardiac endothelial cells regulate reactive oxygen species-induced cardiomyocyte apoptosis through neuregulin-1β/erbB4 signaling. *J. Biol. Chem.* 279:51141–47

57. Fukazawa R, Miller TA, Kuramochi Y, Frantz S, Kim YD, et al. 2003. Neuregulin-1 protects ventricular myocytes from anthracycline-induced apoptosis via erbB4-dependent activation of PI3-kinase/Akt. *J. Mol. Cell. Cardiol.* 35:1473–79

58. Remondino A, Kwon SH, Communal C, Pimentel DR, Sawyer DB, et al. 2003. β-adrenergic receptor–stimulated apoptosis in cardiac myocytes is mediated by reactive oxygen species/c-Jun NH2-terminal kinase–dependent activation of the mitochondrial pathway. *Circ. Res.* 92:136–38

59. Narmoneva D, Davis ME, Vukmirovich R, Kamm RD, Lee RT. 2004. Endothelial cells promote cardiac myocyte survival and spatial reorganization: implications for cardiac regeneration. *Circulation* 110:962–68

60. Edelberg JM, Aird WC, Wu W, Rayburn H, Mamuya WS, et al. 1998. PDGF mediates cardiac microvascular communication. *J. Clin. Invest.* 102:837–43

61. Edelberg JM, Lee SH, Kaur M, Tang L, Feirt NM, et al. 2002. Platelet-derived growth factor-AB limits the extent of myocardial infarction in a rat model: feasibility of restoring impaired angiogenic capacity in the aging heart. *Circulation* 105:608–13

62. Xaymardan M, Zheng JG, Duignan I, Chin A, Holm JM, et al. 2004. Senescent impairment in synergistic cytokine pathways that provide rapid cardioprotection in the rat heart. *J. Exp. Med.* 199:797–804

63. Dallabrida SM, Ismail N, Oberle JR, Himes BE, Rupnick MA. 2005. Angiopoietin-1 promotes cardiac and skeletal myocyte survival through integrins. *Circ. Res.* 96:e8–24

64. Takahashi K, Ito Y, Morikawa M, Kobune M, Huang J, et al. 2003. Adenoviral-delivered angiopoietin-1 reduces the infarction and attenuates the progression of cardiac dysfunction in the rat model of acute myocardial infarction. *Mol. Ther.* 8:584–92

65. Condorelli G, Borello U, De Angelis L, Latronico M, Sirabella D, et al. 2001. Cardiomyocytes induce endothelial cells to trans-differentiate into cardiac muscle: implications for myocardium regeneration. *Proc. Natl. Acad. Sci. USA* 98:10733–38

66. Badorff C, Brandes RP, Popp R, Rupp S, Urbich C, et al. 2003. Transdifferentiation of blood-derived human adult endothelial progenitor cells into functionally active cardiomyocytes. *Circulation* 107:1024–32

67. Yeh ET, Zhang S, Wu HD, Korbling M, Willerson JT, et al. 2003. Transdifferentiation of human peripheral blood CD34+-enriched cell population into cardiomyocytes, endothelial cells, and smooth muscle cells in vivo. *Circulation* 108:2070–73

68. Koyanagi M, Brandes RP, Haendeler J, Zeiher AM, Dimmeler S. 2005. Cell-to-cell connection of endothelial progenitor cells with cardiac myocytes by nanotubes: a novel mechanism for cell fate changes? *Circ. Res.* 96:1039–41

69. Alvarez-Dolado M, Pardal R, Garcia-Verdugo JM, Fike JR, Lee HO, et al. 2003. Fusion of bone-marrow-derived cells with Purkinje neurons, cardiomyocytes and hepatocytes. *Nature* 425:968–73

70. Matsuura K, Wada H, Nagai T, Iijima Y, Minamino T, et al. 2004. Cardiomyocytes fuse with surrounding noncardiomyocytes

and reenter the cell cycle. *J. Cell Biol.* 167:351–63

71. Yue X, Tomanek RJ. 1999. Stimulation of coronary vasculogenesis/angiogenesis by hypoxia in cultured embryonic hearts. *Dev. Dyn.* 216:28–36

72. Lee KW, Lip GY, Blann AD. 2004. Plasma angiopoietin-1, angiopoietin-2, angiopoietin receptor tie-2, and vascular endothelial growth factor levels in acute coronary syndromes. *Circulation* 110:2355–60

73. Lee SH, Wolf PL, Escudero R, Deutsch R, Jamieson SW, et al. 2000. Early expression of angiogenesis factors in acute myocardial ischemia and infarction. *N. Engl. J. Med.* 342:626–33

74. Lee RJ, Springer ML, Blanco-Bose WE, Shaw R, Ursell PC, et al. 2000. VEGF gene delivery to myocardium: deleterious effects of unregulated expression. *Circulation* 102:898–901

75. Kawamoto A, Gwon HC, Iwaguro H, Yamaguchi JI, Uchida S, et al. 2001. Therapeutic potential of ex vivo expanded endothelial progenitor cells for myocardial ischemia. *Circulation* 103:634–37

76. Kawamoto A, Tkebuchava T, Yamaguchi J, Nishimura H, Yoon YS, et al. 2003. Intramyocardial transplantation of autologous endothelial progenitor cells for therapeutic neovascularization of myocardial ischemia. *Circulation* 107:461–68

77. Assmus B, Schachinger V, Teupe C, Britten M, Lehmann R, et al. 2002. Transplantation of progenitor cells and regeneration enhancement in acute myocardial infarction (TOPCARE-AMI). *Circulation* 106:3009–17

78. Kang HJ, Kim HS, Zhang SY, Park KW, Cho HJ, et al. 2004. Effects of intracoronary infusion of peripheral blood stem-cells mobilised with granulocyte-colony stimulating factor on left ventricular systolic function and restenosis after coronary stenting in myocardial infarction: the MAGIC cell randomised clinical trial. *Lancet* 363:751–56

Annu. Rev. Physiol. 2005. 68:67–95
doi: 10.1146/annurev.physiol.68.040104.124611
First published online as a Review in Advance on September 28, 2005

PROTECTING THE PUMP: Controlling Myocardial Inflammatory Responses

Viviany R. Taqueti,[1] Richard N. Mitchell,[2] and Andrew H. Lichtman[2]

[1]Department of Pathology, Brigham and Women's Hospital, and Harvard-MIT Division of Health Sciences and Technology, Harvard Medical School, Boston, Massachusetts 02115; email: taqueti@mit.edu

[2]Department of Pathology, Brigham and Women's Hospital and Harvard Medical School, Boston, Massachusetts 02115; email: rmitchell@rics.bwh.harvard.edu, alichtman@rics.bwh.harvard.edu

Key Words   heart, inflammation, innate immunity, adaptive immunity, myocarditis

■ **Abstract**   Because of the anatomy, function, and nonregenerative nature of the myocardium, inflammation in this tissue is not well tolerated. Nevertheless, various diseases of the heart are characterized by inflammatory responses involving the effector mechanisms of innate and adaptive (lymphocyte-dependent) immunity. The innate immune response to ischemia-reperfusion injury is, by far, the most common cause of myocardial inflammation. Innate responses may have beneficial influences that preserve myocardial function in the short term but may be maladaptive in chronic states. Adaptive responses in the myocardium occur with infection or loss of tolerance, and lead to myocarditis. Given the narrow margin for benefit of cardiac inflammation, special regulatory mechanisms likely raise the threshold, compared to other tissues, for the induction and persistence of adaptive immune responses. These mechanisms include strong central and peripheral T cell tolerance to heart antigens and induction of anti-inflammatory feedback mechanisms involving cytokines such as interferon-γ.

INTRODUCTION

Inflammation is a defensive response to microbial infections or tissue injury and is essential for the survival of higher organisms in contaminated and potentially traumatic environments. Yet, while serving their protective role, inflammatory responses almost invariably contribute to tissue damage. Many diseases are caused by dysregulated inflammatory responses that are initiated in the absence of danger or persist even after resolution of the inciting infection or injury. Inflammation is a common event in many tissues, such as those with epithelial interfaces with the outside world, and these tissues are often able to sustain the damaging effects of acute and chronic inflammation without compromising their essential function. Our ability to tolerate inflammation in various tissues reflects many

0066-4278/06/0315-0067$20.00

factors, including internal redundancy (e.g., the physiologic reserve of the kidney and liver with multiple independently functioning glomeruli or lobules, respectively), external redundancy (i.e., paired organs), and regenerative capacity (e.g., in the skin, gastrointestinal tract, and liver). The heart is an organ without internal or external redundancy and without regenerative capacity. Continuous, uncompromised cardiac function is, of course, required to sustain life. Although inflammatory processes occur in the heart in various disease states, these are not well tolerated, and a healthy myocardium rarely harbors any foci of inflammation. It is therefore reasonable to hypothesize that there are mechanisms that protect the heart from potential inflammatory stimuli and raise the initiation threshold for inflammatory responses above that of most other tissues. This hypothesis serves as the underlying theme for this review of cardiac inflammation.

We discuss a number of pathologic entities where inflammatory responses either cause or contribute to myocardial injury and conclude with a discussion of some of the regulatory pathways that may be particularly important in minimizing inflammatory injury in the heart. Our discussion reflects an immunological perspective in which inflammation is viewed as the combination of multiple effector mechanisms used by the innate and adaptive (lymphocyte-mediated) immune systems to deal with microbes and the damaged self. The discussion focuses on inflammatory processes that primarily involve the myocardium. Dilated cardiomyopathy (DCM), which represents the end stage of functional and anatomic decompensation of damaged myocardium, is a common sequela of myocardial inflammation but does not always have an inflammatory precedent. As such, the pathophysiology of DCM is not discussed. Inflammatory processes extrinsic to the myocardium, including coronary artery atherosclerotic disease and other vasculitidies, pericarditis, and endocarditis, are also not discussed.

IMMUNOLOGICAL QUIESCENCE OF THE NAÏVE MYOCARDIUM: FUNCTIONAL AND ANATOMIC CONSIDERATIONS

In this section, we briefly describe why the physiology and anatomy of the heart limit its capacity to tolerate inflammatory processes. The myocardium is a functional syncytium of myocytes; although individual cells are not fused, they are electrically and physically coupled so that contraction in one location leads to a propagated wave of contraction throughout the heart. Intercalated discs at the end termini of myocytes and a lateral network of gap junctions enable calcium (and other ion) currents to be rapidly propagated among contiguous cells. Pathologic processes such as congestive heart failure lead to significant remodeling of these structures and, as such, materially influence effective contractility. The ability of the heart to function as a syncytium is also influenced by intercellular edema; expansion of the extracellular matrix with fluid (e.g., in inflammatory states) leads to reduced contractility (1). Increased edema may raise coronary vascular resistance and oxygen diffusion distances and result in a functional ischemia. Alternatively, edema may cause extracellular matrix remodeling that results in altered viscoelastic properties of the myocardium or may increase the concentration of an interstitial metabolite that has a negative inotropic effect (1). In any event, the efficient export of excess intercellular fluid from the heart via lymphatics is of obvious importance. Although relatively little has been described concerning myocardial lymphatic drainage, the density and general architecture of these structures within the heart is similar to that of most other solid organs (2).

Several agents (e.g., nitric oxide or superoxides) can affect primary contractility of actin-myosin fibers within myocytes, putatively via effects on calcium storage in the sarcoplasmic reticulum (3). Diminished myofiber contraction significantly impacts cardiac output. Alternatively, reduced oxygen delivery will also reduce ATP production and lead to compromised contractile function.

Normally, the rhythmicity of the heart is maintained by a conduction system driven by the sinoatrial node; regular depolarization of this tissue initiates the wave of electrical activity that passes through the atrioventricular node and subsequently into the bundle of His and Purkinje fibers to activate the apical myocytes and initiate contraction. Injury to this conduction system or the development of ectopic foci of electrical activity can lead to dysrhythmias ranging from profound bradycardia to ventricular fibrillation.

In addition to its central function as a muscular pump, the heart is increasingly appreciated as having endocrine functionality. Some atrial myocytes secrete atrial natriuretic peptide in response to atrial distention; this agent causes vasodilation, natriuresis, and diuresis, putatively intended to restore pressure hemodynamics to the normal range. Similarly, elevated ventricular pressures result in the release of B-type natriuretic peptides by the ventricular myocardium (4). Damage to these endocrine cells would remove an important homeostatic pathway and could conceivably accelerate the time course of congestive failure.

Perhaps the most important feature of the myocardium that underlies the importance of protection from inflammation is that, for all intents and purposes, it is a permanent, nonregenerating tissue. Despite new and exciting data suggesting the existence of either marrow-derived (5) or myocardial-derived (6) stem cells, the ability of these precursors to functionally repopulate myocardium after injury is minimal, at best. Consequently, irreversible damage to the myocardium invariably leads to scarring. Fibrosis is a noncontractile element that severely reduces cardiac output. Moreover, sites of scarring can serve as niduses for arrhythmias.

Although constantly in motion, the unstressed (naïve) heart is a relatively inactive organ from an immunological perspective. Unlike the skin, gut, lung, liver, and spleen, which undergo constant surveillance of the external environment, the heart, in comparison, is secluded from most environmental antigens. As such, the heart is generally not a tissue in which there is a lot of ongoing immune surveillance for foreign antigens. The unstressed heart does not constitutively express proinflammatory cytokines (7–9). Despite evidence that bone marrow–derived, MHC class II–positive dendritic cells (DCs) reside in the interstitial connective tissue of the naïve heart (10–12), these cells are poorly characterized and likely function in a role broader than the classical presentation of environmental antigens (11, 13, 14). In addition to scavenging physiological cell debris, these tissue-resident DCs may serve as key players in maintaining peripheral tolerance to myocardial antigens (discussed below).

MYOCARDIAL INFLAMMATORY CONDITIONS

Our knowledge of how inflammatory responses play out in the heart comes from the study of different human cardiac diseases and their associated animal models. As such, it is useful to consider briefly the range of these various diseases. This review is not meant to be a comprehensive catalog of inflammatory conditions in the heart but rather a consideration of the most common and/or interesting diseases that may illustrate unique features of myocardial inflammation. Cardiac inflammatory processes can be (imperfectly) distinguished as manifestations of either innate immune responses alone or a combination of innate and adaptive immunity. The discussion of human diseases in which cardiac inflammation is a central feature is organized on the basis of that distinction. Table 1 summarizes some features of these diseases that are relevant to this review.

The Role of Innate Immunity: Inflammation Secondary to Ischemia-Reperfusion Injury

The innate immune system responds to molecular patterns common to microbes and to danger signals expressed by injured or infected cells (15). The recognition structures of the innate immune system include soluble proteins in the plasma (e.g., complement proteins and mannose-binding lectins) and cell surface receptors [e.g., scavenger receptors, mannose receptors, and toll-like receptors (TLRs)]. Acute inflammation is a complex response of soluble and cellular factors that together serve as the effector mechanisms of innate immunity. Despite the general immunological quiescence of the heart, occasions exist when the myocardium (like any other organ) becomes involved by innate immune responses and acute inflammation, most commonly as a secondary result of an ischemic insult.

Ischemia-reperfusion injury is a phenomenon in which restoration of blood flow to tissue subjected to an extended period of ischemia results in amplified cellular damage. It was initially recognized in the heart, although it is now well documented in other solid organs such as the kidney, liver, and lung, particularly post-transplantation (16–19). Nearly 50 years ago, Jennings et al. (20) reported on the adverse structural and electrophysiologic changes associated with reperfusion of the ischemic canine heart, and in 1971, Braunwald & Maroko (21, 22) proposed that therapeutic manipulation during the ischemic insult could reduce infarct size following coronary artery occlusion and reperfusion.

Cellular mechanisms underlying ischemic reperfusion injury are complex, involving intrinsic cardiomyocyte contracture in the earliest phase, followed by

TABLE 1   Examples of human diseases characterized by myocardial inflammation

Underlying type of immune response	Category of inciting cause	Example	Characteristics of inflammation
Innate	Myocardial infection	Bacterial, e.g., staphylococcal	Acute inflammation; neutrophilic abscesses; repair with fibrosis (late)
	Ischemia	Infarction	Coagulative necrosis with surrounding acute inflammation (early)-repair and fibrosis (late)
		Reperfusion injury	Acute inflammation (early)-repair with fibrosis (late)
Innate/adaptive	Myocardial infection with autoimmune component	Viral, e.g., coxsackievirus B myocarditis	Lymphocytic infiltration with myocyte necrosis
		Protozoan:*Trypanosoma cruzi* (Chagas disease)	Acute and chronic inflammation with neutrophils, lymphocytes macrophages (early)-repair with fibrosis (late)
	Antimyocardial immune response without myocardial infection	Acute rheumatic carditis s/p Group A streptococcal pharyngitis	T lymphocytes and macrophages around damaged connective tissue and myofibers
		Giant cell myocarditis	Multinucleate giant cells, lymphocytes, eosinophils with myocyte necrosis
		Drug induced hypersensitivity myocarditis	Perivascular lymphocytes, macrophages, and eosinophils
	Alloantigen recognition	Acute allograft rejection	T lymphocytes and macrophages surrounding apoptotic fibers; vascular antibody deposition

activation of endothelial cells and complement, increased vascular permeability, and rapid accumulation of neutrophils and monocytes with subsequent release of reactive oxygen species, elastases and proteases, and arachidonic acid metabolites and platelet-activating factors (23–25). This latter cascade of events, typical of the innate immune response, critically contributes to additional myocyte injury and necrosis, exacerbating the direct ischemic injury and accelerating its onset, e.g., within minutes of reperfusion rather than hours after ischemia (26).

Inhibition of effector mechanisms of innate immunity can reduce ischemic reperfusion injury significantly in many experimental models. For example, administration of soluble complement receptor 1, a C3 convertase inhibitor, dramatically decreased infarct size in a rat myocardial model (27), and pexelizumab, a recombinant monoclonal-antibody fragment against human C5, significantly attenuated postoperative myocardial injury in patients undergoing coronary artery bypass grafting (28). Depletion of neutrophils or inhibition of neutrophil adhesion resulted in a cardioprotective effect in models of experimental canine myocardial reperfusion injury (29–31). In addition, coadministration of antioxidants such as superoxide dismutase and catalase, an O_2 scavenger and H_2O_2-degrading enzyme, respectively, reduced reperfusion injury in canine hearts (32–34). Thus, the inflammatory response constitutes an integral component of the host response to tissue injury and plays a particularly prominent role post myocardial infarction.

Beneficial Effects of Innate Immune Responses in the Heart

The degree of the inflammatory response is a critical determinant of the host's outcome. Recent investigation suggests that the innate immune response may constitute a component of cardiac biology that is actually adaptive in the short term. Central to this theme is the finding that proinflammatory cytokines—which are rapidly elaborated in response to virtually any insult leading to myocardial damage, including hypertension, unstable angina, infarction, reperfusion injury, and heart failure—may constitute a form of ischemic preconditioning in the heart. Proinflammatory cytokines such as tumor necrosis factor (TNF), interleukin-1 (IL-1), and IL-6 are synthesized not only by cells of the immune system but also by cardiac myocytes in response to ischemia or mechanical stretch (35, 36). In the same way that these proinflammatory cytokines serve as effectors of innate immunity (functioning as an early warning system to discriminate self and potential pathogens), expression of these molecules in the heart may allow the myocardium to respond rapidly to tissue injury as part of an early warning system coordinating numerous local homeostatic responses.

This view is supported by studies in which pretreatment of rats with TNFα or IL-1β confers protection for the heart against ischemic reperfusion injury (37–39). This late-phase protection develops within 12–24 h of ischemic stress, lasts for 3–4 days, and is similar to that observed with sublethal stresses such as transient ischemia, hyperthermia, or exercise (40–42). In support of this finding, neutralizing antibodies to both TNFα and IL-1β abolished the protective preconditioning induced by transient cardiac ischemia (43). One potential mechanism involves the free radical scavenger, mitochondrial manganese superoxide dismutase (MnSOD), which was induced in the heart during infusion of these proinflammatory cytokines but was blocked in the neutralization studies. Inhibition of MnSOD using oligodeoxynucleotides abolished the expected cardioprotective effect following ischemic preconditioning (43). More recently, a protective role for the proinflammatory cytokine IL-6 was demonstrated when mice deficient in IL-6 failed to show cardioprotective effects associated with ischemic preconditioning. This loss of the preconditioning effect paralleled decreases in activation of the JAK-STAT pathway and reduced expression of inducible nitric oxide synthase (iNOS) and cyclooxygenase 2 (44). Indeed, the inflammatory cytokines TNFα, IL-1β, and IL-6 promote decreased cardiac contractility (45, 46), which may constitute an adaptive strategy to limit myocardial energy demand in an ischemic setting. They also promote early wound repair, including phagocytosis and resorption of tissue debris, degradation and synthesis of matrix components such as collagens and integrins, angiogenesis and proliferation of myofibroblasts, and to a limited extent, progenitor cell proliferation (47–49). In support of this, anti-IL-1β treatment early postinfarction leads to poor wound healing and delayed collagen deposition (50), as does steroid treatment following myocardial infarction (51).

The TLR4-IRAK1 signaling pathway is another innate immune system signaling pathway related to the IL-1 pathway that may have cardioprotective effects under certain conditions. Lipopolysaccharide (LPS), which signals through TLR4, can protect the myocardium from ischemic injury (52), and TLR4 signaling can protect cardiac myocytes from apoptosis (53). By contrast, TLR4-deficient mice are more resistant to ischemic reperfusion injury in the heart (54). These disparate findings highlight the difficulties in isolating beneficial effects of the innate immune response from the maladaptive effects discussed in the following section.

Maladaptive Effects of Innate Immune Responses in the Heart

Although the transient, self-limited expression of cytokines may serve a fundamentally cardioprotective role in the acute setting by promoting cell survival and wound healing, there is mounting evidence that these inflammatory mediators promote deleterious effects in the heart over time. Sustained expression of proinflammatory cytokines, e.g., as a result of a large infarct or continued robust host responses, can lead to a second wave of cytokine upregulation that may extend to involve remote, noninfarcted regions and contribute to remodeling of the entire myocardium (55–57). Activation of additional inflammatory-cell signaling can contribute to significant myocyte hypertrophy (58), which helps maintain cardiac output following myocyte loss but can chronically decompensate into ventricular dysfunction. This is prominently illustrated in several murine transgenic models of cardiac-specific overexpression of TNFα. Consistently, these animals demonstrate myocardial hypertrophy resulting in inflammatory-cell infiltration, increased interstitial fibrosis, and DCM (59–61).

Increased fibrosis occurs as the infarcted myocardium is replaced by scar tissue. It is important to note that chronic, sustained presence of cytokines, which activate matrix metalloproteinases, also promotes interstitial fibrosis and collagen deposition in noninfarct zones. This leads to accelerated and aberrant extracellular matrix remodeling as the initial degradation gives way to new fiber deposition and redistribution of integrins at the interface between matrix and myocytes (47). Cardiac fibroblasts seem to possess unique properties relative to other fibroblast types, and the mechanisms that govern resolution of acute injury responses versus the transition to chronic activation of cardiac fibroblasts are poorly understood. Interestingly, these cells are particularly responsive to IL-1β, which induces TGFβ1 and iNOS with subsequent promotion of matrix remodeling and other pleiotropic effects in the heart (62). Although adaptive at some level, excessive scar formation critically diminishes myocardial compliance, which contributes to diastolic and eventually systolic dysfunction, disrupts myocyte-to-myocyte electrical connectivity, and further exacerbates the cycle of ischemic stress-inflammation-stress.

Despite intriguing reports suggesting the continuous renewal of a small subpopulation of myocytes derived from differentiating cardiac stem-like cells (5, 6, 63, 64), infarcted myocardium is replaced overwhelmingly with scar tissue. Some reports of positive TUNEL assay staining in hearts following acute myocardial infarction suggested that apoptosis, an immunologically quiet form of cell death, is prominent (65, 66). However, there is very little direct evidence of apoptosis occurring at any stage of infarction, and the premise that apoptosis constitutes a significant form of myocyte death after infarction remains controversial (67, 68). Rather, myocyte necrosis, which can be accompanied by persistent inflammation and fibrosis, likely constitutes the predominant form of cell death in infarcted myocardium. This holds significant implications for understanding the impact of cell death on the ensuing mechanical behavior and structural composition of the heart.

Interestingly, the detrimental cycle of cytokine amplification that contributes to myocardial remodeling can be attenuated significantly with timely intervention of angiotensin-converting enzyme inhibitors, β-blockers, or statins (69–75). These therapies show promise in improving morbidity and mortality from heart failure, potentially related to their increasingly recognized pleiotropic effects as neurohormonal modulators and anti-inflammatory agents (76–78). Since the pioneering description of elevated inflammatory cytokines in patients with heart failure in 1990, there has been growing appreciation of the pathophysiologic consequences of sustained inflammation in the heart. Nevertheless, clinical trials targeting neutralization of TNF in patients with significant heart failure were halted prematurely because of worsening heart failure (79, 80). These findings underscore the complex contributions of the innate immune inflammatory response to cardiac function.

The Role of Adaptive Immunity: Inflammatory Cardiomyopathies

Adaptive immune responses, dependent on antibodies or T lymphocytes, underlie several different inflammatory diseases of the heart (see Table 1). Adaptive immune responses occur when lymphocytes (B or T cells) are activated by antigens. Innate immune responses are usually required to provide danger signals that synergize with signals generated by antigen recognition to activate naïve lymphocytes. For T cells, these danger signals include costimulatory molecules, such as CD80 and CD86, expressed on professional antigen-presenting cells in response to innate immune stimuli. Antigens targeted in adaptive immune responses in different disease states (in the heart as in any other organ) include microbes, products of injured or stressed cells, altered self, and sometimes normal self. Thus, the adaptive immune system plays an important role in infective myocarditis caused by a variety of microbes, autoimmune sequelae of infectious myocarditis, myocarditis due to adaptive immune responses to myocardial antigens in the absence of cardiac infections, and cardiac allograft rejection (see Table 1).

The previous section emphasizes that myocardial inflammation mediated by the innate immune system occurs commonly but secondarily as a result of ischemic heart disease and reperfusion injury. Inflammation mediated by both the innate and adaptive immune systems and specifically triggered by stimuli intrinsic to the myocardium is much less frequent. Myocarditis is defined broadly as inflammation of the heart muscle in which inflammation causes myocardial injury rather than being only a consequence of myocardial injury; myocarditis constitutes the major inflammatory cardiomyopathy (81). As early as 1669, physicians recognized that inflammation of the myocardium is pathologic (82), but not until the latter half of the twentieth century, with the identification of myocarditis in a large number of postmortem studies, has there been renewed interest in the pathophysiology of this condition. Prospective and retrospective studies have identified myocardial inflammation in 1% to 9% of routine postmortem examinations (83–85), although the number of cases in which myocarditis was functionally important is considerably smaller. The number is significant in that it is an extremely low incidence of inflammation compared with the frequent incidence of inflammatory foci in many other tissues.

The most common cause of myocarditis in developed countries is enterovirus infection, particularly by the cardiotropic coxsackievirus B3 (CVB3) (86). Worldwide, the majority of myocarditis is caused by the protozoa *Trypanosoma cruzi*, the agent of Chagas's disease, endemic in rural Central and South America, with myocardial involvement occurring in nearly 80% of affected individuals (87). Other infectious causes include bacteria and helminths. Noninfectious causes of myocarditis include allergic drug hypersensitivities (usually considered clinically insignificant), systemic diseases of immune origin such as rheumatic fever, systemic lupus erythematosus, polymyositis, and, broadly speaking, cardiac transplant rejection (86). Sometimes, as in the case of giant cell myocarditis, the etiology is unclear and may constitute an autoimmune attack by the adaptive immune system. There is evidence that host immune responses contribute to disease in a significant subset of myocarditis patients and in several animal models of myocarditis (88, 89). Infection may lead to the induction of autoimmune responses by activating self-reactive T cells in a breakdown of tolerance or by cross-reaction between microbial and tissue antigens. Interestingly, inflammation persists in many patients despite viral clearance from the heart (89). Evidence of a role for self-reactive T cells in human myocarditis includes the demonstration of myocardial antigen-specific T cells in biopsies from myocarditis patients and the induction of heart disease in severe combined immunodeficiency (SCID) mice that lack B and T cells by transfer of peripheral blood T cells from myocarditis patients (90).

Animal Models of Myocarditis and Adaptive Immune Responses in the Heart

Animal models have been valuable in elucidating the complex interactions between direct viral injury and injury that is mediated by the host immune response. A discussion of three distinct mouse models of myocarditis is illustrative. The first model involves infecting mice with CVB3 to study the induction and effector phases of myocarditis (91, 92). Despite differences in susceptibility to CVB3-induced myocarditis between mouse strains, and probably some mechanistic differences as well, a number of common features are evident. Although virus particles can be detected in the heart during acute myocarditis in this model, the severity of inflammation correlates with increased tissue-specific expression of IL-1β and IL-18 (89, 93). The inflammatory infiltrate is focal and mixed, comprised largely of macrophages, neutrophils, and $CD4^+$ and $CD8^+$ T lymphocytes. Acute inflammation subsides within three weeks postinfection, but a chronic, diffuse lymphocytic infiltrate persists at four weeks. This chronic inflammation eventually becomes associated with regions of necrosis and fibrosis. This later autoimmune phase of the disease is dependent on $CD8^+$ T cells. The progression of the disease involves a complex interaction of T cell subsets. In Balb/c mice, CD1-restricted, γδ-TCR (T cell antigen receptor)-expressing T cells are activated (presumably by a lipid antigen) and secrete interferon (IFN)-γ, which in turn promotes an autoimmune αβ TCR $CD4^+$ T helper 1 (Th1) response (94). The $CD4^+$ Th1 cells provide help for activation of autoimmune $CD8^+$ T cells. The autoimmune phase of CVB3 myocarditis progresses to DCM. Interestingly, this chronic phase develops only in genetically susceptible strains. Although CVB3 replicates at a relatively low level in the heart and no deaths occur during acute viral infection, a persistent challenge remains in distinguishing direct viral damage to the myocardium and injury caused by specific responses to viral antigens and/or autoimmune responses to self-antigens.

A second mouse model of myocarditis, experimental autoimmune myocarditis (EAM), circumvents this limitation by immunizing susceptible mice with cardiac α-myosin heavy chain emulsified in complete Freund's adjuvant. In the model of CVB3 infection previously described, cardiac myosin-specific IgG autoantibodies are found exclusively in mice that develop the late chronic-phase myocarditis (95), supporting the hypothesis of an autoimmune phase of the disease. Immunizing mice with purified myosin in adjuvant induces severe myocarditis that resembles the late-stage disease caused by viral infection (96). The immune responses, histological changes, and genetic susceptibilities observed in EAM closely parallel those of CVB3-induced myocarditis.

These findings raise speculation that human myocarditis may result as a consequence of (*a*) the release of cardiac autoantigens within an adjuvant-enriched environment conducive to robust antigenic presentation inducing an otherwise rare activation of autoreactive lymphocytes or (*b*) antigenic mimicry between epitopes shared by an infectious agent and host cardiac proteins that allows such self-proteins to perpetuate a vigorous lymphocytic response against host tissues (88, 97). Thus far there is no unequivocal evidence to prove either view conclusively. Although common epitopes for IgG antibodies have been described for CVB3 and α-myosin (89), autoantibodies are not required for induction of myocarditis in most mouse strains (98), and immunopathogenic epitopes common to CVB3 and myosin remain elusive. Nevertheless, it is becoming increasingly clear that strong innate immune responses critically contribute to—and likely define—the thresholds required for the development of adaptive immune responses in autoimmune cardiomyopathy. Mechanistic evidence for this hypothesis includes the critical roles of TLR stimulation (93, 99), DC triggering (100, 101), proinflammatory-cytokine expression (102), and complement activation (103, 104) in inducing experimental myocarditis. As described above, cells of innate immunity, such as DCs and macrophages, do not respond to specific antigenic epitopes on pathogens, but rather, they undergo stereotypical responses to classes of pathogens via pattern recognition receptors such as TLRs. As such, innate immune responses critically define the nature of the ensuing adaptive immune response via expression of cytokines and upregulation of molecules important for antigen presentation to lymphocytes.

A third mouse model of myocarditis, developed in our laboratory, focuses on the generation and function of CD8$^+$ effector T cells specific for a cardiac antigen. The model involves adoptive transfer of ovalbumin peptide-specific TCR transgenic CD8$^+$ T cells (OT-I) into a C57/BL6 mouse line (named CMy-mOva) engineered to express membrane-bound ovalbumin exclusively in cardiac myocytes (105). Early in the characterization of CVB3-induced myocarditis models, it was determined that cellular autoimmunity occurs and is mediated by CD8$^+$ cytolytic T lymphocytes (CTL) specifically lysing myocytes (92). Furthermore, these autoreactive CTL effectively transfer myocarditis to recipient mice in the absence of virus (88). In the EAM model, induction of both autoantibodies and autoimmune T cells occurred, yet immunoglobulins were usually not sufficient for transfer of myocarditis, and CD4$^+$ T cells alone (but not CD8$^+$ T cells) transferred disease into SCID or LPS-treated recipients (98). Thus, a limitation of the EAM model is that CD8$^+$ T cell responses, fundamental in the immune response to viruses, may not be efficiently induced by immunization with exogenous protein antigens, which are preferentially processed by the class II MHC pathway of antigen presentation. In contrast, the CMy-mOva model of myocarditis permits investigation of pathogenicity of uniform populations of CD8$^+$ T cells specific for a single antigen expressed in the myocardium. Of relevance to the interdependent roles of innate and adaptive immunity in cardiac inflammation is the finding that strong innate signals are required for the generation of pathogenic (myocarditic) CD8$^+$ CTL from naïve precursors in this model. These signals include in vivo infection with an ovalbumin-expressing virus or in vitro activation of the T cells in the presence of high doses of IL-12 (105).

Cardiac Allograft Rejection: A Special Form of Adaptive Immune–Mediated Myocarditis

Cardiac allograft rejection is a clinically important complication of adaptive immune responses to heart antigens. The increasing frequency of human cardiac transplantation and the extensive experience with experimental models of heart transplantation have improved our understanding of the immunopathology of allograft rejection. Although a detailed consideration of this field is beyond the scope of this review, it is important to note that some of the special features of myocardial inflammation that we discuss are relevant to cardiac allograft rejection and may be distinct from inflammatory processes in other types of allografts. There is some degree of ischemia-reperfusion injury sustained by hearts as a consequence of the transplantation process, and this triggers an innate response that is similar to the innate response seen in conventional ischemic injury of native hearts (106). This innate response is hypothesized to contribute to the activation of alloantigen-specific T cells in the graft recipient, and these T cells then mediate acute allograft rejection. $CD4^+$ and $CD8^+$ effector T cells, as well as T-dependent antibodies, all contribute to acute allograft rejection. Alloantigens expressed on vascular endothelial cells and myocytes are targets of recognition by allospecific T cells and antibodies, and the inflammatory processes that are induced by these immune effectors are similar in nature to those seen in myocarditis. A recently described murine transplant model employs transgenic ovalbumin-expressing hearts and adoptive transfer of ovalbumin-specific $CD8^+$ T cells (OT-I) (107). In this model, OT-I differentiation into CTL that can mediate rejection is dependent on IL-12, which can be produced by DCs interacting with $CD4^+$ T cells. This result is similar to our own observations of the CMy-mOva model of myocarditis and reinforces the concept that DCs and IL-12 bridge pathogenic innate and T cell–mediated immune responses in the heart.

REGULATORY/ANTI-INFLAMMATORY MECHANISMS THAT LIMIT ADAPTIVE IMMUNE RESPONSES IN THE HEART: FOCUS ON T CELL RESPONSES

T cells are involved in most adaptive immune responses, including cell-mediated responses and T-dependent antibody responses. There are several regulatory constraints on the initiation of T cell responses to ensure that aberrant adaptive immune responses to self-antigens are rare events. In light of our hypothesis that the heart is relatively resistant to immune responses, it is of interest to examine how regulation of T cell–mediated immunity plays out in the heart. In this section, we focus on DCs in the initiation of T cell responses to heart antigens, negative regulators of T cell activation that may be particularly important in cardiac tissues, and, finally, the role of IFN-γ and other cytokines in regulating T cell–driven inflammatory responses in the heart.

Dendritic Cells, Lymphatic Drainage, and the Initiation of T Cell–Mediated Responses to Heart Antigens

A first step required for tissue-based T cell–mediated immune responses is the activation and migration of antigen-bearing DCs from tissue into draining lymph nodes. Naïve T cells continually recirculate through lymph nodes and survey the resident DCs for surface peptide-MHC complexes that the T cell may recognize. It is conceivable that the degree of susceptibility of a particular tissue to self-reactive T cell–mediated immune responses correlates with the quantity or phenotype of DCs that are resident in that tissue. As discussed above, DCs readily migrate from bone marrow into the heart (108), and cardiac tissue may be constitutively rich in DCs (11, 12). Bone marrow chimera experiments between different strains of rat demonstrate that bone marrow–derived cells with antigen-presenting function can take up residence in the noninflamed heart (109). These cells have dendritic morphology and express macrophage lineage markers and class II MHC. DC activation is an essential part of innate immune responses and is induced by TLR and cytokine signaling. Adoptive transfer of cardiac-antigen (myosin peptide)-pulsed DCs can cause T cell–mediated myocarditis in mice (110, 111); TLRs or IL-1 receptors are required to activate the DCs so they can initiate myocarditis in this model. The direct trafficking of heart-antigen DCs into regional lymph nodes during the induction phase of inflammatory heart disease has not been demonstrated. In murine cardiac allograft models, it is clear that donor DCs efficiently prime cardiac alloantigen-specific T cell responses and rejection (112, 113), but the migration of DCs and sites of T cell activation in the case of allografts are poorly characterized and differ from the case of surgically unmanipulated hearts.

DCs are phenotypically and functionally heterogeneous, and there is a developing awareness that some DCs may be tolerogenic (114). Tolerogenic DC generated in vitro by treating cell cultures with IL-10 or vitamin D may provide a mechanism for inducing tolerance to cardiac allografts. Whether cardiac resident DCs are enriched for naturally occurring tolerogenic DCs is not clear and is a question worthy of investigative effort.

Draining lymph nodes constitute the anatomic location where DCs interact with naïve T cells. On the basis of studies of immune responses in other tissues, it is highly likely that lymph node–based DC–naïve T cell interactions are essential for initiation of T cell responses to cardiac antigens and for induction of tolerance to these antigens. Nonetheless, it is surprising how little information is available on lymph node–based events in T cell–mediated inflammatory heart disease, such as viral and autoimmune myocarditis. Many anatomic studies have characterized cardiac lymphatic drainage in humans, primates, dogs, and pigs. In all these species, a significant fraction of lymphatic drainage is delivered to the peribronchial/paratracheal lymph nodes (Figure 1) (115–117). Although anatomic studies of lymphatic drainage of the heart have not been performed in mice, we have identified a mouse peribronchial lymph node that is a site of activation of

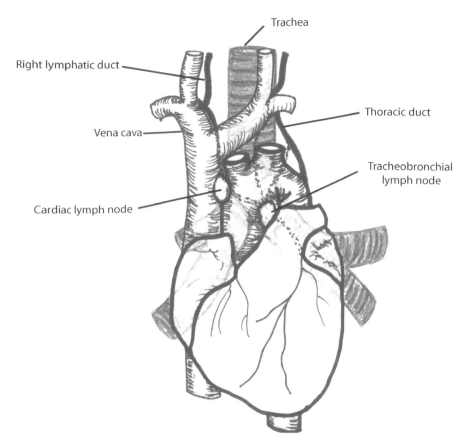

Figure 1   Draining lymph nodes of the myocardium. The locations of two major lymph nodes draining the myocardium, in relation to adjacent anatomical structures, are illustrated. The drawing is based on several studies performed in dogs, pigs, and humans. A tracheobronchial node collects lymphatic drainage mainly from the left ventricle and atrium. A predominant site of right heart drainage is a parabronchial lymph node, often called the cardiac node. As discussed in the text, adaptive immune responses and peripheral tolerance to antigens in the myocardium are likely to depend on events that occur in these lymph nodes. The drawing is adapted from a diagram of dog heart lymphatic drainage (160), includes anatomic information from our own observations of mouse lymph nodes, and is consistent with other studies of pig, monkey, and humans.

heart-antigen-specific CD8$^+$ CTL (105). Because this lymph node is specifically involved in a T cell–mediated cardiac inflammatory disease, its identification provides opportunities to study tolerization or activation of heart-antigen-specific T cells. In our studies of CD8$^+$ T cell–mediated myocarditis, we found that effector CTL migrate into this lymph node and proliferate before they are detectable in the heart (105). Furthermore, we found that T cells with a high CCR7:CXCR3 expression ratio are less pathogenic compared with T cells with lower ratios (105). CXCR3 binds IFN-γ-inducible chemokines found in inflamed tissues, including the heart, and CCR7 binds chemokines that are constitutively expressed in T cell zones of the lymph node.

Regulatory Mechanisms that Prevent Pathologic T Cell–Mediated Responses in the Heart

Several regulatory mechanisms are essential to prevent inappropriate T cell activation to self-antigens, i.e., to maintain self-tolerance, and to downregulate physiologic immune responses. Little is known about central T cell tolerance to heart antigens. Heart-antigen-specific T cell responses contribute to autoimmune myocarditis, but it is not clear if some individuals are susceptible because of a failure of central tolerance or if central tolerance to heart antigens is normally incomplete and disease susceptibility reflects solely a failure of peripheral tolerance. As discussed above, different inbred mouse strains vary in susceptibility to autoimmune myocarditis; the A/J strain is highly susceptible (118), indicating a genetic basis for this disease. Recent genetic mapping studies in recombinant inbred strains identified myocarditis susceptibility loci; these loci regulate apoptosis in thymocytes in the autoimmune-prone NOD mouse (119). Therefore, central tolerance to heart antigens may be defective in susceptible mice, which raises the question of how central tolerance can be maintained to proteins expressed only in the heart, as opposed to ubiquitous self-antigens found in the thymus. The answer may be related to the discovery that many peripheral tissue antigens are expressed in the thymus, several under the control of the transcription factor called autoimmune regulator (AIRE) (120). This affords the opportunity for central deletion or inactivation of developing T cells specific for those antigens. The peripheral tissue antigens known to be expressed by thymic medullary epithelial cells and those expressed in an AIRE-dependent manner in the thymus include some cardiac specific proteins, such as heart-LIM protein and cardiac troponin (121).

Peripheral T cell tolerance requires presentation of self-antigens to T cells under specific conditions that lead to functional inactivation or deletion, or the participation of regulatory T cells, which will suppress self-reactive T cell activation. Adoptive transfer of previously differentiated effector $CD4^+$ or $CD8^+$ T cells specific for myocardial antigens into mice with no cardiac injury or inflammation can rapidly cause myocarditis (105, 109, 122). This demonstrates that myocardial proteins are constitutively processed and presented as peptides on the surface of antigen-presenting cells in the absence of any disease.

The EAM model (discussed above) indicates that naïve cardiac-antigen-specific T cells do emerge from the thymus. Therefore, protection against autoimmune responses to cardiac antigens must involve the prevention of peripheral activation of these naïve T cells. The B7 and CD28 family of molecules play essential roles in the regulation of immune responses, including maintenance of peripheral tolerance (123). CTLA-4 and PD-1 are CD28 family members that are expressed on T cells and which deliver inhibitory signals that block T cell activation. The inhibitory signals involve recruitment and activation of tyrosine phosphatases and are generated when CTLA-4 binds to CD80 or CD86 as well as when PD-1 binds its ligands PD-L1 or PD-L2. The pathologic phenotypes of animals deficient in one or another of these regulatory molecules indicate that they play important roles in suppressing T cell–mediated responses in the heart.

Mice deficient in CTLA-4 die at an early age of a widely distributed lymphoproliferative disease that includes massive infiltration of the heart by both $CD4^+$ and $CD8^+$ T cells (124). Breeding of CTLA-4-deficient mice onto T cell–receptor transgenic backgrounds in which all the T cells are specific for a single nonself-antigen eliminates the disease phenotype (125), which indicates that self-reactive T cells are responsible for the disease seen in CTLA-4-deficient mice with normal T cell repertoires. CTLA-4 is likely involved in tolerizing naïve tissue-antigen-specific T cells when they encounter the antigens in regional lymph nodes. However, effector T cells may also be regulated by CTLA-4. We have observed that CTLA-4-deficient effector CTL are more potent in inducing myocarditis than are control CTL (V. Love & A.H. Lichtman, unpublished observations).

PD-1-deficient Balb/c mice are susceptible to developing DCM (126) mediated by cardiac-troponin-specific autoantibodies that interfere with myocyte function (127). These antibodies are presumably generated in a T-dependent manner. Although the mechanism of disease in this case does not involve a cardiac inflammatory response, the finding does demonstrate the importance of the PD-1 pathway in maintaining B cell tolerance to T-dependent self-antigens in the heart. PD-L1 and PD-L2 are expressed in an inducible manner on a wide variety of cell types, but the heart is a site of particularly abundant expression (128, 129). We have observed PD-L1 expression on heart endothelial cells in vitro and in vivo in our model of $CD8^+$ T cell–mediated myocarditis (130). In vitro, PD-L1 on mouse heart endothelial cells suppresses $CD8^+$ T cell responses to antigens presented by the cardiac endothelial cells (130). The upregulation of PD-L1 in the myocarditis model is dependent on IFN-γ secreted by the infiltrating T cells, and blockade of PD-L1 enhances the intensity for $CD8^+$ T cell–mediated myocarditis (N. Grabie & A.H. Lichtman, unpublished observations). Mice deficient in PD-L1 or PD-L2 do not spontaneously develop autoimmune myocarditis, and PD-L1 and PD-L2 expression is very low in the uninflamed heart (131, 132). These findings are consistent with the hypothesis that the PD-1/PD-L1 pathway serves to suppress the ability of previously differentiated T effector cells to mount responses in the heart. It is possible that INF-γ expressed by DCs and macrophages during innate immune responses may be the essential stimulus for PD-L1 expression and resistance to maladaptive T cell activation in the heart. Data from murine cardiac allograft studies also indicate that the PD-1/PD-L1 pathway can suppress T cell–mediated cardiac inflammatory responses (133, 134).

The Influences of Interferon-γ and Interleukin-12 on T Cell–Dependent Cardiac Inflammation

IFN-γ-secreting CD4$^+$ Th1 T cells have been implicated in many pathologic tissue-antigen-specific immune responses, such as organ-specific autoimmunity [e.g., multiple sclerosis (135), type I diabetes (136) and allograft rejection (137)]. The pleiotropic proinflammatory effects of IFN-γ are well established, and it is convenient to characterize IFN-γ as a purely proinflammatory mediator in these disease processes. Both IFN-γ and IL-12 are expressed by DCs and macrophages during an innate immune response, and they work together to skew T cell differentiation toward a Th1-like, IFN-γ-producing phenotype (138). On the basis of the hypothesis that the heart intrinsically resists the development of inflammatory responses, one might predict that the influences of IFN-γ and IL-12 on this organ are distinct from those in other organs. Experimental approaches that address this question have largely involved mouse models of myocarditis and cardiac allograft rejection.

Myocarditis induced by a cardiotropic virus involves a complex mixture of direct viral cytopathic effects, the local innate response to virus, the innate response to virus-induced cellular injury, and subsequent humoral and cell-mediated autoimmunity (139, 140). Therefore, in viral myocarditis, the proinflammatory/proimmunity effects of IL-12 and IFN-γ may help control viral infection, but these same cytokines may promote collateral tissue damage or autoimmunity. Actually, the net effect of deficiencies in IL-12p35, STAT4, or IFN-γ in CVB3-induced murine myocarditis appears to be increased acute myocarditis (141), and this correlates with inadequate control of viral replication. However, when the chronic phase of CVB3-induced myocarditis is examined, IFN-γ appears to be protective, whereas IL-12 is not (142).

As discussed above, EAM is induced by immunization of rodents with myosin peptide and adjuvants, and the inflammatory disease is largely dependent on myosin-specific CD4$^+$ T cells. This model may not accurately recapitulate the pathogenesis of human myocarditis, but it permits an analysis of the mediators of autoimmune disease in isolation from the effects of direct viral-mediated damage (118). IL-12 and IFN-γ have opposing effects on inflammation in EAM. As demonstrated in studies with cytokine-gene knockout mice and others with blocking anticytokine antibodies, IL-12 has a proinflammatory effect, whereas IFN-γ has an anti-inflammatory/protective effect (143–147). The proinflammatory effects of IL-12 in the heart thus appear to be IFN-γ independent, although the mechanisms for this are not clear. IL-12 contributes to the differentiation of CD8$^+$ T cells that can enter and cause damage to cardiac tissue (105), but it is also likely to have local proinflammatory effects within cardiac tissue.

There are several mechanisms by which IFN-γ may limit inflammatory processes in the heart (Figure 2). These include: (*a*) direct effects on T cells, (*b*) increased iNOS production, and (*c*) upregulation of endothelial PD-L1 expression. IFN-γ has direct antiproliferative and proapoptotic influences on T cells, and these effects have been implicated as important regulators of immune responses to

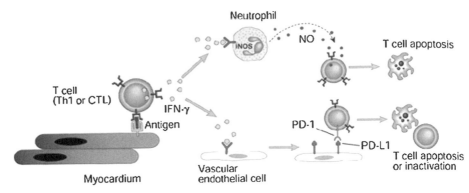

Figure 2   Anti-inflammatory effects of interferon-γ (IFN-γ) in cardiac immune responses. Although IFN-γ has multiple proinflammatory effects and is implicated in immune-mediated injury in a variety of inflammatory and autoimmune diseases, experimental models of myocardial and CNS inflammation suggest that this cytokine exerts early protective anti-inflammatory effects. Two possible mechanisms are depicted. First, T cell–mediated damage to tissues may lead to secondary neutrophilic responses. INF-γ secreted by autoreactive Th1 cells or CTL can upregulate inducible nitric oxide synthase (iNOS) and nitric oxide (NO) production by neutrophils. NO then acts on effector T cells to induce apoptosis or otherwise inhibit their function. This mechanism is based on studies in a mouse model of experimental allergic encephalomyelitis (156). Second, IFN-γ upregulates expression of the negative regulatory molecule PD-L1 on endothelial cells, which bind to PD-1 on T cells. This interaction can inhibit T cell activation (130).

infections (148). The increased severity of EAM seen in the absence of IFN-γ is associated with expansion of a population of apoptosis-resistant CD4$^+$ T cells lacking CD25 expression (149).

Another mechanism by which IFN-γ may be protective in cardiac inflammatory diseases is through its ability to increase nitric oxide (NO) production by upregulating iNOS expression. IFN-γ is responsible for enhanced expression of iNOS and enhanced production of NO in the heart during EAM. In the absence of an IFN-γ receptor, there is a reduction in iNOS and NO and an increase in CD4$^+$ T cell–mediated disease. Furthermore, an iNOS inhibitor enhances EAM disease severity, even in the presence of IFN-γ signaling (147). NO has negative inotropic effects, which may protect the heart during the acute inflammatory phase of myocarditis. There is also evidence that the antiproliferative and proapoptotic effects of IFN-γ on T cells (mentioned above) are related to NO (150). Therefore, NO may act to limit T cell–dependent inflammatory events directly. In this regard, it is instructive to compare EAM with experimental autoimmune encephalomyelitis (EAE), a rodent model of a Th1-mediated autoimmune demyelinating disease that shares many immunological similarities with EAM. In EAE, iNOS-dependent NO generation limits expansion of pathogenic CD4$^+$ T cells by limiting proliferation and promoting apoptosis (151).

IFN-γ-induced PD-L1 expression (discussed above) is another mechanism by which IFN-γ may have anti-inflammatory effects in the heart (Figure 2). In support of this concept, we have found that IFN-γ-deficient CD8$^+$ T cells are as pathogenic as INF-γ-producing CD8$^+$ T cells in an adoptive transfer model of myocarditis (N. Grabie & A.H. Lichtman, unpublished observations).

Studies of murine cardiac allograft rejection also support the concept that IFN-γ has anti-inflammatory effects in the heart. In studies of acute cardiac allograft rejection in mice, the absence of IFN-γ accelerates acute rejection of A/J allografts in C57Bl/6 recipients (152). The inflammatory response in the absence of IFN-γ was characterized by a more severe neutrophilic infiltrate and increased tissue necrosis, as compared with acute rejection in the presence of IFN-γ. In a partial MHC-mismatch model of murine cardiac allograft rejection, the absence of recipient INF-γ also led to enhanced acute parenchymal rejection, although graft arterial disease was markedly diminished (153). At least part of the protective effect of IFN-γ in acute rejection may be attributable to antithrombotic effects in the microcirculation of the cardiac allograft (154).

The Relationship Between Innate and Adaptive Immunity in the Heart: Speculations About T Cells, Neutrophils, and IFN-γ

The paradigm that innate immune responses regulate and influence the phenotype of adaptive immune responses is well accepted. Reverse regulation of innate responses by adaptive immunity is an emerging concept, which is supported by the phenotype of cardiac inflammatory responses both in clinical and experimental settings. Examination of the contributions of neutrophils and lymphocytes to cardiac inflammatory infiltrates provides instructive examples of this bidirectional regulation. As discussed above, the absence of IFN-γ enhances neutrophilic infiltration and cardiac necrosis in murine models of myocarditis and cardiac allograft rejection (141, 152). In our model of CD8$^+$ T cell–mediated myocarditis, depletion of neutrophils diminishes sustained CD8$^+$ CTL infiltration in the heart (155). The immunoregulatory role of neutrophils in T cell–mediated disease has also recently been demonstrated in EAE. Neutrophil infiltration in the CNS is increased in the absence of IFN-γ, and CNS-derived neutrophils can suppress ex vivo T cell responses to CNS antigens. This suppressive effect is dependent on INF-γ production by the T cells and iNOS-mediated NO synthesis by neutrophils (156). The IFN-γ/NO pathway may represent a feedback mechanism that limits amplification of T cell–mediated tissue damage and neutrophilic response to the damaged tissue. It is possible that this regulatory mechanism is particularly important in the heart and CNS but not in other tissues. Another regulatory link between T cells and neutrophils is suggested by the identification of a subset of effector CD4$^+$ T cells that secrete IL-17 but little IFN-γ. The differentiation of this subset, termed Th$_{IL-17}$, depends on IL-23, a cytokine structurally related to IL-12 (157). These IL-23-dependent T cells are highly pathogenic in the murine

EAE model (158). Interestingly, IL-17 drives the proliferation and maturation of neutrophils and induces IL-8 expression, which is the major chemotactic factor for neutrophils. The contribution of IL-17-secreting T cells in myocarditis has not been investigated.

From the perspective of a histopathologist, human cardiac inflammatory diseases tend to obey the rule of mutual exclusion of T cells or neutrophils. The inflammatory infiltrate that surrounds infarcted myocardium is largely neutrophilic, and even in the chronic repair phase post infarction, macrophages are abundant but there are very few lymphocytes. In contrast, and unlike mouse models, acute myocarditis in humans is characterized by T cell infiltrates, and only rarely are significant numbers of neutrophils present, even when there is myocardial necrosis (159). Exceptions to these patterns exist, of course, but the pattern of mixed inflammatory infiltrates that are often found in chronic active inflammation in many other tissues is not common in the heart. The possibility that IFN-γ, or other cytokines elaborated by T cells, inhibits neutrophilic inflammation in human myocardium, and conversely that neutrophils inhibit T cell–mediated inflammation, is worthy of investigation.

SUMMARY AND CONCLUSIONS

Clinical and laboratory investigation has provided an incomplete understanding of the regulation and consequences of inflammation in the heart thus far. The initial hypothesis presented at the beginning of this review proposes that inflammatory responses are not readily initiated in the heart, owing to anatomic and regulatory mechanisms that have evolved to ensure uninterrupted physiologic function. Presumably, failures of those mechanisms can lead to inflammatory diseases and cardiac dysfunction. In light of some of the data discussed here, the validity of our hypothesis appears to be partially correct. Innate immune responses do not frequently occur, largely because environmental pathogens are not readily delivered to the myocardium. However, the intense innate responses to the damaged self occur swiftly and frequently, for example, every time coronary arterial blood flow is obstructed and myocytes die from ischemia. It is unlikely that there has been any evolutionary pressure to minimize the damaging effects of myocardial inflammatory responses to ischemia because this is a disease complication that rarely occurs during reproductive age. By contrast, multiple mechanisms do appear to set a high threshold for the initiation of adaptive immune responses in the myocardium. These mechanisms may reduce risk of autoimmunity but may also increase risk of chronic infection with cardiotropic organisms. As with our knowledge of other organs, our understanding of mechanisms of self-tolerance to myocardial antigens is rudimentary. The emerging recognition of regulatory T cell function and peripheral antigen expression in the thymus will likely lead to insights about tolerance and autoimmunity in the heart. An additional layer of protection against inflammation in the heart (and brain) may result from cross-regulation of innate and adaptive effector cells. Knowledge about how neutrophils inhibit T cells and vice versa may have important therapeutic implications.

ACKNOWLEDGMENTS

Cited studies from the authors' laboratories were supported by the following grants: NIH grant number HL43364 (R.N.M.), Howard Hughes Medical Institute Research Training Fellowship (V.R.T.), and NIH grant numbers AI059610 and HL072056 (A.H.L.).

The *Annual Review of Physiology* is online at
http://physiol.annualreviews.org

LITERATURE CITED

1. Ludwig L, Schertel E, Pratt J, McClure D, Ying A, et al. 1997. Impairment of left ventricular function by acute cardiac lymphatic obstruction. *Cardiovasc. Res.* 33:164–71

2. Geissler H, Bloch W, Forster S, Mehlhorn U, Krahwinkel A, et al. 2003. Morphology and density of initial lymphatics in human myocardium determined by immunohistochemistry. *Thorac. Cardiovasc. Surg.* 51:244–48

3. Bonaventura J, Gow A. 2004. NO and superoxide: opposite ends of the seesaw in cardiac contractility. *Proc. Natl. Acad. Sci. USA* 101:16403–4

4. Sagnella G. 2000. Practical implications of current natriuretic peptide research. *J. Renin Angiotensin Aldosterone Syst.* 1:304–15

5. Orlic D, Kajstura J, Chimenti S, Jakoniuk I, Anderson SM, et al. 2001. Bone marrow cells regenerate infarcted myocardium. *Nature* 410:701–5

6. Beltrami AP, Barlucchi L, Torella D, Baker M, Limana F, et al. 2003. Adult cardiac stem cells are multipotent and support myocardial regeneration. *Cell* 114:763–76

7. Kapadia S, Dibbs Z, Kurrelmeyer K, Kalra D, Seta Y, et al. 1998. The role of cytokines in the failing human heart. *Cardiol. Clin.* 16:645–56

8. Kapadia SR, Oral H, Lee J, Nakano M, Taffet GE, Mann DL. 1997. Hemodynamic regulation of tumor necrosis factor-alpha gene and protein expression in adult feline myocardium. *Circ. Res.* 81:187–95

9. Kapadia S, Lee J, Torre-Amione G, Birdsall HH, Ma TS, Mann DL. 1995. Tumor necrosis factor-α gene and protein expression in adult feline myocardium after endotoxin administration. *J. Clin. Invest.* 96:1042–52

10. Afanasyeva M, Georgakopoulos D, Belardi DF, Ramsundar AC, Barin JG, et al. 2004. Quantitative analysis of myocardial inflammation by flow cytometry in murine autoimmune myocarditis: correlation with cardiac function. *Am. J. Pathol.* 164:807–15

11. Austyn J, Hankins D, Larsen C, Morris P, Rao A, Roake J. 1994. Isolation and characterization of dendritic cells from mouse heart and kidney. *J. Immunol.* 152:2401–10

12. Hart D, Fabre J. 1981. Demonstration and characterization of Ia-positive dendritic cells in the interstitial connective tissues of rat heart and other tissues, but not brain. *J. Exp. Med.* 154:347–61

13. Steiniger B, Klempnauer J, Wonigeit K. 1984. Phenotype and histological distribution of interstitial dendritic cells in

the rat pancreas, liver, heart, and kidney. *Transplantation* 38:169–74

14. Spencer SC, Fabre JW. 1990. Characterization of the tissue macrophage and the interstitial dendritic cell as distinct leukocytes normally resident in the connective tissue of rat heart. *J. Exp. Med.* 171:1841–51

15. Medzhitov R, Janeway CA Jr. 2002. Decoding the patterns of self and nonself by the innate immune system. *Science* 296:298–300

16. Serrick C, Adoumie R, Giaid A, Shennib H. 1994. The early release of interleukin-2, tumor necrosis factor-alpha and interferon-gamma after ischemia reperfusion injury in the lung allograft. *Transplantation* 58:1158–62

17. Oz MC, Liao H, Naka Y, Seldomridge A, Becker DN, et al. 1995. Ischemia-induced interleukin-8 release after human heart transplantation. A potential role for endothelial cells. *Circulation* 92:II428–32

18. Gerlach J, Jorres A, Nohr R, Zeilinger K, Spatkowski G, Neuhaus P. 1999. Local liberation of cytokines during liver preservation. *Transpl. Int.* 12:261–65

19. Lemay S, Rabb H, Postler G, Singh AK. 2000. Prominent and sustained upregulation of gp130-signaling cytokines and the chemokine MIP-2 in murine renal ischemia-reperfusion injury. *Transplantation* 69:959–63

20. Jennings RB, Sommers HM, Smyth GA, Flack HA, Linn H. 1960. Myocardial necrosis induced by temporary occlusion of a coronary artery in the dog. *Arch. Pathol.* 70:68–78

21. Braunwald E, Maroko PR. 1974. The reduction of infarct size—an idea whose time (for testing) has come. *Circulation* 50:206–9

22. Maroko PR, Kjekshus JK, Sobel BE, Watanabe T, Covell JW, et al. 1971. Factors influencing infarct size following experimental coronary artery occlusions. *Circulation* 43:67–82

23. Nian M, Lee P, Khaper N, Liu P. 2004. Inflammatory cytokines and postmyocardial infarction remodeling. *Circ. Res.* 94:1543–53

24. Frangogiannis NG, Smith CW, Entman ML. 2002. The inflammatory response in myocardial infarction. *Cardiovasc. Res.* 53:31–47

25. Piper HM, Meuter K, Schafer C. 2003. Cellular mechanisms of ischemia-reperfusion injury. *Ann. Thorac. Surg.* 75: S644–48

26. Park JL, Lucchesi BR. 1999. Mechanisms of myocardial reperfusion injury. *Ann. Thorac. Surg.* 68:1905–12

27. Weisman HF, Bartow T, Leppo MK, Marsh HC Jr, Carson GR, et al. 1990. Soluble human complement receptor type 1: in vivo inhibitor of complement suppressing post-ischemic myocardial inflammation and necrosis. *Science* 249:146–51

28. Fitch JC, Rollins S, Matis L, Alford B, Aranki S, et al. 1999. Pharmacology and biological efficacy of a recombinant, humanized, single-chain antibody C5 complement inhibitor in patients undergoing coronary artery bypass graft surgery with cardiopulmonary bypass. *Circulation* 100:2499–506

29. Jordan JE, Zhao ZQ, Vinten-Johansen J. 1999. The role of neutrophils in myocardial ischemia-reperfusion injury. *Cardiovasc. Res.* 43:860–78

30. Simpson PJ, Todd RF III, Mickelson JK, Fantone JC, Gallagher KP, et al. 1990. Sustained limitation of myocardial reperfusion injury by a monoclonal antibody that alters leukocyte function. *Circulation* 81:226–37

31. Romson JL, Hook BG, Kunkel SL, Abrams GD, Schork MA, Lucchesi BR. 1983. Reduction of the extent of ischemic myocardial injury by neutrophil depletion in the dog. *Circulation* 67:1016–23

32. Patel BS, Jeroudi MO, O'Neill PG, Roberts R, Bolli R. 1990. Effect of human recombinant superoxide dismutase on canine myocardial infarction. *Am. J. Physiol.* 258:H369–80

33. Jolly SR, Kane WJ, Bailie MB, Abrams GD, Lucchesi BR. 1984. Canine myocardial reperfusion injury. Its reduction by the combined administration of superoxide dismutase and catalase. *Circ. Res.* 54:277–85

34. Bolli R, Jeroudi MO, Patel BS, Aruoma OI, Halliwell B, et al. 1989. Marked reduction of free radical generation and contractile dysfunction by antioxidant therapy begun at the time of reperfusion. Evidence that myocardial "stunning" is a manifestation of reperfusion injury. *Circ. Res.* 65:607–22

35. Mann DL. 2003. Stress-activated cytokines and the heart: from adaptation to maladaptation. *Annu. Rev. Physiol.* 65:81–101

36. Wilson EM, Diwan A, Spinale FG, Mann DL. 2004. Duality of innate stress responses in cardiac injury, repair, and remodeling. *J. Mol. Cell Cardiol.* 37:801–11

37. Wong GH, Goeddel DV. 1988. Induction of manganous superoxide dismutase by tumor necrosis factor: possible protective mechanism. *Science* 242:941–44

38. Brown JM, White CW, Terada LS, Grosso MA, Shanley PF, et al. 1990. Interleukin 1 pretreatment decreases ischemia/reperfusion injury. *Proc. Natl. Acad. Sci. USA* 87:5026–30

39. Eddy LJ, Goeddel DV, Wong GH. 1992. Tumor necrosis factor-alpha pretreatment is protective in a rat model of myocardial ischemia-reperfusion injury. *Biochem. Biophys. Res. Commun.* 184:1056–59

40. Marber MS, Latchman DS, Walker JM, Yellon DM. 1993. Cardiac stress protein elevation 24 hours after brief ischemia or heat stress is associated with resistance to myocardial infarction. *Circulation* 88:1264–72

41. Kuzuya T, Hoshida S, Yamashita N, Fuji H, Oe H, et al. 1993. Delayed effects of sublethal ischemia on the acquisition of tolerance to ischemia. *Circ. Res.* 72:1293–99

42. Hoshida S, Yamashita N, Otsu K, Hori M. 2002. Repeated physiologic stresses provide persistent cardioprotection against ischemia-reperfusion injury in rats. *J. Am. Coll. Cardiol.* 40:826–31

43. Yamashita N, Hoshida S, Otsu K, Taniguchi N, Kuzuya T, Hori M. 2000. The involvement of cytokines in the second window of ischaemic preconditioning. *Br. J. Pharmacol.* 131:415–22

44. Dawn B, Xuan YT, Guo Y, Rezazadeh A, Stein AB, et al. 2004. IL-6 plays an obligatory role in late preconditioning via JAK-STAT signaling and upregulation of iNOS and COX-2. *Cardiovasc. Res.* 64:61–71

45. Goldhaber JI, Kim KH, Natterson PD, Lawrence T, Yang P, Weiss JN. 1996. Effects of TNF-alpha on $[Ca2+]i$ and contractility in isolated adult rabbit ventricular myocytes. *Am. J. Physiol.* 271: H1449–55

46. Yokoyama T, Vaca L, Rossen RD, Durante W, Hazarika P, Mann DL. 1993. Cellular basis for the negative inotropic effects of tumor necrosis factor-alpha in the adult mammalian heart. *J. Clin. Invest.* 92:2303–12

47. Sun M, Opavsky MA, Stewart DJ, Rabinovitch M, Dawood F, et al. 2003. Temporal response and localization of integrins beta1 and beta3 in the heart after myocardial infarction: regulation by cytokines. *Circulation* 107:1046–52

48. Orlic D, Kajstura J, Chimenti S, Limana F, Jakoniuk I, et al. 2001. Mobilized bone marrow cells repair the infarcted heart, improving function and survival. *Proc. Natl. Acad. Sci. USA* 98:10344–49

49. Deten A, Holzl A, Leicht M, Barth W, Zimmer HG. 2001. Changes in extracellular matrix and in transforming growth factor beta isoforms after coronary artery ligation in rats. *J. Mol. Cell Cardiol.* 33:1191–207

50. Hwang MW, Matsumori A, Furukawa Y, Ono K, Okada M, et al. 2001. Neutralization of interleukin-1beta in the acute phase of myocardial infarction promotes

the progression of left ventricular remodeling. *J. Am. Coll. Cardiol.* 38:1546–53

51. Roberts R, DeMello V, Sobel BE. 1976. Deleterious effects of methylprednisolone in patients with myocardial infarction. *Circulation* 53:I204–6

52. Brown JM, Grosso MA, Terada LS, Whitman GJR, Banerjee A, et al. 1989. Endotoxin pretreatment increases endogenous myocardial catalase activity and decreases ischemia-reperfusion injury of isolated rat hearts. *Proc. Natl. Acad. Sci. USA* 86:2516–20

53. Chao W, Shen Y, Zhu X, Zhao H, Novikov M, et al. 2005. Lipopolysaccharide improves cardiomyocyte survival and function after serum deprivation. *J. Biol. Chem.* 280:21997–2005

54. Oyama J-I, Blais C Jr, Liu X, Pu M, Kobzik L, et al. 2004. Reduced myocardial ischemia-reperfusion injury in toll-like receptor 4-deficient mice. *Circulation* 109:784–89

55. Nakamura H, Umemoto S, Naik G, Moe G, Takata S, et al. 2003. Induction of left ventricular remodeling and dysfunction in the recipient heart after donor heart myocardial infarction: new insights into the pathologic role of tumor necrosis factor-alpha from a novel heterotopic transplant-coronary ligation rat model. *J. Am. Coll. Cardiol.* 42:173–81

56. Ono K, Matsumori A, Shioi T, Furukawa Y, Sasayama S. 1998. Cytokine gene expression after myocardial infarction in rat hearts: possible implication in left ventricular remodeling. *Circulation* 98:149–56

57. Irwin MW, Mak S, Mann DL, Qu R, Penninger JM, et al. 1999. Tissue expression and immunolocalization of tumor necrosis factor-alpha in postinfarction dysfunctional myocardium. *Circulation* 99:1492–98

58. Yokoyama T, Nakano M, Bednarczyk JL, McIntyre BW, Entman M, Mann DL. 1997. Tumor necrosis factor-alpha provokes a hypertrophic growth response in adult cardiac myocytes. *Circulation* 95:1247–52

59. Kubota T, McTiernan CF, Frye CS, Demetris AJ, Feldman AM. 1997. Cardiac-specific overexpression of tumor necrosis factor-alpha causes lethal myocarditis in transgenic mice. *J. Card. Fail.* 3:117–24

60. Sivasubramanian N, Coker ML, Kurrelmeyer KM, MacLellan WR, DeMayo FJ, et al. 2001. Left ventricular remodeling in transgenic mice with cardiac restricted overexpression of tumor necrosis factor. *Circulation* 104:826–31

61. Bryant D, Becker L, Richardson J, Shelton J, Franco F, et al. 1998. Cardiac failure in transgenic mice with myocardial expression of tumor necrosis factor-alpha. *Circulation* 97:1375–81

62. Brown RD, Ambler SK, Mitchell MD, Long CS. 2005. The cardiac fibroblast: therapeutic target in myocardial remodeling and failure. *Annu. Rev. Pharmacol. Toxicol.* 45:657–87

63. Kajstura J, Rota M, Whang B, Cascapera S, Hosoda T, et al. 2005. Bone marrow cells differentiate in cardiac cell lineages after infarction independently of cell fusion. *Circ. Res.* 96:127–37

64. Jackson KA, Majka SM, Wang H, Pocius J, Hartley CJ, et al. 2001. Regeneration of ischemic cardiac muscle and vascular endothelium by adult stem cells. *J. Clin. Invest.* 107:1395–402

65. Bardales RH, Hailey LS, Xie SS, Schaefer RF, Hsu SM. 1996. In situ apoptosis assay for the detection of early acute myocardial infarction. *Am. J. Pathol.* 149:821–29

66. Nadal-Ginard B, Kajstura J, Leri A, Anversa P. 2003. Myocyte death, growth, and regeneration in cardiac hypertrophy and failure. *Circ. Res.* 92:139–50

67. Takemura G, Fujiwara H. 2004. Role of apoptosis in remodeling after myocardial infarction. *Pharmacol. Ther.* 104:1–16

68. Primeau AJ, Adhihetty PJ, Hood DA. 2002. Apoptosis in heart and skeletal muscle. *Can. J. Appl. Physiol.* 27:349–95

69. Wei GC, Sirois MG, Qu R, Liu P, Rouleau JL. 2002. Subacute and chronic effects of quinapril on cardiac cytokine expression, remodeling, and function after myocardial infarction in the rat. *J. Cardiovasc. Pharmacol.* 39:842–50

70. Sia YT, Parker TG, Tsoporis JN, Liu P, Adam A, Rouleau JL. 2002. Long-term effects of carvedilol on left ventricular function, remodeling, and expression of cardiac cytokines after large myocardial infarction in the rat. *J. Cardiovasc. Pharmacol.* 39:73–87

71. Gage JR, Fonarow G, Hamilton M, Widawski M, Martinez-Maza O, Vredevoe DL. 2004. Beta blocker and angiotensin-converting enzyme inhibitor therapy is associated with decreased Th1/Th2 cytokine ratios and inflammatory cytokine production in patients with chronic heart failure. *Neuroimmunomodulation* 11:173–80

72. Tatli E, Kurum T. 2005. A controlled study of the effects of carvedilol on clinical events, left ventricular function and proinflammatory cytokines levels in patients with dilated cardiomyopathy. *Can. J. Cardiol.* 21:344–48

73. Deten A, Volz HC, Holzl A, Briest W, Zimmer HG. 2003. Effect of propranolol on cardiac cytokine expression after myocardial infarction in rats. *Mol. Cell Biochem.* 251:127–37

74. Ascer E, Bertolami MC, Venturinelli ML, Buccheri V, Souza J, et al. 2004. Atorvastatin reduces proinflammatory markers in hypercholesterolemic patients. *Atherosclerosis* 177:161–66

75. Holm T, Andreassen AK, Ueland T, Kjekshus J, Froland SS, et al. 2001. Effect of pravastatin on plasma markers of inflammation and peripheral endothelial function in male heart transplant recipients. *Am. J. Cardiol.* 87:815–18, A9

76. Remme WJ. 2003. Pharmacological modulation of cardiovascular remodeling: a guide to heart failure therapy. *Cardiovasc. Drugs Ther.* 17:349–60

77. Scalia R. 2005. Statins and the response to myocardial injury. *Am. J. Cardiovasc. Drugs* 5:163–70

78. Weitz-Schmidt G. 2002. Statins as anti-inflammatory agents. *Trends Pharmacol. Sci.* 23:482–86

79. Mann DL, McMurray JJ, Packer M, Swedberg K, Borer JS, et al. 2004. Targeted anticytokine therapy in patients with chronic heart failure: results of the Randomized Etanercept Worldwide Evaluation (RENEWAL). *Circulation* 109:1594–602

80. Chung ES, Packer M, Lo KH, Fasanmade AA, Willerson JT. 2003. Randomized, double-blind, placebo-controlled, pilot trial of infliximab, a chimeric monoclonal antibody to tumor necrosis factor-alpha, in patients with moderate-to-severe heart failure: results of the Anti-TNF Therapy Against Congestive Heart Failure (ATTACH) trial. *Circulation* 107:3133–40

81. Richardson P, McKenna W, Bristow M, Maisch B, Mautner B, et al. 1996. Report of the 1995 World Health Organization/International Society and Federation of Cardiology task force on the definition and classification of cardiomyopathies. *Circulation* 93:841–42

82. Lower R. 1669. *Tractatus de corde: item de motu & colore sanguinis, et chyli in eum transitu.* Amsterdam: Elsevier. 232 pp.

83. Saphir O. 1941. Myocarditis: a general review, with an analysis of two hundred and forty cases. *Arch. Pathol.* 32:1000–51

84. Gore I, Saphir O. 1947. Myocarditis: a classification of 1402 cases. *Am. Heart J.* 34:827–30

85. Blankenhorn MA, Gall EA. 1956. Myocarditis and myocardosis; a clinico-pathologic appraisal. *Circulation* 13:217–23

86. Feldman AM, McNamara D. 2000. Myocarditis. *N. Engl. J. Med.* 343:1388–98

87. Morris SA, Tanowitz HB, Wittner M, Bilezikian JP. 1990. Pathophysiological

insights into the cardiomyopathy of Chagas' disease. *Circulation* 82:1900–9

88. Huber SA. 1997. Autoimmunity in myocarditis: relevance of animal models. *Clin. Immunol. Immunopathol.* 83:93–102

89. Fairweather D, Kaya Z, Shellam GR, Lawson CM, Rose NR. 2001. From infection to autoimmunity. *J. Autoimmun.* 16:175–86

90. Schwimmbeck PL, Badorff C, Rohn G, Schulze K, Schultheiss HP. 1996. The role of sensitized T-cells in myocarditis and dilated cardiomyopathy. *Int. J. Cardiol.* 54:117–25

91. Wolfgram LJ, Beisel KW, Herskowitz A, Rose NR. 1986. Variations in the susceptibility to coxsackievirus B3-induced myocarditis among different strains of mice. *J. Immunol.* 136:1846–52

92. Huber SA, Lodge PA. 1984. Coxsackievirus B-3 myocarditis in Balb/c mice. Evidence for autoimmunity to myocyte antigens. *Am. J. Pathol.* 116:21–29

93. Fairweather D, Yusung S, Frisancho S, Barrett M, Gatewood S, et al. 2003. IL-12 receptor beta 1 and Toll-like receptor 4 increase IL-1 beta- and IL-18-associated myocarditis and coxsackievirus replication. *J. Immunol.* 170:4731–37

94. Huber SA, Sartini D, Exley M. 2002. Vγ4+ T cells promote autoimmune CD8+ cytolytic T-lymphocyte activation in coxsackievirus B3-induced myocarditis in mice: role for CD4+ Th1 cells. *J. Virol.* 76:10785–90

95. Neu N, Beisel KW, Traystman MD, Rose NR, Craig SW. 1987. Autoantibodies specific for the cardiac myosin isoform are found in mice susceptible to coxsackievirus B3-induced myocarditis. *J. Immunol.* 138:2488–92

96. Neu N, Rose NR, Beisel KW, Herskowitz A, Gurri-Glass G, Craig SW. 1987. Cardiac myosin induces myocarditis in genetically predisposed mice. *J. Immunol.* 139:3630–36

97. Fairweather D, Frisancho-Kiss S, Rose NR. 2005. Viruses as adjuvants for autoimmunity: evidence from coxsackievirus-induced myocarditis. *Rev. Med. Virol.* 15:17–27

98. Eriksson U, Penninger JM. 2005. Autoimmune heart failure: new understandings of pathogenesis. *Int. J. Biochem. Cell Biol.* 37:27–32

99. Lane JR, Neumann DA, Lafond-Walker A, Herskowitz A, Rose NR. 1991. LPS promotes CB3-induced myocarditis in resistant B10.A mice. *Cell. Immunol.* 136:219–33

100. Eriksson U, Kurrer MO, Sonderegger I, Iezzi G, Tafuri A, et al. 2003. Activation of dendritic cells through the interleukin 1 receptor 1 is critical for the induction of autoimmune myocarditis. *J. Exp. Med.* 197:323–31

101. Eriksson U, Ricci R, Hunziker L, Kurrer MO, Oudit GY, et al. 2003. Dendritic cell-induced autoimmune heart failure requires cooperation between adaptive and innate immunity. *Nat. Med.* 9:1484–90

102. Lane JR, Neumann DA, Lafond-Walker A, Herskowitz A, Rose NR. 1992. Interleukin 1 or tumor necrosis factor can promote coxsackie B3-induced myocarditis in resistant B10.A mice. *J. Exp. Med.* 175:1123–29

103. Kaya Z, Afanasyeva M, Wang Y, Dohmen KM, Schlichting J, et al. 2001. Contribution of the innate immune system to autoimmune myocarditis: a role for complement. *Nat. Immunol.* 2:739–45

104. Eriksson U, Kurrer MO, Schmitz N, Marsch SC, Fontana A, et al. 2003. Interleukin-6-deficient mice resist development of autoimmune myocarditis associated with impaired upregulation of complement C3. *Circulation* 107:320–25

105. Grabie N, Delfs MW, Westrich JR, Love VA, Stavrakis G, et al. 2003. IL-12 is required for differentiation of pathogenic CD8+ T cell effectors that cause myocarditis. *J. Clin. Invest.* 111:671–80

106. Land WG. 2005. The role of postischemic

reperfusion injury and other nonantigen-dependent inflammatory pathways in transplantation. *Transplantation* 79:505–14

107. Filatenkov AA, Jacovetty EL, Fischer UB, Curtsinger JM, Mescher MF, Ingulli E. 2005. CD4 T cell-dependent conditioning of dendritic cells to produce IL-12 results in CD8-mediated graft rejection and avoidance of tolerance. *J. Immunol.* 174:6909–17

108. Saiki T, Ezaki T, Ogawa M, Matsuno K. 2001. Trafficking of host- and donor-derived dendritic cells in rat cardiac transplantation: allosensitization in the spleen and hepatic nodes. *Transplantation* 71:1806–15

109. Ratcliffe NR, Wegmann KW, Zhao RW, Hickey WF. 2000. Identification and characterization of the antigen presenting cell in rat autoimmune myocarditis: evidence of bone marrow derivation and nonrequirement for MHC class I compatibility with pathogenic T cells. *J. Autoimmun.* 15:369–79

110. Eriksson U, Ricci R, Hunziker L, Kurrer MO, Oudit GY, et al. 2003. Dendritic cell-induced autoimmune heart failure requires cooperation between adaptive and innate immunity. 9:1484–90

111. Eriksson U, Kurrer MO, Sonderegger I, Iezzi G, Tafuri A, et al. 2003. Activation of dendritic cells through the interleukin 1 receptor 1 is critical for the induction of autoimmune myocarditis. *J. Exp. Med.* 197:323–31

112. Talmage DW, Dart G, Radovich J, Lafferty KJ. 1976. Activation of transplant immunity: effect of donor leukocytes on thyroid allograft rejection. *Science* 191:385–88

113. Lechler RI, Batchelor JR. 1982. Restoration of immunogenicity to passenger cell-depleted kidney allografts by the addition of donor strain dendritic cells. *J. Exp. Med.* 155:31–41

114. Morelli AE, Thomson AW. 2003. Dendritic cells: regulators of alloimmunity and opportunities for tolerance induction. *Immunol. Rev.* 196:125–46

115. Riquet M, Le Pimpec-Barthes F, Hidden G. 2001. Lymphatic drainage of the pericardium to the mediastinal lymph nodes. *Surg. Radiol. Anat.* 23:317–19

116. Palmer AS, Miller AJ, Greene R. 1998. The lymphatic drainage of the left ventricle in the Yucatan minipig. *Lymphology* 31:30–33

117. Miller AJ, DeBoer A, Pick R, Van Pelt L, Palmer AS, Huber MP. 1988. The lymphatic drainage of the pericardial space in the dog. *Lymphology* 21:227–33

118. Afanasyeva M, Georgakopoulos D, Rose NR. 2004. Autoimmune myocarditis: cellular mediators of cardiac dysfunction. *Autoimmun. Rev.* 3:476–86

119. Guler ML, Ligons DL, Wang Y, Bianco M, Broman KW, Rose NR. 2005. Two autoimmune diabetes loci influencing T cell apoptosis control susceptibility to experimental autoimmune myocarditis. *J. Immunol.* 174:2167–73

120. Villasenor J, Benoist C, Mathis D. 2005. AIRE and APECED: molecular insights into an autoimmune disease. *Immunol. Rev.* 204:156–64

121. Gotter J, Brors B, Hergenhahn M, Kyewski B. 2004. Medullary epithelial cells of the human thymus express a highly diverse selection of tissue-specific genes colocalized in chromosomal clusters. *J. Exp. Med.* 199:155–66

122. Smith S, Allen P. 1991. Myosin-induced acute myocarditis is a T cell-mediated disease. *J. Immunol.* 147:2141–47

123. Greenwald RJ, Freeman GJ, Sharpe AH. 2005. The B7 family revisited. *Annu. Rev. Immunol.* 23:515–48

124. Tivol EA, Borriello F, Schweitzer AN, Lynch WP, Bluestone JA, Sharpe AH. 1995. Loss of CTLA-4 leads to massive lymphoproliferation and fatal multiorgan tissue destruction, revealing a critical negative regulatory role of CTLA-4. *Immunity* 3:541–47

125. Oosterwegel MA, Mandelbrot DA, Boyd

SD, Lorsbach RB, Jarrett DY, et al. 1999. The role of CTLA-4 in regulating Th2 differentiation. *J. Immunol.* 163:2634–39

126. Nishimura H, Okazaki T, Tanaka Y, Nakatani K, Hara M, et al. 2001. Autoimmune dilated cardiomyopathy in PD-1 receptor-deficient mice. *Science* 291:319–22

127. Okazaki T, Tanaka Y, Nishio R, Mitsuiye T, Mizoguchi A, et al. 2003. Autoantibodies against cardiac troponin I are responsible for dilated cardiomyopathy in PD-1-deficient mice. *Nat. Med.* 9:1477–83

128. Liang SC, Latchman YE, Buhlmann JE, Tomczak MF, Horwitz BH, et al. 2003. Regulation of PD-1, PD-L1, and PD-L2 expression during normal and autoimmune responses. *Eur. J. Immunol.* 33:2706–16

129. Freeman GJ, Long AJ, Iwai Y, Bourque K, Chernova T, et al. 2000. Engagement of the PD-1 immunoinhibitory receptor by a novel B7 family member leads to negative regulation of lymphocyte activation. *J. Exp. Med.* 192:1027–34

130. Rodig N, Ryan T, Allen JA, Pang H, Grabie N, et al. 2003. Endothelial expression of PD-L1 and PD-L2 down-regulates CD8+ T cell activation and cytolysis. *Eur. J. Immunol.* 33:3117–26

131. Latchman YE, Liang SC, Wu Y, Chernova T, Sobel RA, et al. 2004. PD-L1-deficient mice show that PD-L1 on T cells, antigen-presenting cells, and host tissues negatively regulates T cells. *Proc. Natl. Acad. Sci. USA* 101:10691–96

132. Cai G, Karni A, Oliveira EM, Weiner HL, Hafler DA, Freeman GJ. 2004. PD-1 ligands, negative regulators for activation of naive, memory, and recently activated human CD4+ T cells. *Cell. Immunol.* 230:89–98

133. Ozkaynak E, Wang L, Goodearl A, McDonald K, Qin S, et al. 2002. Programmed death-1 targeting can promote allograft survival. *J. Immunol.* 169:6546–53

134. Ito T, Ueno T, Clarkson MR, Yuan X, Jurewicz MM, et al. 2005. Analysis of the role of negative T cell costimulatory pathways in CD4 and CD8 T cell-mediated alloimmune responses in vivo. *J. Immunol.* 174:6648–56

135. Sospedra M, Martin R. 2004. Immunology of multiple sclerosis. *Annu. Rev. Immunol.* 23:683–747

136. Raz I, Eldor R, Naparstek Y. 2005. Immune modulation for prevention of type 1 diabetes mellitus. *Trends Biotechnol.* 23:128–34

137. Hidalgo LG, Halloran PF. 2002. Role of IFN-gamma in allograft rejection. *Crit. Rev. Immunol.* 22:317–49

138. Berenson LS, Ota N, Murphy KM. 2004. Issues in T-helper 1 development—resolved and unresolved. *Immunol. Rev.* 202:157–74

139. Huber S. 2004. T cells in coxsackievirus-induced myocarditis. *Viral Immunol.* 17:152–64

140. Fairweather D, Kaya Z, Shellam GR, Lawson CM, Rose NR. 2001. From infection to autoimmunity. *J. Autoimmun.* 16:175–86

141. Fairweather D, Frisancho-Kiss S, Yusung SA, Barrett MA, Davis SE, et al. 2005. IL-12 protects against coxsackievirus B3-induced myocarditis by increasing IFN-γ and macrophage and neutrophil populations in the heart. *J. Immunol.* 174:261–69

142. Fairweather D, Frisancho-Kiss S, Yusung SA, Barrett MA, Davis SE, et al. 2004. Interferon-γ protects against chronic viral myocarditis by reducing mast cell degranulation, fibrosis, and the profibrotic cytokines transforming growth factor-β1, interleukin-1β, and interleukin-4 in the heart. *Am. J. Pathol.* 165:1883–94

143. Smith S, Allen P. 1992. Neutralization of endogenous tumor necrosis factor ameliorates the severity of myosin-induced myocarditis. *Circ. Res.* 70:856–63

144. Afanasyeva M, Wang Y, Kaya Z, Park S, Zilliox MJ, et al. 2001. Experimental autoimmune myocarditis in A/J mice is an interleukin-4-dependent disease with a

Th2 phenotype. *Am. J. Pathol.* 159:193–203

145. Afanasyeva M, Wang Y, Kaya Z, Stafford EA, Dohmen KM, et al. 2001. Interleukin-12 receptor/STAT4 signaling is required for the development of autoimmune myocarditis in mice by an interferon-γ-independent pathway. *Circulation* 104:3145–51

146. Eriksson U, Kurrer MO, Sebald W, Brombacher F, Kopf M. 2001. Dual role of the IL-12/IFN-γ axis in the development of autoimmune myocarditis: induction by IL-12 and protection by IFN-γ. *J. Immunol.* 167:5464–69

147. Eriksson U, Kurrer MO, Bingisser R, Eugster HP, Saremaslani P, et al. 2001. Lethal autoimmune myocarditis in interferon-γ receptor-deficient mice: enhanced disease severity by impaired inducible nitric oxide synthase induction. *Circulation* 103:18–21

148. Dalton DK, Haynes L, Chu C-Q, Swain SL, Wittmer S. 2000. Interferon γ eliminates responding CD4 T cells during mycobacterial infection by inducing apoptosis of activated CD4 T cells. *J. Exp. Med.* 192:117–22

149. Afanasyeva M, Georgakopoulos D, Belardi DF, Bedja D, Fairweather D, et al. 2005. Impaired up-regulation of CD25 on CD4+ T cells in IFN-γ knockout mice is associated with progression of myocarditis to heart failure. *Proc. Natl. Acad. Sci. USA* 102:180–85

150. Zettl UK, Mix E, Zielasek J, Stangel M, Hartung H-P, Gold R. 1997. Apoptosis of myelin-reactive T cells induced by reactive oxygen and nitrogen intermediates in vitro. *Cell. Immunol.* 178:1–8

151. Dalton DK, Wittmer S. 2005. Nitric-oxide-dependent and independent mechanisms of protection from CNS inflammation during Th1-mediated autoimmunity: evidence from EAE in iNOS KO mice. *J. Neuroimmunol.* 160:110–21

152. Miura M, El-Sawy T, Fairchild RL. 2003. Neutrophils mediate parenchymal tissue necrosis and accelerate the rejection of complete major histocompatibility complex-disparate cardiac allografts in the absence of Interferon-γ. *Am. J. Pathol.* 162:509–19

153. Nagano H, Mitchell RN, Taylor MK, Hasegawa S, Tilney NL, Libby P. 1997. Interferon-gamma deficiency prevents coronary arteriosclerosis but not myocardial rejection in transplanted mouse hearts. *J. Clin. Invest.* 100:550–57

154. Halloran PF, Miller LW, Urmson J, Ramassar V, Zhu L-F, et al. 2001. IFN-γ alters the pathology of graft rejection: protection from early necrosis. *J. Immunol.* 166:7072–81

155. Grabie N, Hsieh DT, Buono C, Westrich JR, Allen JA, et al. 2003. Neutrophils sustain pathogenic CD8+ T cell responses in the heart. *Am. J. Pathol.* 163:2413–20

156. Zehntner SP, Brickman C, Bourbonniere L, Remington L, Caruso M, Owens T. 2005. Neutrophils that infiltrate the central nervous system regulate T cell responses. *J. Immunol.* 174:5124–31

157. Bettelli E, Kuchroo VK. 2005. IL-12- and IL-23-induced T helper cell subsets: birds of the same feather flock together. *J. Exp. Med.* 201:169–71

158. Langrish CL, Chen Y, Blumenschein WM, Mattson J, Basham B, et al. 2005. IL-23 drives a pathogenic T cell population that induces autoimmune inflammation. *J. Exp. Med.* 201:233–40

159. Aretz HT, Billingham ME, Edwards WD, Factor SM, Fallon JT, et al. 1987. Myocarditis. A histopathologic definition and classification. *Am. J. Cardiovasc. Pathol.* 1:3–14

160. Ullal SR, Kluge TH, Kerth WJ, Gerbode F. 1972. Anatomical studies on lymph drainage of the heart in dogs. *Ann. Surg.* 175(3):305–10

Annu. Rev. Physiol. 2006. 68:97–121
doi: 10.1146/annurev.physiol.68.040104.113828
First published online as a Review in Advance on September 7, 2005

Transcription Factors and Congenital Heart Defects

Krista L. Clark, Katherine E. Yutzey, and D. Woodrow Benson

Divisions of Cardiology and Molecular Cardiovascular Biology, Cincinnati Children's Hospital Medical Center, Cincinnati, Ohio 45229; email: Krista.Clark@cchmc.org, Katherine.Yutzey@cchmc.org, Woody.Benson@cchmc.org

Key Words pediatric heart disease, cardiac anomalies, NKX2.5, GATA4, TBX5

■ **Abstract** Although there have been important advances in diagnostic modalities and therapeutic strategies for congenital heart defects (CHD), these malformations still lead to significant morbidity and mortality in the human population. Over the past 10 years, characterization of the genetic causes of CHD has begun to elucidate some of the molecular causes of these defects. Linkage analysis and candidate-gene approaches have been used to identify gene mutations that are associated with both familial and sporadic cases of CHD. Complementation of the human studies with developmental studies in mouse models provides information for the roles of these genes in normal development as well as indications for disease pathogenesis. Biochemical analysis of these gene mutations has provided further insight into the molecular effects of these genetic mutations. Here we review genetic, developmental, and biochemical studies of six cardiac transcription factors that have been identified as genetic causes for CHD in humans.

INTRODUCTION

Congenital heart defects (CHD) are an important component of pediatric heart disease and constitute a major portion of clinically significant developmental abnormalities (1, 2). Although the clinical literature usually cites an incidence of CHD of ∼1%, to a geneticist it is important to note that the incidence of cardiac malformations is much higher at 3–5% (3). Despite exciting advances in surgical therapy, morbidity and mortality remain high for these important clinical problems. Some malformations are not amenable to surgery, and even defects definitively treated with surgery may result in a shortened life span due to residual heart disease. Until recently, little was known of the causes of these common birth defects.

In a move toward a proactive rather than reactive approach to CHD, molecular genetic studies of human patient populations have been used to identify underlying causes of cardiac developmental anomalies. Both linkage analysis in kindreds and candidate-gene approaches have provided evidence for the contributions of

0066-4278/06/0315-0097$20.00

specific molecular pathways in the pathogenesis of heart malformations. In the past 10 years, researchers have identified a variety of CHD-causing gene mutations, most of which exhibit a dominant pattern of disease transmission. Several of these genes encode transcription factors that regulate various aspects of cardiac development and gene expression. The identification of these human mutations has provided important information for both the developmental biologist and the biochemist. Genetic engineering in model organisms has provided further insights into the biological functions of these genes, particularly their role in normal cardiac development and their contribution to the pathogenesis of human cardiac malformations. Biochemical studies of mutant proteins first identified in patients with CHD have been used to explain how these mutations affect cardiac developmental processes on a molecular level. At the same time, characterization of mutants has been used to identify functional motifs within protein structures. This information has been used in further analyses of the molecular regulation of normal and abnormal heart development.

The complementary nature of genetic, developmental, and biochemical approaches continue to advance our understanding of the molecular basis of CHD in humans. At the core of these molecular and developmental pathways are cardiac transcription factors. These families of proteins play fundamental roles in regulating the pattern and timing of expression of the genes responsible for the cardiac-lineage determination, valvulogenesis, conduction-system development, and heart chamber formation. Here we review advances in the understanding of the pathogenesis of mutations within six cardiac transcription factors that have been identified as genetic causes for CHD in humans: *ZIC3*, *NKX2.5*, *TBX5*, *GATA4*, *TFAP2B*, *TBX1*, and *FOG2*. Because of their role in the orchestration of cardiac development, mutations within these genes result in significant disruption and/or misregulation of downstream gene expression, which leads to cardiovascular malformations. Animal models engineered to express mutations in these genes have been used to reveal the role of these proteins in normal and abnormal heart development. Finally, we review how the biochemical analysis of proteins resulting from mutations within these transcription factor genes has contributed to the evaluation of disease pathogenesis.

IDENTIFICATION OF TRANSCRIPTION FACTORS
INVOLVED IN HUMAN CHD

To date, genes responsible for human disease pathogenesis have been identified using three genetic techniques. All begin with the identification of a disease phenotype in an affected individual, the proband. Determination of phenotype in family members is used to construct a pedigree, and disease transmission is then characterized. Informative kindreds lead to the identification of coinheritance of disease traits with chromosomal markers; this linkage analysis leads to detection of a chromosomal locus, thereby reducing the number of positional candidate genes to those located in a relatively small chromosomal region. This strategy has been successfully used to identify heterozygous mutations in several transcription factors including TBX5, NKX2.5, ZIC3, and TFAP2B (4–7). In other cases, the chromosomal locus is identified by cytogenetic alteration, such as deletion or translocation; this method was used to identify a role for GATA4 in human CHD (8). In recent years, improved understanding of the biology of cardiac development has led to the identification of biological candidate genes. This technique selects a candidate gene on the basis of known protein function and expression patterns during cardiac development. Such approaches have been greatly facilitated by using informatic techniques to mine whole human genome sequence databases. For example, on the basis of the similarity of the cardiac phenotype to tetralogy of Fallot (TOF) in a homozygous deletion mouse, *FOG2* was used as a candidate gene in a cohort of patients with this phenotype (9). As our understanding of cardiac development advances, we can expect the biological candidate approach to be used more often. The discussion below characterizes the genetic approaches leading to the identification of mutations in six cardiac transcription factors and their resulting cardiac phenotypes.

NKX2.5

Also known as CSX (cardiac-specific homeobox), NKX2.5 is a member of the NK2 class of homeodomain transcription factors. An autosomal-dominant disease locus associated with atrioventricular (AV) block and cardiac septal defects was mapped to chromosome 5q35, where the cardiac transcription factor NKX2.5 is encoded (5). Mutational analysis of *NKX2.5* revealed three different heterozygous mutations predicted to disrupt the DNA-binding function of the homeodomain (5). Since the initial report, more than 30 mutations have been identified within *NKX2.5* (Figure 1*a*) (5, 10–17). Heterozygous *NKX2.5* mutations account for ~4% of all CHD; atrial septal defects (ASD) are most common, but ventricular septal defect (VSD), left ventricle noncompaction, TOF, double-outlet right ventricle (DORV), subvalvular aortic stenosis, and Ebstein anomaly of the tricuspid valve have also been observed (Figure 2) (10, 11, 15). The varied manifestations of cardiac defects associated with *NKX2.5* mutation are indicative of its multifunctional capacity during cardiac development. On the basis of the analysis of phenotypes associated with *NKX2.5* mutations, a genotype-phenotype correlation between AV block and specific types of *NKX2.5* mutations has emerged. Nonsense or frameshift mutations at any location or homeodomain missense mutations are associated with an AV conduction block phenotype, whereas individuals with missense mutations in the amino or carboxy terminus do not have this phenotype (18).

GATA4

A member of the zinc-finger transcription factor family, GATA4 is encoded at chromosome 8p23.1. Sixty percent of patients with deletion of this region of chromosome 8p have CHD including ASD, VSD, atrioventricular septal defect

(AVSD), DORV, dextrocardia, and pulmonary stenosis (PS) (Figure 2) (8). More recently, linkage analysis in a large pedigree with an autosomal-dominant transmission pattern of isolated septal defect including ASD or VSD mapped to a locus at 8p22–23, where GATA4 is encoded (19). Using *GATA4* as a candidate gene, missense and nonsense mutations *G296S* and *E359del* were found (19). An additional ASD candidate-gene study identified a missense mutation, *S52F*, which brings the total number of CHD-associated mutations in *GATA4* to three (Figure 1*a*) (20).

TBX5

Mutations in the T-box transcription factor TBX5 cause Holt-Oram syndrome (HOS) (21, 22). First identified in 1960, HOS is a highly penetrant autosomal-dominant disorder characterized by CHD in association with congenital skeletal malformation of the upper limb in the preaxial ray (4, 23, 24). Cardiac anomalies

a)

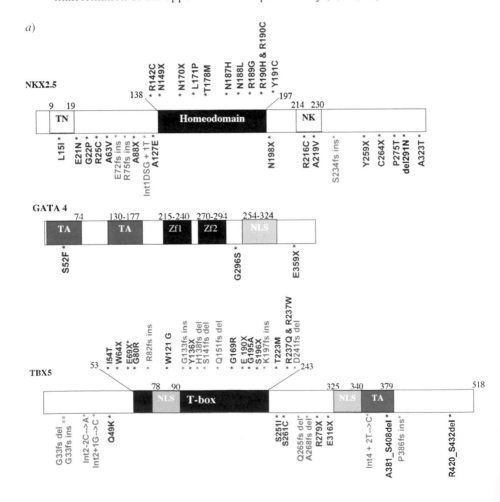

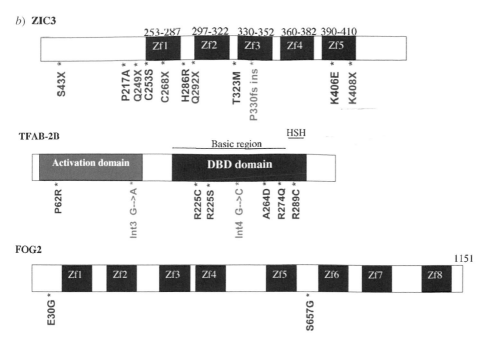

Figure 1 Mutations in human cardiac transcription factors associated with CHD. The structural domains of each protein are shown and labeled above. (*a*) Mutations identified in NKX2.5, GATA4, and TBX5. (*b*) Mutations identified in ZIC3, TFAP2B, and FOG2. Purple: missense; red: nonsense; black: deletions; green: insertions; aqua: intronic mutations. References cited in text.

in these individuals include ASD, VSD, AVSD, and conduction abnormalities (Figure 2). Although the severity and type of cardiac manifestations vary, the upper-limb anomalies are always present (23). The HOS disease locus was initially mapped to the long arm of chromosome 12 in 1994, and positional cloning three years later identified mutations within *TBX5* as the cause of HOS (4, 21, 22). Since the identification of *TBX5* mutation as the basis of HOS, 37 mutations have been identified (Figure 1*a*) (21, 22). To date, nonsyndromic mutations within *TBX5* have not been identified (25).

ZIC3

Heterotaxy is characterized by a variable group of congenital anomalies that include situs inversus or situs ambiguous of visceral organs in conjunction with cardiac anomalies. Associated cardiac defects are varied and include ASD, AVSD, transposition of the great arteries (TGA), PS, and totally anomalous pulmonary venous return (Figure 2) (26). Pedigree analysis of a large kindred with familial heterotaxy established an X-linked pattern of inheritance, and linkage analysis mapped the disease locus to chromosome Xq21–q27 (6). Cytogenetic analysis of

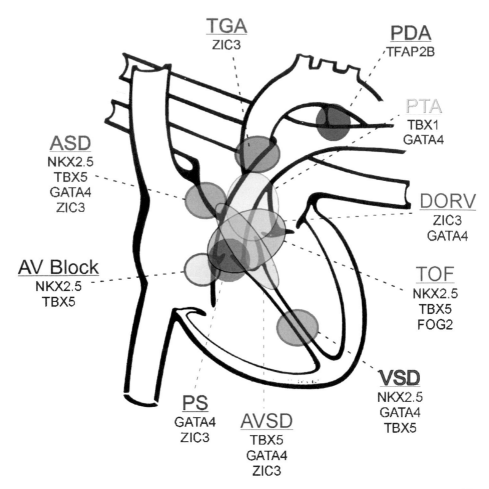

Figure 2   Sites of structural anomalies associated with transcription factor mutations. ASD: atrial septal defect; AV: atrioventricular; AVSD: atrioventricular septal defect; DORV: double-outlet right ventricle; PDA: patent ductus arteriosus; PTA: persistent truncus arteriosus; PS: pulmonary stenosis; TGA: transposition of the great arteries; TOF: tetralogy of Fallot; VSD: ventricular septal defect. References cited in text.

another unrelated male with heterotaxy syndrome indicated a microdeletion at the same chromosomal location (27). Positional cloning of the disease locus identified *ZIC3* as the causative gene. *ZIC3* is a member of the GLI (*gli*oma-associated *o*ncogene) superfamily of zinc-finger transcription factors and is involved in establishing left/right patterning during embryonic development (28). To date, 11 nonsense, frameshift, or missense mutations and two deletions have been identified in male individuals with heterotaxy syndrome (Figure 1*b*) (27–30). Heterozygous female carriers appear to be clinically unaffected (6, 28).

TFAP2B

TFAP2B is a member of the retinoic acid–responsive activator protein (AP) 2 transcription factor family (31). Linkage analysis was used to map the Char syndrome disease locus to chromosome 6p12–p21, where TFAP2B had been previously localized (7, 32). Char syndrome is an autosomal-dominant disease characterized by patent ductus arteriosus (PDA), facial dysmorphism, and hand anomalies (Figure 2) (32). In the initial report, two missense mutations, *A264D* and *R289C*, were identified (33). To date, eight mutations associated with CHD have been identified in the *TFAP2B* gene (Figure 1*b*) (33–35).

TBX1

The *TBX1* gene, which encodes a member of the T-box transcription factor family, is located within a 3-Mb region known as the DiGeorge critical region (DGCR). This interval contains 30 genes that are deleted in patients with 22q11.2 deletion (del22q11) syndrome, which includes the diagnoses of DiGeorge syndrome, velo-cardio-facial syndrome, and conotruncal anomaly face syndrome. The signature cardiovascular manifestations of del22q11.2 syndrome are conotruncal defects, which include TOF, aortic-arch anomalies, persistent truncus arteriosus (PTA), and malaligned VSD (Figure 2) (36). The genes located within the DGCR on 22q11 are almost completely conserved on mouse chromosome 16. Mice with a heterozygous deletion of a 1.5-Mb homologous region of mouse chromosome 16 [Df(16)1] exhibit phenocopy of the aortic-arch, thymic, and parathyroid defects of del22q11.2 syndrome (37, 38). Complementation studies in the *Df(16)1* mice and the generation of *Tbx1*$^{-/-}$ mice identified an essential role for *Tbx1* during normal pharyngeal and cardiac development and made it a candidate gene for the cardiovascular manifestations of del22q11.2 syndrome. These mice, specifically the *Tbx1*$^{-/-}$ mice, appear to exhibit many of the cardiovascular malformations seen in del22q11.2 syndrome patients (39). These observations were used to support *TBX1* as a candidate gene for conotruncal defects in patients without a 22q11.2 deletion (40). However, despite evaluation of large cohorts of deletion-negative patients with conotruncal defects, only three *TBX1* mutations have been identified, and these account for <1% of conotruncal malformations in this population (41–43).

FOG2

FOG2/ZFPM2 is a multitype zinc-finger transcription factor that modulates the activity of the cardiac transcription factor GATA4 (44). The similarity of complex cardiac malformations in *FOG2*-null mice to human TOF led to the analysis of *FOG2* as a candidate gene in 47 unrelated people with this cardiac phenotype (9). Only 2 of these unrelated patients with TOF showed heterozygous missense mutations *S657G* or *E30G* within the *FOG2* gene, indicating its contribution in at least a subset of isolated cases of TOF (Figures 1*b*, 2) (9). To date, familial transmission of CHD associated with *FOG2* mutations has not been analyzed.

TRANSCRIPTIONAL REGULATION OF CARDIAC DEVELOPMENT

In humans, most mutations within the transcription factors that result in CHD are heterozygous alleles. The lack of individuals with homozygous mutations is suggestive of embryonic lethality associated with this genotype. Generation of both hetero- and homozygous deletions of these transcription factors is possible with gene-targeting technologies in mice. Mouse models have provided a substantial amount of knowledge about disease pathogenesis and the function of these transcription factors in normal development. As evidenced by the varying structural and functional cardiac anomalies associated with mutations in these transcription factors, each is involved in multiple aspects of heart development. The study of transcription factor function in the developing heart of murine embryos has also provided important insights into the mechanisms of CHD.

During vertebrate embryonic development, the heart is the first organ to develop and function. The cells that will eventually contribute to the heart are initially located in the anterior lateral plate mesoderm just after gastrulation (45). These progenitors then condense to form two lateral heart primordia that comprise the myocardial and endocardial cell lineages (46). The laterally placed heart primordia then converge along the midline to form a beating heart tube (47). Specification of cardiomyocytes that will contribute to the conduction system also begins during this period (48). During the linear heart tube stage, specific patterns of transcription factor expression begin to differentially define distinct segmental precursors of the outflow tract, atria, and ventricles (45). As cardiac maturation continues, the linear heart tube undergoes rightward looping, which is the first morphologic indication of patterning the right/left axis of the organism (49). Upon completion of looping, septation and morphologic distinction of cardiac chambers and associated valves begins. Coincident with the initiation of valvulogenesis, two extracardiac cell populations begin to invade the developing heart from the neural crest and proepicardium (46). The cardiac neural crest cells migrate from the neural tube and contribute to the septation of the outflow tract into distinct vessels of the aortic and pulmonary arteries. The neural crest cells also contribute to the formation of the aortic-arch arteries and the ductus arteriosus, a connection between the aorta and pulmonary artery. Cells of the proepicardium, a distinct subpopulation of lateral mesoderm, migrate across the surface of the heart to form the epicardium, the outermost cardiac epithelium (46). These cells eventually differentiate into a variety of cell types in the heart including the coronary vessels (46). Cardiac transcription factors associated with human CHD are implicated in each of these processes.

MYOCARDIAL SPECIFICATION, MORPHOGENESIS, SEPTATION, AND CONDUCTION

The first step in the development of the cardiac muscle is the specification of the cardiac lineage. The myogenic progenitors of the heart initially arise from the anterior lateral mesoderm just after gastrulation. Cardiogenic signals induce expression of both *Nkx2.5* and *Gata4*, which are among the earliest markers of cardiomyocytes. In the mouse, both *Nkx2.5* and *Gata4* mRNA are detectable by 7–7.5 dpc in the precardiac mesoderm prior to cardiac differentiation markers, such as atrial naturietic factor (ANF) (2, 50–57). *Tbx5* is also initially expressed throughout the cardiac mesoderm, but it is restricted to the atria and left ventricle at the primitive heart tube stage (58). *Nkx2.5*, *Gata4*, and *Tbx5* are all expressed in cardiac muscle and regulate expression of cardiac contractile protein genes. In addition, *Nkx2.5* and *Tbx5* are both expressed throughout the central conduction system within the heart and may have a cooperative function in the recruitment and maintenance of specialized conduction system myocytes (59–61). In the later stages of heart development, the atria and ventricles undergo septation, a process that is disrupted with mutation of *Nkx2.5*, *Gata4*, or *Tbx5*.

Homozygous null mice have been used to characterize the function of these proteins in normal heart development. In the absence of *Nkx2.5*, the heart tube forms, but abnormal heart morphogenesis ensues by embryonic day 9.5 (E9.5) and embryonic lethality occurs around E10.5 owing to hemodynamic insufficiency (55). A failure of both cardiac looping and molecular differentiation of the ventricles in these mice indicates that *Nkx2.5* is critical in these processes (55). Null mutation of *Gata4* results in abnormal ventral folding of the embryo, failure of the heart primordia to fuse into a single heart tube, and lethality by E10.5 (62, 63). This cardiogenic defect is believed to be due to an earlier defect in the ventral migration of endodermal cells, a process for which *Gata4* is indispensable, thus preventing the subsequent migration of the myocardial cells (48). Mice lacking *Tbx5* do not survive past E10.5 owing to severe defects in heart tube formation with hypoplastic atria (64). Conversely, overexpression of *Tbx5* throughout the heart tube inhibits ventricular chamber maturation, resulting in an "atrialized" heart (65). Together these studies indicate a critical role for *Tbx5* in early delineation of the atrial chambers. In each case, null embryos do not survive beyond midgestation, which suggests that homozygosity of CHD-causing mutant alleles of these transcription factors may never be seen in the human population.

Mice with heterozygous deletion of *Nkx2.5*, *Gata4*, or *Tbx5* provide information about the developmental causes of disease pathogenesis and recapitulate human CHD mutations. Heterozygous mutation of either *Nkx2.5* or *Tbx5* in mice results in ASD and progressive AV block (59, 66). Mice with ventricular muscle-cell restricted loss of *Nkx2.5* show no defects in cardiac morphology but have a gradual conduction atrophy that results in progressive AV block (61). Two transgenic mouse lines, β-MHC-TG (1183P), which expresses a dominant-negative *Nkx2.5*, and α-MHC-TG (DC), which expresses a C-terminal truncation of Nkx2.5, have been made in an attempt to replicate the expression of mutant forms of *Nkx2.5* seen in the human heterozygous condition (67, 69). Both of these mutant mice recapitulate the progressive AV-block phenotype observed in humans (70). *Gata4*$^{+/-}$ mice have been generated and have cardiac defects similar to those found in humans with mutations in *GATA4* including AVSD, DORV, and hypoplastic ventricular myocardium (71). Heterozygous mouse model systems demonstrate a dose dependence for the mutant proteins during development and give insight into their normal functions and disease pathogenesis.

GATA4, TBX5, and NKX2.5 are transcriptional partners, and as a complex they synergize in the induction of downstream cardiac genes including *ANF* (19, 72, 73). ASDs are seen with *TBX5*, *GATA4*, and *NKX2.5* mutations, and both *TBX5* and *NKX2.5* mutations also cause AV block in humans. Similarities in their mutational phenotypes suggest their coordinated function is likely required for these developmental processes. *Tbx5* together with *Nkx2.5* synergistically activate both *ANF* and *connexin 40*, which show markedly decreased expression in *Tbx5*-null mice (64). In vitro analysis of recombinant *Tbx5*-containing missense mutations known to cause ASD in humans indicates these single-nucleotide changes disrupt the interaction with *Gata4* (19). On the basis of these observations and owing to the similarity in the phenotypes of patients with mutations in either *GATA4* or *TBX5*, it is likely these transcriptional partners have a coordinate role in atrial septation. Specific functions for GATA4 and TBX5 in atrial septation and conduction system development have yet to be fully elucidated.

LEFT/RIGHT PATTERNING AND HEART DEVELOPMENT

Left/right patterning of the embryo occurs during gastrulation prior to organ formation (49). Left/right asymmetry is apparent in the positioning of the organs of the chest and abdomen that begin organogenesis in the midline and in accordance with left/right patterning will assume their normal position on one side of the body or the other (45). Heterotaxy syndrome results from inappropriate determination of the left/right axis during embryonic development. Cardiac anomalies associated with heterotaxy are a result of defects with cardiac looping, a process that is governed by left/right axis patterning.

The human genetic studies that led to identification of mutations in *ZIC3* as a cause of human heterotaxy were the first indication that this transcription factor was involved in heart development (6, 28). Although *ZIC3* has no cardiac expression, its presence in the caudal primitive streak at E6.5–E7 at the time of gastrulation, along with the observed phenotypes of the mutants, position it to play a role in left/right axis determination (74). According to Purandare et al. (29) three major steps are involved in establishing the left/right axis: node-dependent symmetry breaking, establishment of the embryonic midline, and node-dependent signaling in the left lateral plate mesoderm. Failure at any step results in the disruption of axis determination, which leads to a hetcrotaxy phenotype. Exactly where ZIC3 is involved in this process remains to be identified.

Homozygous deletion of *Zic3* in mice is embryonically lethal in 50% of mice, with additional neonatal death (29). Examination of stillborn pups shows defects in left/right axis development, the central nervous system, and the axial skeleton. Cardiac defects, found in 68% of stillborn pups, include TGA, aortic-arch anomalies, ASD, and VSD and are similar to those seen in patients with *ZIC3* mutation (29). Examination of embryos at E10.5 showed abnormal cardiac looping and inappropriate positioning of the heart in the left thorax instead of at the midline. One hypothesis for the role of ZIC3 suggests it may modulate the activity of GLI proteins, which are downstream-effector molecules of Sonic hedgehog (Shh) signaling (75). Left/right signaling cascades direct asymmetrical expression of *Shh* and *Nodal* in the lateral mesoderm, and maintenance of their asymmetric expression is critical for proper axial development (45). $Shh^{-/-}$ mice show cardiac defects similar to those seen in $Zic3^{-/-}$ mice, including abnormal cardiac looping and aberrant positioning of the heart (76, 77). *Nodal* is a member of the TGF-β family and is a key player in the initial stages of establishing the left/right axis. Although $Zic3^{-/-}$ mice have normal *Nodal* expression initially, it is not properly maintained and fails to localize asymmetrically to the left side (29). This failed localization results in abnormal signaling downstream of *Nodal*. Future studies are needed to help establish the function of Zic3 in these complex signal transduction pathways.

NEURAL CREST CONTRIBUTION TO THE OUTFLOW TRACT

The mature cardiac outflow track is formed by complex interaction of primary and secondary heart fields and cardiac neural crest. The initial foundations of the outflow track are derived from cells of the anterior heart field (78). Another population of cells, derived from the secondary heart field, is incorporated into the distal portion of the outflow tract and also forms the pharyngeal arch arteries (79). Neural crest cells migrate into the outflow tract and contribute to formation of the ductus arteriosus and septation of the truncus arteriosus into the aortic and pulmonary trunks (47). In the absence of neural crest, as a result of ablation of the premigratory cardiac neural crest in chicken embryos, a multitude of cardiac anomalies are observed. These include PTA, DORV, TOF, and the absence of various pharyngeal arch structures such as aortic-arch arteries and the ductus arteriosus (80). Similar cardiac and craniofacial anomalies are observed in genetically manipulated mice with compromised cardiac neural crest development (10).

In mice, *Tfap2b* is expressed in early premigratory and migrating neural crest cells and has a role in their migration into the aortic arches as they form the ductus arteriosus (33, 81). The ductus arteriosus normally closes shortly after birth. Failure of this closure results in PDA, observed in Char syndrome patients with *TFAP2B* mutations (82). Analysis of the targeted mutation of *TFAP2B* has not been informative as to the disease mechanism because these mice do not recapitulate the Char syndrome phenotype seen in humans. Mice with heterozygous *Tfap2b* deletion mutations are apparently normal, and homozygous deletion induces no apparent cardiac anomalies. Instead, these mice do not survive beyond postnatal day 1 or 2 owing to congenital polycystic kidney disease caused by excessive apoptosis of renal epithelial cells, indicating a possible role for TFAB2B in cell survival during embryogenesis (81). No mechanism has been determined to explain the disparity between the phenotype seen in humans versus that seen in the *Tfap2b*$^{-/-}$ mice. Generation of a mouse model with targeted missense mutations in *Tfap2b* will more accurately represent the genetic lesion seen in humans and may yield a more representative disease model.

To generate a mouse model of del22q11.2 syndrome, *Df(16)1* mice with a heterozygous deletion of a 1.5-Mb region of mouse chromosome 16, which is homologous to the DGCR, were generated (37). The genes located within the DGCR on human 22q11, which include *TBX1*, are mostly conserved within this region of mouse chromosome 16. This mouse model replicates many features of neural crest ablation and has cardiac and craniofacial defects similar to those seen in del22q11.2 syndrome, including aortic-arch anomalies (37). Complementation studies in the *Df(16)1* mice, and the subsequent generation of *Tbx1*$^{-/-}$ mice by three groups, identified an essential role for *Tbx1* during normal pharyngeal and cardiac development (37–40). *Tbx1*$^{+/-}$ mice have interrupted aortic arches, as seen in del22q11.2 syndrome, but rarely have PTA or conotruncal defects. Homozygous deletion of *Tbx1* in mice is embryonically lethal. Examination of the hearts of these *Tbx1*-null mice indicates a number of heart defects affecting formation and growth of the pharyngeal arch arteries, growth and septation of the outflow tract, interventricular septation, and conal alignment (83). The combination of these defects represents all of those seen in del22q11.2 syndrome patients (reviewed in Reference 84). Although the mouse models reiterate the phenotypic abnormalities of del22q11.2 syndrome, mutation analysis of *TBX1* has identified few sequence alterations, indicating that mutations in this gene are not a significant genetic cause of this disease in humans. Because del22q11.2 syndrome in human patients results from a large chromosomal deletion, it is likely the disease phenotype is contributed to by other genes in the deleted region. This gives some explanation to the fact that deletion of *Tbx1* alone in mice does not completely replicate all the phenotype abnormalities seen in human del22q11.2 syndrome.

Because *Tbx1* is expressed in the pharyngeal mesenchymal arch and endodermal pouch but not in neural crest, the neural crest–derived structural defects, specifically PTA, may be due to a non-cell-autonomous mechanism (85, 86; reviewed in 84). This mechanism may involve a diffusible signal regulated by *Tbx1* in the endodermal pouch or pharyngeal mesenchyme that normally provides guidance to the migrating neural crest cells (84). One candidate for this diffusible signal is fibroblast growth factor 8 (Fgf8). *Fgf8* expression overlaps with and is dependent on *Tbx1* expression in the pharyngeal endoderm, and heterozygous deletion of *Fgf8* in *Tbx1*$^{+/-}$ mice increases the penetrance of the cardiovascular defects seen in these animals (83). Using *Tbx1-Cre* mice, Brown et al. (87) developed a mouse with tissue-specific inactivation of *Fgf8*. These mice have congenital cardiovascular defects of the outflow tract similar to those seen in del22q11.2 syndrome, lending support for a role of *Fgf8* downstream of *Tbx1* (87).

EPICARDIAL DEVELOPMENT

In mice and chickens, the epicardium is derived from cells of the septum transversum, also called the proepicardium (88). These cells migrate over the surface of the myocardium to form the epicardium and derivative cell lineages. During normal development, contact with the myocardium induces these proepicardial cells to undergo an epithelial-to-mesenchymal transition (EMT) and migrate into the subepicardial space. A subpopulation of these cells eventually gives rise to the coronary vasculature (88, 89). If the proepicardial cells are prevented from interacting with the myocardium, the developing heart is misaligned and has thin ventricular walls and coronary vessel development is absent (89).

Fog2 is initially expressed in the mouse heart and septum transversum at E8.5 (44). In 2000, *Fog2* deletion mice were reported (44, 90). *Fog2*-null mice die at midgestation with severe cardiac defects reminiscent of TOF (88, 91). This characterization initially led researchers to investigate *FOG2* as a candidate gene for TOF in humans. Developmentally, the TOF phenotype is seen as an outflow tract alignment defect with AVSD. Although the specific role of FOG2 in the formation of TOF has not been determined, evidence suggests that a subpopulation of epicardial cells infiltrate the AV canal and participate in formation of the AV valves and septum (92). *FOG2* mutations may disrupt the migration of this population of cells, resulting in the malaligned outflow tract seen in TOF (92). Epicardial ablation studies in chickens also show VSD and underdevelopment of the outflow tract, similar to TOF in some patients (93). *Fog2* deletion mice also indicate a role for FOG2 in formation of the coronary vasculature, which suggests its downstream effectors may be involved in the signaling between the myocardium and the cells of the migrating epicardium. In addition, FOG2 appears to be required for the EMT of epicardial cells necessary for coronary artery formation (88). FOG2 physically interacts with and regulates the transcriptional function of GATA4 in vitro (44, 94, 95). Interestingly, transgenic mice that express a *Gata4* V217G point mutation, which specifically abolishes its interaction with Fog2, exhibit a phenotype similar to the *Fog2* deletion mutants including lack of coronary vasculature and defective septation. These anomalies support a cooperative role for these two transcription factors in the induction of epicardial EMT (96).

BIOCHEMICAL ANALYSIS OF CARDIAC TRANSCRIPTION FUNCTION

Biochemical and structural analysis of cardiac transcription factors continue to provide information not only about the function of the normal proteins but also regarding how mutations might affect function and/or interactions of a given mutant protein. Many transcription factors discussed in this review physically or functionally interact with each other. Thus, mutations within one protein may interfere with the function of others, suggesting an explanation for the similarity in some of the cardiac phenotypes seen in these patients.

Functionality of a transcription factor can be evaluated on the basis of its nuclear localization, DNA binding, transcriptional activation, or ability to interact with binding partners. Point mutations can result in gain of function, loss of function, or dominant-negative function. Thus, depending on the location of the mutation, various mutations within the same protein may result in different molecular mechanisms of pathogenesis. This may be a molecular basis partially responsible for the variability in phenotypes seen in patients with different mutations within the same protein. Analysis of the biochemical effects of mutations on these transcription factors has provided information about structural requirements for normal function and mechanisms of protein malfunction, resulting in information about disease pathogenesis.

NKX2.5

In addition to the DNA-binding homeodomain [amino acid (aa) 138–197], the NKX2.5 protein contains a tin (TN) domain (aa 9–19), and an NK-2-specific domain (aa 214–230) (Figure 1a) (5, 10, 97–99). Although the function of the TN domain is not known, it is interesting to note that almost all homeodomain proteins containing this domain are expressed in the heart (100). NKX2.5 can function as a homodimer or heterodimer with GATA4 or TBX5. The C terminus, specifically the region after aa 250, is involved in homodimerization (101). The domains involved in heterodimerization are not well characterized and vary for each interacting partner.

Biochemical analyses of *NKX2.5* mutants indicate that missense and nonsense mutations within the homeodomain and nonsense mutations that result in truncated proteins lacking a homeodomain severely impair or abolish DNA binding capabilities. Thus, disease phenotypes seen in patients with these mutations are likely due to happloinsufficiency (10). However, five nonsense C-terminal truncation mutations that do not apparently affect the homeodomain structure have not been characterized, and thus their mechanism of pathology has not been determined.

NKX2.5 has a variety of interacting partners, including but not limited to GATA4, GATA5, dHAND, TBX5, and TBX20 (73, 102–104). The interaction of NKX2.5 with its transcriptional partners appears to be an essential part of the development of the heart as a whole and of the specialized conduction system. Some recombinant mutants of NKX2.5 impair its synergy with GATA4, resulting in decreased activation of a downstream reporter construct (105). In coordination with its various transcriptional partners, NKX2.5 has a multitude of downstream targets including ANF (106, 107), cardiac α-actin (108), A1 adenosine receptor (109), calreticulin (110, 111), cardiac sodium-calcium exchanger 1 gene (112), connexin 40 (64), connexin 43 (70), BMP-10, HOP, MinK, and HCN1 (61). Both inability to bind DNA and decreased/impaired ability to interact with downstream targets would result in a reduction in activation of target genes and contribute to the cardiac phenotypes seen in association with *NKX2.5* mutation. Because GATA4 and TBX5 physically interact with NKX2.5, it is likely that some mutations in each gene affect their interaction, resulting in a mutual inhibition of function that would give a molecular explanation for the similarities in phenotypes seen in humans including ASD and AV block. Variability in cardiac phenotypes seen with different *NKX2.5* mutations may also be explained by the differential ability of the mutant proteins to interact with transcriptional partners.

GATA4

GATA4 is a zinc-finger transcription factor containing an N- and a C-terminal zinc finger. As demonstrated via domain deletion experiments, the C-terminal zinc finger is necessary and sufficient for DNA binding but is stabilized by the N-terminal zinc finger, which does not independently bind DNA (113–115). GATA4 plays a major role in the regulation of several other cardiac promoters including ANF, brain natriuretic peptide, cardiac troponin C, cardiac troponin I, m2 muscarinic acetylcholine receptor, and slow myosin heavy chain 3 (reviewed in Reference 116). Investigators have used in vitro analysis to elucidate GATA4 interactions with a number of cardiac cofactors including FOG2, NKX2.5, and TBX5, which are of specific interest to this review (19, 117). Most protein-protein interactions in which GATA4 participates are via the C-terminal zinc finger, with the exception of FOG2 in which the N-terminal zinc finger is the interaction-mediating motif. The G296S mutation located just outside of the C-terminal zinc finger of GATA4 abrogates its interaction with TBX5 but not NKX2.5, which suggests a possible cause of the pathology seen in patients with this heterozygous mutation (19). Although the E359X mutation found in GATA4 does not disrupt any currently identified motifs within the protein, in vitro analysis indicates this mutant protein is transcriptionally inactive (19).

TBX5

As a member of the T-box family of transcription factors, TBX5 contains a T-box motif that is involved in DNA binding and protein-protein interactions (118). T-box transcription factors can have an activating and/or repressing function, of which TBX5 is an activator. A majority of the 37 *TBX5* mutations that have been identified result in a nonsense mutation predicted to cause early termination of the protein. Thus, the current working model for genetic disease pathology is one of haploinsufficiency, indicating a dose dependence for TBX5 function similar to that seen with NKX2.5 (21, 22, 119). Mutations G80R, R237Q, and R237W within the T-box have dramatically decreased capacity for DNA binding, which leads to reduced activation of transcription and loss of synergistic transcriptional activation with NKX2.5 (120). All mutations within the T-box domain inhibit TBX5 interaction with NKX2.5, resulting in a lack of synergistic activation and reduced expression of downstream targets (120). Owing to the roles of both TBX5 and NKX2.5 in cardiac conduction system development and muscle gene expression, this decreased transcriptional activation likely explains their pathogenesis including ASD and AV block. More recently, the G80R mutation and the R237W mutations have been shown to impair TBX5 interaction with GATA4, again resulting in pathology owing to decreased transcriptional activation (19). Mutations that result in truncation of the protein prior to aa 340 lack the transcriptional activation domain and thus are unlikely to activate transcription properly. Currently, identified missense mutations within the protein affect its ability to interact with DNA and/or other proteins or impair nuclear localization (118). These types of mutations, although they do not cause classic haploinsufficiency with decreased protein concentrations, result in decreased or inhibited molecular function and may provide for a "functional haploinsufficiency," whereby only half of the protein is available to perform its required transcriptional tasks. A 50% reduction in Tbx5 expression in heterozygous deletion mice, which exhibit cardiac defects similar to HOS, causes a marked decrease in expression of ANF and connexin 40 (64). In humans, heterozygous mutations in *TBX5* that yield one nonfunctional copy may affect the levels of activation of downstream targets involved in the development of the myocardium and cells of the conduction system, resulting in the cardiac phenotypes seen in these individuals.

ZIC3

ZIC3 is a zinc-finger transcription factor with five zinc-finger binding domains that shares homology to the GLI family of transcription factors (121) (Figure 1*b*). Independently, ZIC3 is a weak transcriptional activator and thus is believed to function in vivo as a transcriptional coactivator with GLI family transcription factors (121, 122). Downstream targets of ZIC3 have not been identified. In vitro analysis of recombinant versions of ZIC3 mutant proteins demonstrate that mutations within the DNA binding region of the zinc fingers and mutations within the N terminus inhibit their ability to activate transcription. The S43X and Q249X truncation mutations do not result in the production of a functional protein, indicating haploinsufficiency as the likely cause of their pathogenesis (30). Mutant proteins C268X, H286R, and Q292X are inhibited in their nuclear localization and thus are unable to activate transcription (Figure 1*b*). The remaining missense mutations show reduced transcriptional activation, with the exception of the P27A mutation. This mutation has a significant increase in transcriptional activation capability, which may indicate a gain-of-function mechanism of pathogenesis (30). It should be noted that the patient in which this mutation was identified did not present with typical heterotaxy and presented with only ASD and pulmonary stenosis (30).

TFAP2B

Three main functional domains (Figure 1*b*) constitute TFAP2B. The N-terminal portion of the protein contains a transactivation domain with a conserved phosphotyrosine motif. The C terminus contains a highly conserved basic domain, which is necessary for DNA binding, and the helix-span-helix domain, which is involved in DNA binding and dimerization (33, 35) (Figure 1*b*). TFAP2 proteins form both homodimers and heterodimers with other AP2 members. This dimerization is required for initiation of transcription. Eight mutations in TFAP2B have been associated with Char syndrome (Figure 1*b*). Two mutations are intronic and result in abnormal mRNA splice sites likely targeting these mRNAs for nonsense-mediated decay, which then results in haploinsufficiency in these patients (34). The P62R mutation disrupts the conserved phosphotyrosine motif within the transactivation domain, causing reduced transcriptional activation (35, 123). Interestingly, patients with this mutation have a high incidence of PDA phenotype but more mild facial dysmorphisms. They also lack typical hand anomalies associated with Char syndrome, which are not associated with neural crest anomalies (35). Six of the eight mutations, R225C, R225S, A264D, R274Q, and R289C, are missense (33–35). These mutant proteins maintain their ability to dimerize with other AP2 proteins but are unable to bind DNA. Thus, their pathology likely results from a dominant-negative condition, whereby dimerization of mutants with other normal TFAP2 family members adversely affects transactivation (33). The expression of these dominant-negative forms of TFAP2B in the neural crest cells of Char syndrome patients may result in derangement of their migration and disruption of derivative ductal and facial structures (33). It may alter the migration or differentiation of the neural crest cells within the ductus arteriosus, which may prevent its degeneration and/or constriction after birth (33). The downstream effectors of TFAP2B have not been identified.

TBX-1

As a member of the T-box family of transcription factors, TBX1 contains a T-box motif that is involved in DNA binding and protein-protein interactions (Figure 1*b*). Although *TBX1* point mutations cause del22q11.2 syndrome in <1% of the experimental population, in vitro analyses of these isolated point mutants provide the only biochemical information of TBX1 functional domains (42). Three mutations have been identified within TBX1: the nonsense 1223delC mutation and two missense mutations F148Y and G310S (41). The G310S mutation occurs within conserved residues of TBX1, whereas the F148Y mutation occurs within the T-box domain, which suggests the potential for impaired DNA binding and dimerization. But in vitro analyses of these two missense mutations indicate they have normal transcriptional activation. As such, their pathogenesis remains to be identified (42). The 1223delC mutation is predicted to terminate the TBX1 protein after 51 codons. This truncated protein maintains its DNA binding motif but lacks the C-terminal region containing the putative activator and repressor domains (41). In vitro analyses indicate the 1223delC mutant lacks a newly identified nuclear localization site, which suggests improper nuclear localization as a mechanism of pathogenesis (42).

FOG2

FOG2 has eight zinc-finger motifs. Glutathione S-transferase–pull down experiments reveal its interaction with GATA4 and a transcriptional corepressor, C-terminal-binding protein-2 (94). The fifth and sixth zinc-finger of FOG2 are critical for its interaction with the N-terminal zinc finger of GATA4 and for its regulation of GATA4-induced transcription (44, 90). Only two human *FOG2* mutations (*E30G* and *S657G*) have been identified and characterized (9) (Figure 1*b*). Biochemical characterization of these proteins indicates both are still able to bind to GATA4 and repress its activation of downstream targets, although S657G does appear to have a subtle defect in its ability to regulate GATA4 transcriptional activation (9). This decreased GATA4 regulation is likely due to the fact the *S657G* mutation lies directly between zinc finger five and six of FOG2, the domains critical for GATA4 interaction (44, 91). Further molecular analysis of both GATA4 and FOG2 will provide more information about how mutations within their interacting regions may relate to cardiac phenotypes seen in human patients.

SUMMARY

Transcription factors are a significant cause of CHD. Studies to date indicate that transcription factor mutations demonstrate variable expression, i.e., varied phenotypes even from the same mutation, which suggests that other factors including coregulatory elements and epigenetic effects also play a role in the disease manifestations. Identification of human transcription factor mutations has informed developmental biologists and biochemists of the function of these transcription factors. Continued use of these complementary methods (genetics, development, and biochemistry) will provide a more thorough understanding of the molecular function of these transcription factors in cardiac development and disease. Such understanding promises to lead to improved diagnosis and novel therapeutic strategies for these important clinical problems.

ACKNOWLEDGMENTS

We thank Jill Reyna for assistance with illustrations. This work was supported by Awards from the National Institutes of Health (NIH) [HL74728 (D.W.B., K.E.Y.), HL69712 (D.W.B.), HD39946 (D.W.B.), and HL66051 (K.E.Y.)].

The *Annual Review of Physiology* is online at
http://physiol.annualreviews.org

LITERATURE CITED

1. Apergis GA, Crawford N, Ghosh D, Steppan CM, Vorachek WR, et al. 1998. A novel nk-2-related transcription factor associated with human fetal liver and hepatocellular carcinoma. *J. Biol. Chem.* 273:2917–25

2. Azpiazu N, Frasch M. 1993. Tinman and bagpipe: two homeo box genes that determine cell fates in the dorsal mesoderm of Drosophila. *Genes Dev.* 7:1325–40

3. Benson DW. 2002. The genetics of congenital heart disease: a point in the revolution. *Cardiol. Clin.* 20:385–94

4. Basson CT, Cowley GS, Solomon SD, Weissman B, Poznanski AK, et al. 1994. The clinical and genetic spectrum of the Holt-Oram syndrome (heart-hand syndrome). *N. Engl. J. Med.* 330:885–91

5. Schott JJ, Benson DW, Basson CT, Pease W, Silberbach GM, et al. 1998. Congenital heart disease caused by mutations in the transcription factor NKX2-5. *Science* 281:108–11

6. Casey B, Devoto M, Jones KL, Ballabio A. 1993. Mapping a gene for familial situs abnormalities to human chromosome Xq24–q27.1. *Nat. Genet.* 5:403–7

7. Williamson JA, Bosher JM, Skinner A, Sheer D, Williams T, Hurst HC. 1996. Chromosomal mapping of the human and mouse homologues of two new members of the AP-2 family of transcription factors. *Genomics* 35:262–64

8. Pehlivan T, Pober BR, Brueckner M, Garrett S, Slaugh R, et al. 1999. GATA4 haploinsufficiency in patients with interstitial deletion of chromosome region 8p23.1 and congenital heart disease. *Am. J. Med. Genet.* 83:201–6

9. Pizzuti A, Sarkozy A, Newton AL, Conti E, Flex E, et al. 2003. Mutations of ZFPM2/FOG2 gene in sporadic cases of tetralogy of Fallot. *Hum. Mutat.* 22:372–77

10. Benson DW, Silberbach GM, Kavanaugh-McHugh A, Cottrill C, Zhang Y, et al. 1999. Mutations in the cardiac transcription factor NKX2.5 affect diverse cardiac developmental pathways. *J. Clin. Invest.* 104:1567–73

11. Goldmuntz E, Geiger E, Benson DW. 2001. NKX2.5 mutations in patients with tetralogy of fallot. *Circulation* 104:2565–68

12. Gutierrez-Roelens I, Sluysmans T, Gewillig M, Devriendt K, Vikkula M. 2002. Progressive AV-block and anomalous venous return among cardiac anomalies associated with two novel missense mutations in the CSX/NKX2–5 gene. *Hum. Mutat.* 20:75–76

13. Ikeda Y, Hiroi Y, Hosoda T, Utsunomiya T, Matsuo S, et al. 2002. Novel point mutation in the cardiac transcription factor CSX/NKX2.5 associated with congenital heart disease. *Circ. J.* 66:561–63

14. Kasahara H, Lee B, Schott JJ, Benson DW, Seidman JG, et al. 2000. Loss of function and inhibitory effects of human CSX/NKX2.5 homeoprotein mutations associated with congenital heart disease. *J. Clin. Invest.* 106:299–308

15. McElhinney DB, Geiger E, Blinder J, Benson DW, Goldmuntz E. 2003. NKX2.5 mutations in patients with congenital heart disease. *J. Am. Coll. Cardiol.* 42:1650–55

16. Watanabe Y, Benson DW, Yano S, Akagi T, Yoshino M, Murray JC. 2002. Two novel frameshift mutations in NKX2.5 result in novel features including visceral inversus and sinus venosus type ASD. *J. Med. Genet.* 39:807–11

17. Kasahara H, Benson DW. 2001. Biochemical analyses of eight NKX2.5 homeodomain missense mutations causing atrioventricular block and cardiac anomalies. *Cardiovasc. Res.* 64:40–51

18. Benson DW. 2004. Genetics of atrioventricular conduction disease in humans. *Anat. Rec.* 280A:934–39

19. Garg V, Kathiriya IS, Barnes R, Schluterman MK, King IN, et al. 2003. GATA4 mutations cause human congenital heart defects and reveal an interaction with TBX5. *Nature* 424:443–47

20. Hirayama-Yamada K, Kamisago M, Akimoto K, Aotsuka H, Nakamura Y, et al. 2005. Phenotypes with GATA4 or NKX2.5 mutations in familial atrial septal defect. *Am. J. Med. Genet. A* 135:47–52

21. Li QY, Newbury-Ecob RA, Terrett JA, Wilson DI, Curtis AR, et al. 1997. Holt-Oram syndrome is caused by mutations in TBX5, a member of the Brachyury (T) gene family. *Nat. Genet.* 15:21–29

22. Basson CT, Bachinsky DR, Lin RC, Levi T, Elkins JA, et al. 1997. Mutations in human TBX5 [corrected] cause limb and cardiac malformation in Holt-Oram syndrome. *Nat. Genet.* 15:30–35. Erratum. 1997. *Nat. Genet.* 15:411

23. Huang T. 2002. Current advances in Holt-Oram syndrome. *Curr. Opin. Pediatr.* 14:691–95

24. Gruenauer-Kloevekorn C, Froster UG. 2003. Holt-Oram syndrome: a new mutation in the TBX5 gene in two unrelated families. *Ann. Genet.* 46:19–23

25. Goldmuntz E. 2004. The genetic contribution to congenital heart disease. *Pediatr. Clin. North Am.* 51:1721–37

26. Mathias RS, Lacro RV, Jones KL. 1987. X-linked laterality sequence: situs inversus, complex cardiac defects, splenic defects. *Am. J. Med. Genet.* 28:111–16

27. Ferrero GB, Gebbia M, Pilia G, Witte D, Peier A, et al. 1997. A submicroscopic deletion in Xq26 associated with familial situs ambiguus. *Am. J. Hum. Genet.* 61:395–401

28. Gebbia M, Ferrero GB, Pilia G, Bassi MT, Aylsworth A, et al. 1997. X-linked situs abnormalities result from mutations in ZIC3. *Nat. Genet.* 17:305–8

29. Purandare SM, Ware SM, Kwan KM, Gebbia M, Bassi MT, et al. 2002. A complex syndrome of left-right axis, central nervous system and axial skeleton defects in Zic3 mutant mice. *Development* 129:2293–302

30. Ware SM, Peng J, Zhu L, Fernbach S, Colicos S, et al. 2004. Identification and functional analysis of ZIC3 mutations in heterotaxy and related congenital heart defects. *Am. J. Hum. Genet.* 74:93–105

31. Moser M, Imhof A, Pscherer A, Bauer R, Amselgruber W, et al. 1995. Cloning and characterization of a second AP-2 transcription factor: AP-2 beta. *Development* 121:2779–88

32. Satoda M, Pierpont ME, Diaz GA, Bornemeier RA, Gelb BD. 1999. Char syndrome, an inherited disorder with patent ductus arteriosus, maps to chromosome 6p12–p21. *Circulation* 99:3036–42

33. Satoda M, Zhao F, Diaz GA, Burn J, Goodship J, et al. 2000. Mutations in TFAP2B cause Char syndrome, a familial form of patent ductus arteriosus. *Nat. Genet.* 25:42–46

34. Mani A, Radhakrishnan J, Farhi A, Carew KS, Warnes CA, et al. 2005. Syndromic patent ductus arteriosus: evidence for haploinsufficient TFAP2B mutations and identification of a linked sleep disorder. *Proc. Natl. Acad. Sci. USA* 102:2975–79

35. Zhao F, Weismann CG, Satoda M, Pierpont ME, Sweeney E, et al. 2001. Novel TFAP2B mutations that cause Char syndrome provide a genotype-phenotype correlation. *Am. J. Hum. Genet.* 69:695–703

36. Baldini A. 2004. DiGeorge syndrome: an update. *Curr. Opin. Cardiol.* 19:201–4

37. Lindsay EA, Botta A, Jurecic V, Carattini-Rivera S, Cheah YC, et al. 1999. Congenital heart disease in mice deficient for the DiGeorge syndrome region. *Nature* 401:379–83

38. Merscher S, Funke B, Epstein JA, Heyer J, Puech A, et al. 2001. TBX1 is responsible for cardiovascular defects in velo-cardio-facial/DiGeorge syndrome. *Cell* 104:619–29

39. Lindsay EA, Vitelli F, Su H, Morishima M, Huynh T, et al. 2001. Tbx1 haploinsufficieny in the DiGeorge syndrome region causes aortic arch defects in mice. *Nature* 410:97–101

40. Jerome LA, Papaioannou VE. 2001. DiGeorge syndrome phenotype in mice mutant for the T-box gene, Tbx1. *Nat. Genet.* 27:286–91

41. Yagi H, Furutani Y, Hamada H, Sasaki T, Asakawa S, et al. 2003. Role of TBX1 in human del22q11.2 syndrome. *Lancet* 362:1366–73

42. Stoller JZ, Epstein JA. 2005. Identification of a novel nuclear localization signal in Tbx1 that is deleted in DiGeorge syndrome patients harboring the 1223delC mutation. *Hum. Mol. Genet.* 14:885–92

43. Gong W, Gottlieb S, Collins J, Blescia A, Dietz H, et al. 2001. Mutation analysis of TBX1 in non-deleted patients with features of DGS/VCFS or isolated cardiovascular defects. *J. Med. Genet.* 38:E45

44. Svensson EC, Tufts RL, Polk CE, Leiden JM. 1999. Molecular cloning of FOG-2: a modulator of transcription factor GATA-4 in cardiomyocytes. *Proc. Natl. Acad. Sci. USA* 96:956–61

45. Srivastava D, Olson EN. 2000. A genetic blueprint for cardiac development. *Nature* 407:221–26

46. Wessels A, Perez-Pomares JM. 2004. The epicardium and epicardially derived cells (EPDCs) as cardiac stem cells. *Anat. Rec. A Discov. Mol. Cell. Evol. Biol.* 276:43–57

47. Yutzey KE, Kirby ML. 2002. Wherefore heart thou? Embryonic origins of cardiogenic mesoderm. *Dev. Dyn.* 223:307–20

48. Bruneau BG. 2002. Transcriptional regulation of vertebrate cardiac morphogenesis. *Circ. Res.* 90:509–19

49. Brand T. 2003. Heart development: molecular insights into cardiac specification and early morphogenesis. *Dev. Biol.* 258:1–19

50. Bodmer R. 1993. The gene tinman is required for specification of the heart and visceral muscles in Drosophila. *Development* 118:719–29

51. Chen CY, Croissant J, Majesky M, Topouzis S, McQuinn T, et al. 1996. Activation of the cardiac alpha-actin promoter depends upon serum response factor, Tinman homologue, Nkx-2.5, and intact serum response elements. *Dev. Genet.* 19:119–30

52. Fu Y, Yan W, Mohun TJ, Evans SM. 1998. Vertebrate tinman homologues XNkx2–3 and XNkx2–5 are required for heart formation in a functionally redundant manner. *Development* 125:4439–49

53. Grow MW, Krieg PA. 1998. Tinman function is essential for vertebrate heart development: elimination of cardiac differentiation by dominant inhibitory mutants of the tinman-related genes, XNkx2–3 and XNkx2–5. *Dev. Biol.* 204:187–96

54. Lints TJ, Parsons LM, Hartley L, Lyons I, Harvey RP. 1993. Nkx-2.5: a novel murine homeobox gene expressed in early heart progenitor cells and their myogenic descendants. *Development* 119:419–31

55. Lyons I, Parsons LM, Hartley L, Li R, Andrews JE, et al. 1995. Myogenic and morphogenetic defects in the heart tubes of murine embryos lacking the homeo box gene Nkx2–5. *Genes Dev.* 9:1654–66

56. Tanaka M, Wechsler SB, Lee IW, Yamasaki N, Lawitts JA, Izumo S. 1999. Complex modular cis-acting elements regulate expression of the cardiac specifying homeobox gene Csx/Nkx2.5. *Development* 126:1439–50

57. Heikinheimo M, Scandrett JM, Wilson DB. 1994. Localization of transcription factor GATA-4 to regions of the mouse embryo involved in cardiac development. *Dev. Biol.* 164:361–73

58. Chapman DL, Garvey N, Hancock S, Alexiou M, Agulnik SI, et al. 1996.

Expression of the T-box family genes, Tbx1-Tbx5, during early mouse development. *Dev. Dyn.* 206:379–90

59. Moskowitz IP, Pizard A, Patel VV, Bruneau BG, Kim JB, et al. 2004. The T-Box transcription factor Tbx5 is required for the patterning and maturation of the murine cardiac conduction system. *Development* 131:4107–16

60. Thomas PS, Kasahara H, Edmonson AM, Izumo S, Yacoub MH, et al. 2001. Elevated expression of Nkx-2.5 in developing myocardial conduction cells. *Anat. Rec.* 263:307–13

61. Pashmforoush M, Lu JT, Chen H, Amand TS, Kondo R, et al. 2004. Nkx2–5 pathways and congenital heart disease; loss of ventricular myocyte lineage specification leads to progressive cardiomyopathy and complete heart block. *Cell* 117:373–86

62. Molkentin JD, Lin Q, Duncan SA, Olson EN. 1997. Requirement of the transcription factor GATA4 for heart tube formation and ventral morphogenesis. *Genes Dev.* 11:1061–72

63. Kuo CT, Morrisey EE, Anandappa R, Sigrist K, Lu MM, et al. 1997. GATA4 transcription factor is required for ventral morphogenesis and heart tube formation. *Genes Dev.* 11:1048–60

64. Bruneau BG, Nemer G, Schmitt JP, Charron F, Robitaille L, et al. 2001. A murine model of Holt-Oram syndrome defines roles of the T-box transcription factor Tbx5 in cardiogenesis and disease. *Cell* 106:709–21

65. Liberatore CM, Searcy-Schrick RD, Yutzey KE. 2000. Ventricular expression of tbx5 inhibits normal heart chamber development. *Dev. Biol.* 223:169–80

66. Biben C, Weber R, Kesteven S, Stanley E, McDonald L, et al. 2000. Cardiac septal and valvular dysmorphogenesis in mice heterozygous for mutations in the homeobox gene Nkx2–5. *Circ. Res.* 87:888–95

67. Kasahara H, Wakimoto H, Liu M, Maguire CT, Converso KL, et al. 2001. Progressive atrioventricular conduction defects and heart failure in mice expressing a mutant Csx/Nkx2.5 homeoprotein. *J. Clin. Invest.* 108:189–201

68. Wakimoto H, Kasahara H, Maguire CT, Izumo S, Berul CI. 2002. Developmentally modulated cardiac conduction failure in transgenic mice with fetal or postnatal overexpression of DNA non-binding mutant Nkx2.5. *J. Cardiovasc. Electrophysiol.* 13:682–88

69. Wakimoto H, Kasahara H, Maguire CT, Moskowitz IP, Izumo S, Berul CI. 2003. Cardiac electrophysiological phenotypes in postnatal expression of Nkx2.5 transgenic mice. *Genesis* 37:144–50

70. Kasahara H, Ueyama T, Wakimoto H, Liu MK, Maguire CT, et al. 2003. Nkx2.5 homeoprotein regulates expression of gap junction protein connexin 43 and sarcomere organization in postnatal cardiomyocytes. *J. Mol. Cell. Cardiol.* 35:243–56

71. Pu WT, Ishiwata T, Juraszek AL, Ma Q, Izumo S. 2004. GATA4 is a dosage-sensitive regulator of cardiac morphogenesis. *Dev. Biol.* 275:235–44

72. Plageman TF Jr, Yutzey KE. 2004. Differential expression and function of Tbx5 and Tbx20 in cardiac development. *J. Biol. Chem.* 279:19026–34

73. Hiroi Y, Kudoh S, Monzen K, Ikeda Y, Yazaki Y, et al. 2001. Tbx5 associates with Nkx2-5 and synergistically promotes cardiomyocyte differentiation. *Nat. Genet.* 28:276–80

74. Nagai T, Aruga J, Takada S, Gunther T, Sporle R, et al. 1997. The expression of the mouse Zic1, Zic2, and Zic3 gene suggests an essential role for Zic genes in body pattern formation. *Dev. Biol.* 182:299–313

75. Herman GE, El-Hodiri HM. 2002. The role of ZIC3 in vertebrate development. *Cytogenet. Genome Res.* 99:229–35

76. Meyers EN, Martin GR. 1999. Differences in left-right axis pathways in mouse and chick: functions of FGF8 and SHH. *Science* 285:403–6

77. Tsukui T, Capdevila J, Tamura K,

Ruiz-Lozano P, Rodriguez-Esteban C, et al. 1999 Multiple left-right asymmetry defects in Shh$^{-/-}$ mutant mice unveil a convergence of the shh and retinoic acid pathways in the control of Lefty-1. *Proc. Natl. Acad. Sci. USA* 96:11376–81

78. Waldo KL, Kumiski DH, Wallis KT, Stadt HA, Hutson MR, et al. 2001. Conotruncal myocardium arises from a secondary heart field. *Development* 128:3179–88

79. Mjaatvedt CH, Nakaoka T, Moreno-Rodriguez R, Norris RA, Kern MJ, et al. 2001. The outflow tract of the heart is recruited from a novel heart-forming field. *Dev. Biol.* 238:97–109

80. Kirby ML, Waldo KL. 1995. Neural crest and cardiovascular patterning. *Circ. Res.* 77:211–15

81. Moser M, Pscherer A, Roth C, Becker J, Mucher G, et al. 1997. Enhanced apoptotic cell death of renal epithelial cells in mice lacking transcription factor AP-2beta. *Genes Dev.* 11:1938–48

82. Vaughan CJ, Basson CT. 2000. Molecular determinants of atrial and ventricular septal defects and patent ductus arteriosus. *Am. J. Med. Genet.* 97:304–9

83. Vitelli F, Taddei I, Morishima M, Meyers EN, Lindsay EA, Baldini A. 2002. A genetic link between Tbx1 and fibroblast growth factor signaling. *Development* 129:4605–11

84. Yamagishi H, Srivastava D. 2003. Unraveling the genetic and developmental mysteries of 22q11 deletion syndrome. *Trends Mol. Med.* 9:383–89

85. Chieffo C, Garvey N, Gong W, Roe B, Zhang G, et al. 1997. Isolation and characterization of a gene from the DiGeorge chromosomal region homologous to the mouse Tbx1 gene. *Genomics* 43:267–77

86. Garg V, Yamagishi C, Hu T, Kathiriya IS, Yamagishi H, Srivastava D. 2001. Tbx1, a DiGeorge syndrome candidate gene, is regulated by sonic hedgehog during pharyngeal arch development. *Dev. Biol.* 235:62–73

87. Brown CO III, Chi X, Garcia-Gras E,

Shirai M, Feng XH, Schwartz RJ. 2004. The cardiac determination factor, Nkx2-5, is activated by mutual cofactors GATA-4 and Smad1/4 via a novel upstream enhancer. *J. Biol. Chem.* 279:10659–69

88. Tevosian SG, Deconinck AE, Tanaka M, Schinke M, Litovsky SH, et al. 2000. FOG-2, a cofactor for GATA transcription factors, is essential for heart morphogenesis and development of coronary vessels from epicardium. *Cell* 101:729–39

89. Olivey HE, Compton LA, Barnett JV. 2004. Coronary vessel development: the epicardium delivers. *Trends Cardiovasc. Med.* 14:247–51

90. Svensson EC, Huggins GS, Dardik FB, Polk CE, Leiden JM. 2000. A functionally conserved N-terminal domain of the friend of GATA-2 (FOG-2) protein represses GATA4-dependent transcription. *J. Biol. Chem.* 275:20762–6

91. Svensson EC, Huggins GS, Lin H, Clendenin C, Jiang F, et al. 2000. A syndrome of tricuspid atresia in mice with a targeted mutation of the gene encoding Fog-2. *Nat. Genet.* 25:353–56

92. Gittenberger-de Groot AC, Vrancken Peeters MP, Mentink MM, Gourdie RG, Poelmann RE. 1998. Epicardium-derived cells contribute a novel population to the myocardial wall and the atrioventricular cushions. *Circ. Res.* 82:1043–52

93. Gittenberger-de Groot AC, Vrancken Peeters MP, Bergwerff M, Mentink MM, Poelmann RE. 2000. Epicardial outgrowth inhibition leads to compensatory mesothelial outflow tract collar and abnormal cardiac septation and coronary formation. *Circ. Res.* 87:969–71

94. Holmes M, Turner J, Fox A, Chisholm O, Crossley M, Chong B. 1999. hFOG-2, a novel zinc finger protein, binds the co-repressor mCtBP2 and modulates GATA-mediated activation. *J. Biol. Chem.* 274:23491–98

95. Lu JR, McKinsey TA, Xu H, Wang DZ, Richardson JA, Olson EN. 1999. FOG-2, a heart- and brain-enriched cofactor

for GATA transcription factors. *Mol. Cell. Biol.* 19:4495–502

96. Crispino JD, Lodish MB, Thurberg BL, Litovsky SH, Collins T, et al. 2001. Proper coronary vascular development and heart morphogenesis depend on interaction of GATA-4 with FOG cofactors. *Genes Dev.* 15:839–44

97. Damante G, Fabbro D, Pellizzari L, Civitareale D, Guazzi S, et al. 1994. Sequence-specific DNA recognition by the thyroid transcription factor-1 homeodomain. *Nucleic Acids Res.* 22:3075–83

98. Gruschus JM, Tsao DH, Wang LH, Nirenberg M, Ferretti JA. 1997. Interactions of the vnd/NK-2 homeodomain with DNA by nuclear magnetic resonance spectroscopy: basis of binding specificity. *Biochemistry* 36:5372–80

99. Harvey RP. 1996. NK-2 homeobox genes and heart development. *Dev. Biol.* 178:203–16

100. Shiojima I, Komuro I, Mizuno T, Aikawa R, Akazawa H, et al. 1996. Molecular cloning and characterization of human cardiac homeobox gene CSX1. *Circ. Res.* 79:920–29

101. Kasahara H, Usheva A, Ueyama T, Aoki H, Horikoshi N, Izumo S. 2001. Characterization of homo- and heterodimerization of cardiac Csx/Nkx2.5 homeoprotein. *J. Biol. Chem.* 276:4570–80

102. Lee Y, Shioi T, Kasahara H, Jobe SM, Wiese RJ, et al. 1998. The cardiac tissue-restricted homeobox protein Csx/Nkx2.5 physically associates with the zinc finger protein GATA4 and cooperatively activates atrial natriuretic factor gene expression. *Mol. Cell. Biol.* 18:3120–29

103. Stennard FA, Costa MW, Elliott DA, Rankin S, Haast SJ, et al. 2003. Cardiac T-box factor Tbx20 directly interacts with Nkx2–5, GATA4, and GATA5 in regulation of gene expression in the developing heart. *Dev. Biol.* 262:206–24

104. Zang MX, Li Y, Xue LX, Jia HT, Jing H. 2004. Cooperative activation of atrial naturetic peptide promoter by dHAND and MEF2C. *J. Cell. Biochem.* 93:1255–66

105. Zhu W, Shiojima I, Hiroi Y, Zou Y, Akazawa H, et al. 2000. Functional analyses of three Csx/Nkx-2.5 mutations that cause human congenital heart disease. *J. Biol. Chem.* 275:35291–96

106. Durocher D, Chen CY, Ardati A, Schwartz RJ, Nemer M. 1996. The atrial natriuretic factor promoter is a downstream target for Nkx-2.5 in the myocardium. *Mol. Cell. Biol.* 16:4648–55

107. Shiojima I, Komuro I, Oka T, Hiroi Y, Mizuno T, et al. 1999. Context-dependent transcriptional cooperation mediated by cardiac transcription factors Csx/Nkx-2.5 and GATA-4. *J. Biol. Chem.* 274:8231–39

108. Chen JN, Fishman MC. 1996. Zebrafish tinman homolog demarcates the heart field and initiates myocardial differentiation. *Development* 122:3809–16

109. Rivkees SA, Chen M, Kulkarni J, Browne J, Zhao Z. 1999. Characterization of the murine A1 adenosine receptor promoter, potent regulation by GATA-4 and Nkx2.5. *J. Biol. Chem.* 274:14204–9

110. Ruiz-Perez VL, Ide SE, Strom TM, Lorenz B, Wilson D, et al. 2000. Mutations in a new gene in Ellis-van Creveld syndrome and Weyers acrodental dysostosis. *Nat. Genet.* 24:283–86

111. Gu H, Smith FC, Taffet SM, Delmar M. 2003. High incidence of cardiac malformations in connexin40-deficient mice. *Circ. Res.* 93:201–6

112. Muller JG, Thompson JT, Edmonson AM, Rackley MS, Kasahara H, et al. 2002. Differential regulation of the cardiac sodium calcium exchanger promoter in adult and neonatal cardiomyocytes by Nkx2.5 and serum response factor. *J. Mol. Cell. Cardiol.* 34:807–21

113. Whyatt DJ, de Oer E, Grosveld F. 1993. The two zinc finger-like domains of GATA-1 have different DNA binding specificities. *EMBO J.* 12:4993–5005

114. Yang HY, Evans T. 1992. Distinct roles

for the two cGATA-1 finger domains. *Mol. Cell. Biol.* 12:4562–70

115. Morrisey EE, Ip HS, Tang Z, Parmacek MS. 1997. GATA-4 activates transcription via two novel domains that are conserved within the GATA-4/5/6 subfamily. *J. Biol. Chem.* 272:8515–24

116. Charron F, Nemer M. 1999. GATA transcription factors and cardiac development. *Semin. Cell Dev. Biol.* 10:85–91

117. Pikkarainen S, Tokola H, Kerkela R, Ruskoaho H. 2004. GATA transcription factors in the developing and adult heart. *Cardiovasc. Res.* 63:196–207

118. Ghosh TK, Packham EA, Bonser AJ, Robinson TE, Cross SJ, Brook JD. 2001. Characterization of the TBX5 binding site and analysis of mutations that cause Holt-Oram syndrome. *Hum. Mol. Genet.* 10:1983–94

119. Mori AD, Bruneau BG. 2004. TBX5 mutations and congenital heart disease: Holt-Oram syndrome revealed. *Curr. Opin. Cardiol.* 19:211–15

120. Fan C, Liu M, Wang Q. 2003. Functional analysis of TBX5 missense mutations associated with Holt-Oram syndrome. *J. Biol. Chem.* 278:8780–85

121. Mizugishi K, Aruga J, Nakata K, Mikoshiba K. 2001. Molecular properties of Zic proteins as transcriptional regulators and their relationship to GLI proteins. *J. Biol. Chem.* 276:2180–88

122. Koyabu Y, Nakata K, Mizugishi K, Aruga J, Mikoshiba K. 2001. Physical and functional interactions between Zic and Gli proteins. *J. Biol. Chem.* 276:6889–92

123. Wankhade S, Yu Y, Weinberg J, Tainsky MA, Kannan P. 2000. Characterization of the activation domains of AP-2 family transcription factors. *J. Biol. Chem.* 275:29701–8

Annu. Rev. Physiol. 2006. 68:123–58
doi: 10.1146/annurev.physiol.68.040104.124723
First published online as a Review in Advance on October 24, 2005

FROM MICE TO MEN: Insights into the Insulin Resistance Syndromes

Sudha B. Biddinger[1,2] and C. Ronald Kahn[1]

[1]Joslin Diabetes Center and Department of Medicine, Harvard Medical School, Boston, Massachusetts 02215; email: sudha.biddinger@joslin.harvard.edu, c.ronald.kahn@joslin.harvard.edu
[2]Division of Endocrinology, Children's Hospital, Boston, Massachusetts 02215

■ **Abstract** The insulin resistance syndrome refers to a constellation of findings, including glucose intolerance, obesity, dyslipidemia, and hypertension, that promote the development of type 2 diabetes, cardiovascular disease, cancer, and other disorders. Defining the pathophysiological links between insulin resistance, the insulin resistance syndrome, and its sequelae is critical to understanding and treating these disorders. Over the past decade, two approaches have provided important insights into how changes in insulin signaling produce the spectrum of phenotypes associated with insulin resistance. First, studies using tissue-specific knockouts or tissue-specific reconstitution of the insulin receptor in vivo in mice have enabled us to deconstruct the insulin resistance syndromes by dissecting the contributions of different tissues to the insulin-resistant state. Second, in vivo and in vitro studies of the complex network of insulin signaling have provided insight into how insulin resistance can develop in some pathways whereas insulin sensitivity is maintained in others. These data, taken together, give us a framework for understanding the relationship between insulin resistance and the insulin resistance syndromes.

INTRODUCTION

The insulin resistance syndrome is a constellation of findings, including central obesity, glucose intolerance, dyslipidemia, and hypertension, that promotes the development of type 2 diabetes mellitus, cardiovascular disease, cancer, polycystic ovarian disease (PCOS), and nonalcoholic fatty liver disease (Figure 1). Like type 2 diabetes and obesity, the insulin resistance syndrome (also called the metabolic syndrome, the dysmetabolic syndrome, or syndrome X) is increasing in prevalence with alarming rapidity, affecting 27% of adults in the United States (1). Perhaps even more concerning is that up to 50% of severely obese children also have the insulin resistance syndrome (2). Although much is known about the pathogenesis of obesity and type 2 diabetes, the molecular and pathophysiological links between insulin resistance, the various components of the insulin resistance syndrome, and their sequelae are not well understood.

In unraveling the complexity underlying these disorders, it is important to keep two concepts in mind. First, whereas insulin resistance is closely associated with

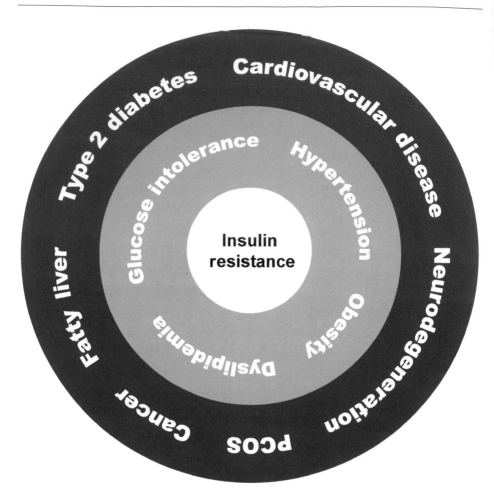

Figure 1   The insulin resistance syndrome. Insulin resistance initially manifests itself as the constellation of symptoms that comprise the insulin resistance syndrome (shown in the *orange circle*). Ultimately, insulin resistance can evolve into a number of diseases (shown in the *red circle*).

alterations in glucose homeostasis, and assessments of insulin resistance usually focus on measurements of glucose or glucose uptake, insulin also regulates the synthesis and storage of fat, protein synthesis, and nonmetabolic processes such as cell growth and differentiation. Thus, alterations in glucose homeostasis are but one aspect of insulin resistance. Second, when we speak of insulin resistance, it is clear that not all insulin-regulated processes and tissues become equally resistant to insulin. Moreover, the hyperinsulinemia that evolves with insulin resistance may lead to increased insulin action in tissues or pathways that are not as resistant as those related to glucose metabolism. For example, hyperinsulinemia acting on the liver, kidney, and ovary leads to hypertriglyceridemia, increased sodium retention and hypertension, and hyperandrogenism, respectively (3).

Over the past decade, two approaches have shed important insights into our understanding of how changes in insulin signaling produce the varied phenotypes associated with insulin resistance syndromes; these form the focus of this review. First, studies using tissue-specific knockouts or tissue-specific reconstitution of the insulin receptor in vivo in mice have enabled us to deconstruct the insulin resistance syndromes by dissecting the contributions of different tissues to the insulin-resistant state. Second, in vivo and in vitro studies of the complex network of insulin signaling have provided insight into how insulin resistance can develop in some pathways while insulin sensitivity is maintained in others. Finally, we discuss how these data impact our thinking about insulin resistance and the insulin resistance syndromes.

TISSUE-SPECIFIC ANALYSIS OF INSULIN ACTION AND INSULIN RESISTANCE

Classically, insulin resistance in muscle, liver, and fat has been viewed as central to the pathogenesis of the insulin resistance syndromes, particularly type 2 diabetes. To study the effects of insulin resistance in a particular tissue, investigators have used various strategies (3a,b; 4; 29; 33). Here we focus on studies in which tissue-specific insulin resistance has been created by conditional knockout of the insulin receptor using Cre-loxP technology (4). In brief, a tissue-specific knockout of the insulin receptor (or other gene) can be produced by introducing two short bacterial DNA sequences called lox (locus of crossing-over) sites into the introns surrounding an essential exon or exons of the relevant gene (exon 4 in the case of the insulin receptor), producing a floxed allele. In the absence of the bacterial recombinase called Cre, this floxed allele behaves similarly to the wild-type allele. However, in the presence of Cre recombinase, these lox sites facilitate the excision of exon 4 of the insulin receptor and produce a missense mRNA with a premature stop codon. The resulting transcript encodes a small fragment of the insulin receptor that is incapable of insulin signaling and is rapidly degraded. By expressing the Cre recombinase in a single tissue, using various tissue-specific promoters to drive Cre recombinase expression, one can create mice with tissue-specific insulin resistance. Such mice have given us the powerful ability to redefine the nature of insulin resistance.

Liver Insulin Receptor Knockout

The liver plays a central role in glucose and lipid metabolism, and hepatic insulin resistance is thought to be largely responsible for the development of fasting hyperglycemia. To better understand the contribution of hepatic insulin resistance to the insulin resistance syndrome, the liver insulin receptor knockout (LIRKO) mouse was created using Cre recombinase driven by the albumin promoter (5).

By two months of age, LIRKO mice exhibit mild fasting hyperglycemia (132 versus 101 mg dl^{-1} in controls) and even more striking hyperglycemia in the fed state (363 versus 135 mg dl^{-1} in controls). These alterations in glucose tolerance appear to stem, in part, from the inability of insulin to suppress hepatic glucose production. Whereas insulin normally suppresses hepatic glucose production by more than 50% during a hyperinsulinemic-euglycemic clamp, this effect is lost in LIRKO mice. Hepatic insulin resistance is also characterized by increased expression of two key enzymes of gluconeogenesis, phosphoenolpyruvate carboxykinase (PEPCK) and glucose-6-phosphatase.

LIRKO mice are also markedly hyperinsulinemic, with insulin levels 20-fold greater than those of control animals. This is due to both the compensatory response of the β-cells of the pancreas, which undergo marked hypertrophy in LIRKO mice, and an impairment in insulin clearance due to the loss of receptor-mediated uptake and degradation in the liver. This leads to insulin resistance in nonhepatic tissues, i.e., tissues in which the insulin receptor has not been knocked out, which can be partially reversed by treating mice with streptozotocin, a toxin that destroys the β-cells of the pancreas (6). The effects of pure hepatic insulin resistance, i.e., the physiology of the LIRKO mouse, are shown in Figure 2.

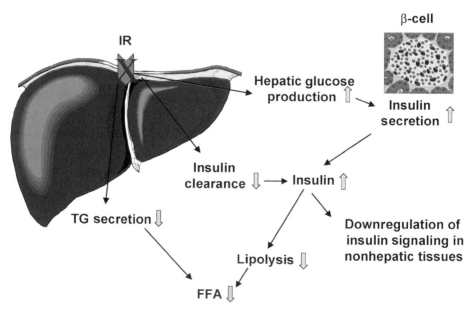

Figure 2   Physiology of pure liver insulin resistance. Knockout of the insulin receptor (IR) in liver is associated with increased hepatic glucose production and decreased insulin clearance, leading to hyperinsulinemia and increased peripheral insulin resistance but decreased triglyceride secretion. This ultimately produces hyperglycemia but decreased triglycerides (TG) and free fatty acids (FFA).

Interestingly, although LIRKO mice exhibit hepatic insulin resistance and severe glucose intolerance, they have decreased levels of circulating free fatty acids and triglycerides (5). This is due in part to the suppression of extrahepatic lipolysis, and hence of the release of free fatty acids, by the high insulin levels in these mice. Additionally, LIRKO mice have decreased rates of hepatic triglyceride secretion (6a), perhaps owing to the inability of hyperinsulinemia to drive triglyceride synthesis in the absence of hepatic insulin signaling. Thus, complete hepatic insulin resistance has beneficial effects on the serum lipid profile, which are not observed in states in which there is only partial insulin resistance.

LIRKO mice also reveal an unexpected role for insulin signaling in maintaining hepatic function. Histological examination of the livers of six-month old LIRKO mice reveals pale nodules with highly vacuolated cells and increased lipid accumulation but no frank fibrosis (5). These histological changes are accompanied by a 50% decrease in serum albumin levels, suggesting a decrease in synthetic function. In addition, as LIRKO mice age, the impaired glucose tolerance resolves and the mice exhibit fasting hypoglycemia, most likely reflecting the decreased capacity of these livers for gluconeogenesis.

Muscle Insulin Receptor Knockouts

Glucose uptake into skeletal muscle accounts for the disposal of 70–90% of an oral glucose load. Furthermore, insulin resistance in muscle is one of the earliest defects in the development of type 2 diabetes (7, 8). To clarify the role of muscle insulin resistance in the development of diabetes and the insulin resistance syndrome, muscle insulin receptor knockout (MIRKO) mice were made using Cre recombinase under the direction of the muscle creatine kinase promoter (4). These mice have >90% of the insulin receptor deleted in all skeletal muscle groups as well as in the heart.

Surprisingly, despite exhibiting complete muscle insulin resistance, MIRKO mice have essentially normal blood glucose and insulin levels and a normal response to glucose tolerance testing. Under hyperinsulinemic-euglycemic clamp conditions, MIRKO muscle has a 75% decrease in insulin-stimulated glucose transport, a 73% decrease in insulin-stimulated glycolysis, and a 78% decrease in glycogen synthesis, indicating muscle insulin resistance (9). MIRKO mice also have a 10–30% decrease in muscle mass, indicating the role of insulin in protein synthesis and/or protection from protein breakdown (9).

Interestingly, adipose tissue from MIRKO mice shows an unexpected threefold increase in insulin-stimulated glucose transport in vivo, relative to controls (9). This increase in fat glucose uptake appears to be due to some unknown circulating factors released from MIRKO muscle because adipocytes isolated from MIRKO mice have rates of glucose uptake similar to controls in vitro. Consistent with this shift of substrates into adipose tissue, MIRKO mice have a 53% increase in epididymal fat pad weight and a 38% increase in whole-body fat content (4, 9, 10). MIRKO mice also have a 16% increase in free fatty acids and a 43% increase in serum triglycerides (9). Thus, MIRKO mice develop several major components of

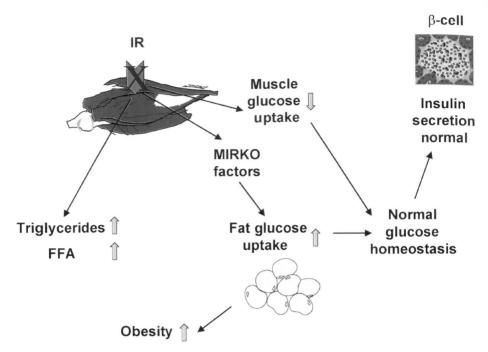

Figure 3 Physiology of pure muscle insulin resistance. Although knockout of the insulin receptor (IR) in muscle decreases muscle glucose uptake, glucose homeostasis overall is normal. This appears to be due to the secretion of muscle factors (MIRKO factors), which increase glucose uptake into fat. The excess flux of nutrients into fat increases adiposity. Also, through unknown mechanisms, MIRKO mice show an increase in serum triglycerides and free fatty acids (FFA). Thus, knockout of the muscle insulin receptor produces obesity and dyslipidemia, but not glucose intolerance.

the insulin resistance syndrome (mild obesity, hypertriglyceridemia, and increased free fatty acids) but do not develop glucose intolerance (Figure 3). MIRKO mice also fail to develop hyperinsulinemia, indicating that muscle insulin resistance alone is not sufficient to produce this compensatory response (4).

Fat Insulin Receptor Knockout

Classically, white adipose tissue has been considered a depot for energy storage, accounting for approximately 5–15% of an oral glucose load. However, the adipocyte is now also recognized as a major endocrine organ, secreting factors that play an important role in appetite regulation and energy homeostasis. To understand the potential role of insulin resistance in the fat in the insulin resistance syndrome, researchers knocked out the insulin receptor using Cre recombinase under the aP2 promoter (11). aP2, a fatty acid–binding protein that is expressed primarily in adipocytes, is one of the most abundant proteins in fat.

In contrast to the muscle and liver forms of insulin resistance, pure fat insulin resistance in the fat insulin receptor knockout (FIRKO) mouse results in slightly improved glucose and lipid homeostasis relative to a normal mouse (11). Although fasted and fed blood sugar levels at two months of age are not significantly different in FIRKO mice as compared to their controls, fasting insulin levels are decreased by ~45%, suggesting improved insulin sensitivity. The responses to glucose and insulin tolerance testing deteriorate by middle age (10 months) in control mice, but not in FIRKO mice. FIRKO mice also have decreased serum triglycerides without a change in serum free fatty acids.

Despite the surprising improvements in glucose homeostasis seen in vivo in FIRKO mice, adipocytes isolated from these mice show the expected defects in glucose and lipid metabolism, with a 90% decrease in insulin-stimulated glucose uptake and a corresponding decrease in insulin-stimulated incorporation of glucose into triglycerides, lactate, and carbon dioxide. Moreover, the ability of insulin to inhibit isoproterenol-induced lipolysis is blunted by approximately 50% (11).

Histological examination of FIRKO fat tissue also reveals an unexpected alteration in adipocyte morphology (11). In wild-type tissue, adipocyte size is normally distributed, with a median diameter between 75 μm and 100 μm. In FIRKO mice, on the other hand, there is a bimodal distribution of adipocytes, with one population of cells having a diameter of <75 μm and the other having a diameter >100 μm. This alteration in fat morphology is not due to differences in knockout efficiency, as both the large and small fat cells show almost complete ablation of insulin receptor expression. It seems instead to stem from some intrinsic heterogeneity in the fat cell response to the loss of insulin signaling. Thus, microarray and proteomics analysis show changes that correlate with both cell size and with changes in insulin signaling. Hence, both large and small FIRKO cells exhibit a decrease in interferon-γ and tumor necrosis factor (TNF)-α as a result of the loss of insulin signaling. By contrast, small adipocytes from both normal and FIRKO mice have decreased expression of enzymes important in lipid and energy metabolism, including fatty acid synthase, and of transcription factors such as C/EBPδ; small FIRKO adipocytes also have a decrease in C/EBPα (12, 13). This suggests that knockout of the insulin receptor unmasks some previously unrecognized intrinsic heterogeneity in white adipose tissue.

One of the most striking features of pure insulin resistance in fat is that rather than leading to development of insulin resistance syndrome, it creates resistance to this disorder (11). Thus, relative to control mice, FIRKO mice eat the same amount and yet have a 50% decrease in both brown and white adipose tissue and a 25% decrease in weight. These differences become even more profound when control and FIRKO mice are put on a high-fat diet or are treated with gold thioglucose, a hypothalamic toxin that causes severe hyperphagia (11, 14).

Leanness and caloric restriction in mammals and defects in insulin signaling in lower organisms are associated with increased longevity (15–19). FIRKO mice are

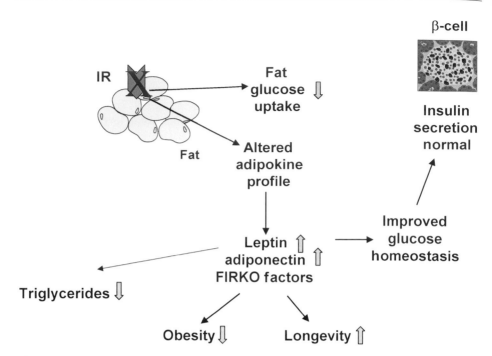

Figure 4   Physiology of pure adipocyte insulin resistance. Knockout of the insulin receptor (IR) in adipocytes decreases glucose uptake into fat and changes adipocyte morphology and gene and protein expression. This produces a more favorable adipokine profile, with increased leptin, adiponectin, and other adipocyte-derived factors (FIRKO factors), which protect against obesity, glucose intolerance, and dyslipidemia while increasing life span.

lean and have insulin signaling defects in fat but are not calorically restricted. Thus, it was important to ask whether these lean mice, which eat more, have normal, increased, or decreased lifespans. In fact, FIRKO mice have an 18% increase in median lifespan compared to control mice (14), and at 30 months of age, when half of the control mice have died, more than 80% of FIRKO mice are still living. This suggests that leanness, even in the absence of reduced caloric intake, can promote longevity. This may also suggest some beneficial effect of insulin resistance that has allowed this trait to survive in evolution.

Although the exact mechanisms that produce these beneficial effects in FIRKO mice remain unknown, one intriguing hypothesis is that insulin resistance in the fat cell alters the complement of hormones elaborated by the adipocyte (Figure 4). For example, fat tissue is known to secrete leptin and adiponectin, hormones that improve insulin sensitivity, as well as resistin and inflammatory cytokines, such as TNF-α and interleukin-6, which promote insulin resistance. In support of this theory, adiponectin and leptin levels are increased and TNF-α expression in fat is decreased in these mice (11, 12).

β-Cell Insulin Receptor Knockout

One of the key features of diabetes is the development of β-cell failure. In type 2 diabetes, this is specifically characterized by a loss of glucose-stimulated insulin secretion as compared to the effects of other secretagogues. To understand the potential role of insulin signaling in this process, β-cell insulin receptor knockout (βIRKO) mice were created using Cre recombinase expression driven by the rat insulin promoter, which produces a near-complete knockout of insulin receptors in β-cells (20).

Prior to weaning, βIRKO mice have normal fasting and fed blood sugar levels (20). However, by two months of age, they develop an abnormal response to glucose tolerance testing, which further worsens with age. This defect is due to a loss of first-phase insulin secretion in response to glucose, whereas insulin secretion in response to arginine is normal. Also, by four months of age, βIRKO islets are significantly smaller than normal and have a 30% decrease in insulin content. This defect in β-cell growth is even more dramatic when βIRKO islets are subjected to an insulin-resistant, hyperinsulinemic state (R.N. Kulkarni, personal communication).

These studies highlight the potential importance of insulin signaling in the β-cell in the pathogenesis of type 2 diabetes. Indeed, the βIRKO mouse recapitulates one of the earliest defects found in patients with type 2 diabetes: impaired first-phase insulin secretion in response to glucose, but normal insulin secretion in response to other secretagogues. Insulin resistance in the β-cell may also contribute to the failure of type 2 diabetics to compensate fully for insulin resistance in other tissues by increasing β-cell mass. Although further studies are needed to determine the role of insulin resistance in the β-cell in humans, it is interesting to note that islets taken from type 2 diabetic individuals show a marked decrease in mRNA for several insulin signaling molecules, suggesting that insulin resistance in the β-cell may indeed be an important component of this disease's pathogenesis (21).

Vascular Endothelium Insulin Receptor Knockout

Vascular disease is a major cause of morbidity associated with the insulin resistance syndrome. Alterations in insulin signaling in the vascular endothelium may contribute to atherogenesis or to abnormal blood vessel proliferation. Also, insulin signaling in the endothelium may play a role in maintaining glucose homeostasis by promoting vasodilation, thus enhancing glucose delivery to muscle (22, 23). Through the use of the Tie2 promoter-enhancer, which is expressed only in vascular endothelial cells and the endocardial cushion, to drive expression of Cre recombinase, the insulin receptor was knocked out of the vascular endothelium (24). VENIRKO mice do not show any abnormalities in glucose homeostasis, and although they tend to have lower heart rates and blood pressure, these changes generally are not significant (24). Surprisingly, however, VENIRKO mice have a 20% decrease in triglycerides as compared to control mice (24, 25). Another unexpected finding is that the retinal vasculature in VENIRKO mice is partially

protected against relative hypoxia in a neonatal oxygen-induced model of retinopathy used to study diabetic retinopathy. Thus, VENIRKO mice exposed to relative
hypoxia have a 50–70% decrease in the size of avascular areas, neovascular tuft
formation, and neovascular nuclei when compared to similarly treated control
mice. Moreover, VENIRKO mice have blunted increases in vascular endothelial
growth factor (VEGF), endothelial nitric oxide synthase (eNOS), and endothelin-1
expression following hypoxia (25). Thus, ablation of the insulin receptor in the vascular endothelium protects against neovascularization and may also have effects
on vascular hemodynamics.

Knockout and Knockdown of the Insulin Receptor
in the Brain

Although the brain is the major site of glucose utilization in the basal state, it is not
classically considered an insulin-sensitive tissue, as insulin does not acutely regulate glucose uptake into the brain. Nonetheless, the presence of insulin receptors
has been demonstrated in a wide variety of brain regions, including the medial and
lateral portions of the arcuate nucleus of the hypothalamus, the olfactory tract, the
cerebellum, and others (26).

To study the effects of insulin resistance in the brain, researchers have used
two approaches. First, neuronal insulin receptor knockout (NIRKO) has been accomplished by expressing the Cre recombinase under the promoter for nestin, an
intermediate-filament protein expressed in neuroepithelial stem cells (27). NIRKO
mice have a >95% decrease in insulin receptor expression in the brain. Second,
the insulin receptor has been knocked down in the hypothalamus using antisense
technology.

KNOCKOUT OF THE NEURONAL INSULIN RECEPTOR   Relative to controls, female
NIRKO mice have a 20% increase in food intake, a 10–15% increase in body
weight, and a twofold increase in fat pad weight (Figure 5) (27). Although male
NIRKO mice are not hyperphagic, they still have a 1.5-fold increase in fat pad
weight relative to controls. On a high-fat diet, both sexes have increased weight
gain relative to controls.

NIRKO mice have normal fasting blood glucose levels and responses to glucose
tolerance testing, but this is in the context of a mild hyperinsulinemia indicating
whole-body insulin resistance (27). NIRKO mice also have a 30% increase in
serum triglycerides. Thus, insulin resistance in the brain produces several components of the insulin resistance syndrome, including mild obesity, hyperinsulinemia,
and hypertriglyceridemia. In addition, insulin action in the brain plays a role in
the control of reproductive function. NIRKO mice of both sexes have a decrease
in fertility due to hypothalamic hypogonadism (27). Female NIRKO mice have
fewer antral follicles and corpora lutea in the ovaries. Similarly, male NIRKO mice
have a 30% decrease in epididymal sperm count and a decrease in the number of
Leydig cells. Both sexes exhibit low LH and FSH levels, and females show a supra-
normal response to the gonadotropin-releasing hormone (GnRH), consistent with

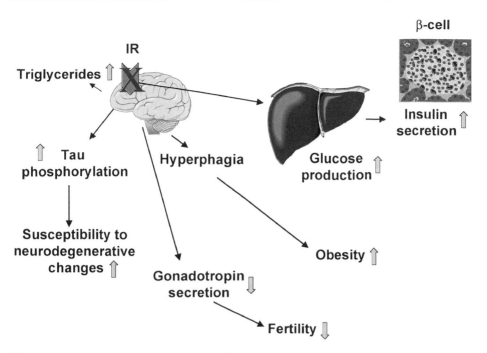

β-cell

IR

Triglycerides ⇑

Insulin ⇑
secretion

⇑ Tau
phosphorylation

Hyperphagia

Glucose ⇑
production

Susceptibility to
neurodegenerative
changes ⇑

Gonadotropin ⇓
secretion

Obesity ⇑

Fertility ⇓

Figure 5 Physiology of pure brain insulin resistance. Knockout of the insulin receptor in the brain results in hyperphagia and therefore obesity. It also leads to decreased gonadotropin secretion and hyperphosphorylation of tau, an early marker of neurodegenerative disease. Lack of insulin signaling in the hypothalamus also impairs the ability of insulin to suppress hepatic glucose production. Hence, neuronal insulin resistance produces obesity, hyperinsulinemia, and dyslipidemia while contributing to reproductive abnormalities and perhaps neurodegenerative disease.

hypothalamic hypogonadism. Although it is well known that there exists an association between insulin resistance and polycystic ovarian disease (PCOS) and that multiple mechanisms may play a role in the development of PCOS, the potential role of neuronal insulin resistance in this syndrome remains unexplored.

Finally, a lack of insulin signaling in the brain produces some changes suggestive of early neurodegenerative disease. This includes a 3.5-fold increase in phosphorylation of the microtubule-associated protein, Tau. Hyperphosphorylated Tau is a major early manifestation in Alzheimer's disease. Moreover, cultured NIRKO neurons are resistant to the protective effects of insulin on apoptosis. Despite this, NIRKO mice have normal learning and memory and do not develop neurofibrillary tangles (28).

KNOCKDOWN OF THE HYPOTHALAMIC INSULIN RECEPTOR   Mindful of the potential importance of the hypothalamus in mediating the central effects of insulin, researchers generated mice with knockdown of the insulin receptor in the hypothalamus (HIRKD) by injecting antisense oligodeoxynucleotides directed against the insulin receptor into the third ventricle (29). HIRKD mice have an 80% decrease of the insulin receptor in the medial arcuate and habenular nuclei as compared to controls in which a scrambled oligodeoxynucleotide is similarly injected.

HIRKD mice show an increase in expression of the orexigenic peptides neuropeptide Y (NPY) and Agouti-related peptide (AgRP), a 50% increase in food intake, and a fourfold increase in subcutaneous fat, but curiously no significant increase in visceral adiposity (29). HIRKD mice have normal serum glucose and insulin levels. However, under hyperinsulinemic-euglycemic clamp conditions, HIRKD mice have a lower glucose infusion rate that can be attributed to a 40% increase in hepatic glucose production, suggesting that insulin signaling in the hypothalamus regulates hepatic glucose output (29). Consistent with this conclusion, it has recently been shown that hepatic vagotomy abolishes the ability of central insulin administration to suppress hepatic glucose production (30).

How Tissues Act Together to Produce the Insulin Resistance Syndrome

Recently, several models have been created to understand how defects in multiple tissues may interact to produce the insulin-resistant phenotype. For example, when insulin receptors are knocked out in both muscle and fat, there are no changes in glucose or insulin levels, indicating that peripheral insulin resistance alone is not sufficient to produce impaired glucose tolerance or the insulin resistance syndrome in the mouse (P.G. Laustsen, S.J. Russell, L. Cui, A. Entingh-Pearsall, M. Holzenberger, et al., manuscript submitted). Likewise, adding insulin resistance in the muscle to insulin resistance in the β-cell improves rather than worsens glucose tolerance. The improvement appears to be due to the ability of MIRKO muscle to enhance insulin secretion from the β-cell and increase glucose uptake in nonmuscle tissues (31). Thus, glucose tolerance in βIRKO-MIRKO double-knockout mice is intermediate between those of βIRKO and MIRKO mice (31). By contrast, when insulin receptors are knocked out simultaneously in the liver and β-cells, mice become markedly diabetic, indicating the importance of the liver and pancreas in the final pathogenesis of the diabetic state (R.N. Kulkarni, personal communication).

A similar conclusion has been reached in a model in which mice with a whole-body knockout of the insulin receptor are rescued by transgenic expression of the insulin receptor in selected tissues. Thus, whereas whole-body insulin receptor knockout mice (IRKO) develop perinatal lethal diabetic ketoacidosis, whole-body IRKO mice in which insulin receptor expression has been reconstituted in the liver, brain, and β-cells are viable, do not develop diabetes (at least in the majority of animals), and have normalized adipose tissue content, lifespan, and reproductive function (32, 33). Whole-body IRKO mice with insulin receptor reconstitution limited to liver and pancreatic β-cells are also rescued from perinatal death but still develop lipoatrophic diabetes and die prematurely (33).

Another combinatorial insulin resistance model has been created by intraperitoneal injection of antisense oligodeoxynucleotides against the insulin receptor (34). In these mice, insulin receptor expression is decreased by 95% in liver and 65% in fat tissue but is unchanged in the hypothalamus and muscle. Such mice have normal glucose, insulin, and leptin levels combined with a decrease in serum free fatty acids and glycerol.

These data, taken together, highlight the importance of insulin signaling in the brain, liver, and β-cells in maintaining glucose homeostasis. It also suggests that insulin resistance in fat and muscle is not detrimental to overall glucose homeostasis and is possibly beneficial in compensating for insulin resistance in the brain, liver, and β-cells.

INSULIN SIGNALING

Although the above studies are helpful in defining the contributions of different tissues to the insulin-resistant state, it is important to remember that the pure insulin resistance associated with ablation of the insulin receptor is only seen in rare, genetic states of insulin resistance, such as leprechaunism and the type A syndrome of insulin resistance (35, 36). Most cases of insulin resistance are due to a combination of factors acting to create partial resistance in some, but not necessarily all, pathways of insulin action. Thus, some pathways may retain sensitivity to insulin, and the degree to which a particular pathway becomes resistant may vary from tissue to tissue. To understand better how different pathways develop varying degrees of insulin resistance, it is necessary to consider the molecular basis of insulin signaling and insulin resistance. Thus, we review below the insulin signaling network (Figure 6) and how knockout of various components of the cascade can impact the insulin-resistant phenotype. We also examine the molecular mechanisms producing insulin resistance as well as the changes in signaling that occur in mouse models and human states of insulin resistance.

Proximal Insulin Signaling

Insulin signaling is triggered by the binding of insulin to its receptor in the plasma membrane of the cell. The insulin receptor is a heterodimeric complex consisting of two α-subunits, capable of insulin binding, and two β-subunits with intrinsic tyrosine kinase activity. Ligand binding of the α-subunits allows the β-subunits to transphosphorylate one another, further increasing their kinase activity (37, 38).

Thus far, more than 10 substrates of the insulin receptor have been identified, of which 4 are structurally related and have been termed the insulin receptor substrate (IRS) proteins (39–43). The IRS family members vary in molecular weight between 60 kDa and 180 kDa. Each contains a pleckstrin homology (PH) domain, a protein tyrosine-binding (PTB) domain, and numerous tyrosine residues that can undergo phosphorylation by the insulin receptor tyrosine kinase. The IRS proteins vary in their tissue distribution and subcellular localization (44–49). Other targets of

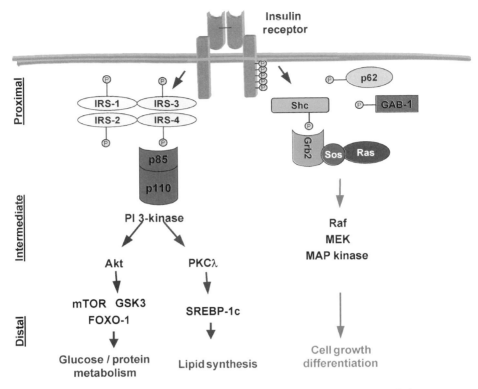

Figure 6   The insulin signaling network. Insulin signaling impacts many cellular processes, including the metabolism of glucose, protein, and lipids, as well as cell growth and differentiation. Thus, the insulin signaling network is broad and complex.

the insulin receptor tyrosine kinase include Shc, Cbl, p62dok, and Gab-1 (50–54). Once phosphorylated, these substrates act as docking molecules for proteins that contain src homology region 2 (SH2) domains, which in turn become activated or associate with other downstream signaling molecules, setting off a complicated cascade of events. Although the details of the insulin signaling network continue to be elucidated, those pathways that are understood better are presented below.

PHOSPHATIDYLINOSITOL 3-KINASE   One of the key intermediates in the insulin signaling network is the class Ia forms of phosphatidylinositol 3-kinase (PI 3-kinase) (reviewed in References 55 and 56). Active PI 3-kinase exists as a heterodimer, consisting of a p110 catalytic subunit, of which there are three isoforms, and a regulatory subunit, of which there are eight known isoforms. The regulatory subunits are encoded by three genes, *Pik3r1*, *Pik3r2*, and *Pik3r3*. *Pik3r1* encodes p85α, the ubiquitously expressed and most abundant regulatory subunit (57). p85α contains multiple domains that may mediate interactions with other proteins, including a src homology region 3 (SH3) domain, a bcr homology domain flanked by two proline-rich regions, and two SH2 domains that flank the p110 catalytic subunit–binding region. The two SH2 regions of PI 3-kinase bind specific phosphotyrosine-containing motifs, pYMXM and pYXXM, found in the IRS proteins. *Pik3r1* can also be alternatively spliced to yield p55α (initially identified as AS53) and p50α (57). These isoforms contain the same two SH2 domains and the p110-binding region; however, they lack the bcr domain, one of the proline-rich regions, and the SH3 domain found in p85α. Instead, p55α and p50α have novel N-terminal sequences of 34 and 6 amino acids, respectively. Additionally, these three isoforms can be alternatively spliced such that aspartate 605, located between the SH2 domains, is replaced by nine amino acids, adding two potential serine phosphorylation sites (58). *Pik3r2* encodes p85β, which is homologous to p85α, whereas *Pik3r3* encodes p55γ (also known as p55PIK), which is similar to p55α.

PI 3-kinase has multiple roles in signal transduction (59). Its most important role in insulin signaling is in the phosphorylation of the phosphoinositides at the 3-position to produce PI-3P, PI-$(3,4)P_2$, and PI-$(3,4,5)P_3$, or PIP_3. These molecules can bind to the PH domains of other signaling molecules, activating them or altering their subcellular location. A major target of the PI 3-phosphates is the AGC family of serine/threonine protein kinases, including PDK1 (PI-dependent kinase 1) and some of the atypical forms of protein kinase C (PKC). The importance of PI 3-kinase in insulin signaling is highlighted by studies showing that virtually all of insulin's effects on glucose transport, lipogenesis, and glycogenesis are abolished by either inhibitors or dominant-negative mutants of PI 3-kinase (55, 60).

TARGETS OF PI 3-KINASE   PI 3-kinase activates the serine/threonine kinase Akt (also known as protein kinase B, or PKB). This appears to occur by two means (55, 56). First, PIP_2 and PIP_3 recruit Akt to the plasma membrane through its PH domain, thus bringing it into proximity with the kinase PDK1, which resides at the membrane. Second, the PI 3-phosphates, particularly PIP_3, activate PDK1. Thus, PDK1 is able to phosphorylate and activate Akt.

Akt has numerous and diverse intracellular targets. It phosphorylates glycogen synthase kinase 3 (GSK3), deactivating it (61). Deactivation of GSK3 leads to derepression, i.e., activation, of glycogen synthase, which is normally inhibited by phosphorylation by GSK3. This results in a stimulation of glycogen synthesis. Akt also phosphorylates the transcription factor FOXO-1. Phosphorylation of FOXO-1 leads to its exclusion from the nucleus, preventing it from activating transcription of various genes, such as phosphoenolpyruvate carboxykinase and glucose 6-phosphatase, two important enzymes in gluconeogenesis (62). Finally, Akt promotes insulin-stimulated glucose uptake, as stable expression of a constitutively active, membrane-bound form of Akt in 3T3-L1 adipocytes results in increased glucose transport and persistent localization of Glut4 to the plasma membrane (63).

PI 3-kinase also activates the atypical forms of PKC, including PKCλ and PKCζ. Like Akt, the atypical PKCs appear to be important in mediating insulin-stimulated glucose uptake. Overexpression of PKCζ or -λ results in increased translocation of the insulin-sensitive glucose transporter, Glut4, to the plasma membrane (64, 65). Conversely, dominant-negative mutants of PKCλ inhibit insulin-stimulated glucose uptake (66). Some recent studies have suggested that these atypical PKCs are more closely linked to IRS-1-mediated signaling in muscle and fat than to IRS-2-mediated signaling, although both IRS proteins activate PI 3-kinase (67). The exact mechanism by which this specificity occurs is unclear.

MAP KINASE PATHWAY   Another major pathway for insulin signaling is the mitogen-activated protein (MAP) kinase pathway. Insulin stimulates tyrosine phosphorylation of the IRS proteins, Gab1, and Shc, allowing them to bind the SH2 domain of the adaptor molecule Grb2. Grb2 recruits the guanyl nucleotide–exchange protein, SOS, to the plasma membrane, where it can activate the G protein Ras. Ras then induces the sequential phosphorylation and activation of the serine/threonine kinases, Raf, MAP kinase–kinase (or MEK), and the MAP kinases ERK 1 and ERK 2. The MAP kinases can phosphorylate substrates in the cytoplasm or translocate to the nucleus to phosphorylate transcription factors such as Elk-1 (68). MAP kinases are associated in particular with the proliferative effects of insulin, as pharmacological inhibitors and dominant-negative mutants of these proteins inhibit the stimulation of cell growth by insulin but have less impact on insulin's metabolic and anabolic effects (69).

SREBP-1c   Whereas many of insulin's effects on the expression of genes in the glucose metabolic pathway are regulated by FOXO1, many of insulin's effects on lipogenesis appear to be mediated via the transcription factor sterol regulatory element–binding protein (SREBP)-1c. The SREBPs are a family of three nuclear transcription factors encoded by two genes (70). SREBP-1c, the dominant isoform in liver and adipose, is capable of activating the entire program of monounsaturated fatty acid synthesis; it also appears to be involved in the regulation of gluconeogenic genes, although to a lesser extent than FOXO1 (71). SREBP-1c also plays an important role in adipocyte differentiation (72).

The SREBPs are subject to complex regulation at the transcriptional and post-translational levels (70). Their transcripts encode membrane-bound precursors, which are retained in the endoplasmic reticulum by Insig proteins. Sterol depletion causes dissociation of the Insig proteins, allowing the SREBPs to proceed to the Golgi apparatus, where the two proteases that cleave the SREBPs are located. Once the N-terminal fragment of SREBP is released from the membrane, it can translocate into the nucleus to activate transcription.

Several lines of evidence suggest that insulin directly regulates SREBP-1c. First, SREBP-1c transcript and nuclear protein are increased by insulin treatment in hepatocytes (73). Second, streptozotocin treatment, which renders mice insulin deficient, results in a decrease in SREBP-1c (73). Similarly, fasting, which also lowers insulin levels, decreases SREBP-1c (74). Conversely, refeeding induces an exaggerated insulin response that is accompanied by an increase in SREBP-1c (74). Third, transgenic overexpression of constitutively active SREBP-1c induces the entire complement of genes necessary for fatty acid synthesis and obviates the need for insulin, whereas dominant-negative forms of SREBP-1c prevent lipogenic gene expression even in the presence of insulin.

Insulin appears to increase SREBP-1c transcription, maturation, and activity. Insulin induces SREBP-1c transcription by activating the transcription factor, liver X receptor (LXR), which is known to bind the SREBP-1c promoter and activate its transcription in an insulin-dependent manner (79). Induction of SREBP-1c transcript appears dependent on PI 3-kinase, through activation of PKCλ, although Akt has also been implicated (75–78). Insulin also suppresses expression of Insig2a, which is the predominant Insig transcript in the livers of fasted animals. Suppression of Insig2a allows SREBP precursors to be escorted to the Golgi apparatus and undergo activation (80). Finally, insulin-stimulated phosphorylation of SREBP-1c mediated by GSK3 and ERK1/ERK 2 may modulate its activity (81, 82).

mTOR   Another important downstream target of Akt that plays a particularly important role in protein synthesis is the protein mTOR (_m_ammalian _t_arget _o_f _r_apamycin). Akt phosphorylates and inhibits TSC2, a component of the tuberous sclerosis heterodimer, which inhibits mTOR (83–85). mTOR, in association with the protein raptor, regulates protein synthesis by phosphorylating and activating the p70 ribosomal S6 kinase (p70S6K). p70S6K phosphorylates the ribosomal S6 protein and thus increases ribosome biosynthesis. mTOR also phosphorylates PHAS-I, also known as eukaryotic translation initiation factor 4E–binding protein 1, enabling it to dissociate from elongation initiation factor 4E and increase translation of mRNAs with a highly structured 5′ untranslated region. Interestingly, it has recently been shown that mTOR, in association with the protein rictor, can activate Akt by phosphorylating Ser473 directly and facilitating phosphorylation of Thr308 by PDK1 (86).

Knockout of Components of the Insulin Signaling Pathway

Over the past decade, the exact roles of many of the signaling components in the insulin signaling network have been studied using genetic knockout strategies as well as knockdown strategies using RNA interference (RNAi) and antisense approaches. Although most of these studies have focused on the role of each of these pathways in the development of glucose intolerance and diabetes, knockout mice also give us some insight into the in vivo role of these proteins in lipid synthesis and cell growth, and help create a picture of how defects at different sites in the insulin signaling network may contribute to the spectrum of the insulin-resistant phenotypes.

KNOCKOUTS OF THE PROXIMAL INSULIN SIGNALING COMPONENTS   Whereas complete knockout of the insulin receptor produces insulin resistance in all pathways,

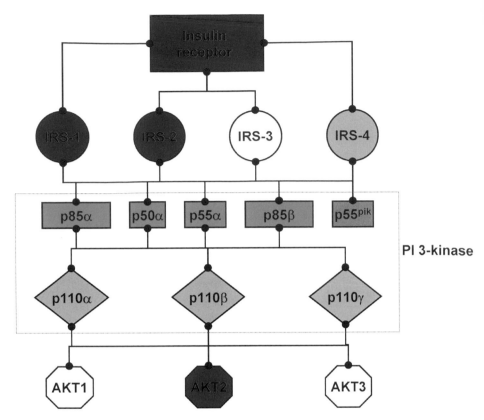

Figure 7   The effects of genetic ablation on glucose homeostasis. The PI 3-kinase portion of the insulin signaling network illustrates the complexity of insulin's metabolic signaling. There are four IRS proteins, five regulatory subunits of PI 3-kinase (with three more produced by alternative splicing), three catalytic units of PI 3-kinase, and three isoforms of Akt. Loss of a particular signaling molecule or isoform can have positive (*green*), slightly negative (*orange*), negative (*red*), or neutral (*white*) effects on glucose homeostasis. Gray denotes that the effects of deletion on glucose homeostasis are unknown.

leading to uncontrolled diabetes and death, knockout and knockdown of the IRS proteins in mice produce varying degrees of glucose intolerance and different types of insulin resistance syndromes (32) (Figure 7). Mice with a complete knockout of IRS-1 exhibit impaired glucose tolerance, increased serum triglycerides (presumably owing to a decrease in triglyceride clearance), and hypertension (87). They also have severe, global growth retardation owing to a decrease in IGF-1-mediated signaling (88). By contrast, IRS-2 knockout mice develop diabetes, which appears to be due to hepatic insulin resistance in addition to a defect in pancreatic β-cell growth and/or development (89, 90). IRS-2 knockouts are also dyslipidemic and hypertensive (91). Moreover, in a mouse model of atherosclerosis, IRS-2 knockout mice show increased neointima formation in response to injury (91). IRS-2 knockout mice do not have global growth retardation but do have a decrease in the size of a few tissues, including certain neuronal cells and the islets of Langerhans (89, 90).

Knockdowns of IRS-1 and IRS-2 using adenoviral-mediated delivery of siRNA, which primarily targets the liver, and not other tissues, produces more specific defects. Decreasing hepatic IRS-1 specifically reduces expression of glucokinase and produces a trend toward hyperglycemia but has no effect on hepatic lipids (92). Knockdown of IRS-2, on the other hand, produces an increase in SREBP-1c and fatty acid synthase, as well as increased hepatic lipid accumulation, but no change in glucose tolerance (92).

The exact role of other substrates in signaling is more difficult to discern. IRS-3 knockout mice appear entirely normal, whereas IRS-4 knockout mice have minimal abnormalities in glucose tolerance (93, 94). By contrast, knockout of hepatic Gab-1, which can mediate signaling to the MAP kinase pathway, actually improves glucose tolerance and signaling through IRS-1 and IRS-2 (95), presumably by preventing inhibitory serine phosphorylation of the IRS proteins by the MAP kinases (see below). In cells in culture, a similar inhibitory effect has been ascribed to IRS-4 (96).

In some cases, combining defects in two proteins increases the severity of the phenotype. For instance, mice with knockdown of both IRS-1 and IRS-2 using adenoviral-mediated delivery of siRNA develop both glucose intolerance and hepatic steatosis (92), and the whole-body double knockout is early-fetal lethal (97). Mice with double knockout of IRS-1 and IRS-3 develop lipoatrophic diabetes and hence a more severe phenotype than either of the single knockouts (98). By contrast, mice with double knockout of IRS-2 and IRS-3 resemble IRS-2 knockout mice and develop diabetes at a similar rate (99). Thus, the degree of functional overlap exhibited by the IRS proteins is variable (99).

KNOCKOUT OF PI 3-KINASE   Disruption of the entire *Pik3r1* gene, and hence of p85α, p55α, and p50α, results in death in mice a few weeks after birth, illustrating the importance of this gene for survival (100). However, knockouts of individual isoforms of the regulatory subunit of PI 3-kinase are viable and paradoxically more insulin sensitive than normal mice. For example, knockout of only p85α (101) or heterozygous knockout of the *Pik3r1* gene (102) results in mice with normal growth and improved insulin sensitivity. In the latter case, the heterozygous knockout can even prevent the development of diabetes in mice with other genetic defects, like those with a combined heterozygous knockout of the insulin receptor and IRS-1 (102). Likewise, selective knockout of p55α and p50α results in mice with normal growth, decreased fat mass, and increased insulin sensitivity (103). Finally, knockout of p85β alone also results in hypoinsulinemia, hypoglycemia, and improved insulin sensitivity (102a).

Although the basis for the improvement in insulin sensitivity has not been fully elucidated, quantitative studies have suggested that it may be due to more efficient signaling of PI 3-kinase. Thus, under normal conditions, the regulatory subunits are in excess of the catalytic subunits (104). As a result, monomeric forms of the regulatory subunit, which are inactive in signaling, can compete with the active heterodimers for binding to the IRS proteins. Thus, reducing the concentration of regulatory subunits improves the ratio of regulatory to catalytic subunits and increases the efficiency of signaling. Conversely, increasing the concentration of regulatory subunits, as appears to occur in pregnancy, actually may result in insulin resistance (105).

Deletion of the catalytic subunits of PI 3-kinase, on the other hand, produces insulin resistance (106). Although homozygous knockout of either the p110α or the p110β catalytic subunit is embryonic lethal, double heterozygous p110$\alpha^{+/-}$p110$\beta^{+/-}$ mice are glucose intolerant, exhibiting mild fasting hyperinsulinemia and decreased IRS-associated PI 3-kinase activity upon insulin stimulation (106–108). Interestingly, mice with knockout of the p110-γ form of the catalytic subunit, which is expressed in pancreatic islets, have a greater hypoglycemic response to insulin than do control mice. However, these mice also are impaired in glucose-stimulated insulin secretion and are thus mildly glucose intolerant (109).

KNOCKOUT OF AKT   There are three isoforms of Akt encoded by three different genes. Knockout of Akt1, the ubiquitously expressed isoform of Akt1, produces a global defect in growth but essentially no alterations in glucose homeostasis (110). Likewise, knockout of Akt3, which is expressed predominantly in the brain and testes, results in a reduction in brain size but no changes in glucose metabolism (111).

By contrast, knockout of Akt2, which is expressed in pancreatic β-cells, skeletal muscle, and brown fat, results in glucose intolerance and hyperinsulinemia. Loss of Akt2 also impairs insulin-stimulated glucose uptake into muscle and decreases the ability of insulin to suppress hepatic glucose output (112). Akt2 knockout mice also have increased serum triglycerides (113) and under some conditions develop β-cell failure and diabetes (113). Interestingly, because Akt2 is expressed in platelets, Akt2 knockout mice also have defects in platelet aggregation and thrombus formation (114).

A role for Akt in insulin secretion has been confirmed by creation of transgenic mice expressing a kinase-dead mutant of Akt specifically in β-cells (115). Such mice have a defect in insulin exocytosis and eventually develop glucose intolerance. Conversely, mice overexpressing a constitutively active form of Akt in the liver become hypoglycemic, hypoinsulinemic, and, by mechanisms that are not clear, hypertriglyceridemic (116).

KNOCKOUT OF THE MAPK PATHWAY   Although the major effects of the MAP kinase pathway focus on cell growth, ERK1 knockout mice have a metabolic phenotype with decreased adiposity, owing to a decrease in the number of adipocytes (117). When challenged with a high-fat diet, these mice, like FIRKO mice, are protected from obesity or glucose intolerance (11). Such findings are consistent with the importance of ERK1 signaling in adipogenesis.

KNOCKOUT OF SREBP-1c   Knockout of SREBP-1c does not alter serum insulin or glucose levels. However, mice lacking SREBP-1c have a 50% decrease in serum triglycerides and a 10% decrease in serum cholesterol (118). On the other hand, transgenic mice expressing a constitutively active form of SREBP-1c develop hepatic steatosis, a finding associated with the insulin resistance syndrome in humans and mice (119–121). Somewhat surprisingly, whereas SREBP-1c promotes adipocyte differentiation in vitro, mice expressing a constitutively active form of SREBP-1c only in adipose tissue have disordered differentiation of white and brown fat (122). The little fat that accumulates in these mice has a histologically immature appearance. Perhaps because of an alteration in adipokine production, they also develop extremely high insulin levels, hepatic steatosis, diabetes, and elevated triglycerides, all of which can be reversed with leptin treatment (122, 123).

Molecular Mechanisms of Insulin Resistance

The insulin signaling cascade is both complex and broad. At each step of the cascade, there appear to be several molecules or isoforms capable of transducing the signal, each with distinct but overlapping functions. For example, in the major pathway leading to the metabolic actions of insulin, the insulin receptor theoretically activates any of the 4 IRS proteins, which in turn activate 1 of 24 possible PI 3-kinase heterodimers, leading to activation of 1 of 3 Akt isoforms. Thus, even in these first three signaling steps in the metabolic pathway of insulin action, there are at least 288 combinations of signaling molecules. But this is not the only known pathway implicated in insulin action; moreover, there may be many additional players and pathways that have yet to be identified. Thus, the number of signaling combinations probably exceeds 1000. Theoretically, diminishing the activity of any single component of the insulin signaling cascade or any combination of components may produce a distinct type of insulin resistance with its own molecular signature.

At the molecular level, insulin resistance can be acquired through multiple mechanisms (Figure 8). First, important components of the insulin signaling cascade can be decreased by either increased degradation or by decreased transcription, as has been described for IRS-1 and IRS-2 (124–127). In this regard, many of the signaling components interact in such a way as to regulate one another. For example, recent studies have shown that SREBP-1c can suppress transcription of IRS-2, suggesting that increased stimulation of SREBP-1c can lead to a decrease in IRS-2 levels (128). Second, as noted above, if the regulatory subunits of PI 3-kinase are further increased, they can act as inhibitors of the normal dimeric form of the enzyme. Third, any of these components can undergo posttranslational modifications that decrease or increase its activity. For example, the insulin receptor and the IRS proteins can undergo serine phosphorylation by PKC, ERK, and the stress kinases, JNK and IKKβ (129–131). Serine phosphorylation of the insulin receptor decreases its kinase activity. Serine phosphorylation of the IRS proteins impairs their ability to be tyrosine phosphorylated and increases their association with the 14-3-3 proteins, removing them from the insulin signaling cascade (132).

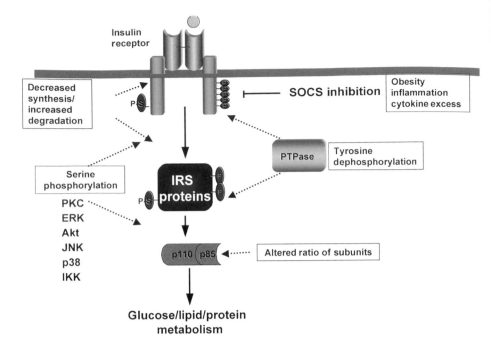

Figure 8   Mechanisms that can produce insulin resistance. Multiple mechanisms exist for downregulating insulin signaling, such as decreased synthesis, increased degradation, inhibitory serine phosphorylation, interaction with inhibitory proteins, and alteration of the ratios of different signaling molecules.

Fourth, insulin resistance can be produced by the interaction of inhibitory proteins with components of the insulin signaling cascade. For example, the suppressors of cytokine signaling (SOCS) proteins, which are induced by inflammatory cytokines, bind to the insulin receptor and block its signaling (133, 134). Finally, insulin resistance can be due to increases in the activity or amount of the enzymes that normally reverse insulin action, including the phosphotyrosine phosphatases, e.g., PTP1b, and the PIP_3 phosphatases, e.g., PTEN and SHIP (135, 136).

Ultimately, the phenotype of insulin resistance will depend on the exact components affected and the exact tissues in which they are affected. For example, insulin resistance caused by downregulation of the insulin receptor itself will decrease insulin signaling at all steps within the cell, whereas alterations in a single IRS protein will alter only those pathways downstream of that specific IRS protein.

Alterations in Insulin Signaling in a Mouse Model of Insulin Resistance

The leptin-deficient *ob/ob* mouse is a well-established model of insulin resistance. The absence of leptin causes these mice to be hyperphagic and have low energy expenditure, resulting in morbid obesity, hyperinsulinemia, and hyperglycemia. *ob/ob* mice also have massive hepatic steatosis and increased very-low-density lipoprotein (VLDL) secretion (137). In this model of obesity, insulin resistance is associated with numerous changes in insulin signaling, although the exact nature of these varies between tissues.

In the *ob/ob* liver, there is a decrease in the expression and phosphotyrosine content of the insulin receptor, IRS-1, and IRS 2 (138, 139). The stoichiometry of the regulatory PI 3-kinase subunits is altered, with a twofold decrease in p85α, the predominant liver isoform; a ninefold increase in p55α; and a twofold increase in p50α. These changes are associated with a decrease in IRS-1, IRS-2, and phosphotyrosine-associated PI 3-kinase activity and Akt phosphorylation (139). Nonetheless, atypical PKC becomes activated to a similar extent (141), and basal ERK1 phosphorylation is greatly increased (140).

SREBP-1c is also increased in the livers of *ob/ob* mice, producing steatosis and increased triglyceride secretion (127, 137, 142). This is in contrast to LIRKO mice, which have a decrease in SREBP-1c, VLDL secretion, and hepatic triglyceride accumulation (6a; S.B. Biddinger & C.R. Kahn, unpublished results). Thus, in the insulin-resistant *ob/ob* liver, there is decreased signaling through PI 3-kinase and Akt, increased activity of SREBP-1c, and dysregulated activity of ERK1.

Unlike the liver, the adipose tissue of *ob/ob* mice becomes resistant to the effects of insulin on lipogenesis (143). Consistent with this, SREBP-1c and lipogenic gene expression is markedly reduced in the adipose tissue of *ob/ob* mice (142). That *ob/ob* mice have large, triglyceride-laden adipocytes suggests the importance of processes other than de novo lipogenesis in determining fat cell size, such as increased uptake and/or decreased release of free fatty acids and triglycerides.

In muscle, *ob/ob* mice also have a decrease in insulin receptor, IRS-1 and IRS-2 protein levels, and phosphotyrosine content (139). They also have decreases in IRS-1, IRS-2, and phosphotyrosine-associated PI 3-kinase activities (139). However, there is no decrease in the activation of atypical PKCs, nor are there any changes in the distribution of the isoforms of the regulatory subunits of PI 3-kinase (139, 141).

By contrast, in retinal vasculature, none of the components of the proximal insulin signaling cascade are downregulated. Thus, there are normal levels of tyrosine-phosphorylated insulin receptor, IRS-1, and IRS-2 (140). Interestingly, Akt phosphorylation is decreased, while ERK1 phosphorylation remains brisk (140). These changes in the retinal vasculature are associated with increased expression of VEGF, which can directly stimulate neovascularization, and decreased expression of the vasodilator, eNOS (140). Similarly, the vasculature of the Zucker fa/fa rat, a related model with a defect in the leptin receptor, also has increased MAPK signaling (144).

In addition, different models of insulin resistance are associated with different defects in insulin signaling. In the high fat–fed rat, for example, IRS-1- and IRS-2-associated PI 3-kinase are increased two- to threefold in liver but decreased in muscle and fat (145). On the other hand, insulin resistance induced by dexamethasone treatment is associated with IRS-1 and PI 3-kinase levels that are increased in liver but decreased in muscle; however, IRS-1-associated PI 3-kinase activity is decreased in both tissues (146).

Alterations in Insulin Signaling in Human Insulin Resistance

It is much more difficult to study insulin signaling in humans, but in both humans and rodents, different pathways do vary in the degree of insulin resistance they manifest.

In obesity, it has been well established that there is a decrease in insulin receptor binding and a decrease in the phosphorylation of the insulin receptor and IRS-1 in muscle (147). There is also a dramatic decrease in insulin-stimulated IRS-1–associated PI 3-kinase activity and glucose uptake (147–149). However, there is no decrease in the activation of Akt or MAPK (147, 149). Even more profound defects in IRS-1 phosphorylation and activation of PI 3-kinase are seen in patients with type 2 diabetes; however, activation of Akt and MAPK is still normal (147, 149). Thus, muscle insulin resistance associated with obesity and type 2 diabetes

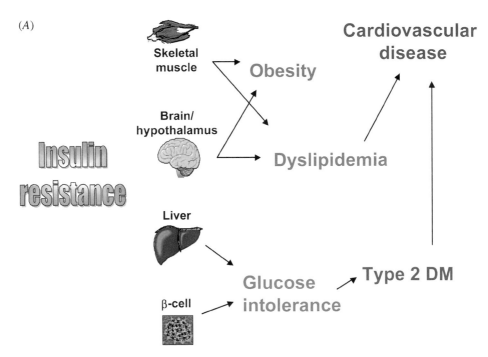

(A)

Skeletal muscle

Brain/ hypothalamus

Insulin resistance

Liver

β-cell

Obesity

Dyslipidemia

Glucose intolerance

Cardiovascular disease

Type 2 DM

Figure 9 The insulin resistance syndrome can be produced by either insulin resistance (A) or excessive insulin action (B) in different tissues. The knockout models reveal that insulin resistance in either muscle or brain produces obesity and dyslipidemia, whereas insulin resistance in the liver and β-cell produces glucose intolerance. By contrast, excessive insulin signaling in the liver or endothelium produces dyslipidemia, whereas excessive insulin action in fat produces obesity, glucose intolerance, and dyslipidemia.

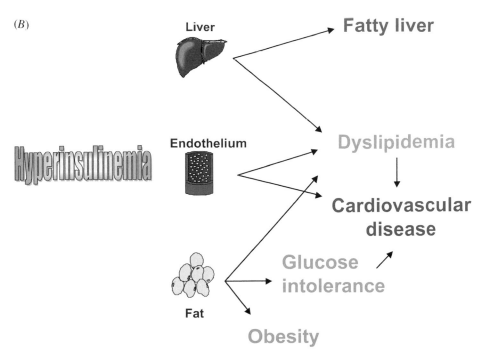

Figure 9   (*Continued*)

is associated with a dramatic decrease in signaling through IRS-1 and PI 3-kinase, while signaling through MAPK is maintained.

In human obesity, adipose tissue shows a decrease in 2-deoxyglucose uptake, both basally and following insulin stimulation (151, 152). Some studies show that adipocytes from patients with obesity or type 2 diabetes also have a decrease in the number of insulin receptors per cell (152–154), whereas others show no change (155, 156). In addition, adipocytes from obese individuals show a decrease in insulin receptor tyrosine kinase activity and IRS-1-associated PI 3-kinase activity; however, IRS-2-associated PI 3-kinase activity is maintained at normal levels (155, 152). In liver, on the other hand, there is essentially no change in the number of insulin receptors per hepatocyte, insulin receptor kinase activity, and the ability of insulin to inhibit glycogenolysis (152, 157).

Implications for Human Disease

The insulin resistance syndrome, a constellation of symptoms including dyslipidemia, hypertension, glucose intolerance, and obesity, can progress to diabetes, cardiovascular disease, cancer, nonalcoholic fatty liver disease, and PCOS. These disorders are all related to insulin resistance, but the pathophysiological links between them are not clear.

Insulin resistance is frequently measured and thought of synonymously with impaired glucose uptake. However, using targeted ablations of the insulin receptor and its downstream signaling molecules, we see that the phenotypic manifestations of the insulin resistance syndrome are not always associated with defects in glucose uptake. For example, insulin resistance in fat greatly diminishes glucose uptake into fat tissue but improves overall glucose homeostasis and the serum lipid profile as well as preventing obesity. Similarly, insulin resistance in brain can produce obesity, hyperinsulinemia, and elevated triglycerides, even though the brain is not a site of insulin-stimulated glucose uptake. Thus, defining insulin resistance in terms of decreased glucose uptake may limit our ability to relate insulin resistance to the various components and sequelae of the insulin resistance syndrome.

Tissue-specific knockouts of the insulin receptor also highlight the fact that either insulin resistance (Figure 9A) or excessive insulin action (Figure 9B) may have detrimental effects. Thus, both insulin resistance in the brain or excessive insulin signaling in fat may produce similar features of the insulin resistance syndrome. Alternatively, mixed insulin sensitivity and resistance in the liver—that is, decreasing activation of Akt and the ability of insulin to suppress glucose production, while increasing activation of SREBP-1c and lipogenesis—may produce glucose intolerance and hepatic steatosis.

Depending on which tissues and pathways are affected, the phenotype of insulin resistance could vary. We could imagine, for example, that if insulin resistance occurred in both the muscle and the fat, the beneficial effects of fat insulin resistance would balance the negative effects of muscle insulin resistance, and disease would be averted. If, instead, insulin resistance in the liver or brain were coupled with a defect in β-cell insulin signaling, diabetes might result. On the other hand, insulin resistance in the brain coupled with continued insulin signaling in the vasculature might promote cardiovascular disease.

Thus, it may be more useful to define and consider insulin resistance in terms of the tissues and pathways in which insulin signaling has been altered. Ultimately, we hope to classify insulin-resistant individuals on the basis of the specific lesions they harbor. By doing so, we may be able to identify better the components and sequelae of the insulin resistance syndrome for which they are most at risk and intervene with more rational therapy. For example, individuals known to have β-cell insulin resistance may be at greater risk for the development of diabetes and may benefit from early institution of insulin therapy. By contrast, therapy for individuals with excess insulin signaling in fat might be directed toward decreasing insulin levels.

CONCLUSIONS

Insulin resistance is a central feature of the insulin resistance syndrome, which manifests itself initially as obesity, glucose intolerance, dyslipidemia, and hypertension but can progress to type 2 diabetes, cardiovascular disease, nonalcoholic fatty liver disease, and other disorders. Mice harboring tissue-specific knockouts of the insulin receptor and targeted ablations of the insulin signaling molecules have given us powerful insights into the nature of insulin resistance. The heterogeneity of phenotypes expressed by these mice suggests that the diverse presentations of insulin resistance in humans may be due to differences in the pathways and tissues that are defective in insulin signaling. These mice also show us that defects in insulin signaling are not always detrimental and may, in fact, be protective against some features of the insulin resistance syndromes. Furthermore, these mice give us new tools with which to deconstruct the insulin resistance syndromes and to distinguish those tissues and processes that have become resistant from those that have remained sensitive. Such insights no doubt will lead to important advances in the understanding and treatment of these disorders.

The *Annual Review of Physiology* is online at
http://physiol.annualreviews.org

LITERATURE CITED

1. Ford ES, Giles WH, Mokdad AH. 2004. Increasing prevalence of the metabolic syndrome among U.S. adults. *Diabetes Care* 27:2444–49

2. Weiss R, Dziura J, Burgert TS, Tamborlane WV, Taksali SE, et al. 2004. Obesity and the metabolic syndrome in children and adolescents. *N. Engl. J. Med.* 350:2362–74

3. Reaven G. 2004. The metabolic syndrome or the insulin resistance syndrome? Different names, different concepts, and different goals. *Endocrinol. Metab. Clin. North Am.* 33:283–303

3a. Chang PY, Benecke H, Le Marchand-Brustel Y, Lawitts J, Moller DE. 1994. Expression of a dominant-negative mutant human insulin receptor in the muscle of transgenic mice. *J. Biol. Chem.* 269:16034–40

3b. Fernandez AM, Kim JK, Yakar S, Dupont J, Hernandez-Sanchez C, et al. 2001. Functional inactivation of the IGF-I and insulin receptors in skeletal muscle causes type 2 diabetes. *Gene Dev.* 15:1926–34

4. Bruning JC, Michael MD, Winnay JN, Hayashi T, Horsch D, et al. 1998. A muscle-specific insulin receptor knockout exhibits features of the metabolic syndrome of NIDDM without altering glucose tolerance. *Mol. Cell* 2:559–69

5. Michael MD, Kulkarni RN, Postic C, Previs SF, Shulman GI, et al. 2000. Loss of insulin signaling in hepatocytes leads to severe insulin resistance and progressive hepatic dysfunction. *Mol. Cell* 6:87–97

6. Fisher SJ, Kahn CR. 2003. Insulin signaling is required for insulin's direct and indirect action on hepatic glucose production. *J. Clin. Invest.* 111:463–68

6a. Hernandez-Ono A, Zhang Y-L, Chiang J, Moon B, Kahn CR, et al. 2005. Lipoprotein metabolism in the liver insulin receptor knockout (LIRKO) mouse: critical role of insulin signaling in the regulation of very low density lipoprotein (VLDL) secretion. *Diabetes* 54(Suppl. 1):A2 (Abstr.)

7. Zierath JR, Wallberg-Henriksson H. 2002. From receptor to effector: insulin signal transduction in skeletal muscle from type II diabetic patients. *Ann. NY Acad. Sci.* 967:120–34

8. Martin BC, Wjkabrsjk CR. 1992. Role

of glucose and insulin resistance in development of type 2 diabetes mellitus: results of a 25-year follow-up study. *Lancet* 340:925–29

9. Kim JK, Michael MD, Previs SF, Peroni OD, Mauvais-Jarvis F, et al. 2000. Redistribution of substrates to adipose tissue promotes obesity in mice with selective insulin resistance in muscle. *J. Clin. Invest.* 105:1791–97

10. Cariou B, Postic C, Boudou P, Burcelin R, Kahn CR, et al. 2004. Cellular and molecular mechanisms of adipose tissue plasticity in muscle insulin receptor knockout mice. *Endocrinology* 145:1926–32

11. Bluher M, Michael MD, Peroni OD, Ueki K, Carter N, et al. 2002. Adipose tissue selective insulin receptor knockout protects against obesity and obesity-related glucose intolerance. *Dev. Cell* 3:25–38

12. Bluher M, Patti M, Kahn C. 2004. Intrinsic heterogeneity in adipose tissue-insulin receptor knockout is associated with differences in metabolic function and in gene expression patterns revealed by microarray analysis. *J. Biol. Chem.* 279:31891–901

13. Bluher M, Wilson-Fritch L, Leszyk J, Laustsen P, Corvera S, Kahn C. 2004. Role of insulin action and cell size on protein expression patterns in adipocytes. *J. Biol. Chem.* 279:31902–9

14. Bluher M, Kahn BB, Kahn CR. 2003. Extended longevity in mice lacking the insulin receptor in adipose tissue. *Science* 299:572–74

15. Kenyon C, Chang J, Gensch E, Rudner A, Tabtiang R. 1993. A *C. elegans* mutant that lives twice as long as wild type. *Nature* 366:461–64

16. Kimura KD, Tissenbaum HA, Liu Y, Ruvkun G. 1997. daf-2, an insulin receptor-like gene that regulates longevity and diapause in *Caenorhabditis elegans*. *Science* 277:942–46

17. Wolkow CA, Kimura KD, Lee MS, Ruvkun G. 2000. Regulation of *C. elegans* life-span by insulinlike signaling in the nervous system. *Science* 290:147–50

18. Tatar M, Kopelman A, Epstein D, Tu MP, Yin CM, Garofalo RS. 2001. A mutant *Drosophila* insulin receptor homolog that extends life-span and impairs neuroendocrine function. *Science* 292:107–10

19. Clancy DJ, Gems D, Harshman LG, Oldham S, Stocker H, et al. 2001. Extension of life-span by loss of CHICO, a *Drosophila* insulin receptor substrate protein. *Science* 292:104–6

20. Kulkarni RN, Bruning JC, Winnay JN, Postic C, Magnuson MA, Kahn CR. 1999. Tissue-specific knockout of the insulin receptor in pancreatic β cells creates an insulin secretory defect similar to that in Type 2 diabetes. *Cell* 96:329–39

21. Gunton JE, Kulkarni RN, Yim S, Okada T, Hawthorne WJ, et al. 2005. Loss of ARNT/HIF1β mediates altered gene expression and pancreatic-islet dysfunction in human type 2 diabetes. *Cell* 122:337–49

22. Etgen GJ Jr, Fryburg DA, Gibbs EM. 1997. Nitric oxide stimulates skeletal muscle glucose transport through a calcium/contraction- and phosphatidylinositol-3-kinase-independent pathway. *Diabetes* 46:1915–19

23. Young ME, Radd GK, Leighton B. 1997. Nitric oxide stimulates glucose transport and metabolism in rat skeletal muscle *in vitro*. *Biochem. J.* 322:223–28

24. Vicent D, Ilany J, Kondo T, Naruse K, Fisher SJ, et al. 2003. The role of endothelial insulin signaling in the regulation of vascular tone and insulin resistance. *J. Clin. Invest.* 111:1373–80

25. Kondo T, Vicent D, Suzuma K, Yanagisawa M, King GL, et al. 2003. Knockout of insulin and IGF-1 receptors on vascular endothelial cells protects against retinal neovascularization. *J. Clin. Invest.* 111:1835–42

26. Havrankova J, Brownstein M, Roth J. 1981. Insulin and insulin receptors in rodent brain. *Diabetologia* 20:268–73

27. Bruning JC, Gautam D, Burks DJ, Gillette J, Schubert M, et al. 2000. Role of brain insulin receptor in control of body weight and reproduction. *Science* 289:2122–25

28. Schubert M, Gautam D, Surjo D, Ueki K, Baudler S, et al. 2004. Role for neuronal insulin resistance in neurodegenerative diseases. *Proc. Natl. Acad. Sci. USA* 101:3100–5

29. Obici S, Feng Z, Karkanias G, Baskin DG, Rossetti L. 2002. Decreasing hypothalamic insulin receptors causes hyperphagia and insulin resistance in rats. *Nat. Neurosci.* 5:566–72

30. Pocai A, Lam TK, Gutierrez-Juarez R, Obici S, Schwartz GJ, et al. 2005. Hypothalamic K_{ATP} channels control hepatic glucose production. *Nature* 434:1026–31

31. Mauvais-Jarvis F, Virkamaki A, Michael MD, Winnay JN, Zisman A, et al. 2000. A model to explore the interaction between muscle insulin resistance and β-cell dysfunction in the development of type 2 diabetes. *Diabetes* 49:2126–34

32. Accili D, Drago J, Lee EJ, Johnson MD, Cool MH, et al. 1996. Early neonatal death in mice homozygous for a null allele of the insulin receptor gene. *Nat. Genet.* 12:106–9

33. Okamoto H, Nakae J, Kitamura T, Park BC, Dragatsis I, Accili D. 2004. Transgenic rescue of insulin receptor-deficient mice. *J. Clin. Invest.* 114:214–23

34. Buettner C, Patel R, Muse ED, Bhanot S, Monia BP, et al. 2005. Severe impairment in liver insulin signaling fails to alter hepatic insulin action in conscious mice. *J. Clin. Invest.* 115:1306–13

35. Taylor SI, Arioglu E. 1998. Syndromes associated with insulin resistance and acanthosis nigricans. *J. Basic Clin. Physiol. Pharmacol.* 9:419–39

36. Kahn CR, Flier JS, Bar RS, Archer JA, Gorden P, et al. 1976. The syndromes of insulin resistance and acanthosis nigricans. Insulin-receptor disorders in man. *N. Engl. J. Med.* 294:739–45

37. Baron V, Kaliman P, Gautier N, Van Obberghen E. 1992. The insulin receptor activation process involves localized conformational changes. *J. Biol. Chem.* 267:23290–94

38. Ablooglu AJ, Kohanski RA. 2001. Activation of the insulin receptor's kinase domain changes the rate-determining step of substrate phosphorylation. *Biochemistry* 40:504–13

39. Sun XJ, Rothenberg PL, Kahn CR, Backer JM, Araki E, et al. 1991. Structure of the insulin receptor substrate IRS-1 defines a unique signal transduction protein. *Nature* 352:73–77

40. Sun XJ, Wang LM, Zhang Y, Yenush L, Myers MG Jr, et al. 1995. Role of IRS-2 in insulin and cytokine signalling. *Nature* 377:173–77

41. Lavan BE, Lane WS, Lienhard GE. 1997. The 60-kDa phosphotyrosine protein in insulin-treated adipocytes is a new member of the insulin receptor substrate family. *J. Biol. Chem.* 272:11439–43

42. Lavan BE, Fantin VR, Chang ET, Lane WS, Keller SR, Lienhard GE. 1997. A novel 160-kDa phosphotyrosine protein in insulin-treated embryonic kidney cells is a new member of the insulin receptor substrate family. *J. Biol. Chem.* 272:21403–7

43. White MF, Yenush L. 1998. The IRS-signaling system: a network of docking proteins that mediate insulin and cytokine action. *Curr. Top. Microbiol. Immunol.* 228:179–208

44. Fantin VR, Lavan BE, Wang Q, Jenkins NA, Gilbert DJ, et al. 1999. Cloning, tissue expression, and chromosomal location of the mouse insulin receptor substrate 4 gene. *Endocrinology* 140:1329–37

45. Sciacchitano S, Taylor SI. 1997.

Cloning, tissue expression, and chromosomal localization of the mouse IRS-3 gene. *Endocrinology* 138:4931–40

46. Inoue G, Cheatham B, Emkey R, Kahn CR. 1998. Dynamics of insulin signaling in 3T3-L1 adipocytes. Differential compartmentalization and trafficking of insulin receptor substrate (IRS)-1 and IRS-2. *J. Biol. Chem.* 273:11548–55

47. Kaburagi Y, Satoh S, Yamamoto-Honda R, Ito T, Ueki K, et al. 2001. Insulin-independent and wortmannin-resistant targeting of IRS-3 to the plasma membrane via its pleckstrin homology domain mediates a different interaction with the insulin receptor from that of IRS-1. *Diabetologia* 44:992–1004

48. Lassak A, Del Valle L, Peruzzi F, Wang JY, Enam S, et al. 2002. Insulin receptor substrate 1 translocation to the nucleus by the human JC virus T-antigen. *J. Biol. Chem.* 277:17231–38

49. Kabuta T, Hakuno F, Asano T, Takahashi S. 2002. Insulin receptor substrate-3 functions as transcriptional activator in the nucleus. *J. Biol. Chem.* 277:6846–51

50. Gustafson TA, He W, Craparo A, Schaub CD, O'Neill TJ. 1995. Phosphotyrosine-dependent interaction of Shc and IRS-1 with the NPEY motif of the insulin receptor via a novel non-SH2 domain. *Mol. Cell Biol.* 15:2500–8

51. Boney CM, Gruppuso PA, Faris RA, Frackelton AR Jr. 2000. The critical role of Shc in insulin-like growth factor-I-mediated mitogenesis and differentiation in 3T3-L1 preadipocytes. *Mol. Endocrinol.* 14:805–13

52. Baumann CA, Ribon V, Kanzaki M, Thurmond DC, Mora S, et al. 2000. CAP defines a second signalling pathway required for insulin-stimulated glucose transport. *Nature* 407:202–7

53. Wick MJ, Dong LQ, Hu D, Langlais P, Liu F. 2001. Insulin receptor-mediated p62dok tyrosine phosphorylation at residues 362 and 398 plays distinct roles for binding GTPase-activating protein

and Nck and is essential for inhibiting insulin-stimulated activation of Ras and Akt. *J. Biol. Chem.* 276:42843–50

54. Lehr S, Kotzka J, Herkner A, Sikmann A, Meyer HE, et al. 2000. Identification of major tyrosine phosphorylation sites in the human insulin receptor substrate Gab-1 by insulin receptor kinase in vitro. *Biochemistry* 39:10898–907

55. Shepherd PR, Withers DJ, Siddle K. 1998. Phosphoinositide 3-kinase: the key switch mechanism in insulin signalling. *Biochem. J.* 333:471–90

56. Vanhaesebroeck B, Alessi DR. 2000. The PI3K-PDK1 connection: more than just a road to PKB. *Biochem. J.* 346(Pt. 3):561–76

57. Inukai K, Funaki M, Ogihara T, Katagiri H, Kanda A, et al. 1997. p85α gene generates three isoforms of regulatory subunit for phosphatidylinositol 3-kinase (PI 3-kinase), p50α, p55α, and p85α, with different PI 3-kinase activity elevating responses to insulin. *J. Biol. Chem.* 272:7873–82

58. Antonetti DA, Algenstaedt P, Kahn CR. 1996. Insulin receptor substrate 1 binds two novel splice variants of the regulatory subunit of phosphatidylinositol 3-kinase in muscle and brain. *Mol. Cell. Biol.* 16:2195–203

59. Carpenter CL, Cantley LC. 1996. Phosphoinositide kinases. *Curr. Opin. Cell Biol.* 8:153–58

60. Cheatham B, Vlahos CJ, Cheatham L, Wang L, Blenis J, Kahn CR. 1994. Phosphatidylinositol 3-kinase activation is required for insulin stimulation of pp70 S6 kinase, DNA synthesis, and glucose transporter translocation. *Mol. Cell. Biol.* 14:4902–11

61. Rommel C, Bodine SC, Clarke BA, Rossman R, Nunez L, et al. 2001. Mediation of IGF-1-induced skeletal myotube hypertrophy by PI(3)K/Akt/mTOR and PI(3)K/Akt/GSK3 pathways. *Nat. Cell. Biol.* 3:1009–13

62. Nakae J, Kitamura T, Silver DL, Accili

D. 2001. The forkhead transcription factor Foxo1 (Fkhr) confers insulin sensitivity onto glucose-6-phosphatase expression. *J. Clin. Invest.* 108:1359–67

63. Kohn AD, Summers SA, Birnbaum MJ, Roth RA. 1996. Expression of a constitutively active Akt Ser/Thr kinase in 3T3-L1 adipocytes stimulates glucose uptake and glucose transporter 4 translocation. *J. Biol. Chem.* 271:31372–78

64. Bandyopadhyay G, Standaert ML, Zhao L, Yu B, Avignon A, et al. 1997. Activation of protein kinase C (α, β, and ζ) by insulin in 3T3/L1 cells. Transfection studies suggest a role for PKC-ζ in glucose transport. *J. Biol. Chem.* 272:2551–58

65. Bandyopadhyay G, Standaert ML, Kikkawa U, Ono Y, Moscat J, Farese RV. 1999. Effects of transiently expressed atypical (ζ, λ), conventional (α, β) and novel (δ, ϵ) protein kinase C isoforms on insulin-stimulated translocation of epitope-tagged GLUT4 glucose transporters in rat adipocytes: specific interchangeable effects of protein kinases C-ζ and C-λ. *Biochem. J.* 337(Pt. 3): 461–70

66. Kotani K, Ogawa W, Matsumoto M, Kitamura T, Sakaue H, et al. 1998. Requirement of atypical protein kinase λ for insulin stimulation of glucose uptake but not for akt activation in 3T3-L1 adipocytes. *Mol. Cell. Biol.* 18:6971–82

67. Sajan MP, Standaert ML, Miura A, Kahn CR, Farese RV. 2004. Tissue-specific differences in activation of atypical protein kinase C and protein kinase B in muscle, liver, and adipocytes of insulin receptor substrate-1 knockout mice. *Mol. Endocrinol.* 18:2513–21

68. Gille H, Kortenjann M, Thomae O, Moomaw C, Slaughter C, et al. 1995. ERK phosphorylation potentiates Elk-1-mediated ternary complex formation and transactivation. *EMBO J.* 14:951–62

69. Reusch JEB, Bhuripanyo P, Carel K, Leitner JW, Hsieh P, et al. 1995. Differ-

ential requirement for p21^{ras} activation in the metabolic signaling by insulin. *J. Biol. Chem.* 270:2036–40

70. Horton JD, Goldstein JL, Brown MS. 2002. SREBPs: activators of the complete program of cholesterol and fatty acid synthesis in the liver. *J. Clin. Invest.* 109:1125–31

71. Yamamoto T, Shimano H, Nakagawa Y, Ide T, Yahagi N, et al. 2004. SREBP-1 interacts with hepatocyte nuclear factor-4 α and interferes with PGC-1 recruitment to suppress hepatic gluconeogenic genes. *J. Biol. Chem.* 279:12027–35

72. Tontonoz P, Kim JB, Graves RA, Spiegelman BM. 1993. ADD1: a novel helix-loop-helix transcription factor associated with adipocyte determination and differentiation. *Mol. Cell. Biol.* 13: 4753–59

73. Shimomura I, Bashmakov Y, Ikemoto S, Horton JD, Brown MS, Goldstein JL. 1999. Insulin selectively increases SREBP-1c mRNA in the livers of rats with streptozotocin-induced diabetes. *Proc. Natl. Acad. Sci. USA* 96:13656–61

74. Horton JD, Bashmakov Y, Shimomura I, Shimano H. 1998. Regulation of sterol regulatory element binding proteins in livers of fasted and refed mice. *Proc. Natl. Acad. Sci. USA* 95:5987–92

75. Azzout-Marniche D, Becard D, Guichard C, Foretz M, Ferre P, Foufelle F. 2000. Insulin effects on sterol regulatory-element-binding protein-1c (SREBP-1c) transcriptional activity in rat hepatocytes. *Biochem. J.* 350(Pt. 2):389–93

76. Matsumoto M, Ogawa W, Teshigawara K, Inoue H, Miyake K, et al. 2002. Role of the insulin receptor substrate 1 and phosphatidylinositol 3-kinase signaling pathway in insulin-induced expression of sterol regulatory element binding protein 1c and glucokinase genes in rat hepatocytes. *Diabetes* 51:1672–80

77. Matsumoto M, Ogawa W, Akimoto K,

Inoue H, Miyake K, et al. 2003. PKCλ in liver mediates insulin-induced SREBP-1c expression and determines both hepatic lipid content and overall insulin sensitivity. *J. Clin. Invest.* 112:935–44

78. Fleischmann M, Iynedjian PB. 2000. Regulation of sterol regulatory-element binding protein 1 gene expression in liver: role of insulin and protein kinase B/cAkt. *Biochem. J.* 349:13–17

79. Chen G, Liang G, Ou J, Goldstein JL, Brown MS. 2004. Central role for liver X receptor in insulin-mediated activation of Srebp-1c transcription and stimulation of fatty acid synthesis in liver. *Proc. Natl. Acad. Sci. USA* 101:11245–50

80. Yabe D, Komuro R, Liang G, Goldstein JL, Brown MS. 2003. Liver-specific mRNA for Insig-2 down-regulated by insulin: implications for fatty acid synthesis. *Proc. Natl. Acad. Sci. USA* 100:3155–60

81. Kotzka J, Muller-Wieland D, Koponen A, Njamen D, Kremer L, et al. 1998. ADD1/SREBP-1c mediates insulin-induced gene expression linked to the MAP kinase pathway. *Biochem. Biophys. Res. Commun.* 249:375–79

82. Kim KH, Song MJ, Yoo EJ, Choe SS, Park SD, Kim JB. 2004. Regulatory role of glycogen synthase kinase 3 for transcriptional activity of ADD1/SREBP1c. *J. Biol. Chem.* 279:51999–2006

83. Inoki K, Li Y, Xu T, Guan KL. 2003. Rheb GTPase is a direct target of TSC2 GAP activity and regulates mTOR signaling. *Genes Dev.* 17:1829–34

84. Inoki K, Li Y, Zhu T, Wu J, Guan KL. 2002. TSC2 is phosphorylated and inhibited by Akt and suppresses mTOR signalling. *Nat. Cell. Biol.* 4:648–57

85. Gao X, Zhang Y, Arrazola P, Hino O, Kobayashi T, et al. 2002. Tsc tumour suppressor proteins antagonize amino-acid-TOR signalling. *Nat. Cell. Biol.* 4:699–704

86. Sarbassov DD, Guertin DA, Ali SM, Sabatini DM. 2005. Phosphorylation and regulation of Akt/PKB by the rictor-mTOR complex. *Science* 307:1098–101

87. Abe H, Yamada N, Kamata K, Kuwaki T, Shimada M, et al. 1998. Hypertension, hypertriglyceridemia, and impaired endothelium-dependent vascular relaxation in mice lacking insulin receptor substrate-1. *J. Clin. Invest.* 101:1784–88

88. Araki E, Lipes MA, Patti ME, Bruning JC, Haag B III, et al. 1994. Alternative pathway of insulin signalling in mice with targeted disruption of the IRS-1 gene. *Nature* 372:186–90

89. Withers DJ, Gutierrez JS, Towery H, Burks DJ, Ren JM, et al. 1998. Disruption of IRS-2 causes type 2 diabetes in mice. *Nature* 391:900–4

90. Kubota N, Tobe K, Terauchi Y, Eto K, Yamauchi T, et al. 2000. Disruption of insulin receptor substrate 2 causes type 2 diabetes because of liver insulin resistance and lack of compensatory β-cell hyperplasia. *Diabetes* 49:1880–89

91. Kubota T, Kubota N, Moroi M, Terauchi Y, Kobayashi T, et al. 2003. Lack of insulin receptor substrate-2 causes progressive neointima formation in response to vessel injury. *Circulation* 107:3073–80

92. Taniguchi CM, Ueki K, Kahn R. 2005. Complementary roles of IRS-1 and IRS-2 in the hepatic regulation of metabolism. *J. Clin. Invest.* 115:718–27

93. Liu SC, Wang Q, Lienhard GE, Keller SR. 1999. Insulin receptor substrate 3 is not essential for growth or glucose homeostasis. *J. Biol. Chem.* 274:18093–99

94. Fantin VR, Wang GE, Lienhard GE, Keller SR. 2000. Mice lacking insulin receptor substrate 4 exhibit mild defects in growth, reproduction, and glucose homostasis. *Am. J. Physiol.* 278:E127–33

95. Bard-Chapeau EA, Hevener AL, Long S, Zhang EE, Olefsky JM, Feng GS. 2005. Deletion of Gab1 in the liver leads to enhanced glucose tolerance and

improved hepatic insulin action. *Nat. Med.* 11:567–71

96. Tsuruzoe K, Emkey R, Kriauciunas KM, Ueki K, Kahn CR. 2001. Insulin receptor substrate 3 (IRS-3) and IRS-4 impair IRS-1- and IRS-2-mediated signaling. *Mol. Cell. Biol.* 21:26–38

97. Withers DJ, Burks DJ, Towery HH, Altamuro SL, Flint CL, White MF. 1999. Irs-2 coordinates Igf-1 receptor-mediated β-cell development and peripheral insulin signalling. *Nat. Genet.* 23:32–40

98. Laustsen PG, Michael MD, Crute BE, Cohen SE, Ueki K, et al. 2002. Lipoatrophic diabetes in Irs1$^{-/-}$/Irs3$^{-/-}$ double knockout mice. *Genes Dev.* 16:3213–22

99. Terauchi Y, Matsui J, Suzuki R, Kubota N, Komeda K, et al. 2003. Impact of genetic background and ablation of insulin receptor substrate (IRS)-3 on IRS-2 knock-out mice. *J. Biol. Chem.* 278:14284–90

100. Fruman DA, Mauvais-Jarvis F, Pollard DA, Yballe CM, Brazil D, et al. 2000. Hypoglycaemia, liver necrosis and perinatal death in mice lacking all isoforms of phosphoinositide 3-kinase p85α. *Nat. Genet.* 26:379–82

101. Terauchi Y, Tsuji T, Satoh S, Minoura H, Murakami K, et al. 1999. Increased insulin sensitivity and hypoglycaemia in mice lacking the p85 α subunit of phosphoinositide 3-kinase. *Nat. Genet.* 21:230–35

102. Mauvais-Jarvis F, Ueki K, Fruman DA, Hirshman MF, Sakamoto K, et al. 2002. Reduced expression of the murine p85α subunit of phosphoinositide 3-kinase improves insulin signaling and ameliorates diabetes. *J. Clin. Invest.* 109:141–49

102a. Ueki K, Yballe CM, Brachmann SM, Vicent D, Watt JM, et al. 2002. Increased insulin sensitivity in mice lacking p85β subunit of phosphoinositide 3-kinase. *Proc. Natl. Acad. Sci. USA* 99:419–24

103. Chen D, Mauvais-Jarvis F, Bluher M, Fisher SJ, Jozsi A, et al. 2004. p50α/p55α phosphoinositide 3-kinase knockout mice exhibit enhanced insulin sensitivity. *Mol. Cell. Biol.* 24:320–29

104. Ueki K, Fruman DA, Brachmann SM, Tseng YH, Cantley LC, Kahn CR. 2002. Molecular balance between the regulatory and catalytic subunits of phosphoinositide 3-kinase regulates cell signaling and survival. *Mol. Cell. Biol.* 22:965–77

105. Barbour LA, Shao J, Qiao L, Leitner W, Anderson M, et al. 2004. Human placental growth hormone increases expression of the p85 regulatory unit of phosphatidylinositol 3-kinase and triggers severe insulin resistance in skeletal muscle. *Endocrinology* 145:1144–50

106. Brachmann SM, Ueki K, Engelman JA, Kahn RC, Cantley LC. 2005. Phosphoinositide 3-kinase catalytic subunit deletion and regulatory subunit deletion have opposite effects on insulin sensitivity in mice. *Mol. Cell. Biol.* 25:1596–607

107. Bi L, Okabe I, Bernard DJ, Nussbaum RL. 2002. Early embryonic lethality in mice deficient in the p110β catalytic subunit of PI 3-kinase. *Mamm. Genome* 13:169–72

108. Bi L, Okabe I, Bernard DJ, Wynshaw-Boris A, Nussbaum RL. 1999. Proliferative defect and embryonic lethality in mice homozygous for a deletion in the p110α subunit of phosphoinositide 3-kinase. *J. Biol. Chem.* 274:10963–68

109. MacDonald PE, Joseph JW, Yau D, Diao J, Asghar Z, et al. 2004. Impaired glucose-stimulated insulin secretion, enhanced intraperitoneal insulin tolerance, and increased β-cell mass in mice lacking the p110γ isoform of phosphoinositide 3-kinase. *Endocrinology* 145:4078–83

110. Cho H, Thorvaldsen JL, Chu Q, Feng F, Birnbaum MJ. 2001. Akt1/PKBα is required for normal growth but dispensable for maintenance of glucose

homeostasis in mice. *J. Biol. Chem.* 276: 38349–52

111. Easton RM, Cho H, Roovers K, Shineman DW, Mizrahi M, et al. 2005. Role for Akt3/protein kinase Bγ in attainment of normal brain size. *Mol. Cell. Biol.* 25:1869–78

112. Cho H, Mu J, Kim JK, Thorvaldsen JL, Chu Q, et al. 2001. Insulin resistance and a diabetes mellitus-like syndrome in mice lacking the protein kinase Akt2 (PKBβ). *Science* 292:1728–31

113. Garofalo RS, Orena SJ, Rafidi K, Torchia AJ, Stock JL, et al. 2003. Severe diabetes, age-dependent loss of adipose tissue, and mild growth deficiency in mice lacking Akt2/PKB β. *J. Clin. Invest.* 112:197–208

114. Woulfe D, Jiang H, Morgans A, Monks R, Birnbaum M, Brass LF. 2004. Defects in secretion, aggregation, and thrombus formation in platelets from mice lacking Akt2. *J. Clin. Invest.* 113:441–50

115. Bernal-Mizrachi E, Fatrai S, Johnson JD, Ohsugi M, Otani K, et al. 2004. Defective insulin secretion and increased susceptibility to experimental diabetes are induced by reduced Akt activity in pancreatic islet β cells. *J. Clin. Invest.* 114:928–36

116. Ono H, Shimano H, Katagiri H, Yahagi N, Sakoda H, et al. 2003. Hepatic Akt activation induces marked hypoglycemia, hepatomegaly, and hypertriglyceridemia with sterol regulatory element binding protein involvement. *Diabetes* 52:2905–13

117. Bost F, Aouadi M, Caron L, Even P, Belmonte N, et al. 2005. The extracellular signal-regulated kinase isoform ERK1 is specifically required for in vitro and in vivo adipogenesis. *Diabetes* 54:402–11

118. Liang G, Yang J, Horton JD, Hammer RE, Goldstein JL, Brown MS. 2002. Diminished hepatic response to fasting/refeeding and liver X receptor agonists in mice with selective deficiency of sterol regulatory element-binding protein-1c. *J. Biol. Chem.* 277:9520–28

119. Marchesini G, Brizi M, Bianchi G, Tomassetti S, Bugianesi E, et al. 2001. Nonalcoholic fatty liver disease: a feature of the metabolic syndrome. *Diabetes* 50:1844–50

120. Yu AS, Keeffe EB. 2002. Nonalcoholic fatty liver disease. *Rev. Gastroenterol. Disord.* 2:11–19

121. Biddinger SB, Almind K, Miyazaki M, Kokkotou E, Ntambi JM, Kahn CR. 2005. Effects of diet and genetic background on sterol regulatory element-binding protein-1c, stearoyl-CoA desaturase 1, and the development of the metabolic syndrome. *Diabetes* 54:1314–23

122. Shimomura I, Hammer RE, Richardson JA, Ikemoto S, Bashmakov Y, et al. 1998. Insulin resistance and diabetes mellitus in transgenic mice expressing nuclear SREBP-1c in adipose tissue: model for congenital generalized lipodystrophy. *Genes Dev.* 12:3182–94

123. Shimomura I, Hammer RE, Ikemoto S, Brown MS, Goldstein JL. 1999. Leptin reverses insulin resistance and diabetes mellitus in mice with congenital lipodystrophy. *Nature* 401:73–76

124. Rui L, Fisher TL, Thomas J, White MF. 2001. Regulation of insulin/insulin-like growth factor-1 signaling by proteasome-mediated degradation of insulin receptor substrate-2. *J. Biol. Chem.* 276:40362–67

125. Zhande R, Mitchell JJ, Wu J, Sun XJ. 2002. Molecular mechanism of insulin-induced degradation of insulin receptor substrate 1. *Mol. Cell. Biol.* 22:1016–26

126. Bar RS, Harrison LC, Muggeo M, Gorden P, Kahn CR, Roth J. 1979. Regulation of insulin receptors in normal and abnormal physiology in humans. *Adv. Int. Med.* 24:23–52

127. Shimomura I, Matsuda M, Hammer RE, Bashmakov Y, Brown MS, Goldstein JL. 2000. Decreased IRS-2 and increased

SREBP-1c lead to mixed insulin resistance and sensitivity in livers of lipodystrophic and ob/ob mice. *Mol. Cell* 6:77–86

128. Ide T, Shimano H, Yahagi N, Matsuzaka T, Nakakuki M, et al. 2004. SREBPs suppress IRS-2-mediated insulin signalling in the liver. *Nat. Cell Biol.* 6:351–57

129. Liu YF, Paz K, Herschkovitz A, Alt A, Tennenbaum T, et al. 2001. Insulin stimulates PKCζ-mediated phosphorylation of insulin receptor substrate-1 (IRS-1). *J. Biol. Chem.* 276:14459–65

130. Hirosumi J, Tuncman G, Chang L, Gorgun CZ, Uysal KT, et al. 2002. A central role for JNK in obesity and insulin resistance. *Nature* 420:333–36

131. Gao Z, Hwang D, Bataille F, Lefevre M, York D, et al. 2002. Serine phosphorylation of insulin receptor substrate 1 (IRS-1) by inhibitor kappa B kinase (IKK) complex. *J. Biol. Chem.* 277:48115–21

132. Craparo A, Freund R, Gustafson TA. 1997. 14-3-3 ε interacts with the insulin-like growth factor I receptor and insulin receptor substrate I in phosphotyrosine-independent manner. *J. Biol. Chem.* 272:11663–70

133. Emanuelli B, Peraldi P, Filloux C, Sawka-Verhelle D, Hilton D, Van Obberghen E. 2000. SOCS-3 is an insulin-induced negative regulator of insulin signaling. *J. Biol. Chem.* 275:15985–91

134. Rui L, Yuan M, Frantz D, Shoelson S, White MF. 2002. SOCS-1 and SOCS-3 block insulin signaling by ubiquitin-mediated degradation of IRS1 and IRS2. *J. Biol. Chem.* 277:42394–98

135. Nakashima N, Sharma PM, Imamura T, Bookstein R, Olefsky J. 2000. The tumor suppressor PTEN negatively regulates insulin signaling in 3T3-L1 adipocytes. *J. Biol. Chem.* 275:12889–95

136. Clement S, Krause U, Desmedt F, Tanti JF, Behrends J, et al. 2001. The lipid phosphatase SHIP2 controls insulin sensitivity. *Nature* 409:92–97

137. Cohen P, Miyazaki M, Socci ND, Hagge-Greenberg A, Liedtke W, et al. 2002. Role for stearoyl-CoA desaturase-1 in leptin-mediated weight loss. *Science* 297:240–43

138. Saad MJA, Araki E, Miralpeix M, Rothenberg PL, White MF, Kahn CR. 1992. Regulation of insulin receptor substrate 1 in liver and muscle of animal models of insulin resistance. *J. Clin. Invest.* 90:1839–49

139. Kerouz NJ, Horsch D, Pons S, Kahn CR. 1997. Differential regulation of insulin receptor substrates-1 and -2 (IRS-1 and IRS-2) and phosphatidylinositol 3-kinase isoforms in liver and muscle of the obese diabetic (ob/ob) mouse. *J. Clin. Invest.* 100:3164–72

140. Kondo T, Kahn CR. 2004. Altered insulin signaling in retinal tissue in diabetic states. *J. Biol. Chem.* 279:37997–8006

141. Standaert ML, Sajan MP, Miura A, Kanoh Y, Chen HC, et al. 2004. Insulin-induced activation of atypical protein kinase C, but not protein kinase B, is maintained in diabetic (ob/ob and Goto-Kakazaki) liver. Contrasting insulin signaling patterns in liver versus muscle define phenotypes of type 2 diabetic and high fat-induced insulin-resistant states. *J. Biol. Chem.* 279:24929–34

142. Yahagi N, Shimano H, Hasty AH, Matsuzaka T, Ide T, et al. 2002. Absence of sterol regulatory element-binding protein-1 (SREBP-1) ameliorates fatty livers but not obesity or insulin resistance in Lep(ob)/Lep(ob) mice. *J. Biol. Chem.* 277:19353–57

143. Bray GA. 1984. Hypothalamic and genetic obesity: an appraisal of the autonomic hypothesis and the endocrine hypothesis. *Int. J. Obes.* 8(Suppl. 1): 119–37

144. Jiang ZY, Lin YW, Clemont A, Feener EP, Hein KD, et al. 1999. Characterization of selective resistance to insulin signaling in the vasculature of obese Zucker

(fa/fa) rats. *J. Clin. Invest.* 104.447–57

145. Anai M, Funaki M, Ogihara T, Kanda A, Onishi Y, et al. 1999. Enhanced insulin-stimulated activation of phosphatidylinositol 3-kinase in the liver of high-fat-fed rats. *Diabetes* 48:158–69

146. Saad MJA, Folli F, Kahn JA, Kahn CR. 1993. Modulation of insulin receptor, insulin receptor substrate-1, and phosphatidylinositol 3-kinase in liver and muscle of dexamethasone-treated rats. *J. Clin. Invest.* 92:2065–72

147. Cusi K, Maezono K, Osman A, Pendergrass M, Patti ME, et al. 2000. Insulin resistance differentially affects the PI 3-kinase- and MAP kinase-mediated signaling in human muscle. *J. Clin. Invest.* 105:311–20

148. Goodyear LJ, Giorgino F, Sherman LA, Carey J, Smith RJ, Dohm GL. 1995. Insulin receptor phosphorylation, insulin receptor substrate-1 phosphorylation, and phosphatidylinositol 3-kinase activity are decreased in intact skeletal muscle strips from obese subjects. *J. Clin. Invest.* 95:2195–204

149. Kim YB, Nikoulina SE, Ciaraldi TP, Henry RR, Kahn BB. 1999. Normal insulin-dependent activation of Akt/protein kinase B, with diminished activation of phosphoinositide 3-kinase, in muscle in type 2 diabetes. *J. Clin. Invest.* 104:733–41

150. Deleted in proof

151. Sinha MK, Taylor LG, Pories WJ, Flickinger EG, Meelheim D, et al. 1987. Long-term effect of insulin on glucose transport and insulin binding in cultured adipocytes from normal and obese humans with and without non-insulin-dependent diabetes. *J. Clin. Invest.* 80:1073–81

152. Caro JF, Dohm LG, Pories WJ, Sinha MK. 1989. Cellular alterations in liver, skeletal muscle, and adipose tissue responsible for insulin resistance in obesity and type II diabetes. *Diabetes Metab. Rev.* 5(8):665–89

153. Freidenberg GR, Henry RR, Klein HH, Reichart DR, Olefsky JM. 1987. Decreased kinase activity of insulin receptors from adipocytes of non-insulin dependent diabetic subjects. *J. Clin. Invest.* 79:240–50

154. Takayama S, Kahn CR, Kubo K, Foley JE. 1988. Alterations in insulin receptor autophosphorylation in insulin resistance: correlation with altered sensitivity to glucose transport and antilipolysis to insulin. *J. Clin. Endocrinol. Metab.* 6:992–99

155. Foley JE. 1988. Mechanisms of impaired insulin action in isolated adipocytes from obese and diabetic subjects. *Diabetes Metab. Rev.* 4:487–505

156. Lockwood DH, Amatruda JM. 1983. Cellular alterations responsible for insulin resistance in obesity and type II diabetes mellitus. *Am. J. Med.* 75:23–31

157. Caro JF, Ittoop O, Pories WJ, Meelheim D, Flickinger EG, et al. 1986. Studies on the mechanism of insulin resistance in the liver from humans with noninsulin-dependent diabetes. *J. Clin. Invest.* 78:249–58

Annu. Rev. Physiol. 2006. 68:159–91
doi: 10.1146/annurev.physiol.68.033104.152158
First published online as a Review in Advance on October 26, 2005

LXRs and FXR: The Yin and Yang of Cholesterol and Fat Metabolism

Nada Y. Kalaany[1] and David J. Mangelsdorf[2]

[1] Whitehead Institute for Biomedical Research, Massachusetts Institute of Technology, Cambridge, Massachusetts 02142; email: kalaany@wi.mit.edu

[2] Howard Hughes Medical Institute, Department of Pharmacology, University of Texas Southwestern Medical Center, Dallas, Texas 75390; email: davo.mango@utsouthwestern.edu

Key Words   nuclear receptors, liver X receptors, farnesoid X receptor, bile acids, lipid metabolism

■ **Abstract**   Liver X receptors (LXRs) and farnesoid X receptor (FXR) are nuclear receptors that function as intracellular sensors for sterols and bile acids, respectively. In response to their ligands, these receptors induce transcriptional responses that maintain a balanced, finely tuned regulation of cholesterol and bile acid metabolism. LXRs also permit the efficient storage of carbohydrate- and fat-derived energy, whereas FXR activation results in an overall decrease in triglyceride levels and modulation of glucose metabolism. The elegant, dual interplay between these two receptor systems suggests that they coevolved to constitute a highly sensitive and efficient system for the maintenance of total body fat and cholesterol homeostasis. Emerging evidence suggests that the tissue-specific action of these receptors is also crucial for the proper function of the cardiovascular, immune, reproductive, endocrine pancreas, renal, and central nervous systems. Together, LXRs and FXR represent potential therapeutic targets for the treatment and prevention of numerous metabolic and lipid-related diseases.

INTRODUCTION

Liver X receptor (LXR) α (NR1H3), LXRβ (NR1H2), and farnesoid X receptor (FXR) (NR1H4) are members of the nuclear hormone receptor superfamily of ligand-activated transcription factors, which regulate diverse physiological processes such as reproduction, development, and overall metabolism (1). These receptors have conserved functional domains, including a DNA-binding domain, a ligand-binding domain, and a C-terminal ligand-dependent activation function domain (AF-2) that mediates receptor activation/repression through binding to coactivator/corepressor complexes. Originally termed "orphan" nuclear receptors, LXRs and FXR are now referred to as "adopted orphans," owing to the identification of their physiological ligands. Whereas LXRs function by binding physiological levels of oxysterols (e.g., 24(S)-hydroxycholesterol and 24(S), 25-epoxycholesterol)

(2, 3), FXR is a bile acid sensor and is activated by binding to endogenous bile acids [e.g., chenodeoxycholic acid (CDCA) and cholic acid (CA)] (4–6). LXRs and FXR function as permissive heterodimers with the retinoid X receptor (RXR), which binds 9-cis retinoic acid and docosahexaenoic acid (7–10). These receptors are thought to be prebound to DNA in a complex with corepressors, such as silencing mediator for retinoic acid and thyroid hormone receptor (SMRT) (11) and nuclear receptor corepressor (N-CoR) (12). Ligand binding to either RXR or its partner releases corepressors and recruits coactivators (13) (e.g., SRC-1, p300, ACS-2, TRRAP, or PGC-1α) (14–20) to initiate transcription. Like other members of the nuclear receptor superfamily, LXRs and FXR bind DNA at specific hormone-response elements, consisting of the minimal core sequence AGGTCA. Whereas direct repeats separated by four nucleotides are preferred response elements for LXRs, FXR preferentially binds inverted repeat elements separated by one nucleotide.[1]

Consistent with their regulation by oxysterols and bile acids, LXRs and FXR are expressed at their highest levels in enterohepatic tissues (i.e., liver and intestine). LXRα is also detected at high levels in macrophages, adipose tissues, lung, spleen, and kidney, and at lower levels in muscle, testes, and adrenals; by contrast, LXRβ is ubiquitously expressed (21). Although recent studies have underscored the important function of LXRs in adipose, brain, pancreas, testes, and skin, the detailed function of LXRs outside of the enterohepatic system and macrophages remains under investigation. Likewise, high levels of FXR have been described not only in liver and intestine but also in kidney and adrenals (21, 22). Lower levels of FXR mRNA have also been reported in the heart, lung, stomach, adipose, thymus, spleen, ovary, testes, and vascular tissue (23–25). However, the physiological function of FXR in these tissues, which are not normally exposed to bile acid circulation, remains controversial. Furthermore, researchers recently have identified four isoforms of the murine (24), hamster, and human (26) FXR (FXRα1, α2, α3, and α4) that differ by their N terminus and by the existence of a four-amino acid insertion in the hinge region. Interestingly, FXR isoforms lacking the four-amino acid insertion bind FXR response elements (FXREs) with higher affinity and display stronger transactivation of the promoters of a number of FXR target genes [e.g., ileal bile acid binding protein (IBABP), syndecan (SDC)-1, fibrinogen, human αA-crystallin] (24, 27–29). Apart from the FXR isoforms, a novel murine nuclear receptor with strong similarity to FXR has been identified and named FXRβ (NR1H5). Whereas it constitutes a pseudogene in humans, FXRβ has been proposed to encode a functional receptor in mice that is expressed at very low levels and can be activated by high concentrations of lanosterol, a cholesterol synthesis intermediate (23). At present, the physiologic relevance of FXRβ remains unclear.

In this review, we describe the transcriptional regulatory cascades of the LXR and FXR receptor systems. In addition to describing the increasing number of LXR and FXR target genes (Tables 1 and 2), we also review the physiological effects

---

[1]For a list of abbreviations, please see the Appendix at the end of this chapter.

TABLE 1 Genes regulated by LXRs in a negative or positive manner

Group[a]	Gene[b]	D/I[c]	LXRE	Regulation[d]	Reference(s)
1	ABCA1	D	DR-4	P	(31, 34)
	ABCG1	D	DR-4	P	(41, 42)
	ABCG5/G8	D	ND	P	(37, 39)
	ApoA5	I		N	(49)
	ApoCI/IV/II	D	IR-1	P	(47)
	ApoD	D	DR-4	P	(48)
	ApoE	D	DR-4	P	(46)
	hCETP	D	DR-4	P	(53)
	mCYP7A1	D	DR-4	P	(3, 30)
	LPL	D	DR-4	P	(50)
	hLXRα	D	DR-4	P	(54)
	PLTP	D	DR-4	P	(51, 52)
	hSR-BI	D	DR-4	P	(59)
2	COX-2	ND		N	(62)
	G-CSF	ND		N	(62)
	IL-1β	ND		N	(62)
	IL-6	ND		N	(62)
	iNOS	ND		N	(62)
	IP-10	ND		N	(62)
	MCP-1	ND		N	(62)
	MCP-3	ND		N	(62)
	MIP-1β	ND		N	(62)
	MMP-9	ND		N	(62)
	SPα[e]	D	DR-4	P	(64)
3	Angptl3	D	DR-4	P	(72, 73)
	FAS	D	DR-4	P	(70)
	SCD-1	ND		P	(79)
	SPOT14	ND		P	(48)
	SREBP-1c	D	DR-4	P	(68, 69)
4	G6Pase	ND		N	(84)
	GK	ND		N	(84)
	GLUT4	D	DR-4	P	(84, 85)
	PEPCK	ND		N	(84, 86, 87)
5	FPPS	D	DR-4	P	(95)
	Fra-1	ND		P	(108)
	IBABP	D	IR-1	P	(186)
	Involucrin	I		P	(107)
	Loricrin	ND		P	(106)
	L-UrdPase	ND		P	(78)
	Profilaggrin	ND		P	(106)
	Renin	D	CNRE	P	(101)
	Transglutaminase-1	ND		P	(107)
	VEGF	D	DR-4	P	(66)

[a]Groups: 1, Cholesterol metabolism; 2, Inflammation/immune response; 3, Fat metabolism; 4, Carbohydrate metabolism; 5, Miscellaneous.

[b]h, human; m, mouse.

[c]D, direct regulation; I, indirect regulation; ND, not determined.

[d]P, positive regulation; N, negative regulation.

[e]LXRα-specific.

TABLE 2   Genes regulated by FXR in a negative or positive manner

Group[a]	Gene[b]	D/I[c]	FXRE	Regulation[d]	Reference(s)
1	CYP7A1	I		N	(32, 116)
	hFGF-19	D	IR-1	P	(124)
	mFGF15	D	IR-1	P	(184)
	mFXR	I		P	(20)
	SHP	D	IR-1	P	(32, 116)
2	BSEP	D	IR-1	P	(127)
	IBABP	D	IR-1	P	(144)
	IBAT	I		N	(142, 143)
	hMDR-3	D	IR-1	P	(130)
	rMDR2	ND		P	(133, 157)
	MRP-2	D	ER-8	P	(137)
	NTCP	I		N	(146)
	OATP1B1	I		N	(150)
	OATP1B3	D	IR-1	P	(151)
	BACS	D	IR-1	P	(152)
	BAT	D	IR-1	P	(152)
	CYP3A4	D	IR-1/ER-8/DR-3	P	(155)
	SULT2A1	D	IR-0	P	(153)
	UGT2B4	D[e]	Half site	P	(154)
3	ApoAI	D[e]	Negative FXRE	N	(175, 176)
	ApoB	I		N	(180)
	ApoCII	D	IR-1	P	(167)
	ApoCIII	I		N	(168)
	ApoE	ND		P	(172)
	C3	D	IR-1	P	(184)
	hHL	I		N	(174)
	MTP	I		N	(180)
	PDK4	ND		P	(185)
	PLTP	D	IR-1	P	(173)
	hPPARα	I		P	(187)
	SREBP-1c	I		N	(179)
	VLDLR	I		P	(182)
4	hαA-crystallin	D	IR-1	P	(29)
	Fibrinogen	D	ND	P	(28)
	hKininogen	D	IR-1	P	(178)
	Syndecan-1	D	DR-1	P	(27)
	VPAC-1	D?	IR-1?	P	(138)

[a]Groups: 1, Bile acid biosynthesis; 2, Bile acid enterohepatic circulation; 3, Triglyceride metabolism; 4, Miscellaneous.

[b]h, human; m, mouse; r, rodent.

[c]D, direct regulation; I, indirect regulation; ND, not determined.

[d]P, positive regulation; N, negative regulation.

[e]Absence of RXR heterodimerization.

of their activation in different tissues, including their coordinated regulation of cholesterol, bile acid, triglyceride, and carbohydrate metabolism.

LXRS MAINTAIN CHOLESTEROL HOMEOSTASIS

The identification of oxysterols as physiological ligands for LXRs initiated extensive research to elucidate their physiological functions (2, 3). In addition to the ligands, the generation of LXR-null mice has provided important tools for uncovering LXR action (30, 31). As detailed below, this research has led to the idea that LXRs function as sterol sensors to enhance transcription of an array of genes involved in the regulation of cholesterol metabolism.

The generation of LXRα-null mice provided the first unequivocal evidence that LXRs are involved in cholesterol homeostasis, as these mice displayed a phenotype of hepatomegaly and accumulation of large amounts of cholesteryl esters in their livers (30). In rodents, LXRs induce cholesterol catabolism by upregulating expression of cytochrome P450 7α-hydroxylase (CYP7A1), which catalyzes the rate-limiting step in the classical pathway of cholesterol conversion into bile acids (3, 30). LXR-dependent regulation of bile acid synthesis is due to the presence of a potent LXRE in the promoter of the rodent CYP7A1 gene (3); this LXRE is not conserved in humans. LXR-mediated transcription of CYP7A1 also requires binding of the liver receptor homolog-1 (LRH-1, NR5A2) to the CYP7A1 promoter (32). LRH-1 is a nuclear receptor that is highly expressed in the enterohepatic tissues, where it functions as a competence factor for the transcription of several genes involved in bile acid and cholesterol homeostasis (33).

LXRs also contribute significantly to the enhancement of reverse cholesterol transport through upregulation of the expression of another target gene, ATP-binding cassette transporter A1 (ABCA1) (31, 34), the genetic mutations of which are the cause of Tangier disease and familial hypoalphalipoproteinemia (35). ABCA1 plays a key role in the ApoAI-mediated efflux of cholesterol and phospholipids from lipid-laden macrophages back to the liver. Although oxysterol-induction of ABCA1 expression was observed in the mouse intestine (31) and to a lesser extent in the liver (36), the physiological role of ABCA1 in these tissues remains unclear. Researchers later discovered that the oxysterol-induced, LXR-mediated decrease in sterol absorption is due mainly to the intestinal upregulation of two other ABC transporter superfamily members, the ABCG5 and ABCG8 half-transporters (37). Mutations in these ABC half-transporters lead to the development of sitosterolemia, a disease characterized by premature atherosclerosis (38). ABCG5 and ABCG8 encode proteins that play major roles in the LXR-mediated decrease of dietary sterol absorption in the intestine (37, 39) and the promotion of sterol excretion from liver into bile (40). An additional ABC half-transporter, ABCG1, is also a target of LXRs in macrophages (41, 42), in which it mediates net cholesterol efflux through a mode of function that appears to differ from that of ABCA1. HDL-2 and HDL-3 particles serve as ABCG1 cholesterol acceptors, rather than the lipid-poor apolipoproteins (e.g., ApoAI), as is the case for ABCA1 (43, 44). Owing to the significant contribution of HDL-2 and HDL-3 particles to the total HDL fraction in plasma, ABCG1 may have an antiatherosclerotic effect, through modulation of HDL cholesterol levels. Further in vivo studies have supported this hypothesis (45).

In addition to the ABC transporters that mediate cholesterol efflux, several apolipoproteins and lipid-modulating enzymes involved in reverse cholesterol transport are also transcriptional targets for LXRs. These include apoliprotein (Apo) E in macrophages and adipose (46), the entire ApoCI/ApoCIV/ApoCII gene cluster in macrophages (47), ApoD in adipose (48), ApoAV in liver (49), lipoprotein lipase (LPL) in liver and macrophages (50), phospholipid transfer protein (PLTP) (51, 52), and human cholesteryl ester transfer protein (CETP) (53). In human macrophages, the LXR-induced transcriptional regulation of cholesterol efflux genes in response to lipid loading is thought to be further amplified through autoregulation of the LXRα promoter (54). The significance of the tissue-specific regulation of ApoE expression by LXRs was studied in ApoE-null mice that expressed ApoE only in macrophages, as well as in wild-type mice specifically lacking ApoE in macrophages. Whereas the former were protected against atherosclerosis, the latter showed enhanced formation of atherosclerotic lesions (55–57). Although ApoCI, ApoCII, and ApoCIV can all serve as cholesterol acceptors from ABCA1, and ApoD may be involved in lipid transfer or reverse cholesterol transport, further studies are needed to understand their functions in vivo. ApoAV represents the only apolipoprotein described so far to be downregulated by LXR activation. This regulation, which requires SREBP-1c induction, may contribute to LXR-induced increases in plasma triglyceride levels (49). The functions of the lipoprotein modifying enzymes LPL, CETP, and PLTP are more complex, but all are involved in cholesterol transport and are directly regulated by LXRs (58). For instance, LPL seems to be antiatherogenic in liver, but not in macrophages. Human CETP can enhance cholesterol reverse transport by transferring cholesteryl esters from HDL to LDL particles. However, its overexpression in ApoE-null mice markedly decreases HDL and increased lesion formation. By contributing to the generation of preβ HDL, PLTP can enhance cholesteryl ester clearance, but it also enhances VLDL secretion from the liver (reviewed in Reference 58). Furthermore, LXR activation induces the expression of human SR-BI (scavenger receptor-BI) in hepatoma cell lines and 3T3-L1 adipocytes. Because SR-BI mediates the transfer of HDL cholesterol to hepatocytes, this regulation probably contributes to the overall LXR-mediated enhancement of reverse cholesterol transport (59).

The potential antiatherogenic function of LXRs through enhancing reverse cholesterol transport has been investigated in atherosclerotic mouse models. ApoE-null mice that received a bone marrow transplant from LXRα,β double-knockout mice and that hence had LXR deficient macrophages had increased atherosclerotic lesions (60). By contrast, administration of an LXR synthetic agonist resulted in a 50% decrease in lesion formation in both ApoE-null and LDL receptor–null mice (61). This decrease was accompanied by a simultaneous induction in the expression of ABCA1 and ABCG1 in the atherosclerotic lesions. Further support for the antiatherogenic function of LXRs has come from the finding that an array of proinflammatory genes is repressed by LXR agonists in activated macrophages (62). These genes include inducible nitric oxide synthase, cyclooxygenase, granulocyte colony–stimulating factor, interleukins IL-1β and IL-6, matrix metalloproteinase MMP-9, and the chemokines monocyte chemoattractant protein-1 and -3, macrophage inflammatory protein-1β, and interferon-inducible protein-10. Although the mechanisms underlying the anti-inflammatory function of LXRs are not yet understood, in vivo studies have supported these findings. LXRα,β-deficient mice exhibit enhanced stimuli-induced inflammatory responses, whereas administration of LXR agonists decreases inflammation in murine models of contact dermatitis in an LXRα- and LXRβ-dependent manner (62, 63). Interestingly, common LXR-mediated regulatory pathways may govern lipid homeostasis and innate immunity (64). Bone marrow transplant studies in mice demonstrated that loss of LXRs results in enhanced susceptibility to bacterial infection owing to markedly decreased macrophage survival and pathogen clearance. LXRα, whose own expression increases following bacterial infection, may play a specific and predominant role in innate immune responses, partly by upregulating the expression of SPα (64, 65). Spα (also called AIM) is a scavenger receptor cysteine-rich protein family member known to antagonize apoptosis in macrophages and to be involved in regulating immune responses following bacterial infection (65). Furthermore, LXRs can directly induce the transcription of vascular endothelial growth factor in macrophages and adipose, independent of hypoxic or inflammatory stimuli. However, the significance of the induction of this major stimulator of angiogenesis by LXRs remains to be investigated (66).

LXRS AS MASTER REGULATORS OF FAT METABOLISM

In addition to their roles in cholesterol metabolism, the LXRs have profound effects on hepatic fat metabolism. These effects are mediated in part through the transcriptional regulation of SREBP-1c, the master regulator of fatty acid and triglyceride synthesis (67, 68). LXRs bind two functional LXREs in the promoter region of the SREBP-1c gene, which are required for both the basal and inducible expression of SREBP-1c protein (68, 69). In wild-type mice, cholesterol feeding or administration of a synthetic ligand for LXR and its heterodimeric partner RXR resulted in increased SREBP-1c expression, nuclear SREBP-1c protein levels, and fatty acid synthesis. These effects were abrogated in LXR-null mice, which are refractory to both cholesterol- and insulin-induced expression of SREBP-1c (68). In addition to SREBP-1c, other lipogenic genes, such as fatty acid synthase (FAS), have been reported as direct LXR targets (70). Expression of angiopoietin-like protein 3 (Angptl3), a liver-secreted protein that raises plasma lipid levels, also is regulated by LXRs and therefore is believed to be responsible for LXR-induced hypertriglyceridemia (71–73). Angptl3 increases plasma triglycerides and free fatty acids by inhibiting LPL activity in different tissues and activating lipolysis in adipocytes, respectively (74, 75).

In addition to the regulation of lipogenic gene expression, other lines of evidence implicate LXRs in the regulation of fat metabolism, particularly as it relates to the diet. For example, a wild-type mouse requires LXRs to process and store dietary fat properly (76). When fed a high-fat, high-cholesterol "Western"-style diet, LXR-null mice are resistant to obesity and demonstrate better glucose tolerance. Instead of storing dietary fat, LXR-null mice burn it off in their peripheral tissues and display significantly increased metabolic rates and markedly suppressed hepatic fatty acid synthesis. Interestingly, the same study found that resistance to diet-induced obesity depends on the presence of cholesterol in the diet; this finding uncovered an essential role for LXRs in regulating the balance between fat storage and oxidation (76). This study and others revealed the existence of an SREBP-1c-independent regulatory pathway for governing hepatic fat metabolism and demonstrated that LXRs are required for both the basal (30, 68) and insulin-induced expression of SREBP-1c (77). Furthermore, LXRs induce the transcription of a novel, liver-specific uridine phosphorylase gene (L-UrdPase), suggesting a possible role for LXRs in diet-induced clearance of plasma uridine and the maintenance of uridine homeostasis. Nevertheless, because β-alanine, a precursor to fatty acid synthesis, is a major product of uridine catabolism, LXRs induction of L-UrdPase may reinforce the important role of LXRs in lipogenesis (78).

From a pharmacological point of view, the finding that LXR agonists can regulate fat metabolism has been a thorn in the development of drugs such as those for reducing body cholesterol levels, enhancing insulin sensitivity, and protecting against atherosclerosis. Indeed, oral administration of the potent synthetic agonist T0901317 in mice and hamsters leads to increased hepatic fatty acid synthesis, hepatic steatosis, and increased VLDL triglycerides (79, 80). When administered to hypercholesterolemic ApoE$^{-/-}$ or LDL receptor$^{-/-}$ mice, the LXR ligand also causes a 12-fold increase in plasma triglycerides (80). Importantly, a majority of these detrimental lipid responses are due to activation of the SREBP-1c pathway (76, 81), whereas activation of ABC transporters results in the beneficial therapeutic effects of lowering cholesterol. It is of interest to note that SREBP-1c and the ABCA1 genes exhibit a striking difference in the ways in which LXRs regulate their promoters (82). For example, the absence of LXRs in mice causes derepression of basal ABCA1 expression in macrophages and the intestine; by contrast, basal SREBP-1c levels are decreased significantly. This is thought to result from the LXR-dependent, differential recruitment of the corepressors NCoR and SMRT to the ABCA1 and SREBP-1c promoters. Therefore, the development of an LXR partial agonist that is able to enhance basal expression of ABCA1 without affecting SREBP-1c expression would have potential therapeutic value. The rationale for such pharmacological agents has been established with other nuclear receptor targets, such as the estrogen receptor (83).

From a physiological point of view, LXRs appear to function as upsteam master regulators of SREBP-1c-dependent and -independent lipogenic pathways that are induced in response to increased sterol as well as insulin levels. Hence, the physiological function of LXRs points to a survival response by the body to efficiently store carbohydrate- and high-fat diet–derived energy in the form of body fat.

LXRS AND GLUCOSE METABOLISM

A role for LXRs in the regulation of glucose homeostasis is currently emerging. Administration of the synthetic LXR agonist GW3965 to a mouse model of obesity and insulin resistance (obese, high-fat diet–fed C57Bl/6 mice) improves glucose tolerance (84). This was attributed to a coordinate transcriptional regulation of genes involved in glucose metabolism in both the liver and adipose tissue. In liver, activation of LXRs leads to a suppression of the gluconeogenic program through a decrease in peroxisome proliferator–activated receptor-γ coactivator-1 (PGC-1), phosphoenolpyruvate carboxykinase (PEPCK), and glucose-6-phosphatase (G6Pase) expression. This is accompanied by an induction of hepatic glucokinase (GK) expression, which enhances glucose utilization. In white adipose tissue, expression of the insulin-sensitive glucose transporter (GLUT4), which promotes glucose uptake and utilization, is upregulated following LXR activation (84). Two separate groups have identified GLUT4 as a direct target gene of LXRs in white adipose (84, 85). Additionally, GLUT4 expression correlates to that of LXRα during adipocyte differentiation, and although activation of either LXRα or LXRβ can induce GLUT4 expression, only LXRα affects GLUT4 basal expression. Moreover, two other groups confirmed the suppression of gluconeogenesis through LXR activation. One study demonstrated that treatment of diabetic insulin-resistant Zucker (fa/fa) rats with the synthetic LXR ligand T0901317 improves their insulin sensitivity (86). This results mainly from a dramatic reduction in glucose output due to the inhibited expression of several gluconeogenic genes, including PEPCK. Another study showed that whereas T0901317 oral administration to db/db diabetic mice leads to severe lipogenesis, it simultaneously suppresses PEPCK expression (87). Interestingly, a novel role for LXR in the pancreas was recently described. Addition of the LXR agonist T0901317 to pancreatic islets or insulin-secreting cells results in enhanced glucose-dependent insulin secretion (88). On the other hand, absence of LXRβ in mice leads to an impairment in such function and to an accumulation of lipid droplets, which is attributed to a possible disturbance in cholesterol efflux due to decreased expression of ABC transporters (89).

LXR ACTION OUTSIDE ENTEROHEPATIC TISSUES AND THE VASCULATURE

It is becoming increasingly evident that LXR function is crucial for the homeostatic integrity of several tissues in which LXR-mediated regulation of lipid and cholesterol metabolism has remained underexplored. These include the brain, skin, kidney, adrenals, and testes.

Disturbances in cholesterol homeostasis, including increased cholesterol turnover across the brain, are correlated with enhanced susceptibility to neurodegenerative diseases such as Alzheimer's disease (AD) (90). A role for the sterol sensor LXR in maintaining lipid homeostasis in the brain became evident as aging LXRα,β double-knockout mice (more than 1 year old) displayed gross brain abnormalities, including closed lateral ventricles resulting from excessive lipid deposition (91). More recently, researchers revealed a critical role for LXRβ in maintaining normal brain function. Mice lacking LXRβ displayed, at a relatively younger age (7 months old), impaired motor coordination resulting from loss of motor neurons and other features reminiscent of the symptoms of amyotrophic lateral sclerosis, a neurodegenerative disease (92). Interestingly, several reports point to potential therapeutic effects of LXR activation in neuronal cells. For instance, treatment of primary astrocyte cultures with LXR agonists leads to increased expression of ABCA1 and increased protein levels and secretion of ApoE, resulting in enhanced cholesterol efflux (93, 94). In these studies, LXR regulation of genes involved in maintaining cholesterol homesostasis was reproduced in vivo in mice, in which expression of ABCA1, ABCG1, and SREBP-1 was upregulated in the hippocampus, cerebellum, and that of ApoE in the hippocampus and cerebral cortex. A separate report demonstrated that farnesyl pyrophosphate synthase (FPPS)—an enzyme important in the formation of farnesyl pyrophosphate, an intermediate in cholesterol synthesis—is a direct target gene for LXR in the brain. Enhanced FPPS expression by LXR may decrease sterol synthesis by increasing levels of farnesol, which is known to specifically enhance degradation of HMG-CoA reductase (95). Increased LXR target gene expression in brain cells following LXR activation leads to decreased secretion of amyloid β, a peptide whose deposition in the brain is a hallmark of AD. Amyloid β formation is the result of a sequential cleavage by β- and then γ-secretase activities of the type I integral membrane amyloid precursor protein (APP). Indeed, LXR activation leads to ABCA1-dependent reduction in β-secretase as well as γ-secretase activities, in addition to enhancing ApoAI- and ApoE-mediated cholesterol efflux (96–98). Administration of a synthetic LXR agonist to a mouse model of AD (APP-23 mice) confirmed the beneficial effect of LXR activation, as it led to a marked decrease in amyloidogenic processing of APP, and pointed to the potential use of LXRs as therapeutic targets in the battle against neurodegenerative diseases (98).

In addition to the brain, other tissues such as the kidney and testes seem to require normal LXR function for the prevention of excessive lipid accumulation, which may result in glomerular atherosclerosis, or even infertility. In the kidney, inflammation increases lipid accumulation, partially by inhibiting LXR-induced ABCA1 expression in human kidney mesangial cells (MCs) (99), whereas treatment of rabbit MCs with T0901317 markedly enhances basal and ApoAI-mediated cholesterol efflux and ABCA1 expression (100). Interestingly, LXRs were shown to mediate the adrenergic regulation of the renin/angiotensin system in the kidney (101). Indeed, LXRs colocalize with renin in juxtaglomerular cells, in which LXRα is specifically enriched. As previously has been shown for LXRα (102),

LXRβ also binds a noncanonical, negative cAMP response element in the renin promoter, and both LXRs positively regulate renin basal expression in vitro and in mice in vivo. However, only LXRα has been shown to be crucial for the β-adrenergic/cAMP-inducible upregulation of renin expression, which is totally abrogated in LXRα-null mice (101).

In aging LXRα,β-null mice, the caput epididymides are severely disrupted owing to abnormal accumulation of neutral lipids, resulting in sperm fragililty and infertility (103). LXRβ is the predominant LXR transcript in the testes, and LXRβ-null mice exhibit excessive cholesterol accumulation in Sertoli cells, abnormal spermatogenesis, and decreased fertility (104), features reminiscent of the phenotype described in ABCA1-null mice (105).

Finally, LXRβ activation is now known to be important in the skin, where it mediates oxysterol-induced keratinocyte differentiation via increased transcription of several proteins involved in the formation of the cornified envelope (e.g., transglutaminase-1, involucrin, loricrin, and profilaggrin) (106–108). LXR-mediated upregulation of the expression of AP-1 protein complex components (e.g., Fra-1, Jun-D, and c-Fos) mediates this effect on the involucrin promoter (108). However, the role of LXR in skin and other tissues such as heart, lung, spleen, kidney, and adrenals remains underexplored and awaits further investigation.

FXR MAINTAINS BILE ACID HOMEOSTASIS

Excessive accumulation of cholesterol in the body can lead to deleterious effects resulting in the onset of cardiovascular (109) as well as neurodegenerative diseases (90, 110). Efficient elimination of this potentially toxic sterol largely occurs in the liver through conversion of cholesterol to bile acids. Bile acid biosynthesis follows two main pathways: a classical neutral pathway accounting for 90% of bile acid produced and an alternate acidic pathway, both of which result in the production of the primary bile acids CA and CDCA in humans. The generation of small, amphipathic bile acid molecules from cholesterol allows for the solubilization of cholesterol in the liver and its excretion into the bile. In addition, bile acids facilitate the digestion and absorption of cholesterol, lipid nutrients, and fat-soluble vitamins in the small intestine. Constant availability of bile salts in the body is secured by a highly efficient enterohepatic circulation process, which allows bile acid reabsorption by enterocytes and their recycling back to the liver. This results in the conservation of more than 95% of the bile acids secreted daily, limiting compensatory de novo bile acid synthesis to less than 5% (111). Despite their beneficial physiological function, bile acids, like cholesterol, are intrinsically toxic. Increased intracellular bile acid concentrations can cause liver damage, leading to the development of cholestasis and its associated pathologies (112). Therefore, both the synthesis and enterohepatic circulation of bile acids undergo extensive dynamic regulation in order to insure the maintenance of a precise balanced amount of cholesterol and bile acids in the body. In this respect, the bile acid sensor FXR plays a central role in the maintenance of bile acid homeostasis (113). FXR induces transcriptional feedback repression and feedforward regulatory loops, resulting in the suppression of bile acid synthesis, the covalent modification of bile acids into less toxic molecules, and the induction of hepatic bile acid efflux. In this way, the FXR-induced response prevents hepatic toxicity.

FXR and Bile Acid Biosynthesis

The long-known bile acid–induced feedback repression of its own biosynthesis occurs at the transcriptional level, causing the downregulated expression of cytochrome P450 hydroxylase enzymes, which are important in bile acid synthesis (114, 115). These include CYP7A1, the microsomal cholesterol 7α-hydroxylase, which regulates the first and rate-limiting step in the major classic pathway of bile acid synthesis, and CYP8B1, the microsomal sterol 12α-hydroxylase enzyme, which modulates bile acid hydrophobicity by increasing the production of CA (111). This transcriptional repression is mediated, at least in part, by the bile acid sensor FXR through upregulation of small heterodimer partner (SHP, NR0B2) expression (32, 116). SHP is an atypical member of the nuclear hormone receptor superfamily that lacks a DNA-binding domain and is capable of mediating the repression of several basic helix-loop-helix transcription factor family members and a diverse number of nuclear hormone receptors, including LRH-1 and hepatocyte nuclear factor-4α (HNF-4α, NR2A1), both of which bind to bile acid–response element regions (BAREs) in the CYP7A1 and CYP8B1 promoters (32, 116, 117). Interestingly, SHP harbors an LRH-1 element in its own promoter and can therefore feedback inhibit its own expression, providing a mechanism for fine-tuning of bile acid homeostasis at the biosynthesis level (32). Unlike rodents, humans lack positive oxysterol regulation of the CYP7A1 promoter. Instead, LXR activation induces direct upregulation of SHP expression through LXREs overlapping the previously identified bile acid response region, causing CYP7A1 transcriptional repression (118). The differential regulation of human CYP7A1 expression, which results in decreased bile acid synthesis in response to oxysterol overload, may help maintain cholesterol homeostasis by decreasing its absorption rather than inducing its catabolism, as is the case in rodents.

The generation of transgenic mouse models harboring deletions in the SHP or FXR genes helped underscore the physiological relevance of the FXR-SHP repression of CYP7A1. However, these mouse models also revealed the existence of SHP-independent, as well as FXR-independent, redundant parallel pathways for CYP7A1 and CYP8B1 repression by bile acids. Activation of the xenobiotic nuclear hormone receptor pregnane X receptor (PXR, NR1I2) constitutes one alternative FXR-independent pathway for CYP7A1 repression in SHP-null mice, although the mechanism of such a repression remains to be investigated (119–121). This activation probably results from the metabolism by the gut flora of the 1%

CA supplemented in the diet to secondary toxic bile acids (e.g., lithocholic acid), known potent activators of PXR (119, 120).

However, in vivo evidence in mice suggested the existence of PXR-independent pathways for CYP7A1 repression that involve the stress-induced c-Jun N-terminal kinase (JNK) signaling pathway (122). The use of a JNK activation inhibitor in wild-type mice demonstrated increased CYP7A1 expression in the absence or presence of CA (119). CYP7A1 repression is further blunted in SHP knockout mice, indicating a possible cross talk between the SHP-dependent and the JNK-induced pathways. Studies performed in primary rat hepatocytes suggested such a cross talk may occur through JNK-induced upregulation of SHP expression (123). More recently, the fibroblast growth factor-19 (FGF-19), a known upstream activator of JNK signaling pathway, was identified as a bona fide direct target gene of FXR. This discovery reinforced the importance of FXR in the feedback repression of bile acid synthesis and defined a whole new signaling cascade for CYP7A1 repression downstream of FXR activation (124). Although SHP is thought to perform its repressive function by competing with coactivator binding to transcription factors or through direct repression, possibly involving the recruitment of histone deacetylases, the mechanism of FGF-19-JNK-mediated repression still needs dissection. Nevertheless, cell culture studies performed on the human HepG2 cell line implicate the stress-induced MAP kinase module, which can also induce JNK activation, in the inhibition of HNF-4α activity on the BARE of the CYP7A1 promoter (125). Another FXR-independent pathway was recently described in the same human cell line. This pathway involved the disruption by bile acids of the recruitment of PGC-1α and CBP (cAMP response element–binding protein–binding protein) coactivators to HNF-4α on the CYP7A1 promoter (126). Whether this pathway requires JNK or MAP kinase activity is currently unknown, and further molecular studies are needed to elucidate the complex interconnected and parallel pathways for repression of the CYP7A1 promoter by bile acids.

FXR and Bile Salt Enterohepatic Circulation

Following their synthesis, the majority of bile acids undergo enzymatic conjugation with taurine or glycine, resulting in their ionization at physiological pH and their transformation into anionic bile salts. Owing to their increased hydrophilicity, conjugated bile salts are less permeable to the cell membrane than unconjugated bile salts. Therefore, they require the coordinated and regulated activities of several transporters expressed in the liver and the small intestine to facilitate their enterohepatic circulation (111). In preventing hepatic bile acid accumulation, FXR regulates the enterohepatic circulation of bile salts at different levels, including hepatic bile salt efflux, absorption by enterocytes, and reuptake by hepatocytes (112).

At the hepatic bile salt efflux level, FXR induces the transcription of two ABC transporters that are expressed on the canalicular membrane of hepatocytes: the bile salt export pump (BSEP, also called ABCB11 or sister of P-glycoprotein)

(127–129) and the multidrug resistance gene (MDR) [MDR-3 in humans and MDR-2 in mice (ABCB4)] (130). BSEP activity, which controls the rate-limiting step in bile formation, accounts for the majority of conjugated bile salt excretion into bile, whereas MDR-3 is an ATP-dependent flippase that translocates phospholipids, mostly phosphatidylcholine, across canalicular membranes of hepatocytes. The activities of BSEP and MDR-3 are crucial to the liver because genetic mutations in these genes result in the development of progressive familial intrahepatic cholestasis type II and type III, respectively (131, 132). Furthermore, the coordinated activities of BSEP and MDR-3 are thought to maintain a proper balance of biliary bile salts and phospholipids, both of which are crucial for the solubilization of cholesterol in bile. A recent study in mice highlighted the physiological relevance of FXR-mediated regulation of BSEP and MDR-3 expression. Feeding of a lithogenic diet to FXR-null mice caused biliary cholesterol precipitation resulting from increased indices of bile hydrophobicity and cholesterol saturation. By contrast, pharmacological activation of FXR by administration of the synthetic agonist GW4064 prevented gallstone formation in susceptible wild-type mice that recapitulate the human disease (133).

FXR activation also affects bile flow by inducing the transcription of the multidrug resistance-associated protein family member, MRP-2. MRP-2, which is mutated in the Dubin-Johnson syndrome, encodes a protein that induces canalicular excretion of glutathione, glucuronidated and sulfated bile salts, and a variety of drug substrates (134–136). FXR mediates transactivation of the MRP-2 promoter through binding to an atypical FXRE capable of also binding PXR and constitutive androstane receptor (CAR, NR1I3) (137). Interestingly, FXR can induce transcription of vasoactive intestinal peptide receptor-1, which is expressed predominantly in the gallbladder, and functions as a high-affinity receptor of vasoactive intestinal peptide, an important regulator of bile secretion (138). It is thus that FXR activation may have protective effects in gallstone disease.

At the intestinal lumen of the terminal ileum, FXR plays a key role in decreasing bile salt absorption by mediating species-specific bile acid–induced repression of the major apical sodium-dependent bile salt transporter (ASBT, SLC10A2), also known as the ileal bile acid transporter (IBAT). Mutations in IBAT, whose function is crucial for efficient hydrophilic and conjugated bile salt reuptake by the ileum, result in marked suppression of bile salt reabsorption in mice as well as in humans (139, 140). Whereas IBAT promoter activity in rats does not seem affected by changes in bile acid concentrations (141), the mechanism for FXR-induced IBAT repression in mice and humans involves differential SHP-mediated interference with transactivation factors on the IBAT promoter. These include LRH-1 in mice (142) and RXR/RAR, or possibly GR, in humans (113, 143). The activity of ileal bile acid–binding protein (IBABP), a small soluble protein, facilitates the intestinal uptake of bile salts and their intracellular trafficking to the basolateral side, where they undergo efflux into the portal blood (4, 144). FXR positively regulates the expression of IBABP (4, 144). Although IBABP basal expression levels are strongly downregulated in FXR knockout mice, these mice undergo efficient intestinal bile salt absorption, suggesting the existence of FXR-independent pathways for the regulation of bile salt enterohepatic circulation (145).

The last step in enterohepatic circulation that is essential for bile formation involves the hepatocellular uptake of bile salts from the portal circulation. Two types of transporters, expressed on the basolateral membrane of hepatocytes, achieve this important task: the sodium-dependent sodium taurocholate cotransporting polypeptide (NTCP, SLC10A1) and the sodium-independent organic anion transporting polypeptides (OATP, SLCO). NTCP activity accounts for more than 80% of conjugated but less than 50% of unconjugated bile salt hepatic reabsorption. As with IBAT, FXR negatively regulates NTCP expression through SHP-mediated repression. In rats, SHP interferes with retinoic acid activation of the NTCP promoter (146) and may interfere in a similar manner with HNF-4α, which also binds and transactivates the NTCP promoter from a DNA element overlapping that of the retinoic acid response (113, 147). However, other SHP-independent pathways for NTCP repression seem to exist, at least in mice (119), and may involve JNK signaling and phosphorylation-induced decreased binding of RXR and retinoic acid receptor (RAR) to the NTCP promoter (123, 148). Although neither RXR/RAR nor HNF-4α can transactivate the human NTCP promoter, SHP-mediated NTCP repression is conserved and may function through interference with GR-mediated upregulation of human NTCP expression (113). In addition to repressing the sodium-dependent bile salt uptake, FXR also downregulates the expression of OATP1B1 (SLCO1B1), the major contributor to sodium-independent bile salt uptake in humans. Through induction of SHP expression, FXR mediates a repressive effect on HNF-4α, whose activity is needed for the expression of HNF-1α, an effective transcriptional activator of OATP1B1 (149, 150). However, FXR also directly upregulates the expression of another OATP family member, OATP1B3, which shares high sequence identity with OATP1B1 and is also predominantly expressed in liver. Because both OATP1B1 and OATP1B3 have overlapping substrate specificities, upregulation of OATP1B3 may maintain the hepatic extraction of harmful molecules, such as xenobiotics, under cholestatic conditions (151).

Overall, activation of FXR seems to fulfill two different functions: the maintenance of bile acid homeostasis and the prevention of bile acid–induced liver toxicity under normal as well as cholestatic conditions. In support of the latter function of FXR, several bile acid modifying enzymes, known to decrease the hydrophobicity and hence toxicity of bile acids, are direct targets of FXR. Although increased expression of many of these genes requires pharmacological doses of FXR agonists and are not easily explained mechanistically, these effects may explain the redundancy the liver has evolved to respond to toxic conditions. Such genes encode the enzymes responsible for bile acid conjugation to taurine and glycine, bile acid-CoA synthetase (BACS), and bile acid-CoA: amino acid N-acetyltransferase (BAT) (152). Other potential FXR targets include UDP-glucuronosyltransferase UGT2B4 and sulfotransferase SULT2A1 enzymes, which transform hydrophobic bile acids into more soluble derivatives and facilitate their excretion. Whereas

SULT2A1 promoter is activated by the binding of FXR to a functional IR-0 capable of also binding PXR, CAR, and vitamin D receptor (VDR, NR1I1) (153), activation of UGT2B4 expression appears to involve a controversial mechanism involving FXR binding to a single hexameric element in its promoter (154). FXR also induces the transcription of human αA-crystallin, an abundant ocular lens protein, whose expression was also detected in liver (29). Because αA-crystallin possesses chaperone-like activities, its upregulated expression by FXR may protect the liver from toxicity under cholestatic conditions.

The role of FXR in preventing bile acid–induced toxicity is supported further by findings that primary bile acids, such as CDCA, and the FXR synthetic agonist GW4064 induce CYP3A4, which is the predominant CYP450 detoxifying enzyme in humans and a well-known target gene of PXR. In vivo studies in FXR and PXR transgenic mouse models and in vitro cell culture studies revealed the presence of two functional FXREs in the 5'-flanking region of CYP3A4: an ER-8 and an ER-1 overlapping the DR-3 element necessary for PXR binding (155). Moreover, FXR activation by CA feeding in mice leads to an increase in bile acid output and enhances the excretion of more toxic, unconjugated bile acids (156). Furthermore, administration of GW4064 to rat models of intrahepatic and extrahepatic cholestasis results in significantly decreased liver damage, reduced necrosis, inflammation, and bile duct proliferation as compared to control vehicle–treated cholestatic rats (157).

FXR AND TRIGLYCERIDE METABOLISM

The generation of FXR knockout mice shed light on an interesting and important role for this nuclear receptor in the regulation of systemic lipid metabolism (158). In addition to their increased serum bile acid levels, FXR-null mice have a distinctive proatherogenic lipid profile characterized by increased hepatic and serum triglyceride and cholesterol levels. The increase in total serum cholesterol in FXR-null mice was reflected in increased levels of VLDL, LDL (158), and HDL cholesterol; increased synthesis of ApoB-containing lipoproteins; and reduced clearance rate of HDL cholesteryl ester (159). Such a correlation between bile acid and lipid metabolism was described more than two decades ago, although the mechanism of its action was largely unknown (160–166). CDCA, which has been used in the past to treat cholesterol gallstone patients, caused a concomitant decrease in plasma triglyceride levels and was therefore suggested as a potential drug for the treatment of hypertriglyceridemia (160–162, 166). In parallel, treatment of dyslipidemic patients with bile-sequestering resins, such as cholestyramine, led to an undesirable increase in plasma triglyceride and VLDL production levels (163–165). In view of the dyslipidemic phenotype of the FXR-null mouse, FXR is currently thought to mediate much of the bile acid modulation of lipid metabolism, although the molecular details of its action remain to be elucidated. This hypothesis is supported by reports identifying an array of lipid-modulating proteins as direct or indirect downstream targets of FXR. For instance, FXR activation induces the expression of ApoCII through direct binding to two functional IR-1 elements in its hepatic control regions (167). On the other hand, FXR suppresses the expression of ApoCIII (168), the concentration of which is positively correlated with plasma triglyceride levels in humans (169). Because ApoCII is an obligate cofactor for LPL, whereas ApoCIII can inhibit LPL activity (170, 171), FXR modulation of these apolipoproteins may reduce plasma triglyceride levels by promoting LPL-mediated triglyceride hydrolysis in VLDL and chylomicrons. Interestingly, addition of GW4064 onto a HepG2 cell line stably infected with FXR markedly enhances the uptake and degradation of labeled LDL particles (27). This effect correlates with increased expression of syndecan (SDC)-1, a transmembrane heparan sulfate proteoglycan involved in binding and internalization of extracellular ligands. SDC-1 was identified as a FXR target gene, harboring a DR-1 element in its proximal promoter that is necessary and sufficient for activation by FXR (27).

FXR has also been suggested to govern hepatic expression of ApoE, a member of the same gene cluster as ApoCII (172), which would be expected to promote the hepatic clearance of atherogenic remnant lipoproteins and reduce plasma cholesterol levels. Consistent with this finding, researchers showed that FXR induces PLTP expression (173) and represses the expression of hepatic lipase (HL) in humans (174). Taken together, these studies would imply a role for FXR in increasing plasma HDL levels. However, other reports describe a role for FXR in mediating bile acid–induced downregulation of expression of ApoAI, which encodes the major lipoprotein of HDL, through an unclear mechanism that appears to involve binding of FXR as a monomer to a negative element in the ApoAI promoter (175, 176). Because it is controversial whether bile acids (and hence FXR activation) can affect HDL levels in humans (162, 166, 177), these studies remain confusing. Indeed, two genes encoding liver-secreted proteins involved in opposite coagulatory functions have been identified as positive FXR targets: (*a*) human kininogen, involved in the induction of anticoagulation, vasodilation, and inflammation (178), and (*b*) fibrinogen, whose cleavage by thrombin is critical for blood coagulation (28).

In spite of these latter reports, the well-documented, bile acid–induced decrease in triglyceride levels has been shown to involve FXR in several ways, including through the indirect modulation of triglyceride synthesis and secretion as well as of fatty acid uptake and oxidation. In particular, FXR represses expression of SREBP-1c, the master regulator of fatty acid and triglyceride synthesis (179). The mechanism of SREBP-1c repression is not known but may involve interference with LXR activation, which, as described above, is crucial for the expression of SREBP-1c and its downstream target genes (68, 76). In other experiments in cultured human hepatic cells, FXR also repressed expression of microsomal triglyceride transfer protein (MTP), a liver- and intestine-specific protein that, along with ApoB, is crucial for the assembly and secretion of chylomicrons and VLDL particles (180, 181). In the human and mouse liver, FXR also controls the expression of VLDL receptor (VLDLR) (182). Although the role of VLDLR in liver is not well known, its function in peripheral tissues seems to correlate with increased uptake of circulating VLDL triglycerides (183). Complement C3 promoter has also been identified as a FXR target gene in human cells; C3 protein levels are increased in mice and rats following administration of the FXR agonist GW4064 (184). This result led to the speculation that subsequent C3 conversion to acylation stimulating protein (ASP) may decrease plasma triglyceride levels by promoting free fatty acid uptake and triglyceride synthesis in adipocytes (184). However, such a hypothesis requires further investigation.

Increasing fatty acid oxidation represents another means for FXR-mediated reduction of plasma triglyceride levels. This may occur via FXR-mediated up-regulation of pyruvate dehydrogenase kinase (PDK4), which promotes utilization of fat rather than glucose as an energy source (185). In addition, FXR activation induces the expression of peroxisome proliferator–activated receptor α (PPARα, NR1C1), a nuclear receptor that plays a pivotal role in promoting hepatic fatty acid oxidation (186). This regulation appears to be species-specific, because only human PPARα expression has been induced by the binding of FXR to a nonconserved DR-5 element (187). Again, such studies have yet to be validated in vivo. Nevertheless, elevated levels of primary bile acids have previously been shown to antagonize PPARα activation of downstream fatty acid oxidation genes, evidently by impairing coactivator recruitment (188).

Finally, there is evidence implicating FXR in governing PGC-1α modulation of triglyceride metabolism, although the mechanistic details remain obscure (18–20). The most convincing data come from a study showing that the decrease in triglyceride levels known to occur following fasting and the induction of PGC-1α expression does not take place in FXR-null mice but is correlated with upregulation of FXR in wild-type mice (20). In vitro cell culture studies further suggested that PGC-1α coactivates PPARγ and HNF-4α on the FXR gene promoter, and even FXR itself, to enhance transcription of its target genes (20). However, the mechanism reported in these studies still remains controversial (18, 19).

FXR AND GLUCOSE METABOLISM

The involvement of FXR in the regulation of carbohydrate metabolism has been the subject of ongoing investigation. At present, results from different studies appear contradictory. One study showed that bile acid feeding to mice represses the expression of gluconeogenic genes following fasting (189). In vitro cell culture studies suggest that this effect can be mediated via SHP competition with the coactivator, CBP, which enhances the activity of HNF-4α and FOXO1 on the gene promoters for PEPCK, fructose 1,6-bis phosphatase, and G6Pase (189). SHP also antagonizes PGC-1α coactivation of the glucocorticoid receptor (GR) and inhibits both GR and HNF-4α transactivation of the PEPCK promoter (190). However, in another study, addition of the synthetic FXR agonist GW4064 to primary human or rat cells led to an induction rather than suppression of PEPCK expression and stimulated glucose output (191). When GW4064 was administered to mice that were not fasted prior to sacrifice, a similar PEPCK upregulation was observed as a result of Foxo1 activation (191). These apparently contradicting results may reflect differential effects of bile acids and GW4064 under different in vivo fasting/refeeding metabolic conditions. Indeed, absence of FXR in mice accentuates the fasting/refeeding response by enhancing the expression of glycolytic, lipogenic genes and suppressing that of gluconeogenic genes (192). In addition, studies in primary rat hepatocytes using an FXR agonist showed attenuation of the glucose-induced lipogenic response (192). FXR expression itself was induced by glucose levels in these cells, whereas it was significantly decreased in diabetic rats (193).

CONCLUSIONS

The phylogeny of the nuclear hormone receptor superfamily places LXRα, LXRβ, and FXR within a subfamily of closely related genes (194, 195) that we now know regulate common physiologic pathways. From an evolutionary point of view, it is tempting to speculate that by acquiring properties to bind different physiologic metabolites, LXRs and FXR have coevolved to allow for the congruous regulation of overlapping and intricately interconnected metabolic pathways governing sterol and lipid homeostasis. This hypothesis is supported by the above-described physiological functions of these receptors and by an increasing number of reports underscoring their delicately balanced regulation of mammalian sterol, lipid, and carbohydrate metabolism.

Although several examples of genes that can be transcriptionally regulated in a similar manner by LXRs or FXR have been shown to exist under in vitro or artificial conditions, e.g., ApoCII (47, 167), ApoE, PLTP (172), and IBABP (196), the general picture depicts a well-coordinated yin and yang regulatory interplay between LXRs and FXR at several levels (Figure 1). For instance, in addition to their well-described opposing roles in mediating cholesterol metabolism and bile acid biosynthesis, a less-appreciated interplay occurs at the level of regulation of enterohepatic excretion, circulation, and absorption of cholesterol, lipids, and bile acids. In addition to enhancing cholesterol conversion into bile acids through upregulation of CYP7A1 expression, oxysterol-mediated activation of LXRs induces the transcription of ABCG5 and ABCG8, the half-transporters crucial for cholesterol secretion into bile. As bile acid concentrations build up, FXR activation causes repression of bile acid synthesis and simultaneous increases in biliary bile salt and phospholipid secretion through upregulation of BSEP and MDR-3. This effect enhances cholesterol solubility, preventing cholesterol gallstone formation.

Another example of the LXR/FXR interplay also occurs at the general regulatory level of triglyceride metabolism. Activation of LXRs promotes fatty acid and triglyceride synthesis and storage, increasing liver and plasma triglyceride levels, whereas FXR activation results in a consistent and marked decrease in hepatic

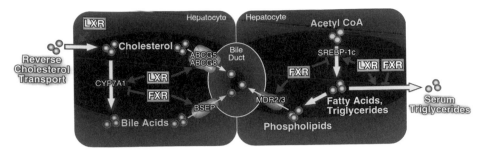

Figure 1   A yin and yang relationship between LXRs and FXR maintains cholesterol, bile acid, and triglyceride homeostasis. Shown are the key pathways and target genes of the LXR and FXR enterohepatic regulatory cascade. Oxysterol activation of LXRs induces cholesterol conversion to bile acids, enhances biliary cholesterol secretion, and increases hepatic and plasma triglyceride levels. Bile acid activation of FXR prevents excessive accumulation of bile acids, through feedback regulation of bile acid biosynthesis and the conversion of bile acids into less toxic compounds. FXR also enhances biliary bile acid and phospholipid secretion, solubilizing biliary cholesterol, and preventing gallstone formation. In addition, FXR activation counteracts the effects of LXRs on fat metabolism by markedly decreasing hepatic and plasma triglyceride levels, preventing hypertriglyceridemia. See text for further details.

as well as plasma triglyceride levels (160–162, 166, 179). The LXR/FXR regulatory interplay reinforces the hypothesis that these receptors evolved to govern the organismal metabolic balance at different levels and help maintain overall lipid homeostasis. In this respect, pharmaceutical design of tissue-specific LXR or FXR modulators holds great therapeutic promise. Indeed, the combinatorial administration of selective modulators of LXRs and/or FXR may lead to greater efficacy in the fight against an increasing number of lipid-related and metabolic diseases.

SUMMARY POINTS

1. LXRs and FXR are nuclear receptors that function as lipid sensors for cholesterol and bile acids. Together these two receptor systems coordinately govern transcription of genes involved in cholesterol, bile acid, lipid, and carbohydrate homeostasis.

2. In the enterohepatic system, LXRs and FXR often oppose each other's action by diametrically regulating transcription of the same target genes. This yin and yang relationship maintains a normal balance of sterols, lipids, and glucose.

3. As nuclear receptors, LXRs and FXR represent promising therapeutic targets for drug discovery. Such drugs may have utility in treating or preventing

diseases such as atherosclerosis, cardiovascular disease, cholesterol gall-stone disease, nonalcoholic steatohepatitis, and cholestasis, as well as other metabolic syndrome disorders.

FUTURE ISSUES TO BE RESOLVED

1. The function of LXRs and FXR in tissues other than the enterohepatic system and macrophages. Examples of such tissues include adipose tissue and brain (for LXRs) and adrenal and kidney (for LXRs and FXR).
2. Generation of conditional knockouts of LXRs and FXR in mice as models to study tissue- and temporal-specific receptor activities.
3. Development of LXR-selective modulators that retain the ability to induce reverse cholesterol transport but that do not cause hypertriglyceridemia.
4. The utility of ligands for LXRs and FXR in humans.

APPENDIX

Abbreviations used: ABC, ATP-binding cassette transporter; AD, Alzheimer's disease; Angptl3, angiopoietin-like protein 3; Apo, apoliprotein; APP, amyloid precursor protein; ASBT, apical sodium-dependent bile salt transporter; ASP; acylation stimulating protein; BACS, bile acid-CoA synthetase; BARE, bile acid response element; BAT, bile acid-CoA: amino acid N-acetyltransferase; BSEP, bile salt export pump; CA, cholic acid; CAR, constitutive androstane receptor; CBP, cAMP response element–binding protein-binding protein; CDCA, chenodeoxycholic acid; CETP, cholesteryl ester transfer protein; CYP, cytochrome P450; FAS, fatty acid synthase; FGF, fibroblast growth factor; FPPS, farnesyl pyrophosphate synthase; FXR, farnesoid X receptor; FXRE, FXR response element; G6Pase, glucose-6-phosphatase; GK, glucokinase; GLUT, glucose transporter; HDL, high-density lipoprotein; HL, hepatic lipase; HNF, hepatocyte nuclear factor; IBABP, ileal bile acid–binding protein; IBAT, ileal bile acid transporter; JNK, c-Jun N-terminal kinase; LDL, low-density lipoprotein; LPL, lipoprotein lipase; LRH-1, liver receptor homolog-1; L-UrdPase, liver-specific uridine phosphorylase gene; LXR, liver X receptor; LXRE, LXR response element; MDR, multidrug resistance gene; MRP, multidrug resistance-associated protein; MTP, microsomal triglyceride transfer protein; N-CoR, nuclear receptor corepressor; NTCP, sodium-dependent sodium taurocholate cotransporting polypeptide; OATP, sodium-independent organic anion–transporting polypeptides; PDK4, pyruvate dehydrogenase kinase; PEPCK, phosphoenolpyruvate carboxykinase; PGC-1, peroxisome proliferator–activated receptor-gamma coactivator-1; PLTP, phospholipid transfer protein; PPAR, peroxisome proliferator–activated receptor; PXR, pregnane X receptor; RAR, retinoic acid receptor; RXR, retinoid X receptor; SDC, syndecan; SHP, small heterodimer partner; SMRT, silencing mediator for retinoic acid and thyroid hormone receptor; SR-BI, scavenger receptor-BI; SREBP-1c, sterol regulatory element–binding protein-1c; SULT, sulfotransferase; UGT, UDP-glucuronosyltransferase; VDR, vitamin D receptor; VLDL, very-low-density lipoprotein; VLDLR, VLDL receptor.

DISCLOSURE STATEMENT

D.J.M. is a consultant for Exelixis, Inc.

The *Annual Review of Physiology* is online at
http://physiol.annualreviews.org

LITERATURE CITED

1. Chawla A, Repa JJ, Evans RM, Mangelsdorf DJ. 2001. Nuclear receptors and lipid physiology: opening the X-files. *Science* 294:1866–70

2. Janowski BA, Willy PJ, Devi TR, Falck JR, Mangelsdorf DJ. 1996. An oxysterol signalling pathway mediated by the nuclear receptor LXR α. *Nature* 383:728–31

3. Lehmann JM, Kliewer SA, Moore LB, Smith-Oliver TA, Oliver BB, et al. 1997. Activation of the nuclear receptor LXR by oxysterols defines a new hormone response pathway. *J. Biol. Chem.* 272:3137–40

4. Makishima M, Okamoto AY, Repa JJ, Tu H, Learned RM, et al. 1999. Identification of a nuclear receptor for bile acids. *Science* 284:1362–65

5. Parks DJ, Blanchard SG, Bledsoe RK, Chandra G, Consler TG, et al. 1999. Bile acids: natural ligands for an orphan nuclear receptor. *Science* 284:1365–68

6. Wang H, Chen J, Hollister K, Sowers LC, Forman BM. 1999. Endogenous bile acids are ligands for the nuclear receptor FXR/BAR. *Mol. Cell.* 3:543–53

7. Heyman RA, Mangelsdorf DJ, Dyck JA, Stein RB, Eichele G, et al. 1992. 9-cis retinoic acid is a high affinity ligand for the retinoid X receptor. *Cell* 68:397–406

8. Levin AA, Sturzenbecker LJ, Kazmer S, Bosakowski T, Huselton C, et al. 1992. 9-cis retinoic acid stereoisomer binds and activates the nuclear receptor RXR α. *Nature* 355:359–61

9. Mangelsdorf DJ, Borgmeyer U, Heyman RA, Zhou JY, Ong ES, et al. 1992. Characterization of three RXR genes that mediate the action of 9-cis retinoic acid. *Genes Dev.* 6:329–44

10. de Urquiza AM, Liu S, Sjoberg M, Zetterstrom RH, Griffiths W, et al. 2000. Docosahexaenoic acid, a ligand for the retinoid X receptor in mouse brain. *Science* 290:2140–44

11. Chen JD, Evans RM. 1995. A transcriptional co-repressor that interacts with nuclear hormone receptors. *Nature* 377:454–57

12. Horlein AJ, Naar AM, Heinzel T, Torchia J, Gloss B, et al. 1995. Ligand-independent repression by the thyroid hormone receptor mediated by a nuclear receptor co-repressor. *Nature* 377:397–404

13. Glass CK, Rosenfeld MG. 2000. The coregulator exchange in transcriptional functions of nuclear receptors. *Genes Dev.* 14:121–41

14. Huuskonen J, Fielding PE, Fielding CJ. 2004. Role of p160 coactivator complex in the activation of liver X receptor. *Arterioscler. Thromb. Vasc. Biol.* 24:703–8

15. Kim SW, Park K, Kwak E, Choi E, Lee S, et al. 2003. Activating signal cointegrator

2 required for liver lipid metabolism mediated by liver X receptors in mice. *Mol. Cell. Biol.* 23:3583–92

16. Unno A, Takada I, Takezawa S, Oishi H, Baba A, et al. 2005. TRRAP as a hepatic coactivator of LXR and FXR function. *Biochem. Biophys. Res. Commun.* 327:933–38

17. Oberkofler H, Schraml E, Krempler F, Patsch W. 2003. Potentiation of liver X receptor transcriptional activity by peroxisome-proliferator-activated receptor γ co-activator 1 α. *Biochem. J.* 371: 89–96

18. Kanaya E, Shiraki T, Jingami H. 2004. The nuclear bile acid receptor FXR is activated by PGC-1α in a ligand-dependent manner. *Biochem. J.* 382:913–21

19. Savkur RS, Thomas JS, Bramlett KS, Gao Y, Michael LF, Burris TP. 2005. Ligand-dependent coactivation of the human bile acid receptor FXR by the peroxisome proliferator-activated receptor γ coactivator-1α. *J. Pharmacol. Exp. Ther.* 312:170–78

20. Zhang Y, Castellani LW, Sinal CJ, Gonzalez FJ, Edwards PA. 2004. Peroxisome proliferator-activated receptor-γ coactivator 1α (PGC-1α) regulates triglyceride metabolism by activation of the nuclear receptor FXR. *Genes Dev.* 18:157–69

21. Lu TT, Repa JJ, Mangelsdorf DJ. 2001. Orphan nuclear receptors as eLiXiRs and FiXeRs of sterol metabolism. *J. Biol. Chem.* 276:37735–38

22. Forman BM, Goode E, Chen J, Oro AE, Bradley DJ, et al. 1995. Identification of a nuclear receptor that is activated by farnesol metabolites. *Cell* 81:687–93

23. Otte K, Kranz H, Kober I, Thompson P, Hoefer M, et al. 2003. Identification of farnesoid X receptor β as a novel mammalian nuclear receptor sensing lanosterol. *Mol. Cell. Biol.* 23:864–72

24. Zhang Y, Kast-Woelbern HR, Edwards PA. 2003. Natural structural variants of the nuclear receptor farnesoid X receptor affect transcriptional activation. *J. Biol. Chem.* 278:104–10

25. Bishop-Bailey D, Walsh DT, Warner TD. 2004. Expression and activation of the farnesoid X receptor in the vasculature. *Proc. Natl. Acad. Sci. USA* 101:3668–73

26. Huber RM, Murphy K, Miao B, Link JR, Cunningham MR, et al. 2002. Generation of multiple farnesoid-X-receptor isoforms through the use of alternative promoters. *Genetics* 290:35–43

27. Anisfeld AM, Kast-Woelbern HR, Meyer ME, Jones SA, Zhang Y, et al. 2003. Syndecan-1 expression is regulated in an isoform-specific manner by the farnesoid-X receptor. *J. Biol. Chem.* 278:20420–28

28. Anisfeld AM, Kast-Woelbern HR, Lee H, Zhang Y, Lee FY, Edwards PA. 2005. Activation of the nuclear receptor FXR induces fibrinogen expression: a new role for bile acid signaling. *J. Lipid Res.* 46:458–68

29. Lee FY, Kast-Woelbern HR, Chang J, Luo G, Jones SA, et al. 2005. α-A-crystallin is a target gene of the farnesoid X-activated receptor (FXR) in human livers. *J. Biol. Chem.* 280:31792–800

30. Peet DJ, Turley SD, Ma W, Janowski BA, Lobaccaro JM, et al. 1998. Cholesterol and bile acid metabolism are impaired in mice lacking the nuclear oxysterol receptor LXR α. *Cell* 93:693–704

31. Repa JJ, Turley SD, Lobaccaro JA, Medina J, Li L, et al. 2000. Regulation of absorption and ABC1-mediated efflux of cholesterol by RXR heterodimers. *Science* 289:1524–29

32. Lu TT, Makishima M, Repa JJ, Schoonjans K, Kerr TA, et al. 2000. Molecular basis for feedback regulation of bile acid synthesis by nuclear receptors. *Mol. Cell.* 6:507–15

33. Fayard E, Auwerx J, Schoonjans K. 2004. LRH-1: an orphan nuclear receptor involved in development, metabolism and steroidogenesis. *Trends Cell Biol.* 14:250–60

34. Costet P, Luo Y, Wang N, Tall AR.

2000. Sterol-dependent transactivation of the ABC1 promoter by the liver X receptor/retinoid X receptor. *J. Biol. Chem.* 275:28240–45

35. Singaraja RR, Brunham LR, Visscher H, Kastelein JJ, Hayden MR. 2003. Efflux and atherosclerosis: the clinical and biochemical impact of variations in the ABCA1 gene. *Arterioscler. Thromb. Vasc. Biol.* 23:1322–32

36. Singaraja RR, Bocher V, James ER, Clee SM, Zhang LH, et al. 2001. Human ABCA1 BAC transgenic mice show increased high density lipoprotein cholesterol and ApoAI-dependent efflux stimulated by an internal promoter containing liver X receptor response elements in intron 1. *J. Biol. Chem.* 276:33969–79

37. Berge KE, Tian H, Graf GA, Yu L, Grishin NV, et al. 2000. Accumulation of dietary cholesterol in sitosterolemia caused by mutations in adjacent ABC transporters. *Science* 290:1771–75

38. Salen G, Shefer S, Nguyen L, Ness GC, Tint GS, Shore V. 1992. Sitosterolemia. *J. Lipid Res.* 33:945–55

39. Repa JJ, Berge KE, Pomajzl C, Richardson JA, Hobbs H, Mangelsdorf DJ. 2002. Regulation of ATP-binding cassette sterol transporters ABCG5 and ABCG8 by the liver X receptors α and β. *J. Biol. Chem.* 277:18793–800

40. Yu L, Hammer RE, Li-Hawkins J, Von Bergmann K, Lutjohann D, et al. 2002. Disruption of Abcg5 and Abcg8 in mice reveals their crucial role in biliary cholesterol secretion. *Proc. Natl. Acad. Sci. USA* 99:16237–42

41. Venkateswaran A, Repa JJ, Lobaccaro JM, Bronson A, Mangelsdorf DJ, Edwards PA. 2000. Human white/murine ABC8 mRNA levels are highly induced in lipid-loaded macrophages. A transcriptional role for specific oxysterols. *J. Biol. Chem.* 275:14700–7

42. Sabol SL, Brewer HB Jr, Santamarina-Fojo S. 2005. The human ABCG1 gene: identification of LXR response elements that modulate expression in macrophages and liver. *J. Lipid Res.* 46:2151–67

43. Klucken J, Buchler C, Orso E, Kaminski WE, Porsch-Ozcurumez M, et al. 2000. ABCG1 (ABC8), the human homolog of the *Drosophila* white gene, is a regulator of macrophage cholesterol and phospholipid transport. *Proc. Natl. Acad. Sci. USA* 97:817–22

44. Wang N, Lan D, Chen W, Matsuura F, Tall AR. 2004. ATP-binding cassette transporters G1 and G4 mediate cellular cholesterol efflux to high-density lipoproteins. *Proc. Natl. Acad. Sci. USA* 101:9774–79

45. Kennedy MA, Barrera GC, Nakamura K, Baldan A, Tarr P, et al. 2005. ABCG1 has a critical role in mediating cholesterol efflux to HDL and preventing cellular lipid accumulation. *Cell. Metab.* 1:121–31

46. Laffitte BA, Repa JJ, Joseph SB, Wilpitz DC, Kast HR, et al. 2001. LXRs control lipid-inducible expression of the apolipoprotein E gene in macrophages and adipocytes. *Proc. Natl. Acad. Sci. USA* 98:507–12

47. Mak PA, Laffitte BA, Desrumaux C, Joseph SB, Curtiss LK, et al. 2002. Regulated expression of the apolipoprotein E/C-I/C-IV/C-II gene cluster in murine and human macrophages. A critical role for nuclear liver X receptors α and β. *J. Biol. Chem.* 277:31900–8

48. Hummasti S, Laffitte BA, Watson MA, Galardi C, Chao LC, et al. 2004. Liver X receptors are regulators of adipocyte gene expression but not differentiation: identification of apoD as a direct target. *J. Lipid Res.* 45:616–25

49. Jakel H, Nowak M, Moitrot E, Dehondt H, Hum DW, et al. 2004. The liver X receptor ligand T0901317 down-regulates APOA5 gene expression through activation of SREBP-1c. *J. Biol. Chem.* 279:45462–69

50. Zhang Y, Repa JJ, Gauthier K, Mangelsdorf DJ. 2001. Regulation of lipoprotein lipase by the oxysterol receptors, LXRα and LXRβ. *J. Biol. Chem.* 276:43018–24

51. Laffitte BA, Joseph SB, Chen M, Castrillo A, Repa J, et al. 2003. The phospholipid transfer protein gene is a liver X receptor target expressed by macrophages in atherosclerotic lesions. *Mol. Cell. Biol.* 23:2182–91

52. Cao G, Beyer TP, Yang XP, Schmidt RJ, Zhang Y, et al. 2002. Phospholipid transfer protein is regulated by liver X receptors in vivo. *J. Biol. Chem.* 277:39561–65

53. Luo Y, Tall AR. 2000. Sterol upregulation of human CETP expression in vitro and in transgenic mice by an LXR element. *J. Clin. Invest.* 105:513–20

54. Laffitte BA, Joseph SB, Walczak R, Pei L, Wilpitz DC, et al. 2001. Autoregulation of the human liver X receptor α promoter. *Mol. Cell. Biol.* 21:7558–68

55. Bellosta S, Mahley RW, Sanan DA, Murata J, Newland DL, et al. 1995. Macrophage-specific expression of human apolipoprotein E reduces atherosclerosis in hypercholesterolemic apolipoprotein E-null mice. *J. Clin. Invest.* 96:2170–79

56. Linton MF, Atkinson JB, Fazio S. 1995. Prevention of atherosclerosis in apolipoprotein E-deficient mice by bone marrow transplantation. *Science* 267:1034–37

57. Fazio S, Babaev VR, Murray AB, Hasty AH, Carter KJ, et al. 1997. Increased atherosclerosis in mice reconstituted with apolipoprotein E null macrophages. *Proc. Natl. Acad. Sci. USA* 94:4647–52

58. Tontonoz P, Mangelsdorf DJ. 2003. Liver X receptor signaling pathways in cardiovascular disease. *Mol. Endocrinol.* 17:985–93

59. Malerod L, Juvet LK, Hanssen-Bauer A, Eskild W, Berg T. 2002. Oxysterol-activated LXRα/RXR induces hSR-BI-promoter activity in hepatoma cells and preadipocytes. *Biochem. Biophys. Res. Commun.* 299:916–23

60. Tangirala RK, Bischoff ED, Joseph SB, Wagner BL, Walczak R, et al. 2002. Identification of macrophage liver X receptors as inhibitors of atherosclerosis. *Proc. Natl. Acad. Sci. USA* 99:11896–901

61. Joseph SB, McKilligin E, Pei L, Watson MA, Collins AR, et al. 2002. Synthetic LXR ligand inhibits the development of atherosclerosis in mice. *Proc. Natl. Acad. Sci. USA* 99:7604–9

62. Joseph SB, Castrillo A, Laffitte BA, Mangelsdorf DJ, Tontonoz P. 2003. Reciprocal regulation of inflammation and lipid metabolism by liver X receptors. *Nat. Med.* 9:213–19

63. Fowler AJ, Sheu MY, Schmuth M, Kao J, Fluhr JW, et al. 2003. Liver X receptor activators display anti-inflammatory activity in irritant and allergic contact dermatitis models: liver-X-receptor-specific inhibition of inflammation and primary cytokine production. *J. Invest. Dermatol.* 120:246–55

64. Joseph SB, Bradley MN, Castrillo A, Bruhn KW, Mak PA, et al. 2004. LXR-dependent gene expression is important for macrophage survival and the innate immune response. *Cell* 119:299–309

65. Arai S, Shelton JM, Chen M, Bradley MN, Castrillo A, et al. 2005. A role for the apoptosis inhibitory factor AIM/Spa/Api6 in atherosclerosis development. *Cell Metab.* 1:201–13

66. Walczak R, Joseph SB, Laffitte BA, Castrillo A, Pei L, Tontonoz P. 2004. Transcription of the vascular endothelial growth factor gene in macrophages is regulated by liver X receptors. *J. Biol. Chem.* 279:9905–11

67. Horton JD, Goldstein JL, Brown MS. 2002. SREBPs: activators of the complete program of cholesterol and fatty acid synthesis in the liver. *J. Clin. Invest.* 109:1125–31

68. Repa JJ, Liang G, Ou J, Bashmakov Y, Lobaccaro JM, et al. 2000. Regulation of mouse sterol regulatory element-binding protein-1c gene (SREBP-1c) by oxysterol receptors, LXRα and LXRβ. *Genes Dev.* 14:2819–30

69. Yoshikawa T, Shimano H, Amemiya-Kudo M, Yahagi N, Hasty AH, et al. 2001.

Identification of liver X receptor retinoid X receptor as an activator of the sterol regulatory element-binding protein 1c gene promoter. *Mol. Cell. Biol.* 21:2991–3000

70. Joseph SB, Laffitte BA, Patel PH, Watson MA, Matsukuma KE, et al. 2002. Direct and indirect mechanisms for regulation of fatty acid synthase gene expression by liver X receptors. *J. Biol. Chem.* 277:11019–25

71. Koishi R, Ando Y, Ono M, Shimamura M, Yasumo H, et al. 2002. Angptl3 regulates lipid metabolism in mice. *Nat. Genet.* 30:151–57

72. Kaplan R, Zhang T, Hernandez M, Gan FX, Wright SD, et al. 2003. Regulation of the angiopoietin-like protein 3 gene by LXR. *J. Lipid Res.* 44:136–43

73. Inaba T, Matsuda M, Shimamura M, Takei N, Terasaka N, et al. 2003. Angiopoietin-like protein 3 mediates hypertriglyceridemia induced by the liver X receptor. *J. Biol. Chem.* 278:21344–51

74. Shimizugawa T, Ono M, Shimamura M, Yoshida K, Ando Y, et al. 2002. ANGPTL3 decreases very low density lipoprotein triglyceride clearance by inhibition of lipoprotein lipase. *J. Biol. Chem.* 277:33742–48

75. Shimamura M, Matsuda M, Kobayashi S, Ando Y, Ono M, et al. 2003. Angiopoietin-like protein 3, a hepatic secretory factor, activates lipolysis in adipocytes. *Biochem. Biophys. Res. Commun.* 301: 604–9

76. Kalaany NY, Gauthier KC, Zavacki AM, Mammen PP, Kitazume T, et al. 2005. LXRs regulate the balance between fat storage and oxidation. *Cell. Metab.* 1:231–44

77. Chen G, Liang G, Ou J, Goldstein JL, Brown MS. 2004. Central role for liver X receptor in insulin-mediated activation of Srebp-1c transcription and stimulation of fatty acid synthesis in liver. *Proc. Natl. Acad. Sci. USA* 101:11245–50

78. Zhang Y, Repa JJ, Inoue Y, Hayhurst GP, Gonzalez FJ, Mangelsdorf DJ. 2004.

79. Schultz JR, Tu H, Luk A, Repa JJ, Medina JC, et al. 2000. Role of LXRs in control of lipogenesis. *Genes Dev.* 14:2831–38

80. Grefhorst A, Elzinga BM, Voshol PJ, Plosch T, Kok T, et al. 2002. Stimulation of lipogenesis by pharmacological activation of the liver X receptor leads to production of large, triglyceride-rich very low density lipoprotein particles. *J. Biol. Chem.* 277:34182–90

81. Liang G, Yang J, Horton JD, Hammer RE, Goldstein JL, Brown MS. 2002. Diminished hepatic response to fasting/refeeding and liver X receptor agonists in mice with selective deficiency of sterol regulatory element-binding protein-1c. *J. Biol. Chem.* 277:9520–28

82. Wagner BL, Valledor AF, Shao G, Daige CL, Bischoff ED, et al. 2003. Promoter-specific roles for liver X receptor/corepressor complexes in the regulation of ABCA1 and SREBP1 gene expression. *Mol. Cell. Biol.* 23:5780–89

83. Osborne CK, Zhao H, Fuqua SA. 2000. Selective estrogen receptor modulators: structure, function, and clinical use. *J. Clin. Oncol.* 18:3172–86

84. Laffitte BA, Chao LC, Li J, Walczak R, Hummasti S, et al. 2003. Activation of liver X receptor improves glucose tolerance through coordinate regulation of glucose metabolism in liver and adipose tissue. *Proc. Natl. Acad. Sci. USA* 100:5419–24

85. Dalen KT, Ulven SM, Bamberg K, Gustafsson JA, Nebb HI. 2003. Expression of the insulin-responsive glucose transporter GLUT4 in adipocytes is dependent on liver X receptor α. *J. Biol. Chem.* 278:48283–91

86. Cao G, Liang Y, Broderick CL, Oldham BA, Beyer TP, et al. 2003. Antidiabetic action of a liver x receptor agonist

Identification of a liver-specific uridine phosphorylase that is regulated by multiple lipid-sensing nuclear receptors. *Mol. Endocrinol.* 18:851–62

mediated by inhibition of hepatic gluco-neogenesis. *J. Biol. Chem.* 278:1131–36

87. Chisholm JW, Hong J, Mills SA, Lawn RM. 2003. The LXR ligand T0901317 induces severe lipogenesis in the db/db diabetic mouse. *J. Lipid Res.* 44:2039–48

88. Efanov AM, Sewing S, Bokvist K, Gromada J. 2004. Liver X receptor activation stimulates insulin secretion via modulation of glucose and lipid metabolism in pancreatic β-cells. *Diabetes* 53(Suppl. 3):S75–78

89. Gerin I, Dolinsky VW, Shackman JG, Kennedy RT, Chiang SH, et al. 2005. LXRβ is required for adipocyte growth, glucose homeostasis, and β cell function. *J. Biol. Chem.* 280:23024–31

90. Dietschy JM, Turley SD. 2004. Thematic review series: brain lipids. Cholesterol metabolism in the central nervous system during early development and in the mature animal. *J. Lipid. Res.* 45:1375–97

91. Wang L, Schuster GU, Hultenby K, Zhang Q, Andersson S, Gustafsson JA. 2002. Liver X receptors in the central nervous system: from lipid homeostasis to neuronal degeneration. *Proc. Natl. Acad. Sci. USA* 99:13878–83

92. Andersson S, Gustafsson N, Warner M, Gustafsson JA. 2005. Inactivation of liver X receptor β leads to adult-onset motor neuron degeneration in male mice. *Proc. Natl. Acad. Sci. USA* 102:3857–62

93. Whitney KD, Watson MA, Collins JL, Benson WG, Stone TM, et al. 2002. Regulation of cholesterol homeostasis by the liver X receptors in the central nervous system. *Mol. Endocrinol* 16:1378–85

94. Liang Y, Lin S, Beyer TP, Zhang Y, Wu X, et al. 2004. A liver X receptor and retinoid X receptor heterodimer mediates apolipoprotein E expression, secretion and cholesterol homeostasis in astrocytes. *J. Neurochem.* 88:623–34

95. Fukuchi J, Song C, Ko AL, Liao S. 2003. Transcriptional regulation of farnesyl pyrophosphate synthase by liver X receptors. *Steroids* 68:685–91

96. Sun Y, Yao J, Kim TW, Tall AR. 2003. Expression of liver X receptor target genes decreases cellular amyloid β peptide secretion. *J. Biol. Chem.* 278:27688–94

97. Koldamova RP, Lefterov IM, Ikonomovic MD, Skoko J, Lefterov PI, et al. 2003. 22R-hydroxycholesterol and 9-cis-retinoic acid induce ATP-binding cassette transporter A1 expression and cholesterol efflux in brain cells and decrease amyloid β secretion. *J. Biol. Chem.* 278:13244–56

98. Koldamova RP, Lefterov IM, Staufenbiel M, Wolfe D, Huang S, et al. 2005. The liver X receptor ligand T0901317 decreases amyloid β production in vitro and in a mouse model of Alzheimer's disease. *J. Biol. Chem.* 280:4079–88

99. Ruan XZ, Moorhead JF, Fernando R, Wheeler DC, Powis SH, Varghese Z. 2004. Regulation of lipoprotein trafficking in the kidney: role of inflammatory mediators and transcription factors. *Biochem. Soc. Trans.* 32:88–91

100. Wu J, Zhang Y, Wang N, Davis L, Yang G, et al. 2004. Liver X receptor-α mediates cholesterol efflux in glomerular mesangial cells. *Am. J. Physiol. Renal Physiol.* 287:F886–95

101. Morello F, de Boer RA, Steffensen KR, Gnecchi M, Chisholm JW, et al. 2005. Liver X receptors α and β regulate renin expression in vivo. *J. Clin. Invest.* 115:1913–22

102. Tamura K, Chen YE, Horiuchi M, Chen Q, Daviet L, et al. 2000. LXRα functions as a cAMP-responsive transcriptional regulator of gene expression. *Proc. Natl. Acad. Sci. USA* 97:8513–18

103. Frenoux JM, Vernet P, Volle DH, Britan A, Saez F, et al. 2004. Nuclear oxysterol receptors, LXRs, are involved in the maintenance of mouse caput epididymidis structure and functions. *J. Mol. Endocrinol.* 33:361–75

104. Robertson KM, Schuster GU, Steffensen KR, Hovatta O, Meaney S, et al. 2005. The liver X receptor-β is essential for

maintaining cholesterol homeostasis in the testis. *Endocrinology* 146:2519–30

105. Selva DM, Hirsch-Reinshagen V, Burgess B, Zhou S, Chan J, et al. 2004. The ATP-binding cassette transporter 1 mediates lipid efflux from Sertoli cells and influences male fertility. *J. Lipid Res.* 45:1040–50

106. Komuves LG, Schmuth M, Fowler AJ, Elias PM, Hanley K, et al. 2002. Oxysterol stimulation of epidermal differentiation is mediated by liver X receptor-β in murine epidermis. *J. Invest. Dermatol.* 118:25–34

107. Hanley K, Ng DC, He SS, Lau P, Min K, et al. 2000. Oxysterols induce differentiation in human keratinocytes and increase Ap-1-dependent involucrin transcription. *J. Invest. Dermatol.* 114:545–53

108. Schmuth M, Elias PM, Hanley K, Lau P, Moser A, et al. 2004. The effect of LXR activators on AP-1 proteins in keratinocytes. *J. Invest. Dermatol.* 123:41–48

109. Carleton RA, Dwyer J, Finberg L, Flora J, Goodman DS, et al. 1991. Report of the expert panel on population strategies for blood cholesterol reduction. A statement from the National Cholesterol Education Program, National Heart, Lung, and Blood Institute, National Institutes of Health. *Circulation* 83:2154–232

110. Miller LJ, Chacko R. 2004. The role of cholesterol and statins in Alzheimer's disease. *Ann. Pharmacother.* 38:91–98

111. Russell DW. 2003. The enzymes, regulation, and genetics of bile acid synthesis. *Annu. Rev. Biochem.* 72:137–74

112. Pauli-Magnus C, Stieger B, Meier Y, Kullak-Ublick GA, Meier PJ. 2005. Enterohepatic transport of bile salts and genetics of cholestasis. *J. Hepatol.* 43:342–57

113. Eloranta JJ, Kullak-Ublick GA. 2005. Coordinate transcriptional regulation of bile acid homeostasis and drug metabolism. *Arch. Biochem. Biophys.* 433:397–412

114. Shefer S, Hauser S, Mosbach EH. 1968. 7-α hydroxylation of cholestanol by rat liver microsomes. *J. Lipid Res.* 9:328–33

115. Danielsson H, Einarsson K, Johansson G. 1967. Effect of biliary drainage on individual reactions in the conversion of cholesterol to taurochlic acid. Bile acids and steroids 180. *Eur. J. Biochem.* 2:44–49

116. Goodwin B, Jones SA, Price RR, Watson MA, McKee DD, et al. 2000. A regulatory cascade of the nuclear receptors FXR, SHP-1, and LRH-1 represses bile acid biosynthesis. *Mol. Cell.* 6:517–26

117. Chiang JY. 2004. Regulation of bile acid synthesis: pathways, nuclear receptors, and mechanisms. *J. Hepatol.* 40:539–51

118. Goodwin B, Watson MA, Kim H, Miao J, Kemper JK, Kliewer SA. 2003. Differential regulation of rat and human CYP7A1 by the nuclear oxysterol receptor liver X receptor-α. *Mol. Endocrinol.* 17:386–94

119. Wang L, Lee YK, Bundman D, Han Y, Thevananther S, et al. 2002. Redundant pathways for negative feedback regulation of bile acid production. *Dev. Cell* 2:721–31

120. Xie W, Radominska-Pandya A, Shi Y, Simon CM, Nelson MC, et al. 2001. An essential role for nuclear receptors SXR/PXR in detoxification of cholestatic bile acids. *Proc. Natl. Acad. Sci. USA* 98:3375–80

121. Staudinger JL, Goodwin B, Jones SA, Hawkins-Brown D, MacKenzie KI, et al. 2001. The nuclear receptor PXR is a lithocholic acid sensor that protects against liver toxicity. *Proc. Natl. Acad. Sci. USA* 98:3369–74

122. Kerr TA, Saeki S, Schneider M, Schaefer K, Berdy S, et al. 2002. Loss of nuclear receptor SHP impairs but does not eliminate negative feedback regulation of bile acid synthesis. *Dev. Cell* 2:713–20

123. Gupta S, Stravitz RT, Dent P, Hylemon PB. 2001. Down-regulation of cholesterol 7α-hydroxylase (CYP7A1) gene expression by bile acids in primary rat hepatocytes is mediated by the c-Jun

N-terminal kinase pathway. *J. Biol. Chem* 276:15816–22

124. Holt JA, Luo G, Billin AN, Bisi J, Mc-Neill YY, et al. 2003. Definition of a novel growth factor-dependent signal cascade for the suppression of bile acid biosynthesis. *Genes Dev.* 17:1581–91

125. De Fabiani E, Mitro N, Anzulovich AC, Pinelli A, Galli G, Crestani M. 2001. The negative effects of bile acids and tumor necrosis factor-α on the transcription of cholesterol 7α-hydroxylase gene (CYP7A1) converge to hepatic nuclear factor-4: a novel mechanism of feedback regulation of bile acid synthesis mediated by nuclear receptors. *J. Biol. Chem.* 276:30708–16

126. De Fabiani E, Mitro N, Gilardi F, Caruso D, Galli G, Crestani M. 2003. Coordinated control of cholesterol catabolism to bile acids and of gluconeogenesis via a novel mechanism of transcription regulation linked to the fasted to-fed cycle. *J. Biol. Chem.* 278:39124–32

127. Ananthanarayanan M, Balasubramanian N, Makishima M, Mangelsdorf DJ, Suchy FJ. 2001. Human bile salt export pump promoter is transactivated by the farnesoid X receptor/bile acid receptor. *J. Biol. Chem.* 276:28857–65

128. Plass JR, Mol O, Heegsma J, Geuken M, Faber KN, et al. 2002. Farnesoid X receptor and bile salts are involved in transcriptional regulation of the gene encoding the human bile salt export pump. *Hepatology* 35:589–96

129. Schuetz EG, Strom S, Yasuda K, Lecureur V, Assem M, et al. 2001. Disrupted bile acid homeostasis reveals an unexpected interaction among nuclear hormone receptors, transporters, and cytochrome P450. *J. Biol. Chem.* 276:39411–18

130. Huang L, Zhao A, Lew JL, Zhang T, Hrywna Y, et al. 2003. Farnesoid X receptor activates transcription of the phospholipid pump MDR3. *J. Biol. Chem.* 278:51085–90

131. Strautnieks SS, Bull LN, Knisely AS, Ko-shis SA, Dahl N, et al. 1998. A gene encoding a liver-specific ABC transporter is mutated in progressive familial intrahepatic cholestasis. *Nat. Genet.* 20:233–38

132. de Vree JM, Jacquemin E, Sturm E, Cresteil D, Bosma PJ, et al. 1998. Mutations in the MDR3 gene cause progressive familial intrahepatic cholestasis. *Proc. Natl. Acad. Sci. USA* 95:282–87

133. Moschetta A, Bookout AL, Mangelsdorf DJ. 2004. Prevention of cholesterol gallstone disease by FXR agonists in a mouse model. *Nat. Med.* 10:1352–58

134. Borst P, Elferink RO. 2002. Mammalian ABC transporters in health and disease. *Annu. Rev. Biochem.* 71:537–92

135. Keppler D, Konig J. 2000. Hepatic secretion of conjugated drugs and endogenous substances. *Semin. Liver Dis.* 20:265–72

136. Konig J, Nies AT, Cui Y, Leier I, Keppler D. 1999. Conjugate export pumps of the multidrug resistance protein (MRP) family: localization, substrate specificity, and MRP2-mediated drug resistance. *Biochim. Biophys. Acta* 1461:377–94

137. Kast HR, Goodwin B, Tarr PT, Jones SA, Anisfeld AM, et al. 2002. Regulation of multidrug resistance-associated protein 2 (ABCC2) by the nuclear receptors pregnane X receptor, farnesoid X-activated receptor, and constitutive androstane receptor. *J. Biol. Chem.* 277:2908–15

138. Chignard N, Mergey M, Barbu V, Finzi L, Tiret E, et al. 2005. VPAC1 expression is regulated by FXR agonists in the human gallbladder epithelium. *Hepatology* 42:549–57

139. Dawson PA, Haywood J, Craddock AL, Wilson M, Tietjen M, et al. 2003. Targeted deletion of the ileal bile acid transporter eliminates enterohepatic cycling of bile acids in mice. *J. Biol. Chem.* 278:33920–27

140. Oelkers P, Kirby LC, Heubi JE, Dawson PA. 1997. Primary bile acid malabsorption caused by mutations in the ileal sodium-dependent bile acid transporter

gene (SLC10A2). *J. Clin. Invest.* 99: 1880–87

141. Arrese M, Trauner M, Sacchiero RJ, Crossman MW, Shneider BL. 1998. Neither intestinal sequestration of bile acids nor common bile duct ligation modulate the expression and function of the rat ileal bile acid transporter. *Hepatology* 28:1081–87

142. Chen F, Ma L, Dawson PA, Sinal CJ, Sehayek E, et al. 2003. Liver receptor homologue-1 mediates species- and cell line-specific bile acid-dependent negative feedback regulation of the apical sodium-dependent bile acid transporter. *J. Biol. Chem.* 278:19909–16

143. Neimark E, Chen F, Li X, Shneider BL. 2004. Bile acid-induced negative feedback regulation of the human ileal bile acid transporter. *Hepatology* 40:149–56

144. Grober J, Zaghini I, Fujii H, Jones SA, Kliewer SA, et al. 1999. Identification of a bile acid-responsive element in the human ileal bile acid-binding protein gene. Involvement of the farnesoid X receptor/9-cis-retinoic acid receptor heterodimer. *J. Biol. Chem.* 274:29749–54

145. Kok T, Hulzebos CV, Wolters H, Havinga R, Agellon LB, et al. 2003. Enterohepatic circulation of bile salts in farnesoid X receptor-deficient mice: efficient intestinal bile salt absorption in the absence of ileal bile acid-binding protein. *J. Biol. Chem.* 278:41930–37

146. Denson LA, Sturm E, Echevarria W, Zimmerman TL, Makishima M, et al. 2001. The orphan nuclear receptor, shp, mediates bile acid-induced inhibition of the rat bile acid transporter, ntcp. *Gastroenterology* 121:140–47

147. Jung D, Hagenbuch B, Fried M, Meier PJ, Kullak-Ublick GA. 2004. Role of liver-enriched transcription factors and nuclear receptors in regulating the human, mouse, and rat NTCP gene. *Am. J. Physiol. Gastrointest. Liver Physiol.* 286:G752–61

148. Li D, Zimmerman TL, Thevananther S, Lee HY, Kurie JM, Karpen SJ. 2002.

Interleukin-1 β-mediated suppression of RXR:RAR transactivation of the Ntcp promoter is JNK-dependent. *J. Biol. Chem.* 277:31416–22

149. Jung D, Hagenbuch B, Gresh L, Pontoglio M, Meier PJ, Kullak-Ublick GA. 2001. Characterization of the human OATP-C (SLC21A6) gene promoter and regulation of liver-specific OATP genes by hepatocyte nuclear factor 1 α. *J. Biol. Chem.* 276:37206–14

150. Jung D, Kullak-Ublick GA. 2003. Hepatocyte nuclear factor 1 α: a key mediator of the effect of bile acids on gene expression. *Hepatology* 37:622–31

151. Jung D, Podvinec M, Meyer UA, Mangelsdorf DJ, Fried M, et al. 2002. Human organic anion transporting polypeptide 8 promoter is transactivated by the farnesoid X receptor/bile acid receptor. *Gastroenterology* 122:1954–66

152. Pircher PC, Kitto JL, Petrowski ML, Tangirala RK, Bischoff ED, et al. 2003. Farnesoid X receptor regulates bile acid-amino acid conjugation. *J. Biol. Chem.* 278:27703–11

153. Song CS, Echchgadda I, Baek BS, Ahn SC, Oh T, et al. 2001. Dehydroepiandrosterone sulfotransferase gene induction by bile acid activated farnesoid X receptor. *J. Biol. Chem.* 276:42549–56

154. Barbier O, Torra IP, Sirvent A, Claudel T, Blanquart C, et al. 2003. FXR induces the UGT2B4 enzyme in hepatocytes: a potential mechanism of negative feedback control of FXR activity. *Gastroenterology* 124:1926–40

155. Gnerre C, Blattler S, Kaufmann MR, Looser R, Meyer UA. 2004. Regulation of CYP3A4 by the bile acid receptor FXR: evidence for functional binding sites in the CYP3A4 gene. *Pharmacogenetics* 14:635–45

156. Miyata M, Tozawa A, Otsuka H, Nakamura T, Nagata K, et al. 2005. Role of farnesoid X receptor in the enhancement of canalicular bile acid output and excretion of unconjugated bile acids: a

mechanism for protection against cholic acid-induced liver toxicity. *J. Pharmacol. Exp. Ther.* 312:759–66

157. Liu Y, Binz J, Numerick MJ, Dennis S, Luo G, et al. 2003. Hepatoprotection by the farnesoid X receptor agonist GW4064 in rat models of intra- and extrahepatic cholestasis. *J. Clin. Invest.* 112:1678–87

158. Sinal CJ, Tohkin M, Miyata M, Ward JM, Lambert G, Gonzalez FJ. 2000. Targeted disruption of the nuclear receptor FXR/BAR impairs bile acid and lipid homeostasis. *Cell* 102:731–44

159. Lambert G, Amar MJ, Guo G, Brewer HB Jr, Gonzalez FJ, Sinal CJ. 2003. The farnesoid X-receptor is an essential regulator of cholesterol homeostasis. *J. Biol. Chem.* 278:2563–70

160. Angelin B, Einarsson K, Hellstrom K, Leijd B. 1978. Effects of cholestyramine and chenodeoxycholic acid on the metabolism of endogenous triglyceride in hyperlipoproteinemia. *J. Lipid. Res.* 19:1017–24

161. Bateson MC, Maclean D, Evans JR, Bouchier IA. 1978. Chenodeoxycholic acid therapy for hypertriglyceridaemia in men. *Br. J. Clin. Pharmacol.* 5:249–54

162. Carulli N, Ponz de Leon M, Podda M, Zuin M, Strata A, et al. 1981. Chenodeoxycholic acid and ursodeoxycholic acid effects in endogenous hypertriglyceridemias. A controlled double-blind trial. *J. Clin. Pharmacol.* 21:436–42

163. Crouse JR 3rd. 1987. Hypertriglyceridemia: a contraindication to the use of bile acid binding resins. *Am. J. Med.* 83:243–48

164. Beil U, Crouse JR, Einarsson K, Grundy SM. 1982. Effects of interruption of the enterohepatic circulation of bile acids on the transport of very low density-lipoprotein triglycerides. *Metabolism* 31: 438–44

165. Bard JM, Parra HJ, Douste-Blazy P, Fruchart JC. 1990. Effect of pravastatin, an HMG CoA reductase inhibitor, and cholestyramine, a bile acid sequestrant,

on lipoprotein particles defined by their apolipoprotein composition. *Metabolism* 39:269–73

166. Leiss O, von Bergmann K. 1982. Different effects of chenodeoxycholic acid and ursodeoxycholic acid on serum lipoprotein concentrations in patients with radiolucent gallstones. *Scand. J. Gastroenterol.* 17:587–92

167. Kast HR, Nguyen CM, Sinal CJ, Jones SA, Laffitte BA, et al. 2001. Farnesoid X-activated receptor induces apolipoprotein C-II transcription: a molecular mechanism linking plasma triglyceride levels to bile acids. *Mol. Endocrinol.* 15:1720–28

168. Claudel T, Inoue Y, Barbier O, Duran-Sandoval D, Kosykh V, et al. 2003. Farnesoid X receptor agonists suppress hepatic apolipoprotein CIII expression. *Gastroenterology* 125:544–55

169. Stocks J, Holdsworth G, Galton D. 1979. Hypertriglyceridaemia associated with an abnormal triglyceride-rich lipoprotein carrying excess apolipoprotein C-III-2. *Lancet* 2:667–71

170. Ginsberg HN, Le NA, Goldberg IJ, Gibson JC, Rubinstein A, et al. 1986. Apolipoprotein B metabolism in subjects with deficiency of apolipoproteins CIII and AI. Evidence that apolipoprotein CIII inhibits catabolism of triglyceride-rich lipoproteins by lipoprotein lipase in vivo. *J. Clin. Invest.* 78:1287–95

171. Jong MC, Rensen PC, Dahlmans VE, van der Boom H, van Berkel TJ, Havekes LM. 2001. Apolipoprotein C-III deficiency accelerates triglyceride hydrolysis by lipoprotein lipase in wild-type and apoE knockout mice. *J. Lipid Res.* 42:1578–85

172. Mak PA, Kast-Woelbern HR, Anisfeld AM, Edwards PA. 2002. Identification of PLTP as an LXR target gene and apoE as an FXR target gene reveals overlapping targets for the two nuclear receptors. *J. Lipid Res.* 43:2037–41

173. Urizar NL, Dowhan DH, Moore DD. 2000. The farnesoid X-activated receptor

mediates bile acid activation of phospholipid transfer protein gene expression. *J. Biol. Chem.* 275:39313–17

174. Sirvent A, Verhoeven AJ, Jansen H, Kosykh V, Darteil RJ, et al. 2004. Farnesoid X receptor represses hepatic lipase gene expression. *J. Lipid Res.* 45:2110–15

175. Claudel T, Sturm E, Duez H, Torra IP, Sirvent A, et al. 2002. Bile acid-activated nuclear receptor FXR suppresses apolipoprotein A-I transcription via a negative FXR response element. *J. Clin. Invest.* 109:961–71

176. Srivastava RA, Srivastava N, Averna M. 2000. Dietary cholic acid lowers plasma levels of mouse and human apolipoprotein A-I primarily via a transcriptional mechanism. *Eur. J. Biochem.* 267:4272–80

177. Albers JJ, Grundy SM, Cleary PA, Small DM, Lachin JM, Schoenfield LJ. 1982. National Cooperative Gallstone Study: the effect of chenodeoxycholic acid on lipoproteins and apolipoproteins. *Gastroenterology* 82:638–46

178. Zhao A, Lew JL, Huang L, Yu J, Zhang T, et al. 2003. Human kininogen gene is transactivated by the farnesoid X receptor. *J. Biol. Chem.* 278:28765–70

179. Watanabe M, Houten SM, Wang L, Moschetta A, Mangelsdorf DJ, et al. 2004. Bile acids lower triglyceride levels via a pathway involving FXR, SHP, and SREBP-1c. *J. Clin. Invest.* 113:1408–18

180. Hirokane H, Nakahara M, Tachibana S, Shimizu M, Sato R. 2004. Bile acid reduces the secretion of very low density lipoprotein by repressing microsomal triglyceride transfer protein gene expression mediated by hepatocyte nuclear factor-4. *J. Biol. Chem.* 279:45685–92

181. Hussain MM, Shi J, Dreizen P. 2003. Microsomal triglyceride transfer protein and its role in apoB-lipoprotein assembly. *J. Lipid Res.* 44:22–32

182. Sirvent A, Claudel T, Martin G, Brozek J, Kosykh V, et al. 2004. The farnesoid X receptor induces very low density lipoprotein receptor gene expression. *FEBS Lett.* 566:173–77

183. Tacken PJ, Teusink B, Jong MC, Harats D, Havekes LM, et al. 2000. LDL receptor deficiency unmasks altered VLDL triglyceride metabolism in VLDL receptor transgenic and knockout mice. *J. Lipid Res.* 41:2055–62

184. Li J, Pircher PC, Schulman IG, Westin SK. 2005. Regulation of complement C3 expression by the bile acid receptor FXR. *J. Biol. Chem.* 280:7427–34

185. Savkur RS, Bramlett KS, Michael LF, Burris TP. 2005. Regulation of pyruvate dehydrogenase kinase expression by the farnesoid X receptor. *Biochem. Biophys. Res. Commun.* 329:391–96

186. Evans RM, Barish GD, Wang YX. 2004. PPARs and the complex journey to obesity. *Nat. Med.* 10:355–61

187. Pineda Torra I, Claudel T, Duval C, Kosykh V, Fruchart JC, Staels B. 2003. Bile acids induce the expression of the human peroxisome proliferator-activated receptor α gene via activation of the farnesoid X receptor. *Mol. Endocrinol.* 17:259–72

188. Sinal CJ, Yoon M, Gonzalez FJ. 2001. Antagonism of the actions of peroxisome proliferator-activated receptor-α by bile acids. *J. Biol. Chem.* 276:47154–62

189. Yamagata K, Daitoku H, Shimamoto Y, Matsuzaki H, Hirota K, et al. 2004. Bile acids regulate gluconeogenic gene expression via small heterodimer partner-mediated repression of hepatocyte nuclear factor 4 and Foxo1. *J. Biol. Chem.* 279:23158–65

190. Borgius LJ, Steffensen KR, Gustafsson JA, Treuter E. 2002. Glucocorticoid signaling is perturbed by the atypical orphan receptor and corepressor SHP. *J. Biol. Chem.* 277:49761–66

191. Stayrook KR, Bramlett KS, Savkur RS, Ficorilli J, Cook T, et al. 2005. Regulation of carbohydrate metabolism by the farnesoid X receptor. *Endocrinology* 146:984–91

192. Duran-Sandoval D, Cariou B, Percevault F, Hennuyer N, Grefhorst A, et al. 2005. The farnesoid X receptor modulates hepatic carbohydrate metabolism during fasting/refeeding transition. *J. Biol. Chem.* 280:29971–79

193. Duran-Sandoval D, Mautino G, Martin G, Percevault F, Barbier O, et al. 2004. Glucose regulates the expression of the farnesoid X receptor in liver. *Diabetes* 53:890–98

194. Robinson-Rechavi M, Escriva Garcia H, Laudet V. 2003. The nuclear receptor superfamily. *J. Cell. Sci.* 116:585–86

195. Nucl. Recept. Nomencl. Comm. 1999. A unified nomenclature system for the nuclear receptor superfamily. *Cell* 97:161–63

196. Landrier JF, Grober J, Demydchuk J, Besnard P. 2003. FXRE can function as an LXRE in the promoter of human ileal bile acid-binding protein (I-BABP) gene. *FEBS Lett.* 553:299–303

Annu. Rev. Physiol. 2006. 68:193–221
doi: 10.1146/annurev.physiol.68.040104.105418
First published online as a Review in Advance on October 24, 2005

DESIGN AND FUNCTION OF SUPERFAST MUSCLES:
New Insights into the Physiology of Skeletal Muscle

Lawrence C. Rome

*Department of Biology, University of Pennsylvania, Philadelphia, Pennsylvania 19104;
The Marine Biological Laboratory, Woods Hole, Massachusetts 02543;
email: lrome@sas.upenn.edu*

Key Words   parvalbumin, Ca^{2+} release, Ca^{2+} uptake, cross-bridges, adaptation, sound production

■ **Abstract**   Superfast muscles of vertebrates power sound production. The fastest, the swimbladder muscle of toadfish, generates mechanical power at frequencies in excess of 200 Hz. To operate at these frequencies, the speed of relaxation has had to increase approximately 50-fold. This increase is accomplished by modifications of three kinetic traits: (*a*) a fast calcium transient due to extremely high concentration of sarcoplasmic reticulum (SR)-Ca^{2+} pumps and parvalbumin, (*b*) fast off-rate of Ca^{2+} from troponin C due to an alteration in troponin, and (*c*) fast cross-bridge detachment rate constant (*g*, 50 times faster than that in rabbit fast-twitch muscle) due to an alteration in myosin. Although these three modifications permit swimbladder muscle to generate mechanical work at high frequencies (where locomotor muscles cannot), it comes with a cost: The high *g* causes a large reduction in attached force-generating cross-bridges, making the swimbladder incapable of powering low-frequency locomotory movements. Hence the locomotory and sound-producing muscles have mutually exclusive designs.

INTRODUCTION: THE FUNCTION
OF SUPERFAST MUSCLES

Our perception of speed is greatly influenced by movement we can see. Most animal movement, such as the running of a squirrel or mouse, is macroscopic. These movements seem fast until one watches the wings of a hovering humming bird. It is a natural assumption that these fast movements are powered by the fastest vertebrate muscles, but in fact this is not the case. The fastest muscles are used to produce sounds, and the movements occur inside the body, and usually, but not always (e.g., rattlesnake), no external movement can be detected. These superfast muscles produce sound from approximately 90 Hz to approximately 250 Hz.

There has been considerable research through the years concerning the behavior of sound production, the neural control of muscle function, and the morphology of the muscle (1–3). This review focuses exclusively on more recent work, begun less than a decade ago, on the physiology and mechanics of superfast muscles (4). The recent progress in understanding how these muscles function has been based on integrative and comparative approaches. Below, we discuss these general approaches before discussing the particulars of superfast muscle.

THE IMPORTANCE OF INTEGRATIVE APPROACHES TO THE STUDY OF MUSCLE FUNCTION

Vertebrates perform a tremendous variety of motor tasks, ranging from making explosive movements taking less than ~50 ms, to swimming thousands of miles in the ocean, to producing sound at several hundred Hertz for communication. These activities require different outputs from the muscles (5). Accordingly, research over the past several decades has revealed that muscle tissue has tremendous diversity, perhaps more than any other tissue.

Although it has been clear for many years that contractile properties of the muscles themselves are key in understanding muscular system design, over the past 15 years it has become equally clear that one cannot study the contractile properties in isolation. Rather, the whole muscular system must be examined because one cannot appreciate the contractile properties of the fibers without examining the other components (e.g., joints, muscle moments, masses, kinematics of movement) to which they are matched. Hence the field has moved toward comprehensive integrative studies (6–9). This movement represents a turning point in the field of integrative muscle physiology and biomechanics.

This current revolution in integrative muscle physiology stems from both intellectual and technological improvements in the fields of biophysics, quantitative physiology, quantitative biochemistry, biomechanics, and musculoskeletal modeling. However, these advances would not have been possible without the recognition of the need for, and the development of, exceptional experimental animal models. Although today, few would consider using the mechanics and energetics of frog muscle (10) to try to understand the mechanics and energetics of running lizards (11), such practice was considered at the cutting edge of experimentation as recently as 20–25 years ago (reviewed in 12). Hence, by providing exceptional experimental models in which all pertinent measurements could be made in a single species, comparative physiology has played a major role in understanding muscle function from the molecular to the whole animal level.

COMPARATIVE PHYSIOLOGY: PURE AND EXTREME MODELS PROVIDE A LINCHPIN OF MODERN BIOLOGY

The importance of comparative physiology as a central approach underlying much of the modern study of biology is not fully appreciated. Perhaps nowhere is this better illustrated than the study of superfast muscle, in which comparative physiology has been the enabler of a large number of approaches from other fields. In this case of superfast muscles, comparative physiology provides an opportunity to utilize a pure tissue with extreme modifications. As discussed below, these combine with great effect.

The approaches used in studying superfast muscles represent the confluence of many fields, including cellular and molecular physiology, biophysics, integrative physiology, quantitative physiology, quantitative biochemistry, quantitative ultrastructure, modeling, and neuroethology. The linchpin to this interdisciplinary approach, however, is the experimental models provided by comparative physiology—in this case, fish muscles that express a single, pure fiber type. The purity of the tissue permits the dissection of bundles of fibers literally of any size, which, in turn, permits us to perform diverse studies ranging from single- (or split-) fiber biophysics and mechanics (using micrograms of tissue) to intact muscle energetics (with milligrams of tissue) to biochemical purification and quantification of muscle proteins (grams to kilograms of tissue).

Aside from providing information about the workings of superfast muscles, studying these extremes also provides insight into the way in which normal muscle works. For instance, different muscle types from our experimental model, the toadfish, vary 50-fold in speed and include the fastest and among the slowest in the whole vertebrate subphylum. Understanding how each of these extremes functions can give an important perspective absent when examining relatively small differences in function (e.g., between IIx and IIa fibers in mammals). As we also see below, looking at extremes can permit us to see functional differences clearly, whereas the smaller differences in more normal muscles can be easily lost in the noise. Finally, extreme designs may require very high concentrations of some compounds [e.g., parvalbumin (PARV)], which permits the development of new types of techniques that can elucidate new mechanisms.

TOADFISH (*OPSANUS TAU*) SWIMBLADDER MUSCLE: THE KEY MODEL

In general, fish provide excellent models for studying muscular system design because their different fiber types are anatomically separated; hence (a) different-sized bundles of fibers pure in fiber type can be isolated for various experiments and (b) the usage of different fibers during normal activity can be monitored easily by electromyography. Toadfish is exceptional because its slow-twitch red muscle, used for swimming at <1 Hz, is among the slowest twitch fibers known (twitch half-width ≈ 500 ms at $15°C$), whereas the superfast toadfish swimbladder muscle used for sound production at 100 Hz (at $15°C$; >200 Hz at $25°C$) is the fastest vertebrate muscle ever measured (twitch half-width ≈ 10 ms). Therefore, in this one species, it is possible to examine the function of three muscle fiber types (including the intermediate-speed fast-twitch white muscle) whose twitch speed and speed of operation in vivo vary ~ 50-fold, spanning nearly the full range found in vertebrates.

THE ACTIVATION-RELAXATION CYCLE:
WHAT IS NEEDED FOR SPEED?

Molecular Basis of Activation and Relaxation in Toadfish

The importance of muscle activation and relaxation in muscle design has been appreciated for 20 years (13). Workloop experiments on locomotory and sound-producing muscles (7, 9, 14–16) have demonstrated that the kinetics of activation and relaxation are critical for in vivo performance. Until approximately 10 years ago, research dealt with the kinetics of activation and relaxation phenomenologically. This approach was necessary because, although the basic mechanisms of activation and relaxation were known for some time, a quantitative understanding of which processes set twitch speed or how Ca^{2+} cycles during normal contraction had been lacking. The kinetics of activation and relaxation are complex and require the integration of many components for which values remain largely unknown. Activation is usually thought to consist of Ca^{2+} release from the sarcoplasmic reticulum (SR) (Figure 1, step 1) and Ca^{2+} binding to troponin C (TNC) (Figure 1, step 2). Measurements and models show that this is rapid (i.e., 10 ms) (17). However, for purposes of understanding muscle function during locomotion or other motor behaviors, maximal activation is not achieved until cross-bridges attach and generate force (or undergo transitions between states) (Figure 1, step 3). This takes considerably longer and lags well behind formation of the Ca^{2+}-TNC complex. Relaxation is even more complex. During and following Ca^{2+} release, Ca^{2+} is sequestered by the SR (Figure 1, step 4) and also becomes bound to PARV, which is in relatively high concentration in frog muscle and fish white muscle. Myoplasmic $[Ca^{2+}]$ drops, causing Ca^{2+} to dissociate from TNC (Figure 1, step 5), cross-bridges to detach (Figure 1, step 6), and force to drop to zero.

Although the phenomenological approach was a necessity, it was limiting because it neither explained what molecular properties must be changed to alter the kinetics of the muscle to appropriate values nor how different components are related. Hence, we applied a molecular approach to gain a better understanding of these issues. The rate of relaxation varies tremendously among different types of muscle (reviewed in Reference 18). Despite the many experiments that examined the individual components involved in muscle relaxation, little was known about how relaxation speed was set. One reason for this was that measurements of the different components were made on different species. Furthermore, sufficient information to develop an integrative understanding of relaxation had been collected for only one species (frog) (17, 19–22); thus, there was no other muscle type to which this could be compared.

Hence, the main goal was to determine the principal factors that set the kinetics of twitches in three fiber types of toadfish whose twitch duration varied 50-fold. Because twitch duration is set primarily by relaxation, we measured the kinetics of the three systems involved in relaxation (Figure 1): (*a*) Ca^{2+} reuptake, (*b*) Ca^{2+} off-rate (k_{off}) from troponin (indicated as a shift in force-pCa relationship),

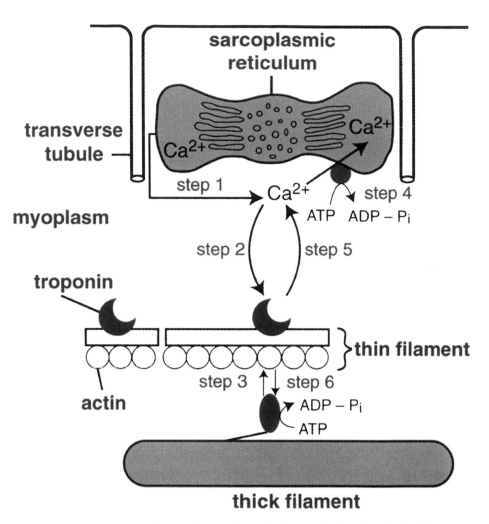

Figure 1   The major kinetic steps in muscle activation and relaxation. Activation occurs in steps 1–3, and relaxation in steps 4–6. (*Step 1*) Ca^{2+} is released from the sarcoplasmic reticulum into the myoplasm. (*Step 2*) Ca^{2+} binds to troponin, releasing inhibition of the thin filament. (*Step 3*) Cross-bridges attach and generate force. (*Step 4*) Ca^{2+} is resequestered from the myoplasm by the Ca^{2+} pumps. (*Step 5*) Ca^{2+} comes off troponin, thereby preventing subsequent cross-bridge attachment. (*Step 6*) Cross-bridges detach. For a muscle to relax rapidly, steps 4–6 all must occur rapidly.

TABLE 1   Cross-bridge kinetic constants, force, and concentration of constituent protein in toadfish muscle[*]

Fiber	$f(s^{-1})$	$g(s^{-1})$	$f/(f+g)$	Force[a] (kN/m²)	MHC (μM)[b]	TNC (μM)	PARV (mM)[c]
Red	6.6	2.8	70	244	117	—	
White	16.2	10.3	61	192	167	—	
Swim	12.7	108	10.5	24	69	35	1.5

[*]All data are from Reference 23 except for the TNC and PARV concentrations (B. Tikunov, personal communication). MHC, myosin heavy chain; TNC, troponin C; PARV, parvalbumin.

[a]Force in intact fibers. Note that swimbladder has only about 50% myofibrils and hence a lower myosin heavy chain (MHC) concentration.

[b]This is an abbreviation for μmol kg^{-1} of muscle.

[c]This is an abbreviation for mmol kg^{-1} of muscle.

and (c) cross-bridge detachment. To a first approximation, these processes occur in series. Thus, as twitch speed increased in different fiber types, so did the rate of each one of the processes (Figures 2–5; Table 1) (4, 23).

Figure 2 shows that there was a systematic acceleration of the Ca^{2+} transient as the twitch speed increased. The Ca^{2+} transient in the sonic muscles is the fastest ever measured for any fiber type (a half-width of ~3.4 ms at 16°C and 1.5 ms at 25°C) (4). The importance of the Ca^{2+} transient duration in setting the twitch duration is evident in Figure 2, which shows that between the slow-twitch red fibers and the superfast-twitch swimbladder fibers, the half-widths of the Ca^{2+} transient and the twitch sped up in parallel (by ~50-fold).

The significance of a fast Ca^{2+} transient is most apparent during repetitive stimulation. During stimulation of slow red muscle at a modest 3.5 Hz (Figure 3), the time course of Ca^{+2} uptake is so slow that $[Ca^{2+}]$ does not have time to return to baseline between stimuli. Even the lowest myoplasmic $[Ca^{2+}]$ between stimuli was above the threshold required for force generation in this fiber type, thus resulting in a partially fused tetanus. By contrast, the swimbladder muscle's Ca^{2+} transient is so rapid that even with 67 Hz stimulation, the return of $[Ca^{2+}]$ to baseline between stimuli is complete (Figure 3, *right*; note the 50-times faster time base). In addition, $[Ca^{2+}]$ is below the threshold for force generation for more than half of the time in all but the first stimulus. Hence, the Ca^{2+} transient is sufficiently rapid to permit the oscillation in force required for sound production.

Even though $[Ca^{2+}]$ returned rapidly to baseline, the swimbladder fiber could not relax quickly unless its troponin rapidly released bound Ca^{2+} (Figure 1, step 5). Indeed, kinetic modeling (figure 3 in Reference 4) indicates that if the swimbladder troponin had the k_{off} for Ca^{2+} estimated for fast-twitch fibers of frog ($115 \ s^{-1}$), then Ca^{2+} would not unbind from troponin with sufficient rapidity to permit the observed rapid fall in force. Figure 4 shows a systematic right shift in the force-pCa relationship with increasing twitch speed. The threefold right shift of the force versus $[Ca^{2+}]$ relationship of swimbladder fibers with respect to frog

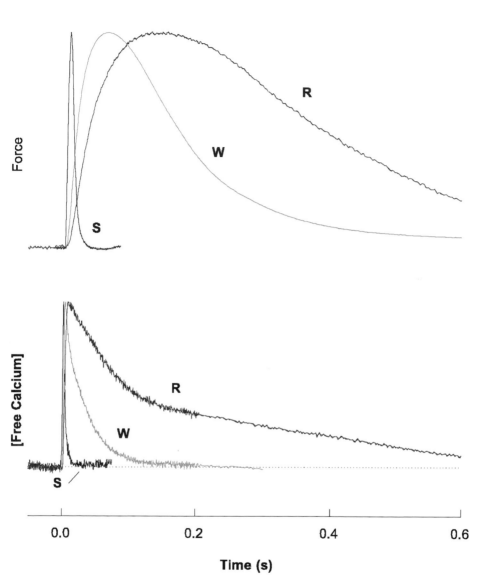

Figure 2   Twitch tension (*top*) and calcium transients (*bottom*) of three fiber types from toadfish at 15°C. In each case, the force and the calcium records have been normalized to their maximum value. The twitch and calcium transient become briefer going from the slow-twitch red fiber (R, *red*), to the fast-twitch white fiber (W, *orange*), to the superfast-twitch swimbladder fiber (S, *blue*). Based on Reference 4.

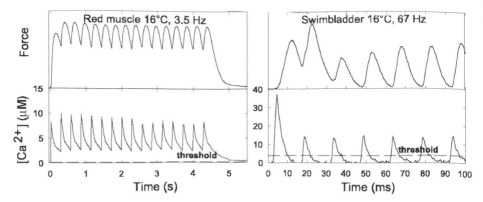

Figure 3   Force production (*top*) and calcium transients (*bottom*) during repetitive stimulation. (*Left*) Slow-twitch red fiber (*red*) stimulated at 3.5 Hz. The threshold [Ca^{2+}] for force generation was derived by force-pCa experiment on skinned fibers (from Figure 4) and is shown with a light blue dashed line. (*Right*) Swimbladder fiber (*blue*) stimulated at 67 Hz. The [Ca^{2+}] threshold for force production is much higher for the swimbladder than for the red fiber. Note the different time and [Ca^{2+}] scales. Based on Reference 4.

fibers was interpreted as a threefold higher k_{off} in swimbladder TNC (345 s^{-1}) as the on-rate (k_{on}) is fixed by diffusion. With this higher k_{off}, the rate of troponin deactivation is no longer limiting (Figure 3 in Reference 4).

The final requirement for force to drop quickly following the dissociation of Ca^{2+} from troponin is a fast cross-bridge detachment rate (Figure 1, step 6). The maximum velocity of shortening (V_{max}), thought to be determined by the cross-bridge detachment rate, showed a systematic increase in different fiber type with increasing twitch speed: The V_{max} of the swimbladder muscle (~12 ML s^{-1}) is exceptionally fast (5- and 2.5-fold faster than toadfish red and white muscle, respectively) (Figure 5). Even more striking is the systematic 50-fold increase in the cross-bridge detachment rate constant, g (Table 1) (23). The cross-bridge attachment rate constant (f) and g were calculated from measurements of steady-state relative stiffness and cross-bridge ATP utilization per myosin head (isometric ATP utilization/myosin heavy chain concentration). We found that f increased 2–3-fold between the red and white fibers and then remained fairly constant between the fast-twitch and superfast-twitch muscles (approximately 15 s^{-1}). Similarly, g increased approximately 3-fold between the slow-twitch red and the fast-twitch white muscles. However, between the fast-twitch white muscle and the superfast-twitch swimbladder, g increased another 10-fold to an extraordinary 108 s^{-1}, by far the fastest ever measured. By comparison, the best-studied mammalian fiber type, the fast-twitch psoas fiber of rabbit, has a g of only 2 s^{-1} (24). This fast g appears necessary to enable the rapid relaxation observed (see below) (25).

In short, we found that at 15°C the swimbladder muscle had (*a*) the fastest Ca^{2+} transient ever measured, (*b*) the most right-shifted force versus [Ca^{2+}] curve

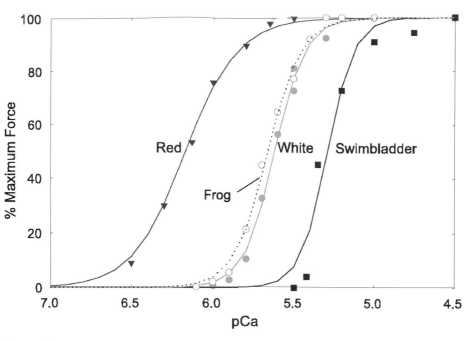

Figure 4   Force-pCa relationship for three muscle fiber types in toadfish (red fiber, *red*; white fiber, *orange*; swimbladder fiber, *blue*). A frog (*Rana temporaria*) fiber (*dotted black line*) is shown for comparison. The force from swimbladder fibers rose more sharply than the fitted curve at forces below 50% and more gradually than the fitted curve at forces above 80%. Based on Reference 4.

(interpreted as the fastest k_{off}), and (*c*) the fastest detachment rate constant ever measured. This suggests that all three rate processes are important in setting the kinetics of this superfast fiber.

Further analysis of the Ca^{2+} transient and cross-bridge kinetics suggests that the process that ultimately limits relaxation rate may differ between fiber types. Figure 6 compares the detachment rate constant *g* with the relaxation rate of intact muscle as twitch force falls from 95% to 80% of maximum tension [i.e., within this range, the fiber is thought to remain isometric, and the decline in force is not sped up artifactually by the collapsing sarcomere structure (21)]. There is a remarkable concordance between *g* and the relaxation rate for the swimbladder muscle ($108 \ s^{-1}$ versus $104 \ s^{-1}$) and for white muscle ($10.3 \ s^{-1}$ versus $10.6 \ s^{-1}$). Furthermore, at the point of the twitch at which force had fallen below 95%, the free $[Ca^{2+}]$ had already dropped below the threshold level for force generation and thus could not be limiting relaxation. Hence, for white and swimbladder muscles, between which speed varies ~10-fold, we conclude it is likely that *g* limits the speed of muscle relaxation. (However, as k_{off} of TNC has not been measured, we cannot discount the possibility that it may have contributed as well.) By contrast, red

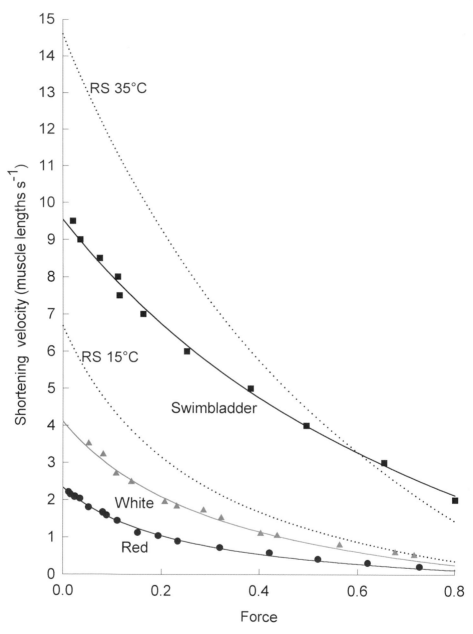

Figure 5 Force-velocity curves of toadfish and rattlesnake shaker fibers. The force-velocity curves of the toadfish red (*red*) and white (*orange*) and swimbladder fibers (*blue*) are shown with solid symbols and lines. The swimbladder V_{max} (12 muscle lengths s^{-1}) was determined from the slack test, and its force-velocity curve is shown only for comparison. The force-velocity curves of the rattlesnake shaker fibers (*RS*) are shown at 16°C and 35°C with dotted black curves. Based on Reference 4.

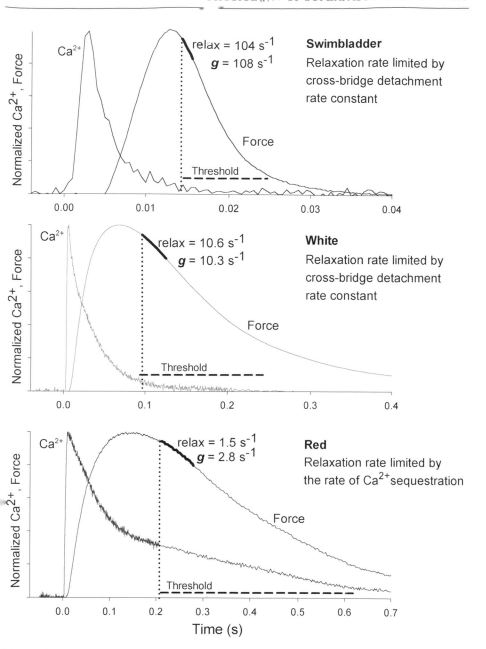

Figure 6   Comparison of muscle relaxation and the detachment rate constant g. For each muscle fiber (swimbladder, *top*; white, *middle*; red, *bottom*), the normalized Ca^{2+} transient and force are shown. Relaxation rate was calculated over the time when force was reduced from 95 to 80% of isometric force (*thick black trace*). The threshold Ca^{2+} for force generation determined from force-pCa (i.e., Figure 4) is denoted with a dashed horizontal line. Based on Reference 23.

muscle yields a different result. The value of g (2.8 s^{-1}) was substantially faster than the initial relaxation rate (1.5 s^{-1}). Furthermore, the myoplasmic [Ca^{2+}], although falling, remained above threshold during much of the decline in force (Figure 6). This suggests that, in red muscle, Ca^{2+} reuptake may be the rate-limiting step in relaxation.

It is important to note that although only one step appears to be limiting in each fiber type, all three adaptations are needed for high speed. For instance, a muscle with a g of the swimbladder muscle and Ca^{2+} transient of white muscle would relax slowly (now Ca^{2+} reuptake would become rate limiting).

Trading Force for Speed

Measurements of cross-bridge kinetics have elucidated another remarkable specialization. In locomotory muscle, $f \gg g$, and thus the number of attached cross-bridges [proportional to $f/(f + g)$] is high. In the red and white muscles, this calculation suggests that 60–70% of the cross-bridges are attached during isometric contraction. In the swimbladder muscle, however, because such a high g is required for rapid relaxation, $g \gg f$, and thus there should have been few (approximately 1/5 that of locomotory muscle) attached cross-bridges. This observation suggests that these muscles would generate very low forces, which they do. Swimbladder muscle generates only approximately 1/10 of the force of normal locomotory muscle (Table 1). Hence in superfast fibers there is a tradeoff of force for speed! This may represent a new design rule with significant implications for other very fast muscles (e.g., mammalian eye muscles). Although most of the reduction in force is due to cross-bridge kinetics, some is attributable to the fact that myofibrils make up only ~50% of the cell volume (most of the remainder is taken up by a large SR) (18).

The Downside of Speed: Mutually Exclusive Designs

It is generally thought that, although fast-twitch muscle can generate more power than can slow-twitch muscle at any given velocity (based on generating more force at any given velocity), animals must also have slow-twitch muscles because slow-twitch muscles can generate power more efficiently at low shortening velocities (6). The low force generated by the swimbladder muscle led us to reevaluate this concept (14). We performed optimized workloop experiments in which the muscle was driven under sinusoidal length changes of various magnitudes and frequencies while the duration and phase of the stimulus were adjusted to obtain maximum power output (13). Figure 7 shows that red and white muscles cannot generate power above 2 and 12 Hz, respectively. This result is expected on the basis of the slow cross-bridge and Ca^{2+} kinetics of these muscle types. However, we also found that the swimbladder muscle could not generate the power at the low frequencies required to drive swimming. Because of low forces, the mass-specific power generated by the swimbladder muscle at the low frequencies toadfish use for swimming (0.8–4 Hz) is a small fraction of that generated by the swimming muscles (i.e., 1/

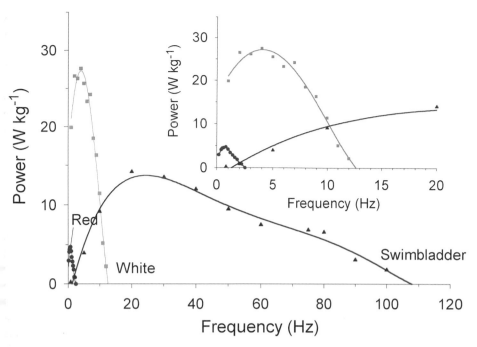

Figure 7   A plot of red muscle (*red*), white muscle (*orange*), and swimbladder muscle (*blue*) mean power output (W kg^{-1}) versus cycle frequency (Hz). Values were determined from optimized workloop experiments. Based on Reference 14.

and 1/20 of that generated by white and red muscles, respectively). To generate the same total power required for swimming, the volume of the swimbladder muscle would have to be 6-fold and 20-fold greater than that of white and red muscles, respectively, and simply could not fit into the fish! Hence, the different fiber types in toadfish have mutually exclusive designs and the locomotory red and white fibers can operate only at low frequencies, whereas the swimbladder muscle can operate effectively only at high frequencies. Thus the modifications made for speed in the swimbladder muscle are so extreme that it can no longer perform the function of powering swimming but is highly specialized for working at high frequencies.

Quantitative Energetics of Swimbladder Muscle: Superfast Contraction Without Superfast Energetics

There is a well-accepted relationship between steady-state ATP usage and the twitch speed of muscle (26). As the twitch speed of the muscle increases, so must the kinetics of the ATP-utilizing cross-bridges and SR-Ca^{2+} pumps, resulting in an increase in the steady-state rate of ATP utilization. We hypothesized that, because of the great twitch speed of swimbladder muscle, it should use ATP extraordinarily quickly. We measured the energetics of swimbladder muscle (i.e., the total rate at which skinned fibers utilized ATP). Surprisingly, despite performing twitches eight times faster than did frog fast-twitch fibers, the swimbladder did not use ATP any faster than did frog muscle fibers during steady-state contractions.

Because the overall rate of ATP usage by fibers is low, ATP utilization by cross-bridges and by Ca^{2+} pumps must both be low. We used standard partitioning techniques (i.e., knocking out the Ca^{2+} pumps as discussed in Reference 27) and also exploited the right-shifted force-pCa curve to measure SR-Ca^{2+} ATP utilization directly (i.e., at pCa below which force generation starts) to partition and measure directly the rate of ATP usage by the cross-bridges and SR-Ca^{2+} pumps (28). The means by which the swimbladder achieves the superfast g needed for rapid relaxation with a modest ATP utilization per myosin head can be understood by the mechanism described above: The swimbladder muscle has very few attached cross-bridges at any one time [$=f/(f + g)$]; thus the cross-bridge cycling rate (ATPase = number of attached bridges $\times$ g) is modest. However, the means by which the swimbladder muscle could have a superfast Ca^{2+} transient with a modest Ca^{2+} pumping rate remains a puzzle.

CALCIUM CYCLING IN VIVO

A Simple Model of Calcium Movement: Quantitative Neuroethology and Quantitative Biochemistry

To be able to develop an expectation of the Ca^{2+} fluxes during calling, one has to know the biomechanics of the call as well as the concentrations of constituent proteins in toadfish muscle. If we assume that the amount of Ca^{2+} released and taken up during every contraction is equal to that which just saturates troponin binding sites, then the precise TNC concentration must be known to predict Ca^{2+} flux. Using refined approaches (29), we found that the TNC concentration in swimbladder muscle is only 35 μmoles per kg muscle.[1] Hence swimbladder muscle has approximately 70 μmol kg^{-1} of Ca^{2+} binding sites. Thus to saturate and desaturate TNC, 70 μmol kg^{-1} of Ca^{2+} would need to be released (and pumped back into the SR) during each twitch.

Toadfish calls range from ~80 Hz at low temperatures to ~250 Hz at high temperatures (30). To obtain a more precise understanding of what the muscle must do requires further information on the number and frequencies of the sound pulses that comprise the call. Using hydrophones we recorded and analyzed the sound pulses (31) and found the calls could be divided temporally into two phases. Phase II represents the majority of the call and is characterized by extremely regular sound pulses with interpulse intervals of a constant value (Figure 8). By contrast,

---

[1]Although many authors (17) provide concentrations in terms of muscle water, we report values for the total muscle mass.

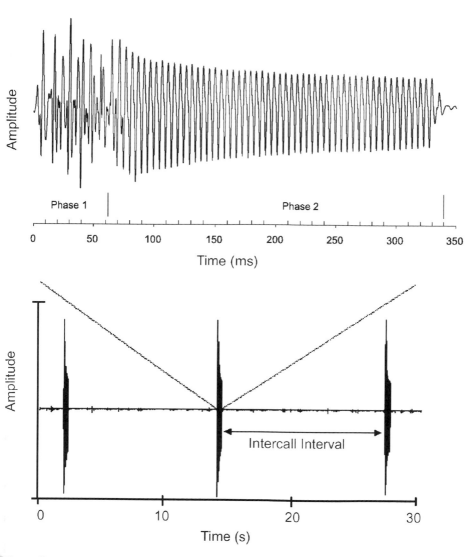

Figure 8 The boatwhistle mating call of toadfish. The bottom trace shows calls over a slow time base. Note that the calls are low duty cycle and that there can be long intercall intervals that appear to be obligatory. The upper trace shows an expanded view of one call. Whereas the first short portion of the call (Phase I) is irregular, the majority of the call (Phase II) is very regular. Based on Reference 31.

Phase I represents the short, initial portion of the call, when interpulse intervals are more variable. Whether Phase I represents the motor pattern with which the swimbladder is driven or is an emergent property of the whole system is unclear. We focused on Phase II because of its regularity. From this work, we determined the necessary pattern with which to stimulate the muscle. We concentrated our work at 15°C, used 80 and 100 Hz to simulate the muscle stimulation frequency, and used a call duration of 400 ms.

Hence, if during each twitch, 70 μmol kg^{-1} of Ca^{2+} is released, saturates troponin, and then is pumped back into the SR before the next stimulus, then calling at 100 Hz would require a steady state pumping rate of 7 mmol of Ca^{2+} kg^{-1} s^{-1}. If we assume a stoichiometry of two Ca^{2+} per ATP (32), our measurement of ATPase rate associated with Ca^{2+} pumping (1 mmol kg^{-1}s^{-1}) is equivalent to a Ca^{2+} pumping rate[2] of 2 mmol kg^{-1} s^{-1} (Figure 9). This value is three- to fourfold slower than that which appears necessary for the swimbladder muscle to twitch continuously at the high frequencies (100 Hz) used for calling at 15°C. Hence the SR-Ca^{2+} pumps simply cannot keep up in real time. So how does swimbladder muscle generate rapid Ca^{2+} transients with a relatively low Ca^{2+} pumping rate?

Two Mechanisms and the Need for New Approaches

The classical description of the activation-relaxation process is that the SR releases sufficient Ca^{2+} to saturate troponin (or considerably more should there be additional buffers such as PARV), and during the relaxation process all the Ca^{2+} is pumped back into the SR. Two mechanisms can explain the apparent paradox we describe above:

> Mechanism 1: During calling there may not be complete saturation and desaturation of troponin during each twitch; rather, the troponin saturation level may rise slightly and fall slightly while operating on a steep portion of the force-versus-troponin saturation curve. This would provide an oscillation of force with relatively little exchange of Ca^{2+}, and thus considerable reduction in the average Ca^{2+} release and reuptake per stimulus. Evidence to support this mechanism is the observation that, when held isometrically and stimulated at 100 Hz, the swimbladder muscles do not fully relax between stimuli.

> Mechanism 2: Most of the Ca^{2+} released during a call temporarily binds to PARV and is released from PARV and pumped back into the SR during the long intercall interval (i.e., only after the muscle has relaxed).

To test Mechanism 1, we needed to know how much Ca^{2+} is released per twitch in an intact muscle. However, there was no technique available to measure

---

[2] An approximate value was found (28) prior to the development of new techniques utilizing N-benzyl-p-toluene sulfonoaminde. N-benzyl-p-toluene sulfonoaminde usage permitted a more accurate measurement of the maximum rate (Figure 9) (33).

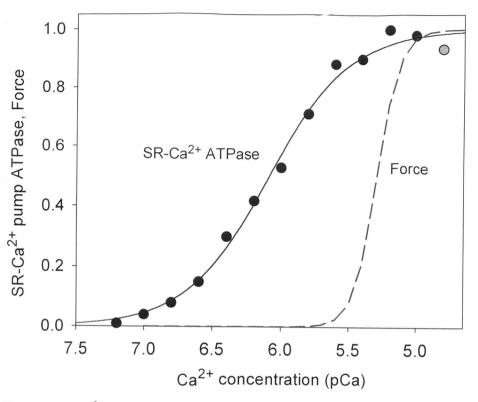

Figure 9   SR-Ca^{2+} pump ATP utilization as a function of free [Ca^{2+}] in skinned fibers. All experiments were performed in the presence of 25 μM N-benzyl-p-toluene sulfonoaminde, which removed more than 95% of the cross-bridge ATP utilization and 5 mM caffeine to prevent back-inhibition of the SR-Ca^{2+} pumps. ATP utilization (*blue*) reached a maximum at approximately pCa 5.2 and then declined slightly with higher [Ca^{2+}] (*shown as gray circle*; not included in curve fit). Force-pCa plot (*purple*) from another swimbladder experiment is added for comparison. Based on Reference 33.

this. Although it is often assumed that the Ca^{2+} transient describes the amount of Ca^{2+} released, such an assumption blurs a crucial distinction between free [Ca^{2+}] and total Ca^{2+}. Ca^{2+} transient durations provide information about the speed of changes in free [Ca^{2+}]; however, they do not by themselves provide information about the magnitude and time course of the total amount of Ca^{2+} released, its reuptake, or its binding to different proteins in the cell. The vast majority of Ca^{2+} released in a muscle cell is bound during contraction (i.e., to TNC and PARV), and there are significant kinetic delays between (*a*) the drop of free [Ca^{2+}] and (*b*) Ca^{2+} unbinding from PARV and being pumped back into the SR (34). Hence it is difficult to equate the kinetics of the Ca^{2+} transient (or muscle relaxation) with the amount and time course of Ca^{2+} release and reuptake (17, 33, 35).

Thus, we had to develop new techniques for measuring the amounts of Ca^{2+} released and pumped back into the SR per twitch. New techniques were also necessary to test Mechanism 2 (see below). It became apparent that in solving the riddle posed by the swimbladder muscle, we had to answer general questions about calcium cycling and PARV function. Hence, here we depart from discussing just superfast muscle and explain in more detail the approach and findings, which are generalizable to all muscles.

The new approaches to answering these questions are based on two facts: First, Ca^{2+} pumping requires the utilization of ATP in a fixed stoichiometry of two Ca^{2+} per ATP. Thus the total amount of ATP utilized by the SR-Ca^{2+} pumps can be used as a proxy for the amount of Ca^{2+} pumped. Second, we have been able to show that a new myosin II inhibitor, N-benzyl-p-toluene sulfonoaminde (BTS), can be used to knock out cross-bridge ATPase (as well as cross-bridge force) and thereby unmask SR-Ca^{2+} pump ATP utilization. Although highly effective SR-Ca^{2+} pump inhibitors [e.g., 2,5-di-(tert-butyl)-1,4-benzohydroquinone (TBQ)] have been available for some time, there has been a strong need for a specific and potent cross-bridge inhibitor that does not affect Ca^{2+} pumping rate. Using combinatorial chemistry and high throughput screening, Mitchison and colleagues (36) found that BTS appeared to block myosin ATPase from fast-twitch fibers and knocked out force generation while seeming to have little effect on the Ca^{2+} transient in frog fibers.

We recognized that BTS could be used not only to distinguish between the two mechanisms but more broadly to help answer a series of questions that had not been accessible by previously available techniques. We performed a series of experiments to determine whether (*a*) BTS could completely block the cross-bridge ATPase and (*b*) it affects SR-Ca^{2+} pump ATPase at that concentration. First, we found that at 25 μM, BTS reduced the cross-bridge ATPase to only 3% of its original value, and force fell to a very low value indistinguishable from zero. Second, we then exploited the right-shifted force-pCa relationship in swimbladder muscle, which permitted us to measure SR-Ca^{2+} ATP utilization both with and without a cross-bridge blocker up to $[Ca^{2+}]$ of pCa 5.8 (i.e., at lower $[Ca^{2+}]$, the cross-bridges do not generate force or utilize ATP). At 25 μM, the BTS had no apparent effect on the SR-Ca^{2+} pump ATPase.[3] Hence, we were then able to measure SR-Ca^{2+} pump ATPase over a wide range of $[Ca^{2+}]$. We found that the SR-Ca^{2+} pump ATPase increased with increasing $[Ca^{2+}]$ with a relatively low Hill coefficient ($\sim$1.45) and that the 50% pumping rate occurred at a $[Ca^{2+}]$ of $\sim$0.83 μM (pCa = 6.08). By using SR-Ca^{2+} pumping as a proxy for Ca^{2+} pumping, we determined the maximal steady-state pumping rate of 2 mmol kg^{-1} s^{-1}.

Most importantly, we found that BTS not only works on skinned fibers but also can be used in large intact bundles suitable for energetics. As in skinned fibers,

---

[3]Note that because of its right-shifted force-pCa curve, the swimbladder is the only muscle known in which a direct test of BTS could have been performed. In all other muscles, cross-bridge ATPase completely overlaps SR-Ca^{2+} pump ATPase.

exposure of intact bundles to 25 μM BTS knocks out more than 95% of the cross-bridge ATP utilization, thereby unmasking the SR-Ca^{2+} pump ATP utilization. The other important property of intact muscle is that, because it is a closed system, if a muscle is given sufficient time to recover from contraction, all of the Ca^{2+} that was released will eventually be pumped back into the SR. Thus Ca^{2+} pumped equals Ca^{2+} released.

We used recovery oxygen consumption experiments to measure the high-energy phosphate breakdown, enabling us to determine how much Ca^{2+} was released per twitch (37). Preliminary evidence shows that in the first stimulus, 240 μmol kg^{-1} (or three and one-half times the amount of TNC sites) is released, but this amount declines dramatically with increasing stimuli. After approximately the tenth twitch, Ca^{2+} release declines to a level (48 μmol kg^{-1}) below that required to saturate and desaturate troponin. This observation provides strong evidence for the reduction in Ca^{2+} release during calling.

Two processes are known to modulate the magnitude of Ca^{2+} release. First, Chandler and colleagues (38) showed in cut frog fibers (in which the amount of Ca^{2+} in the SR could be manipulated by changing [Ca^{2+}] in the bathing solution) that the absolute quantity of Ca^{2+} release per stimulus depends on the amount of Ca^{2+} remaining in the SR (39). In the swimbladder, the amount of Ca^{2+} in the SR should decline over the period of the call (because much of the Ca^{2+} may be bound to PARV in the myoplasm). Thus it is critical to know how much Ca^{2+} is left in the SR prior to any given stimulus. By the tenth stimulus, the Ca^{2+} release per stimulus had declined fivefold, even though preliminary results suggest that only 20% of total SR-Ca^{2+} had been released. Hence, dwindling SR Ca^{2+} levels play only a minor role in the decline of Ca^{2+} released per stimulus.

The second process that modulates Ca^{2+} release is Ca^{2+} release channel inactivation. Paired stimuli experiments have shown that the size of the Ca^{2+} transient in intact fibers (and the amount of Ca^{2+} release in cut fibers) declines dramatically in the second stimulus if it is given in close succession to the first (39). It is likely that inactivation plays a large part in the reduction of Ca^{2+} release from swimbladder muscle. Hollingworth & Baylor (40) showed that, similar to frog muscle, there is considerable inactivation in swimbladder muscle with a time constant of 25 ms.

The Role of Parvalbumin

Although the amount of Ca^{2+} released per twitch drops below that needed to saturate the troponin, this is not quantitatively large enough to provide the observed 3.5-fold shortfall in the rate of Ca^{2+} pumping. By exclusion, this suggests that PARV must play an important role. Ca^{2+} sequestration is equal to the sum of that associated with PARV and that associated with the SR-Ca^{2+} pumps. The quantitative role of PARV in setting the Ca^{2+} transient and twitch duration, however, is controversial. Because PARV is found only in fast muscles, it is generally associated with speed. In fact, there have been a number of reports in which twitch speed is positively correlated with PARV concentrations (41, 42), and there are several direct demonstrations of PARV's ability to speed relaxation (43, 44). Thus, one interpretation is that high PARV concentration is a requisite of fast contraction and hence is a major determinant of the Ca^{2+} transient. Seemingly in confirmation of this notion, toadfish swimbladder muscle, the fastest-contracting vertebrate muscle, also has the highest concentration of PARV measured (Table 1) (18).

Although it is clear that PARV can accelerate the Ca^{2+} transient and relaxation in some fiber types, kinetic and binding measurements suggest PARV may not be as effective in accelerating the Ca^{2+} transient in superfast muscle like the swimbladder. Most Ca^{2+} binding sites on PARV are thought to be filled by Mg^{2+} in the resting muscle (17, 21, 35, 45). Protecting these sites is a necessity to prevent PARV from competing with TNC during Ca^{2+} release. Although Ca^{2+} has a 10,000-fold higher affinity and can outcompete Mg^{2+} for PARV binding sites, the absolute rate at which sites become available in the muscle depends on the product of the kinetic off-rate of Mg^{2+} from PARV, which is slow [2 s^{-1} at 10°C (21, 45)], and [PARV]. This would suggest that the potential effectiveness of PARV may diminish as the speed of the Ca^{2+} transient increases.

Another unknown is the function PARV plays during normal activity. As noted above, PARV increases the speed of muscle relaxation for short stimuli, but what functional advantage does this bestow on the animal during normal activity? For instance, it is generally assumed that PARV would be useless to muscles involved in sustained activities because it would quickly become saturated (21, 45). Consistent with this assumption, PARV is not found in cardiac muscle and very little is found in slow-twitch muscle (which is the most-often recruited). However, Metzger and colleagues (44) found that the addition of PARV (by injection of cDNA) not only accelerates the relaxation rate in cardiac muscle but also improves cardiac function during normal continuous heart pumping. This surprising result underscores our incomplete understanding of PARV's in vivo function.

Toward a Quantitative Measure of Parvalbumin Usage During In Vivo Function

A measurement of Ca^{2+} binding of PARV during contraction in a well-defined, naturally occurring experimental system is necessary to understand the function of PARV. By using BTS and the ATP utilization by the SR-Ca^{2+} pumps as a proxy, we have been able to track the movements of Ca^{2+} onto and off PARV.

Toadfish call for many hours, which may suggest PARV would become quickly saturated, and hence ineffective at sequestering Ca^{2+}. However, a more detailed examination of the calling pattern suggests that PARV could play a large role. Even though the toadfish can call for many hours, the duty cycle of the call [call duration/(call duration + intercall interval)] is small. A typical call is 400 ms, but the intercall interval is 5–15 s. Hence, the duty cycle is approximately 2.5–7.5%. Given this time course of calling (low duty cycle and long intercall interval) and the large [PARV], PARV could contribute significantly to the sequestration of Ca^{2+} during calling. During the actual call, much of the released Ca^{2+} could bind to PARV. This Ca^{2+} would be subsequently pumped back into the SR during the relatively long intercall intervals. Furthermore, the high concentration of Ca^{2+} binding sites associated with PARV (3 mM) (18) seems to be sufficient to sequester much of the Ca^{2+} released during a contraction. Electron microprobe studies on frog muscle (34) suggest that considerable Ca^{2+} can be left bound in the myoplasm at the point of relaxation and that this is sequestered during the postrelaxation phase accompanied by hydrolysis of ATP (46, 47). Hence, toadfish seem to have both the need and the opportunity to utilize PARV. Thus the swimbladder muscle represents a compelling model with which to examine the role of PARV during in vivo function.

We developed a method by which we can track Ca^{2+} unbinding from PARV. At the end of the 400-ms call, we anticipated that the PARV would be nearly saturated with Ca^{2+} and that during the intercall interval (i.e., postrelaxation), the Ca^{2+} would unbind from PARV and be pumped into the SR by the SR-Ca^{2+} pumps. Thus we can use the ATP breakdown by the SR-Ca^{2+} pumps during the intercall interval to assess the time course and magnitude of the Ca^{2+} unbinding from PARV (48).

Muscle bundles preincubated in BTS were stimulated for 400 ms (duration of calls), and some were frozen immediately after the muscle relaxed with a robotic muscle-freezing device. Other bundles were stimulated for the same 400 ms; however, we delayed freezing them for 100–5000 ms following relaxation. To prevent high-energy phosphate regeneration during the interim, glycolysis was poisoned with iodoacetic acid, and oxidative phosphorylation was prevented by incubating the muscle bundles in nitrogen-saturated Ringers solution.

To obtain several duplicates at different time points, 20–25 experimental bundles and controls were necessary. Fish again represented a particularly good model because 20–30 bundles could be dissected from a single swimbladder. By contrast, such measurements in frogs would have required a minimum of 20–30 individuals.[4]

Under these conditions there was an increase in Pi, a fall in CrP, and little change in ATP. Because the Pi measurement was the most repeatable, we focused our analysis on those data. The ATP that was utilized during the 400-ms contraction was the difference between values from bundle stimulated for 400 ms (and then frozen immediately) and the resting Pi concentration obtained from unstimulated bundles. These preliminary data show that the amount of ATP used, and thus the amount of Ca^{2+} pumped during the call, was small. To determine the magnitude and time course of the postrelaxation ATP utilization, we subtracted the average value obtained from the muscle frozen immediately after relaxation from values obtained from muscle frozen at different times. Preliminary results show that between 1.5–2 mmol kg^{-1} of high-energy phosphate were broken down following relaxation. This greatly outweighed the amount of ATP broken down during contraction.

---

[4]Frogs exhibit a large interindividual variation, and the muscle from one leg must be used as a control for the experimental muscle on the other leg. Hence, one obtains only a single experimental datum per frog. In toadfish, four bundles serve as controls for all the rest.

Thus approximately 80% of the ATP utilization by the SR-Ca^{2+} pumps occurs following relaxation. This represents a fundamentally different way of thinking about how muscle is turned on and off but supports our prediction of possible PARV function based on our quantitative analysis of SR-Ca^{2+} pumping.

Evidence is strong that the large postrelaxation SR-Ca^{2+} pump ATPase is due to Ca^{2+} coming off PARV and being pumped back into the SR:

1. The results of a sister set of experiments show that the magnitude of postrelaxation ATP utilization is saturable, which would be expected if it reflected a buffer of fixed size like PARV.

2. This magnitude is appropriate for PARV. Toadfish swimbladder has the highest PARV concentration ever measured (1.5 mM), corresponding to 3-mM Ca^{2+} binding sites. Assuming two Ca^{2+} pumped per ATP, our results are consistent with all of the 3-mM Ca^{2+} sites binding Ca^{2+} during the call, followed by the Ca^{2+} being pumped back into the SR.

3. The kinetics appear to be too fast for PARV. The rate constant we observed is 3 s^{-1}, whereas the off-rate of Ca^{2+} from frog PARV is 1 s^{-1} (45). However, we found that toadfish possess different PARV isoforms. We have purified PARV from swimbladder, and preliminary results from stopped-flow kinetics on the predominant isoform in the same-size fish used in this experiment show a rate of approximately 3 s^{-1} (J. Davis, B.A. Tikunov, J. Rall, and L.C. Rome, personal communication).

4. Finally, the animal's calling behavior is consistent with this mechanism. It takes approximately 3–5 s for Ca^{2+} to be completely resequestered, and toadfish always leave 5 s between the end of one call and the beginning of the next.

Thus, by using this unique animal model, we have been able to track calcium cycling and PARV usage in muscle undergoing its natural stimulation pattern. We showed that two factors enable toadfish swimbladder muscle to power fast Ca^{2+} transients with moderate Ca^{2+} pumping: (*a*) reduced Ca^{2+} release per twitch and, more importantly, (*b*) PARV binding Ca^{2+} during the call (and slowly releasing it during the long intercall interval).

THE STUDY OF SUPERFAST MUSCLES IN THE MIDSHIPMAN

Midshipman (*Porichthys notatus*) calls for its mate by using its swimbladder muscle. The communication system as well as the ultrastructure of its muscle have been beautifully studied by Bass and colleagues (49). Males call at ∼100 Hz at 12–15°C. Preliminary experiments show that the males' muscles have fast twitches (L.C. Rome, A.A. Klimov, B.A. Tikunov, and A.H. Bass, personal communication). Furthermore, they have a right-shifted force-pCa curve (neither the Ca^{2+}

transient noi the cross-bridge detachment rate constant has been measured). Finally, we have found that, as in toadfish, the muscle generates very low forces. Although showing that the modifications made in the superfast muscle from another species are similar to those found in the toadfish is important to verify the necessity and generality of these traits, there are four additional reasons to study midshipman:

1. Unlike toadfish that call with a 2.5–7.5% duty cycle, midshipman call with a 100% duty cycle (i.e., they call continuously for hours at a time).

2. Unlike toadfish, which show only minor differences between male and female (18), midshipman exhibit profound sexual dimorphism in their swimbladder muscle (49). The Type 1 (calling) males have a large muscle (~3–7 mm thick) with a very high aerobic capacity (i.e., red in color). By contrast, the females, which do not call, have only a thin wisp of white muscle (several fibers thick).

3. Perhaps the most striking sexual dimorphism occurs at the ultrastructural level: In the calling males, the z-bands of the swimbladder muscle are 1-μm thick or ~20 times longer than in other vertebrates. By contrast, the males' locomotory muscle as well as the z-band of the female swimbladder have normal thicknesses of ~0.05 μm. Because the other myofilaments in the calling male have normal lengths, the resting sarcomere length in the swimbladder muscle is 3.2 μm, compared with 2.2 μm for all other midshipman skeletal muscles.

4. The dimorphisms we discuss above appear to be based on function rather than just sex. Although the calling male (also called Type 1) has a different muscle ultrastructure than has the female, some male midshipman that do not call (Type 2) have exactly the same muscle structure as the female. These "sneaker males" stay in the vicinity of a calling male's nest and try to mate with the female that has been attracted by the Type 1 male's call (50).

Two questions must be answered about midshipman. First, how do they cycle Ca^{2+} without being able to use PARV as a temporary buffer? If no intercall interval exists, then PARV would become saturated quickly, and there would be no time period during which Ca^{2+} could unbind from PARV and be pumped back into the SR. Preliminary experiments show that the SR-Ca^{2+} pumping rate is no higher in the swimbladder of Type 1 males than it is in toadfish swimbladder (J. Marx, B.A. Tikunov, A.H. Bass, and L.C. Rome, personal communication). Second, what is the functional significance of the long z-bands in midshipman swimbladder muscle? We have hypothesized that, at a minimum, they act as an inert spacer (so that the muscle could span a given distance with fewer sarcomeres). As the z-bands do not contain cross-bridges or SR-Ca^{2+} pumps, they would not use ATP during contractions, which would result in an overall reduction in ATP usage. However, the long z-bands may form an elastic structure (51) that in some way permits the midshipman swimbladder to oscillate force with little exchange of Ca^{2+}.

TEMPERATURE EFFECTS: WHAT MAKES A FAST MUSCLE SUPERFAST?

One interesting question regards the physiological, biochemical, and morphological criteria used to categorize a fast muscle as superfast. From a physiological/biomechanical viewpoint, superfast muscle must be able to perform mechanical work at high frequencies as measured by workloop experiments, and there is strong evidence that this requires fast kinetics of the three sequential mechanisms involved in relaxation. However, fast kinetics can be obtained in two ways: (a) by altering the myosin and troponin isoforms and the quantity of SR pumps (as found in swimbladder) or (b) by simply raising the temperature. Temperature has a large effect on kinetic processes involved in muscle relaxation. At very low temperatures, SR-Ca^{2+} pumps have a Q_{10} of nearly 5, and V_{max} (which is a manifestation of cross-bridge kinetics) has a Q_{10} of 3. Although these Q_{10}s moderate at higher temperatures, they are still large. This raises the question: Can a muscle that is only "fast" at low temperatures become "superfast" at high temperatures? As an example, at 35°C the rattlesnake shaker muscle's Ca^{2+} transient, V_{max}, and twitch are faster than those of the toadfish swimbladder at 15°C, and it can generate mechanical power at frequencies exceeding 90 Hz (4). However, at 15°C, compared to swimbladder, the shaker muscle[5] has a 2–3-ms slower Ca^{2+} transient, a considerably slower V_{max}, and a much slower twitch and can generate mechanical power up to only approximately 30 Hz. Although, based solely on the kinetics, the results from 35°C suggest that the shaker muscle is certainly a superfast muscle, the values obtained at 15°C are within the realm of the fastest locomotory muscles at high temperatures (e.g., hummingbird at 40°C). However, taking into account that 30 Hz is obtained at 15°C puts the shaker muscle into the superfast category. Furthermore, all superfast muscle appear to have high SR volumes (~30%) and an isoform of myosin different from that of the locomotory muscles. Therefore, from a morphological and biochemical viewpoint, the shaker muscle would be categorized as superfast (25, 52).

As superfast muscles have probably evolved multiple times [perhaps even in birds[6] (53) for vocalization], it is unlikely that each muscle has precisely the

---

[5]Maximum isometric force was not measured in these rattlesnake shaker muscles. An indication of little shift in the force-pCa curve at 15°C was also found.
[6]The criteria used to categorize the bird muscle as superfast (it did not tetanize until stimulated at more than 200 Hz, and twitch force was very low) are not sufficient to conclude that this muscle is superfast (i.e., this property could be the result of the membrane properties). It must be demonstrated that the muscle is capable of generating power at high frequencies by the workloop technique.

same changes in myosin or troponin. Because we believe we have identified three functional traits that are necessary for power production at high frequencies, understanding the evolution of these traits (all three of which appear necessary) is a fascinating area for future research on superfast muscles.

CONCLUSION

From one viewpoint, the work reviewed here provides information about the function of superfast muscles, but in a larger sense it provides both an improved understanding of muscle physiology in general and a paradigm for studying it. This approach represents a confluence of cellular and molecular physiology, biophysics, integrative physiology, quantitative physiology, quantitative biochemistry, quantitative ultrastructure, and modeling as well as biomechanics and neuroethology. All these approaches have contributed novel insights to our understanding. First, integrative physiology is important because it is through the process of integration that gaps in our understanding become apparent. Integrating function frequently reveals crucial experiments that have never been performed and also provides the motivation for performing them. Integration is not only important across higher levels of organization but also necessary within a cell (i.e., cross-bridge cycling and calcium pumping). Furthermore, only through quantitative approaches (e.g., quantitative physiology, quantitative biochemistry, and modeling) can hypotheses concerning integrated physiological function be proven. As we have seen, quantitative bookkeeping often fails (e.g., Ca^{2+} pumps cannot keep up with Ca^{2+} flux out of the SR), leading to the assessment of other mechanisms.

The use of the comparative approach has a number of important advantages that have been explored in this review. First, the advantage of using an experimental model with pure muscle fibers (fish) is incalculable and can be best understood if one envisions trying similar experiments in mammals. As reviewed in Reference 54, the main limitation of mammalian experimental models is that the muscles are typically heterogeneous (i.e., slow-twitch muscle fibers are interspersed with various types of fast-twitch muscle fibers). This fact leads to two nearly insurmountable problems preventing the type of analysis used for the swimbladder muscle. First, electromyography electrodes implanted in the mammalian muscles receive signals from fibers of all types, making it nearly impossible to discriminate which fiber type(s) are powering a particular movement. Second, any given bundle of fibers dissected from a mammalian muscle necessarily will contain more than one fiber type, making it nearly impossible to differentiate the properties of one fiber type from those of another (12). Hence, in mammals, measuring function in a pure fiber type requires going to the single-cell level, making it very difficult to perform the energetics and quantitative biochemistry experiments used in toadfish (i.e., larger preparations are needed). Furthermore, mechanics experiments in mammalian fibers require skinning, which destroys the sarcolemma membrane and precludes measurements of some important steps in the activation-relaxation cycle (e.g., Ca^{2+} cycling).

A second advantage of the comparative approach is that one can study muscles at the extremes of function. On the one hand, this provides perspective on the function and mechanisms in mammalian muscle (i.e., in toadfish, we studied muscles whose speed varied by more than 50-fold, which represents close to the total speed range among vertebrates). On the other hand, studying extreme muscles can elucidate mechanisms that may otherwise be difficult to discover. For instance, by studying the fastest vertebrate muscle, we were able to show that force generation is greatly reduced in superfast muscle. The underlying mechanism appeared to be fewer attached cross-bridges. Although there had been several observations of reduced force in the muscles associated with the mammalian eye (23), these declines were small, and no mechanism had been proposed, in part because experimental studies were confounded by the heterogeneity of the muscle, but also because the mammalian eye muscles are not as fast as the swimbladder muscle and hence do not appear to have as large a reduction in force. The 10-fold reduction in force in swimbladder muscle helped us elucidate the mechanism, thereby providing evidence that the previous observations from mammalian eye muscles were not due to muscle damage or incomplete activation but may instead conform to a general design principle.

A third advantage of studying extreme function is that often it is a consequence of extreme modification to the muscle, making it technically possible to observe function that is unmeasurable in normal muscle. In the case of PARV, we devised new techniques to take advantage of the high concentration of PARV, which we predicted would provide a large postrelaxation ATP utilization signal that could be tracked. The extreme modifications for speed and synchronized function also provided the high densities of Ca^{2+} release sites as well as regular T-tubule disposition, which aided in the discovery of the molecular structure of the SR triad (55).

Finally, and perhaps most importantly, using lower vertebrate animal models has facilitated the establishment of paradigms that should be used when addressing issues of muscular system design and function in any animal system. By being able to bring to bear so many quantitative approaches on toadfish muscle, we have learned a good deal about superfast muscle function and muscle function in general. We believe that this type of approach is necessary, but because of the technical difficulties associated with working on mammals (and particularly humans), it will not be possible to obtain all the data required by the paradigm in the near term. Nonetheless, the paradigm still serves two important functions: (*a*) When certain information is not available, investigators can appreciate better the limitations required in interpreting the data they have obtained, and (*b*) the paradigm describes the type of data that is necessary to obtain and hence provides an impetus to develop new techniques to obtain this information.

ACKNOWLEDGMENTS

This work was made possible by support from NIH grants AR38404 and AR46125 as well as the University of Pennsylvania Research Foundation. I thank Drs. Boris

Iikunov, James Marx, and Richard Essner for their helpful comments on the manuscript.

The *Annual Review of Physiology* is online at
http://physiol.annualreviews.org

LITERATURE CITED

1. Skoglund CR. 1961. Functional analysis of swim-bladder muscle engaged in sound production by toadfish. *J. Biophys. Biochem. Cytol.* 10:187–200

2. Bennett NT, Weiser M, Baker R, Bennett MVL. 1985. Toadfish sonic motor system: I. Physiology. *Biol. Bull.* 169:546

3. Fine ML, Mosca PJ. 1989. Anatomical study of innervation pattern of the sonic muscle of the oyster toadfish. *Brain Behav. Evol.* 34:265–72

4. Rome LC, Syme DA, Hollingworth S, Lindstedt SL, Baylor SM. 1996. The whistle and the rattle: the design of sound producing muscles. *Proc. Natl. Acad. Sci. USA* 93:8095–100

5. Rome LC, Lindstedt SL. 1997. Mechanical and metabolic design of the muscular system in vertebrates. In *Handbook of Physiology. Section 13, Comparative Physiology*, ed. W Dantzler, pp. 1587–651. New York: Oxford Univ. Press

6. Rome LC, Funke RP, Alexander RM, Lutz GJ, Aldridge HDJN, et al. 1988. Why animals have different muscle fibre types. *Nature* 355:824–27

7. Rome LC, Swank D, Corda D. 1993. How fish power swimming. *Science* 261:340–43

8. Lutz GJ, Rome LC. 1994. Built for jumping: the design of frog muscular system. *Science* 263:370–72

9. Marsh RL, Olson JM, Guzik SK. 1992. Mechanical performance of scallop adductor muscle during swimming. *Nature* 357:411–13

10. Rome LC, Kushmerick MJ. 1983. The energetic cost of generating isometric force as a function of temperature in isolated frog muscle. *Am. J. Physiol. Cell Physiol.* 244:C100–9

11. Rome LC. 1982. The energetic cost of running with different muscle temperatures in Savannah monitor lizards. *J. Exp. Biol.* 97:411–26

12. Rome LC. 1998. Some advances in integrative muscle physiology. *Comp. Biochem. Physiol. B* 120:51–72

13. Josephson RK. 1985. Mechanical power output from striated muscle during cyclic contraction. *J. Exp. Biol.* 114:493–512

14. Young IS, Rome LC. 2001. Mutually exclusive muscle designs: power output of locomotory and sonic muscles of the oyster toadfish (*Opsanus tau*). *Proc. R. Soc. London Ser. B* 268:1975–80

15. Johnson TP, Johnston IA. 1991. Power output of fish muscle fibres performing oscillatory work: effects of acute and seasonal temperature change. *J. Exp. Biol.* 157:409–23

16. Josephson RK. 1985. The mechanical power output of a tettigoniid wing muscle during singing and flight. *J. Exp. Biol.* 117:357–68

17. Baylor SM, Hollingworth S. 1988. Fura-2 calcium transients in frog skeletal muscle fibres. *J. Physiol.* 403:151–92

18. Appelt D, Shen V, Franzini-Armstrong C. 1991. Quantitation of Ca ATPase, feet and mitochondria in super fast muscle fibers from the toadfish, *Opsanus tau. J. Muscle Res. Cell Motil.* 12:543–52

19. Baylor SM, Hollingworth S. 1998. Model of sarcomeric Ca^{2+} movements, including ATP Ca^{2+} binding and diffusion, during activation of frog skeletal muscle. *J. Gen. Physiol.* 112:297–316

20. Jiang Y, Johnson JD, Rall JA. 1996. Parval-bumin relaxes frog skeletal muscle when the sarcoplasmic reticulum Ca-ATPase is inhibited. *Am. J. Physiol. Cell Physiol.* 270:C411–17

21. Hou TT, Johnson JD, Rall JA. 1991. Parval-bumin content and Ca²⁺ and Mg²⁺ dissoci-ation rates correlated with changes in relax-ation rate of frog muscle fibres. *J. Physiol.* 441:285–304

22. Wahr PA, Johnson JD, Rall JA. 1998. De-terminants of relaxation rate in skinned frog skeletal muscle fibers. *Am. J. Physiol.* 274:C1608–15

23. Rome LC, Cook C, Syme DA, Con-naughton MA, Ashley-Ross M, et al. 1999. Trading force for speed: Why su-perfast crossbridge kinetics leads to super-low forces. *Proc. Natl. Acad. Sci. USA* 96: 5826–31

24. Brenner B. 1988. Effect of Ca²⁺ on cross-bridge turnover kinetics in skinned single rabbit psoas fibers: implications for regu-lation of muscle contraction. *Proc. Natl. Acad. Sci. USA* 85:3265–69

25. Rome LC, Lindstedt SL. 1998. The quest for speed: muscles built for high frequency contractions. *News Physiol. Sci.* 13:261–68

26. Barany M. 1967. ATPase activity of myosin correlated with speed of muscle shortening. *J. Gen. Physiol.* 50:197–218

27. Stienen GJM, Zaremba R, Elzinga G. 1995. ATP utilization of calcium uptake and force production in skinned muscle fibres of *Xenopus laevis. J. Physiol.* 482:109–22

28. Rome LC, Klimov AA. 2000. Superfast contractions without superfast energetics: ATP usage by SR-Ca²⁺ pumps and cross-bridges in the toadfish swimbladder mus-cle. *J. Physiol.* 526:279–98

29. Tikunov BA, Sweeney HL, Rome LC. 2001. Quantitative electrophoretic analysis of myosin heavy chains in single muscle fibers. *J. Appl. Physiol.* 90:1927–35

30. Fine ML. 1978. Seasonal and geographical variation of the mating call of the oyster toadfish *Opsanus tau* L. *Oecologia* 36:45–57

31. Edds-Walton P, Mangiamele L, Rome LC. 2002. Boatwhistles from oyster toadfish (*Opsanus tau*) around Waquoit Bay, Mas-sachusetts. *J. Bioacoust.* 13:153–73

32. Martonosi AN, Beeler TJ. 1983. Mecha-nism of Ca²⁺ transport by sarcoplasmic reticulum. In *Handbook of Physiology,* ed. LD Peachy, pp. 417–85. Bethesda, MD: Am. Physiol. Soc.

33. Young IS, Harwood CL, Rome LC. 2003. Cross-bridge blocker BTS permits the di-rect measurement of Ca²⁺ pump ATP uti-lization in skinned toadfish swimbladder muscle fibers. *Am. J. Physiol. Cell Phys-iol.* 285:C781–87

34. Somlyo AV, McClellan G, Gonzalez-Serratos H, Somlyo AP. 1985. Electron probe X-ray microanalysis of post-tetanic Ca²⁺ and Mg²⁺ movements across the sar-coplasmic reticulum *in situ. J. Biol. Chem.* 260:6801–7

35. Baylor SM, Chandler WK, Marshall MW. 1983. Sarcoplasmic reticulum calcium re-lease in frog skeletal muscle fibres ex-timated from Arsenazo III calcium tran-sients. *J. Physiol.* 344:625–66

36. Cheung A, Dantzig JA, Hollingworth S, Baylor SM, Goldman YE, et al. 2002. A small-molecule inhibitor of skeletal muscle myosin II. *Nat. Cell Biol.* 4:83–88

37. Harwood CL, Young IS, Rome LC. 2001. Cheap twitches in superfast toadfish swim-bladder (SB) muscle. *Biophys. J.* 80:A271 (Abstr.)

38. Jong DS, Pape PC, Geibel J, Chandler WK. 1996. Sarcoplasmic reticulum calcium re-lease in frog cut muscle fibers in the pres-ence of a large concentration of EGTA. *Soc. Gen. Physiol. Ser.* 51:255–68

39. Pape PC, Jong DS, Chandler WK. 1995. Calcium release and its voltage dependence in frog cut muscle fibers equilbrated with 20 mM EGTA. *J. Gen. Physiol.* 106:259–336

40. Hollingworth S, Baylor SM. 1996. Sar-coplasmic reticulum (SR) calcium release in intact superfast toadfish swimbladder

(TSB) and fast frog twitch fibers. *Biophys. J.* 70:A235 (Abstr.)

41. Heizman CW, Berchtold MW, Rowlerson AM. 1982. Correlations of parvalbumin concentration with relaxation speed in mammalian muscles. *Proc. Natl. Acad. Sci. USA* 79:7243–47

42. Klug GA, Leberer E, Leisner E, Simoneau J, Pette D. 1988. Relationship between parvalbumin content and the speed of relaxation in chronically stimulated rabbit fast-twitch muscle. *Pflugers Arch.* 411:126–31

43. Muntener M, Kaser L, Weber J, Berchtold MW. 1995. Increase of skeletal muscle relaxation speed by direct injection of parvalbumin cDNA. *Proc. Natl. Acad. Sci. USA* 92:6504–8

44. Szatkowski ML, Westfall MV, Gomez CA, Wahr PA, Michele DE, et al. 2001. In vivo acceleration of heart relaxation performance by parvalbumin gene delivery. *J. Clin. Invest.* 107:191–98

45. Hou TT, Johnson JD, Rall JA. 1992. Effect of temperature on relaxation rate and Ca^{2+}, Mg^{2+} dissociation rates from parvalbumin of frog fibres. *J. Physiol.* 449:399–410

46. Homsher E, Lacktis J, Yamada T, Zohman G. 1987. Repriming and reversal of the isometric unexplained enthalpy in frog skeletal muscle. *J. Physiol.* 393:157–70

47. Rall JA. 1989. Relationship of isometric unexplained energy production to parvalbumin content in frog skeletal muscle. In *Muscle Energetics*, ed. R.J. Paul, G. Elzinga, K. Yamada, pp. 117–26. New York: A.R. Liss, Inc.

48. Tikunov BA, Klimov AA, Rome LC. 2003. *Biophys. J.* 84(2):A1892 (Abstr.)

49. Bass AH, Marchaterre MA. 1989. Sound-generating (sonic) motor system in teleost fish (*Porichthys notatus*): sexual polymorphism in the ultrastructure of myofibrils. *J. Comp. Neurol.* 286:141–53

50. Bass AH, Grober MS. 2001. Social and neural modulation of sexual plasticity in teleost fish. *Brain Behav. Evol.* 57:293–300

51. Lewis MK, Nahirney PC, Chen V, Adhikari BB, Wright J, et al. 2003. Concentric intermediate filament lattice links to specialized z-band junctional complexes in sonic muscle fibers of the type I male midshipman fish. *J. Struct. Biol.* 143:56–71

52. Schaeffer PJ, Conley KE, Lindstedt SL. 1996. Structural correlates of speed and endurance in skeletal muscle: the rattlesnake tailshaker muscle. *J. Exp. Biol.* 199:351–58

53. Elemans CP, Spierts IL, Muller UK, van Leeuwen JL, Goller F. 2004. Bird song: Superfast muscles control dove's trill. *Nature* 431:146

54. Rome LC. 2002. The design of vertebrate muscular systems: comparative and integrative approaches. *Clin. Orthop.* 403S:S59–76

55. Block BA, Imagawa T, Campbell KP, Franzini-Armstrong C. 1988. Structural evidence for direct interaction between the molecular components of the transverse tubule/sarcoplasmic reticulum junction in skeletal muscle. *J. Cell Biol.* 107:2587–600

Annu. Rev. Physiol. 2006. 68:223–51
doi: 10.1146/annurev.physiol.68.040104.105739
First published online as a Review in Advance on October 19, 2005

THE COMPARATIVE PHYSIOLOGY OF FOOD DEPRIVATION: From Feast to Famine

Tobias Wang,[1] Carrie C.Y. Hung,[2] and David J. Randall[2]

[1]Department of Zoophysiology, Aarhus University, 8000 Aarhus C, Denmark;
email: tobias.wang@biology.au.dk
[2]Department of Biology and Chemistry, City University of Hong Kong, Kowloon Tong,
Hong Kong PRC; email: carriehung2@yahoo.com, bhrand@cityu.edu.hk

Key Words   feeding, fasting, starvation, metabolism, atrophy, digestion, gastrointestinal organs, specific dynamic action, phenotypic plasticity

■ **Abstract**   The ability of animals to survive food deprivation is clearly of considerable survival value. Unsurprisingly, therefore, all animals exhibit adaptive biochemical and physiological responses to the lack of food. Many animals inhabit environments in which food availability fluctuates or encounters with appropriate food items are rare and unpredictable; these species offer interesting opportunities to study physiological adaptations to fasting and starvation. When deprived of food, animals employ various behavioral, physiological, and structural responses to reduce metabolism, which prolongs the period in which energy reserves can cover metabolism. Such behavioral responses can include a reduction in spontaneous activity and a lowering in body temperature, although in later stages of food deprivation in which starvation commences, activity may increase as food-searching is activated. In most animals, the gastrointestinal tract undergoes marked atrophy when digestive processes are curtailed; this structural response and others seem particularly pronounced in species that normally feed at intermittent intervals. Such animals, however, must be able to restore digestive functions soon after feeding, and these transitions appear to occur at low metabolic costs.

INTRODUCTION

All animals supply the energy required for basal metabolism, physical activity, growth, and reproduction from their food. When food is not available, animals must use internal energy stores to fuel these activities. Starvation resistance reflects an animal's ability to store energy and control its allocation during extreme resource limitation. Many animals live in environments in which food abundance and quality vary drastically over time, and periods of starvation are common. Bacteria can lead a feast-or-famine existence (1, 2), many planktonic species live in a resource-limited world (3), and many fish overwinter with little or no food. Sommer (4) suggested organisms can be grouped according to their responses to variations in food supplies. "Velocity specialists" are those species with high, maximum population-growth rates that respond to sudden increases in resource abundance with rapid increases in population density; many bacteria fall into this category. "Affinity specialists" can maintain growth at low-food-intake levels; this may describe many tropical fish. Lastly, "storage specialists" such as hibernating mammals create large internal stores that increase the chances of survival when resource abundance is extremely low.

Many of the animals that inhabit environments with fluctuating food availability are adapted to consume very large meals when prey is available. Classic examples are sit-and-wait predators, such as snakes; such species feed only once or a few times a year. Very large meals followed by low rates of energy expenditure, typical of crocodiles as well as a number of other reptiles and various fish, allow for days or weeks between feeding. When the lack of food is due to seasonal changes in temperature or water availability, animals may enter into dormancy, during which digestive processes are curtailed. With the exception of birds and some large mammals, which migrate to more desirable areas, dormancy is common in vertebrates living in temperate or arctic environments that hibernate during cold periods of the year. In tropical areas, many vertebrates, with the notable exception of birds, enter into a dormant estivating state during dry, and typically warm, periods. Fasting may also occur when animals engage in activities that compete with or even preclude feeding or the search for food. Such activities include migration, moulting, or the care of eggs or young (e.g., 5–7). Examples of voluntary anorexia exist in all major groups of vertebrates and may be very prolonged. Some species of penguins, for example, do not eat for several months when tending to eggs (8, 9); Pacific salmon do not feed during their upstream migration, which can be more than 1000 km in length (10); and eels do not feed during their migration across the Atlantic to spawn.

The ability of animals to survive food deprivation is clearly of considerable survival value. Unsurprisingly, therefore, all animals exhibit adaptive biochemical and physiological responses to the lack of food. These responses prolong survival when food is not available. Equally important, however, these responses also help animals to (a) preserve physiological functions so that behaviors, such as physical activity to avoid predators or to seek food, can be maintained and (b) ensure that the animals can resume digestive and metabolic processes when food becomes available again.

OVERALL FASTING TOLERANCE IS DETERMINED BY ENERGY STORES RELATIVE TO USAGE

When animals experience food deprivation, they must derive the energetic costs for basal metabolism, physical activity, growth, and reproduction from the internal energy available at the onset of fasting. Large energy stores at the onset of fasting, therefore, obviously aid in prolonging starvation tolerance; animals commonly gain weight before dormancy. A reduction in metabolism also prolongs starvation tolerance, and various biochemical and physiological responses to food deprivation contribute to the efficient use of resources. Animals reduce metabolism in diverse ways. Many animals inhibit reproduction and reduce both activity and body temperature. In fact, many animals breed only when food supplies are readily available, for example, in the springtime in temperate regions, and they do not breed during the winter, when food supplies are limited. The response to starvation is integrated at all levels of organization and is directed toward the survival of the species.

Starvation has been more extensively studied in birds and mammals than in invertebrates and other vertebrates. Birds and mammals are special among vertebrates because they normally must eat at regular intervals owing to their high metabolic rates relative to their body stores. Most definitions and ideas of fasting come from the human literature. The extent to which our knowledge of fasting and starvation in birds and mammals can be transferred to other animals is not clear, as we discuss below.

The Phases of Fasting and Starvation in Mammals and Birds

The responses to absolute food deprivation in birds and mammals proceed in stages, culminating in death. The initial period involves fasting, and the later stages starvation. The demarcation between these two states is rarely appreciated, perhaps owing to lack of definition. In humans, fasting often refers to abstinence from food, whereas starvation is used for a state of extreme hunger resulting from a prolonged lack of essential nutrients. In other words, starving is a state in which an animal, having depleted energy stores, normally would feed to continue normal physiological processes. The metabolic transitions of food deprivation in birds and mammals have been divided into three phases. Most investigators probably would agree that the transition from fasting to starvation occurs by the end of phase II or the start of phase III. Alternatively, and not in conflict with this view, others have argued that the transition between fasting and starvation may occur whenever animals opt to abort voluntary anorexia (sensu Reference 5). In our view, fasting should denote voluntary anorexia. Thus, salmon moving upstream and eels crossing the Atlantic to spawn are fasting rather than starving. The stages of starvation, if they exist in fish, amphibians, and reptiles, span a much more extended time frame, and the distinction between fasting and starvation becomes somewhat esoteric. In any event, the severity of the food deprivation and time required before an animal enters actual starvation vary among species, and the transition to starvation will depend on individual responses and nutritional status as well as a number of environmental conditions. Thus, whereas a small endothermic animal may be starving within a day of lacking food, it would take much longer for large ectothermic animals to undergo the same transition. Future studies should address questions such as whether large predatory ectotherms, such as sharks, that have not fed for a few days, are starving or fasting while waiting for the next meal. Similarly, is a hibernating ground squirrel starving or fasting?

The different phases of response to food deprivation, which are defined according to the progressive metabolic changes that occur, were initially characterized for humans and other mammals (e.g., 11–14). These stages have been successfully applied to birds (8, 9, 15), and although not thoroughly investigated in ectothermic vertebrates, the overall progression in metabolic adaptation appears similar for most vertebrates. As a major difference, however, resting and maximal metabolism are much lower in ectothermic vertebrates than in endotherms (e.g., 16). For a given body composition and amount of energy stores, ectothermic vertebrates therefore can maintain normal metabolic functions for much longer than can endothermic animals, thereby deferring the detrimental consequences of food deprivation. Thus many ectothermic vertebrates can tolerate more lengthy starvation than can endotherms. Eels, for example, can migrate many thousands of kilometers over almost a year without feeding and may survive lack of food for many years (17, 18), whereas a similar-sized mammal would die from starvation within a few days or a week. Likewise, small animals are much more susceptible to food deprivation than are larger animals.

In mammals, the three metabolic phases during food deprivation are characterized as follows on the bases of the primary fuel available for use and the associated changes in overall body mass:

Phase I. The postabsorptive phase is the initial phase of fasting immediately after the last meal has been absorbed from the gastrointestinal tract. During this period, which normally lasts for hours, metabolism is largely fueled by glycogenolysis, or glycogen depletion of liver stores, which maintain constant blood sugar levels. In addition, fatty acids are liberated from adipose depots, and the availability of plasma fatty acids allows for some tissues, such as skeletal muscle, to spare the overall use of glucose.

Phase II. When liver glycogen stores are depleted, gluconeogenesis becomes necessary to supply the requirements of glucose-requiring organs such as the brain. In humans, the initial fuel for gluconeogenesis is amino acids from proteolysis of muscle protein, but this contribution falls markedly as increased amounts of glycerol, another substrate for gluconeogenesis, is liberated from adipose tissues. Increased oxidation of fatty acids leads to an elevated production of ketone bodies, which can be used as an oxidative fuel in many tissues including the brain. As phase II progresses, protein degradation is rather slow, and degradation of adipose tissue fuels most bodily metabolism. In owls, lipid contributes more than 90% of the energy consumption in phase II, and approximately 2.5% of the energy consumption is derived from protein (19). In humans, this state can be maintained for several weeks and has often been referred to as a period of adapted starvation. Because of the high energy content of lipid, weight loss is rather slow during this state.

Phase III. If starvation continues until the adipose stores are depleted, muscle is rapidly degraded for gluconeogenesis. The rapid loss in muscle mass cannot be sustained for long and eventually kills the animal.

Figure 1 presents these three phases in rats. It shows how, as rats enter phase III, the rate of body mass loss along with nitrogenous waste production and its excretion increase as a result of protein degradation.

Although similar metabolic changes and the transitions between fasting and starvation remain to be studied in detail for ectothermic vertebrates, numerous

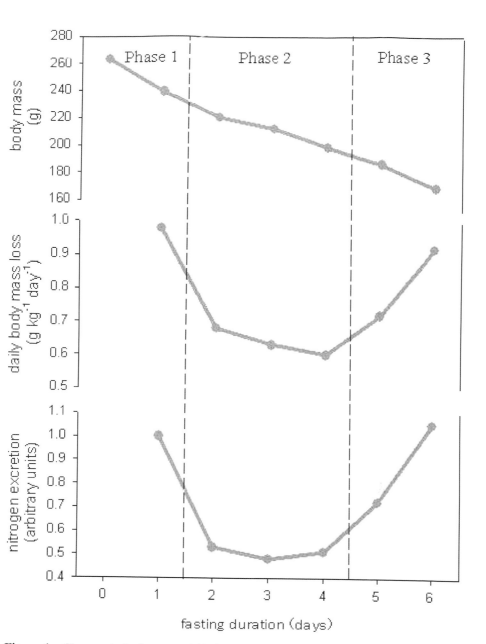

Figure 1   Changes in body mass, daily loss of body mass, and excretion of nitrogenous waste in rats during the three phases of starvation (modified from Reference 48).

studies have reported on the gradual but slow decrease in body weight and somatic indices as food is withheld. Most species studied utilize fat before protein is degraded (e.g., 10, 20–24). Other reports suggest that in ectotherms, as in mammals, glycogen is utilized even before lipid or protein (25, 26). Table 1 shows expression profiles of the main energy-generating pathways that are related to the three phases of the mammalian response to starvation. Mammals have to utilize energy reserves much earlier than fish in response to starvation. Genes that encode protein products in lipolysis and protein turnover were induced after 24 and 48 h of starvation in mice and rats. Figure 2 shows that starvation did not trigger significant changes in gene expression in carp until after at least 16 days of food deprivation. Lipolysis genes, such as β-oxidation, remained unchanged in carp liver throughout the six weeks of food deprivation, reflecting the fact that, unlike in mice and rats, hepatic lipid utilization was not enhanced in carp. Carp hepatic ubiquitin-proteasome genes were upregulated by approximately 1.3-fold after 28 days of starvation but did not trigger a significant decline in total hepatic protein content. In fact, hepatic protein appears to be well conserved during starvation in carp (27–30). Of course, before mobilizing hepatic reserves, carp use other lipid sources such as visceral lipid, as do rainbow trout (31). Carp contained a large amount of visceral lipids, although they were not quantified in the experiment. Carp hepatic glycogen, on the other hand, was mobilized during the first four days of starvation and then declined again after six weeks, which coincided with an increase in glycolytic gene expression. Hepatic glycogen was not exhausted completely in carp after 100 days of starvation (32). Early mobilization of hepatic glycogen in carp may be related to glucagon release during the initial phase of starvation; this has been observed in teleosts and may be a response to stress rather than starvation (24, 33). Migrating salmon utilize lipid and spare protein until later in the migration phase, when lipid stores are almost completely depleted (10).

REDUCTIONS IN ENERGY EXPENDITURE DURING FOOD DEPRIVATION

Responses to starvation occur at the behavioral, physiological, biochemical, and molecular levels. In general, the time to reach starvation-induced death increases with body mass (34–37), reflecting larger animals' greater abilities to lower specific metabolic rate and increase stores of energy. Reductions in energy expenditure can also occur via a reduction in body temperature, which reduces metabolic rate (the Q_{10} or Arrhenius effect). Animals also can decrease energy expenditure during starvation by reducing locomotor activity as well as other behavioral and physiological functions such as reproduction and care for young. Many of these activities, although interconnected, are often studied in isolation. Protein synthesis is decreased, and expression of many metabolic genes is also reduced. Whether gene expression is reduced in response to decreased energy expenditure or vice versa is not clear.

TABLE 1 Gene expression of mammals and carp liver in response to starvation. Upregulation of genes involved in lipolysis and proteolysis is observed only in starved mammals after 1 to 2 days of starvation, but it is not observed in common carp even after weeks of food deprivation. For details of gene lists, please refer to original publications.

Animal model	Methods	Duration of starvation	Upregulated pathways	Downregulated pathways	Probable phase	Reference
Male 129/sv mice (8–15 weeks old)	cDNA microarray	24 h	Lipolysis: β-oxidation genes Urea cycle genes Amino metabolism S-adenosylmethionine (SAM) cycle	Lipogenesis Cholesterol synthesis and DHEA metabolism	II	148
		48 h	Stronger induction of genes involved in the above pathways	Stronger suppression of genes involved in the above pathways	II, III transition	
Male Sprague-Dawley rats (145–155 g)	Suppression subtraction hybridization (SSH)*	48 h	Lipolysis: β-oxidation Gluconeogenesis Protein turnover		II	149
Cyprinus carpio (200–300 g)	cDNA microarray	4 days 8 days 16 days 28 days	Glycolysis (2 genes) Tricarboxylic (TCA) cycle Ubiquitin-proteasome pathway (1.3-fold) Glycolysis	Vitellogenin 16 to 42 days: Lipid biosynthesis Gluconeogenesis	?	26
		42 days	Sustained expression of the above genes	Varied expressions of gluconeogenic genes	?	

*SSH technique identifies only induced but not suppressed genes.

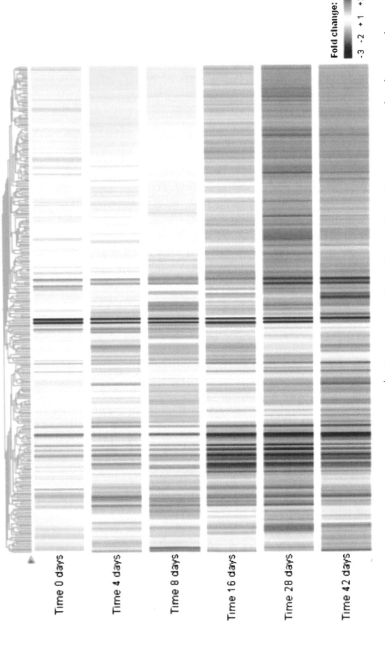

Figure 2 Expression profiles of statistically significant genes of carp liver indicate that gene expression does not change until after 16 days of starvation. Genes that have similar expression profiles are clustered together.

Basal and Resting Metabolic Rates

Humans and other mammals decrease resting metabolism during fasting and starvation (38). A very pronounced example is that of the golden spiny mouse, *Acomys russatus*, which inhabits dry deserts in the Middle East. Within one day, this small rodent apparently reduces oxygen uptake to half of the normal value and maintains this low metabolism when kept on restricted food availability for two weeks (39). This reduction takes places without changes in body temperature and may represent sympathetic control of energy-requiring processes (39). However, a recent study on the same species showed that food restriction elicits a reduction in both body temperature and metabolism that resembles torpor in other mammalian species (40). Nevertheless, although most animals do seem to lower body temperature when food is limited, substantial reductions in basal metabolic rate may occur. In salmon, for example, oxygen uptake decreased gradually over approximately two months when food was withheld (22).

A reduction in basal metabolic rate, with no attendant decline in body temperature, requires that some energy-requiring processes be reduced at the cellular level. The changes in cellular metabolism responsible for a reduction in basal metabolism in fasting or food-restricted animals have not been studied, but the response may involve some of the same mechanisms as those occurring during the metabolic reduction observed during hypoxia. Many vertebrates respond to lack of oxygen by lowering protein synthesis, and a lower membrane permeability decreases the demand for active ion transport. Cell cycle may arrest, and cell proliferation may decrease, leading to the observed reduction in growth (41). Certainly food restriction and therefore the lower rate of intestinal nutrient uptake should also decrease protein synthesis, cell proliferation, and growth. It is difficult to envision, however, that changes in membrane properties would occur without an associated loss of function.

Measuring basal metabolic rate while controlling for changes in spontaneous activity or alertness and sleep is difficult. Also, as described in more detail below, digestion also involves specific dynamic action and a rise in metabolism of animals seemingly at basal conditions. Fasting or starvation may therefore be associated with an apparent decline in basal metabolism that actually should be ascribed to a gradual cessation of digestive processes. The mammalian digestive tract displays great morphological and functional changes in response to starvation. Epithelial cell renewal and cell migration from crypt to villi tips were both reduced in starved mice (42) and rats (43). Total intestinal and jejunal mucosal mass was decreased by a factor of two in fasting rats, accompanied by reductions in jejunal crypt size and villi size and numbers as well as increases in villi tip cell apoptosis (44). Another recent study on rats, however, reports only minor changes in intestinal villus apoptosis (45). In any event, these studies concur that during phase III, apoptosis decreases and cell proliferation and migration increase (44, 45). Altogether, this upregulation of cellular events and presumably of absorptive capacity may reflect preparation for future feeding (45). The restoration of the intestine during phase III of starvation may be related to a "refeeding signal," which has been described in penguins (46), and/or to behavioral changes in food hunting that mammals display during this phase (47). Intestinal restoration may also be related to increased expression of genes encoding orexigenic, hypothalamic peptides such as neuropeptide Y, agouti-related protein, and pro-opiomelanocortin (48). Rapid restoration of intestinal structures was observed during refeeding in mammals (49). This restoration took place as quickly as 30 minutes after refeeding following phase II starvation (43). By contrast, starvation had no significant effect on the intestinal tract of the common carp. After 42 days of starvation, intestinal mucosal thickness of carp was not affected, and cellular events (e.g., apoptosis and cell proliferation) of the gut remained active. The expression of many digestive genes—including those for chemotrypsin A and B, elastases, trypsins, carboxypeptidase A and B, proproteinase E, and amylase 3—were suppressed greatly during starvation in carp, with most of the downregulation occurring after 16 days of starvation.

The acquisition and processing of food are expensive processes; their cessation is manifested as a reduction in basal metabolic rate. The large reduction in gut mucosal surface area during starvation probably results from a large reduction in energy expenditure in maintaining the gut. This reduction in gut energy expenditure constitutes a part of the overall bodily reduction in energy expenditure during starvation. Several studies reporting on metabolic depression of fasting animals did not report body temperature, and some of the decline in metabolism may stem from hypothermia. Thus, part of the alleged reduction in standard metabolic rate during food deprivation may be ascribed to factors that lead to a reduction in cellular metabolism.

Body Temperature

In both birds and mammals, fasting and the associated depletion of energy reserves are important physiological cues to initiate torpor, which is a reduction in body temperature during inactive parts of the diurnal cycle (50). Thus, in some species, torpor occurs only when energy stores have reached a certain threshold and can be prevented by artificial administration of nutrients such as glucose (e.g., 36, 40, 51–53). The gradual depletion of energy stores may also explain why the hypothermic response is enhanced as fasting is prolonged (e.g., 54). Torpor reduces energy usage by the direct effect of temperature on metabolism and because the metabolism of activity is negligible. Hypothermia is more pronounced in small as compared to large mammals and birds (e.g., 55–57). Some hummingbirds, for example, may decrease body temperature by as much as 30°C. Small animals may benefit from this body-size effect because of their higher mass-specific metabolism and greater ease for heat transfer due to their large surface area relative to body mass. However, body mass alone does not explain the occurrence and patterns of torpor. Many small birds, such as passerines, rarely reduce body temperature by more than 3–5°C whereas some larger birds can undergo much larger changes (e.g., 57–59). Torpor appears to be more pronounced in animals that inhabit areas in which large fluctuations in temperature and food availability are common, such as deserts. However, torpor also occurs in laboratory rats (e.g., 60) and may therefore be a rather common and widespread response.

An entrance into torpor in response to food deprivation has been described in various animals. Nocturnal reductions in peripheral temperatures, associated with lower heart rate and presumably reduced metabolism, occur in large mammals such as red deer and reindeer (e.g., 61, 62). Torpor also occurs in primates; for example, torpor in response to food deprivation has recently been documented in the gray mouse lemur, *Microcebus murinu* (63). Also, barnacle geese and Puerto Rican todies undergoing long-term migration without feeding reduce body temperature by several degrees (64, 65). However, as Schleucher (57) points out, neither food supply nor energetic stress per se appear to be the ultimate factor determining the hypothermic response in these species, as the response is more pronounced in fatter premigratory birds. Torpor, therefore, may be a strategy to reduce energy expenditure during accumulation of fat stores. Thus, in addition to fasting and energy status of the individual, diverse ecological, morphological, and physiological variables, breeding, or migration periods, as well as physical parameters such as weather and annual cycles, are likely to influence the extent to which different endotherms utilize torpor (e.g., 57).

Ectothermic animals rely on external heat sources and appropriate behavior to regulate body temperature, and when provided with these opportunities, they maintain remarkably constant and well-regulated body temperatures. The effects of food deprivation have been studied in a few species freely selecting body temperature in laboratory settings. Several of these studies on fish and lizards have shown reductions in the preferred body temperature by a few degrees, which develops progressively as food is withheld (e.g., 66, 67).

Physical Activity

Decreasing physical activity and allowing body temperature to decline are likely to contribute more to energy sparing, and thereby to tolerance of starvation, than do reductions in basal metabolic rate, which are comparatively small. When food is not available, however, animals may search more actively for food, at the expense of increased energy usage, or decrease activity so as to reduce energy expenditure. An animal's use of these alternatives depends on its foraging mode, the causes of food deprivation, and many other aspects of the animal's natural history. In general, sit-and-wait predators are likely to reduce activity when food is not available, whereas active hunters and grazers are more likely to increase activity as they search for food. Furthermore, although many animals reduce physical activity during the initial phases of fasting, many other animals exhibit a marked stimulation of activity during the later and more critical phases of starvation.

In captive rats, food deprivation leads to reduced physical activity during the initial phases of food deprivation (68), followed by a marked hyperactivity when the animals enter phase III of starvation (47; see also Reference 69). Similar events ensue in captive emperor penguins, in which the transition from phase II to III and the associated depletion of fat stores coincide with increased activity and escape behaviors. Teleonomically, these responses appear beneficial, as the transition to phase III of starvation signifies that existing resources are limited and that need for food is acute.

Some [but not all (23, 66)] fasting fish and amphibians reduce activity. Mendez & Wieser (21) proposed that the behavioral response of fish to starvation consists of three phases, which has some resemblance to the biochemical changes outlined above. The first phase is short lasting (approximately 24 h) and involves the increased activity of food searching. A transition phase, in which the fish gradually reduce swimming activity and thereby lower energy expenditure, then follows. The third and final phase, adaptation, is characterized by low activity and metabolism, which persist until the fish are presented with the possibility of food. Van Dijk et al. (66) did not observe the stress phase in fasting roach (*Rutilus rutilus*), and it is quite likely that the specific responses will vary among fish with different behaviors and with the experimental setting.

Reproduction

Reproduction is energy expensive and requires either large energy stores in the mother or a ready source of food for both parents and offspring. Reproduction in many animals coincides with a high probability of food. Starvation is an inhibitor of reproduction in vertebrates; for example, most anorexic human females do not menstruate and cannot conceive (38). Female hamsters generally fail to ovulate and show little interest in sex if deprived of food for one or two estrous cycles. Ovulatory failure in these animals is related to an absence of an ovulatory gonadotropin surge and a set of immature follicles (70). Starvation for three days suppresses sexual receptivity in female rats, and this is associated with a reduction in the estrogenic response in the ventromedial nuclei of the hyopthalamus, critical for some reproductive behaviors (71). Sexually mature zebrafish spawn daily, but when they are starved, the number of eggs they produce per day drops rapidly (Figure 3) (26). The prompt decline in egg production is associated with decreased expression of CYP19a, an enzyme that converts testosterone into estrogen in female zebrafish (Figure 4).

Not all animals exhibit inhibition of reproduction during starvation. Tessier et al. (72) suggested that reproduction during starvation may be advantageous for short-lived species. Some rotifer species maintain or even increase egg production during starvation (73), whereas other rotifer species reduce reproduction and survive starvation for longer than the reproducing rotifer species. Those rotifers that increase reproduction when energy supplies are limited invest resources in their offspring, which presumably have a better chance of surviving starvation. This is at the expense of the parent's survival, possibly because of the accompanying benefits of predator avoidance, reduced energy requirements of the young, and/or increased chance of moving to a resource-rich environment.

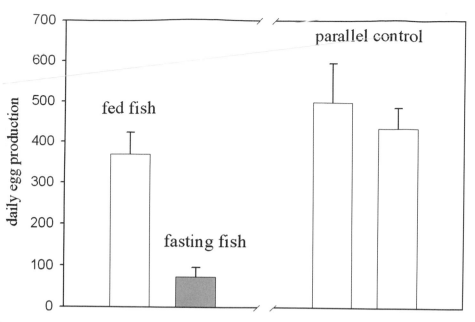

Figure 3 Daily egg production of zebrafish when fed continuously for 10 days and after 11 days of food deprivation (fasting). A parallel experiment on continuously fed fish, which served as a control group, is shown on the right side of the figure. Each group represents results from 26 pairs of zebrafish, sex ratio 1:1, and presented as mean ± S.E. Food-deprived fish produced significantly lower numbers of eggs produced than did fed fish (modified from Reference 26).

Some vertebrates starve during reproduction; the Pacific salmon and eel are classic examples. Yellow eels feed and grow in freshwater, but they stop feeding when they become silver and start their migration across the Atlantic to spawn in the Sargasso Sea. Eels that have been starved in both seawater (74) and freshwater (75) for up to three to four years have survived. During this time, these eels lost

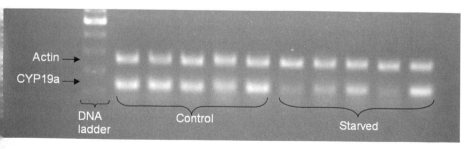

Figure 4 RT-PCR analysis of CYP19a mRNA expression in female gonad of control and zebrafish starved for 11 days.

between 70 and 80% of their body weight. Interestingly, starvation was associated with an increase in growth hormone (GH) brain-cell hypertrophy and plasma GH levels (75). GH binds predominately to liver cell membrane receptors, but the increase in plasma GH during starvation is not associated with an increase in liver growth, as these liver GH receptors are downregulated.

Atlantic salmon and trout species spawn repeatedly, whereas Pacific salmon die after their first spawning. The lifestyles of Pacific salmon vary between species. Sockeye salmon, *Oncorhynchus nerka*, are born in freshwater and enter the ocean as one- or two-year olds, weighing between 4 and 15 g. They return to their natal stream to spawn after two or three years at sea, weighing between 1.6 and 3 kg (76). These animals do not feed once they have entered the river. The upstream migration may be more than 1000 km, depending on the river. Not only do these animals swim such large distances but they also produce numerous eggs. The gonad of the female reaches 14% of the pre-spawning total body mass. The starving migrating Pacific salmon use fat to fuel both their upstream migration and egg production; at death, both sexes will have expended more than 95% of their fat reserves (10). The fat reserves when the fish enters the river are, in general, proportional to the distance to be traveled. Female pink salmon spend less energy on migration than do males but more on gonad production such that, in the end, total energy expenditure is approximately the same for both sexes. When fat reserves start to exhaust, first protein (from white muscle, but not heart or red muscle) and then carbohydrates are utilized (77). When the fish reaches the spawning ground, the calorific content of the fish is reduced to less than half of that which existed when the fish entered the river (10). There is increased interrenal activity and cortisol production, presumably directing some of the metabolic changes. The fish spawn and are usually in a moribund condition associated with energy depletion. All of the fish die, but they are not all moribund, and the cause of death is not clear although the depletion of energy reserves must be an important component. Other salmonids, such as trout and Atlantic salmon, are repeat spawners.

Immediately after reproduction, survival of the parent rather than the offspring is usually favored; this is particularly true for vertebrates. In penguins that fast on the ice during incubation of their eggs, the transition to phase III also leads to increased activity, and parents will abandon their eggs to secure their own survival at the expense of successful reproduction (46). There are, however, exceptions. Some species of octopus continually ventilate their developing embryos; feeding behavior is inhibited during this time, and these octopi can starve to death in the process. Adult cuttlefish, *Sepia officinalis*, migrate toward coastal waters to spawn and then die a month later. They also stop feeding and age rapidly during this period, with a marked deterioration in long-term memory. The cessation of feeding is associated with defects in visuomotor coordination as a result of degenerative changes in the central nervous system (78). There may be selection for genes that cause the rapid death of the postreproductive, or even just older and larger sized, individuals within the population. This may be the case in spawning salmon and lampreys. Nothing is known of such death genes, but if they exist, they may be functionally similar to genes observed in Sepia (78), in which degenerative changes in the central nervous system lead to a loss in prey capture ability and, as a consequence, death by starvation.

PHENOTYPIC PLASTICITY OF THE VISCERAL ORGANS IN RESPONSE TO DIGESTIVE STATE

The size and functional capacity of most visceral organs and muscle change in response to the physiological demands that are placed upon them (e.g., 79, 80). This phenotypic plasticity is pronounced for the gastrointestinal organs, which undergo a marked structural and functional reduction during fasting. The gastrointestinal organs are very metabolically active and have been estimated to account for as much as 40% of basal metabolic rate (e.g., 81). Thus, a reduction in organ size during fasting may confer a significant energetic savings, which may contribute to a marked reduction in the basal metabolic rate of fasting animals.

Within nonmammalian vertebrates, the effects of food deprivation on gastrointestinal organs have predominantly been studied in snakes. This group of reptiles has attracted particular interest because they tolerate very prolonged fasting periods—in some cases up to several years—and because they can ingest very large meals. Thus, under natural conditions, some snakes may eat only a few times a year, but when they do eat, they can consume prey items of 50% of their own body mass or more (e.g., 82, 83). When snakes such as pythons or rattlesnakes eat these large meals after fasting for a few weeks or longer, the mass of the small intestine increases drastically within the first 12–24 h after ingestion (84–86). The gut wall of reptiles, like that of other vertebrates, consists of an outer muscular coat and an inner mucosal layer with an epithelial lining toward the gut lumen (80, 87, 88). The mucosa in particular increases in mass upon feeding (Figure 5) (85, 89), and is attended by a many-fold increase in the transport capacities for various amino acids and glucose (85). This rise in nutrient transport capacity likely reflects that the length of the intestinal microvilli increases almost fivefold within 24 h (Figure 6) (90).

In ectothermic vertebrates, whether the nutrient transport proteins are being synthesized de novo as the enterocytes expand and microvilli lengthen, or whether the increased nutrient transport capacity merely reflects the increased surface area and that more transport proteins are exposed to the lumen, is unknown. When expressed relative to the mass of the intestine, nutrient transport capacity actually increases in mammals during hibernation (91–93). Thus, although the mucosa atrophies, the intestinal transport proteins are well preserved, and the mRNA levels of the transporter protein SGLT1 does not change during hibernation in ground squirrels (94). This is also the case for the activity and mRNA levels of Na^+, K^+ ATPase (94). The membrane potential of the enterocytes actually increases slightly during hibernation, which enhances the Na^+ gradient that drives many of the intestinal nutrient transporters (92). It is not clear how this hyperpolarization

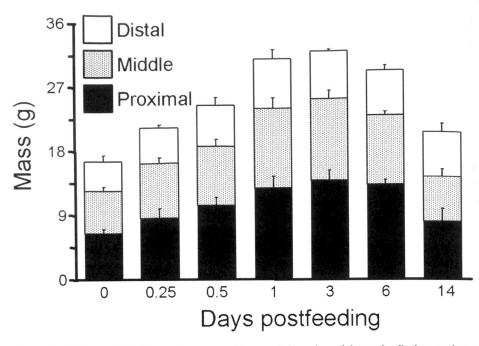

Figure 5 Effects of feeding on the mass of the small intestine of the snake *Python molurus* before and after ingestion of a meal equalling 25% of the snake's body mass. Intestinal mass is shown for fasting snakes (time 0). Each bar represents the total mass of the intestine and is divided into three parts representing the three parts of the intestine: proximal (*black*), middle (*gray*), and distal (*white*). The data are modified from Reference 90.

of the enterocytes affects cellular metabolism, and it certainly would be of interest to perform similar studies in ectothermic vertebrates in which cellular functions could be compared at similar temperatures in fasting and digesting animals.

In contrast to that of the mucosa, the thickness of the intestinal muscle layer appears unchanged (85, 89). Furthermore, digestive status does not affect the total number of gut neurons in the intestinal muscle layer, spontaneous activity of the muscle layer in vitro, or the motility responses of isolated intestinal preparations when exposed to various excitatory neuropeptides (87). The mass and nutrient transport capacity of the gastrointestinal system undergo progressive reductions during subsequent food deprivation. Although other species of nonmammalian vertebrates have received much less attention than have snakes, phenotypic plasticity both in terms of mass of the organs and their functional correlates, is seemingly universal, albeit less pronounced, in species with a more continuous feeding pattern, where prolonged periods of food deprivation are less common (e.g., 95). Thus, a progressive reduction in the intestinal epithelium during fasting, which is rapidly reversed after feeding, occurs in all other major groups of ectothermic vertebrates (e.g., 95–98, 150, 151).

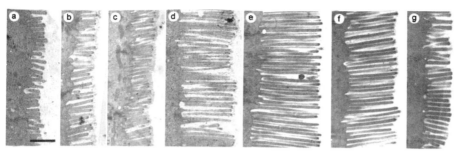

Figure 6   Electron micrographs of proximal intestinal microvilli of *Python molurus* during fasting (*a*) and at 0.25 (*b*), 0.5 (*c*), 1 (*d*), 3 (*e*), 6 (*f*), and 14 (*g*) days after ingestion of a meal equalling 25% of the snake's body mass. Note the immediate lengthening of the microvilli and the subsequent regression (bar = 1 µm). Modified from Reference 90.

Organ growth can occur by increased cell size (hypertrophy) or cell proliferation (hyperplasia). A two- or threefold increase in mucosal mass through hyperplasia would require extremely high rates of cell division and mitotic activities and presumably would be energetically expensive. Several recent studies on snakes and other ectothermic vertebrates have shown that the feeding-induced rise in intestinal mass is due predominantly to increased size of the individual enterocytes (84, 89, 90, 99–101), suggesting that hypertrophy is the major mechanism. Thus, although cell proliferation may start early in the digestive phase, cell division reaches its maximal rate rather late in the digestive process. Therefore, the cells that have been "worn down" during digestion may be replaced. In this manner, the fully functional gut can be rapidly restored when food becomes available again (89). In all ectothermic vertebrates, the epithelium—which in fasting animals is pseudostratified, with folded cell membranes of neighboring cells—may be rapidly unfolded after feeding and converted to a single layer of cells with stretched membranes as the enterocytes expand (reviewed in Reference 80). Each enterocyte appears to swell owing to a very rapid incorporation of lipid droplets (89, 90, 100), and there is also a small increase in fluid content as relative wet mass of the intestine increases (87). Although evidence is still inconclusive, the lipid droplets likely come from the ingested food; alternatively, some of the lipids may stem from fat bodies in the body of the predator. Although incorporation of the lipid droplets certainly must account for a major part of enterocytic expansion, Starck & Beese (89) also have suggested that increased lymphatic pressure contributes. The increased water content of the enterocytes, however, cannot be caused by lymph pressure per se but would require movements of osmolytes such that osmotically obliged water is dragged along. Clearly, this aspect needs further experimental clarification.

The structure and function of the intestines of birds and mammals are also flexible. However, mammals and birds normally feed on a much more regular basis than do ectothermic vertebrates, and the former generally have a constant renewal of the gut epithelium. Structural and functional changes occur rapidly after food deprivation in small mammals, whose high metabolism places extra premium on energy-saving mechanisms. For frequent feeders that are not adjusted to long periods of fasting, prolonged food deprivation or actual starvation may be more destructive to the gut, as compared to animals that normally experience long periods of fasting. The reduction in intestinal mass is due to atrophy, and the restoration of the gut upon subsequent feeding is accomplished by hyperplasia, although hypertrophy also contributes (e.g., 49, 102–105). This is also the case in hibernating mammals, although lower body temperature and metabolism greatly extend starvation tolerance (91–93, 106). The evolution of endothermy, which occurred independently in birds and mammals and has required much higher rates of nutrient uptake across the gut because of the high metabolism requirements (107), seemingly has led to a structure for which gastrointestinal and digestive plasticity is energetically more expensive.

The signals that elicit the growth of the gastrointestinal organs are not well understood in nonmammalian vertebrates. In mammals, gastrointestinal growth can be elicited through luminal, hormonal, neural, and secretory pathways (e.g., 108). Although these regulatory pathways appear phylogenetically old and conserved (e.g., 88), few studies have experimentally investigated the respective roles of the individual mechanisms. Secor et al. (109) performed systematic infusions of nutrients into the intestine of fasting animals. Infusion of amino acids or protein directly into the intestine increased intestinal mass and transport capacity, whereas infusion of glucose, lipid, or bile had no effect (109). However, only infusion of homogenized rats caused a structural and functional response equivalent to that elicited by a normal meal (109). Cephalic responses, investigated by allowing the snake to constrict a prey item, followed by its removal, did not affect the intestine (109). Luminal signals predominantly from protein, therefore, appear sufficient for intestinal expansion and rise in transport capacity during digestion. However, the mucosal mass and transport capacity of surgically isolated portions of the intestine (Thiry-Vella loops) increase in voluntarily eating snakes (101), so hormonal and/or neural pathways also seem to contribute to gut expansion. Circulating levels of a number of regulatory peptides, some released from various gastrointestinal organs, increase dramatically during digestion (110); some of these peptides may serve as trophic factors.

THE METABOLIC RESPONSE TO DIGESTION: SPECIFIC DYNAMIC ACTION

Lavoisier first showed that metabolism increases in response to digestion, and this metabolic response, documented in all animals investigated, now represents a general phenomenon. The postprandial rise in metabolism, normally referred to as the specific dynamic action of food (SDA), includes the energetic costs associated with the ingestion, digestion, absorption, and assimilation of the food. Thus the physiological mechanisms that underlie the SDA response may vary among different animals depending on feeding habits, food composition, temperature, and other factors. In the past ten years, SDA in carnivorous reptiles has received much attention, owing both to its magnitude and its potential to elucidate the large structural and functional changes in the gut. Also, in the animals that exhibit a large SDA response, the costs of digestion may account for a large proportion of the total energy budget, and the metabolic response to feeding becomes ecologically relevant (e.g., 111, 112).

The SDA response is normally characterized as the factorial rise in oxygen uptake and by its duration. Another useful manner of expressing the response is via the SDA coefficient, which is the integrated metabolic response, calculated in caloric equivalents, relative to the energy consumed. Although some have criticized this parameter (113), it provides information on the energetic costs of digestion and allows therefore for a quantitative comparison between digestive responses under different environmental parameters or among different types or amounts of food ingested. There does not appear to be an anaerobic contribution to the SDA response, and the entire response therefore is reflected in the rate of oxygen uptake (114–117). The causes and determinants of the SDA response in vertebrates is beyond the scope of the present review, and this area has been summarized elsewhere (95, 118–120). Also, the respiratory and cardiovascular correlates of the high metabolic rate during digestion have been reviewed recently (121, 122).

The effects of fasting duration on the SDA response can be a particularly insightful example of the phenotypic plasticity of the gastrointestinal organs. Thus, Overgaard et al. (123) studied the effects of the previous fasting duration on SDA response. Upon feeding, animals exhibit elevated intestinal mass and function for many days, suggesting that if the expansion of the gut is energetically expensive, then a second meal, ingested while intestinal function is still elevated, should elicit a SDA response smaller than the first response. Overgaard et al. (123) showed that the SDA coefficient does not change with a fasting duration between 3–60 days (Figure 7) and that intestinal growth does not constitute a major contributor to SDA response. Fasting duration does not affect the SDA coefficient in skinks or rattlesnakes either (124, 125). A small contribution of intestinal growth was also suggested for turtles (126). Collectively, these findings are consistent with the proposal that intestinal expansion is structurally simple and energetically cheap (89). Secor (127) subsequently estimated that gastrointestinal upregulation contributes only 5% of the SDA response in pythons. A recent study of frogs, nevertheless, shows that the rate of digestion of the first meal following three months of estivation is slower than for subsequent meals and that reconstitution of the gut may account for this delay (98). The efficiencies of accumulation of various nutrients, however, were not affected (98).

A recent study on snakes has implied that the stomach and the secretion of acid and digestive enzymes are the main contributors to the SDA response (97). In this study, the SDA response to a meal of 25% of the snake's body weight was reduced by more than half when digesting a liquid meal. The study further showed that the response was a third of its normal value when the liquid meal was infused

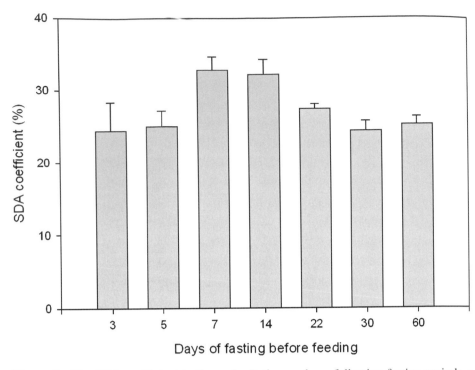

Figure 7   The SDA coefficient in the snake *Python molurus* following fasting periods o various duration. The SDA coefficient does not change with the duration of the previou fasting duration, indicating that structural and functional upregulation of the intestine occur at a low energetic cost (modified from Reference 123).

directly into the small intestine. It was estimated therefore that gastric function contribute 55% of the SDA response and that the stomach operates on a "pay before-pumping" principle, in which the snakes must spend endogenous energ: sources to initiate acid production and other digestive processes before ingeste◄ nutrients can be absorbed and used for metabolic pathways. To investigate thi possibility further, we recently used another strategy: tying off of the pylorus which is the anatomical connection between the stomach and the intestine, s◄ that the chyme was unable to enter the intestine from the stomach. In the thus operated animals, the SDA response was completely abolished, whereas sham operated animals had a normal response (Figure 8). Visual inspection of the pre◄ items clearly indicated that gastric functions had started digestion, and these dat therefore suggest that secretion of acid and digestive enzymes can proceed at relatively low energetic cost. That gastric acid secretion has a low energetic co◄ is further supported by the observation that treatment with omeprazole, a specifi inhibitor of the H^+, K^+ ATPase that drives gastric acid secretion, does not affec the SDA response in another snake species, *Boa constrictor* (128).

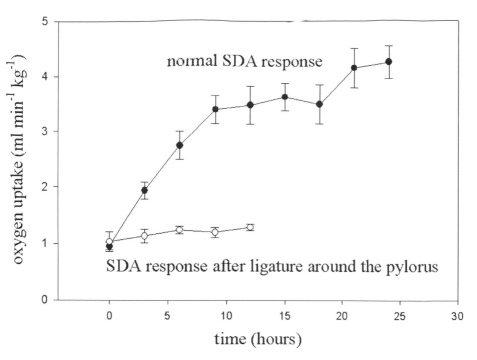

Figure 8 The metabolic response to digestion in snakes (*Python molurus*) in which the pylorus has been ligated to prevent chime from entering the intestine from the stomach (M. Andersen, H. Cueto, & T. Wang, unpublished data).

In most animals studied, the SDA response elicited by a given food type increases proportionally with meal size, and both the maximal oxygen uptake during digestion and the duration of the response increase as meal size increases (e.g., 129–133). In most cases, the SDA coefficient is unaffected by meal size, indicating that the costs of digestion are proportional to the amount of food ingested. Although these data are often interpreted to reflect that it is merely the caloric content of the food that determines the SDA, numerous studies have documented that protein-rich meals elicit larger metabolic changes than do diets composed of fat or carbohydrates. Thus, force-feeding reptiles with fat or carbohydrates elicits almost no metabolic response (e.g., 134–138). This would indicate that stimulation of protein synthesis in response to high circulating levels of amino acids (139) is a major contributor to the SDA response (140). The role of protein metabolism in the SDA response (140, 141) is pivotal in fasting catfish, toads, alligators, and pythons in which either systemic infusion of amino acids, or infusion of protein or amino acids directly into the stomach, leads to a rise in metabolism that is comparable to that observed during normal feeding (138, 141–144). In catfish and pythons, inhibition of protein synthesis with cyclohexamide completely abolishes the SDA response (138, 142, 143). If increased protein synthesis is indeed the major contributor to the SDA response, metabolism of all organs should increase during the postprandial period, a reasonable suggestion in light of the very high growth efficiency of snakes in which some 40–60% of ingested food is directed to growth (123, 145, 146). Obviously, the resulting rates of growth must require protein synthesis in all organs.

SUMMARY AND FUTURE DIRECTIONS

Digestive status affects virtually all physiological and behavioral responses, and selective pressure to enhance feeding strategies and digestive processes must be significant. The ectothermic vertebrates, with their lower metabolic rates, can endure prolonged periods of fasting, and many of these species exhibit much more pronounced changes in gastrointestinal organs than are normal in healthy mammals (see also Reference 147). The extreme structural and functional changes in their dynamic guts make ectothermic vertebrates useful models to explore largely unresolved issues regarding the interaction and prioritization of physiological functions among organ systems. These issues are of basic physiological importance. Such studies may contribute to our understanding of the mechanisms that enable organs to adapt to physiological demands. They also may help us to understand the factors that in humans can promote intestinal repair following either intestinal resections or diseases such as colitis and Crohn's disease in which there is inflammatory destruction.

ACKNOWLEDGMENTS

The authors are supported by the Danish Research Council as well as the Research Grants Council of Hong Kong Special Administrative Region, People's Republic of China (project number: CityU RGC1224/02M).

The *Annual Review of Physiology* is online at
http://physiol.annualreviews.org

LITERATURE CITED

1. Koch AL. 1971. The adaptive responses of *Escherichia coli* to a feast or famine existence. *Adv. Microb. Physiol.* 6:147–217
2. Morita RY. 1993. Bioavailability of energy and the starvation state. In *Starvation in Bacteria*, ed. S Kjelleberg, pp. 1–53. New York: Plenum
3. McCauley E, Murdoch WW, Nisbet RM. 1990. Growth, reproduction, and mortal-

ity of daphnia-pulex leydig—life at low food. *Funct. Ecol.* 4:505–14
4. Sommer U. 1984. The paradox of the plankton: Fluctuations of phosphorus availability maintain diversity of phytoplankton in flow-through cultures. *Limn Oceanogr.* 29:633–36
5. Mrosovsky N, Sherry DF. 1980. Animal anorexias. *Science* 207:837–42
6. Robin JP, Frain M, Sardet C, Groscolas R

Le Maho Y. 1988. Protein and lipid utilization during long-term fasting in emperor penguins. *Am. J. Physiol.* 254:R61–68

7. Cherel Y, Robin J-P, Le Maho Y. 1988. Physiology and biochemistry of long-term fasting birds. *Can. J. Zool.* 66:159–66

8. Le Maho Y. 1984. Metabolic adaptations to prolonged fasting in birds. *J. Physiol.* 79:113–19

9. Groscolas R, Robin J-P. 2001. Long-term fasting and re-feeding in penguins. *Comp. Biochem. Physiol.* 128A:645–55

10. Brett JR. 1995. Energetics. In *Physiological Ecology of Pacific Salmon*, ed. C Groot, L Margolis, WC Clarke, pp. 1–68. Vancouver: UBC Press

11. Cahill GF Jr, Herrera MG, Morgan AP, Soeldner JS, Steinke J, et al. 1966. Hormone-fuel interrelationships during fasting. *J. Clin. Invest.* 45:1751–69

12. Cahill GF Jr. 1976. Starvation in man. *Clin. Endocrinol. Metab.* 5:397–415

13. Henry CJK. 1990. Body mass index and the limits of human survival. *Eur. J. Clin. Nutr.* 44:329–35

14. Owen OE, Tappy L, Mozzoli MA, Smalley KJ. 1990. Acute starvation. In *The Metabolic and Molecular Basis of Acquired Disease*, ed. RD Cohen, B Lewis, KGMM Alberti, AM Denman, pp. 550–70. London: Bailliere Tinall

15. Cherel Y, Groscolas R. 1999. Relationships between nutrient storage and nutrient utilisation in long-term fasting birds and mammals. In *Proc. 22nd Int. Ornithol. Congr., Durban*, eds. NJ Adams, RH Slotow, pp. 17–34. Johannsesburg: BirdLife South Africa

16. Nagy KA, Girard IA, Brown TK. 1999. Energetics of free-ranging mammals, reptiles, and birds. *Annu. Rev. Nutr.* 19:247–77

17. Schmidt J. 1923. Breeding places and migration of the eel. *Nature* 111:51–54

18. Van Ginneken VJT, Antonissen E, Müller UK, Booms R, Eding E, et al. 2005. Eel migration to the Sargasso: remarkably high swimming efficiency and low energy costs. *J. Exp. Biol.* 208:1329–35

19. Thouzeau C, Robin J-P, Le Maho Y, Handrich Y. 1999. Body reserve dynamics and energetics of barn owls during fasting in the cold. *J. Comp. Physiol.* 169B:612–20

20. Jobling M. 1980. Effects of starvation on proximate chemical composition and energy utilization of plaice, *Pleuronectes platessa* L. *J. Fish Biol.* 17:325–34

21. Mendez G, Wieser W. 1993. Metabolic responses to food deprivation and refeeding in juveniles of *Rutilus rutilus* (Teleostei: Cyprinidae). *Environ. Biol. Fishes* 36:73–81

22. Cook JT, Sutterlin AM, McNiven MA. 2000. Effect of food deprivation on oxygen consumption and body composition of growth-enhanced transgenic Atlantic salmon *Salmo salar*. *Aquaculture* 188:47–63

23. Hervant F, Mathieu J, Durand J. 2001. Behavioural, physiological and metabolic responses to long-term starvation and refeeding in a blind cave-dwelling (*Proteus anguinus*) and a surface-dwelling (*Euproctus asper*) salamander. *J. Exp. Biol.* 204:269–81

24. Figueiredo-Garutti M, Navarro LI, Capilla E, Souza RH, Moraes G, et al. 2002. Metabolic changes in *Brycon cephalus* (Teleostei, Characidae) during post-feeding and fasting. *Comp. Biochem. Physiol* 132A:467–76

25. Navarro I, Gutierrez J. 1995. Fasting and starvation. In *Biochemistry and Molecular Biology of Fishes*, Vol. 4, ed. PW Hochachka, TP Mommsen, pp. 393–434. Amsterdam: Elsevier Science B.V.

26. Hung CY. 2005. *Survival strategies of common carp*, Cyprinus carpio, *to prolonged starvation and hypoxia*. PhD thesis. City Univ. Hong Kong. 254 pp.

27. Takeuchi T, Watanabe T. 1982. The effects of starvation and environmental temperature on proximate and fatty acid

compositions of carp and rainbow trout. *Bull. Jap. Soc. Sci. Fish* 48:1307–16

28. Shimeno S, Kheyyali D, Takeda M. 1990. Metabolic adaptation to prolonged starvation in carp. *Nippon Suisan Gakkaishi* 56:35–41

29. Shimeno S, Saida Y, Tabata T. 1996. Response of hepatic NAD- and NADP-isocitrate dehydrogenase activities to several dietary conditions in fishes. *Nippon Suisan Gakkaishi* 62:642–46

30. Shimeno S, Shikata T. 1993. Effects of acclimation temperature and feeding rate on carbohydrate-metabolizing enzyme activity and lipid content of common carp. *Nippon Suisan Gakkaishi* 59:661–66

31. Jezierska B, Hazel JR, Gerking SD. 1982. Lipid mobilization during starvation in the rainbow trout, *Salmo gairdneri* R., with attention to fatty acids. *J. Fish Biol.* 21:681–92

32. Nagai M, Ikeda S. 1971. Carbohydrate metabolism in fish—I. Effects of starvation and dietary composition on the blood glucose level and the hepatopancreatic glycogen and lipid content in carp. *Bull. Jap. Soc. Scient. Fish* 37:404–409

33. Moon TW, Foster GD. 1995. Tissue carbohydrate metabolism, gluconeogenesis and hormonal and environmental influences. In *Biochemistry and Molecular Biology of Fishes*, Vol. 4, ed. PW Hochachka, TP Mommsen, pp. 254–96. Amsterdam: Elsevier Science B.V.

34. Threlkeld ST. 1976. Starvation and the size structure of zooplankton communities. *Freshw. Biol.* 6:489–96

35. Peters RH. 1983. *The Ecological Implications of Body Size*. Cambridge, MA: Cambridge Univ. Press

36. Calder WA. 1994. When do hummingbirds use torpor in nature? *Physiol. Zool.* 67:1051–76

37. Øritsland NA. 1990. Starvation survival and body composition in mammals with particular reference to *Homo sapiens*. *Bull. Math. Biol.* 52:643–55

38. Keys A, Brozek J, Hennschel A,

Michelsen O, Taylor HL. 1950. *The Biology of Human Starvation*. Minneapolis: Univ. Minn. Press

39. Merkt JR, Taylor CR 1994. "Metabolic switch" for desert survival. *Proc. Natl. Acad. Sci. USA* 91:12313–16

40. Ehrhardt N, Heldmaier G, Exner C. 2005. Adaptive mechanisms during food restriction in *Acomys russatus*: the use of torpor for desert survival. *J. Comp. Physiol. B* 175: 193–200

41. Poon WL. 2005. *In vivo changes in common carp (*Cyprinus carpio L.*) liver during hypoxia at the molecular and cellular levels.* PhD thesis. City Univ. Hong Kong. 182 pp.

42. Brown HO, Levine ML, Lipkin M. 1963. Inhibition of intestinal epithelial cell renewal and migration induced by starvation. *Am. J. Physiol.* 205:868–872

43. Habold C, Chevalier C, Dunel-Erb S, Foltzer-Jourdainne C, Le Maho Y, Lignot J-H. 2004. Effects of fasting and refeeding on jejunal morphology and cellular activity in rats in relation to depletion of body stores. *Scand. J. Gastroenterol.* 39:531–39

44. Iwakiri R, Gotoh Y, Noda T, Sugihara H, Fujimoto K, et al. 2001. Programmed cell death in rat intestine: effect of feeding and fasting. *Scand. J. Gastroenterol.* 36:39–47

45. Habold C, Foltzer-Jourdainne C, Le Maho Y, Lignot J-H. 2005. Intestinal apoptotic changes linked to metabolic status in fasted and refed rats. *Eur. J. Physiol.* In press

46. Robin J-P, Boucentet L, Chillet P, Griscolas R. 1998. Behavioral changes in fasting emperor penguins: evidence for a "refeeding signal" linked to metabolic shifts. *Am. J. Physiol.* 43:R746–53

47. Koubi HE, Robin JP, Dewasmes G, Le Maho Y, Frutoso J, Minaire Y. 1991. Fasting-induced rise in locomotor activity in rats coincides with increased protein utilization. *Physiol. Behav.* 50:337–43

48. Bertile F, Oudart H, Criscuolo F, Le Maho

Y, Raclot T. 2003. Hypothalamic gene expression in long-term fasted rats: relationship with body fat. *Biochem. Biophys. Res. Commun.* 303.1106–13

49. Dunel-Erb S, Chevalier C, Laurent P, Bach A, Decrock F, Le Maho Y. 2001. Restoration of the jejunal mucosa in rats refed after prolonged fasting. *Comp. Biochem. Physiol.* 129A:933–47

50. Geiser F. 2004. Metabolic rate and body temperature reduction during hibernation and daily torpor. *Annu. Rev. Physiol.* 66:239–74

51. Lovegrove BG, Raman J, Perrin MR. 2001. Daily torpor in elephant shrews (*Macroscelidea: Elephantulus spp.*) in response to food deprivation. *J. Comp. Physiol.* 171B:11–21

52. Bech C, Abe AS, Steffensen JF, Berger M, Bicudo JEPW. 1997. Torpor in three species of Brazilian hummingbirds under semi-natural conditions. *Condor* 99:780–88

53. Powers DR, Brown AR, Van Hook JA. 2003. Influence of normal daytime fat deposition on laboratory measurements of torpor use in territorial versus nonterritorial hummingbirds. *Physiol. Biochem. Zool.* 76:389–97

54. Prinzinger R, Schleucher E, Preßmar A. 1992. Long-term telemetry of body temperature with synchronous measurement of metabolic rate in torpid and non-torpid blue naped mousebirds (*Urocolius macrourus*). *J. Ornithol.* 133:446–50

55. Geiser F, Ruf T. 1995. Hibernation versus daily torpor in mammals and birds: physiological variables and classification of torpor patterns. *Physiol. Zool.* 68:935–66

56. Geiser F. 1998. Evolution of daily torpor and hibernation in birds and mammals: importance of body size. *Clin. Exp. Pharmacol. Physiol.* 25:736–40

57. Schleucher E. 2004. Torpor in birds: taxonomy, energetics, and ecology. *Physiol. Biochem. Zool.* 77:942–49

58. Graf R, Krishna S, Heller HC. 1989. Reg-

ulated nocturnal hypothermia induced in pigeons by food deprivation. *Am. J. Physiol.* 256:R733–38

59. McKechnie AE, Lovegrove BG. 2002. Avian facultative hypothermic responses: a review. *Condor* 104:704–24

60. Severinsen T, Munch IC. 1999. Body core temperature during food restriction in rats. *Acta. Physiol. Scand.* 165:299–305

61. Mesteig K, Tyler NJ, Blix AS. 2000. Seasonal changes in heart rate and food intake in reindeer (*Rangifer tarandus tarandus*). *Acta. Physiol. Scand.* 170:145–51

62. Arnold W, Ruf T, Reimoser S, Tataruch F, Onderscheka K, Schober F. 2004. Nocturnal hypometabolism as an overwintering strategy of red deer (*Cervus elaphus*). *Am. J. Physiol.* 286:R174–81

63. Genin F, Perret M. 2003. Daily hypothermia in captive grey mouse lemurs (*Microcebus murinus*): effects of photoperiod and food restriction. *Comp. Biochem. Physiol.* 136B:71–81

64. Merola-Zwartjes M, Ligon JD. 2000. Ecological energetics of the Puerto Rican tody: heterothermy, torpor and intraisland variation. *Ecology* 81:990–1002

65. Butler PJ, Woakes AJ. 2001. Seasonal hypothermia in a large migrating bird: saving energy or fat deposition? *J. Exp. Biol.* 204:1361–67

66. Van Dijk PLM, Staaks G, Hardewig I. 2002. The effect of fasting and refeeding on temperature preference, activity and growth of roach, *Rutilus rutilus*. *Oecologia* 130:496–504

67. Brown RP, Griffin S. 2005. Lower selected body temperatures after food deprivation in the lizard *Anolis carolinensis*. *J. Thermal Biol.* 30:79–83

68. Sclafani A, Rendel A. 1978. Food deprivation-induced activity in dietary obese, dietary lean and normal-weight rats. *Behav. Biol.* 24:220–28

69. Pierre PJ, Skjoldager P, Bennett AJ, Renner MJ. 2001. A behavioral characterization of the effects of food deprivation on food and nonfood object interaction: an

investigation of the information-gathering functions of exploratory behavior. *Physiol. Behav.* 72:189–97

70. Morin LP. 1986. Environment and hamster reproduction: responses to phase-specific starvation during estrous cycle. *Am. J. Physiol.* 251:R663–69

71. Jones JE, Wade GN. 2002. Acute fasting decreases sexual receptivity and neural estrogen receptor-α in female rats. *Physiol. Behav.* 77:19–25

72. Tessier AJ, Henry LL, Goulden CE, Durand MW. 1983. Starvation in daphnia: energy reserves and reproductive allocation. *Limnol. Oceanogr.* 28:667–76

73. Kirk KL. 1997. Life-history responses to variable environments: starvation and reproduction in planktonic rotifers. *Ecology* 78:434–41

74. Boetius I, Boetius J. 1985. Lipid and protein content in *Anguilla anguilla* during growth and starvation. *Dana* 4:1–17

75. Olivereau M, Olivereau JM. 1997. Long-term starvation in the European eel: general effects and responses of pituitary growth hormone-(GH) and somatolactin-(SL) secreting cells. *Fish Physiol. Biochem.* 17:261–69

76. Weatherley AH, Gill HS. 1985. Dynamics of increase in muscle fibers in fishes in relation to size and growth. *Experientia* 41:353–54

77. French CJ, Hochachka PW, Mommsen TP. 1983. Metabolic organization of liver during spawning migration of sockeye salmon. *Am. J. Physiol.* 245:R827–30

78. Chichery MP, Chichery R. 1992. Behavioural and neurohistological changes in aging *Sepia*. *Brain Res.* 574:77–84

79. Piersma T, Lindstrom A. 1997. Rapid reversible changes in organ size as a component of adaptive behaviour. *Trends Ecol. Evol.* 12:134–38

80. Starck JM. 2005. Structural flexibility of the digestive system of tetrapods—patterns and processes at the cellular and

tissue levels. In *Physiological and Ecological Adaptations to Feeding in Vertebrates*, ed. JM Starck, T Wang, pp. 175–200. New Delhi: Sci. Publ. Inc.

81. Cant JP, McBide BW, Croom WJ. 1996. The regulation of intestinal metabolism and its impact on whole animal energetics. *J. Anim. Sci.* 74:2541–53

82. Greene HW. 1983. Dietary correlates of the origin and radiation of snakes. *Am. Zool.* 23:431–41

83. Shine R, Harlow PS, Keogh JS, Boeadi 1998. The influence of sex and body size on food habits of a giant tropical snake *Python reticulatus*. *Func. Ecol.* 12:248–58

84. Secor SM, Stein ED, Diamond J. 1994. Rapid upregulation of snake intestine in response to feeding: a new mode of intestinal adaptation. *Am. J. Physiol.* 29:G695–705

85. Secor SM, Diamond J. 1995. Adaptive responses to feeding in Burmese pythons: pay before pumping. *J. Exp. Biol.* 198:1313–25

86. Secor SM, Diamond J. 1998. A vertebrate model of extreme physiological regulation. *Nature* 395:659–62

87. Holmberg A, Joanna K, Persson A, Jensen J, Wang T, Holmgren S. 2002. Effects of digestive status on the reptilian gut. *Comp. Biochem. Physiol.* 133A:499–18

88. Holmgren S, Holmberg A. 2005. Control of gut motility and secretion in fasting and fed non-mammalian vertebrates. In *Physiological and Ecological Adaptations to Feeding in Vertebrates*, ed. JM Starck, T Wang, pp. 325–62. New Delhi: Sci. Publ. Inc.

89. Starck JM, Beese K. 2001. Structural flexibility of the intestine of Burmese python in response to feeding. *J. Exp. Biol.* 204:325–35

90. Lignot J-H, Helmstetter C, Secor SM. 2005. Postprandial morphological response of the intestinal epithelium of the Burmese python (*Python molurus*). *Comp. Biochem. Physiol.* 141A:280–91

91. Carey HV. 1990. Seasonal changes in mucosal structure and function in ground squirrel intestine. *Am. J. Physiol.* 259:R385–92

92. Carey HV. 1995. Gut feelings about hibernation. *News Physiol. Sci.* 10:55–61

93. Carey HV. 2005. Gastrointestinal responses to fasting in mammals: lessons from hibernators. In *Physiological and Ecological Adaptations to Feeding in Vertebrates*, ed. JM Starck, T Wang, p. 229–254. New Delhi: Sci. Publ. Inc.

94. Carey HV, Martin SL. 1996. Preservation of intestinal gene expression during hibernation. *Am. J. Physiol.* 271:804–13

95. Secor SM. 2001. Regulation of digestive performance: a proposed adaptive response. *Comp. Biochem. Physiol.* 128A:563–75

96. McLeese JM, Moon TW. 1989. Seasonal changes in the intestinal mucosa of the winter flounder, *Pseudopleuronectes americanus* (Wallbaum) from Passamquoddy Bay, New Brunswick. *J. Fish Biol.* 35:381–93

97. Secor SM. 2005. Physiological responses to feeding, fasting and estivation for anurans. *J. Exp. Biol.* 208:2595–609

98. Cramp RL, Franklin CE. 2003. Is refeeding efficiency compromised by prolonged starvation during aestivation in the green striped burrowing frog, *Cyclorana alboguttata*? *J. Exp. Zool.* 300:126–32

99. Jackson K, Perry G. 2000. Changes in intestinal morphology following feeding in the brown treesnake, Boiga irregularis. *J. Herpetol.* 34:459–62

100. Starck JM, Beese K. 2002. Structural flexibility of the small intestine and liver of Garter snakes in response to feeding and fasting. *J. Exp. Biol.* 205:1377–88

101. Secor SM, Whang EE, Lane JS, Ashley SW, Diamond J. 2000. Luminal and systemic signals trigger intestinal adaptation in the juvenile python. *Am. J. Physiol.* 279:G1177–87

102. Altmann GG. 1972. Influence of starvation and refeeding on mucosal size and epithelial renewal in the rat small intestine. *Am. J. Anat.* 133:391–400

103. Starck JM, Kloss E. 1995. Structural responses of Japanese quail intestine to different diets. *Dtsch. Tierarztl. Wochenschr.* 102:146–50

104. Hume ID, Biebach H. 1996. Digestive tract function in the long distance migratory garden warbler, *Sylvia borin*. *J. Comp. Physiol. B* 166:388–95

105. Ferrari RP, Carey HV. 2000. Intestinal transport during fasting and malnutrition. *Annu. Rev. Nutr.* 20:195–19

106. Hume ID, Beiglbock C, Ruf T, Frey-Ross F, Bruns U, Arnold W. 2002. Seasonal changes in morphology and function of the gastrointestinal tract of free-living alpine marmots (*Marmota marmota*). *J. Comp. Physiol. B* 172:197–207

107. Karasov WH, Diamond JM. 1985. Digestive adaptations for fueling the cost of endothermy. *Science* 228:202–4

108. Johnson LR. 1997. *Gastrointestinal Physiology*. St. Louis, MO: Mosby. 1023 pp.

109. Secor SM, Lane JS, Whang EE, Ashley SW, Diamond J. 2002. Luminal nutrient signals for intestinal adaptation in pythons. *Am. J. Physiol.* 283:G1298–1309

110. Secor SM, Fehsenfeld D, Diamond J, Adrian TE. 2001. Responses of python gastrointestinal regulatory peptides to feeding. *Proc. Natl. Acad. Sci. USA* 98: 13637–42

111. Peterson CC, Walton BM, Bennett AF. 1999. Metabolic costs of growth in free-living garter snakes and the energy budgets of ectotherms. *Func. Ecol.* 13:500–507

112. McCue MD, Lillywhite HB. 2002. Oxygen consumption and the energetics of island-dwelling Florida cottonmouth snakes. *Physiol. Biochem. Zool.* 75:165–78

113. Beaupre SJ. 2005. Ratio representations of specific dynamic action (mass-specific SDA and SDA coefficient) do not standardize for body mass and meal size. *Physiol. Biochem. Zool.* 78:126–31

114. Andersen JB, Wang T. 2003. Cardiorespiratory effects of forced activity and digestion in toads. *Physiol. Biochem. Zool.* 76:459–70

115. Overgaard J, Busk M, Hicks JW, Jensen FB, Wang T. 1999. Respiratory consequences of feeding in the snake *Python molorus. Comp. Biochem. Physiol.* 124A:361–67

116. Busk M, Jensen FB, Wang T. 2000. The effects of feeding on blood gases, acid-base parameters and selected metabolites in the bullfrog *Rana catesbeiana. Am. J. Physiol.* 278:R185–95

117. Busk M, Overgaard J, Hicks JW, Bennett AF, Wang T. 2000. Effects of feeding on arterial blood gases in the American alligator, *Alligator mississippiensis. J. Exp. Biol.* 203:3117–24

118. Jobling M. 1981. The influences of feeding on the metabolic rate of fishes: a short review. *J. Fish Biol.* 18:385–400

119. Wang T, Zaar M, Arvedsen S, Vedel C, Overgaard J. 2002. Effects of temperature on the metabolic response to feeding in *Python molurus. Comp. Biochem. Physiol.* 133A:519–27

120. Andrade DV, Abe AS, Cruz-Neto AP, Wang T. 2005. Specific dynamic action in ectothermic vertebrates: a general review on the determinants of the metabolic responses to digestion in fish, amphibians and reptiles. In *Adaptations in Physiological and Ecological Adaptations to Feeding in Vertebrates*, ed. JM Starck, T Wang, pp. 305–24. New Delhi: Sci. Publ. Inc.

121. Wang T, Busk M, Overgaard J. 2001. The respiratory consequences of feeding in amphibians and reptiles. *Comp. Biochem. Physiol.* 128A:533–47

122. Wang T, Andersen J, Hicks JW. 2005. Effects of digestion on the respiratory and cardiovascular physiology of amphibians and reptiles. In *Adaptations in Physiological and Ecological Adaptations to Feeding in Vertebrates*, ed. JM Starck, T Wang, pp. 279–303. New Delhi: Sci. Publ. Inc.

123. Overgaard J, Andersen JB, Wang T. 2002. The effects of fasting duration on the metabolic response to feeding in *Python molurus*: an evaluation of the energetic costs associated with gastrointestinal growth and upregulation. *Physiol. Biochem. Zool.* 75:360–68

124. Iglesias S, Thompson MB, Seebacher F. 2003. Energetic cost of a meal in a frequent feeding lizard. *Comp. Biochem. Physiol.* 135A:377–82

125. Zaidan F, Beaupre SJ. 2003. Effects of body mass, meal size, fast length, and temperature on specific dynamic action in the timber rattlesnake (*Crotalus horridus*). *Physiol. Biochem. Zool.* 76:447–58

126. Hailey A. 1998. The specific dynamic action of the omnivorous tortoise *Kinixys spekii* in relation to diet, feeding pattern, and gut passage. *Physiol. Zool.* 71:57–66

127. Secor SM. 2003. Gastric function and its contribution to the postprandial metabolic response of the Burmese python *Python molurus. J. Exp. Biol.* 206:1621–30

128. Andrade DV, Toledo LP, Abe AS, Wang T. 2004. Ventilatory compensation of the alkaline tide during digestion in the snake *Boa constrictor. J. Exp. Biol.* 207:1379–85

129. Soofiani NM, Hawkins AD. 1982. Energetic costs at different levels of feeding in juvenile cod, *Gadus morhua* L. *J. Fish Biol.* 21:577–92

130. Andrade DV, Cruz-Neto AP, Abe AS. 1997. Meal size and specific dynamic action in the rattlesnake *Crotalus durissus* (Serpentes: Viperidae). *Herpetologica* 53:485–93

131. Boyce SJ, Clarke A. 1997. Effect of body size and ration on specific dynamic action in the Antarctic plunderfish, *Harpagifer antarcticus* Nybelin 1947. *Physiol. Zool.* 70:679–90

132. Secor SM, Diamond J. 1997. Effects of meal size on postprandial responses in juvenile Burmese pythons (*Python molurus*). *Am. J. Physiol.* 272:R902–12

133. Secor SM, Faulkner AC. 2002. Effects of meal size, meal type, body temperature, and body size on the specific dynamic action of the marine toad, *Bufo marinus*. *Physiol. Biochem. Zool.* 75:557–71

134. Benedict FG. 1932. *The Physiology of Large Reptiles with Special Reference to the Heat Production of Snakes, Tortoises, Lizards, and Alligators*. Washington, DC: Carnegie Inst. Publ. 254 pp.

135. Jobling M, Davies PS. 1980. Effects of feeding on metabolic rate, and the specific dynamic action in plaice, *Pleuronectes platessa* L. *J. Fish Biol.* 16:629–38

136. Coulson RA, Hernandez T. 1983. Alligator metabolism: studies on chemical reactions in vivo. *Comp. Biochem. Physiol.* 74A:1–182

137. Somanath B, Palavesam A, Lazarus S, Ayyapan M. 2000. Influence of nutrient source on specific dynamic action of pearl spot, *Etroplus suratensis* (Bloch). *Naga* 23:15–17

138. McCue MD, Bennett AF, Hicks JW. 2005. The effect of meal composition on specific dynamic action in burmese pythons (*Python molurus*). *Physiol. Biochem. Zool.* 78:182–92

139. Houlihan DF. 1991. Protein turnover in ectotherms and its relationships to energetics. *Adv. Comp. Physiol. Biochem.* 7:1–43

140. Ashworth A. 1969. Metabolic rates during recovery from protein-calorie malnutrition: the need for a new concept of specific dynamic action. *Nature* 223:407–409

141. Coulson RA, Hernadez T. 1979. Increase in metabolic rate of the alligator fed proteins or amino acids. *J. Nutr.* 109:538–50

142. Brown CR, Cameron JN. 1991. The relationship between specific dynamic action (SDA) and protein synthesis rates in the channel catfish. *Physiol. Zool.* 64:298–309

143. Brown CR, Cameron JN. 1991. The induction of specific dynamic action in channel catfish by infusion of essential amino acids. *Physiol. Zool.* 64:276–97

144. Wang T, Burggren WW, Nobrega E. 1995. Metabolic, ventilatory, and acid-base responses associated with specific dynamic action in the toad *Bufo marinus*. *Physiol. Zool.* 68:192–205

145. Vinegar A, Hutchison VH, Dowling HG. 1970. Metabolism, energetics and thermoregulation during brooding of snakes of the genus Python (Reptilia, Boidae). *Zoologica* 55:19–48

146. Secor SM, Diamond J. 1997. Determinants of the postfeeding metabolic response of burmese pythons, *Python molurus*. *Physiol. Zool.* 70:202–12

147. Pennisi E. 2005. The dynamic gut. *Science* 307:1896–99

148. Bauer M, Hamm AC, Bonaus M, Jacob A, Jaekel J, et al. 2004. Starvation response in mouse liver shows strong correlation with life-span-prolonging processes. *Physiol. Genomics* 17:230–44

149. Zhang J, Underwood LE, D'Ercole AJ. 2001. Hepatic mRNAs up-regulated by starvation: an expression profile determined by suppression subtractive hybridization. *FASEB J.* 15:1261–63

150. Rios FS, Kalinin AL, Fernandes MN, Rantin FT. 2004. Changes in gut gross morphology of traira, *Hoplias malabaricus* (Teleostei, Erythrinidae) during long-term starvation and after refeeding. *Braz. J. Biol.* 64:683–89

151. Van Dijk PLM, Hardewig I, Holker F. 2005. Energy reserves during food deprivation and compensatory growth in juvenile roach: the importance of season and temperature. *J. Fish Biol.* 66:167–181

Annu. Rev. Physiol. 2006. 68:253–78
doi: 10.1146/annurev.physiol.68.040104.110001
Copyright © 2006 by Annual Reviews. All rights reserved
First published online as a Review in Advance on August 25, 2005

Oxidative Stress in Marine Environments:
Biochemistry and Physiological Ecology

Michael P. Lesser

*Department of Zoology and Center for Marine Biology, University of New Hampshire,
Durham, NH 03824; email: mpl@unh.edu*

Key Words   reactive oxygen species, antioxidants, superoxide dismutase,
apoptosis, superoxide radicals

■ **Abstract**   Oxidative stress—the production and accumulation of reduced oxygen
intermediates such as superoxide radicals, singlet oxygen, hydrogen peroxide, and hydroxyl radicals—can damage lipids, proteins, and DNA. Many disease processes of
clinical interest and the aging process involve oxidative stress in their underlying etiology. The production of reactive oxygen species is also prevalent in the world's oceans,
and oxidative stress is an important component of the stress response in marine organisms exposed to a variety of insults as a result of changes in environmental conditions
such as thermal stress, exposure to ultraviolet radiation, or exposure to pollution. As
in the clinical setting, reactive oxygen species are also important signal transduction
molecules and mediators of damage in cellular processes, such as apoptosis and cell
necrosis, for marine organisms. This review brings together the voluminous literature
on the biochemistry and physiology of oxidative stress from the clinical and plant
physiology disciplines with the fast-increasing interest in oxidative stress in marine
environments.

INTRODUCTION

Early History of Oxygen

The geological record provides convincing evidence of the long history of life on
Earth, starting in the Archean as far back as 3.8 Gyr (1). The atmosphere of Earth
was originally highly reduced and dominated by microbes (2), but by the mid-
to-early Archean cyanobacteria capable of oxygenic photosynthesis had evolved
(1, 2). With an abundance of carbon dioxide (CO_2), water (H_2O) as a reductant, and
solar radiation, oxygenic photosynthesis by cyanobacteria spread and evolved into
other taxa by serial endosymbioses (3). As a result, molecular oxygen, or dioxygen
(O_2), appeared in significant amounts in the Earth's atmosphere $\sim$2.5 Gyr and
accumulated in the upper atmosphere. The accumulation of O_2 changed terrestrial
and shallow oceanic habitats and provided strong selective pressures on anaerobic
life forms existing at the end of the Archean. The evolution of aerobic respiration,

with its greater efficiency and higher yields of energy, is believed to have been critical to the development of complex multicellular eukaryotic organisms.

Theory of Oxygen Toxicity

It has only been fifty years since it was proposed that free radicals are responsible for the toxic effects of oxygen (4). Atmospheric O_2 in its ground state is distinctive among the elements because it has two unpaired electrons (and thus is known as a biradical) (5–8). This property makes O_2 paramagnetic, which significantly limits its ability to interact with organic molecules unless it is "activated." The univalent reduction of molecular oxygen produces reactive intermediates such as the superoxide radical (O_2^-), singlet oxygen ($^1O_2$), hydrogen peroxide (H_2O_2), hydroxyl radical ($HO^\bullet$), and finally water (H_2O) (5–8). Often biologists label all of the reduction products of oxygen as free radicals. As defined above, however, a free radical is an atom or molecule with an unpaired electron. It therefore is more appropriate to refer to the intermediate reduction products of oxygen as activated and not as free radicals, but for consistency I use reactive oxygen species (ROS) throughout and include H_2O_2 in that definition.

All photosynthetic and respiring cells produce ROS, including O_2^- via the univalent pathway; H_2O_2 is formed by the continued reduction of O_2^-; and eventually $HO^\bullet$ is formed and then reduced to the hydroxyl ion and water (5–8). For biological systems the production of ROS is directly and positively related to the concentration of O_2 (9). Oxidative stress, the production and accumulation of ROS beyond the capacity of an organism to quench these reactive species, can damage lipids, proteins, and DNA, but ROS can also act in signal transduction (5–8). The central purposes of antioxidant defenses in biological systems are to quench $^1O_2$ at the site of production and to quench or reduce the flux of reduced oxygen intermediates such as O_2^- and H_2O_2 to prevent the production of $HO^\bullet$, the most damaging of the ROS (5–8).

Reactive Oxygen Species

SINGLET OXYGEN   In biological systems $^1O_2$ is produced through several photochemical and chemical pathways. Singlet oxygen is often produced by photosensitization reactions in which molecules absorb light of a specific wavelength and are raised to a higher energy state. The energy can then be passed to O_2 and forms $^1O_2$ while the sensitizing molecule returns to its ground state. The lifetime of $^1O_2$ is ~ 3.7 μs in aqueous media. Its high reactivity with cellular components is controlled primarily by diffusion, whose mean distance has been estimated to be ~ 82 nm; therefore, site-specific effects in biological systems are likely to occur with this ROS (5–8).

SUPEROXIDE RADICALS O_2^- can act as either an oxidant or a reductant in biological systems. The dismutation of O_2^-, leading to the formation of H_2O_2, occurs spontaneously or is catalyzed by the antioxidant enzyme superoxide dismutase with a rate constant of $2 \times 10^9 \, mol^{-1} \, s^{-1}$ (5). In its protonated state (pKa = 4.8) O_2^- forms the perhydroxyl radical ($^\bullet OOH$), which is a powerful oxidant (5–8), but its biological relevance is probably minor because of its low concentration at physiological pH. Within the aprotic interiors of biological membranes, such as mitochondrial or thylakoid membranes, O_2^- is stable (10, 11) and can diffuse across the membrane in a concentration-dependent manner but at extremely slow rates (2.1×10^{-6} cm s^{-1}) (5–8). Although SOD reduces the steady-state concentration of O_2^- by several orders of magnitude, O_2^- still has significant, and independent, damaging potential (12) with a lifetime of 50 μs and a diffusion distance of $\sim$320 nm (5–8).

HYDROGEN PEROXIDE   Because hydrogen peroxide is uncharged, it readily diffuses across biological membranes. H_2O_2 causes significant damage because it is not restricted to its point of synthesis in the cell and can enter into numerous other reactions. Exposure to H_2O_2 can damage many cellular constituents directly, such as DNA and enzymes involved in carbon fixation (5–8). H_2O_2 is also involved in pathways such as programmed cell death, or apoptosis (8). If H_2O_2 is further reduced, it can produce $HO^\bullet$. One source of electrons for that reduction in biological systems is transition metals via so-called Fenton chemistry, such as the conversion of Fe from its ferrous to ferric form (5–8).

HYDROXYL RADICAL   The $HO^\bullet$ is the most reactive oxygen radical. It has tremendous potential for biological damage because it attacks all biological molecules in a diffusion-controlled fashion, with a lifetime of 10^{-7} s and mean diffusion distance of 4.5 nm. It also tends to initiate free radical chain reactions, can oxidize membrane lipids, and causes proteins and nucleic acids to denature (5–8). The production of $HO^\bullet$ in biological systems is regulated by the availability of ferrous iron. Any recycling of iron from the ferric to the ferrous form by a reducing agent can maintain an ongoing Fenton reaction, leading to the generation of $HO^\bullet$. One excellent reducing agent is O_2^-, which participates in the metal-catalyzed Haber-Weis reaction (6–8). Metals other than iron (e.g., copper) may also participate in these electron transfer reactions by cycling between the oxidized and reduced states.

REACTIVE NITROGEN SPECIES   Many cells also produce nitric oxide, or nitrogen monoxide ($NO^\bullet$), a molecule implicated initially in neurotransmission but now a known participant in diverse processes involving oxidative stress (13). Nitric oxide synthase produces $NO^\bullet$, which can react with O_2^- to form the peroxynitrite anion ($ONOO^-$), a potent oxidant (12). Because the solubility of $NO^\bullet$ is similar to that of H_2O, the former can readily diffuse across biological membranes. It can then react at near-diffusion-limited rates with free radicals, especially O_2^-, to form $ONOO^-$, which can diffuse across biological membranes at rates 400 times greater than does O_2^- (14, 15). The half-life of $ONOO^-$ is <0.1 s at physiological pH, mostly because of its high reactivity with organic molecules, especially lipids.

The high concentrations of NO$^\bullet$ may create significant competition between NO$^\bullet$ and SOD for O_2^-. This balance between the competition for O_2^- may be a major determinant of oxidative stress in many organisms. Many investigators are now re-evaluating the role of O_2^- in oxidative stress because of these new insights and because many of the observed in vitro effects ascribed to O_2^- may in fact be mediated by ONOO$^-$ (8).

Cellular Sites of ROS Production

CHLOROPLASTS   Chloroplasts, because of their photosynthetic nature, are hyperoxic, produce ROS, and are susceptible to oxidative stress. ROS in the chloroplast may damage photosystem (PS) II, primarily through oxidative degradation of the D1 protein (5, 16–19), and also inhibit the repair of damage to PS II (20). In addition to $^1O_2$ (21), O_2^- and HO$^\bullet$ also are produced in the PS II reaction center (22). The reducing side of PS I can reduce O_2 to O_2^- by the Mehler reaction and is the most significant site of O_2^- production in the chloroplast (5, 16, 23). The production of O_2^- increases under stressful conditions, such as exposure to xenobiotics or pollutants, high visible irradiances, exposure to ultraviolet radiation (UVR), and/or exposure to thermal stress. This elevated production can overwhelm antioxidant defenses to produce damage to both PS II and to the carbon fixation process (5, 16).

MITOCHONDRIA   Two main sites of O_2^- generation in the inner mitochondrial membrane are (a) NADH dehydrogenase at complex I and (b) the interface between ubiquinone and complex III. Once generated, the O_2^- is then converted to H_2O_2 by spontaneous dismutation or by SOD (24). The integrity of the inner membrane and the associated complexes is essential to oxidative phosphorylation. The inner membrane is also permeable to H^+. Although it causes energy loss, this H^+ leakage can be beneficial because it reduces ROS production. The loss of energy and the production of ROS via H^+ leakage can also be regulated by specific uncoupling proteins, which themselves are upregulated by the production of ROS (25).

ENDOPLASMIC RETICULUM   The endoplasmic reticulum of animals, plants, and some bacteria contain cytochromes collectively known as cytochrome P-450. Cytochrome P-450 is involved in several detoxification processes, including hydroxylations, dealkylations, deaminations, dehalogenations, and desaturations that involve the reduction of O_2 (8). These mixed-function oxygenase (MFO) reactions add an O_2 atom to an organic substrate using NADPH as the electron donor. Superoxide can be produced by microsomal NADPH-dependent electron transport involving cytochrome P-450.

MICROBODIES   Peroxisomes and glyoxysomes are subcellular organelles that contain enzymes involved in the β-oxidation of fatty acids and in photorespiration, such as glycolate oxidase, catalase, and several peroxidases. Found in both animals and plants, these organelles were initially believed to be involved in detoxification reactions and the quenching of H_2O_2. The H_2O_2 synthesized by these microbodies may also contribute to the pool of signal transduction molecules (26). In the glyoxisomes of plants, glycolate oxidase produces H_2O_2 in a two-electron transfer from glycolate to oxygen (26). In addition to H_2O_2, glyoxisomes produce O_2^- via a xanthine oxidase reaction with purines (26, 27). The dismutation of O_2^- to H_2O_2 also occurs via SOD in both peroxisomes and glyoxisomes (28).

THE DUAL ROLE OF REACTIVE OXYGEN SPECIES

Oxidative Damage

Reactive oxygen species are both agents of disease and cellular damage, but they are also participants in many normal cellular functions. Below I briefly describe some of the major sites of damage and pathways in which ROS plays an important regulatory role.

OXIDATIVE DAMAGE TO LIPIDS   The reaction of ROS, especially of HO•, with lipids is one of the most prevalent mechanisms of cellular injury and is dependent on the degree of membrane fluidity, which in turn is a function of the saturation state of the lipid bilayer (8). The degradation products of lipid peroxidation are aldehydes, such as malondialdehyde, and hydrocarbons, such as ethane and ethylene (29, 30). Lipid peroxidation in mitochondria is particularly cytotoxic, with multiple effects on enzyme activity and ATP production as well as on the initiation of apoptosis (31).

OXIDATIVE DAMAGE TO PROTEINS   Oxidative attack on proteins results in site-specific amino acid modifications, fragmentation of the peptide chain, aggregation of cross-linked reaction products, altered electrical charge, and increased susceptibility to removal and degradation. The amino acids in a peptide differ in their susceptibility to attack, and the various forms of ROS also differ in their potential reactivity. The primary, secondary, and tertiary structure of a protein determines the susceptibility of each amino acid to attack by ROS (8, 30). For many enzymes, the oxidation by O_2^- of iron-sulphur centers inactivates enzymatic function (30, 32), and other amino acids, such as histidine, lysine, proline, arginine, and serine, form carbonyl groups when oxidized (33). A wide range of proteins and their amino acid building blocks is damaged or degraded by ROS (34), and the accumulation of these proteins in cells has been hypothesized to be part of the aging process (35).

OXIDATIVE DAMAGE TO DNA   The generation of ROS can induce numerous lesions in DNA that cause deletions, mutations, and other lethal genetic effects. Both the sugar and the base moieties are susceptible to oxidation, causing base degradation, single-strand breakage, and cross-linking to proteins (36, 37). In vitro,

H_2O_2 or O_2^- cannot by themselves cause strand breaks under normal physiological conditions, and therefore, their toxicity in vivo is most likely the result of Fenton reactions in the presence of a transition metal (36, 37). Both prokaryotic and eukaryotic cells have DNA repair enzymes; for a cell with DNA damage, it is the balance between damage and repair that determines the fate of that cell (38).

Signal Transduction

ROS are also produced for specific cellular functions, and it has been proposed that the antioxidant systems of cells regulate intracellular levels of ROS so that they can function as second messengers (39). ROS as second messengers are important for the expression of several transcription factors and other signal transduction molecules such as heat shock–inducing factor, nuclear factor, the cell-cycle gene $p53$, mitogen-activated protein kinase, and $oxyR$ gene products (40, 41).

Oxidative stress also plays a role in apoptosis through several cell-cycle genes (42). Two apoptotic pathways, the death-receptor and the mitochondrial pathways, have been described. The mitochondrial pathway is commonly associated with DNA damage and upregulation or activation of $p53$ (43). Exposure to UVR also causes ROS production in the electron transport chain of mitochondria (44) and can lead to apoptosis through the activation of caspases (45). Both cellular necrosis and apoptosis, which have overlapping features, can result from oxidative stress and lead to cell death (42). Whereas high levels of oxidative stress cause cell necrosis, lower levels either cause DNA damage and cell-cycle arrest or initiate apoptosis (8, 41).

Exposure to ROS and subsequent apoptosis are also common in higher plants, and many caspases homologous to animal caspases have been identified (46, 47). ROS are also an important component of plant defense systems against pathogens (48); O_2^- is directly involved in the apoptotic hypersensitive reaction of higher plants against pathogens (49). Interestingly, caspases have also been identified in unicellular photoautotrophic eukaryotes (i.e., phytoplankton) as well as in simple metazoans, and when compared to more derived plant and metazoan caspases, they are regulated in a similar fashion during experimentally induced apoptosis (50, 51).

One of the most interesting signal transduction roles for ROS is the mediation of morphogenic events associated with the onset of mutualistic symbiotic associations. The symbiosis of the serpiolid squid (*Euprymna scolopes*) light organ and the bioluminescent bacterium *Vibrio fisheri* is one of the best understood systems in terms of the attraction, initiation, and ultimate establishment of a symbiotic association that involves dramatic changes in host morphology to accommodate the symbionts (52). When the *V. fisheri* cells enter the eventual light organ, they first encounter the hostile environment of the ducts and then the crypt space, which includes epithelial cells that line the crypt (52). Potential symbionts associate themselves with the microvilli of the crypt epithelial cells and induce changes in the light-organ crypt epithelial cells that help maintain this unique symbiosis (52). Macrophages are abundant in the light-organ crypt and apparently patrol this space for nonspecific bacteria, while leaving symbiotic bacteria unharmed (53). Additionally, the epithelial cells apparently secrete a halide peroxidase that produces bacteriocidal hypohalous acid from H_2O_2, which presumably comes from the phagocytic activity of the macrophages (52, 53). Colonization of the crypt by *V. fisheri* in this hostile environment requires the removal of H_2O_2 using a catalase enzyme that is required for bacterial competency to successfully establish the symbiotic association (52–54). In addition, the luciferase enzyme of *V. fisheri* is a MFO that utilizes molecular oxygen (54). The luciferase enzyme is expressed in large quantities and, combined with bacterial respiration, can maintain a low pO_2 that subsequently results in lower ROS production (9) and an environment conducive to the successful maintenance of the symbiosis (54).

ANTIOXIDANT DEFENSES

Enzymatic Antioxidants

SUPEROXIDE DISMUTASE   Superoxide dismutase (SOD) (EC 1.15.1.1), originally discovered by McCord & Fridovitch (55), occurs as different metalloproteins with different cellular distributions. The Cu/Zn SOD is principally a cytosolic enzyme in eukaryotes but is also found in chloroplasts, bacteria, and peroxisomes, and as an extracellular enzyme (8, 28). The Mn form of SOD is principally found in mitochondria and bacteria, and the Fe SOD is found in chloroplasts and bacteria (5–8). The prokaryotic Mn SOD and Fe SOD and the eukaryotic Cu/Zn SOD are dimers, whereas the Mn SODs of mitochondria are tetramers, with each subunit consisting of 151 amino acids (5–8). All forms of the SOD are nuclear-encoded and are targeted to their respective subcellular compartments by an amino-terminal targeting sequence (56). SOD is an efficient catalyst and can keep the steady-state concentration of O_2^- at 10^{-10} mol liter^{-1} (5–8). With SOD concentrations at 10^{-5} mol liter^{-1}, any molecule of O_2^- is more likely to encounter a molecule of SOD then another O_2^- (5–8). At a rate constant (k_2) of 2×10^9 mol liter^{-1} s^{-1}, the lifetime of O_2^- is significantly shortened by SOD (5–8).

Prokaryotic cells and many eukaryotic algae contain the Mn SOD and Fe SOD enzymes, which are believed to be more ancient forms of SOD, whereas some phytoplankton also contain a Ni metalloprotein (57). Protein sequence data clearly show two distinct evolutionary paths for the Cu/Zn and the Fe/Mn SODs, and within the Cu/Zn SOD clade there is a varying degree of conservation in the protein sequences (5–8). The evolution of the Mn and Fe forms of SOD, most likely from a common ancestral protein, is attributed to the availability of the Mn and Fe metal cofactors under conditions when O_2 was four orders of magnitude lower than it is today (58). The Cu/Zn form has been reported to be present only in higher plants and animals and in the Charophycean alga, *Spirogyra* sp. (59). These data suggest that the divergence of the chloroplast and cytosolic forms of the Cu/Zn SOD occurred very early in the evolution of the protein (59). There is, however, evidence that the Cu/Zn SOD exists in unicellular eukaryotic algae, specifically dinoflagellates (60–62), whose evolutionary history extends back to the early Jurassic and which contain plastids derived from an ancestral red alga by secondary symbiosis (3).

CATALASE   Catalase (EC 1.11.1.6) is a heme-containing enzyme that catalyzes the conversion of H_2O_2 to H_2O and O_2. The enzyme is a tetramer with molecular weights in excess of 220 kD and has a high K_m for H_2O_2, which makes it most efficient at scavenging high concentrations of H_2O_2 (5–8). An unusual feature of catalase is its sensitivity to light and rapid turnover, which may result from light absorption by the heme group. Conditions that reduce the rate of protein turnover, such as osmotic, heat, or cold stress, can lower catalase activity (5–8, 63). For photoautotrophs, this feature of the catalase enzyme may affect their ability to tolerate oxidative stress when exposed to environmental perturbations. Phylogenetically, catalases from plants and animals are unique and divergent from those of bacteria and fungi, with bacteria containing several separate lineages (64).

PEROXIDASES   Peroxidases, like catalase, catalyze the reduction of H_2O_2 to H_2O, but they require a source of electrons that subsequently becomes oxidized. Ascorbate peroxidase (EC 1.11.1.11) is a heme-containing monomeric enzyme with a molecular mass of 30 kD (5). It has a significantly lower K_m for H_2O_2 than does catalase and uses a large pool (10–20 mM) of ascorbate as its specific electron donor to reduce H_2O_2 to H_2O in the stroma and on the thylakoids of chloroplasts (5).

Glutathione peroxidase (EC 1.11.1.9) is a tetrameric enzyme with a molecular weight of 84 kD. The enzyme is found in both selenium-containing and selenium-independent forms in the cytosol and mitochondria of animal tissues, but not in plants (8). This enzyme catalyzes the oxidation of glutathione, a low-molecular-weight tripeptide thiol compound, with H_2O_2 (8). Glutathione is very abundant in animal tissues through the action of glutathione reductase, which regenerates reduced glutathione (8).

Nonenzymatic Antioxidants

ASCORBIC ACID   L-ascorbic acid, or vitamin C, is an essential vitamin in animals and is abundant in plant tissues. All plants and animals, except humans, can synthesize ascorbate de novo; animals also can obtain vitamin C through their diet. Ascorbate functions as a reductant source for many ROS, thereby minimizing the damage caused by oxidative stress. Ascorbate scavenges not only H_2O_2 but also O_2^-, $HO^\bullet$, and lipid hydroperoxides without enzyme catalysts (5–8), and it can indirectly scavenge ROS by recycling α-tocopherol to its reduced form. Ascorbate has been found in plant cell chloroplasts and cytosol, where it also acts as a substrate for ascorbate peroxidase.

GLUTATHIONE   Glutathione (GSH) is a tripeptide (Glu-Cys-Gly) found in animals and plants. It forms a thiyl radical that reacts with a second oxidized glutathione, forming a disulphide bond (GSSG) when oxidized (8). The ratio of GSH/GSSG is often used as an indicator of oxidative stress in cells, and glutathione functions as an antioxidant in many ways by reacting with $^1O_2$, O_2^-, and $HO^•$. Glutathione can also act as a chain-breaker of free radical reactions and is an essential substrate for glutathione peroxidase (8). The maintenance of GSH levels, and therefore the reducing environment of cells, is crucial in preventing damage to cells exposed to conditions that promote oxidative stress.

TOCOPHEROL   The tocopherols, specifically α-tocopherol (vitamin E), are lipid-soluble antioxidants that scavenge ROS (5–8). This phenolic antioxidant is found in both animals and plants. α-tocopherol, due to its hydrophobic nature, is located exclusively within the bilayers of cell membranes. α-tocopherol is generally considered to be the most active form of the tocols. Plants synthesize α-tocopherol in chloroplasts, with the aromatic ring formed by the shikimic acid pathway—the same pathway that produces UVR-absorbing compounds, the mycosporine-like amino acids (MAAs; see below), in many marine algae. By contrast, animals must acquire tocopherol through their diet. The antioxidant properties of tocopherol are the result of its ability to quench both $^1O_2$ and peroxides (5–8). A marine-derived tocopherol known as α-tocomonoenol has been isolated from salmon eggs and provides enhanced antioxidant protection because of its ability to diffuse in viscous lipids and prevent lipid peroxidation (65).

CAROTENOIDS   Carotenoids are lipid-soluble molecules that protect both plants and animals against oxidative damage. Photoautotrophs produce carotenoids de novo, whereas animals must acquire carotenoids dietarily. In photosynthetic organisms, some carotenoids function as accessory pigments in light harvesting, whereas others specifically quench ROS produced as a result of overexcitation of the photosynthetic apparatus by light (5–8, 66). β-carotene can quench both excited triplet-state chlorophyll and $^1O_2$ because they have highly conjugated double bonds. Carotenoids can also dissipate excess excitation energy through the xanthophyll cycle (5–8, 67), a process, also known as dynamic photoinhibition, that prevents the overexcitation of the photosynthetic apparatus. Many carotenoids also serve as effective quenchers of ROS and can prevent lipid peroxidation in marine animals (68).

SMALL-MOLECULE ANTIOXIDANTS   Uric acid, a product of purine metabolism, can quench both $^1O_2$ and $HO^•$ (69). It is found in high concentrations in marine invertebrates, in which it can be a potent antioxidant (70). Another group of small-molecule antioxidants is compatible solutes (71). In particular, mannitol can quench $HO^•$ and prevent damage to critical carbon-fixing enzymes in photoautotrophs (72). Dimethylsulfide (DMS) is an important component of global sulfur cycles and a significant contributor to aerosol fractions in the atmosphere

(73). Many species of marine macrophytes and phytoplankton produce DMS from dimethylsulphoniopropionate (DMSP), whose primary function had been assumed to be as an osmolyte (73). Both DMS and DMSP have been shown to quench HO$^\bullet$. DMS can diffuse through biological membranes and act as an effective antioxidant in any cellular compartment (74).

Mycosporine-like amino acids are UVR-absorbing compounds with broadband absorption from 310–360 nm. They have been extensively studied in a wide variety of marine organisms (75). Some MAAs have antioxidant activity (76–78). These compounds are synthesized de novo by the shikimic acid pathway in photoautotrophs but are acquired by animals through their diet (75). Mycosporine-glycine can quench $^1O_2$ (79), whereas other MAAs can quench O_2^- (78). In reef-forming corals, mycosporine-glycine concentrations decline significantly upon exposure to prolonged high-temperature stress, while antioxidant enzymes increase (80). The concentration of MAAs in corals declines with increasing depth in proportion to photooxidative potential caused by exposure to UVR and hyperoxia due to photosynthesis (75).

OXIDATIVE STRESS IN THE MARINE ENVIRONMENT

Reactive Oxygen Production in Seawater

In marine systems, the absorption of solar radiation, and especially of its UVR wavelengths, by dissolved organic matter in seawater leads to the photochemical production of diverse reactive transients, including ROS (81). These ROS may have deleterious effects on bacteria and phytoplankton by affecting cell membranes or inhibiting photosynthesis. Hydrogen peroxide has the longest lifetime in seawater and the highest steady-state concentrations (10^{-7} M) and can readily pass through biological membranes (5–8, 81).

Hydrothermal vents also produce ROS (82). The abundance of hydrogen sulfide (H_2S) and O_2 near vents leads to the oxidation of H_2S in seawater and the production of both oxygen- and sulfur-centered radicals (82). In particular, electron paramagnetic resonance spin-trapping has shown convincingly that sulfide oxidation produces O_2^- (82). High concentrations of O_2^- near vents probably leads to H_2O_2 production by the dismutation of O_2^- and to subsequent oxidative stress for vent fauna. As may be expected, vent worms (*Riftia pachyptila*), vent clams (*Calyptogena magnifica*), and their bacterial symbionts all express SOD and exhibit peroxidase activity (83).

Oxidative Stress in Marine Organisms

Marine organisms are exposed to and adjust to a wide variety of environmental factors on varying temporal and spatial scales, from polar to tropical and from hourly to seasonal, in order to maintain homeostasis and growth and to reproduce. It should be no surprise that, just like any other metabolic pathway, those processes that lead to the production of ROS vary significantly over large gradients in many environmental factors, and adjustments in antioxidant defenses are required in order to maintain the steady-state concentration of ROS at low levels and thus prevent oxidative stress and cellular damage.

Although antioxidant protection is almost always associated with aerobic organisms, there exists a wide spectrum of oxygen tolerance in anaerobic organisms, principally prokaryotes (8, 37). Specialized anaerobic bacteria from hydrothermal vent environments have evolved novel enzymes to quench O_2^- without producing oxygen, which would also be toxic (84). *Pyrococcus furiosus*, a hyperthermophilic anaerobic bacterium, contains a superoxide reductase that reduces O_2^- to H_2O_2, which is then reduced to H_2O by peroxidases (84). The superoxide reductase maintains its activity at $25°C$, which is far below the growth optimum of $100°C$ for this bacterium but may be adaptive, as these free-living bacteria in the hydrothermal fluids are mixed with the surrounding cold water (84).

Many bacteria contain Fe and Mn SOD, but several also contain Cu/Zn SOD (8). These Cu/Zn SODs are distinct from and may be the evolutionary precursor to eukaryotic Cu/Zn SODs (8). One unique example of a prokaryotic Cu/Zn SOD is from the bacterium *Photobacterium leiognathi*, a bioluminescent bacterium symbiotic with pony fish (85). Originally believed to arise from horizontal gene transfer from eukaryotes to prokaryotes, differences in the gene sequence of the *P. leiognathi* and other prokaryotic Cu/Zn SODs, as well as important differences in gene structure and function, actually support a prokaryotic origin for the Cu/Zn SODs (86). One of the important attributes of these bacterial symbioses is the use of bioluminescence as a mechanism of signaling between con-specific hosts. The luciferase enzyme that produces bioluminescence is a MFO that utilizes molecular oxygen. Several recent studies strongly support the hypothesis that the original selective pressure for the evolution of luciferase was to prevent oxidative stress and that it then was co-opted for its bioluminescent characteristic, originally a by-product of its antioxidant activities that can still be experimentally demonstrated in *V. harveyi* (87–90).

The production of ROS is a consistent feature of photoautotrophs, and marine algae are no exception. In unicellular eukaryotic algae, especially dinoflagellates, all three metalloproteins of SOD have been identified (60–62, 91). Many of these cells exhibit a daily cycling of maximum SOD activities and other antioxidant enzymes that are associated with peak midday irradiances and the production of ROS (61, 62, 91, 92). At least one study has demonstrated that this daily rhythm is under transcriptional control and that new SOD protein is produced on a daily basis (91). Other studies have reported distinct seasonal regulation of antioxidant enzymes based on total daily irradiance in addition to daily rhythms (62, 92). Some of these species are toxic to bacteria and fish owing to production of extracellular ROS (93, 94).

Green, brown, and red macrophytes are conspicuous components of many marine ecosystems, but especially of rocky intertidal systems, in which many species of attached seaweed are dominant members of the community. These algae withstand some of the harshest environmental conditions known, including freezing, desiccation, carbon limitation, and heat stress. These environmental extremes are conducive to the formation of ROS and contribute to the photoinhibition of photosynthesis observed in these ecologically important marine algae. The production of ROS in the brown alga *Fucus evanescens* has been detected with fluorescent dyes (95) and is enhanced in freezing, high light, and desiccation stress (96–98). The increase in ROS production also causes an increase in lipid peroxidation and a decrease in the quantum yield of PSII fluorescence (96). Additionally, species vary in susceptibility to oxidative stress (97) and in seasonal acclimatization of antioxidant defenses to changes in temperature-induced oxidative stress (98). Two red algae, *Mastocarpus stellatus* and *Chondrus crispus*, exhibit zonational patterns in temperate rocky intertidal ecosystems that reflect their ability to resist freezing and the accompanying oxidative stress (99, 100). The activities of enzymatic and nonenzymatic antioxidants increase with tidal height, as does, therefore, daily exposure to air temperatures, for *M. stellatus*, which always has greater antioxidant capabilities than *C. crispus*, which is found in the lower intertidal zone (99). Seasonal acclimatization to irradiance, both its visible and ultraviolet components, also occurs in macrophytes at high latitudes, where the amplitude in the changes in seawater temperature is low (101). Macrophytes exposed to increased visible radiation and UVR during the breakup of sea ice increase SOD, catalase, and MAAs, all of which prevent oxidative stress and its subsequent effects on photosynthesis (101).

Many marine invertebrates produce ROS. Bivalve molluscs produce ROS in response to xenobiotics (102) and changes in temperature, especially heat stress (103). ROS are also important in the cell-mediated immune response of molluscs to both prokaryotic and eukaryotic pathogens (104). Interestingly, many bivalve molluscs are euryoxic and survive fluctuations between hypoxia/anoxia and normoxia with each tidal cycle. During anoxic-normoxic transitions, euryoxic species produce far less ROS and therefore avoid oxidative stress (105).

Sponges (Phylum: Porifera) with symbiotic cyanobacteria undergo elevated pO_2 in their tissues from photosynthetically produced O_2 (106, 107). Exposure to summertime highs in seawater temperature result in the highest values of total oxidative scavenging capacity and catalase, which is attributed to the production of H_2O_2 (107). Similar temperature-related increases in prooxidant pressure have been observed in the eurythermal lug worm, *Arenicola marina* (Annelida: Polychaeta) (108). The increase in ROS production is associated with an increase in mitochondrial substrate oxidation and higher rates of proton leakage in summer animals as compared to winter animals (108). Oxybiotic meiofauna also can experience photochemically generated ROS in intertidal pools (109). H_2O_2 can diffuse across the redoxcline, and annelid worms such as *Nereis diversicolor* respond by increasing their activities of catalase (109). Worms maintained in anoxic conditions also increase SOD activities, which is, again, a physiological adaptation to withstand the transition from anoxia to normoxia and the subsequent burst in the production of ROS (109). Thiobiotic meiofauna, including gastrotrichs and turbellarians, living in anoxia and exposed to H_2S also have higher activities of antioxidant enzymes than their oxybiotic counterparts and may be exposed to oxygen- and sulfur based radicals like their hydrothermal vent cousins (110).

Marine arthropods (i.e., crabs, lobsters, and shrimp) vary in their antioxidant defenses according to their level of aerobic metabolism, exposure to chronically cold environments, or exposure to UVR (111–113). Surprisingly, marine arthropods lack Cu/Zn SOD (114, 115). Instead, these animals use a copper-dependent hemocyanin for oxygen transport and have an unusual cytosolic Mn SOD lacking the signal transit peptide that would otherwise direct it to the mitochondrion (115). This occurs in all Crustacea that use a copper-dependent oxygen transport system and is believed to be evolutionarily linked to the fluctuation in copper metabolism induced by the use of copper-dependent oxygen transport systems (115).

Studies on oxidative stress in echinoderms (i.e., sea stars, sea urchins, sea cucumbers, and crinoids) are few. To prevent polyspermy, sea urchins create a physical barrier to multiple fertilizations of a single ovum. They accomplish this by hardening the vitelline membrane to create the fertilization membrane, which raises and hardens as a result of covalently cross-linked products of the oxidation of tyrosyl residues released by the cortical granules. This reaction requires extracellular H_2O_2, which is formed by a membrane-bound NADPH oxidase during a respiratory burst upon fertilization (116). The excess H_2O_2 is quenched by a secreted peroxidase and ovothiol C, a nonenzymatic scavenger of H_2O_2 (116). Although the H_2O_2 produced is extracellular, this H_2O_2, if not scavenged in the extracelluar space, may diffuse back into the now newly fertilized zygote. Recently, major yolk proteins from sea urchin eggs, which were believed to be vitellogenin, have been identified as transferrin-like, iron-chelating proteins (117) that could potentially be very useful in preventing H_2O_2 from participating in Fenton chemistry within the zygote.

Many echinoderms are important broadcast-spawning members of benthic marine communities, and their planktonic embryos and larvae may therefore be susceptible to the detrimental effects of UVR. Although total exposure to UVR is dependent on the stability and optical properties of the water column, planktonic larvae can be easily advected into surface waters, where irradiances of UVR are higher. Sea urchin embryos exposed to UVR irradiances equivalent to shallow temperate coastal environments show symptoms of oxidative stress, as indicated by elevated concentrations of SOD protein, DNA damage, and apoptosis (118). Field exposures of embryos at fixed depths reveal similar results down to a depth of 8 m (M.P. Lesser, unpublished data). Both laboratory and field experiments also show abnormal embryonic morphology typical of that seen in apoptosis (118).

Marine vertebrates are not immune from oxidative stress. Fish in particular, including those from the Antarctic, have several well-characterized Cu/Zn SODs (119–121) that have evolved to maintain catalytic function over a wide range of temperatures (121–122). Fish also respond to prooxidant pressure due to differences in metabolic rates (123), pollution (124), and exposure to UVR (125). As in sea urchin embryos, the larvae of Atlantic cod (*Gadus morhua*) exposed to UVR show significant increases in SOD activity and in expression of the cell-cycle gene *p*53 (125).

Physiological Extremes

A relatively new area of investigation is oxidative stress in polar, especially Antarctic, environments (122). Because solubility of O_2 is high in the constant, $-1.8°C$, seawater temperatures of Antarctica, polar ectotherms potentially experience

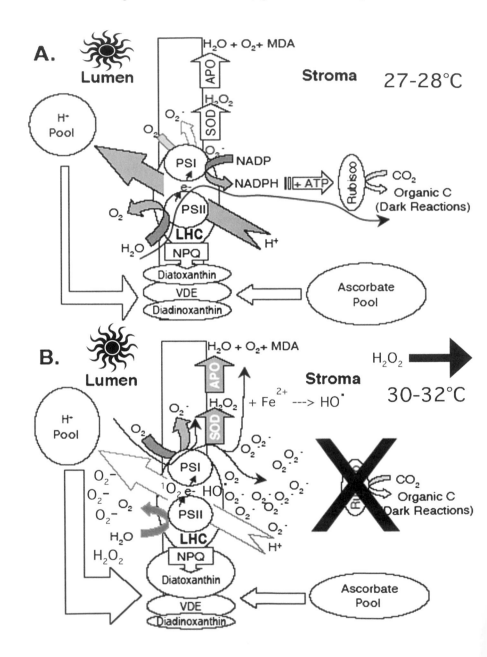

increased prooxidant pressure and metabolic costs associated with antioxidant defenses. But low temperatures also reduce the conductance of O_2 because of changes in tissue viscosity. Increases in mitochondrial volume density and lipid stores may potentially compensate for this decreased conductance (122). The increased unsaturation of membranes in Antarctic ectotherms can also promote lipid peroxidation by ROS unless antioxidant defenses are available. Indeed, the Antarctic bivalve, *Laternula elliptica*, exhibits a greater potential for lipid peroxidation than does the temperate species, *Mya arenaria*, with similar total lipid concentrations (126). But *L. elliptica* also has higher concentrations of α-tocopherol and β-carotene, both lipid-soluble antioxidants known for their lipid peroxidation chain–breaking capabilities (5–8, 66). Polar invertebrates that may be predisposed to oxidative stress also contain higher activities of antioxidant enzymes. The Antarctic scallop, *Adamussium colbecki*, has significantly higher activities of SOD in its gills as compared to the Mediterranean scallop, *Pecten jacobaeus* (127). Consistent with the chronically cold environment in which it lives, *A. colbecki* also exhibits seasonally invariant antioxidant capacities except during the austral spring phytoplankton bloom or during reproduction (128). Studies to date, mostly involving measurements of enzyme activity, generally support that antioxidant enzymes compensate for exposure to chronically cold seawater temperatures (129, 130). Whether these compensation strategies are quantitative or qualitative, *sensu* Hochachka & Somero (131), is unknown.

One of the best-understood marine invertebrate systems, as relating to oxidative stress, is that of cnidarians (i.e., sea anemones, corals, and jellyfish) with symbiotic zooxanthellae. In particular, reef-forming corals are important members of

---

Figure 1   Detail of events leading to oxidative stress on the thylakoid membrane of the chloroplast of zooxanthellae. (*A*) During normal temperatures and irradiances, light is absorbed by the light-harvesting complex (LHC) and photochemistry in PSI and PSII produce ATP and NADPH for the dark reactions, in which CO_2 is fixed by the enzyme Rubisco. The efficiency of photochemistry is regulated by the interconversion of the two pigments diatoxanthin and diadinoxanthin, which is part of the xanthophyll cycle used to protect the photosystems from overexcitation. Superoxide dismutase (SOD) and ascorbate peroxidase enzymes in the chloroplast degrade reactive oxygen species (ROS). (*B*) During heat stress, membrane fluidity changes (154) result in the production of ROS. Subsequently, the simultaneous overreduction of photosynthetic electron transport and the decreased fixation of CO_2 (i.e., sink limitation) result in the overexcitation of the photosystems and the flow of excitation energy primarily through PSI. The excess absorbed energy cannot be dissipated by the xanthophyll cycle (NPQ, nonphotochemical quenching). More ROS are formed than can be quenched by the available enzymatic and nonenzymatic antioxidants, and some species (e.g., H_2O_2) can be exported from the chloroplast (*bold horizontal arrow*). APO, ascorbate peroxidase; LHC, light-harvesting complex; VDE, violaxanthin de-epoxidase. Adapted from Jones et al. (150) and Hoegh-Guldberg (132).

this group. Global climate change, principally the emission of greenhouse gases (e.g., CO_2 and CH_4), and the subsequent effects on seawater temperature are the primary causes of "coral-bleaching" events around the world (132, 133). Seawater temperatures of $2-3°C$ above long-term average summer temperatures result in a stress response, known as bleaching in corals, in which they lose their zooxanthellae (132, 133). Both field and laboratory studies on bleaching in corals and other symbiotic cnidarians have established a causal link between temperature stress and bleaching (132, 133). The extent of coral-bleaching, the extent of subsequent mortality, and the underlying mechanism(s) that cause bleaching are related to the magnitude of temperature elevation and the duration of exposure for any individual event.

Although thermal stress is seen as the principal cause of coral bleaching, other environmental factors, including those that are affected by anthropogenic influences, act synergistically by effectively lowering the threshold temperature at which coral bleaching occurs. The abiotic factor that has the most significant influence on the severity of thermally induced coral bleaching is solar radiation, both its visible and ultraviolet components (UVB: 290–320 nm; UVA: 320–400 nm) (133, 134). Exposure to UVR is particularly important during the hyperoxic conditions (135, 136) that occur intracellularly in corals during photosynthesis, and leads to the photodynamic production of ROS (5–8). An important response of corals during exposure to UVR is the synthesis of MAAs and enzymes involved in the protection of both the host and symbiont from oxidative stress (60, 75, 137, 138).

Exposure to elevated temperatures alone (139), UVR alone (60), or in combination (137, 140) can result in photoinhibition of photosynthesis in zooxanthellae. Photoinhibition occurs as a result of the reduction in photosynthetic electron transport combined with the continued high absorption of excitation energy and the production of ROS. ROS have many cellular targets, including photosystem II and the primary carboxylating enzyme, Rubisco, in zooxanthellae (Figure 1) (60, 137). Elevated temperatures functionally lower the set point for light-induced photoinhibition. Enzymic defenses in the cnidarian host occur in proportion to the potential for photooxidative damage in symbiotic cnidarians (141, 142). However, high fluxes of ROS in the host (141, 143) or zooxanthellae (60, 137, 144) can overwhelm the protective enzymatic response and result in hydroxyl radical production via the Fenton reaction (5–8). Both the cnidarian host and zooxanthellae express Cu/Zn and Mn SODs (60, 138, 145), whereas zooxanthellae also express an Fe SOD (146).

Oxidative stress has been proposed as a unifying mechanism for several environmental insults that cause bleaching (137, 138). Oxidative stress can lead to bleaching of corals by zooxanthellae exocytosis from coral host cells (140, 147, 148) or by apoptosis (138, 147–149). A cellular model of bleaching in symbiotic cnidarians has been developed (Figure 2) that includes oxidative stress, PSII damage, sink limitation, DNA damage, and apoptosis as underlying processes (137, 138, 140, 147, 150, 151). This model is consistent with biomarker proteins expressed in corals during thermal stress (152, 153). Recent findings can also be included

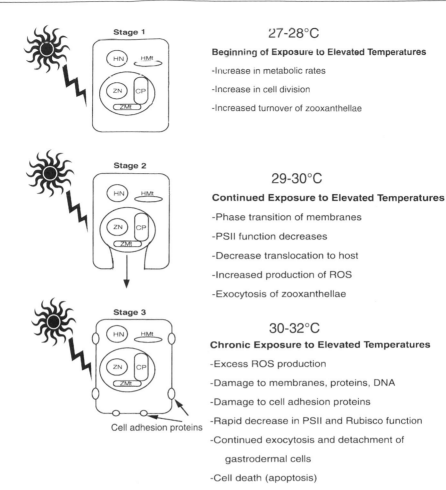

Stage 1

27-28°C

Beginning of Exposure to Elevated Temperatures

-Increase in metabolic rates

-Increase in cell division

-Increased turnover of zooxanthellae

Stage 2

29-30°C

Continued Exposure to Elevated Temperatures

-Phase transition of membranes

-PSII function decreases

-Decrease translocation to host

-Increased production of ROS

-Exocytosis of zooxanthellae

Stage 3

30-32°C

Chronic Exposure to Elevated Temperatures

-Excess ROS production

-Damage to membranes, proteins, DNA

-Damage to cell adhesion proteins

-Rapid decrease in PSII and Rubisco function

-Continued exocytosis and detachment of

   gastrodermal cells

-Cell death (apoptosis)

Cell adhesion proteins

Figure 2   Model of coral bleaching caused by oxidative stress that incorporates photoinhibition (150) and apoptosis (137, 138, 147). The chloroplast and mitochondria (both host and zooxanthellae) are major sources of ROS. Continued exposure to elevated seawater temperatures, concurrent with the increase in ROS production, causes a progression from Stage 1 (27–28°C) to Stage 3 (30–32°C) and apoptosis or cell necrosis. HN, host nucleus; ZN, zooxanthellae nucleus; HMt, host mitochondria; CP, chloroplast; ZMt, zooxanthellae mitochondria.

in this model and include differential sensitivity of zooxanthellae thylakoid membranes to thermal stress (154) and the presence of nitric oxide synthase activity (155, 156), which produces NO$^\bullet$ and in turn reacts with O_2^- to form $ONOO^-$, whose ability to diffuse through membranes is much greater than that of O_2^-.

UVR and thermal stress also damage DNA in corals (169). DNA damage can also lead to apoptosis if not repaired. The expression pattern of a putative *p53* protein in *Montastraea faveolata* after exposure to thermal stress and high irradiances of solar radiation is consistent with DNA damage (138). Morphological evidence indicates both apoptosis and necrosis in host and algal cells of thermally stressed symbiotic sea anemones. Similarly, in thermally stressed symbiotic cnidarians, ROS-mediated apoptosis and possibly necrosis are consistent with morphological evidence and the upregulation of a putative $p53$ protein. Apoptosis and cell necrosis are extremes in a range of cellular responses of corals to oxidative stress caused by thermal stress, with and without the synergistic effects of solar radiation (133).

CONCLUSIONS AND FUTURE DIRECTIONS

Increasingly, ecologists/physiologists are examining oxidative stress. Additionally, oxidative stress is emerging as a common theme in connection with the impact of global climate change (e.g., global warming and ozone depletion) on natural ecosystems at all trophic levels. Responses of various marine taxa to this impact both mitigate protein damage (e.g., heat shock proteins and ubiquination) and, by quenching ROS, limit damage to DNA, proteins, and lipids.

The future for integrated studies includes molecular genetics, microarrays, proteomics, RNAi assays, knockouts, and marine model organisms, combined with a quantitative organismal approach. Methods routinely used (e.g., electron paramagnetic resonance, enzyme assays, and fluorochromes) to assess the level of oxidative stress should be applied to a variety of marine taxa. These techniques should be combined with assessments of (*a*) costs to respond to and repair the damage from oxidative stress and (*b*) sublethal impacts on growth and reproduction. Biomarker development should also be integrated more extensively into an ecological setting. Few antioxidant genes have been sequenced for marine organisms or have had their expression quantified. Some research groups have established EST libraries, which will facilitate the development of microarrays for stress genes and genes of intermediate metabolism, which can assess stress and energetic costs for marine taxa under diverse environmental conditions.

The *Annual Review of Physiology* is online at
http://physiol.annualreviews.org

LITERATURE CITED

1. Nisbet EG, Sleeo NH. 2001. The habitat and nature of early life. *Nature* 409:1083–91
2. Kasting JF, Siefert JL. 2002. Life and the evolution of Earth's atmosphere. *Science* 296:1066–68
3. Falkowski PG, Katz ME, Knoll AH, Quigg A, Raven JA, et al. 2004. The evolution of modern eukaryotic phytoplankton. *Science* 305:354–60
4. Gerschman R, Gilbert DL, Nye SW, Dwyer P, Fenn WO. 1954. Oxygen

poisoning and X-irradiation: a mechanism in common. *Science* 119:4562–70
5. Asada K, Takahashi M. 1987. Production and scavenging of active oxygen in photosynthesis. In *Photoinhibition*, ed. DJ Kyle, CB Osmond, CJ Arntzen, pp. 228–87. Elsevier: Amsterdam
6. Cadenas E. 1989. Biochemistry of oxygen toxicity. *Annu. Rev. Biochem.* 58:79–110
7. Fridovich I. 1998. Oxygen toxicity: a radical explanation. *J. Exp. Biol.* 201:1203–9
8. Halliwell B, Gutteridge JMC. 1999. *Free Radicals in Biology and Medicine*. New York: Oxford Univ. Press. 936 pp.
9. Jamieson D, Chance B, Cadenas E, Boveris A. 1986. The relation of free radical production to hyperoxia. *Annu. Rev. Physiol.* 48:703–19
10. Takahashi M, Asada K. 1988. Superoxide production in aprotic interior of chloroplast thylakoids. *Arch. Biochem. Biophys.* 267:714–22
11. Shiraishi T, Takahashi M, Asada K. 1994. Generation of superoxide anion radicals and hydroxyl radicals in chloroplast thylakoids. In *Frontiers of Reactive Oxygen Species in Biology and Medicine*, ed. K Asada, T Yoshikawa, pp. 31–32. Amsterdam: Elsevier Sci.
12. Fridovich I. 1986. Biological effects of the superoxide radical. *Arch. Biochem. Biophys.* 247:1–11
13. Fang FC. 2004. Antimicrobial reactive oxygen and nitrogen species: concepts and controversies. *Nat. Rev. Microbiol.* 2:820–32
14. Marla SS, Lee J, Groves JT. 1997. Peroxynitrite rapidly permeates phopholipid membranes. *Proc. Natl. Acad. Sci. USA* 94:14243–48
15. Denicola A, Souza JM, Radi R. 1998. Diffusion of peroxynitrite across erythrocyte membranes. *Proc. Natl. Acad. Sci. USA* 95:3566–71
16. Asada K. 1994. Mechanisms for scavenging reactive molecules generated in chloroplasts under light stress. In *Photoin-*

hibition of Photosynthesis, ed. NR Baker, JB Boyer, pp. 129–42. Oxford, UK: BIOS
17. Falkowski PG, Raven JA. 1997. *Aquatic Photosynthesis*. Malden, MA: Blackwell Sci. 375 pp.
18. Richter M, Rüle W, Wild A. 1990. Studies on the mechanism of photosystem II photoinhibition II. The involvement of toxic oxygen species. *Photosynth. Res.* 24:237–43
19. Lupínkova L, Komenda J. 2004. Oxidative modification of the photosystem II D1 protein by reactive oxtgen species: From isolated protein to cyanobacterial cells. *Photochem. Photobiol.* 79:152–62
20. Nishiyama Y, Yamamoto H, Allakhverdiev SI, Inaba M, Yokota A, et al. 2001. Oxidative stress inhibits the repair of photodamage to the photosynthetic machinery. *EMBO J.* 20:5587–94
21. Macpherson AN, Telfer A, Barber J, Truscott GT. 1993. Direct detection of singlet oxygen from isolated photosystem II reaction centers. *Biochem. Biophys. Acta* 1143:301–9
22. Liu K, Sun J, Song Y, Liu B, Xu Y, et al. 2004. Superoxide, hydrogen peroxide, and hydroxyl radical in D1/D2/cytochrome *b-559* photosystem II reaction center complex. *Photosynth. Res.* 81:41–47
23. Asada K. 1999. The water-water cycle in chloroplasts: scavenging of active oxygens and dissipation of excess photons. *Annu. Rev. Plant Physiol. Mol. Biol.* 50:601–39
24. Brookes PS. 2005. Mitochondrial H^+ leak and ROS generation: An odd couple. *Free Radic. Biol. Med.* 38:12–23
25. Brand MD, Affouritit C, Esteves TC, Green K, Lamber AJ, et al. 2005. Mitochondrial superoxide: production, biological effects, and activation of uncoupling proteins. *Free Radic. Biol. Med.* 37:755–67
26. Corpas FJ, Barroso JB, del Rio LA. 2001. Peroxisomes as a source of reactive oxygen species and nitric oxide signal

molecules in plant cells. *Trends Plant Sci.* 6:145–50

27. Sandalio LM, Fernández VM, Rupérez FL, del Rio LA. 1988. Superoxide free radicals are produced in glyoxisomes. *Plant Physiol.* 87:1–7

28. Sandalio LM, del Rio LA. 1988. Intraorganellar distribution of superoxide dismutase in plant peroisomes (glyoxisomes and leaf peroxisomes). *Plant Physiol.* 88:1215–18

29. Gutteridge JMC, Halliwell B. 1990. The measurement and the mechanism of lipid peroxidation in biological systems. *Trends Biochem. Sci.* 15:129–35

30. Freeman BA, Crapo JD. 1982. Biology of disease, free radicals and tissue injury. *Lab. Invest.* 47:412–26

31. Green DR, Reed JC. 1998. Mitochondria and apoptosis. *Science* 281:1309–12

32. Hyslop PA, Hinshaw DB, Halsey WA Jr, Schraufsätter IU, Sauerheber RD, et al. 1988. Mechanisms of oxidant-mediated cell injury. *J. Biol. Chem.* 263:1665–75

33. Stadtman ER. 1986. Oxidation of proteins by mixed-function oxidation systems: implication in protein turnover, aging and neutrophil function. *Trends Biochem. Sci.* 11:11–12

34. Davies KJA. 1987. Protein damage and degradation by oxygen radicals. *J. Biol. Chem.* 262:9895–901

35. Dean RT, Gieseg S, Davies MJ. 1993. Reactive species and their accumulation on radical-damaged proteins. *Trends Biochem. Sci.* 18:437–41

36. Imlay JA, Linn S. 1988. DNA damage and oxygen radical toxicity. *Science* 240:1302–9

37. Imlay JA. 2003. Pathways of oxidative damage. *Annu. Rev. Microbiol.* 57:395–418

38. Beyer W, Imlay J, Fridovich I. 1991. Superoxide dismutases. *Prog. Nucleic Acid Res.* 40:221–53

39. Schrek R, Baeuerle PA. 1991. A role for oxygen radicals as second messengers. *Trends Cell Biol.* 1:39–42

40. Zheng M, Åslund F, Storz G. 1998. Activation of the OxyR transcription factor by reversible disulfide bond formation. *Science* 279:1718–21

41. Martindale JL, Holbrook NJ. 2002. Cellular response to oxidative stress: signaling for suicide and survival. *J. Cell. Physiol.* 192:1–15

42. Johnson TM, Yu Z, Ferrans VJ, Lowenstein RA, Finkel T. 1996. Reactive oxygen species are downstream mediators of p53-dependent apoptosis. *Proc. Natl. Acad. Sci. USA* 93:11848–52

43. Hengartner MO. 2000. The biochemistry of apoptosis. *Nature* 407:770–76

44. Gniadecki R, Thorn T, Vicanova J, Petersen A, Wulf HC. 2000. Role of mitochondria in ultraviolet-induced oxidative stress. *J. Cell. Biochem.* 80:216–22

45. Pourzand C, Tyrell RM. 1999. Apoptosis, the role of oxidative stress and the example of solar UV radiation. *Photochem Photobiol.* 70:380–90

46. Korthout HAAJ, Berecki G, Bruin W, van Duijn B, Wang M. 2000. The presence and subcellular localization of caspase 3 like proteinases in plant cells. *FEBS Lett* 475:139–44

47. Lam E, del Pozo O. 2000. Caspase-like protease involvement in the control of plant cell death. *Plant Mol. Biol.* 44:417–28

48. Mehdy MC. 1994. Active oxygen specie in plant defense against pathogens. *Plant Physiol.* 105:467–72

49. Jabs T, Dietrich RA, Dang JL. 1996. Initiation of runaway cell death in an *Arabidopsis* mutant by extracellular superoxide. *Science* 273:1853–56

50. Cikala M, Wilm B, Hobmayer E, Böttger A, David CN. 1999. Identification of caspases and apoptosis in the simple metazoan *Hydra. Curr. Biol.* 9:959–62

51. Segovia M, Haramaty L, Berges JA, Falkowski PG. 2003. Cell death in the unicellular chlorophyte *Dunaliella tertiolecta.* A hypothesis on the evolution o

apoptosis in higher plants and metazoans. *Plant Physiol.* 132:99–105

52. McFall-Ngai MJ. 1999. Consequences of evolving with bacterial symbionts: insights from the squid-*Vibrio* associations. *Annu. Rev. Ecol. Syst.* 30:235–56

53. Visick KL, McFall-Ngai MJ. 2000. An exclusive contract: specificity in the *Vibrio fisheri-Euprymna scolopes* partnership. *J. Bacteriol.* 182:1779–87

54. Ruby EG, McFall-Ngai MJ. 1999. Oxygen-utilizing reactions and symbiotic colonization of the squid light organ by *Vibrio fisheri. Trends Microbiol.* 7:414–20

55. McCord JM, Fridovich I. 1969. Superoxide dismutase, an enzymatic function for erythrocuprein. *J. Biol. Chem.* 244:6049–55

56. Bannister JV, Bannister WH, Rotils G. 1987. Aspects of the structure, function and applications of superoxide dismutase. *CRC Crit. Rev. Biochem.* 22:110–80

57. Wolfe-Simon F, Grzebyk D, Schofield O, Falkowski PG. 2005. The role and evolution of superoxide dismutases in algae. *J. Phycol.* 41:453–65

58. Asada K, Kanematsu S, Okada S, Hayakawa T. 1980. Phylogenetic distribution of three types of superoxide dismutase in organisms and in cell organelles. In *Chemical and Biochemical Aspects of Superoxide and Superoxide Dismutase*, ed. JV Bannister, HAO Hill, pp. 136–53. Amsterdam: Elsevier Sci.

59. Kanematsu S, Asada K. 1989. CuZn-superoxide dismutases from the fern *Equisetum arvense* and the green alga *Spirogyra* sp.: occurrence of chloroplast and cytosol types of enzyme. *Plant Cell Physiol.* 30:717–27

60. Lesser MP, Shick JM. 1989. Effects of irradiance and ultraviolet radiation on photoadaptation in the zooxanthellae of *Aiptasia pallida*: primary production, photoinhibition, and enzymic defenses against oxygen toxicity. *Mar. Biol.* 102:243–55

61. Hollnagel HC, di Mascio P, Asano CS, Okamoto OK, Stringer CG, et al. 1996. The effect of light on the biosynthesis of β-carotene and superoxide dismutase activity in the photosynthetic alga *Gonyaulax polydra. Brazil. J. Med. Biol. Res.* 29:105–10

62. Butow BJ, Wynne D, Tel-Or E. 1997. Superoxide dismutase activity in *Peridinium gatunense* in Lake Kinneret: effect of light regime and carbon dioxide concentration. *J. Phycol.* 33:787–93

63. Hertwig B, Steb P, Feierabend J. 1992. Light dependence of catalase synthesis and degradation in leaves and the influence of interfering stress conditions. *Plant Physiol.* 100:1547–53

64. Klotz MG, Klassen GR, Loewen PC. 1997. Phylogenetic relationships among prokaryotic and eukaryotic catalases. *Mol. Biol. Evol.* 14:951–58

65. Yamamoto Y, Fujisawa A, Hara A, Dunlap WC. 2001. An unusual vitamin E constituent (α-tocomonoenol) provides enhanced antioxidant protection in marine organisms adapted to cold-water environments. *Proc. Natl. Acad. Sci. USA* 98: 13144–48

66. Demmig-Adams B, Adams WW. 1993. The xanthophyll cycle. In *Antioxidants in Higher Plants*, ed. RG Alscher, JL Hess, pp. 59–90. Boca Raton, FL: CRC Press

67. Krinsky NI. 1989. Antioxidant functions of carotenoids. *Free Radic. Biol. Med.* 7:617–35

68. Miki W, Otaki N, Shimidzu N, Yokoyama A. 1994. Carotenoids as free radical scavengers in marine animals. *J. Mar. Biotechnol.* 2:35–37

69. Ames BN, Cathcart R, Schwiers E, Hochstein P. 1981. Uric acid provides an antioxidant defense in humans against oxidant- and radical-caused aging and cancer: a hypothesis. *Proc. Natl. Acad. Sci. USA* 78:6858–62

70. Dunlap WC, Shick JM, Yamamoto Y. 2000. UV protection in marine organisms. I. Sunscreens, oxidative stress and

antioxidants. In *Free Radicals in Chemistry, Biology and Medicine*, ed. T Yoshikawa, S Toyokuni, Y Yamamoto, Y Naito, pp. 200–14. London: OICA Int.

71. Smirnoff N, Cumbes QJ. 1989. Hydroxyl radical scavenging activity of compatible solutes. *Phytochemistry* 28:1057–60

72. Shen B, Jensen RG, Bohnert HJ. 1997. Mannitol protects against oxidation by hydroxyl radicals. *Plant Physiol.* 115:527–32

73. Broadbent AD, Jones GB, Jones RJ. 2002. DMSP in corals and benthic algae from the Great Barrier Reef. *Estuar. Coast. Shelf. Sci.* 55:547–55

74. Sunda W, Kleber DJ, Klene RP, Huntsman S. 2002. An antioxidant function for DMSP and DMS in marine algae. *Nature* 418:317–20

75. Shick JM, Dunlap WC. 2000. Mycosporine-like amino acids and related gadusols: biosynthesis, accumulation, and UV-protective functions in aquatic organisms. *Annu. Rev. Physiol.* 64:223–62

76. Dunlap WC, Yamamoto Y. 1995. Small-molecule antioxidants in marine organisms: antioxidant activity of mycosproine-glycine. *Comp. Biochem. Physiol. B* 112:105–14

77. Conde FR, Churio MS, Previtali CM. 2000. The photoprotector mechanism of mycosporine-like amino acids. Excited-state properties and photostability of porphyra-334 in aqueous solution. *J. Photochem. Photobiol. B* 56:139–44

78. Kim CS, Lim WA, Cho YC. 2001. Mycosporine-like amino acids as the UV sunscreen with oxygen radical scavenging activity. *Bull. Natl. Fish. Res. Dev. Inst. Korea* 60:65–71

79. Suh H, Lee H, Jung J. 2003. Mycosporine glycine protects biological systems against photodynamic damage by quenching singlet oxygen with a high efficiency. *Photochem. Photobiol.* 78:109–13

80. Yakoleva I, Bhagooli R, Takemura A, Hidaka M. 2004. Differential susceptibility to oxidative stress of two scleractinian corals: antioxidant functioning of mycosporine-glycine. *Comp. Biochem. Physiol. B* 139:721–30

81. Mopper K, Kieber DJ. 2000. Marine photochemistry and its impact on carbon cycling. In *The Effects of UV Radiation in the Marine Environment*, ed. S De Mora, S Demers, M Vernet, pp. 101–30. Cambridge, UK: Cambridge Univ. Press

82. Tapley DW, Buettner GR, Shick JM. 1999. Free radicals and chemiluminescence as products of the spontaneous oxidation of sulfide in seawater, and their biological implications. *Biol. Bull.* 196:52–56

83. Blum J, Fridovich I. 1984. Enzymatic defenses against oxygen toxicity in the hydrothermal vent animals *Riftia pachyptila* and *Calyptogena magnifica*. *Arch. Biochem. Biophys.* 228:617–20

84. Jenny FE Jr, Verhagen MFJM, Cui X, Adams MWW. 1999. Anaerobic microbes: oxygen detoxification withot superoxide dismutase. *Science* 286:306–9

85. Bannister JV, Parker MW. 1985. The presence of a copper/zinc superoxide dismutase in the bacterium *Photobacterium leiognathi*: a likely case of gene transfer from eukaryotes to prokaryotes. *Proc. Natl. Acad. Sci. USA* 82:149–52

86. Bourne Y, Redford SM, Steinman HM, Lepock JR, Tainer JA, et al. 1996. Novel dimeric interface and electrostatic recognition in bacterial Cu, Zn superoxide dismutase. *Proc. Natl. Acad. Sci. USA* 93:12774–79

87. Timmins GS, Jackson SK, Swartz HM. 2001. The evolution of bioluminescent oxygen consumption as an ancient oxygen detoxification mechanism. *J. Mol. Evol.* 52:321–32

88. Rees J-F, de Wergifosse B, Noiset O, Dubuisson M, Janssens B, et al. 1998. The origins of marine bioluminescence: turning oxygen defense mechanisms into deep-sea communication tools. *J. Exp. Biol.* 201:1211–21

89. Szpilewska H, Czyz A, Wegrzyn G. 2003. Experimental evidence for the physiological role of bacterial luciferase in the protection of cells against oxidative stress. *Curr. Microbiol.* 47:379–82

90. Wergifosse B, Dubuisson M, Marchand-Brynaert J, Trout A, Rees J-F. 2004. Coelenterazine: a two stage antioxidant in lipid micells. *Free Radic. Biol. Med.* 36:278–87

91. Okamato OK, Robertson DL, Fagan TF, Hastings JW, Colepicolo P. 2001. Different regulatory mechanisms modulate the expression of a dinoflagellate iron-superoxide dismutase. *J. Biol. Chem.* 276:19989–93

92. Butow BJ, Wynne D, Tel-Or E. 1997. Antioxidative protection of *Peridinium gatunense* in Lake Kinneret: seasonal and daily variation. *J. Phycol.* 33:780–86

93. Yang CZ, Albright LJ, Yousif AN. 1995. Oxygen-radical-mediated effects of the toxic phytoplankter *Heterosigma carterae* on juvenile rainbow trout *Oncorhynchus mykiss*. *Mar. Ecol. Prog. Ser.* 23:101–8

94. Oda T, Ishimatsu A, Shimada M, Takeshita S, Muramatsu T. 1992. Oxygen-radical-mediated toxic effects of the red tide flagellate *Chattonella marina* on *Vibrio alginolyticus*. *Mar. Biol.* 112:505–9

95. Collén J, Davison IR. 1997. *In vivo* measurement of active oxygen production in the brown alga *Fucus evanescens* using 2',7'-dichlorohydrofluorescein diacetate. *J. Phycol.* 33:643–48

96. Collén J, Davison IR. 1999. Reactive oxygen production and damage in intertidal *Fucus* spp. (Phaeophyceae). *J. Phycol.* 35:54–61

97. Collén J, Davison IR. 1999. Reactive oxygen metabolism in intertidal *Fucus* spp. (Phaeophyceae). *J. Phycol.* 35:62–69

98. Collén J, Davison IR. 2001. Seasonal and thermal acclimation of reactive oxygen metabolism in *Fucus vesiculosus* (Phaeophyceae). *J. Phycol.* 37:474–81

99. Collén J, Davison IR. 1999. Stress tolerance and reactive oxygen metabolism in the intertidal red seaweeds *Mastocarpus stellatus* and *Chondrus crispus*. *Plant Cell Environ.* 22:1143–51

100. Lohrmann NL, Logan BA, Johnson AS. 2004. Seasonal acclimatization of antioxidants and photosynthesis in *Chondrus crispus* and *Mastocarpus stellatus*, two co-occurring red algae with differing stress tolerances. *Biol. Bull.* 207:225–32

101. Aguilera J, Bischof K, Karsten U, Hanelt D, Wiencke C. 2002. Seasonal variation in ecophysiological patterns in macroalgae from an Arctic fjord. II. Pigment accumulation and biochemical defence systems against high light stress. *Mar. Biol.* 140:1087–95

102. Winston GW, Livingstone DR, Lips F. 1990. Oxygen reduction metabolism by the digestive gland of the common marine mussel, *Mytilus edulis* L. *J. Exp. Zool.* 255:296–308

103. Abele D, Heise K, Pörtner HO, Puntarulo S. 2002. Temperature-dependence of mitochondrial function and production of reactive oxygen species in the intertidal mud clam *Mya arenaria*. *J. Exp. Biol.* 205:1831–41

104. Adema CM, Van der Knaap WPW, Sminia T. 1991. Molluscan hemocyte-mediated cytotoxicity: the role of reactive oxygen intermediates. *Rev. Aquat. Sci.* 4:201–23

105. Dykens JA, Shick JM. 1988. Relevance of purine catabolism to hypoxia and recovery in euryoxic and stenoxic marine invertebrates, particularly bivalve molluscs. *Comp. Biochem. Physiol. C* 91:35–41

106. Regoli F, Cerrano C, Chierici E, Bompadre S. 2000. Susceptibility to oxidative stress of the Mediterranean demosponge *Petrosia ficiformis*: role of endosymbionts and solar irradiance. *Mar. Biol.* 137:453–61

107. Regoli F, Cerrano C, Chierici E, Chiantore MC, Bavestrello G. 2004. Seasonal variability of prooxidant pressure and antioxidant adaptation to symbiosis in the Mediterranean demosponge *Petrosia*

ficiformis. Mar. Ecol. Prog. Ser. 275:129–37

108. Keller M, Sommer AM, Pörtner HO, Abele D. 2004. Seasonality of energetic functioning and production of reactive oxygen species by lugworm (*Arenicola marina*) mitochondria exposed to acute temperature changes. *J. Exp. Biol.* 207:2529–38

109. Abele-Oeschger D, Oeschger R, Theede H. 1994. Biochemical adaptations of *Nereis diversicolor* (Polychaeta) to temporarily increased hydrogen peroxide levels in intertidal sandflats. *Mar. Ecol. Prog. Ser.* 106:101–10

110. Morrill AC, Powell EN, Bidigare RR, Shick JM. 1988. Adaptations to life in the sulfide system: a comparison of oxygen detoxifying enzymes in thiobiotic and oxybiotic meiofauna (and freshwater planarians). *J. Comp. Physiol. B* 158:335–44

111. Maciel FE, Rosa CE, Santos EA, Monserrat JM, Nery LEM. 2004. Daily variations in oxygen consumption, antioxidant defenses, and lipid peroxidation in the gills and hepatopancreas of an estuarine crab. *Can. J. Zool.* 82:1871–77

112. Camus L, Gulliksen B. 2005. Antioxidant defense properties of Arctic amphipods: comparison between deep-, sublittoral and surface species. *Mar. Biol.* 146:355–62

113. Gouveia GR, Marques DS, Cruz BP, Geracitano LA, Nery LEM, et al. 2005. Antioxidant defenses and DNA damage induced by UV-A and UV-B radiation in the crab *Chasmagnathus granulata* (Decopoda, Brachyura). *Photochem. Photobiol.* 81:398–403

114. Brouwer M, Brouwer TH, Grater W, Enghild JJ, Thogersen IB. 1997. The paradigm that all oxygen-respiring eukaryotes have cytolosolic CuZn-suproxide dismutase and that Mn-superoxide dismutase is localized to mitochondria does not apply to a large group of marine arthropods. *Biochemistry* 36:13381–88

115. Brouwer M, Brouwer TH, Grater W, Brown-Peterson N. 2003. Replacement of a cytosolic copper/zinc superoxide dismutase by a novel cytosolic manganese superoxide dismutase in crustaceans that use copper (haemocyanin) for oxygen transport. *Biochem. J.* 374:219–28

116. Shapiro BM. 1991. The control of oxidant stress at fertilization. *Science* 252:533–36

117. Brooks JM, Wessel GM. 2002. The major yolk protein in sea urchins is a transferrin-like, iron binding protein. *Dev. Biol.* 245:1–12

118. Lesser MP, Kruse VA, Barry TM. 2003. Exposure to ultraviolet radiation causes apoptosis in developing sea urchin embryos. *J. Exp. Biol.* 206:4097–103

119. Ken C-F, Lin C-T, Shaw JF, Wu JL. 2003. Characterization of fish Cu/Zn-superoxide dismutase and its protection from oxidative stress. *Mol. Biotechnol.* 5:167–73

120. Natoli G, Calabrese L, Capo C, O'Neil P, di Prisco G. 1990. Icefish (*Chaenocephalus aceratus*) Cu,Zn superoxide dismutase conservation of the enzyme properties in extreme adaptation. *Comp. Biochem. Physiol. B* 95:29–33

121. Cassini A, Favero M, de Laureto P, Albergon V. 1997. Cu-Zn suproxide dismutase from *Pagothenia bernacchii*: catalytic and molecular properties. In *Antarctic Communities, Species, Structure, and Survival*, ed. B Battaglia, J Valencia DWH Walton, pp. 266–71. Cambridge UK: Cambridge Univ. Press

122. Abele D, Puntarulo S. 2004. Formation of reactive species and induction of antioxidant defence systems in polar and temperate marine invertebrates and fish. *Comp Biochem. Physiol. A* 138:405–15

123. Janssens B, Childress JJ, Baguet F, Rees J-F. 2000. Reduced enzymatic antioxidative defenses in deep-sea fish. *J. Exp. Biol* 203:3717–25

124. Bacanskas LR, Whitaker J, di Giulio RT 2004. Oxidative stress in two population

of killifish (*Fundulus heteroclitus*) with differing contaminant histories. *Mar. Environ. Res.* 58:597–601

125. Lesser MP, Farrell JH, Walker CW. 2001. Oxidative stress, DNA damage, and p53 expression in the larvae of Atlantic cod (*Gadus morhua*) exposed to ultraviolet (290–400 nm) radiation. *J. Exp. Biol.* 204:157–64

126. Estevez MS, Abele D, Puntarulo S. 2002. Lipid radical generation in polar (*Laternula elliptica*) and temperate (*Mya arenaria*) bivalves. *Comp. Biochem. Physiol. B* 132:729–37

127. Viarengo A, Canesi L, Martinez PG, Peters LD, Livingstone DR. 1995. Prooxidant processes and antioxidant defence systems in the tissues of the Antarctic scallop (*Adamussium colbecki*) compared with the Mediteranean scallop (*Pecten jacobaeus*). *Comp. Biochem. Physiol. B* 111:119–26

128. Regoli F, Nigro M, Chiantore M, Winston GM. 2002. Seasonal variations of susceptibility to oxidative stress in *Adamussium colbecki*, a key bioindicator species for the Antarctic marine environment. *Sci. Total Environ.* 289:205–11

129. Abele D, Tesch C, Wencke P, Pörtner HO. 2001. How does oxidative stress relate to thermal tolerance in the Antarctic bivalve *Yolida eightsi*. *Antarct. Sci.* 13:11–18

130. Viarengo A, Abele-Oeschger D, Burlando B. 1998. Effects of low temperature on prooxidant processes and antioxidant defence systems in marine organisms. In *Cold Ocean Physiology*, ed. HO Pörtner, RC Playle, pp. 212–35. Cambridge, NY: Cambridge Univ. Press

131. Hochachka PW, Somero GN. 2002. *Biochemical Adaptation: Mechanism and Process in Physiological Evolution*. New York: Oxford Univ. Press. 478 pp.

132. Hoegh-Guldberg O. 1999. Climate change, coral bleaching and the future of the world's coral reefs. *Mar. Freshw. Res.* 50:839–66

133. Lesser MP. 2004. Experimental coral reef biology. *J. Exp. Mar. Biol. Ecol.* 300:217–52

134. Shick JM, Lesser MP, Jokiel P. 1996. Effects of ultraviolet radiation on corals and other coral reef organisms. *Glob. Change Biol.* 2:527–45

135. Dykens JA, Shick JM. 1982. Oxygen production by endosymbiotic algae controls superoxide dismutase activity in their animal host. *Nature* 297:579–80

136. Kühl M, Cohen Y, Dalsgaard T, Jørgensen BB, Revsbech NP. 1995. Microenvironment and photosynthesis of zooxanthellae in scleractinian corals studied with microsensors for O_2, pH, and light. *Mar. Ecol. Prog. Ser.* 117:159–72

137. Lesser MP. 1996. Exposure of symbiotic dinoflagellates to elevated temperatures and ultraviolet radiation causes oxidative stress and inhibits photosynthesis. *Limnol. Oceanogr.* 41:271–83

138. Lesser MP, Farrell J. 2004. Solar radiation increases the damage to both host tissues and algal symbionts of corals exposed to thermal stress. *Coral Reefs* 23:367–77

139. Iglesias-Prieto R, Matta JL, Robins WA, Trench RK. 1992. Photosynthetic response to elevated temperature in the symbiotic dinoflagellate *Symbiodinium microadriaticum* in culture. *Proc. Natl. Acad. Sci. USA* 89:10302–5

140. Lesser MP. 1997. Oxidative stress causes coral bleaching during exposure to elevated temperatures. *Coral Reefs* 16:187–92

141. Dykens JA, Shick JM, Benoit C, Buettner GR, Winston GW. 1992. Oxygen radical production in the sea anemone *Anthopleura elegantissima*: and its symbiotic algae. *J. Exp. Biol.* 168:219–41

142. Richier S, Furla P, Plantivaux A, Merle P-L, Allemand D. 2005. Symbiosis-induced adaptation to oxidative stress. *J. Exp. Biol.* 208:277–85

143. Nii CM, Muscatine L. 1997. Oxidative stress in the symbiotic sea anemone *Aiptasia pulchella* (Calgren, 1943): Contribution of the animal to superoxide ion

production at elevated temperature. *Biol. Bull.* 192:444–56

144. Matta JL, Trench RK. 1991. The enzymatic response of the symbiotic dinoflagellate *Symbiodinium microadriaticum* (Freudenthal) to growth *in vitro* under varied oxygen tensions. *Symbiosis* 11:31–45

145. Plantivaux A, Furla P, Zoccola D, Garello G, Forcioli D, et al. 2004. Molecular characterization of two CuZn-superoxide dismutases in a sea anemone. *Free Radic. Biol. Med.* 37:1170–81

146. Matta JL, Govind NS, Trench RK. 1992. Ployclonal antibodies against iron-superoxide dismutase from *Escherichia coli* B cross react with superoxide dismutases from *Symbiodinium microadriaticum* (Dinophyceae). *J. Phycol.* 28:343–46

147. Gates RD, Baghdasarian G, Muscatine L. 1992. Temperature stress causes host cell detachment in symbiotic cnidarians: implications for coral bleaching. *Biol. Bull.* 182:324–32

148. Franklin DJ, Hoegh-Guldberg O, Jones RJ, Berges JA. 2004. Cell death and degeneration in the symbiotic dinoflagellates of the coral *Stylophora pistillata* during bleaching. *Mar. Ecol. Prog. Ser.* 272: 117–30

149. Dunn SR, Bythell JC, Le Tessier DA, Burnett WJ, Thomason JC. 2002. Programmed cell death and necrosis activity during hyperthermic stress-induced bleaching of the symbiotic sea anemone *Aiptasia* sp. *J. Exp. Mar. Biol. Ecol.* 272: 29–53

150. Jones RJ, Hoegh-Guldberg O, Larkum AWD, Schreiber U. 1998. Temperature-induced bleaching of corals begins with impairment of the CO_2 fixation mechanism in zooxanthellae. *Plant Cell Environ.* 21:1219–30

151. Warner ME, Fitt WK, Schmidt GW. 1999. Damage to photosystem II in symbiotic dinoflagellates: a determinant of coral bleaching. *Proc. Natl. Acad. Sci. USA* 96: 8007–12

152. Downs CA, Mueller E, Phillips S, Fauth JE, Woodley CM. 2000. A molecular biomarker system for assessing the health of coral (*Montastraea faveolata*) during heat stress. *Mar. Biotechnol.* 2:533–44

153. Downs CA, Fauth JE, Halas JC, Dustan P, Bemiss J, Woodley CM. 2002. Oxidative stress and seasonal coral bleaching. *Free Radic. Biol. Med.* 33:533–43

154. Tchernov D, Gorbunov MY, de Vargas C, Yadav SN, Milligan AJ, et al. 2004. Membrane lipids of symbiotic algae are diagnostic of sensitivity to thermal bleaching in corals. *Proc. Natl. Acad. Sci. USA* 101:13531–35

155. Trapido-Rosenthal H, Zielke S, Owen R, Buxton L, Boeing B, et al. 2005. Increased zooxanthellae nitric oxide synthase activity is associated with coral bleaching. *Biol. Bull.* 208:3–6

156. Morrall CE, Galloway TS, Trapido-Rosenthal HG, Depledge MH. 2000. Characterization of nitric oxide synthase activity in the tropical sea anemone *Aiptasia pallida*. *Comp. Biochem. Physiol. B* 125:483–91

Annu. Rev. Physiol. 2006. 68:279–305
doi: 10.1146/annurev.physiol.68.040504.094635
Copyright © 2006 by Annual Reviews. All rights reserved
First published online as a Review in Advance on September 28, 2005

BRAINSTEM CIRCUITS REGULATING GASTRIC FUNCTION

R. Alberto Travagli, Gerlinda E. Hermann, Kirsteen N. Browning, and Richard C. Rogers

Department of Neuroscience, Pennington Biomedical Research Center, Louisiana State University System, Baton Rouge, Louisiana 70808; email: travagra@pbrc.edu, hermange@pbrc.edu, brownik@pbrc.edu, rogersrc@pbrc.edu

Key Words   dorsal motor nucleus of the vagus, nucleus tractus solitarius, gastrointestinal reflexes, plasticity

Abstract   Brainstem parasympathetic circuits that modulate digestive functions of the stomach are comprised of afferent vagal fibers, neurons of the nucleus tractus solitarius (NTS), and the efferent fibers originating in the dorsal motor nucleus of the vagus (DMV). A large body of evidence has shown that neuronal communications between the NTS and the DMV are plastic and are regulated by the presence of a variety of neurotransmitters and circulating hormones as well as the presence, or absence, of afferent input to the NTS. These data suggest that descending central nervous system inputs as well as hormonal and afferent feedback resulting from the digestive process can powerfully regulate vago-vagal reflex sensitivity. This paper first reviews the essential "static" organization and function of vago-vagal gastric control neurocircuitry. We then present data on the opioidergic modulation of NTS connections with the DMV as an example of the "gating" of these reflexes, i.e., how neurotransmitters, hormones, and vagal afferent traffic can make an otherwise static autonomic reflex highly plastic.

OVERVIEW

The gastrointestinal (GI) tract possesses an intrinsic nervous plexus that allows the intestine to have a considerable degree of independent neural control. The stomach and the esophagus, however, are almost completely dependent upon extrinsic nervous inputs arising from the central nervous system (CNS) (1–6). CNS control over the smooth and coordinated digestive functions of the stomach is mediated by parasympathetic and sympathetic pathways that either originate in, or are controlled by, neural circuits in the caudal brainstem.

Sympathetic control of the stomach stems from cholinergic preganglionic neurons in the intermediolateral column of the thoracic spinal cord (T6 through T9 divisions), which impinge on postganglionic neurons in the celiac ganglion, of which the catecholaminergic neurons provide the stomach with most of its sympathetic supply. Sympathetic regulation of motility primarily involves inhibitory

1066-4278/06/0315-0279$20.00

presynaptic modulation of vagal cholinergic input to postganglionic neurons in the enteric plexus. The magnitude of sympathetic inhibition of motility is directly proportional to the level of background vagal efferent input (7). Recognizing that the GI tract is under the dual control of the sympathetic and parasympathetic nervous systems, we refer the reader to other comprehensive reviews on the role of the sympathetic control of gastric function (7–9). The present review focuses on the functionally dominant parasympathetic control of the stomach via the dorsal motor nucleus of the vagus (DMV) and the means by which this input is modulated, i.e., plasticity of vago-vagal reflexes.

ORGANIZATION OF THE BRAINSTEM VAGAL CIRCUITS CONTROLLING THE STOMACH

Visceral Afferent Inputs to Brainstem Gastric Control Circuits

Vagal afferent fibers carry a large volume of information about the physiological status of the gut directly to brainstem circuits regulating gastric function. The cell bodies of origin for this afferent sensory pathway are contained in the nodose ganglion. Neurons in the nodose ganglia are the cranial equivalent of dorsal root ganglion cells in that they are bipolar and connect the gut directly with nucleus tractus solitarius (NTS) neurons in the brainstem with no intervening synapse (reviewed recently in References 10–13).

There are several types of visceral receptors and afferent fibers (7, 9, 14–31). Independent of their functions or modalities, however, all vagal afferents use glutamate as the primary neurotransmitter to transfer information to the NTS (32–42).

Electrophysiological experiments have shown that glutamate released onto presumptive GI NTS neurons activates both NMDA and non-NMDA receptors (35, 41, 43), as do putative cardiovascular NTS neurons (36, 44, 45). The release of glutamate at this interface between peripheral perception (sensory afferent fibers) and central integration (NTS neurons) is open to modulation by a wide variety of transmitter and hormonal agonists, including glutamate itself (46–51), cholecystokinin (CCK) (52–54), leptin (55, 56), tumor necrosis factor (57), and ATP (58), to name but a few. These data strongly suggest that vago-vagal reflex functions are subject to pre- as well as postsynaptic modulation.

Nucleus Tractus Solitarius

The general hypothesis of a vago-vagal reflex connecting vagal afferent input brainstem NTS neurons, and vagal efferent projections to the stomach was formulated some years ago and has been reviewed extensively (6, 10, 38, 59–66). The early models presented a framework to explain many of the salient features of brainstem reflex control of the stomach. Simply put, reflex action is initiated by stimulation of sensory vagal afferent pathways, which activate second-order NTS neurons via glutamate action on NMDA and non-NMDA receptors. These

NTS neurons can use several different neurotransmitters to control the output from DMV cells, which, in turn, control gastric functions and complete the vago-vagal loop (Figure 1).

Visceral sensory afferents are organized in an overlapping topographic manner within the NTS subnuclei. Terminal fields from the intestine are represented in the subnuclei commissuralis and medialis, the stomach sends its afferent inputs to the subnuclei medialis and gelatinosus, and the esophagus to the subnucleus centralis (66–70).

Although sensory inputs from distinct visceral areas, such as, for example, the aortic branches or gastric branches, do not seem to converge on single NTS neurons (71), the same subnucleus receives sensory information from more than one peripheral organ. Afferent terminals involved in arterial baroreflex circuits as well as terminals involved in vago-vagal gastric reflexes, for example, impinge upon neurons similarly located in the subnucleus medialis of the NTS (17, 38, 67, 70, 72–76). This overlap of the terminal fields from different organs within the NTS makes the anatomical and functional identification of individual neurons difficult.

The subnucleus centralis (cNTS), however, provides a convenient exception in that it seems to receive inputs from vagal afferent fibers originating almost exclusively in the esophagus (17, 38, 66, 67, 77), making it an excellent model for the study of a population of second-order neurons controlling esophageal-mediated reflexes. The cNTS of the rat is located between the tractus solitarius and the lateral third of the DMV at a level of approximately 0.5 to 1.5 mm rostral to the calamus scriptorius (70). Note that although the cNTS projects to the DMV as part of a classic example of a vago-vagal reflex, it also projects to areas that are not related to CNS control of gastric functions. Indeed, the cNTS projects to the nucleus ambiguus (NAmb), the reticular formation, and the ependymal layer of the fourth ventricle, as well as other NTS subnuclei (66). Thus, although cNTS neurons receive sensory information almost exclusively from the esophagus, these neurons control outputs related to gastric, esophageal, cardiovascular, deglutitive, and respiratory functions. Neurons of the cNTS may also be subject to plastic changes in function that translate into dramatic shifts in the reflex responses to afferent input dependent upon, and appropriate to, the feeding status of the animal.

Dorsal Motor Nucleus of the Vagus

The parasympathetic motor supply to the stomach is provided by the efferent vagus nerve originating from neurons located in the NAmb and DMV. The NAmb contains the soma of cells contributing to the vagal control of the esophagus, upper esophageal sphincter, and cardiorespiratory system. We refer the reader to recent reviews dealing with the NAmb (6, 8, 63, 77, 78).

The cell bodies for the great majority of parasympathetic efferent fibers that project to the upper GI tract originate along the whole rostro-caudal extent of the DMV (79–81). The DMV is a paired structure in the dorsal caudal medulla adjacent to the central canal, the majority of whose cells project to neurons in the myenteric plexus or onto interstitial cells of Cajal (ICC) of the upper GI tract

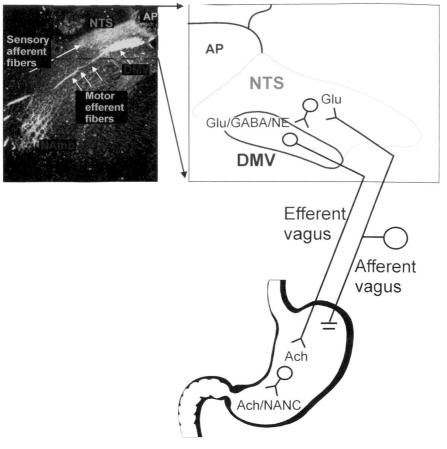

Figure 1 Cytoarchitecture of the dorsal vagal complex. Darkfield photomicrograph of a coronal section of rat brainstem at the level of the area postrema/intermediate level following application of HRP crystals to the subdiaphragmatic vagus. Note the intense labeling of the nucleus tractus solitarius (NTS), dorsal motor nucleus of the vagus (DMV), and nucleus ambiguus (NAmb). An area of the photomicrograph (*dotted line*) has been expanded into cartoon form to allow a more detailed illustration of the brainstem circuitry of gastrointestinal (GI) vago-vagal reflexes. Note that vagal afferent neurons, whose cell bodies lie in the nodose ganglion, receive sensations from the GI tract. The central terminals of these afferent fibers enter the brainstem via the tractus solitarius and terminate within the NTS, utilizing principally glutamate as their neurotransmitter. These afferent signals are integrated by neurons of the NTS that project to, among other areas, the adjacent DMV using, mainly, glutamate, GABA, or NE as neurotransmitters. Neurons of the DMV are the preganglionic parasympathetic motoneurons that provide the motor output to the GI tract via the efferent vagus, where they release acetylcholine onto their postganglionic target. Postganglionic parasympathetic neurons are either excitatory [cholinergic (ACh)] or inhibitory [nonadrenergic, noncholinergic (NANC)].

(19, 82–85), with the highest density of efferent fibers terminating in the stomach (87).

Retrograde tracing experiments have determined that the DMV is organized in medio-lateral columns spanning its rostro-caudal extent (79–81). Efferent projections originating from soma located in the medial portions of the nucleus form both the dorsal and ventral gastric vagal branches. Neurons in the lateral portions of the DMV send axons to the vagal celiac and accessory celiac branches, whereas scattered neurons in the left DMV provide axons to the hepatic branch. The medio-lateral organization of the columns is not maintained when the target organs of the various vagal branches are considered. In fact, the two gastric branches innervate the stomach, a portion of the proximal duodenum, and some visceral structures such as the pancreas; the celiac branches innervate the GI tract from the duodenum to the transverse colon; and the hepatic branch innervates parts of the stomach, liver, and proximal duodenum (87, 88).

Although the DMV may not be organized in a rigid organotypic manner, evidence suggests that the vagal innervation of the stomach is segregated within the DMV by function. For example, descending vagal pathways responsible for causing gastric contractions versus gastric relaxation [i.e., the nonadrenergic, noncholinergic (NANC) pathway] appear to be localized in different regions of the DMV. That is, putative NANC-pathway neurons appear to be located in the caudomedial and rostrolateral divisions of the DMV, whereas the gastroexcitatory neurons are located in the more rostral and medial divisions of the DMV (66, 89).

The DMV is comprised of neuronal populations that are nonhomogeneous with respect to morphological features such as soma size and shape, number of dendritic branches, and extent of dendritic arborization (90–96). The significance of these morphological differences is not well understood, although a possible explanation is discussed below.

Huang and colleagues (95) have hypothesized that neuronal DMV subgroups in the human may form functional units innervating specific organs; our studies in the rat suggest a similar organization. Combined retrograde tracing, whole cell patch clamp, and postrecording neuronal reconstruction techniques show a functional and morphological correlation between DMV neurons and the peripheral target organs that they innervate (92). Specifically, in the rat, DMV neurons projecting to the gastric fundus have a smaller soma size and fewer dendritic branches than neurons projecting to the corpus, duodenum, or cecum; neurons that project to the cecum have the largest soma out of all these. Additionally, there is evidence to support a relationship between structure and function of neurons in the rat DMV; e.g., neurons responsive to gastric or intestinal distension can be distinguished into separate morphological groups (96). Furthermore, these different DMV subgroups produce different profiles of extracellular recorded action potentials. This last observation suggests that DMV neurons engaged in different functions may possess different membrane properties.

Indeed, several investigators have reported a large array of unevenly distributed membrane currents in the DMV (92, 94, 97–100). Using the current clamp configuration, Browning et al. (92) showed that gastric-projecting DMV neurons can be easily distinguished from intestinal-projecting DMV neurons by the charac teristic shape and duration of the after-hyperpolarization (AHP). That is, intestinal projecting neurons have a much larger and slower AHP, in contrast to the smaller and faster AHP of gastric-projecting neurons. This characteristic implies that gastric DMV neurons are more prone to change membrane potential in response to synaptic inputs than larger DMV neurons that project to other targets. This is of particular relevance, as one of the distinguishing features of DMV neurons is that they maintain a spontaneous, slow (1–2 pulses sec^{-1}) pacemaker-like activity (65, 101, 102). This pacemaker activity implies that afferent inputs altering the membrane potential of these neurons by even a few mV, in either direction, cause dramatic changes to the vagal motor output (97, 103). One immediate implication is that the stomach, at rest, is controlled by a tonic vagal efferent outflow provided by small gastric-projecting DMV cells, the vagal motor output of which is continuously sculpted by impinging inputs. Conversely, larger intestinal DMV neurons need more robust synaptic inputs to achieve the same membrane displacement. This size principle, widely accepted for spinal motoneuron activation and recruitment (104), states that neurons with the smallest cell bodies have the lowest threshold for synaptic activation and, therefore, can be activated by weaker synaptic inputs. The resting membrane potential of gastric smooth muscle is nonuniform, with a gradient from approximately -48 mV in the proximal stomach to -70 mV in the distal stomach (105). The threshold for smooth muscle contraction is approximately -50 mV (106). The myenteric plexus of the stomach essentially serves as a follower of vagal efferent input (107): There is a high degree of fidelity for activation of gastric myenteric neurons by vagal efferent inputs (108). As a result, the proximal stomach is highly sensitive to minute changes in input from either the cholinergic excitatory branch or the NANC inhibitory branch of the DMV, and slight changes in DMV activity in either division translate into dramatic effects on gastric function.

Synaptic Connections Between the Nucleus Tractus Solitarius and the Dorsal Motor Nucleus of the Vagus

The majority of NTS neurons do not possess pacemaker activity; their inputs onto DMV neurons must be driven and are modulated by synaptic activity, either from the afferent vagus, from other CNS areas, or via circulating hormones (10, 41, 53, 57, 65, 109–118). By being subject to a vast array of modulatory activity, the synaptic connections between NTS and DMV by implication play a major role in shaping the vagal efferent output.

There are numerous NTS neuronal phenotypes that can potentially contribute input to the DMV and induce potent effects on vagally mediated gastric function (17, 119–134). Despite the presence of this vast array of neurotransmitters in various NTS subnuclei, electrophysiological data show that the NTS primarily controls the DMV through glutamatergic (37, 61, 101, 135, 136), GABAergic (61, 101, 137), and catecholaminergic (138–140) inputs. Electrical stimulation of the

NTS subnucleus commissuralis, for example, evokes noradrenergic α2-mediated inhibitory potentials in approximately 10% of the DMV neurons (140), whereas an α1-mediated current can be evoked by stimulation of the A2 area between the subnuclei medialis and cNTS of the NTS (K.N. Browning & R.A. Travagli, unpublished data). Additionally, stimulation of other portions of the NTS evokes both inhibitory GABA-mediated or/and excitatory glutamate-mediated currents in DMV neurons (61, 101, 103, 141–144). Pharmacological and immunocytochemical experiments have determined that GABA released onto DMV neurons activates GABA-A receptor subtypes (101, 135, 146–148).

Most of the NTS-induced inhibition of the DMV is mediated by GABA, interacting with the GABA-A receptor; conversely, most of the excitation delivered to the DMV by the NTS is mediated by glutamate interacting with both NMDA and non-NMDA receptors (101, 144, 146, 149–152). Recent data from our laboratory show a major functional role of catecholamines (likely originating from the A2 area) in both excitatory and inhibitory control of DMV. Application of glutamate or catecholamine agonists in the dorsal vagal complex (DVC) has profound effects on gastric motility and tone (114, 138, 152–154). Glutamate injected into the medial portion of the DMV causes a brisk gastric excitation, whereas norepinephrine injections cause gastroinhibition, perhaps owing to the simultaneous activation of the NANC inhibitory pathway and inhibition of the cholinergic gastroexcitatory pathway (66, 138, 139).

Microinjections of glutamate and catecholamine antagonists in the same area, however, do not induce noticeable effects on gastric motility and tone unless GABAergic transmission is blocked (37, 135, 138). Conversely, microinjections of GABA antagonists into the DVC induce profound excitatory effects on esophageal motility and on gastric motility and secretion (135, 147, 148). Taken together, these data suggest that the DMV output is restrained by a tonic GABAergic input arising, most likely, from the NTS. Furthermore, they suggest that factors modulating GABAergic inputs from the NTS to the DMV may have a significant impact on vagal reflex control of the stomach. By contrast, glutamatergic and catecholaminergic inputs do not seem to play a major role in setting tonic vagal output to the GI tract but rather seem to be invoked phasically by specific reflexes. Thus, NTS neurons that display some type of spontaneous activity use GABA as their main neurotransmitter. Glutamatergic and adrenergic NTS neurons are probably silent unless activated by afferent input.

Vagal Efferent (Motor) Control

To better appreciate the role of the efferent vagus in controlling GI functions, one has to consider the short-term effects of vagotomy (reviewed recently in References 9 and 155). Acute vagotomy induces an increase in fundic tone and a decrease in antral motility, impairing the reservoir function of the stomach. The conflicting effects of vagotomy suggested to Pavlov that the vagus nerve controlled gastric motility through both inhibitory and excitatory pathways, a concept that has received ample support over the past 100 years.

The vast majority (>95%) of DMV neurons are cholinergic, i.e., choline acetyl-transferase immunoreactive (156). Some DMV neurons also express immunoreactivity for nitric oxide synthase (NOS) or catecholamines [e.g., tyrosine hydroxylase (TH)] (89, 156–159). Although the projections of NOS- (158) or TH- (159) positive DMV neurons target selective areas of the stomach, the physiological significance of this detail is not clear. Interestingly, in both rodent (66, 89) as well as in feline (160) models, these markers for alternate neurotransmitter synthesis (i.e., TH and NOS) tend to coincide with the location of neurons involved in the inhibitory effect of vagal input to the stomach. For example, NOS-containing DMV neurons are found in the extreme caudal and rostrolateral portions of the DMV. Stimulation of these areas evoke gastric relaxation (66, 89, 160). This is far from establishing a connection between NOS- (or TH-) positive neurons and gastroinhibition; the correlation, however, is quite intriguing.

Using an intact vagus-gastric myenteric plexus in vitro preparation, Schemann & Grundy (107) demonstrated that stimulation of vagal preganglionic fibers, likely originating from DMV somata, induces exclusively cholinergic nicotinic potentials. These data indicate that acetylcholine, interacting with nicotinic receptors, is the principal neurotransmitter released from vagal efferent terminals and that preganglionic vagal fibers excite only enteric neurons. Because activation of vagal efferent fibers can induce both excitatory as well as inhibitory effects on gastric smooth muscles (18, 38, 62, 66, 160–180), it is clear that both excitatory and inhibitory postganglionic neuroeffectors are released from enteric neurons in response to excitatory vagal input. Note that the inhibitory vagal action on the stomach is directed toward the control of motility and tone but not of gastric secretion (181–185).

The principal excitatory postganglionic neurotransmitter is acetylcholine acting on muscarinic receptors in gastric smooth muscle, ICC, and parietal cells. Activation of muscarinic receptors depolarize smooth muscle and the ICC to augment the smooth-muscle activity that drives peristalsis and tone (1, 2, 4, 26, 186–190).

Inhibitory postganglionic neurons comprise the NANC path between the nicotinic preganglionic vagal efferent fibers and gastric smooth muscle and ICCs. The two most likely candidates mediating this connection are nitric oxide and vasoactive intestinal polypeptide, although other mediators such as adenosine and/or serotonin (5HT) have also been implicated (84, 162, 164, 165, 169–171, 176, 179, 191–194). Activation of this vagal NANC path produces a profound relaxation of the proximal stomach and depresses motility in the antrum. Together, these excitatory and inhibitory vagal efferent control mechanisms provide the brainstem with the tools to exert a very fine control of gastric motility (10).

Vago-Vagal Reflexes

Afferent input to the NTS is organized in a distinct viscerotopic manner: The intestine is represented in the subnuclei commissuralis and medialis, the stomach in the subnuclei medialis and gelatinosus, and the esophagus in the cNTS (67–69). Elucidation of this viscerotopic organization of afferent information, together with the knowledge that the DMV may be organized in a more functional manner, led to the hypothesis that vago-vagal reflex functions are organized in a sort of functional matrix (82). According to this hypothesis, contributions of vagal afferent input from different regions of the gut lead to different patterns of gut-directed vagal efferent outflow. Accordingly, afferents serving one gut region synapse on a subset of NTS neurons that, in turn, terminate on an appropriate collection of DMV neurons, completing the reflex. Medial NTS neurons (i.e., the cells that receive vagal afferent input from the stomach and intestine) possess long mediolaterally oriented dendrites ($\sim$600 μm) that extend across the terminal zones of all gastrointestinal afferent inputs (195). A priori, such an arrangement tends to favor a convergence of vagal afferent input of different modalities and different visceral loci onto select populations of NTS neurons. Selected NTS neurons provide dominant local synaptic control over the appropriate DMV projections to the stomach (61, 82) and determine a specific pattern of gastric control (59, 60, 108). In general, stimulation of the proximal GI (i.e., esophagus, fundic stomach) vagal afferents activates some gastric-projecting vagal fibers while inhibiting others. By contrast, activation of vagal afferents from more distal sites in the antrum and intestine results in inhibition of practically all vagal outflow to the stomach (see reviews in References 10, 61, 62, and 108). In combination with other physiological studies, we see that vagal afferent input integrated by the NTS ultimately evokes gastroinhibition by either the withdrawal of cholinergic tone (i.e., generated by the pacemaker activity of DMV neurons), the activation of the vagal NANC pathway, or a combination of both, as described above.

This model provides the framework for the hardware responsible for coordinating vago-vagal responses. In its static form, one may presume that activation of any given afferent elicits the same hard-wired efferent response. Recent studies have, however, revealed a high degree of plasticity in available responses, such as in the vago-vagal reflex control over the stomach. One agonist signal may "gate" another, and the tonic effects of vagal afferent input may "gate" agonist responses. As mentioned above, DMV neurons are spontaneously active and highly sensitive to inputs from the NTS. This means that influences modulating NTS input to the DMV can have potent effects on the regulation of gastric motility. Although many examples of presynaptic modulation of the NTS are available, we focus below on the regulation of opioidergic presynaptic signaling, as it is perhaps the best-developed example.

Agonist and Afferent "Gating" of Opiate Effects on Connections Between the NTS and DMV

CNS opioid release is proportional to the anticipated quality of a reward (196–198). With respect to feeding, opioid release is associated with the expected quality of food to be consumed. It is clear that activation of central opioidergic mechanisms is associated with the consumption of large amounts of palatable food. In evolutionary terms, rapid consumption of scarce, palatable food was desirable, in that palatability is an excellent predictor of caloric density or metabolic usefulness.

There is some evidence suggesting that opiates can act in the dorsal vagal complex to augment feeding as well as to coordinate gastric function in anticipation of large (and in evolution, scarce) palatable meals (199–201). It would be advantageous, under these circumstances, to have a "cephalic-phase" mechanism by which opioids could augment digestive processes in order to assimilate high-quality food as quickly as possible. It would be especially convenient if the effects of opioids to accelerate digestion could be gated or regulated by the presence of other agonists whose release would signal either imminent feeding or that feeding is taking place. In this way, the effects of opioids on digestion would work best while taking food rewards, but not during the taking of other rewards.

Activation of central opioid receptors, abundant also in the DVC, may signal the consumption of palatable food and elicit gastric relaxation (199, 200, 202–208). Enhancement of proximal gastric relaxation during feeding would assist in the assimilation of large, palatable meals. Teleologically speaking, such a mechanism would explain why celebrants at Thanksgiving dinner may complain that they can't eat another bite of green beans, yet mysteriously "find room" for pie and ice cream. Although this laboratory group did not set out specifically to investigate the hedonic mechanism of gastric relaxation in response to palatable meals, our basic neurophysiological results do suggest that opioid effects within the DVC to provoke gastric relaxation are gated by factors associated with the taking of meals.

When we analyzed the cellular mechanisms of μ-opioid actions in the DVC, our group first showed that endogenous opioid peptides inhibit all of the glutamatergic but none of the GABAergic currents between the NTS and the DMV (141). The selectivity of the effects of opioids on the brainstem circuitry was explained by the presence of μ-opioid receptors on the surface of glutamatergic, but not GABAergic, profiles impinging on DMV neurons (141) (Figure 2). It is possible, however, to induce an inhibitory effect of opioids on approximately 60% of the GABAergic currents by activating the cAMP-PKA pathway either via forskolin or via pretreatment with hormones such as thyrotropin-releasing hormone (TRH) or CCK. TRH and CCK are well-known markers of cephalic-phase digestive functions (10, 209) and the consumption of fatty meals, respectively (210).

The increase in cAMP-PKA activity in NTS neurons promotes fast (within five minutes) but short-lasting (approximately 60 minutes) trafficking of μ-opioid receptors from intracellular compartments to the outer membrane of the GABAergic terminals. Once on the surface of the synaptic terminal, opioid receptors are available for activation by the μ-agonist, resulting in a decrease in the net charge of the GABAergic current (211). This mechanism may explain, in part, feeding-related changes in the effectiveness of central gastroinhibitory action of opiates. Opioid receptors are normally available for interaction with their endogenous ligand on a limited population of glutamatergic synapses impinging on DMV neurons of the cholinergic (gastroexcitatory) pathway. Inhibition of this glutamatergic input induces a partial cholinergic withdrawal in the stomach and, by consequence, a modest inhibition of gastric motility. Activation of the cAMP-PKA pathway in the NTS, for example by peptides such as TRH or CCK, induces trafficking of

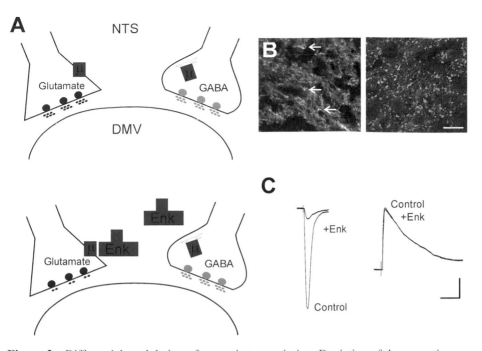

Figure 2   Differential modulation of synaptic transmission. Depiction of the synaptic connections between the NTS and the DMV. (*A*) Under control conditions, the μ-opioid receptor
(μ) is expressed on the nerve terminal of glutamatergic NTS profiles apposing DMV neurons;
by contrast, on GABAergic nerve terminals, the μ-opioid receptor is internalized and associated with the Golgi apparatus (*upper panel*). Opioid agonists (e.g., Enk) can act to inhibit
glutamate synaptic transmission, but not GABAergic transmission (*lower panel*). (*B*) High-
power photomicrographs of the cloned μ-opioid receptor (MOR1)-immunoreactivity (-IR;
TRITC filters, *red*) and γ-glutamyl glutamate-IR (Glu-IR; FITC filters, *green*; *left panel*)
used as a marker for glutamate nerve terminals, and glutamic acid decarboxylase-IR (GAD-
IR; FITC filters, *green*; *right panel*) used as a marker for GABAergic nerve terminals in
the rat DVC. Note that MOR1-IR is colocalized with glutamate nerve terminals (*yellow*;
arrows) but not with GABA nerve terminals. Scale bar: 15 μm. (*C*) Representative traces of
evoked excitatory postsynaptic currents (eEPSCs) (*left panel*) and evoked inhibitory postsynaptic currents (eIPSCs) (*right panel*) in gastric-projecting DMV neurons voltage-clamped at
–60 mV and –50 mV, respectively, evoked by electrical stimulation of the NTS. Perfusion
with Enk inhibits the eEPSCs but not eIPSCs. Scale bar: 50 pA and 30 ms.

μ-opioid receptors on GABA terminals synapsing on DMV neurons that comprise
the NANC pathway (211). Opioid agonists, by reducing the GABAergic input to
these DMV neurons, increase the vagal output of NANC neurotransmitters to the
target organ and, by consequence, further enhance its inhibition. By this means, the
effectiveness of opiates to provoke gastroinhibition can be greatly amplified by the
action of other agonists.

Agonist modulation of receptor trafficking may also be responsible for the dramatic and very different effects neuropeptide Y (NPY) has on gastric motility in the fed versus fasted state. Briefly, a vagally mediated increase in gastric motility and secretion is observed when NPY is administered centrally to fasted rats (212, 213), but gastric motility is reduced when NPY is administered to fed rats (213, 214). The cellular explanation of this intriguing difference in gastric responses to central administration of NPY is probably similar to those explanations regarding the effects of opioids at the level of the GABAergic synapse between the NTS and the DMV (62, 141, 143, 211, 215). Our group has shown recently that the GABAergic NTS synapse with DMV neurons is subject to receptor trafficking modulation by a number of neurotransmitter and hormone agonists, including TRH, CCK, 5HT, norepinephrine, pancreatic polypeptides, and opiates (103, 211, 216–221).

In another hypothesis, an additional type of short-term plasticity, determined by the state of activation of vagal afferent fibers, also influences receptor availability at the level of the DMV-NTS GABAergic synaptic connection. This, in turn, allows the vago-vagal reflex circuits to adapt their responses to the immediate physiological needs of the animal, i.e., to different phases of the digestive process. As mentioned above, this hypothesis is quite appealing because it suggests that multiple factors regulating the release of glutamate from vagal afferents (e.g., CCK, leptin, tumor necrosis factor, and glutamate itself) may modulate basic reflex functions.

Preliminary data from our group are starting to provide explanations for the cellular and structural mechanisms that underlie the capability of this circuit to adapt its responses to the visceral afferent state of activation. Although this state of activation is reflected in the levels of activity of the cAMP-PKA pathway in the GABAergic NTS nerve terminals impinging on DMV neurons, it is set, principally, by vagal afferent fiber inputs onto NTS neurons and/or terminals. We came to this conclusion by observing that following vagal sensory deafferentation, either by perivagal capsaicin or by selective sensory rhizotomy, opioids reduced the GABAergic currents between the NTS and the DMV, but without the need for increasing the activity of the cAMP-PKA pathway (215). These data imply that vagal sensory afferent inputs chronically dampen the level of activity of the cAMP-PKA pathway at the level of the GABAergic synapses between the NTS and the DMV, making the synapse unavailable for modulation. As mentioned above, glutamate is the main neurotransmitter of vagal afferent fibers impinging on brainstem circuits (32–35, 37–42, 44, 45). Following its release, glutamate activates both ionotropic and metabotropic receptors (mGluRs). All metabotropic glutamate receptor subunits identified to date have been found in the DVC (50). The presence of group II and III mGluRs, both of which play a role in the modulation of brainstem vagal circuits (47–49, 222, 223), is of particular interest, because their activation decreases the levels of cAMP (224). We have recently investigated whether these receptor groups are involved in setting the levels of cAMP-PKA activation in GI brainstem circuits and, by consequence, in determining the availability of the NTS-DMV GABAergic synapse to modulation.

Our data indicate that group II mGluRs, but not group III mGluRs, are involved in this type of modulation of the GI brainstem circuits. In fact, pretreatment with selective mGluR group II antagonists "primes" the μ-opioid receptors on GABAergic NTS-DMV synapses, making it possible for opioids to modulate the inhibitory currents without the need to pharmacologically increase the activity of the cAMP-PKA pathway (K.N. Browning, Z.L. Zheng, & R.A. Travagli, submitted manuscript).

In summary, sensory vagal afferent fibers use group II mGluRs to dampen the activity of the cAMP-PKA pathway in GABAergic NTS nerve terminals. While held in check, the low levels of cAMP-PKA activity keep the μ-opioid receptors in inaccessible intracellular compartments such that enkephalins do not have a modulatory effect on the NTS-DMV GABAergic currents. The use of glutamate both to activate the sensory vagal pathways and at the same time keep the GABAergic NTS neurons in a nonmodulatory state is an apparent conundrum. In fact, it implies that glutamate released by sensory vagal afferent fibers exerts opposite functions via simultaneously acting at ionotropic receptors (which carry information about the visceral organs and prepare the circuit for a sophisticated level of modulation to allow the appropriate response) and at mGluRs (which dampen the brainstem GABAergic circuitry and prevent its modulation by neurotransmitters such as pancreatic polypeptides, opioids, indolamines, and catecholamines).

We have to keep in mind, however, that group II mGluRs may be located perisynaptically (225) and may be activated by the release of glutamate that leaks from vagal afferent terminals on NTS neurons in an action potential–independent manner. If this is the case, then mGluRs and ionotropic glutamate receptors may coexist without interfering with each others' functions; furthermore, peptides impinging on NTS neurons, such as TRH and CCK, that increase the activity of the cAMP-PKA pathway may overcome the tonic activation of group II mGluRs, increase the levels of cAMP, and induce altered sensitivity of neural circuits to a variety of inputs. The role of glutamate activation of group II mGluR, then, may be to change the synaptic state of the NTS-DMV GABAergic circuit. The widespread use by different neurotransmitters of the cAMP-PKA pathway argues in favor of its utilization in a general manner in the control of vago-vagal circuits. Also, depending on whether the affected GABAergic NTS-DMV synapse controls cholinergic excitatory pathways or NANC inhibitory pathways, a decrease in synaptic transmission may result in an increased or decreased vagal motor output, respectively. The physiological correlate of these experimental conditions is modulation of the NTS-DMV GABAergic synapse by neurotransmitters that are coupled to adenylate cyclase and that generate cAMP. Alterations in sensory inputs from the GI tract, such as those following activation of vago-vagal reflexes or changes in the feeding status, for example, are expected to modify the ability of tonic GABAergic inputs controlling the vagal brainstem circuits to be modulated. For example, the constant perception of ongoing GI tract activity exerts a tonic inhibition of cAMP-PK pathways in the DVC. Activation of vago-vagal reflexes or changes in the state of activation (i.e., from fasted to fed or vice versa) may change the

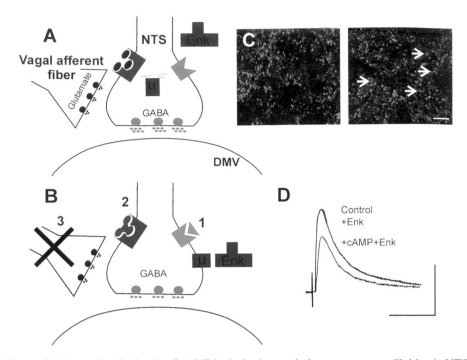

Figure 3   Increasing the levels of cAMP in the brainstem induces receptor trafficking in NTS nerve terminals. (*A*) In control conditions, i.e., when the levels of cAMP are low either because the tonic release of glutamate from vagal afferent fibers activates group II metabotropic glutamate receptors (*left*) or because Gαs-coupled receptors are not activated (*right*), the terminals of GABAergic neurons in the NTS store μ-opioid receptors (μ) in internal compartments associated with the Golgi complex. In this situation, μ-opioid agonists (e.g., Enk) cannot modulate the release of GABA onto DMV neurons. (*B*) Following increases in cAMP levels within the GABAergic nerve terminal, for example, by (*1*) activation of a receptor coupled to Gαs e.g., TRH or CCK, (*2*) antagonism of group II metabotropic glutamate receptors, or (*3*) removal of tonic vagal afferent input, μ-opioid receptors are released from the Golgi apparatus and translocated to the nerve terminal membrane, where opioid agonists can inhibit GABA synaptic transmission between the NTS and the DMV. (*C*) High-power photomicrographs of the cloned μ-opioid receptor (MOR1)-immunoreactivity (-IR; TRITC filters, *red*) and glutamic acid decarboxylase-IR (GAD-IR; FITC filters, *green*) used as a marker for GABAergic nerve terminals in the rat DVC. In control conditions (*left panel*), note the absence of MOR- and GAD-IR colocalized profiles. Following vagal afferent rhizotomy (*right panel*), many nerve terminals show MOR- and GAD-IR colocalized profiles (*yellow; arrows*). Images represent three-dimensional reconstructions from Z-stack image series. Scale bar: 10 μm. (*D*) Representative traces of eIPSCs evoked in a gastric-projecting DMV neuron voltage-clamped at −50 mV. Perfusion with Enk does not affect the amplitude of IPSCs evoked by electrical stimulation of the NTS. Following five minutes' perfusion with substances that increase the cAMP levels, however, reapplication of Enk reduces the amplitude of the eIPSCs. Scale bar: 50 ms and 200 pA.

levels of cAMP and influence the ability of circulating hormones or locally released neurotransmitters to modulate inhibitory synaptic transmission between the NTS and the DMV (Figure 3).

SUMMARY

GI brainstem circuits that are part of vago-vagal reflexes are comprised of sensory afferent fibers whose terminals impinge on NTS neurons. The NTS neurons then project to DMV cells, which in turn provide the preganglionic efferent fibers controlling cholinergic excitatory and NANC inhibitory postganglionic cells. The strategic location outside the blood-brain barrier of portions of brainstem vagal circuits makes them accessible to a multitude of circulating hormones, cytokines, and chemokines that can dramatically alter vago-vagal reflex responsiveness. Add to this the dense and phenotypically diverse descending projections from anterior forebrain structures to the dorsal vagal complex (8), and it becomes clear that the number of mechanisms by which vago-vagal reflex selectivity and sensitivity can be modulated is virtually limitless.

The presynaptic modulation, and hence plasticity, of these circuits allows for a finely tuned control over gastric functions. It is now apparent that these circuits are not static entities devoted to the relay of unprocessed information between the brain and the gut. Indeed, by controlling the levels of activity of the cAMP-PKA pathway in synaptic terminals, circulating molecules, descending neurotransmitter inputs, and primary afferent traffic can act on brainstem circuits to affect the sensitivity of the synapses to modulation. If variations of the cAMP levels are determined, for example by the substances circulating in different situations such as during fasting or following a meal or by differing levels of visceral afferent traffic, then the net effect of activation of a particular circuit can vary according to the cAMP levels within the synapse itself. This implies, then, that gross differences in the performance of gastric control reflexes that occur at different times of the digestive process may result from variations in the cAMP levels at GABAergic NTS-DMV synapses.

In conclusion, we are just starting to understand the cellular mechanisms that underlie the exquisite coordination of digestive processes with ongoing and anticipated changes in behavior as a consequence of the diversity of input. The work to be conducted in the years to come must investigate the circulating factors that can cause rapid changes in brain-gut control through their direct action on brainstem vagal reflex control circuits and the cellular entities involved in this control. A multitude of possible modulatory mechanisms exist within these circuits to guarantee speed, precision, and flexibility in the control of digestive processes.

ACKNOWLEDGMENTS

The authors would like to thank the NIH (grants nos. DK-55530, DK-56373, and DK-52142) and NSF (grant no. IBN-0456291) for their support. We also thank

Cesare M. Travagli, Hans Hermann, and Lois and Richard F. Rogers for their support and encouragement.

The *Annual Review of Physiology* is online at
http://physiol.annualreviews.org

LITERATURE CITED

1. Wood JD. 1987. Physiology of the enteric nervous system. In *Physiology of the Gastrointestinal Tract*, ed. LR Johnson, pp. 67–110. New York: Raven
2. Goyal RK, Hirano I. 1996. The enteric nervous system. *N. Engl. J. Med.* 334: 1106–15
3. Costa M, Brookes SJH, Hennig GW. 2000. Anatomy and physiology of the enteric nervous system. *Gut* 47:15–19
4. Bornstein JC, Costa M, Grider JR. 2004. Enteric motor and interneuronal circuits controlling motility. *Neurogastroenterol. Motil.* 16(Suppl. 1):34–38
5. Goyal RK, Paterson WG. 1989. Esophageal motility. In *Handbook of Physiology—The Gastrointestinal System I*, ed. J.D. Wood, pp. 865–908. Bethesda, MD: Am. Phys. Soc.
6. Chang HY, Mashimo H, Goyal RK. 2003. Musings on the wanderer: What's new in our understanding of vago-vagal reflex?: IV. Current concepts of vagal efferent projections to the gut. *Am. J. Physiol. Gastrointest. Liver Physiol.* 284:G357–66
7. Roman C, Gonella J. 1987. Extrinsic control of digestive tract motility. In *Physiology of the Gastrointestinal Tract*, ed. LR Johnson, pp. 507–54. New York: Raven
8. Blessing WW. 1997. *The Lower Brainstem and Bodily Homeostasis*. New York: Oxford Univ. Press
9. Rogers RC, Hermann GE, Travagli RA. 2005. Brainstem control of the gastric function. In *Physiology of the Gastrointestinal Tract*, ed. LR Johnson. San Diego, CA: Elsevier. In press
10. Rogers RC, McTigue DM, Hermann GE. 1995. Vagovagal reflex control of diges-

tion: afferent modulation by neural and "endoneurocrine" factors. *Am. J. Physiol.* 268:G1–10
11. Browning KN. 2003. Excitability of nodose ganglion cells and their role in vagovagal reflex control of gastrointestinal function. *Curr. Opin. Pharmacol.* 3:613–17
12. Browning KN, Mendelowitz D. 2003. Musings on the wanderer: what's new in our understanding of vago-vagal reflexes?: II. Integration of afferent signaling from the viscera by the nodose ganglia. *Am. J. Physiol. Gastrointest. Liver Physiol.* 284:G8–14
13. Zhuo H, Ichikawa H, Helke CJ. 1997. Neurochemistry of the nodose ganglion. *Prog. Neurobiol.* 52:79–107
14. Berthoud HR, Kressel M, Raybould HE, Neuhuber WL. 1991. Vagal sensors in the rat duodenal mucosa: distribution and structure as revealed by in vivo DiI-tracing. *Anat. Embryol.* 191:203–12
15. Berthoud HR, Powley TL. 1992. Vagal afferent innervation of the rat fundic stomach: morphological characterization of the gastric tension receptor. *J. Comp. Neurol.* 319:261–76
16. Neuhuber WL, Kressel M, Stark A, Berthoud HR. 1998. *J. Auton. Nerv. Syst.* 70:92–102 (Abstr.)
17. Berthoud HR, Neuhuber WL. 2000. Functional and chemical anatomy of the afferent vagal system. *Auton. Neurosci.* 85:1–17
18. Mayer EA. 1994. The physiology of gastric storage and emptying. In *Physiology of the Gastrointestinal Tract*, ed. LR Johnson, pp. 929–76. New York: Raven

19. Berthoud HR, Patterson LM, Zheng H. 2001. Vagal-enteric interface: vagal activation-induced expression of c-Fos and p-CREB in neurons of the upper gastrointestinal tract and pancreas. *Anat. Rec.* 262:29–40

20. Sengupta JN. 2000. An overview of esophageal sensory receptors. *Am. J. Med.* 108 (Suppl. 4a):87S–9S

21. Berthoud HR, Blackshaw LA, Brookes SJ, Grundy D. 2004. Neuroanatomy of extrinsic afferents supplying the gastrointestinal tract. *Neurogastroenterol. Motil.* 16 (Suppl. 1):28–33

22. Zagorodnyuk V, Chen B, Brookes S. 2001. Intraganglionic laminar endings are mechano-transduction sites of vagal tension receptors in the guinea-pig stomach. *J. Physiol.* 534:255–68

23. Blackshaw LA, Grundy D, Scratcherd T. 1987. Involvement of gastrointestinal mechano- and intestinal chemoreceptors in vagal reflexes: an electrophysiological study. *J. Auton. Nerv. Syst.* 18:225–34

24. Phillips RJ, Powley TL. 2000. Tension and stretch receptors in gastrointestinal smooth muscle: re-evaluating vagal mechanoreceptor electrophysiology. *Brain Res. Rev.* 34:1–26

25. Zagorodnyuk VP, Brookes SJH. 2000. Transduction sites of vagal mechanoreceptors in the guinea pig esophagus. *J. Neurosci.* 20:6249–55

26. Grundy D, Schemann M. 1993. The interface between the enteric and central nervous system. In *Innervation of the Gut: Pathophysiological Implications*, ed. Y Tache, DL Wingate, TF Burks, pp. 157–66. Boca Raton: CRC

27. Powley TL, Phillips RJ. 2002. Musings on the wanderer: what's new in our understanding of vago-vagal reflexes? I. Morphology and topography of vagal afferents innervating the GI tract. *Am. J. Physiol. Gastrointest. Liver Physiol.* 283:G1217–25

28. Zagorodnyuk VP, Chen BN, Costa M, Brookes SJ. 2002. 4-aminopyridine- and dendrotoxin-sensitive potassium channels influence excitability of vagal mechanosensitive endings in guinea-pig oesophagus. *Br. J. Pharmacol.* 137:1195–206

29. Grundy D. 2002. Neuroanatomy of visceral nociception: vagal and splanchnic afferent. *Gut* 51 (Suppl. 1):i2–i5

30. Zagorodnyuk VP, Chen BN, Costa M, Brookes SJ. 2003. Mechanotransduction by intraganglionic laminar endings of vagal tension receptors in the guinea-pig oesophagus. *J. Physiol.* 553:575–87

31. Kirkup AJ, Brunsden AM, Grundy D. 2001. Receptors and transmission in the brain-gut axis: potential for novel therapies. 1. Receptors on visceral afferents. *Am. J. Physiol. Gastrointest. Liver Physiol.* 280:G787–94

32. Kawai Y, Senba E. 1996. Organization of excitatory and inhibitory local networks in the caudal nucleus of tractus solitarius of rats revealed in in vitro slice preparation. *J. Comp. Neurol.* 373:309–21

33. Smith BN, Dou P, Barber WD, Dudek FE. 1998. Vagally evoked synaptic currents in the immature rat nucleus tractus solitarii in an intact in vitro preparation. *J. Physiol.* 512:149–62

34. Lu WY, Bieger D. 1998. Vagal afferent transmission in the NTS mediating reflex responses of the rat esophagus. *Am. J. Physiol.* 274:R1436–45

35. Glatzer NR, Hasney CP, Bhaskaran MD, Smith BN. 2003. Synaptic and morphologic properties in vitro of premotor rat nucleus tractus solitarius neurons labeled transneuronally from the stomach. *J. Comp. Neurol.* 464:525–39

36. Andresen MC, Yang M. 1990. Non-NMDA receptors mediate sensory afferent synaptic transmission in medial nucleus tractus solitarius. *Am. J. Physiol.* 259:H1307–11

37. Hornby PJ. 2001. Receptors and transmission in the brain-gut axis. II. Excitatory amino acid receptors in the brain-gut axis. *Am. J. Physiol. Gastrointest. Liver Physiol.* 280:G1055–60

38. Jean A. 2001. Brainstem control of swallowing: neuronal network and cellular mechanisms. *Physiol. Rev.* 81:929–69

39. Lachamp P, Balland B, Tell F, Crest M, Kessler JP. 2003. Synaptic localization of the glutamate receptor subunit GluR2 in the rat nucleus tractus solitarii. *Eur. J. Neurosci.* 17:892–96

40. Yen JC, Chan JY, Chan SHH. 1999. Differential roles of NMDA and non-NMDA receptors in synaptic responses of neurons in nucleus tractus solitarii of the rat. *J. Neurophysiol.* 81:3034–43

41. Baptista V, Zheng ZL, Coleman FH, Rogers RC, Travagli RA. 2005. Characterization of neurons of the nucleus tractus solitarius pars centralis. *Brain Res.* 1052:139–46

42. Tell F, Jean A. 1991. Activation of N-methyl-D-aspartate receptors induces endogenous rhythmic bursting activities in nucleus tractus solitarii neurons: an intracellular study on adult rat brainstem slices. *Eur. J. Neurosci.* 3:1353–65

43. Wang YT, Bieger D, Neuman RS. 1991. Activation of NMDA receptors is necessary for fast information transfer at brainstem vagal motoneurons. *Brain Res.* 567:260–66

44. Aylwin ML, Horowitz JM, Bonham AC. 1997. NMDA receptors contribute to primary visceral afferent transmission in the nucleus of the solitary tract. *J. Neurophysiol.* 77:2539–48

45. Andresen MC, Kunze DL. 1994. Nucleus tractus solitarius—gateway to neural circulatory control. *Annu. Rev. Physiol.* 56:93–116

46. Jones NM, Monn JA, Beart PM. 1998. Type I and II metabotropic glutamate receptors regulate the outflow of [3H]D-aspartate and [14C]γ-aminobutyric acid in rat solitary nucleus. *Eur. J. Pharmacol.* 353:43–51

47. Chen CY, Ling Eh EH, Horowitz JM, Bonham AC. 2002. Synaptic transmission in nucleus tractus solitarius is depressed by Group II and III but not Group I presynaptic metabotropic glutamate receptors in rats. *J. Physiol.* 538:773–86

48. Chen CY, Bonham AC. 2005. Glutamate suppresses GABA release via presynaptic metabotropic glutamate receptors at baroreceptor neurones in rats. *J. Physiol.* 562:535–51

49. Page AJ, Young RL, Martin CM, Umaerus M, O'Donnell TA, et al. 2005. Metabotropic glutamate receptors inhibit mechanosensitivity in vagal sensory neurons. *Gastroenterology* 128:402–10

50. Hay M, McKenzie H, Lindsley K, Dietz N, Bradley SR, et al. 1999. Heterogeneity of metabotropic glutamate receptors in autonomic cell groups of the medulla oblongata of the rat. *J. Comp. Neurol.* 403:486–501

51. Hay M, Hoang CJ, Pamidimukkala J. 2001. Cellular mechanisms regulating synaptic vesicle exocytosis and endocytosis in aortic baroreceptor neurons. *Ann. NY Acad. Sci.* 940:119–31

52. Simasko SM, Ritter RC. 2003. Cholecystokinin activates both A- and C-type vagal afferent neurons. *Am. J. Physiol. Gastrointest. Liver Physiol.* 285:G1204–13

53. Appleyard SM, Bailey TW, Doyle MW, Jin YH, Smart JL, et al. 2005. Proopiomelanocortin neurons in nucleus tractus solitarius are activated by visceral afferents: regulation by cholecystokinin and opioids. *J. Neurosci.* 25:3578–85

54. Baptista V, Zheng Z, Coleman FH, Rogers RC, Travagli RA. 2005. Cholecystokinin octapeptide increases spontaneous glutamatergic synaptic transmission to neurons of the nucleus tractus solitarius centralis. *J. Neurophysiol.* 94:2763–71

55. Peters JH, McKay BM, Simasko SM, Ritter RC. 2005. Leptin-induced satiation mediated by abdominal vagal afferents. *Am. J. Physiol Regul. Integr. Comp. Physiol.* 288:R879–84

56. Kaplan JM, Moran TH. 2004. Gastrointestinal signaling in the control of food intake. In *Neurobiology of Food and Fluid Intake, Vol. 14 of Handbook of*

Behavioral Neurobiology, ed. E Stricker, S Woods, pp. 275–305. New York: Kluwer Acad./Plenum Publ.

57. Emch GS, Hermann GE, Rogers RC. 2000. TNF-α activates solitary nucleus neurons responsive to gastric distention. Am. J. Physiol. Gastrointest. Liver Physiol. 279:G582–86

58. Jin YH, Bailey TW, Li BY, Schild JH, Andresen MC. 2004. Purinergic and vanilloid receptor activation releases glutamate from separate cranial afferent terminals in nucleus tractus solitarius. J. Neurosci. 24:4709–17

59. Miolan JP, Roman C. 1984. The role of oesophageal and intestinal receptors in the control of gastric motility. J. Auton. Nerv. Syst. 10:235–41

60. Miolan JP, Roman C. 1978. Discharge of efferent vagal fibers supplying gastric antrum: indirect study by nerve suture technique. Am. J. Physiol. 235:E366–73

61. Travagli RA, Rogers RC. 2001. Receptors and transmission in the brain-gut axis: potential for novel therapies. V. Fast and slow extrinsic modulation of dorsal vagal complex circuits. Am. J. Physiol. Gastrointest. Liver Physiol. 281:G595–601

62. Travagli RA, Hermann GE, Browning KN, Rogers RC. 2003. Musings on the wanderer: What's new in our understanding of vago-vagal reflexes?: III. Activity-dependent plasticity in vago-vagal reflexes controlling the stomach. Am. J. Physiol. Gastrointest. Liver Physiol. 284:G180–87

63. Goyal RK, Padmanabhan R, Sang Q. 2001. Neural circuits in swallowing and abdominal vagal afferent-mediated lower esophageal sphincter relaxation. Am. J. Med. 111 (Suppl. 8A):95S–105S

64. Bieger D. 1993. The brainstem esophagomotor network pattern generator: a rodent model. Dysphagia 8:203–8

65. McCann MJ, Rogers RC. 1992. Impact of antral mechanoreceptor activation on the vago-vagal reflex in the rat: func-tional zonation of responses. J. Physiol. 453:401–11

66. Rogers RC, Hermann GE, Travagli RA. 1999. Brainstem pathways responsible for oesophageal control of gastric motility and tone in the rat. J. Physiol. 514:369–83

67. Altschuler SM, Bao X, Bieger D, Hopkins DA, Miselis RR. 1989. Viscerotopic representation of the upper alimentary tract in the rat: sensory ganglia and nuclei of the solitary and spinal trigeminal tracts. J. Comp. Neurol. 283:248–68

68. Altschuler SM, Ferenci DA, Lynn RB, Miselis RR. 1991. Representation of the cecum in the lateral dorsal motor nucleus of the vagus nerve and commissural subnucleus of the nucleus tractus solitarii in rat. J. Comp. Neurol. 304:261–74

69. Altschuler SM, Escardo J, Lynn RB, Miselis RR. 1993. The central organization of the vagus nerve innervating the colon of the rat. Gastroenterology 104:502–9

70. Barraco R, El-Ridi M, Parizon M, Bradley D. 1992. An atlas of the rat subpostremal nucleus tractus solitarius. Brain Res. Bull. 29:703–65

71. Paton JFR, Li Y-W, Deuchars J, Kasparov S. 2000. Properties of solitary tract neurons receiving inputs from the sub-diaphragmatic vagus nerve. Neurosci. 95:141–53

72. Chan RKW, Peto CA, Sawchenko PE. 2000. Fine structure and plasticity of barosensitive neurons in the nucleus solitary tract. J. Comp. Neurol. 422:338–51

73. Bradley RM, Grabauskas G. 1998. Excitation, inhibition and synaptic plasticity in the rostral gustatory zone of the nucleus of the solitary tract. Ann. NY Acad. Sci. 855:467–74

74. Mifflin SW, Felder RB. 1990. Synaptic mechanisms regulating cardiovascular afferent inputs to solitary tract nucleus. Am. J. Physiol. Heart Circ. Physiol. 28:H653–61

75. Loewy AD, Spyer KM. 1990. Vagal preganglionic neurons. In Central Regulation of Autonomic Functions, ed. AD Loewy,

KM Spyer, pp. 68–87. New York/Oxford: Oxford Univ. Press

76. Spyer KM. 1981. Neural organisation and control of the baroreceptor reflex. *Rev. Physiol. Biochem. Pharmacol.* 88:24–124

77. Broussard DL, Altschuler SM. 2000. Brainstem viscerotopic organization of afferents and efferents involved in the control of swallowing. *Am. J. Med.* 108:79S–86S

78. Mendelowitz D. 1999. Advances in parasympathetic control of heart rate and cardiac function. *News Physiol. Sci.* 14:155–61

79. Fox EA, Powley TL. 1985. Longitudinal columnar organization within the dorsal motor nucleus represents separate branches of the abdominal vagus. *Brain Res.* 341:269–82

80. Shapiro RE, Miselis RR. 1985. The central organization of the vagus nerve innervating the stomach of the rat. *J. Comp. Neurol.* 238:473–88

81. Norgren R, Smith GP. 1988. Central distribution of subdiaphragmatic vagal branches in the rat. *J. Comp. Neurol.* 273:207–23

82. Powley TL, Berthoud HR, Fox EA, Laughton W. 1992. The dorsal vagal complex forms a sensory-motor lattice: the circuitry of gastrointestinal reflexes. In *Neuroanatomy and Physiology of Abdominal Vagal Afferents*, ed. S Ritter, RC Ritter, CD Barnes, pp. 55–79. Boca Raton, FL: CRC

83. Beckett EA, McGeough CA, Sanders KM, Ward SM. 2003. Pacing of interstitial cells of Cajal in the murine gastric antrum: neurally mediated and direct stimulation. *J. Physiol.* 553:545–59

84. Berthoud HR. 1995. Anatomical demonstration of vagal input to nicotinamide acetamide dinucleotide phosphate diaphorase-positive (nitrergic) neurons in rat fundic stomach. *J. Comp. Neurol.* 358:428–39

85. Zheng H, Berthoud HR. 2000. Functional vagal input to gastric myenteric plexus as assessed by vagal stimulation-induced Fos expression. *Am. J. Physiol. Gastrointest. Liver Physiol.* 279:G73–81

86. Deleted in proof

87. Berthoud HR, Carlson NR, Powley TL. 1991. Topography of efferent vagal innervation of the rat gastrointestinal tract. *Am. J. Physiol.* 260:R200–7

88. Berthoud HR. 2004. The caudal brainstem and the control of food intake and energy balance. In *Neurobiology of Food and Fluid Intake, Vol. 14 of Handbook of Behavioral Neurobiology*, ed. E Stricker, S Woods, pp. 195–240. New York: Kluwer Acad./Plenum Publ.

89. Krowicki ZK, Sharkey KA, Serron SC, Nathan NA, Hornby PJ. 1997. Distribution of nitric oxide synthase in rat dorsal vagal complex and effects of microinjection of NO compounds upon gastric motor function. *J. Comp. Neurol.* 377:49–69

90. Fox EA, Powley TL. 1992. Morphology of identified preganglionic neurons in the dorsal motor nucleus of the vagus. *J. Comp. Neurol.* 322:79–98

91. Jarvinen MK, Powley TL. 1999. Dorsal motor nucleus of the vagus neurons: a multivariate taxonomy. *J. Comp. Neurol.* 403:359–77

92. Browning KN, Renehan WE, Travagli RA. 1999. Electrophysiological and morphological heterogeneity of rat dorsal vagal neurones which project to specific areas of the gastrointestinal tract. *J. Physiol.* 517:521–32

93. Martinez-Pena y Valenzuela IM, Browning KN, Travagli RA. 2004. Morphological differences between planes of section do not influence the electrophysiological properties of identified rat dorsal motor nucleus of the vagus neurons. *Brain Res.* 1003:54–60

94. Browning KN, Coleman FH, Travagli RA. 2005. Characterization of pancreas-projecting rat dorsal motor nucleus of the vagus neurons. *Am. J. Physiol. Gastrointest. Liver Physiol.* 288:G950–55

95. Huang X, Tork I, Paxinos G. 1993. Dorsal motor nucleus of the vagus nerve: a cyto- and chemoarchitectonic study in the human. *J. Comp. Neurol.* 330:158–82

96. Fogel R, Zhang X, Renehan WE. 1996. Relationships between the morphology and function of gastric and intestinal distention-sensitive neurons in the dorsal motor nucleus of the vagus. *J. Comp. Neurol.* 364:78–91

97. Travagli RA, Gillis RA. 1994. Hyperpolarization-activated currents I_H and I_{KIR}, in rat dorsal motor nucleus of the vagus neurons in vitro. *J. Neurophysiol.* 71:1308–17

98. Sah P. 1996. Ca^{2+}-activated K^+ currents in neurones: types, physiological roles and modulation. *Trends Neurosci.* 19:150–54

99. Dean JB, Huang R-Q, Erlichman JS, Southard TL, Hellard DT. 1997. Cell-cell coupling occurs in dorsal medullary neurons after minimizing anatomical-coupling artifacts. *Neuroscience* 80:21–40

100. Pedarzani P, Kulik A, Muller M, Ballanyi K, Stocker M. 2000. Molecular determinants of Ca^{2+}-dependent K^+ channel function in rat dorsal vagal neurones. *J. Physiol.* 527:283–90

101. Travagli RA, Gillis RA, Rossiter CD, Vicini S. 1991. Glutamate and GABA-mediated synaptic currents in neurons of the rat dorsal motor nucleus of the vagus. *Am. J. Physiol.* 260:G531–36

102. Marks JD, Donnelly DF, Haddad GG. 1993. Adenosine-induced inhibition of vagal motoneuron excitability: receptor subtype and mechanisms. *Am. J. Physiol.* 264:L124–32

103. Browning KN, Travagli RA. 2001. The peptide TRH uncovers the presence of presynaptic 5-HT1A receptors via activation of a second messenger pathway in the rat dorsal vagal complex. *J. Physiol.* 531:425–35

104. Henneman E, Somjen G, Carpenter DO. 1965. Functional significance of cell size in spinal motoneurons. *J. Neurophysiol.* 28:560–80

105. Hunt JN. 1983. Mechanisms and disorders of gastric emptying. *Annu. Rev. Med.* 34:219–29

106. Szurszewski JH. 1977. Modulation of smooth muscle by nervous activity: a review and a hypothesis. *Fed. Proc.* 36:2456–61

107. Schemann M, Grundy D. 1992. Electrophysiological identification of vagally innervated enteric neurons in guinea pig stomach. *Am. J. Physiol.* 263:G709–18

108. Grundy D, Schemann M. 2002. Motor control of the stomach. In *Innervation of the Gastrointestinal Tract*, ed. S Brookes, M Costa, pp. 57–102. London: Taylor and Francis

109. Champagnat J, Denavit-Saubie M, Grant K, Shen KF. 1986. Organization of synaptic transmission in the mammalian solitary complex, studied in vitro. *J. Physiol.* 381:551–73

110. Fortin G, Champagnat J. 1993. Spontaneous synaptic activities in rat nucleus tractus solitarius neurons in vitro: evidence for re-excitatory processing. *Brain Res.* 630:125–35

111. Schild JH, Khushalani S, Clark JW, Andresen MC, Kunze DL, Yang M. 1993. An ionic current model for neurons in the rat medial nucleus tractus solitarii receiving sensory afferent input. *J. Physiol.* 469:341–63

112. Fan W, Andresen MC. 1998. Differential frequency-dependent reflex integration of myelinated and nonmyelinated rat aortic baroreceptors. *Am. J. Physiol.* 275:H632–40

113. Andresen MC, Doyle MW, Bailey TW, Jin YH. 2004. Differentiation of autonomic reflex control begins with cellular mechanisms at the first synapse within the nucleus tractus solitarius. *Braz. J. Med. Biol. Res.* 37:549–58

114. Gillis RA, Quest JA, Pagani FD, Norman WP. 1989. Control centers in the central

nervous system for regulating gastrointestinal motility. In *Handbook of Physiology. Section 6: The Gastrointestinal System Motility and Circulation Volume 1, Part 2*, ed. JD Wood, pp. 621–83. Bethesda, MD: Am. Physiol. Soc.

115. Dekin MS, Getting PA. 1984. Firing pattern of neurons in the nucleus tractus solitarius: modulation by membrane hyperpolarization. *Brain Res.* 324:180–84

116. Dekin MS, Getting PA. 1987. In vitro characterization of neurons in the ventral part of the nucleus tractus solitarius. II. Ionic basis for repetitive firing patterns. *J. Neurophysiol.* 58:215–29

117. Haddad GG, Getting PA. 1989. Repetitive firing properties of neurons in the ventral region of nucleus tractus solitarius. In vitro studies in adult and neonatal rat. *J. Neurophysiol.* 62:1213–24

118. Paton JFR, Rogers WT, Schwaber JS. 1991. Tonically rhythmic neurons within a cardiorespiratory region of the nucleus tractus solitarii of the rat. *J. Neurophysiol.* 66:824–38

119. Maley BE. 1996. Immunohistochemical localization of neuropeptides and neurotransmitters in the nucleus solitarius. *Chem. Senses* 21:367–76

120. Ritchie TC, Westlund KN, Bowker RM, Coulter JD, Leonard RB. 1982. The relationship of the medullary catecholamine containing neurones to the vagal motor nuclei. *Neuroscience* 7:1471–82

121. Kalia M, Fuxe K, Hokfelt T, Johansson O, Lang R, et al. 1984. Distribution of neuropeptide immunoreactive nerve terminals within the subnuclei of the nucleus of the *tractus solitarius* of the rat. *J. Comp. Neurol.* 222:409–44

122. Takagi H, Kubota Y, Mori S, Tateishi K, Hamaoka T, Tohyama M. 1984. Fine structural studies of cholecystokinin-8-like immunoreactive neurons and axon terminals in the nucleus of *tractus solitarius* of the rat. *J. Comp. Neurol.* 227:369–79

123. Kalia M, Fuxe K, Goldstein M. 1985. Rat medulla oblongata. II. Dopaminergic, noradrenergic (A1 and A2) and adrenergic neurons, nerve fibers, and presumptive terminal processes. *J. Comp. Neurol.* 233:308–32

124. Harfstrand A, Fuxe K, Terenius L, Kalia M. 1987. Neuropeptide Y-immunoreactive perikarya and nerve terminals in the rat medulla oblongata: relationship to cytoarchitecture and catecholaminergic cell groups. *J. Comp. Neurol.* 260:20–35

125. Thor KB, Hill KM, Harrod C, Helke CJ. 1988. Immunohistochemical and biochemical analysis of serotonin and substance P colocalization in the nucleus tractus solitarii and associated ganglia of the rat. *Synapse* 2:225–31

126. Pickel VM, Chan J, Massari VJ. 1989. Neuropeptide Y-like immunoreactivity in neurons of the solitary tract nuclei: vesicular localization and synaptic input from GABAergic terminals. *Brain Res.* 476:265–78

127. Caccatelli S, Seroogy KB, Millhorn DE, Terenius L. 1992. Presence of a dynorphin-like peptide in a restricted subpopulation of catecholaminergic neurons in rat nucleus tractus solitarii. *Brain Res.* 589:225–30

128. Ohta A, Takagi H, Matsui T, Hamai Y, Iida S, Esumi H. 1993. Localization of nitric oxide synthase-immunoreactive neurons in the solitary nucleus and ventrolateral medulla oblongata of the rat: their relation to catecholaminergic neurons. *Neurosci. Lett.* 158:33–35

129. Lynn RB, Hyde TM, Cooperman RR, Miselis RR. 1996. Distribution of bombesin-like immunoreactivity in the nucleus of the solitary tract and dorsal motor nucleus of the rat and human colocalization with tyrosine-hydroxylase. *J. Comp. Neurol.* 369:552–70

130. Simonian SX, Herbison AE. 1996. Localization of neuronal nitric oxide synthase immunoreactivity within sub-population of noradrenergic A1 and A2 neurons in the rat. *Brain Res.* 732:247–52

131. Merchenthaler I, Lane M, Shughrue P. 1999. Distribution of pre-pro-glucagon and glucagon like peptide-1 receptor messenger RNAs in the rat central nervous system. *J. Comp. Neurol.* 403:261–80

132. Atkinson L, Batten TFC, Deuchars J. 2000. P2X$_2$ receptor immunoreactivity in the dorsal vagal complex and area postrema of the rat. *Neuroscience* 99:683–96

133. Lin L-H, Talman WT. 2000. N-methyl-D-aspartate receptors on neurons that synthetize nitric oxide in rat nucleus tractus solitarius. *Neuroscience* 100:581–88

134. Yao ST, Barden JA, Lawrence AJ. 2001. On the immunohistochemical distribution of ionotropic P2X receptors in the nucleus tractus solitarius of the rat. *Neuroscience* 108:673–85

135. Sivarao DV, Krowicki ZK, Hornby PJ. 1998. Role of GABA$_A$ receptors in rat hindbrain nuclei controlling gastric motor function. *Neurogastroenterol. Motil.* 10:305–13

136. Derbenev AV, Stuart TC, Smith BN. 2004. Cannabinoids suppress synaptic input to neurones of the rat dorsal motor nucleus of the vagus nerve. *J. Physiol.* 559:923–38

137. Williforl DJ, Ormsbee HS, Norman WP, Harmon JW, Garwey TQ, et al. 1981. Hindbrain GABA receptors influence parasympathetic outflow to the stomach. *Science* 214:193–94

138. Rogers RC, Travagli RA, Hermann GE. 2003. Noradrenergic neurons in the rat solitary nucleus participate in the esophageal-gastric relaxation reflex. *Am. J. Physiol. Regul. Integr. Comp. Physiol.* 285:R479–89

139. Martinez-Pena YVI, Rogers RC, Hermann GE, Travagli RA. 2004. Norepinephrine effects on identified neurons of the rat dorsal motor nucleus of the vagus. *Am. J. Physiol. Gastrointest. Liver Physiol.* 286:G333–39

140. Fukuda A, Minami T, Nabekura J, Oomura Y. 1987. The effects of noradrenaline on neurones in the rat dorsal motor nucleus of the vagus, *in vitro. J. Physiol.* 393:213–31

141. Browning KN, Kalyuzhny AE, Travagli RA. 2002. Opioid peptides inhibit excitatory but not inhibitory synaptic transmission in the rat dorsal motor nucleus of the vagus. *J. Neurosci.* 22:2998–3004

142. Lewis MW, Hermann GE, Rogers RC, Travagli RA. 2002. In vitro and in vivo analysis of the effects of corticotropin releasing factor on rat dorsal vagal complex. *J. Physiol.* 543:135–46

143. Browning KN, Travagli RA. 2003. Neuropeptide Y and peptide YY inhibit excitatory synaptic transmission in the rat dorsal motor nucleus of the vagus. *J. Physiol.* 549:775–85

144. Willis A, Mihalevich M, Neff RA, Mendelowitz D. 1996. Three types of postsynaptic glutamatergic receptors are activated in DMNX neurons upon stimulation of NTS. *Am. J. Physiol.* 271:R1614–19

145. Deleted in proof

146. Broussard DL, Li H, Altschuler SM. 1997. Colocalization of GABA A and NMDA receptors within the dorsal motor nucleus of the vagus nerve (DMV) of the rat. *Brain Res.* 763:123–26

147. Feng HS, Lynn RB, Han J, Brooks FP. 1990. Gastric effects of TRH analogue and bicuculline injected into dorsal motor vagal nucleus in cats. *Am. J. Physiol.* 259:G321–26

148. Washabau RJ, Fudge M, Price WJ, Barone FC. 1995. GABA receptors in the dorsal motor nucleus of the vagus influence feline lower esophageal sphincter and gastric function. *Brain Res. Bull.* 38:587–94

149. Corbett EK, Saha S, Deuchars J, McWilliam PN, Batten TF. 2003. Ionotropic glutamate receptor subunit immunoreactivity of vagal preganglionic neurones projecting to the rat heart. *Auton. Neurosci.* 105:105–17

150. Laccasagne O, Kessler JP. 2000. Cellular

and subcellular distribution of the amino 3-hydroxy-5-metyl-4-isoxazole proprionate receptor subunit GluR2 in the rat dorsal vagal complex. *Neurosci.* 99:557–63

151. Nabekura J, Ueno T, Katsurabayashi S, Furuta A, Akaike N, Okada M. 2002. Reduced NR2A expression and prolonged decay of NMDA receptor-mediated synaptic current in rat vagal motoneurons following axotomy. *J. Physiol.* 539:735–41

152. Sivarao DV, Krowicki ZK, Abrahams TP, Hornby PJ. 1999. Vagally-regulated gastric motor activity: evidence for kainate and NMDA receptor mediation. *Eur. J. Pharmacol.* 368:173–82

153. Pagani FD, Dimicco JA, Hamilton BL, Souza JD, Schmidt B, Gillis RA. 1987. Stress-induced changes in the function of the parasympathetic nervous system are mimicked by blocking GABA in the CNS of the cat. *Neuropharmacology* 26:155–60

154. Ferreira M Jr, Sahibzada N, Shi M, Panico W, Niedringhaus M, et al. 2002. CNS site of action and brainstem circuitry responsible for the intravenous effects of nicotine on gastric tone. *J. Neurosci.* 22:2764–79

155. Li Y, Owyang C. 2003. Musings on the wanderer: What's new in our understanding of vago-vagal reflexes? V. Remodeling of vagus and enteric neural circuitry after vagal injury. *Am. J. Physiol. Gastrointest. Liver Physiol.* 285:G461–69

156. Armstrong DM, Manley L, Haycock JW, Hersh LB. 1990. Co-localization of choline acetyltransferase and tyrosine hydroxylase within neurons of the dorsal motor nucleus of the vagus. *J. Chem. Neuroanat.* 3:133–40

157. Kalia M, Fuxe K, Goldstein M, Harfstrand A, Agnati LF, Coyle JT. 1984. Evidence for the existence of putative dopamine-, adrenaline- and noradrenaline-containing vagal motor neurons in the brainstem of the rat. *Neurosci. Lett.* 50:57–62

158. Zheng ZL, Rogers RC, Travagli RA. 1999. Selective gastric projections of nitric oxide synthase-containing vagal brainstem neurons. *Neuroscience* 90:685–94

159. Guo JJ, Browning KN, Rogers RC, Travagli RA. 2001. Catecholaminergic neurons in rat dorsal motor nucleus of vagus project selectively to gastric corpus. *Am. J. Physiol. Gastrointest. Liver Physiol.* 280:G361–67

160. Rossiter CD, Norman WP, Jain M, Hornby PJ, Benjamin SB, Gillis RA. 1990. Control of lower esophageal sphincter pressure by two sites in dorsal motor nucleus of the vagus. *Am. J. Physiol.* 259:G899–906

161. Abrahamsson H. 1973. Studies on the inhibitory nervous control of gastric motility. *Acta Physiol. Scand.* 390 (Suppl.): 1–38

162. Abrahamsson H. 1973. Vagal relaxation of the stomach induced from the gastric antrum. *Acta Physiol. Scand.* 89:406–14

163. Andrews PLR, Grundy D, Scratcherd T. 1980. Reflex excitation of antral motility induced by gastric distension in the ferret. *J. Physiol.* 298:79–84

164. Goyal RK, Rattan S. 1980. VIP as a possible neurotransmitter of non-cholinergic non-adrenergic inhibitory neurons. *Nature* 288:378–80

165. Abrahamsson H. 1986. Non-adrenergic non-cholinergic nervous control of gastrointestinal motility patterns. *Arch. Int. Pharmacodyn. Ther.* 280:50–61

166. Azpiroz F, Malagelada JR. 1986. Vagally mediated gastric relaxation induced by intestinal nutrients in the dog. *Am. J. Physiol.* 251:G727–35

167. Ito S, Ohga A, Ohta T. 1988. Gastric relaxation and vasoactive intestinal peptide output in response to reflex vagal stimulation in the dog. *J. Physiol.* 404:683–93

168. Azpiroz F, Malagelada JR. 1990. Perception and reflex relaxation of the stomach in response to gut distention. *Gastroenterol.* 98:1193–98

169. Desai KM, Zembowicz A, Sessa WC,

Vane JR. 1991. Nitroxergic nerves mediate vagally induced relaxation in the isolated stomach of the guinea pig. *Proc. Natl. Acad. Sci. USA* 88:11490–94

170. Desai KM, Sessa WC, Vane JR. 1991. Involvement of nitric oxide in the reflex relaxation of the stomach to accommodate food or fluid. *Nature* 351:477–79

171. Lefebvre RA, Hasrat J, Gobert A. 1992. Influence of N^G-nitro-L-arginine methyl ester on vagally induced gastric relaxation in the anaesthetized rat. *Br. J. Pharmacol.* 105:315–20

172. Lefebvre RA, Baert E, Barbier AJ. 1992. Influence of N^G-nitro-L-arginine on non-adrenergic non-cholinergic relaxation in the guinea-pig gastric fundus. *Br. J. Pharmacol.* 106:173–79

173. Meulemans AL, Helsen LF, Schuurkes AJ. 1993. Role of NO in vagally-mediated relaxations of guinea-pig stomach. *Naunyn-Schmiedebergs Arch. Pharmacol.* 347:225–30

174. Barbier AJ, Lefebvre RA. 1993. Involvement of the L-arginine: Nitric oxide pathway in nonadrenergic noncholinerigic relaxation of the cat gastric fundus. *J. Pharmacol. Exp. Ther.* 266:172–78

175. Meulemans AL, Eelen JG, Schuurkes JA. 1995. NO mediates gastric relaxation after brief vagal stimulation in anesthetized dogs. *Am. J. Physiol.* 269:G255–61

176. Takahashi T, Owyang C. 1995. Vagal control of nitric oxide and vasoactive intestinal polypeptide release in the regulation of gastric relaxation in rat. *J. Physiol.* 484:481–92

177. Curro D, Volpe AR, Preziosi P. 1996. Nitric oxide synthase activity and nonadrenergic non-cholinergic relaxation in the rat gastric fundus. *Br. J. Pharmacol.* 117:717–23

178. Krowicki ZK, Hornby PJ. 1996. Contribution of acetylcholine, vasoactive intestinal polypeptide and nitric oxide to CNS-evoked vagal gastric relaxation in the rat. *Neurogastroenterol. Motil.* 8:307–17

179. Takahashi T, Owyang C. 1997. Characterization of vagal pathways mediating gas tric accomodation reflex in rats. *J. Physiol.* 504:479–88

180. Krowicki ZK, Sivarao DV, Abrahams TP, Hornby PJ. 1999. Excitation of dorsal motor vagal neurons evokes non-nicotinic receptor-mediated gastric relaxation. *J. Auton. Nerv. Syst.* 77:83–89

181. Schubert ML. 2003. Gastric secretion. *Curr. Opin. Gastroenterol.* 19:519–25

182. Konturek PC, Konturek SJ, Ochmanski W. 2004. Neuroendocrinology of gastric H^+ and duodenal HCO^{3-} secretion: the role of brain-gut axis. *Eur. J. Pharmacol.* 499:15–27

183. Tache Y, Yang H. 1993. Role of medullary TRH in the vagal regulation of gastric function. In *Innervation of the Gut: Pathophysiological Implications*, ed. Y Tache, DL Wingate, TF Burks, pp. 67–80. Boca Raton: CRC

184. Rogers RC, Hermann GE. 1985. Vagal afferent stimulation-evoked gastric secretion suppressed by paraventricular nucleus lesion. *J. Auton. Nerv. Syst.* 13:191–99

185. Raybould HE, Tache Y. 1989. Capsaicin-sensitive vagal afferent fibers and stimulation of gastric acid secretion in anesthetized rats. *Eur. J. Pharmacol.* 167:237–43

186. Galligan JJ. 2002. Pharmacology of synaptic transmission in the enteric nervous system. *Curr. Opin. Pharmacol.* 2:623–29

187. Horowitz B, Ward SM, Sanders KM. 1999. Cellular and molecular basis for electrical rhythmicity in gastrointestinal muscles. *Annu. Rev. Physiol.* 61:19–43

188. Kunze WAA, Furness JB. 1999. The enteric nervous sytem and regulation of intestinal motility. *Annu. Rev. Physiol.* 61:117–42

189. Hirst GD, Ward SM. 2003. Interstitial cells: involvement in rhythmicity and neural control of gut smooth muscle. *J. Physiol.* 550:337–46

190. Sanders KM, Koh SD, Ordog T, Ward SM. 2004. Ionic conductances involved in generation and propagation of electrical slow waves in phasic gastrointestinal muscles. *Neurogastroenterol. Motil.* 16 (Suppl. 1):100–5

191. Abrahamsson H. 1974. Reflex adrenergic inhibition of gastric motility elicited from the gastric antrum. *Acta Physiol. Scand.* 90:14–24

192. Lefebvre RA, De Vriese A, Smits GJM. 1992. Influence of vasoactive intestinal polypeptide and N^G-nitro-L-arginine methyl ester on cholinergic neurotransmission in the rat gastric fundus. *Eur. J. Pharmacol.* 221:235–42

193. Bojo L, Lefebvre RA, Nellgard P, Cassuto J. 1993. Involvement of vasoactive intestinal polypeptide in gastric reflex relaxation. *Eur. J. Pharmacol.* 236:443–48

194. Boeckxstaens GE, Pelckmans PA, Bogers J, Bult H, De Man JG, et al. 1991. Release of nitric oxide upon stimulation of noadrenergic noncholinergic nerves in the rat gastric fundus. *J. Pharmacol. Exp. Ther.* 256:441–47

195. Rogers RC, McCann MJ. 1993. Intramedullary connections of the gastric region in the solitary nucleus: a biocytin histochemical tracing study in the rat. *J. Auton. Nerv. Syst.* 42:119–30

196. Dum J, Herz A. 1984. Endorphinergic modulation of neural reward systems indicated by behavioral changes. *Pharmacol. Biochem. Behav.* 21:259–66

197. Levine AS, Kotz CM, Gosnell BA. 2003. Sugars and fats: the neurobiology of preference. *J. Nutr.* 133:831S–4S

198. Fullerton DT, Getto CJ, Swift WJ, Carlson IH. 1985. Sugar, opioids and binge eating. *Brain Res. Bull.* 14:673–80

199. Giraudo SQ, Kotz M, Billington CJ, Levine AS. 1998. Association between the amygdala and the nucleus of the solitary tract in μ-opioid induced feeding in the rat. *Brain Res.* 802:184–88

200. Kotz CM, Billington CJ, Levine AS. 1997.

201. Opioids in the nucleus of the solitary tract are involved in feeding in the rat. *Am. J. Physiol.* 272:R1028–32

201. Kotz CM, Glass MJ, Levine AS, Billington CJ. 2000. Regional effect of naltrexone in the nucleus of the solitary tract in blockade of NPY-induced feeding. *Am. J. Physiol. Regulatory Integrative Comp. Physiol.* 278:R499–503

202. Burks TF, Galligan JJ, Hirning LD, Porreca F. 1987. Brain, spinal cord and peripheral sites of action of enkephalins and other endogenous opioids on gastrointestinal motility. *Gastroenterol. Clin. Biol.* 11:44B–51B

203. Del Tacca M, Bernardini C, Corsano E Soldani G, Roze C. 1987. Effects of morphine on gastric ulceration, barrier mucus and acid secretion in pylorus-ligated rats *Pharmacology* 35:174–80

204. Fox DA, Burks TF. 1988. Roles of central and peripheral μ, δ, and κ opioid receptors in the mediation gastric acid secretory effects in the rat. *J. Pharmacol. Exp. Ther.* 244:456–62

205. Farkas E, Jansen ASP, Loewy AD. 1997. Periaqueductal gray matter projection to vagal preganglionic neurons and the nucleus tractus solitarius. *Brain Res* 764:257–61

206. Liubashina O, Jolkkonen E, Pitkanen A. 2000. Projections from the central nucleus of the amygdala to the gastric related area of the dorsal vagal complex: a *Phaseolus vulgaris*-leucoagglutinin study in rats *Neurosci. Lett.* 291:85–88

207. Moriwaki A, Wang JB, Svingos A, Van Bockstaele EJ, Cheng P, et al. 1996 μ-opiate receptor immunoreactivity in rat central nervous system. *Neurochem. Res* 21:1315–31

208. Pickel VM, Colago EEO. 1999. Presence of μ-opioid receptors in targets of efferent projections from the central nucleus of the amygdala to the nucleus of the solitary tract. *Synapse* 33:141–52

209. Martinez V, Barrachina MD, Ohning G Tache Y. 2002. Cephalic phase of acid

secretion involves activation of medullary TRH receptor subtype 1 in rats. *Am. J. Physiol. Gastrointest. Liver Physiol.* 283:G1310–19

210. Moran TH. 2004. Gut peptides in the control of food intake: 30 years of ideas. *Physiol. Behav.* 82:175–80

211. Browning KN, Kalyuzhny AE, Travagli RA. 2004. μ-opioid receptor trafficking on inhibitory synapses in the rat brainstem. *J. Neurosci.* 24:9344–52

212. Yoneda M, Yokohama S, Tamori K, Sato Y, Nakamura K, Makino I. 1997. Neuropeptide Y in the dorsal vagal complex stimulates bicarbonate-dependent bile secretion in rats. *Gastroenterology* 112: 1673–80

213. Chen CH, Stephens RL Jr, Rogers RC. 1997. PYY and NPY control of gastric motility via action on Y1 and Y2 receptors in the DVC. *Neurogastroenterol. Motil.* 9:109–16

214. Matsuda M, Aono M, Moriga M, Okuma M. 1993. Centrally administered NPY inhibits gastric emptying and intestinal transit in the rat. *Dig. Dis. Sci.* 38:845–50

215. Browning KN, Kalyuzhny AE, Travagli RA. 2003. *State dependent activation of brainstem circuits by opioids.* Presented at Neurosci. Meet., 33rd, New Orleans.

216. Travagli RA, Browning KN. 2001. *Electrophysiological properties of identified pancreatic neurons of the rat dorsal motor nucleus of the vagus.* Presented at Neurosci. Meet., 31st, San Diego.

217. Browning KN, Travagli RA. 2001. *NPY in the rat dorsal motor nucleus of the vagus (DMV) inhibit excitatory but not inhibitory synaptic transmission.* Presented at Dig. Dis. Week, Atlanta.

218. Browning KN, Travagli RA. 2003. *State dependent activation of brainstem circuits by opioids.* Presented at Dig. Dis. Week, Orlando, FL.

219. Browning KN, Gillis RA, Travagli RA. 2003. *State dependent activation of brainstem circuits by opioids.* Presented at Neurosci. Meet., 33rd, New Orleans.

220. Browning KN, Travagli RA. 2005. *Modulation of brainstem circuits by pancreatic polypeptides.* Presented at Int. Soc. Auton. Neurosci., Marseille.

221. Travagli RA, Browning KN. 2005. *Short term receptor trafficking in the dorsal vagal complex.* Presented at Int. Soc. Auton. Neurosci., Marseille.

222. Jin YH, Bailey TW, Andresen MC. 2004. Cranial afferent glutamate heterosynaptically modulates GABA release onto second-order neurons via distinctly segregated metabotropic glutamate receptors. *J. Neurosci.* 24:9332–40

223. Browning KN, Travagli RA. 2005. *Differential distribution of metabotropic glutamate receptors within gastric vagal brainstem circuits.* Presented at Dig. Dis. Week, Chicago.

224. Conn PJ, Pin JP. 1997. Pharmacology and functions of metabotropic glutamate receptors. *Annu. Rev. Pharmacol. Toxicol.* 37:205–37

225. Cartmell J, Schoepp DD. 2000. Regulation of neurotransmitter release by metabotropic glutamate receptors. *J. Neurochem.* 75:889–907

Annu. Rev. Physiol. 2006. 68:307–343
doi: 10.1146/annurev.physiol.68.040504.094718
Copyright © 2006 by Annual Reviews. All rights reserved
First published online as a Review in Advance on October 24, 2005

INTERSTITIAL CELLS OF CAJAL AS PACEMAKERS IN THE GASTROINTESTINAL TRACT

Kenton M. Sanders, Sang Don Koh, and Sean M. Ward

Department of Physiology and Cell Biology, University of Nevada School of Medicine, Reno, Nevada 89557; email: kent@unr.edu, sdk@unr.edu, sean@unr.edu

Key Words   smooth muscle, electrical rhythmicity, slow wave

■ **Abstract**   In the gastrointestinal tract, phasic contractions are caused by electrical activity termed slow waves. Slow waves are generated and actively propagated by interstitial cells of Cajal (ICC). The initiation of pacemaker activity in the ICC is caused by release of Ca^{2+} from inositol 1,4,5-trisphosphate (IP_3) receptor–operated stores, uptake of Ca^{2+} into mitochondria, and the development of unitary currents. Summation of unitary currents causes depolarization and activation of a dihydropyridine-resistant Ca^{2+} conductance that entrains pacemaker activity in a network of ICC, resulting in the active propagation of slow waves. Slow wave frequency is regulated by a variety of physiological agonists and conditions, and shifts in pacemaker dominance can occur in response to both neural and nonneural inputs. Loss of ICC in many human motility disorders suggests exciting new hypotheses for the etiology of these disorders.

INTRODUCTION

The gastrointestinal (GI) tract has the tasks of ingesting food, digesting and absorbing nutrients, and removing waste products. To accomplish these critical bodily functions, sophisticated motor patterns have developed in GI organs to maximize transit from one chamber to the next or to retain food in one region until particle size is reduced, digestion and absorption are accomplished, fluids and electrolytes are recovered, or a conscious decision for defecation has been received. Control of motor patterns comes from multiple layers of regulatory mechanisms that start with the basic properties of smooth muscle cells that permit contractile behavior or precondition cells for contraction or relaxation when other stimuli are overlaid. Like other motor systems in the body, GI muscles are controlled by a variety of neural and nonneural factors. For example, spontaneous electrical activity, intrinsic to many smooth muscle tissues, is generated by interstitial cells of Cajal (ICC) that are electrically coupled to the smooth muscle cells. In addition to this level of control, inputs from the enteric nervous system, hormonal influences, and paracrine factors regulate motor activity during normal physiological responses. Inflammatory mediators contribute additionally during the course of pathophysiological conditions. Smooth muscle cells integrate the inputs from all

0066-4278/06/0315-0307$20.00

levels of control and respond in normal individuals with appropriate contractile responses.

An interesting level of control in GI motility comes from the ongoing discharge of pacemaker activity by ICC. Pacemaker activity, and the slow waves that result, provide background conditioning of the smooth muscle syncytium, organizing the open probability of voltage-dependent Ca^{2+} channels in smooth muscle cells to generate phasic contractile activity in many regions of the GI tract (see Table 1 for definition of slow wave). Slow waves are both generated and actively propagated within ICC networks. ICC networks extend along and around the organs of the GI tract to all points involved in phasic contractions. ICC loss or disruption of ICC networks may be an important, causative factor in a number of GI motility disorders. Work on this interesting cell type has led to novel concepts for the mechanism of pacemaker activity, the means of slow wave propagation, and the role of nerves in regulating pacemaker activity. This review describes some of the most recent progress in this field.

ELECTROPHYSIOLOGY OF GASTROINTESTINAL MUSCLES

Muscles of the GI tract exhibit a variety of electrical patterns, from quiescent or tonic membrane potentials that change slowly in response to excitatory and inhibitory neural inputs to muscles with intrinsic rhythmicity (1). The latter are referred to as "phasic muscles," and these muscles are commonly paced by slow depolarization/repolarization events known as slow waves (Figure 1). The function of slow waves is to change membrane potential from a state of low open probability for voltage-dependent Ca^{2+} channels (e.g., resting potential or the maximum potential between slow waves of -80 to -55 mV) to one of potentials, when there is elevated probability of channel opening (i.e., peak depolarization to approximately -40 to -25 mV). The oscillation in Ca^{2+} channel openings results in periodic Ca^{2+} entry, and by this mechanism, the electrical activity organizes excitation-contraction coupling into a pattern of phasic contractions. Slow wave shape, frequency, amplitude, and duration vary in different species and in different parts of the GI tract (Figure 1), but slow waves always take membrane potentials of smooth muscle cells into a range of potentials in which there is a steep relationship between activation of Ca^{2+} channels and membrane potential. Thus, small changes in potential at the peak of slow waves can greatly affect Ca^{2+} entry and contraction. Some authors have referred to the membrane potentials reached during slow waves as the "mechanical threshold" (2, 3). The mechanical threshold defines the degree of depolarization at which sufficient Ca^{2+} entry occurs to accomplish excitation-contraction coupling.

In addition to the variability in slow wave shape, the response of smooth muscle cells to the depolarization caused by slow waves also varies considerably. In some cases muscles respond to slow waves with a generalized increase in open probability of Ca^{2+} channels during most of the duration of the slow wave.

TABLE 1   Terminology used to describe the intrinsic electrical events in GI smooth muscles and interstitial cells of Cajal (ICC)

Term	Definition or synonym used in literature
Action potential	Excitable event due to activation of dihydropyridine-sensitive Ca^{2+} channels in smooth muscle cells. Also called "spikes" and electrical response activity[a] (ERA). These events are blocked by μM concentrations of dihydropyridines. Large contractions are associated with the occurrence of action potentials (see Reference 5[b]).
Slow wave[c]	Spontaneous activity of interstitial cells of Cajal; not blocked by dihydropyridines (13). Slow waves conduct to smooth muscle cells and cannot be actively propagated by smooth muscle. Also called basic electrical rhythm[a] (BER), electrical control activity[a] (ECA), pacemaker potentials[a], slow wave–like action potentials[a] (SWAP), and action potentials[a] by various authors in the older literature. This term has also been used to describe slow waves recorded specifically from circular muscle cells in intact muscles (e.g., Reference 63).
Upstroke depolarization	The initial fast depolarization characteristic of slow waves recorded from ICC or smooth muscle cells in intact muscles. In recordings from smooth muscle cells, there is often a partial repolarization before the development of the plateau potential. The upstroke is due to the activation of dihydropyridine-resistant Ca^{2+} channels (70).
Plateau depolarization	The second component of slow waves; distinct in recordings from smooth muscle cells in intact muscles. Owing to summation of unitary potentials but may be augmented and lengthened by activation of ion channels in smooth muscle cells or in ICC-IM.
Unitary current	Basic pacemaker events in ICC. Conducts to smooth muscle cells and can be recorded in these cells as membrane potential fluctuations (see details in this review).
Pacemaker current	Inward current through main pacemaker conductance; summation of currents through these channels produces unitary currents (78).
Unitary potential	Voltage response to unitary currents (59).
Regenerative potentials[d]	The smooth muscle contribution to slow waves recorded from circular muscles (see Reference 5). These events are thought to result from summation of unitary potentials in small bundles of smooth muscle and are due to the currents generated by ICC-IM. Regenerative potentials have similar properties as the plateau potentials of slow waves and typically do not display prominent fast upstroke components (64). Also called slow potentials (66).
Pacemaker potential	Slow wave recorded from ICC (63).
Driving potential	Slow wave recorded from ICC (63).
Follower potentials	Slow wave recorded in longitudinal muscle cells (63).

[a]Older terms for action potentials and slow waves that have not had popular usage in recent years.

[b]References given are illustrative of papers in which the terms have been used in the recent literature.

[c]Some authors have chosen to use different terms for slow waves recorded from different compartments (i.e., ICC, circular muscle, and longitudinal muscle). In this review we use "slow wave" to refer to activity recorded from all three compartments.

[d]Regenerative potentials are likely to be similar to plateau potentials.

The small magnitude of inward current during the several second periods of slow waves is sufficient to accomplish excitation-contraction coupling (see Reference 4). In other cases, slow wave depolarization elicits regenerative Ca^{2+} action potentials in smooth muscle cells. Ca^{2+} action potentials result in brief but intense periods of Ca^{2+} entry, resulting in strong contractile responses. Some muscles have a relatively low threshold for Ca^{2+} action potentials (e.g., taenia coli) (5), whereas other muscles require tens of millivolts of depolarization before threshold is reached. The response of smooth muscle cells to slow wave depolarization is a function of ion channel expression and availability. Preconditioning of the input resistance of the smooth muscle syncytium by enteric neural, hormonal, and

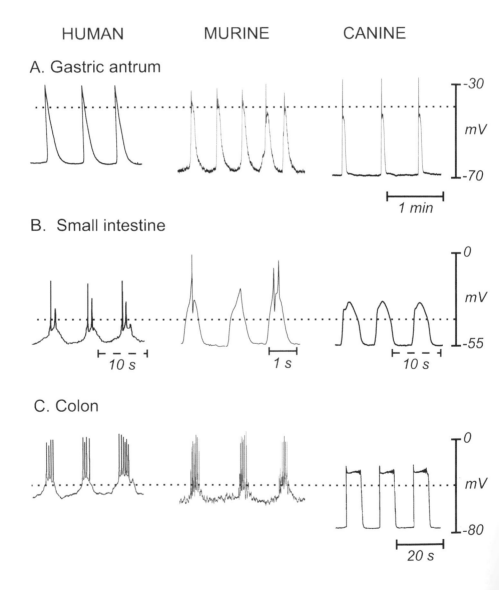

HUMAN   MURINE   CANINE

A. Gastric antrum

B. Small intestine

C. Colon

paracrine inputs is also a major determinant of the smooth muscle response to each slow wave cycle.

Slow wave activity in GI muscles arises from defined pacemaker regions in each organ. Unlike the heart, GI muscles do not have point sources (or nodes) for pacemaker activity. Instead, the phasic regions of the GI tract have a continuous network of electrically coupled pacemaker cells. Thus, every region of phasic muscle has intrinsic pacemaker capability. In the stomach, for example, pacemaker activity can be measured in small, isolated muscle strips taken from any region from the orad corpus to the pylorus (1). In vivo these pacemakers function as a unit, and the pacemaker cells along the greater curvature of the corpus usually provide the dominant frequency because these cells pace at the most rapid frequency (6). There is a proximal-to-distal pacemaker frequency gradient in the stomach, and the intrinsic frequency at more distal sites is much slower than the corpus frequency. Thus, there is time for an event generated in the corpus to propagate around the stomach and along the stomach to the pylorus before a distal pacemaker event occurs. This organizational feature of gastric slow waves is the basis for gastric peristaltic waves, which begin in the corpus and spread at the rate of slow wave propagation to the pyloric canal.

Slow waves do not propagate from the stomach to the small intestine (7, 8). There is a region of muscle at the junction between the pyloric sphincter and the

---

Figure 1 Examples of the diverse electrical activity recorded from GI muscles of three species. Short segments of intracellular electrophysiological records from human, murine, and canine stomach (gastric antrum) (*A*), small intestine (*B*), and colon (*C*) are shown. There are both differences and similarities in the electrical events recorded from the different species and gut regions. Slow wave frequencies and the maximal level of polarization between slow waves (resting potential) vary significantly in different GI muscles (note scales differ somewhat in each panel). Slow wave amplitude and coupling to generation of action potentials (see Table 1 for definition of action potential) also differ. The dotted line denotes the -40 mV point in each panel. This potential was chosen as a highlight because the Ca^{2+} current can typically be resolved in isolated GI smooth muscle cells with depolarization to approximately -40 mV (e.g., References 99–101). At potentials in the range of -40 mV, the voltage-dependent activation properties of L-type Ca^{2+} channels in smooth muscle cells (CaV 1.2 channels) become steeply dependent upon membrane potential. Thus, the mechanical threshold is reached in most GI smooth muscles at potentials near -40 mV. Note that each slow wave approaches or achieves potentials positive to this line. Action potential complexes always overshoot the mechanical threshold. L-type Ca^{2+} channels in smooth muscles also experience a "window-current phenomenon" in which there is incomplete inactivation of the channels with sustained depolarization (100, 102, 103). The window current is also in the range of potentials near -40 mV, so the duration of time that cells remain depolarized positive to -40 mV also influences the amount of Ca^{2+} entry and the degree of excitation-contraction coupling (4, 103).

duodenum that is electrically quiescent. Slow waves from the stomach or small bowel decay and die out as they conduct into this region (7). Recent evidence has suggested a discontinuity in the pacemaker cell network in this region (8). The small intestine has intrinsic pacemaker activity from the duodenum to the terminal ileum, and there is also a proximal-to-distal pacemaker frequency gradient in this organ (9, 10). The frequency of small intestinal pacemakers is higher than gastric pacemaker frequency. The higher frequency and the long length of the small bowel make slow wave propagation from duodenum to the ileum impossible. In fact, long distance propagation of slow waves (and peristaltic contractions) would tend to create rapid movement of contents and interfere with the main tasks of the small bowel (i.e., digestion and absorption of nutrients). Calculations based on pacemaker frequency and propagation rates suggest that slow waves can propagate only a few centimeters before colliding with a slow wave generated in an adjacent segment. Thus, pacemaker frequencies and propagation properties of slow waves in the small bowel naturally organize motility into segmental contractions, which are typical of the postprandial motor pattern in this region of the gut.

The colon displays far more varied activity and greater species differences in terms of electrical rhythmicity. In the stomach and small bowel, the circular and longitudinal muscles pace together (i.e., the same dominant pacemaker in each local region paces both muscle layers). But different frequencies and electrical activities occur in the circular and longitudinal muscles of the colon (11). Some animals, like the dog, display well-defined slow wave activity in the colon; however, others, e.g., human and mouse, display rather small depolarization cycles that may be considered slow waves, and these events elicit clusters of Ca^{2+} action potentials. The details of colonic slow wave propagation are rather vague, but many texts suggest a distal-to-proximal slow wave frequency gradient (see Reference 12).

Within each of the organs discussed above, there are well-defined regions within the thickness of the tunica muscularis that generates the dominant pacemaker activity. Dissection experiments in the 1980s carefully detailed the source and spread of slow waves from pacemaker regions in the stomach, small bowel, and colon (reviewed in detail in Reference 13). These studies showed that (a) pacemaker activity was generated and actively propagated through defined planes in GI muscles and (b) the removal of cells in the pacemaker regions blocked both spontaneous slow wave generation and active propagation of slow waves (see Reference 14). In the stomach the dominant pacemaker lies in the area between the circular and longitudinal muscle layer (15, 16), although cells within the muscle layers also display some intrinsic rhythmicity (16–19). The dominant pacemaker also exists between the circular and longitudinal muscle layers in the small intestine (20, 21). However, in larger animals, cells near the submucosal surface of the circular muscle layer (in the region of the deep muscular plexus) are also capable of pacemaker activity (see Reference 22). The colon displays two pacemaker regions, one at the submucosal surface of the circular muscle layer and the other between the circular and longitudinal muscle layers (11).

The terminology used to describe the electrical activity in the GI tract is rather complicated because various authors have coined their own terms for the events they have observed. Table 1 provides definitions of the various terms found in the old and new literature on slow waves and electrical rhythmicity in the gut.

INTERSTITIAL CELLS OF CAJAL ARE PACEMAKERS IN GASTROINTESTINAL MUSCLES

Interstitial Cells of Cajal Populate Pacemaker Regions

Pacemaker areas in GI muscle contain a variety of cell types, including enteric neurons, glial cells, smooth muscle cells, ICC, immune cells, and fibroblast-like cells. It is well established that slow waves are generated in the absence of neural inputs, and this activity is typically referred to in texts as "myogenic." For example, in glial cell line–derived neurotrophic factor (GDNF) knockout mice, the enteric nervous system fails to develop. These mice display normal slow wave activity (23). Morphological studies have shown that each of the pacemaker regions in the GI tract contains ICC (see References 15, 20, 24, and 25). To anatomists, the morphological features of ICC suggested that ICC may serve as pacemaker cells (see References 25 and 26; see also the older references within these papers). Physiological evidence for this hypothesis has been collected over the past 16 years. ICC between the circular and longitudinal muscle layers lie in the plane of the myenteric plexus and are usually referred to as ICC-MY or ICC-MP (ICC-AP for Auerbach's plexus has also been used). There are also ICC in the small intestine near the submucosal surface of the circular muscle layer in close association with the nerve terminals in the deep muscular plexus (DMP). These cells are called ICC-DMP. Pacemaker cells in the colon along the submucosal surface of the circular muscle layer are referred to as ICC-SM. The gastric antrum and pylorus also have populations of ICC-SM (16, 27), but the physiological roles of these ICC have not been described. ICC also lie in septa between smooth muscle bundles, and these cells are called ICC-SEP. Additionally, colon and stomach have ICC intermixed with smooth muscle fibers. These intramuscular ICC are referred to as ICC-IM. These cells are critical for inputs from enteric motor neurons (24, 28). Although the main function of ICC-IM is to transduce inputs from enteric motor neurons, these cells also have intrinsic pacemaker activity and can mediate changes in slow wave frequency in response to neural inputs and stretch. In fact, neural drive may cause ICC-IM to emerge as the dominant pacemakers under some circumstances (29–31). A generalized naming scheme for the various types of ICC and their functions is given in a previous review (32).

Interstitial Cells of Cajal Generate Slow Waves in Intact Muscles

It was difficult to deduce the functions of ICC in intact muscles before current labeling techniques (see Reference 21 and review in Reference 32) because it was hard to identify these cells in living tissues. One study showed loss of slow wave activity in tissues exposed to methylene blue (33), although the specificity of methylene blue as a selective poison for ICC was questioned (34). Rhodamine 123, a dye that accumulates in mitochondria, also inhibited slow wave activity (35). ICC contain an abundance of mitochondria; however, all cells in the tunica muscularis contain these organelles, so the specificity of rhodamine 123 may also be questionable. Even with direct recording of activity from ICC (as claimed in Reference 36), it was not possible to deduce the pacemaker function of ICC with a single point of recording. ICC are electrically coupled to smooth muscle cells, and both cell types reflect the activity of the other. To determine properly the sequence of events in ICC versus smooth muscle cells, it is necessary to perform double electrode recordings, with one site of recording in an ICC and the other in a nearby smooth muscle cell. Hirst and colleagues (15, 37) performed such elegant experiments, showing that (*a*) ICC and smooth muscle cells in the stomach are electrically coupled in situ, and (*b*) slow waves originate in ICC and conduct to smooth muscle cells (Figure 2). Kito & Suzuki (38) have made similar recordings from the small intestine. Optical experiments using fluorescent Ca^{2+} dyes have also confirmed the temporal sequence of activation in ICC and smooth muscle cells (39).

---→

Figure 2   ICC and smooth muscle cells are electrically coupled. Hirst and colleagues (37) have provided direct evidence for the electrical coupling between smooth muscle cells and ICC. (*A*) Electrotonic responses to injected current in pairs of cells in the guinea pig antrum. Currents were passed via one electrode into the first cell and voltage responses were recorded by another electrode in a second nearby cell. Circular muscle cells were well coupled, but as evidenced by the decreasing amplitude of the electrotonic potentials, circular muscle cells and ICC-MY and circular-to-longitudinal muscle cells were not as well coupled, resulting in large voltage drops. Hirst et al. (37) also demonstrated that there was no rectification in the junctions between these cells. Simultaneous recordings of slow waves in ICC (*B*) and nearby circular smooth muscle cells (*C*). Slow waves occur in a 1:1 fashion in both cell types. The large drop in amplitude in the smooth muscle cell, however, reflects the relatively weak coupling between these cells and the conduction of slow waves within the smooth muscle syncytium [rather than active propagation as occurs in ICC networks; (63)]. (*D*) Sections of the records in *B* and *C* marked by black bars (*a* and *b*, respectively) that have been expanded and superimposed. Events begin in ICC and spread (with decrement) to the smooth muscle cell. There is an inflection in the records from smooth muscle cells during the upstroke. The portion of the event above this inflection is referred to as the "regenerative potential" and is thought to be due to the summation of unitary potentials in ICC-IM (see Table 1 for definitions of unitary potential and regenerative potential). Figures are redrawn from the original publications (37, 63) with permission.

A. Electrotonic responses

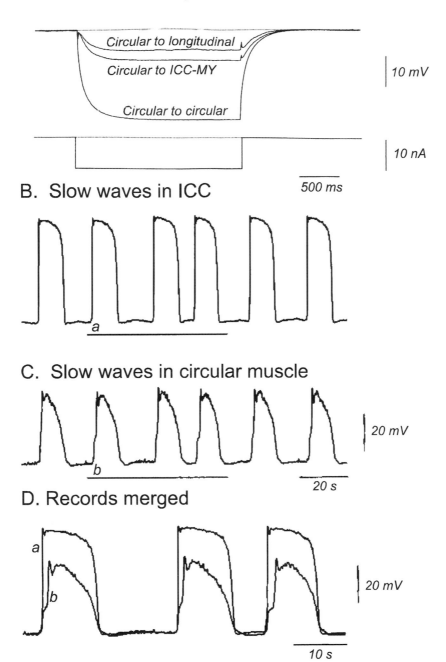

10 mV

10 nA

B. Slow waves in ICC

500 ms

a

C. Slow waves in circular muscle

20 mV

b

20 s

D. Records merged

a

b

20 mV

10 s

Isolated and Cultured Interstitial Cells of Cajal Generate Slow Wave Activity

The first direct test of the pacemaker function of ICC came when these cells were isolated from the submucosal pacemaker area of the canine colon (40). ICC, identified at that time by electron microscopy, were spontaneously active and generated slow wave–type depolarizations with properties much like the slow waves recorded in situ. Nearly a decade later, murine ICC were shown to retain a rhythmic phenotype in short-term cultures (41, 42). The slow wave activity of cultured ICC is very similar to slow waves recorded from intact muscle strips, and these cells have been useful for studies of the mechanism of electrical rhythmicity. Studies on isolated and cultured ICC have shown that ICC express the intracellular apparatus and plasma membrane conductances required for the generation and propagation of slow waves. These are specialized mechanisms in ICC, and GI smooth muscle cells or smooth muscle tissues lacking ICC cannot generate or propagate slow waves.

Slow Waves Are Lost in Animals Devoid of Specialized Populations of Interstitial Cells of Cajal

The most convincing test of the hypothesis that ICC are pacemakers came from studies in which ICC were removed from GI muscles. Maeda and coworkers (43) treated newborn animals with a neutralizing antibody for Kit, a receptor tyrosine kinase that is important in hematopoiesis. The authors noted GI pathology in treated animals that included a distended stomach full of milk (evidence of a gastric emptying disorder) and a pronounced intestinal ileus. They also measured contractile activity of intestinal muscles and found an irregular pattern of phasic contractions. We found that animals treated with the Kit neutralizing antibody lost ICC-MY in the small intestine and slow waves could not be recorded from the affected muscles (44). Kit is encoded by the *dominant white spotting* (W) locus in mice, and loss-of-function mutations in this locus resulted in the loss of specific ICC populations, including ICC-MY in the small intestine. The W mutant mice also failed to generate slow wave activity (Figure 3) (21, 45). In another study we used organotypic cultures to show that a neutralizing anti-Kit antibody also caused loss of ICC and slow wave activity in the stomach (46). There have also been a variety of developmental, genetic, and pathophysiological studies since these initial studies that have confirmed that loss of ICC in pacemaker regions causes loss of slow wave activity. Space here is not sufficient to review these experiments in full, and the reader is referred to another review (Reference 32) for a more complete description.

In summary, most investigators now agree that ICC are pacemakers in GI muscles. This conclusion is based on the following points: (*a*) Morphological studies place ICC in the correct anatomical positions for pacemaker activity and demonstrate electrical coupling between these cell types. (*b*) ICC are spontaneously rhythmic when isolated, and the phenotype of these cells is retained in culture.

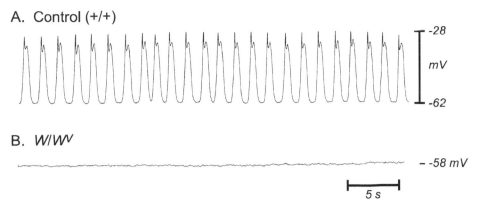

Figure 3   Slow waves are lost in the small intestine of mice with compromised Kit signaling. The Kit signaling pathway is essential for development of ICC and maintenance of the ICC phenotype after birth. Slow waves were recorded from wild-type control muscles (*A*). After treatment with neutralizing antibodies for Kit (44) or in mutant animals, such as *W/W*^{V} mice (21), ICC were lost from the pacemaker region in the small intestine (compare Figure 4*A,B*). There was no slow wave activity in mutant (*B*) or treated muscles (*not shown*). Copied with permission from Reference 21.

(*c*) Simultaneous electrical and optical recordings from ICC and nearby smooth muscle cells show initiation of slow waves in ICC and conduction of these events to smooth muscle cells. (*d*) Muscles without ICC lack rhythmicity, and slow waves cannot propagate through regions of muscle without ICC.

STRUCTURAL FEATURES IMPORTANT FOR PACEMAKER ACTIVITY IN INTERSTITIAL CELLS OF CAJAL

As space is insufficient to review ICC morphology thoroughly, the reader is referred to a variety of excellent reviews (e.g., References 25, 26, and 47–50). There are a variety of morphological features of ICC that are thought to facilitate pacemaker activity. Our knowledge of ICC structure is still developing, but we now understand many important features, such as (*a*) the importance of ICC organization into electrically coupled networks at the multicellular level, (*b*) the basis for electrical coupling between ICC and smooth muscle cells, (*c*) essential ultrastructural features of ICC, and (*d*) the types of transport proteins that are necessary for generation and propagation of slow waves.

The connectivity that ICC develop with other ICC and with smooth muscle cells is critical to pacemaker function. ICC that serve as pacemakers typically have fusiform cell bodies and multiple thin processes that contact other ICC (Figure 4) and form intricate, electrically coupled networks via gap junctions. ICC express a variety of gap junction proteins, including connexins 40, 43, and 45 (51–53). Some investigators have argued on the basis of morphological studies that there is little structural evidence for gap junction coupling between pacemaker ICC and smooth muscle cells (see Reference 54 for review). It is admittedly difficult to find obvious gap junction structures between these cells. However, ICC-MY and ICC-SM do express gap junction proteins and occasionally form small gap junctions with neighboring smooth muscle cells (16, 55). Electrical coupling has also been demonstrated by passing currents between smooth muscle cells and ICC (37). Relatively weak coupling between ICC and smooth muscle cells may be necessary for pacemaker activity so that the smooth muscle syncytium, which represents a low-impedance sink to the minority population of ICC, does not drain off all pacemaker current before the threshold for entrainment of pacemaker currents can be attained (see Table 1 for definition of pacemaker current). Gap junctions between pacemaker ICC and smooth muscle cells are relatively rare and small in size, possibly to limit the degree of coupling between these networks of cells. However, even this weak coupling appears to be sufficient to enable the conduction of slow waves into the smooth muscle syncytium.

ICC-MY and ICC-SM also have membrane caveolae that often serve in cellular signaling processes and provide scaffolding for localization of transport and signaling proteins (see morphological reviews of ICC and Reference 56 for review of general properties of caveolae). L-type Ca^{2+} channels, plasma membrane Ca^{2+} pump protein, and Na^+/Ca^{2+} exchanger type 1 were found to be colocalized with caveolin-1 in ICC (57), however, the full significance of caveolae in ICC is poorly understood at present. ICC also express an abundance of mitochondria and prominent smooth endoplasmic reticulum (SER) (Figure 4E,F). Mitochondria are often packed and tightly wound into convoluted structures within the perinuclear regions of ICC and often closely associated with SER. The mitochondria/SER

$\longrightarrow$

Figure 4   ICC networks and ultrastructural features of ICC in the murine and human gastrointestinal tracts. (A, B) ICC networks in the murine small intestine, labeled with an anti-Kit antibody. (A) Both ICC-MY (arrows) and ICC-DMP (arrowheads) are visible in wild-type animals. (B) In W/W^V mutant animals, ICC-DMP networks (arrowheads) are present, but ICC-MY are mostly lost. The loss of ICC-MY in W/W^V mutants is associated with a loss of electrical slow waves (see Figure 3, above), although enteric motor neurotransmission in these animals appears normal. (C) ICC-MY networks (arrows) in the human stomach, labeled with an anti-Kit. (D) Ultrastructural features of ICC-MY between the circular (CM) and longitudinal (LM) muscle layers in the murine intestine. ICC-MY possess an abundance of mitochondria (M) and a prominent endoplasmic reticulum (ER). (E) ICC-MY also form gap junctional complexes (arrows) with neighboring circular smooth muscle cells. (F) The ultrastructural components of the pacemaking unit in ICC-MY. Numerous mitochondria (M) are found wrapped in sheaths of endoplasmic reticulum (arrows) that also often come into close apposition with the plasma membrane. Scale bars are as indicated on each panel.

complex lies in close proximity (within 10 20 nm) to the plasma membrane (Figure 4F). Based on pharmacological and mechanistic experiments conducted on ICC (see next section), we have formulated the idea that mitochondria, SER, and the cytoplasmic volume between these organelles and the plasma membrane form "pacemaker units" in ICC (58). Pacemaker units, via Ca^{2+} handling mechanisms between compartments, generate pacemaker activity in ICC. Figure 5 illustrates

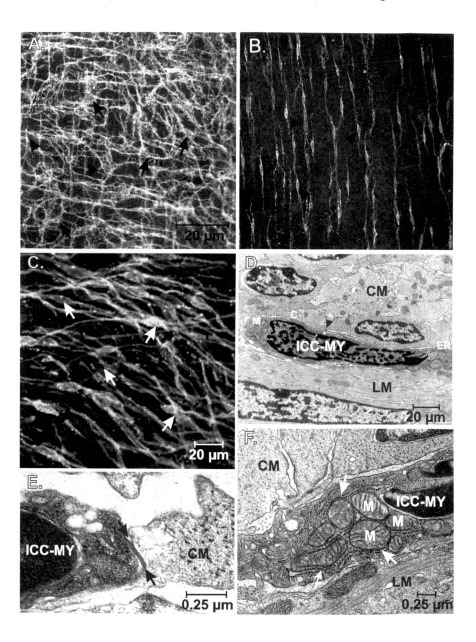

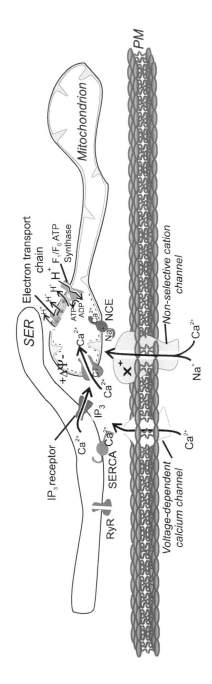

Pacemaker unit

Figure 5 Basic essential structural and molecular features of pacemaker function in ICC. This figure depicts an idealized "pacemaker unit" in ICC that is composed of smooth endoplasmic reticulum (SER), a mitochondrion (M), and ion channels in the plasma membrane (PM). These structures are abundant in the perinuclear regions of ICC and closely associated with the PM. The cytoplasmic volume between the SER, M, and PM is likely to be a subcompartment in ICC and is very small. Consequently, small numbers of ions moving into or out of this volume can significantly affect the ionic activity in the cytoplasmic volume of the pacemaker unit. Ca^{2+} handling by various transport proteins [e.g., sarco-endoplasmic reticulum Ca^{2+}-ATPase (SERCA) pumps; inositol 1,4,5-trisphosphate (IP_3) receptors] in the SER and in mitochondria [e.g., Ca^{2+} uptake transporter, electron transport complex, and probably mitochondrial Na^+/Ca^{2+} exchanger (NCE)] is essential for the regulation of Ca^{2+} concentration within the cytoplasmic volume. Ryanodine receptors (RYR) are likely to be present but do not appear to have a direct role in pacemaker activity. Our theory is that a drop in Ca^{2+} caused by mitochondrial uptake gates opens nonselective cation channels in the plasma membrane. These channels are responsible for unitary potentials and pacemaker activity. The dynamics of pacemaker activity are shown in Figure 8.

the essential basic components of pacemaker units in ICC. There are many sites that meet the anatomical characteristics of pacemaker units. Thus, it might be assumed that individual ICC contain many spontaneously active regions of membrane and that the currents generated by different areas of membrane summate to produce cellular pacemaker activity. In other words, the ICC themselves are not the basic pacemakers in GI muscles; rather, subcellular structures within the ICC serve this basic function. In sections below, descriptions are given about the basic pacemaker event (unitary current; see Table 1 for definition), the ways in which the pacemaker units generate unitary currents, and the voltage-dependent mechanisms that may entrain the activities of multiple and distant pacemaker units to produce and propagate slow waves.

UNITARY CURRENTS: THE BASIS FOR PACEMAKER ACTIVITY

Small strips of GI muscles, ICC-MY in situ, or cultured ICC display small amplitude fluctuations in membrane potential (Figure 6) (24, 38, 59–61). These events have been referred to as unitary potentials and result from small inward-directed unitary currents generated by the intrinsic pacemaker activity of ICC. The inward currents responsible for unitary potentials appear to be the fundamental pacemaker events underlying slow waves. There is controversy about the conductance responsible for unitary currents. Some investigators attribute these events to the activation of Ca^{2+}-activated Cl^- channels (e.g., 62), and we believe that a nonselective cation conductance is responsible for unitary currents and the plateau phase of slow waves (see Spontaneous Electrical Rhythmicity of Interstitial Cells of Cajal section, below). Both ICC-MY and ICC-IM are capable of generating unitary currents; however, there are differences in the pharmacology and properties of the unitary potentials in different regions of the GI tract. Thus, different channel species may be responsible for unitary currents in different types of ICC.

Because individual ICC show morphological evidence for pacemaker units (see section above), many pacemaker current discharge sites may exist in each cell. Unitary potentials result from a stochastic activation of pacemaker channels (63), so a given cell may be capable of generating unitary currents at random intervals at many sites along the plasma membrane. Multiply this activity by the number of ICC within a small strip of muscle, and the result would seem to explain the ongoing noisy discharge of unitary potentials observed with sharp electrode recordings in small bundles of muscle (Figure 6C) (59). The discharge of unitary currents has been analyzed by noise analysis, and power spectrum plots provide important information about the properties and frequency components of unitary potentials (59). Comparison of spectral densities before and after experimental treatments has shown the effects of various drugs and conditions on the ability of ICC to generate unitary currents.

A. Unitary currents single ICC (V=-80)

50 pA

2 s

B. Unitary potentials single ICC (I=0)

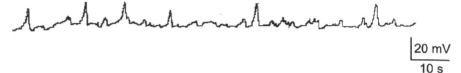

20 mV

10 s

C. Unitary potentials muscle bundle(I=0)

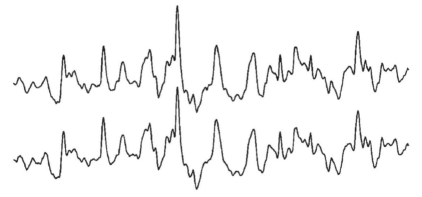

D. Unitary potentials BAPTA-AM (20 µM)

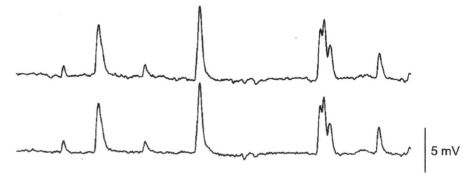

5 mV

2 s

The discharge of unitary currents is regulated by a variety of biophysical and pharmacological conditions. For example, intracellular Ca^{2+} appears to be important for unitary current generation because treatment of muscles with the acetoxymethyl ester of bis-(aminophenoxy)ethane-N,N,N',N'-tetraacetic acid (BAPTA AM) greatly reduced the frequency of unitary potentials (Figure 6D) (59). Unitary potentials also depend upon the loading of Ca^{2+} stores as unitary potentials are blocked by thapsigargan and cyclopiazonic acid (60). Caffeine also inhibits the generation of unitary potentials, but the mechanism for this action has not been determined (e.g., Reference 59). We have also found that caffeine inhibits unitary currents in cultured ICC (K.M. Sanders & N.G. Jin, unpublished observations), and it may be the release of Ca^{2+} by caffeine and enhanced Ca^{2+} activity in the pacemaker unit domain that inhibits unitary current discharge. Depolarization of cells (or relieving of hyperpolarization—i.e., anode break) increases the frequency of unitary potentials. Thus, a voltage sensor is linked to the discharge of unitary currents. This may be the same voltage sensor that is responsible for entrainment of unitary currents and the active propagation of slow waves (see Mechanism of Propagation of Slow Waves section, below).

Summation of unitary potentials (in response to depolarization, anode break, or nerve stimulation) results in events called regenerative potentials (see References 59, 60, and 64). These events display a threshold phenomenon in which the latency between depolarization and activation of the regenerative potential depends upon the magnitude and duration of the depolarization. Regenerative potentials also display long (many-second) absolute and relative refractory periods during which events either could not be initiated or were of reduced amplitude. The threshold properties and long absolute and relative refractory periods are similar to the properties of slow waves in muscle strips (65). We found that the upstroke potentials of slow waves recovered rapidly but that the plateau phase of slow waves (which is likely to be due to summation of unitary potentials) was well-described by the refractory properties of regenerative potentials. Another interesting similarity between the summed unitary potentials of regenerative potentials and slow waves is the reduction in relative refractory period caused by muscarinic stimulation

---

←

Figure 6   Unitary potentials in ICC. (A) Unitary currents from a single ICC from the murine colon after two days in culture (61). The cell was voltage clamped to -80 mV. (B) Unitary potentials resulting from the unitary currents. The recording was made under current clamp with no applied current ($I = 0$). (C) Simultaneous recordings from two cells in a small bundle of guinea pig antrum (59). There is a constant discharge of membrane potential fluctuations and a very high degree of coupling between the two cells (records are essentially identical from two sites of recording). (D) Waveforms of single unitary potentials isolated after adding the acetoxymethyl ester of bis-(aminophenoxy)ethane-N,N,N',N'-tetraacetic acid (BAPTA AM), which reduces the frequency of unitary potentials (59). Figure is a composite redrawn from References 59 and 61 (with permission).

(65, 66). Publicover & Sanders (66) attribute this form of regulation to activation of protein kinase C.

Several important questions remain concerning the refractory properties of unitary potentials, regenerative potentials, and slow waves. These include the following:

1. What is the cause of the many-second refractory periods for regenerative potentials and slow waves? Voltage-dependent ion channels typically recover from inactivation on a much faster time scale than from the refractory periods of regenerative potentials and slow waves. Because unitary potentials and slow waves require release of Ca^{2+} from stores (see Mechanisms Responsible for Pacemaker Currents, below), it is likely that the long refractory periods are related to reloading of Ca^{2+} stores in pacemaker units. After the firing of a unitary potential (or many unitary potentials to accomplish regenerative potentials or slow waves), time is required to reload stores before another event can be elicited.

2. What is the refractory period for a single pacemaker unit? If it is true that the time required for reloading of stores is the rate-limiting step responsible for the long refractory periods, then the reloading period of the Ca^{2+} store in a single pacemaker should be statistically similar to the refractory periods for regenerative potentials and slow waves.

3. What signaling pathways in ICC are responsible for regulating the refractory period, and are these mechanisms similar to mechanisms that produce chronotropic effects in ICC? The latter issue is discussed in Chronotropic Mechanisms in the Interstitial Cells of Cajal, below.

The rhythmicity of networks of ICC includes a mechanism that coordinates the discharge of pacemaker units and thus the occurrence of unitary currents. The proposed mechanism to accomplish this task is described in Mechanism of Propagation of Slow Waves, below. Briefly, this mechanism includes activation of a voltage sensor that can organize the discharge of many pacemaker units within ICC networks. Thus, the physiological means of coordinating the discharge of pacemaker units may be simulated by depolarization (or anode break). It is possible that this mechanism is also responsible for the initiation of "autonomous currents," which are likely to be the cause of regenerative potentials, in freshly dispersed ICC (67). We believe the voltage sensor responsible for coordination of unitary currents is a voltage-dependent, dihydropyridine-resistant Ca^{2+} conductance expressed by ICC (61, 68), and voltage-dependent Ca^{2+} entry is the primary factor that organizes the discharge of pacemaker units. Several studies have now shown that blockade of voltage-dependent Ca^{2+} entry via the dihydropyridine-resistant Ca^{2+} conductance inhibits slow wave propagation (69) and the ability of pacemaker potentials in ICC to entrain (see Table 1 for definition of pacemaker potential) (38, 70). Reducing the ability of slow waves to entrain and propagate results in a decrease in slow wave frequency and resolution of unitary potentials (70), suggesting that the mechanism of coordination of unitary currents is blocked by inhibitors of the voltage-dependent Ca^{2+} entry.

MECHANISMS RESPONSIBLE FOR PACEMAKER CURRENTS

Model Systems of Electrical Rhythmicity

The complexities of the ICC syncytium, along with the fact that ICC are a minority component of a greater electrical syncytium that includes smooth muscle cells, make it difficult to perform mechanistic studies on ICC in situ. Under such circumstances, every experimental manipulation can have consequences on a variety of cell types, and these consequences can easily affect the behaviors of other electrically coupled cells. Thus, a reductionist approach, coupled with pharmacological and genetic manipulations, has been required to obtain mechanistic information. Freshly dispersed ICC would seem the ideal tool for such an investigation, but fresh cells have been difficult to identify in dispersions of GI muscle tissues (extracellular epitopes of Kit protein appear to be destroyed during enzymatic digestion). Some investigators have attempted to identify ICC by the gross structural characteristics of isolated cells (40), but this form of identification needs to be verified by other time-intensive criteria (i.e., ultrastructure, molecular expression, etc.). Therefore, few studies of freshly dispersed ICC appear in the literature.

It has been possible to culture ICC from cell dispersions, and under these conditions the phenotype of electrical rhythmicity is maintained for two to three days (41, 42). These cells express Kit and can be distinguished either as individual cells or as small networks of interconnected cells (Figure 7A–D). Some investigators have expressed concerns about the fidelity of cultured ICC as a model for slow wave activity. Certainly, the phenotype of some cells changes dramatically after separation from the extracellular matrix. ICC also change during the first week of cell culture, and the expression of muscle-like genes increases as a function of time in these cells (71). We have compared the electrical events of cultured ICC with the slow wave events recorded from source muscles and found remarkable similarity between the activities of the two preparations (Figure 7E,F) (42). In addition, much of the pharmacology known to affect various steps responsible for electrical rhythmicity in cultured ICC has similar consequences on slow waves in situ (e.g., Reference 28). Thus, cultured ICC will be an important tool for investigation of the mechanism of rhythmicity until efficient techniques are developed to identify freshly dispersed ICC.

Initial evidence suggested that the spontaneous activity in cultured cells was due to a voltage-independent, nonselective cation conductance, but there was criticism of this conclusion based on the relatively poor space clamp obtained in voltage-clamp studies of ICC networks. Recent studies, however, with better space clamp have observed a nonselective, "autonomous" conductance in freshly

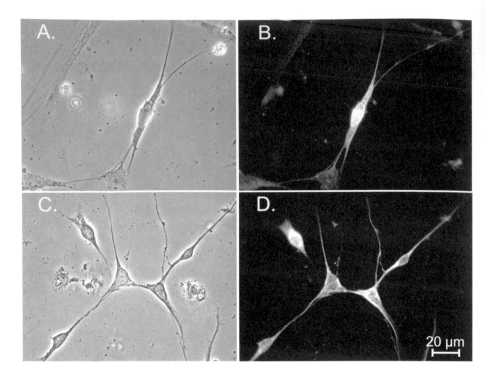

E. Slow waves in cultured ICC

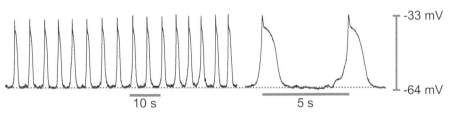

F. Slow waves in ICC in situ

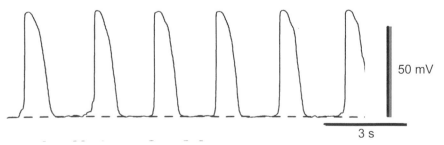

isolated ICC that is also attributed to activation of a nonselective cation conductance (67).

Possible Mechanisms for the Spontaneous Electrical Rhythmicity Observed in the Interstitial Cells of Cajal

All investigators seem to agree that release of Ca^{2+} from inositol 1,4,5-trisphosphate (IP_3) receptor–operated stores is the fundamental cellular event that initiates pacemaker current. Uptake of Ca^{2+} (and cycle-to-cycle maintenance of Ca^{2+} levels in IP_3 receptor–operated stores), therefore, is critical preparation for Ca^{2+} release events. Ca^{2+} pumps (SERCA), inhibited by thapsigargan and cyclopiazonic acid (CPA), are required for this process (28, 73). The specific isoforms of SERCA pumps active in ICC have not been identified. The importance of Ca^{2+} release from IP_3 receptor–operated stores in slow waves has been demonstrated with pharmacological experiments using cultured ICC (28), experiments on mutant mice lacking the IP_3R1 isoform of IP_3 receptors (74), and pharmacological studies of intact muscle strips (28, 73).

However, there is controversy about the ionic conductance(s) necessary for pacemaker activity and the steps between Ca^{2+} release and activation of pacemaker current. From studies of intact muscle preparations using pharmacological tools, other investigators have suggested that a Ca^{2+}-activated Cl^- conductance is responsible for pacemaker currents in ICC (62, 72). These investigators have suggested that Ca^{2+} released from IP_3 receptor–operated stores directly activates Ca^{2+}-activated Cl^- channels in the plasma membrane. However, a major problem with this concept is the lack of evidence for a Ca^{2+}-activated Cl^- conductance in ICC. One report suggested that ICC from the murine small intestine generate spontaneous Ca^{2+}-activated Cl^- currents (75). However, this conclusion was based on pharmacological tests using a Cl^- channel–blocking drug (4-acetoamido-4-isothiocyanat-ostilbene-2,2'-disulphonic acid; SITS) that may have nonspecific effects at the concentrations used. In addition, the complexities of ion redistributions caused by altering the Cl^- gradient in intact muscles also bring the authors' conclusion into question. Direct measurements of conductances available in either freshly dispersed ICC (67, 68) or cultured ICC (61) have not identified Ca^{2+}-activated Cl^- conductances. In cells that express such a conductance,

←

Figure 7 Spontaneous slow wave activity in ICC. (*A–D*) Short-term cultured ICC from the murine small intestine (42). (*A,C*) Phase contrast images. (*B,D*) The same fields as in *A* and *C*, respectively, labeled with an anti-Kit antibody to identify the cells as ICC. (*E*) Spontaneous slow wave activity recorded from a network of ICC under current clamp. (*F*) Slow waves recorded from ICC of the murine small intestine in situ (70). Although the waveforms in *E* and *F* are somewhat different, the slow wave activities in the two preparations are remarkably similar. The figure is redrawn from images in References 42 and 70 with permission.

depolarization to 0 mV activates Ca^{2+} entry and the development of a Ca^{2+}-dependent outward current. Repolarization typically results in a prominent and slowly decaying inward tail current. These signatures of Ca^{2+}-activated Cl^- conductances, (a) Ca^{2+} entry generating outward current positive to E_{Cl}, and (b) current reversal as membrane potential repolarizes negative to E_{Cl}, have not been observed in ICC. Although voltage-dependent Ca^{2+} currents are expressed by ICC, the presence of a Ca^{2+}-activated Cl^- conductance is not detected during depolarization, Ca^{2+} entry, or repolarization (e.g., see References 61 and 68). Another problem is that caffeine, a drug that releases Ca^{2+} from intracellular stores in many cells and activates large Ca^{2+}-activated Cl^- currents in cells that express these channels, blocks pacemaker currents in ICC (63, 72). Two additional papers have reported other species of Cl^- conductances in ICC (76, 77). However, the currents were not Ca^{2+} dependent, and no mechanism was provided to link these conductances to pacemaker activity.

We have never been able to resolve Ca^{2+}-activated inward currents with spontaneous pacemaker activity in ICC (78). Reduction in cytoplasmic Ca^{2+} initiates a large inward, nonselective cation current in ICC. The conductance responsible for this current is not a typical store-depletion or calcium release–activated calcium current (I_{CRAC}) conductance and is not activated by thapsigargin (28). The pacemaker conductance in ICC appears to be gated open by reduced Ca^{2+}. Our data suggest that uptake of Ca^{2+} from the cytoplasmic space between IP_3 receptor–operated stores, mitochondria, and the plasma membrane is the signal for activation of the pacemaker conductance (28). As we describe in Structural Features Important for Pacemaker Activity in the Interstitial Cells of Cajal (above), we refer to the close association of these cellular organelles and the plasma membrane as the pacemaker unit. The restricted cytoplasmic volume associated with the pacemaker units is small but poorly defined and may be quite variable. Estimates of the cytoplasmic volume associated with a pacemaker unit suggest that very small fluxes of Ca^{2+} ions into and out of this space may have significant impact on local Ca^{2+} activity. A careful morphometric study of this region in ICC, however, is needed to appreciate fully the extent and details of the pacemaker unit and to improve estimates of Ca^{2+} fluxes and concentration in this space during pacemaker activity. The dynamics of Ca^{2+} exchange in the volume of the pacemaker unit may be difficult to monitor directly with fluorescent Ca^{2+} dyes and current imaging techniques because the dimensions of the cytoplasmic volume of the pacemaker unit are well below the wavelengths of visible light.

The essential step of reduction in Ca^{2+} near the plasma membrane appears to be accomplished by mitochondrial uptake because (a) activation of pacemaker current is preceded by a rise in mitochondrial Ca^{2+} and (b) a variety of drugs known to inhibit mitochondrial Ca^{2+} uptake block mitochondrial Ca^{2+} transients and pacemaker currents (28). The following sequence of events has been suggested from experimental evidence to initiate pacemaker currents (see 28, 58) (Figure 8):

1. Puffs of Ca^{2+} released from IP_3 receptor–operated stores gate open Ca^{2+} uptake channels in mitochondria, which are in close apposition to the release sites in the endoplasmic (or sarcoplasmic) reticulum (the presumed Ca^{2+} store in pacemaker units.

2. The mitochondria is not in equilibrium with the cytoplasmic space, so opening a Ca^{2+} uptake pathway into the mitochondria removes more Ca^{2+} from the pacemaker unit than just the Ca^{2+} released in the puff.

3. As Ca^{2+} falls in the space between the plasma membrane and the Ca^{2+} store/mitochondria complex, nonselective cation channels in the plasma membrane are activated.

4. The inward current resulting from the transient dip in Ca^{2+} is the primary pacemaker current in ICC and the conductance responsible for unitary potentials in the pacemaker cells.

The mechanism above suggests the expression of channels in the plasma membrane of ICC that have intrinsically high open probability but are gated closed by physiological levels of intracellular Ca^{2+}. A Ca^{2+}-inhibited nonselective cation conductance has been observed in murine ICC. This conductance, which is thought to be gated closed by binding of Ca^{2+}/calmodulin (78), is likely to be the primary pacemaker conductance responsible for unitary potentials and the plateau phase of slow waves. Oscillations in Ca^{2+} within the cytoplasmic space in pacemaker units appear to regulate the open probabilities of these channels. Buffering of cytoplasmic Ca^{2+} to low levels by dialysis of ICC with BAPTA-AM activates a nonselective cation conductance that generates inward current. Single channels with a unitary conductance of approximately 13 pS are also activated in ICC by Ca^{2+} buffering. Channels of the same conductance are also activated when on-cell single channel recordings are made from ICC (78), and the oscillatory openings of clusters of these channels occur at the frequency of spontaneous slow wave activity. Excision of patches from rhythmically active cells reveals the presence of nonselective cation channels (13.5 pS) (Figure 9A) (78). Increasing Ca^{2+} concentration from 10^{-7} to 10^{-6} M reduces the open probability of these channels (Figure 9B). Some nonselective cation channels, such as those encoded by *transient receptor potential channel* (*TRPC*) genes are negatively gated by Ca^{2+}/calmodulin (79). In ICC, calmidazolium and W-7 (both calmodulin inhibitors) increased the activity of the 13.5 pS nonselective cation channels in ICC in both on-cell and off-cell recording conditions, suggesting that inhibition of channel gating occurs via Ca^{2+}/calmodulin binding. Expressed TRPC4 channels display properties similar to the native pacemaker conductance (80) such as (*a*) similar single channel conductance (17 versus 13.5 pS), (*b*) negative regulation by Ca^{2+}, and (*c*) activation by calmodulin inhibitors. Pacemaker currents appear not to be dependent upon the expression of TRPC4 channels, however, because $TRPC4^{-/-}$ mice (a gift of Professor V. Flockerzi) displayed normal slow waves and pacemaker currents in ICC (S.M. Ward, S.D. Koh, and K.M. Sanders, unpublished observations).

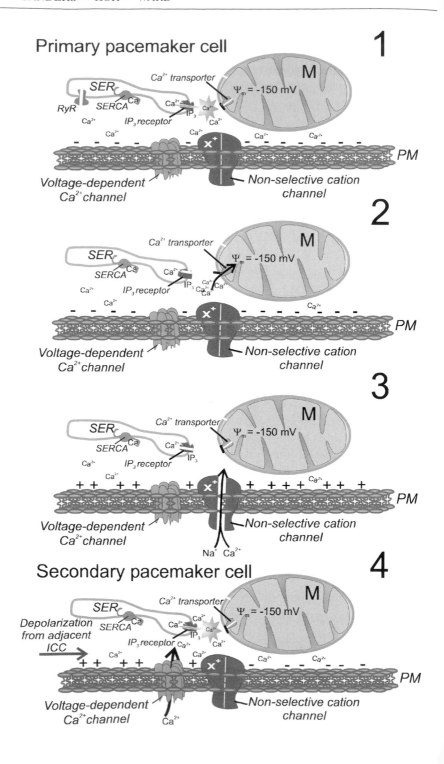

A study of freshly dispersed ICC from the murine intestine has also reported the presence of a nonselective cation conductance that produced a so-called autonomous current (67). ICC in enzymatically treated intestinal muscles were identified by Kit immunoreactivity. A large inward, nonselective cation current with a time course of 500 ms was activated by depolarization of ICC under voltage clamp. Removal of extracellular Ca^{2+} blocked the activation of the autonomous current. Although there are similarities between the autonomous current and the nonselective current in cultured ICC, more studies are needed to determine whether these are due to the same conductance(s).

MECHANISM OF SLOW WAVE PROPAGATION

Electrical slow waves can propagate for many centimeters in GI muscles (e.g., the mechanism underlying gastric peristalsis requires the propagation of slow waves from the dominant corpus pacemaker to the pylorus). At the macroscopic level, propagation occurs as a smooth activity front at rates of 5–40 mm s^{-1}. This rate of propagation is greater than would be expected from a purely diffusional mechanism of entrainment (i.e., Ca^{2+} waves that spread at rates that are at least two orders of magnitude less than the propagation rates of slow waves). Thus, an electrical mechanism is necessary for synchronization of pacemaker events and active

---

Figure 8   Phases in pacemaker activation and active propagation of slow waves. This figure shows a sequence for the proposed mechanism of activation of pacemaker current in ICC (*phases 1–3*) and how activation of a voltage-dependent, dihydropyridine-resistant Ca^{2+} conductance participates in active propagation (*phase 4*). Each panel is a single frame in time illustrating the functions of the various elements of the pacemaker unit. (*Phase 1*) IP_3 receptors in smooth endoplasmic reticulum (SER) discharge, releasing a puff of Ca^{2+} into the space between the Ca^{2+} store and the mitochondrion (M). (*Phase 2*) The Ca^{2+} gates open a Ca^{2+} transporter in the mitochondrial membrane, causing Ca^{2+} entry down the steep electrochemical gradient across the mitochondrial inner membrane due to the electron transport chain. More Ca^{2+} than the small amount released in the puff is taken up by the mitochondrion. Consequently, the Ca^{2+} in the cytoplasmic volume between the mitochondrion and the plasma membrane is temporarily reduced (*phase 3*). The drop in cytoplasmic Ca^{2+} gates opens a nonselective cation conductance in the plasma membrane (PM). Inward current depolarizes membrane potential in the pacemaker unit. (*Phase 4*) Cells near the primary pacemaker (secondary pacemaker cells) are subsequently depolarized, resulting in activation of the pacemaker apparatus (as in *phases 1–3*) in these cells. The primary pacemaker unit (or cell) serves as the dominant pacemaker for this cycle, causing sequential activation of the pacemaker apparatus in the secondary pacemaker cells and propagation of the slow wave. Information for this figure was obtained from References 28, 61, 74, and 78.

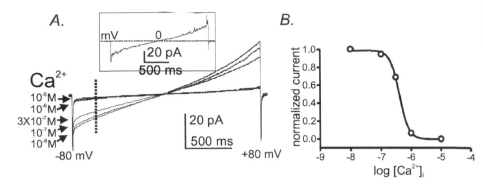

Figure 9 Relationship between cytoplasmic Ca^{2+} and open probability of nonselective cation channels in ICC. Openings of the nonselective cation channels were flickery, and it was difficult to obtain open probability measurements in the standard manner. (A) Ramp potentials applied to an excised patch containing nonselective cation channels. The response to a single ramp is shown in the inset. The patch was repetitively ramped from -80 to $+80$ mV, and the currents elicited were summed to produce smooth current traces. Ramps were passed after decreasing the Ca^{2+} concentration at the cytoplasmic surface from 10^{-5} M to 10^{-8} M. (B) Summary of the effects of Ca^{2+} on current. Reducing Ca^{2+} increased the current density of the nonselective current.

propagation. Active propagation of slow waves occurs in ICC networks. Smooth muscle cells do not possess the mechanism necessary either to generate or regenerate slow waves (81). Thus, slow waves cannot actively propagate over long distances in GI muscles lacking ICC. Dissection experiments and genetic and pathophysiological manipulation of ICC networks have demonstrated that when ICC are lost from a region of muscle, active propagation fails through that region, and slow waves decay in amplitude as a function of the cable properties of the remaining smooth muscle syncytium (e.g., References 14 and 46). Simultaneous recording of slow waves in ICC and neighboring smooth muscle cells clearly demonstrates that slow waves conduct passively to neighboring smooth muscle cells (15, 37). These findings have important implications in pathophysiological conditions in which portions of ICC networks are damaged. In such circumstances, slow wave propagation over long distances becomes impossible, and regions retaining functional ICC that are normally coupled or driven by other regions become independent. Thus, loss of ICC or even heterogeneous lesions in ICC may result in aberrant motility patterns.

Networks of ICC in pacemaker areas consist of thousands of spontaneously active cells. How the activity of this multitude of cells can be coordinated is an extremely important question. For regenerative propagation of slow waves to take place, the opening of the channels responsible for pacemaker current in a network of ICC must be coordinated so that the myriad of normally spontaneous pacemakers activate in a sequential manner. Voltage-dependent sequential activation (as occurs with Na^+ channels in skeletal and cardiac muscles) of pacemaker channels cannot be the basis for active propagation in ICC networks, as the ionic conductances hypothesized to be responsible for pacemaker currents (i.e., either the nonselective cation conductance or a Ca^{2+}-activated Cl^- conductance) are not voltage dependent. Because release of Ca^{2+} from IP_3 receptor–operated Ca^{2+} stores is the initiating event in pacemaker activity, voltage-dependent coordination of Ca^{2+} release may be a mechanism for the sequential activation of pacemaker channels in ICC networks. Some authors have suggested that ICC possess a voltage sensor that controls the synthesis of IP_3 or another second messenger that regulates IP_3 receptor–operated Ca^{2+} release (37, 82). However, voltage-dependent regulation of IP_3 synthesis in ICC has never been demonstrated.

Regulation of the Ca^{2+} concentration in the cytoplasmic space within pacemaker units may be another way to provide voltage-dependent coordination of Ca^{2+} release from IP_3 receptors. The open probability of IP_3 receptors displays a bell-shaped dependence toward cytoplasmic Ca^{2+} (e.g., 83). At basal levels of IP_3, small increases in Ca^{2+} concentration can enhance the open probability of IP_3 receptors. Thus, IP_3 receptors exhibit the property of Ca^{2+}-induced Ca^{2+} release. We have suggested that voltage-dependent Ca^{2+} entry that increases Ca^{2+} activity in pacemaker units near IP_3 receptors may be responsible for coordination of Ca^{2+} release events and entrainment of unitary currents within a network of ICC (61).

We have demonstrated a voltage-dependent Ca^{2+} conductance in ICC that appears to be responsible for the upstroke depolarization of slow waves and for active slow wave propagation (61, 69, 70) (see Figure 10). Our hypothesis of how this current coordinates the discharge of pacemaker units and the generation of unitary currents is as follows:

1. Unitary potentials occur randomly in ICC as a result of stochastic activation of pacemaker units (17).

2. Summation of unitary potentials causes depolarization and activates a voltage-dependent Ca^{2+} current. A substantive part of this current is due to dihydropyridine-resistant Ca^{2+} channels (61) because dihydropyridines only slightly reduce the upstroke velocities of slow waves and have no effect on unitary potentials.

3. Ca^{2+} entry through these channels increases Ca^{2+} concentration near the IP_3 receptors in pacemaker units that have not discharged. Owing to the cable properties of the ICC/smooth muscle syncytium, Ca^{2+} channels closest to the source of depolarization experience the largest increase in open probability. A relatively small influx of Ca^{2+} ions may significantly increase local Ca^{2+} concentration in the small volumes of pacemaker units.

4. The rise in Ca^{2+} synchronizes the release of Ca^{2+} from IP_3 receptors in many pacemaker units and initiates the sequence of events responsible for activation of the pacemaker conductance in these cells (see Mechanisms Responsible for Pacemaker Currents, above).

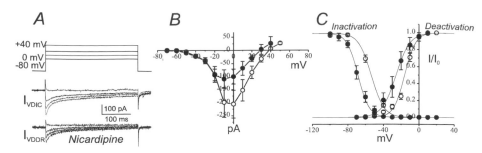

Figure 10 Voltage-dependent, dihydropyridine-resistant Ca^{2+} current in ICC. Single ICC studied under voltage clamp were stepped to potentials from -70 mV to $+50$ mV. (A) Currents from four potentials are compared before (*middle traces*) and after (*bottom traces*) addition of nicardipine (1 μM). Nicardipine blocked approximately half of the inward current. (B) Summary I–V curves from 15 experiments before (*open circles*) and after (*closed circles*) nicardipine. (C) The shift in the voltage dependence of activation and inactivation before (*open circles*) and after (*closed circles*) nicardipine. A low-threshold, dihydropyridine-resistant current is present in these cells. Data are copied from Reference 61 with permission.

5. Synchronized activation of many pacemaker units is responsible for the plateau phase of slow waves.

The conductance responsible for pacemaker entrainment is not blocked by dihydropyridines because these compounds do not block slow waves in GI muscles. Dihydropyridine-resistant Ca^{2+} conductances have been described in freshly dispersed ICC from the canine colon and in cultured ICC from murine small intestine (Figure 10) (61, 68). The dihydropyridine-resistant currents are due to a Ca^{2+} selective conductance that is activated by depolarization and reduced by Ni^{2+} (100 μM) and mibefradil (<1 μM). Block of the dihydropyridine-sensitive currents in ICC caused the voltage dependence of activation and inactivation of the remaining Ca^{2+} current to shift toward more negative potentials. The reversal potential of the Ca^{2+} current remaining after nicardipine also shifted 14 mV toward more negative potentials (Figure 10). At present the molecular species responsible for the dihydropyridine-resistant Ca^{2+} current in ICC has not been identified.

Recent studies on ICC in situ and on intact muscles have confirmed the importance of a dihydropyridine-resistant, voltage-dependent Ca^{2+} conductance in slow wave propagation. Kito and colleagues (38, 70) showed that dihydropyridines did not block slow waves recorded from ICC in situ in the murine small intestine. Reduced external Ca^{2+}, elevated external K$^+$, and Ni^{2+} or mibefradil, however, caused concentration-dependent reduction in the velocity of the upstroke depolarization of slow waves recorded from ICC (see Table 1 for definition of upstroke depolarization). At sufficient concentration, these treatments blocked slow waves. At concentrations less than required for complete block, these treatments also reduced slow wave frequency and uncoupled pacemaker activity, reducing the ability of unitary potentials to entrain and generate slow waves. The latter was apparent by the increased resolution of unitary potentials in ICC treated with drugs and conditions to reduce voltage-dependent Ca^{2+} currents. Experiments on intact muscles have been performed using a partitioned chamber apparatus to facilitate selective application of drugs to specific regions of muscles (69). Under control conditions slow waves actively propagate from one region of muscle to the next over many centimeters. Drugs and conditions expected to block or reduce the availability of dihydropyridine-resistant, voltage-dependent Ca^{2+} currents—i.e., reduced external Ca^{2+} (to reduce driving force), depolarization with elevated K^+ (to produce inactivation), hyperpolarization (to inhibit activation), or addition of Ni^{2+} or mibefradil (to directly block the current)—caused failure of slow wave propagation through a treated section of muscle. Taken together, these findings suggest that slow wave propagation is dependent upon the dihydropyridine-resistant, voltage-dependent Ca^{2+} channels in ICC.

CHRONOTROPIC MECHANISMS IN INTERSTITIAL CELLS OF CAJAL

It is well recognized that slow wave frequency is regulated by a variety of physiological agonists and conditions. Slow wave frequency may be regulated as a feature of normal physiological activity, but there are also instances in which frequency is affected abnormally, leading to motility dysfunction. As discussed above, there are gradients in slow wave frequency in some GI organs that permit pacemaker dominance of one region over another. For example, in the stomach slow wave frequency is higher in the corpus than in the antrum, and the corpus pacemaker dominates in gastric peristalsis (6). This gradient must be maintained for normal gastric peristalsis and gastric emptying, but pathological conditions exist in which antral pacemaker frequency is elevated to the point at which the corpus can no longer drive antral activity. The latter has been termed tachygastria, which can lead to delayed gastric emptying (84, 85). Elevation in antral frequency need only approach the corpus frequency for functional uncoupling between the corpus and antrum to occur. Physiological substances such as prostaglandin E_2 (PGE_2) can participate in the generation of pathologically high slow wave frequency in the antrum and serve as a cause of tachygastria (86). Thus, it is extremely important to understand the mechanisms responsible for physiological and pathophysiological regulation of pacemaker activity.

On the basis of our model for pacemaker activity and propagation (see Figure 8), it is now possible to predict which agonists may affect slow wave frequency. Some of these agonists have been tested on GI muscles and ICC. In general, agonists that enhance synthesis (or impede breakdown) of IP_3 should have positive chronotropic effects because IP_3 levels within the pacemaker unit should affect the tendency for Ca^{2+} release events from IP_3 receptor–operated stores. Agonists or conditions that increase Ca^{2+} flux through voltage-dependent Ca^{2+} channels should also increase pacemaker frequency because increased Ca^{2+} entry should make it easier to reach threshold for an upstroke depolarization and to entrain pacemaker activity. These agonists will also tend to increase the rate of slow wave propagation. Agonists that block Ca^{2+} entry through dihydropyridine-resistant, voltage-dependent Ca^{2+} channels in ICC should reduce slow wave frequency and reduce propagation rates (as described above). This may include drugs that affect membrane potential because either hyperpolarization or depolarization may negatively affect the availability of voltage-dependent Ca^{2+} channels. Finally, mechanisms that affect intracellular Ca^{2+} handling, by either IP_3 receptor–operated stores or mitochondrial function, may have effects on pacemaker frequency.

Muscarinic stimulation has been shown to have significant positive chronotropic effects on slow waves in gastric antral muscles (87), and muscarinic agonists also greatly enhance the frequency of pacemaker currents in ICC (88). The chronotropic effects of muscarinic stimulation are mediated by M_3 cholinoreceptors, which are coupled to enhanced production of IP_3. PGE_2 also has profound positive chronotropic effects on antral muscles (86, 89, 90), and these effects are mediated by prostoglandin EP_3–type receptors (90). Many other compounds produced within the body or given as therapeutic agents may affect slow wave frequency, and some of the GI side effects of drugs may be mediated by their effects on slow wave frequency.

ROLE OF INTRAMUSCULAR INTERSTITIAL CELLS OF CAJAL IN PACEMAKING

In the first studies using W/W^V mice to elucidate the physiological roles of ICC there appeared to be a division of labor between the intramuscular ICC and the ICC in pacemaker areas. Loss of ICC-MY in the small intestine resulted in loss of slow wave activity but no obvious defect in neurotransmission (28, 45). ICC-DMP are not apparently affected in the small intestines of W/W^V mice (see Figure 4), and it was surmised that ICC-MY generated slow waves and ICC-DMP were involved in neurotransmission. Loss of ICC-IM in the stomach resulted in the opposite circumstances: Slow waves were preserved, but both nitrergic and cholinergic motor neurotransmission were greatly reduced (24, 27, 91). Thus, it was suggested that ICC-MY (and ICC-SM in the colon) serve as pacemaker cells and that ICC-IM (and ICC-DMP in the small intestine) mediate neurotransmission (13). When this was proposed, there were two caveats that since have developed into major arguments against the division-of-labor concept. First, in small intestinal muscle with the entire myenteric region (and ICC-MY) removed, electrical rhythmicity can be recorded from smooth muscle cells near the DMP (22). Gastric muscle lacking ICC-MY (such as the fundus and small strips dissected from the circular muscle layer of the antrum) produced unitary potentials (see Unitary Currents: The Basis for Pacemaker Activity, above) (17, 24). In the distal stomach, unitary potentials can be stimulated to generate slow wave, plateau-like events (regenerative potentials) (17). Unitary potentials and the ability to develop regenerative potentials were abolished in stomachs of W/W^V mice. Muscles of the gastric antrum separated from the myenteric region are also able to generate slow wave activity, albeit at a reduced rate from intact muscles (16). Thus, it appears that ICC-IM and ICC-DMP also express a pacemaker mechanism. To date, there is little or no evidence, however, that ICC-MY are innervated or involved in neurotransmission.

ICC-IM can even become the dominant pacemaker under some circumstances. For example, Hirst and colleagues showed that when cholinergic nerves are stimulated in the gastric antrum, premature slow waves can be initiated. These events can drive events in ICC-MY (31). Inputs from enteric motor neurons are directed at ICC-IM, not ICC-MY (24, 91). Thus, it is logical that neural regulation of slow waves must be mediated via ICC-IM. Recent studies have shown that it is possible to pace slow waves via stimulation of intrinsic neurons and that this effect is lost in gastric muscles of W/W^V mutants, which lack ICC-IM (29). During periods of sustained cholinergic nerve stimulation in gastric antral muscles, stimulation of muscarinic (M_3) receptors on ICC-IM causes these cells to become rhythmic and generate slow waves at a rate higher than the intrinsic ICC-MY frequency (30). Thus, during sustained cholinergic stimulation, as occurs during the gastric phase of digestion, ICC-IM become the dominant pacemaker in antral muscles. Through this mechanism, inputs from enteric motor neurons can drive the normally dominant ICC-MY pacemaker cells.

Another means of activating ICC-IM is stretch. Stretching of antral muscles by approximately 25% with precise length ramps caused depolarization and increased slow wave frequency by a nonneural mechanism (92). The magnitude of the stretch responses depended upon the rate of stretch, and the chronotropic effects were reduced by addition of Ni^{2+} or reduction of external Ca^{2+}. The responses to stretch were similar to effects caused by EP_3 receptor agonists (90), and indomethacin blocked the stretch-dependent effects. Responses to stretch were missing muscles from cyclo-oxygenase II (COX II) knockout mice. This suggests that production of eicosanoids, such as PGE_2, by COX II may be responsible for the chronotropic and membrane potential effects of stretch. Responses to stretch were absent in antral muscles of W/W^V mice. Thus, it is also likely that ICC-IM, and not ICC-MY, are the site of mechanosensitive chronotropic effects. The mechanosensitive transducer in ICC-IM has not been identified. Conditions of stretch, therefore, are another example of ICC-IM emerging as the dominant pacemaker in gastric muscles.

INTEGRATED RESPONSES AND THE LINK BETWEEN INTERSTITIAL CELLS OF CAJAL AND GASTROINTESTINAL PATHOLOGIES

As we discuss in this review, ICC provide several important functions in the GI tract: (*a*) generation of electrical slow wave activity, (*b*) coordination of pacemaker activity and active propagation of slow waves, (*c*) transduction of motor neural inputs from the enteric nervous system, and (*d*) mechanosensation to stretch of GI muscles. These are important functions that contribute to both the regulation of excitation-contraction coupling in the gut and to connectivity between smooth muscle cells and the motor output of the enteric nervous system. Motility patterns require timed contractions to create ring-like peristaltic and segmental contractions, and as discussed above, slow waves provide oscillations in the open probability of Ca^{2+} channels of smooth muscle cells that facilitates phasic contractions (103). It would not be possible to generate the motor programs stored in the enteric nervous system without the patterned electrical activity and synaptic connectivity provided by ICC. A variety of motility disorders, such as gastroparesis, pseudo-obstruction, chronic constipation, and posttraumatic or postinfectious dysfunction, have been associated with either profound or partial loss of ICC (see reviews in References 32 and 93). The important physiological roles of ICC also suggest the possibility of a causative role for ICC loss in a variety of motor disorders. In addition, the labeling of ICC with anti-Kit antibodies provided a basis for the realizations that (*a*) gastrointestinal stromal tumors (GIST) may be derived from ICC, and (*b*) Kit mutations may result in cell transformation from ICC to malignant cells (94). Blocking the tyrosine kinase activity of Kit receptors has provided successful new approaches for the therapeutic control of GIST (95–98). The discoveries listed above are excellent examples of the benefits of physiological research for mankind. The work toward understanding the role of ICC in basic physiological processes has provided new insights and exciting new hypotheses about the causes of some GI diseases, leading to novel potential therapies for these disorders.

The *Annual Review of Physiology* is online at
http://physiol.annualreviews.org

LITERATURE CITED

1. Szurszewski JH. 1987. Electrical basis for gastrointestinal motility. In *Physiology of the Gastrointestinal Tract*, ed. LR Johnson, pp. 383–422. New York: Raven
2. Morgan KG, Szurszewski JH. 1980. Mechanism of phasic and tonic actions of pentagastrin on canine gastric smooth muscle. *J. Physiol. London* 301:229–42
3. Morgan KG, Muir TC, Szurszewski JH. 1981. The electrical basis for contraction and relaxation in canine fundal smooth muscle. *J. Physiol. London* 311:475–88
4. Ozaki H, Stevens RJ, Blondfield DP, Publicover NG, Sanders KM. 1991. Simultaneous measurement of membrane potential, cytosolic Ca^{2+}, and tension in

intact smooth muscles. *Am. J. Physiol.* 260:C917–25
5. Tomita T. 1981. Electrical activity (spikes and slow waves) in gastrointestinal smooth muscles. In *Smooth Muscle: An Assessment of Current Knowledge*, ed. E Bulbring, AF Brading, AW Jones, T Tomita, pp. 171–97. London: Edward Arnold
6. Kelly KA, Code CF. 1971. Canine gastric pacemaker. *Am. J. Physiol.* 220:112–18
7. Sanders KM, Vogalis F. 1989. Organization of electrical activity in the canine pyloric canal. *J. Physiol. London* 416:49–66
8. Wang XY, Lammers WJ, Bercik P, Huizinga JD. 2005. Lack of pyloric

interstitial cells of Cajal explains distinct peristaltic motor patterns in stomach and small intestine. *Am. J. Physiol. Gastrointest. Liver. Physiol.* 289:G539–49

9. Diamant NE, Bortoff A. 1969. Effects of transection on the intestinal slow-wave frequency gradient. *Am. J. Physiol.* 216:734–43

10. Diamant NE, Bortoff A. 1969. Nature of the intestinal slow-wave frequency gradient. *Am. J. Physiol.* 216:301–7

11. Smith TK, Reed JB, Sanders KM. 1987. Interaction of two electrical pacemakers in muscularis of canine proximal colon. *Am. J. Physiol.* 252:C290–99

12. Christensen J. 1975. Myoelectric control of the colon. *Gastroenterology* 68:601–9

13. Sanders KM. 1996. A case for interstitial cells of Cajal as pacemakers and mediators of neurotransmission in the gastrointestinal tract. *Gastroenterology* 111:492–515

14. Sanders KM, Stevens R, Burke E, Ward SW. 1990. Slow waves actively propagate at submucosal surface of circular layer in canine colon. *Am. J. Physiol.* 259:G258–63

15. Dickens EJ, Hirst GD, Tomita T. 1999. Identification of rhythmically active cells in guinea-pig stomach. *J. Physiol. London* 514:515–31

16. Horiguchi K, Semple GS, Sanders KM, Ward SM. 2001. Distribution of pacemaker function through the tunica muscularis of the canine gastric antrum. *J. Physiol. London* 537:237–50

17. Edwards FR, Hirst GD, Suzuki H. 1999. Unitary nature of regenerative potentials recorded from circular smooth muscle of guinea-pig antrum. *J. Physiol. London* 519:235–50

18. van Helden DF, Imtiaz MS, Nurgaliyeva K, von der Weid P, Dosen PJ. 2000. Role of calcium stores and membrane voltage in the generation of slow wave action potentials in guinea-pig gastric pylorus. *J. Physiol. London* 524:245–65

19. van Helden DF, Imtiaz MS. 2003. Ca^{2+}

phase waves: a basis for cellular pacemaking and long-range synchronicity in the guinea-pig gastric pylorus. *J. Physiol. London* 548:271–96

20. Suzuki N, Prosser CL, Dahms V. 1986. Boundary cells between longitudinal and circular layers: essential for electrical slow waves in cat intestine. *Am. J. Physiol.* 250:G287–94

21. Ward SM, Burns AJ, Torihashi S, Sanders KM. 1994. Mutation of the proto-oncogene c-*kit* blocks development of interstitial cells and electrical rhythmicity in murine intestine. *J. Physiol. London* 480:91–97

22. Jimenez M, Cayabyab FS, Vergara P, Daniel EE. 1996. Heterogeneity in electrical activity of the canine ileal circular muscle: interaction of two pacemakers. *Neurogastroenterol. Motil.* 8:339–49

23. Ward SM, Ordog T, Bayguinov JR, Horowitz B, Epperson A, et al. 1999. Development of interstitial cells of Cajal and pacemaking in mice lacking enteric nerves. *Gastroenterology* 117:584–94

24. Burns AJ, Lomax AEJ, Torihashi S, Sanders KM, Ward SM. 1996. Interstitial cells of Cajal mediate inhibitory neurotransmission in the stomach. *Proc. Natl. Acad. Sci. USA.* 93:12008–13

25. Thuneberg L. 1982. Interstitial cells of Cajal: intestinal pacemaker cells? *Adv. Anat. Embryol. Cell Biol.* 71:1–130

26. Faussone Pellegrini MS, Cortesini C, Romagnoli P. 1977. Ultrastructure of the tunica muscularis of the cardial portion of the human esophagus and stomach, with special reference to the so-called Cajal's interstitial cells. *Arch. Ital. Anat. Embriol.* 82:157–77

27. Ward SM, Morris G, Reese L, Wang XY, Sanders KM. 1998. Interstitial cells of Cajal mediate enteric inhibitory neurotransmission in the lower esophageal and pyloric sphincters. *Gastroenterology* 115:314–29

28. Ward SM, Ordog T, Koh SD, Baker SA, Jun JY, et al. 2000. Pacemaking in

interstitial cells of Cajal depends upon calcium handling by endoplasmic reticulum and mitochondria. *J. Physiol. London* 525:355–61

29. Beckett EA, McGeough CA, Sanders KM, Ward SM. 2003. Pacing of interstitial cells of Cajal in the murine gastric antrum: neurally mediated and direct stimulation. *J. Physiol. London* 553:545–59

30. Forrest AS, Ördög T, Sanders KM. 2005. Neural regulation of slow wave frequency in the murine gastric antrum. *Am. J. Physiol. Gastrointest. Liver Physiol.* In press

31. Hirst GD, Dickens EJ, Edwards FR. 2002. Pacemaker shift in the gastric antrum of guinea-pigs produced by excitatory vagal stimulation involves intramuscular interstitial cells. *J. Physiol. London* 541:917–28

32. Sanders KM, Ordog T, Koh SD, Torihashi S, Ward SM. 1999. Development and plasticity of interstitial cells of Cajal. *Neurogastroenterol. Motil.* 11:311–38

33. Thuneberg L, Johansen V, Rumenssen JJ, Andersen BG. 1983. Interstitial cells of Cajal (ICC): Selective uptake of methylene blue inhibits slow wave activity. In *Gastrointestinal Motility*, ed. C Roman, pp. 495–502. Lancaster: MTP Press Ltd.

34. Sanders KM, Burke EP, Stevens RJ. 1989. Effects of methylene blue on electrical rhythmicity of the canine colon. *Am. J. Physiol.* 256:G779–84

35. Ward SW, Burke EP, Sanders KM. 1990. Use of rhodamine 123 to label and lesion interstitial cells of Cajal in canine colonic circular muscle. *Anat. Embryol.* 182:215–24

36. Barajas-Lopez C, Berezin I, Daniel EE, Huizinga JD. 1989. Pacemaker activity recorded in interstitial cells of Cajal of the gastrointestinal tract. *Am. J. Physiol.* 257:C830–35

37. Cousins HM, Edwards FR, Hickey H, Hill CE, Hirst GD. 2003. Electrical coupling between the myenteric interstitial cells of Cajal and adjacent muscle layers in the guinea-pig gastric antrum. *J. Physiol. London* 550:829–44

38. Kito Y, Suzuki H. 2003. Properties of pacemaker potentials recorded from myenteric interstitial cells of Cajal distributed in the mouse small intestine. *J. Physiol. London* 553:803–18

39. Hennig GW, Hirst GD, Park KJ, Smith CB, Sanders KM, et al. 2004. Propagation of pacemaker activity in the guinea-pig antrum. *J. Physiol. London* 556:585–99

40. Langton P, Ward SM, Carl A, Norell MA, Sanders KM. 1989. Spontaneous electrical activity of interstitial cells of Cajal isolated from canine proximal colon. *Proc. Natl. Acad. Sci. USA* 86:7280–84

41. Thomsen L, Robinson TL, Lee JC, Farraway LA, Hughes MJ, et al. 1998. Interstitial cells of Cajal generate a rhythmic pacemaker current. *Nat. Med.* 4:848–51

42. Koh SD, Sanders KM, Ward SM. 1998. Spontaneous electrical rhythmicity in cultured interstitial cells of cajal from the murine small intestine. *J. Physiol. London* 513:203–13

43. Maeda H, Yamagata A, Nishikawa S, Yoshinaga K, Kobayashi S, et al. 1992. Requirement of c-kit for development of intestinal pacemaker system. *Development* 116:369–75

44. Torihashi S, Ward SM, Nishikawa S-I, Nishi K, Kobayashi S, Sanders KM. 1995. c-kit-dependent development of interstitial cells and electrical activity in the murine gastrointestinal tract. *Cell Tissue Res.* 280:97–111

45. Huizinga JD, Thuneberg L, Kluppel M, Malysz J, Mikkelsen HB, Bernstein A. 1995. W/kit gene required for interstitial cells of Cajal and for intestinal pacemaker activity. *Nature* 373:347–49

46. Ordog T, Ward SM, Sanders KM. 1999. Interstitial cells of cajal generate electrical slow waves in the murine stomach. *J. Physiol. London* 518:257–69

47. Faussone-Pellegrini MS. 1992. Histogenesis, structure and relationships of interstitial cells of Cajal (ICC): from morpholog

to functional interpretation. *Eur. J. Morphol.* 30:137–48

48. Faussonc-Pellegrini MS, Thuneberg L. 1999. Guide to the identification of interstitial cells of Cajal. *Microsc. Res. Tech.* 47:248–66

49. Komuro T, Tokui K, Zhou DS. 1996. Identification of the interstitial cells of Cajal. *Histol. Histopathol.* 11:769–86

50. Rumessen JJ, Thuneberg L. 1996. Pacemaker cells in the gastrointestinal tract: interstitial cells of Cajal. *Scand. J. Gastroenterol. Suppl.* 216:82–94

51. Seki K, Komuro T. 2001. Immunocytochemical demonstration of the gap junction proteins connexin 43 and connexin 45 in the musculature of the rat small intestine. *Cell Tissue Res.* 306:417–22

52. Seki K, Zhou DS, Komuro T. 1998. Immunohistochemical study of the c-kit expressing cells and connexin 43 in the guinea-pig digestive tract. *J. Auton. Nerv. Syst.* 68:182–87

53. Wang YF, Daniel EE. 2001. Gap junctions in gastrointestinal muscle contain multiple connexins. *Am. J. Physiol. Gastrointest. Liver. Physiol.* 281:G533–43

54. Daniel EE. 2004. Communication between interstitial cells of Cajal and gastrointestinal muscle. *Neurogastroenterol. Motil.* 16(Suppl. 1):118–22

55. Horiguchi K, Sanders KM, Ward SM. 2003. Enteric motor neurons form synaptic-like junctions with interstitial cells of Cajal in the canine gastric antrum. *Cell Tissue Res.* 311:299–313

56. Cohen AW, Hnasko R, Schubert W, Lisanti MP. 2004. Role of caveolae and caveolins in health and disease. *Physiol. Rev.* 84:1341–79

57. Cho WJ, Daniel EE. 2005. Proteins of interstitial cells of Cajal and intestinal smooth muscle, colocalized with caveolin-1. *Am. J. Physiol. Gastrointest. Liver. Physiol.* 288:G571–85

58. Sanders KM, Ordog T, Koh SD, Ward SM. 2000. A novel pacemaker mechanism drives gastrointestinal rhythmicity. *News Physiol. Sci.* 5:291–98

59. Edwards FR, Hirst GD, Suzuki H. 1999. Unitary nature of regenerative potentials recorded from circular smooth muscle of guinea-pig antrum. *J Physiol. London* 519:235–50

60. Kito Y, Suzuki H, Edwards FR. 2002. Properties of unitary potentials recorded from myenteric interstitial cells of Cajal distributed in the guinea-pig gastric antrum. *J. Smooth Muscle Res.* 38:165–79

61. Kim YC, Koh SD, Sanders KM. 2002. Voltage-dependent inward currents of interstitial cells of Cajal from murine colon and small intestine. *J. Physiol. London* 541:797–810

62. Hirst GD, Bramich NJ, Teramoto N, Suzuki H, Edwards FR. 2002. Regenerative component of slow waves in the guinea-pig gastric antrum involves a delayed increase in [Ca^{2+}]$_i$ and Cl$^-$ channels. *J. Physiol. London* 540:907–19

63. Hirst GD, Edwards FR. 2001. Generation of slow waves in the antral region of guinea-pig stomach—a stochastic process. *J. Physiol. London* 535:165–80

64. Suzuki H, Hirst GD. 1999. Regenerative potentials evoked in circular smooth muscle of the antral region of guinea-pig stomach. *J. Physiol. London* 517:563–73

65. Publicover NG, Sanders KM. 1986. Effects of frequency on the waveform of propagated slow waves in canine gastric antral muscle. *J. Physiol. London* 371:179–89

66. Kito Y, Fukuta H, Yamamoto Y, Suzuki H. 2002. Excitation of smooth muscles isolated from the guinea-pig gastric antrum in response to depolarization. *J. Physiol. London* 543:155–67

67. Goto K, Matsuoka S, Noma A. 2004. Two types of spontaneous depolarizations in the interstitial cells freshly prepared from the murine small intestine. *J. Physiol. London* 559:411–22

68. Lee HK, Sanders KM. 1993. Comparison

of ionic currents from interstitial cells and smooth muscle cells of canine colon. *J. Physiol. London* 460:135–52

69. Ward SM, Dixon RE, de Faoite A, Sanders KM. 2004. Voltage-dependent calcium entry underlies propagation of slow waves in canine gastric antrum. *J. Physiol. London* 561:793–810

70. Kito Y, Ward SM, Sanders KM. 2005. Pacemaker potentials generated by interstitial cells of Cajal in the murine intestine. *Am. J. Physiol. Cell Physiol.* 288:C710–20

71. Epperson A, Hatton WJ, Callaghan B, Doherty P, Walker RL, et al. 2000. Molecular markers expressed in cultured and freshly isolated interstitial cells of Cajal. *Am. J. Physiol.* 279:C529–39

72. Kito Y, Fukuta H, Suzuki H. 2002. Components of pacemaker potentials recorded from the guinea pig stomach antrum. *Pflugers Arch.* 445:202–17

73. Malysz J, Donnelly G, Huizinga JD. 2001. Regulation of slow wave frequency by IP_3-sensitive calcium release in the murine small intestine. *Am. J. Physiol.* 280:G439–48

74. Suzuki H, Takano H, Yamamoto Y, Komuro T, Saito M, et al. 2000. Properties of gastric smooth muscles obtained from mice which lack inositol trisphosphate receptor. *J. Physiol. London* 525:105–11

75. Tokutomi N, Maeda H, Tokutomi Y, Sato D, Sugita M, et al. 1995. Rhythmic Cl^- current and physiological roles of the intestinal c-kit-positive cells. *Pflugers Arch.* 431:169–77

76. Huizinga JD, Zhu Y, Ye J, Molleman A. 2002. High-conductance chloride channels generate pacemaker currents in interstitial cells of Cajal. *Gastroenterology* 123:1627–36

77. Zhu Y, Mucci A, Huizinga JD. 2005. Inwardly rectifying chloride channel activity in intestinal pacemaker cells. *Am. J. Physiol. Gastrointest. Liver. Physiol.* 288:G809–21

78. Koh SD, Jun JY, Kim TW, Sanders KM. 2002. A Ca^{2+}-inhibited non-selective cation conductance contributes to pacemaker currents in mouse interstitial cell of Cajal. *J. Physiol. London* 540:803–14

79. Zhang Z, Tang J, Tikunova S, Johnson JD, Chen Z, et al. 2001. Activation of Trp3 by inositol 1,4,5-trisphosphate receptors through displacement of inhibitory calmodulin from a common binding domain. *Proc. Natl. Acad. Sci. USA* 98:3168–73

80. Walker RL, Koh SD, Sergeant GP, Sanders KM, Horowitz B. 2002. TRPC4 currents have properties similar to the pacemaker current in interstitial cells of Cajal. *Am. J. Physiol. Cell Physiol.* 283:C1637–45

81. Horowitz B, Ward SM, Sanders KM. 1999. Cellular and molecular basis for electrical rhythmicity in gastrointestinal muscles. *Annu. Rev. Physiol.* 61:19–43

82. Edwards FR, Hirst GD. 2005. An electrical description of the generation of slow waves in the antrum of the guinea-pig. *J. Physiol. London* 564:213–32

83. Mak DO, McBride S, Foskett JK. 1998. Inositol 1,4,5-trisphosphate activation of inositol trisphosphate receptor Ca^{2+} channel by ligand tuning of Ca^{2+} inhibition *Proc. Natl. Acad. Sci. USA* 95:15821–25

84. You CH, Chey WY. 1984. Study of electromechanical activity of the stomach in humans and in dogs with particular attention to tachygastria. *Gastroenterology* 86:1460–68

85. Koch KL, Stern RM, Stewart WR, Vasey MW. 1989. Gastric emptying and gastric myoelectrical activity in patients with diabetic gastroparesis: effect of long-term domperidone treatment. *Am. J. Gastroenterol.* 84:1069–75

86. Sanders KM. 1984. Role of prostaglandins in regulating gastric motility *Am. J. Physiol.* 247:G117–26

87. el-Sharkawy TY, Szurszewski JH. 1978. Modulation of canine antral circula

smooth muscle by acetylcholine, noradrenaline and pentagastrin. *J. Physiol. London* 279:309–20

88. Kim TW, Koh SD, Ordog T, Ward SM, Sanders KM. 2003. Muscarinic regulation of pacemaker frequency in murine gastric interstitial cells of Cajal. *J. Physiol.* 546:415–25

89. Sanders KM, Szurszewski JH. 1981. Does endogenous prostaglandin affect gastric antral motility? *Am. J. Physiol.* 241:G191–95

90. Kim TW, Beckett EA, Hanna R, Koh SD, Ordog T, et al. 2002. Regulation of pacemaker frequency in the murine gastric antrum. *J Physiol. London* 538:145–57

91. Ward SM, Beckett EAH, Wang X-Y, Baker F, Khoyi M, Sanders KM. 2000. Interstitial cells of Cajal mediate cholinergic neurotransmission from enteric motor neurons. *J. Neurosci.* 20:1393–403

92. Won KJ, Sanders KM, Ward SM. 2005. Interstitial cells of Cajal mediate mechanosensitive responses in the stomach. *Proc. Natl. Acad. Sci. USA* 102: 14913–18

93. Vanderwinden JM, Rumessen JJ. 1999. Interstitial cells of Cajal in human gut and gastrointestinal disease. *Microsc. Res. Technol.* 47:344–60

94. Hirota S, Isozaki K, Moriyama Y, Hashimoto K, Nishida T, Ishiguro S, et al. 1998. Gain-of-function mutations of c-kit in human gastrointestinal stromal tumors. *Science* 279:577–80

95. Candelaria M, de la Garza J, Duenas-Gonzalez A. 2005. A clinical and biological overview of gastrointestinal stromal tumors. *Med. Oncol.* 22:1–10

96. Kitamura Y, Hirotab S. 2004. Kit as a human oncogenic tyrosine kinase. *Cell Mol. Life Sci.* 61:2924–31

97. Corless CL, Fletcher JA, Heinrich MC. 2004. Biology of gastrointestinal stromal tumors. *J. Clin. Oncol.* 22:3813–25

98. Kitamura Y, Hirota S, Nishida T. 2003. Gastrointestinal stromal tumors (GIST): a model for molecule-based diagnosis and treatment of solid tumors. *Cancer Sci.* 94:315–20

99. Sims SM. 1992. Calcium and potassium currents in canine gastric smooth muscle cells. *Am. J. Physiol.* 262:G859–67

100. Langton PD, Burke EP, Sanders KM. 1989. Participation of Ca^{2+} currents in colonic electrical activity. *Am. J. Physiol.* 257:C451–60

101. Ward SM, Sanders KM. 1992. Dependence of electrical slow waves of canine colonic smooth muscle on calcium current. *J. Physiol. London* 455:307–19

102. Cohen NM, Lederer WJ. 1987. Calcium current in isolated neonatal rat ventricular myocytes. *J. Physiol. London* 391:169–91

103. Vogalis F, Publicover NG, Hume JR, Sanders KM. 1991. Relationship between calcium current and cytosolic calcium concentration in canine gastric smooth muscle cells. *Am. J. Physiol.* 260:C1012–18

Annu. Rev. Physiol 2006. 68:345–74
doi: 10.1146/annurev.physiol.68.040504.094707
First published online as a Review in Advance on September 22, 2005

SIGNALING FOR CONTRACTION AND RELAXATION IN SMOOTH MUSCLE OF THE GUT

Karnam S. Murthy

*Department of Physiology, Virginia Commonwealth University Medical Center,
Richmond, Virginia 23298; email: skarnam@hsc.vcu.edu*

Key Words MLC_{20} phosphorylation, MLC phosphatase, Rho kinase, ZIP kinase, integrin-linked kinase

■ **Abstract**   Phosphorylation of Ser^{19} on the 20-kDa regulatory light chain of myosin II (MLC_{20}) by Ca^{2+}/calmodulin-dependent myosin light-chain kinase (MLCK) is essential for initiation of smooth muscle contraction. The initial $[Ca^{2+}]_i$ transient is rapidly dissipated and MLCK inactivated, whereas MLC_{20} and muscle contraction are well maintained. Sustained contraction does not reflect Ca^{2+} sensitization because complete inhibition of MLC phosphatase activity in the absence of Ca^{2+} induces smooth muscle contraction. This contraction is suppressed by staurosporine, implying participation of a Ca^{2+}-independent MLCK. Thus, sustained contraction, as with agonist-induced contraction at experimentally fixed Ca^{2+} concentrations, involves (*a*) G protein activation, (*b*) regulated inhibition of MLC phosphatase, and (*c*) MLC_{20} phosphorylation via a Ca^{2+}-independent MLCK. The pathways that lead to inhibition of MLC phosphatase by $G_{q/13}$-coupled receptors are initiated by sequential activation of $G\alpha_q/\alpha_{13}$, RhoGEF, and RhoA, and involve Rho kinase–mediated phosphorylation of the regulatory subunit of MLC phosphatase (MYPT1) and/or PKC-mediated phosphorylation of CPI-17, an endogenous inhibitor of MLC phosphatase. Sustained MLC_{20} phosphorylation is probably induced by the Ca^{2+}-independent MLCK, ZIP kinase. The pathways initiated by G_i-coupled receptors involve sequential activation of $G\beta\gamma_i$, PI 3-kinase, and the Ca^{2+}-independent MLCK, integrin-linked kinase. The last phosphorylates MLC_{20} directly and inhibits MLC phosphatase by phosphorylating CPI-17. PKA and PKG, which mediate relaxation, act upstream to desensitize the receptors ($VPAC_2$ and NPR-C), inhibit adenylyl and guanylyl cyclase activities, and stimulate cAMP-specific PDE3 and PDE4 and cGMP-specific PDE5 activities. These kinases also act downstream to inhibit (*a*) initial contraction by inhibiting Ca^{2+} mobilization and (*b*) sustained contraction by inhibiting RhoA and targets downstream of RhoA. This increases MLC phosphatase activity and induces MLC_{20} dephosphorylation and muscle relaxation.

INTRODUCTION

Smooth muscle of the gut possesses distinct regional and functional properties that distinguish it from other types of visceral and vascular smooth muscle. Smooth muscle of the proximal stomach, sphincters, and gall bladder exhibits sustained tone, whereas smooth muscle of the distal stomach, small intestine, and colon exhibits variable tone on which are superimposed rhythmic contractions driven by cycles of membrane depolarization and repolarization known as slow waves (1–5). These cycles originate in pacemaker cells located at the boundaries and in the substance of the inner, circular muscle layer, from which they spread to the outer, longitudinal muscle layer (4, 5). The depolarization phase primarily reflects activation of voltage-gated Ca^{2+} channels, resulting in Ca^{2+} entry and contraction. Concurrent stimulation of rhythmic smooth muscle by excitatory neurotransmitters, chiefly acetylcholine and tachykinins, causes further depolarization and Ca^{2+} entry and activates signaling cascades that result in a transient Ca^{2+} release (1, 6–8). Stimulation of tonic smooth muscle by excitatory neurotransmitters causes transient Ca^{2+} release and contraction followed by a sustained Ca^{2+}-independent contraction. The molecular mechanisms underlying these processes in tonic smooth muscle are the subject of this review.

SIGNALING FOR CONTRACTION IN SMOOTH MUSCLE OF THE GUT

Receptors and G Proteins

The apparatus of signal transduction in smooth muscle cells consists of three types of membrane-bound proteins: (*a*) a membrane-spanning receptor, (*b*) a GTP-binding protein that couples to the receptor, and (*c*) effector enzymes capable of generating cascades of intracellular regulatory signals. The ligands for these receptors in smooth muscle of the gut are derived mainly from enteric excitatory motor neurons (e.g., acetylcholine, tachykinins), from smooth muscle cells [e.g., sphingosine-1-phosphate (S1P), lysophosphatidic acid (LPA), endocannabinoids], and to lesser extents from adventitious cells and via the circulation (6–10). A large variety of receptors and receptor subtypes have been identified as capable of mediating smooth muscle contraction. Among these are receptors for peptides (e.g., tachykinins, endothelin, motilin), amines (histamine, 5-hydroxytryptamine), pyrimidines/purines (UTP and ATP), and lipids (S1P, LPA) (11–18). Muscarinic m2 and m3 receptors are the predominant receptor types, although only m3 receptors mediate contraction (19, 20) (Table 1). All of these receptors are uniformly distributed in smooth muscle cells of the outer, longitudinal muscle layer and the densely innervated, inner circular muscle layer, with the exception of somatostatin, opioid, and neuropeptide Y (NPY) Y_2 and Y_4 receptors, which are absent from smooth muscle cells of the longitudinal muscle layer (21–23).

TABLE 1 Diversity of signaling by G protein–coupled receptors in smooth muscle of the gut*

Ligand	Receptor	G protein	Effector	Messenger
ACh	m3	$G\alpha_q$	PLC-β1 ↑	IP$_3$/DAG ↑
		$G\alpha_{13}$	RhoA ↑	ROK/DAG ↑
Endothelin	ET$_A$	$G\alpha_q$	PLC β1 ↑	IP$_3$/DAG ↑
		$G\alpha_{13}$	RhoA ↑	ROK/DAG ↑
	ET$_B$	$G\alpha_q$	PLC-β1 ↑	IP$_3$/DAG ↑
S1P	S1P$_2$	$G\alpha_q$	PLC-β1 ↑	IP$_3$/DAG ↑
		$G\beta\gamma_i$	PLC-β3 ↑	IP$_3$/DAG ↑
		$G\alpha_{13}$	RhoA ↑	ROK/DAG ↑
		$G\alpha_{i1/2/3}$	AC ↓	cAMP ↓
ATP/UTP	P$_{2Y2}$	$G\alpha_q$	PLC-β1 ↑	IP$_3$/DAG ↑
		$G\beta\gamma_{i3}$	PLC-β3 ↑	IP$_3$/DAG ↑
		$G\alpha_{i3}$	AC ↓	cAMP ↓
NPY/PP	Y$_2$/Y$_4$	$G\alpha_q$	PLC-β1 ↑	IP$_3$/DAG ↑
		$G\alpha_{i2}$	AC ↓	cAMP ↓
Opioid	δ, μ, κ	$G\beta\gamma_{i2}$	PLC-β3 ↑	IP$_3$/DAG ↑
		$G\beta\gamma_{i2}$	PI 3-K	
		$G\alpha_{i2}$	AC ↓	cAMP ↓
ACh	m2	$G\beta\gamma_{i3}$	PLC-β3 ↑	IP$_3$/DAG ↑
		$G\beta\gamma_{i3}$	PI 3-K	
		α_{i3}	AC ↓	cAMP ↓
Anandamide	CB$_1$	$G\alpha_{i2}$	AC ↓	cAMP ↓

*The pattern of signaling by m3 receptors involving G$_q$ and G$_{13}$ is also seen with endothelin ET$_A$ and motilin receptors. G protein–coupling of serotonin 5-HT$_{2c}$, histamine H$_1$, and tachykinin NK$_1$/NK$_2$ receptors is probably similar to that of m3 receptors, although activation of G$_{13}$ has not yet been examined. ET$_B$ receptors are coupled to G$_q$, but not G$_{13}$, and do not activate RhoA. Sphingosine-1 phosphate (S1P$_2$) receptors and purine/pyrimidine (P$_{2Y2}$) receptors are coupled to G$_q$/G$_{13}$ and to various isoforms of G$_i$. Y$_2$ and Y$_4$ receptors are coupled to G$_q$ and G$_{i2}$ but do not signal via G$\beta\gamma_i$. Opioid μ,δ, κ; somatostatin sstr$_3$; and adenosine A$_1$ receptors are coupled to various isoforms of G$_i$ and signal in similar fashion via Gα_i and G$\beta\gamma_i$. Muscarinic m2 receptors are coupled to the same G proteins as A$_1$ receptors, but they signal atypically, as detailed in the text. Cannabinoid CB$_1$ receptors are coupled to an atypical G protein complex (Gα_{i2}-Gβ_5-RGS6) and signal via Gα_{i2} only.

Effector Enzymes

PHOSPHOLIPASE C-β1 AND PLC-β3   Receptors mediating contraction in smooth muscle of the gut are coupled to G$_q$, G$_i$, or both. For example, m3, ET$_A$ and ET$_B$, and 5-HT$_2$ receptors are coupled to G$_q$, somatostatin sstr$_3$, opioid μ, δ, and κ; adenosine A$_1$ receptors are coupled to various isoforms of G$_i$; and S1P$_2$ and P2Y$_2$ receptors are coupled to both G$_q$ and G$_i$ (12, 15–17, 20–24). All of the receptors stimulate phosphoinositide (PI) hydrolysis by activating phospholipase C (PLC)-β1 via Gα_q and/or PLC-β3 via G$\beta\gamma_i$. Although all four PLC-β isozymes are expressed in smooth muscle cells of the gut, neither PLC-β2 nor PLC-β4 is activated

(Table 1). PLC-β1 is activated via binding of its COOH-terminal tail, a characteristic feature of PLC-β isozymes, to Gα_q, whereas PLC-β3 is activated via binding of its NH$_2$-terminal pleckstrin homology (PH) domain to G$\beta\gamma_i$ (25). In circular smooth muscle cells, as in most cell types, phosphatidylinositol 4,5-bisphosphate (PIP$_2$) is the predominant phosphoinositide substrate hydrolyzed by PLC-β1 or PLC-β3, resulting in generation of diacylglycerol (DAG) and the diffusible Ca^{2+}-mobilizing messenger, inositol 1,4,5-trisphosphate (IP$_3$) (26). In longitudinal smooth muscle cells, however, the predominant phosphoinositide substrate hydrolyzed by PLC-β1 or PLC-β3 is phosphatidylinositol 4-phosphate, yielding inositol 1,4-bisphosphate (IP$_2$), an ineffective Ca^{2+}-mobilizing product, and DAG (26). In both cell types, activation of PLC-β is transient (<2 min) and occurs mainly during the initial phase of contraction (26, 27).

PHOSPHOLIPASE D   G$_q$/G$_{13}$-coupled receptors mediating contraction also activate phosphatidylcholine (PC)-specific PLC and phospholipase D (PLD). Activation of these enzymes occurs mainly, but not exclusively, during the sustained phase of contraction. In smooth muscle cells of the gut, m3, ET$_A$, and S1P$_2$ receptors activate PLD sequentially via G$_{13}$ and the monomeric G protein/GTPase RhoA (12, 17, 20, 28, 29). Phosphatidic acid, the primary product of PLD activity, is dephosphorylated to DAG, which causes sustained activation of protein kinase C (PKC). Formation of DAG and activation of PKC are biphasic, with an early peak coinciding with PI hydrolysis followed by a sustained increase reflecting PC-PLC and PLD activities (28).

PLC-δ1   G$_i$-coupled receptors do not activate G$_{13}$, RhoA, or PLD in smooth muscle cells of the gut yet can cause transient PI hydrolysis by activating PLC-β3 via G$\beta\gamma_i$ as well as delayed PI hydrolysis by activating PLC-δ1, the most widely expressed PLC-δ isozyme (30). Although the PH domain of PLC-δ binds G$\beta\gamma$, the binding is too weak to stimulate enzyme activity. PLC-δ, which is highly sensitive to Ca^{2+} and is typically activated by intracellular Ca^{2+} concentrations in the range of 0.1–1 μM, appears to be preferentially activated by Ca^{2+} influx via capacitative Ca^{2+} channels (30, 31). PLC-δ1 activity in smooth muscle of the gut is minimally inhibited by blockade of voltage-gated Ca^{2+} channels but is strongly inhibited by blockade of store-operated Ca^{2+} channels. Capacitative Ca^{2+} influx induced by G$_q$-coupled receptors does not lead to activation of PLC-δ1 because concurrent activation of RhoA blocks PLC-δ1 activation. Inactivation of RhoA by the *Clostridium botulinum* C3 exoenzyme or by expression of a dominant-negative mutant of RhoA unmasks Ca^{2+}-stimulated PLC-δ1 activity (30). The functional significance of PLC-δ1 activity and resultant IP$_3$ formation in the context of depleted sarcoplasmic Ca^{2+} stores is unclear. Furthermore, as noted below, the concurrent activation of PKC via PLC-δ1 is not involved in the sustained contraction mediated by G$_i$-coupled receptors.

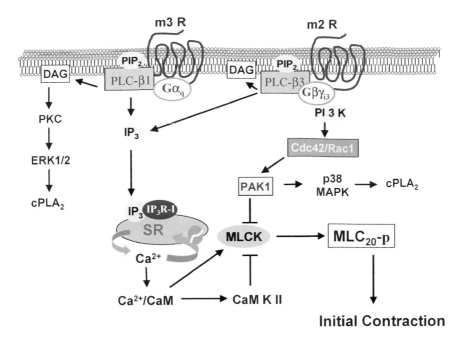

Figure 1 Signaling by muscarinic m3 and m2 receptors during the initial phase of contraction. Initial contraction and MLC_{20} (20-kDa regulatory light chain of myosin II) phosphorylation in circular smooth muscle are mediated exclusively by m3 receptors; m2 receptors do not cause contraction because they initiate a parallel pathway involving sequential activation of PI 3-kinase, Cdc42/Rac1, and PAK1, which results in PAK1-mediated inactivation of myosin light-chain kinase (MLCK). This inhibitory pathway is specific to m2 receptors and is not shared by other G_i-coupled receptors that cause initial contraction (e.g., δ-opioid, A_1, $sstr_3$). Both m3 and m2 receptors also activate $cPLA_2$ via distinct pathways involving ERK1/2 and p38 MAP kinase (p38 MAPK), respectively. A distinct mechanism for Ca^{2+} mobilization in longitudinal smooth muscle involving activation of $cPLA_2$ is described in the text. Signaling by other receptors is summarized in Table 1. CaM K II, calcium/calmodulin-activated kinase II; IP_3R-I, Type 1 IP_3 receptor; PI 3K, PI 3-kinase; SR, sarcoplasmic reticulum.

CYTOPLASMIC PLA_2 G_q- and G_i-coupled receptors activate cytoplasmic PLA_2 ($cPLA_2$) via different mechanisms, as evidenced in the signaling cascades initiated by m3 and m2 receptors. Sustained stimulation of $cPLA_2$ activity via m3 receptors involves sequential activation of $G\alpha_q \rightarrow$ PLC-β1 $\rightarrow$ PKC $\rightarrow$ ERK1/2 $\rightarrow cPLA_2$, whereas sustained stimulation via m2 receptors involves sequential activation of $G\beta\gamma_i \rightarrow$ PI 3-kinase $\rightarrow$ Cdc42/Rac1 $\rightarrow$ PAK1 (*p*21-*a*ctivated protein *k*inase 1) $\rightarrow$ p38 MAP kinase $\rightarrow cPLA_2$. Both pathways result in phosphorylation of $cPLA_2$ at Ser^{505}, and their effects are additive (32) (Figure 1).

CA^{2+} MOBILIZATION IN CIRCULAR AND LONGITUDINAL SMOOTH MUSCLE

The mechanisms of Ca^{2+} mobilization in smooth muscle cells of the circular and longitudinal muscle layers vary. In muscle cells from both layers, G protein–coupled agonists initiate contraction by increasing cytosolic Ca^{2+}, or $[Ca^{2+}]_i$. Initial contraction of smooth muscle cells and the increase in $[Ca^{2+}]_i$ are not affected by Ca^{2+} channel blockers or by withdrawal of extracellular Ca^{2+} in circular smooth muscle cells but are abolished in longitudinal smooth muscle cells. This implies that Ca^{2+} mobilization in longitudinal muscle is dependent upon an obligatory step involving Ca^{2+} influx (33–35).

IP$_3$-Induced Ca^{2+} Release via IP$_3$R-I Receptors/Ca^{2+} Channels in Circular Smooth Muscle

Circular smooth muscle cells exhibit preferential PIP$_2$ hydrolysis and rapid IP$_3$ formation. Their sarcoplasmic Ca^{2+} stores contain high-affinity IP$_3$ receptors/Ca^{2+} channels and release Ca^{2+} in response to low cytosolic concentrations of IP$_3$ (33). Two IP$_3$ receptors, IP$_3$R-I and IP$_3$R-III, are expressed in these cells, but only IP$_3$R-I mediates Ca^{2+} release (33, 36). Agonist-stimulated increase in $[Ca^{2+}]_i$ is abolished by inhibitors of PI hydrolysis; upon depletion of sarcoplasmic Ca^{2+} stores in intact muscle cells; and by treatment with Gα_q antibodies, PLC-β antibodies, or IP$_3$ receptor inhibitors in permeabilized muscle cells (33, 36) (Figure 1).

Ca^{2+}- and Cyclic ADP Ribose–Induced Ca^{2+} Release via Ryanodine Receptors/Ca^{2+} Channels in Longitudinal Smooth Muscle

Longitudinal smooth muscle cells, which exhibit preferential hydrolysis of PIP, contain small amounts of PIP$_2$ and generate minimal amounts of IP$_3$. Their sarcoplasmic stores are devoid of high-affinity IP$_3$ receptors and exhibit minimal sensitivity to IP$_3$, but they possess high-affinity ryanodine receptors (K$_d$ = 7 nM) and exhibit high sensitivity to Ca^{2+} (100–500 nM) and cyclic ADP ribose (EC$_{50}$ = 1 nM) (37, 38). Muscle cells from this layer possess membrane-bound ADP ribosyl cyclase, which can be activated in a concentration-dependent fashion to yield cyclic ADP ribose (38).

The search for a messenger that triggers Ca^{2+} influx and results in Ca^{2+}- and cyclic ADP ribose–induced Ca^{2+} release led to arachidonic acid, a product of PC hydrolysis by cPLA$_2$ (39). An early transient peak in cPLA$_2$ activity, which precedes sustained cPLA$_2$ activity, is observed only in longitudinal smooth muscle cells. Blockade of peak activity with cPLA$_2$ inhibitors (or with GDPβS in reversibly permeabilized muscle cells) abolishes Ca^{2+} mobilization. Low concentrations of arachidonic acid applied exogenously induce membrane depolarization and the opening of voltage-gated Ca^{2+} channels (39, 40). The entry of Ca^{2+}

stimulates cyclic ADP ribose formation and induces synergistic Ca^{2+}- and cyclic ADP ribose–induced Ca^{2+} release via ryanodine receptors/Ca^{2+} channels.

INITIAL AND SUSTAINED MLC_{20} PHOSPHORYLATION AND CONTRACTION

Overview of Signaling Pathways Mediating Contraction

Phosphorylation of Ser^{19} on the 20-kDa regulatory light chain of myosin II (MLC_{20}) by Ca^{2+}/calmodulin-dependent myosin light-chain kinase (MLCK) is essential for activation of actin-activated myosin ATPase and interaction of actin and myosin, which initiates smooth muscle contraction (41, 42) (Figure 1). The $[Ca^{2+}]_i$ transient responsible for the initial contraction is rapidly and efficiently dissipated by Ca^{2+} extrusion from the cell and uptake into sarcoplasmic Ca^{2+} stores. This causes an equally rapid decline in MLCK activity, which is intensified by phosphorylation of MLCK via Ca^{2+}/calmodulin-dependent protein kinase II and p21-activated protein kinase 1 (PAK1) (Figure 1) (20, 43, 44). The decline in Ca^{2+} mobilization reflects the transient natures of PLC-β activity in circular smooth muscle and of $cPLA_2$ activity in longitudinal smooth muscle. Despite the declines in $[Ca^{2+}]_i$ and MLCK activity, MLC_{20} phosphorylation and muscle contraction are relatively well maintained. This pattern led to a search for an alternative process that was postulated to sensitize a Ca^{2+}-dependent contraction. It was soon demonstrated that an agonist-mediated, G protein–dependent process augments contractile activity at fixed Ca^{2+} concentrations (45, 46). Subsequent studies, however, showed that contraction does not reflect Ca^{2+} sensitization because complete inhibition of MLC phosphatase activity with microcystin-LR in the absence of Ca^{2+} induces smooth muscle contraction that staurosporine, a nonspecific kinase inhibitor, can suppress (47, 48). Agonist-induced contraction at experimentally fixed Ca^{2+} concentrations involves three elements: agonist-induced activation of a G protein, regulated inhibition of MLC phosphatase, and MLC_{20} phosphorylation via a Ca^{2+}-independent MLCK. The linkage between these elements is evident physiologically in the signaling pathways that regulate the sustained phase of agonist-induced MLC_{20} phosphorylation and muscle contraction.

Ca^{2+}-Dependent Initial Contraction and MLC_{20} Phosphorylation

Initial MLC_{20} phosphorylation and contraction induced by G_q-coupled receptor agonists in circular smooth muscle are mediated by Ca^{2+}/calmodulin-dependent MLCK and are blocked by $G\alpha_q$ or PLC-β1 antibodies in permeabilized muscle cells and by expression of a $G\alpha_q$ C-terminal minigene in intact muscle cells. By contrast, MLC_{20} phosphorylation and contraction induced by G_i-coupled receptor agonists are blocked by $G\beta$ or PLC-β3 antibodies in permeabilized muscle cells and by expression of the $G\alpha_i$ C-terminal minigene in intact muscle cells (13, 17,

20, 33). The pathways for all receptor types converge downstream, so that MLC_{20} phosphorylation and contraction are blocked by IP_3 receptor inhibitors, calmodulin antagonists, MLCK inhibitors, and depletion of sarcoplasmic Ca^{2+} stores (49).

Because of their obligatory dependence on an initial step involving arachidonic acid–mediated Ca^{2+} influx, MLC_{20} phosphorylation and contraction in longitudinal smooth muscle are blocked by voltage-gated Ca^{2+} channel blockers and inhibitors of $cPLA_2$, calmodulin, and MLCK (49).

Ca^{2+}-Independent Sustained Contraction and MLC_{20} Phosphorylation

A critical difference between initial and sustained smooth muscle contraction is the balance between MLCK and MLC phosphatase activities. Although Ca^{2+}-independent MLCK(s) are much less potent than Ca^{2+}-activated MLCK(s), the former appear to derive their effectiveness from the concurrent inhibition of MLC phosphatase during the sustained phase of contraction (47, 48). The pathways that lead to inhibition of MLC phosphatase by $G_{q/13}$-coupled receptors are initiated by sequential activation of $G\alpha_q/\alpha_{13}$, RhoGEF, and RhoA and involve Rho kinase–mediated phosphorylation of the regulatory subunit of MLC phosphatase, MYPT1, and/or PKC-mediated phosphorylation of CPI-17, an endogenous inhibitor of MLC phosphatase (20, 41, 50–54). Sustained MLC_{20} phosphorylation is induced by a Ca^{2+}-independent kinase, most likely zipper-interacting protein (ZIP) kinase (47, 55–58) (Figure 2). The pathways initiated by G_i-coupled receptors involve sequential activation of $G\beta\gamma_i$, PI 3-kinase, and integrin-linked kinase (ILK); the last of these phosphorylates MLC_{20} directly and indirectly inhibits MLC phosphatase by phosphorylating CPI-17 (PK_C_-_p_otentiated _i_nhibitor _17_-kDa protein) (59).

G PROTEIN–MEDIATED ACTIVATION OF RHOA AND RHO KINASE   Agonists that activate G_q also activate G_{13}; both G proteins are capable of activating RhoA, but their involvement in smooth muscle of the gut is receptor-specific. Muscarinic m3 and CCK-A receptors activate RhoA via $G\alpha_{13}$ only, whereas $S1P_2$ and motilin receptors activate RhoA via both $G\alpha_q$ and $G\alpha_{13}$ (13, 17, 20, 29) (Table 1). RhoA, in contrast to other RhoA family G proteins such as Cdc42 and Rac1, is inactivated via ADP ribosylation by the _C. botulinum_ C3 exoenzyme (60–61). The inactive form of RhoA (RhoA.GDP) is present in the cytosol bound to a guanine dissociation inhibitor (GDI). Activation of RhoA by $G\alpha_q$ and/or $G\alpha_{13}$ is mediated by various Rho-specific guanine nucleotide exchange factors (RhoGEFs), which promote the exchange of GDP for GTP. The RhoGEF family of proteins includes p115RhoGEF, PDZ-RhoGEF, and LARG (_l_eukemia-_a_ssociated _R_ho_G_EF). These proteins share common structural motifs that include RGS (_r_egulator of _G_ protein _s_ignaling), Dbl homology (DH), and pleckstrin homology (PH) domains (60–61). The DH domain is responsible for the exchange of GDP for GTP. In the GTP-bound conformation, RhoA interacts with and stimulates the activity of downstream effectors including Rho kinase and PLD (29, 62).

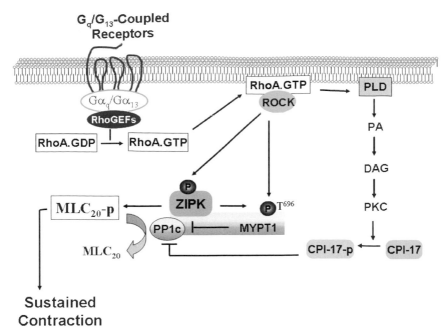

Figure 2   Signaling by G_q- and/or G_{13}-coupled receptors during the sustained phase of contraction. Sustained MLC_{20} phosphorylation and muscle contraction are initiated by $G_{q/13}$-dependent activation of RhoA via RhoGEFs. Translocation of activated RhoA to the membrane initiates two pathways involving Rho kinase (ROCK) and PKC, respectively. PKC is activated by PLD via dephosphorylation of the primary product of PLD, phosphatidic acid (PA), to DAG. Both pathways result in inhibition of MLC phosphatase. ROCK phosphorylates the regulatory MYPT1 at Thr^{696}, causing its dissociation from and inhibition of the catalytic subunit of MLC phosphatase ($PPIc\delta$). PKC phosphorylates an endogenous inhibitor of MLC phosphatase, CPI-17, at Thr^{38}, greatly enhancing its ability to inhibit MLC phosphatase activity. MLC_{20} is phosphorylated by ZIP kinase (ZIPK), a Ca^{2+}-independent MLCK associated with MYPT1. Sustained MLC_{20} phosphorylation reflects the activity of ZIPK and concurrent inhibition of MLC phosphatase. Some receptors (e.g., m3) are coupled to RhoA via G_{13} only, whereas others (e.g., SIP_2, motilin) are coupled to RhoA via both G_q and G_{13} (Table 1). G_i-coupled receptors initiate a distinct pathway involving sequential activation of $G\beta\gamma_i$, PI 3-kinase, and ILK (*not shown*); the latter acts as a Ca^{2+}-independent MLCK and inhibits MLC phosphatase by phosphorylating CPI-17 at Thr^{38}.

The Rho kinases, ROCK I and ROCK II, the latter of which is predominantly expressed in smooth muscle, belong to a family of serine/threonine protein kinases that includes _m_yotonic _d_ystrophy _p_rotein kinase (DMP kinase), Cdc42-binding kinase, and citron kinase (61). All of these kinases possess an NH_2-terminal kinase domain, a coiled-coil region, and other functional motifs at the COOH terminus. In the inactive form, the PH and Rho-binding domains of Rho kinase bind to the catalytic NH_2-terminal region, forming an autoinhibitory loop. Activated RhoA (RhoA.GTP) binds to the Rho-binding domain, causing the enzyme to unfold and freeing its catalytic activity (61).

INHIBITION OF MLC PHOSPHATASE VIA RHO KINASE–MEDIATED PHOSPHORYLATION OF MYPT1   Rho kinase phosphorylates the regulatory subunit of MLC phosphatase and inhibits catalytic activity (50, 51). The holoenzyme is a heterotrimer with (*a*) a 110- to 130-kDa regulatory subunit [myosin phosphatase target subunit 1 (MYPT1)] that is widely expressed and is present in high concentrations in smooth muscle, (*b*) a 37-kDa catalytic subunit of type 1 phosphatase ($PP1c\delta$), and (*c*) a 20-kDa subunit of unknown function (51). MYPT1 contains a PP1c-binding motif (K/R-I/V-X-F/W) adjacent to seven NH_2-terminal ankyrin repeats. Interaction of $PP1c\delta$ with MYPT1 alters substrate specificity and enhances catalytic activity. Phosphorylation of Thr^{696} in MYPT1 by Rho kinase or by ZIP kinase (which is bound to MYPT1 and is also known as MYPT1-associated kinase) fosters dissociation and inhibits activity of the catalytic subunit (Figure 2) (50, 51, 63–65). Phosphorylation of an adjacent Ser^{695} by cAMP- or cGMP-dependent protein kinase (PKA and PKG, respectively) blocks the ability of Rho kinase to phosphorylate Thr^{696} and is one of several mechanisms by which PKA and PKG restore MLC phosphatase activity (66). Phosphorylation of Thr^{853} by Rho kinase has also been reported, but its functional significance remains uncertain.

INHIBITION OF MLC PHOSPHATASE VIA PKC-MEDIATED PHOSPHORYLATION OF CPI-17   PLD, like Rho kinase, is activated downstream of RhoA. As noted above, dephosphorylation of phosphatidic acid, the primary product of phosphatidylcholine hydrolysis by PLD, yields DAG, leading to sustained activation of PKC (Figure 2). The involvement of PKC in sustained, but not initial, contraction and MLC_{20} phosphorylation has been amply demonstrated and reflects inhibition of MLC phosphatase via a Rho-dependent, PKC-mediated pathway (i.e., CPI-17) distinct from that involving Rho kinase and MYPT1 (49, 52–54). Sustained contraction and MLC_{20} phosphorylation are blocked by PKC inhibitors (e.g., bisindolylmaleimide or calphostin C) (20, 49). The involvement of PKC isozymes appears to be receptor-specific, with PKC-δ and PKC-ε mediating the effects of G protein–coupled receptor agonists and PKC-α mediating the effects of growth factors and phorbol esters (49, 67).

The target of PKC is CPI-17, an endogenous phosphorylation-dependent inhibitor of the catalytic $PP1c\delta$ subunit of MLC phosphatase. CPI-17 is preferentially expressed in vascular and visceral smooth muscle, whereas its orthologue, phosphatase holoenzyme inhibitor (PHI)-1, is more widely expressed (52–54, 68). CPI-17 and PHI-1 are the only endogenous PP1c inhibitors regulated by PKC. The mechanism of inhibition reflects a large increase ($\sim$1000-fold) in the binding affinities of CPI-17 and PHI-1 to $PP1c\delta$ following PKC-mediated phosphorylation of CPI-17 at Thr^{38} or of PHI-1 at Thr^{57} (52–54, 68). Several kinases, including Rho kinase, protein kinase N, PAK1, and ILK, phosphorylate CPI-17 at Thr^{38}

in vitro (64, 69, 70). Of these, only Rho kinase has been shown to participate in CPI-17 phosphorylation in vivo (20, 67, 71). For example, CPI-17 phosphorylation at Thr^{38} in response to acetylcholine or histamine is abolished by PKC inhibitors and partially inhibited by Rho kinase inhibitors (20, 67). Phosphorylation of CPI-17 in response to contractile agonists is correlated with sustained MLC_{20} phosphorylation and muscle contraction (52–54). Relaxant agonists inhibit CPI-17 phosphorylation at Thr^{38}; the mechanism has not been determined but may reflect suppression of sustained PKC activity via inhibitory phosphorylation of RhoA at Ser^{188} by PKA or PKG (72–74). As noted below, CPI-17 can be dephosphorylated via PP2A and PP2C but is a poor substrate for dephosphorylation by PP1 (75).

RECEPTOR-SPECIFIC INVOLVEMENT OF CPI-17 AND MYPT1 IN INHIBITION OF MLC PHOSPHATASE   Although both pathways (Rho kinase/MYPT1 and PKC/CPI-17) are downstream of RhoA, their relative involvements in inhibition of MLC phosphatase appear to be receptor-specific. Most $G_{q/13}$-coupled receptors (e.g., m3, $S1P_2$, motilin) engage both pathways, whereas ET_A receptors engage only Rho kinase/MYPT1 and LPA_3 receptors engage only PKC/CPI-17 (12, 13, 17, 18). The initial response to endothelin is mediated by both ET_A and ET_B receptors, whereas the sustained response is mediated only by ET_A receptors and reflects Rho kinase–mediated phosphorylation of MYPT1 at Thr^{696} (12). Although PKC is activated by endothelin, CPI-17 is not phosphorylated and thus does not contribute to inhibition of MLC phosphatase. The absence of CPI-17 phosphorylation by PKC reflects active dephosphorylation of CPI-17 by PP2A via a pathway involving ET_B-dependent sequential activation of p38 MAP kinase and PP2A (12, 76). ET_A-induced PKC-dependent phosphorylation of CPI-17 can be unmasked by blocking either p38 MAP kinase or PP2A activity.

MLC_{20} Phosphorylation by Ca^{2+}-Independent MLCKs

Agonist-mediated sustained MLC_{20} phosphorylation and contraction are not affected by inhibitors of Ca^{2+}/calmodulin-dependent MLCK, implying that MLC_{20} phosphorylation during this phase may be mediated by a Ca^{2+}-independent MLCK (20, 47, 48). The complete suppression of MLC phosphatase activity during this phase by itself cannot account for MLC_{20} phosphorylation. As noted above, contraction of arterial smooth muscle induced by the phosphatase inhibitor microcystin-LR in Ca^{2+}-free medium is abolished by the nonspecific kinase inhibitor staurosporine, implying the participation of a Ca^{2+}-independent kinase (57). Diphosphorylation of MLC_{20} at both Ser^{19} and Thr^{18} by a Ca^{2+}-independent MLCK is a characteristic feature of MLC_{20} phosphorylation during agonist-mediated sustained contraction and under conditions designed to elicit Ca^{2+} sensitization (41, 47, 48, 77–79).

ZIPPER-INTERACTING PROTEIN KINASE   The search for a Ca^{2+}-independent kinase led to the discovery of ZIP kinase. Addition of the purified form of this enzyme to permeabilized arterial smooth muscle in the absence of Ca^{2+} or phosphatase inhibitors causes contraction and MLC_{20} phosphorylation at both Ser^{19} and Thr^{18} (47, 55, 57). ZIP kinase, a serine/threonine kinase expressed in various tissues including smooth muscle, is a member of the death-associated protein (DAP) kinase family, which includes DAP kinase and DAP kinase–related protein 1 (80). ZIP kinase cloned from rat aorta is homologous (49%) to smooth muscle Ca^{2+}-dependent MLCK, but unlike MLCK, it has neither an autoinhibitory nor a calmodulin-binding region (55). In intestinal smooth muscle, purified ZIP kinase phosphorylates MYPT1 at Thr^{696}, and its potency in vitro ($K_m = 3~\mu M$) is greater than its potency in phosphorylating MLC_{20} ($K_m = 53~\mu M$), suggesting that MYPT1 may be the preferred substrate (65). ZIP kinase is also more effective than Rho kinase in phosphorylating full-length MYPT1 and is insensitive to inhibition by the Rho kinase inhibitor Y27632 (57). Recent evidence suggests that ZIP kinase can also phosphorylate CPI-17 at Thr^{38} in vitro (64).

The pattern of ZIP kinase activities suggests that in addition to its role in the ultimate phosphorylation of MLC_{20}, ZIP kinase may mediate Rho kinase–dependent phosphorylation of MYPT1 and inhibition of MLC phosphatase. That Rho kinase does not phosphorylate ZIP kinase in vitro suggests that phosphorylation of ZIP kinase by Rho kinase may be indirect via one or more kinases. ZIP kinase is activated by contractile agonists and is subject to both autophosphorylation and phosphorylation by DAP kinase at six COOH-terminal sites (57, 81). Activated plasma membrane–bound Rho kinase may be linked to MYPT1 via myofilament-bound ZIP kinase. The functional linkage between Rho kinase, ZIP kinase, and MYPT1 is supported by recent studies in HEK 293 cells in which expression of constitutively active RhoA increases the association of ZIP kinase and MYPT1 (55).

INTEGRIN-LINKED KINASE   A distinct Ca^{2+}-independent MLCK mediates sustained MLC_{20} phosphorylation and contraction mediated by G_i-coupled receptors (59). These receptors are not coupled to G_{13} or RhoA and do not activate Rho kinase or PKC in smooth muscle of the gut (30). Instead, they activate PI 3-kinase via $G\beta\gamma_i$ and various pathways downstream of PI 3-kinase. As noted above, one pathway involves sequential activation of Cdc42/Rac1, PAK1, and p38 MAP kinase; another involves activation of c-Src; and a third involves PIP_3-dependent activation of ILK (32, 59). ILK, a serine/threonine kinase, was first discovered as a protein bound to the cytoplasmic domain of integrin β subunits and involved in cytoskeletal regulation via interaction with actin-binding proteins. Walsh and co-workers (48, 78) recently isolated a cellular fraction of ILK bound to myofilaments and identified ILK as a Ca^{2+}-independent MLCK.

In addition to phosphorylating MLC_{20}, ILK phosphorylates CPI-17 at Thr^{38} and PHI-1 at Thr^{57} (68). CPI-17 and PHI-1 thio-phosphorylated by ILK inhibit MLC phosphatase activity in vitro and induce contraction at subthreshold concentrations of Ca^{2+} upon being added to permeabilized rat arterial smooth muscle. Recent studies show that ILK is associated with the MLC phosphatase holoenzyme and phosphorylates MYPT1 at three sites, including functionally active Thr[695], in the chicken gizzard (82). Phosphorylation of MYPT1 by ILK in vitro produces only moderate inhibition of MLC phosphatase in comparison with phosphorylation by Rho kinase; the functional significance of ILK in this context is uncertain. Thus, ILK, like ZIP kinase, can induce smooth muscle contraction by direct phosphorylation of MLC_{20} combined with CPI-17-dependent inhibition of MLC phosphatase.

Recent studies on intestinal smooth muscle cells have shown that three G_i-coupled receptors (G_{i1}-coupled sstr3, G_{i2}-coupled δ-opioid, and G_{i3}-coupled adenosine A_1 receptors) initiate contraction by activating PLC-β3 and induce sustained contraction and MLC_{20} phosphorylation via preferential activation of a PI 3-kinase/ILK pathway (59). Both MLC_{20} and CPI-17 (but not MYPT1) are phosphorylated, and both phosphorylations are inhibited by the PI 3-kinase inhibitor, LY-294002, and by expression of dominant-negative ILK(R211A) or siRNA for ILK, providing a clear demonstration of the role of ILK as a Ca^{2+}-independent MLCK and inhibitor of MLC phosphatase in vivo.

In contrast to sstr3, δ-opioid, and A_1 receptors, G_{i3}-coupled m2 receptors do not induce sustained contraction and preferentially activate Cdc42/Rac1, PAK1, and p38 MAP kinase (32). The last inhibits ILK activity and prevents ILK-induced sustained contraction and MLC_{20} phosphorylation; inhibition of p38 MAP kinase restores contraction and MLC_{20} phosphorylation (K.S. Murthy, unpublished observations). Thus, activation of PAK1 and p38 MAP kinase by m2 receptors prevents activation of MLCK and ILK during the initial and sustained phases of contraction, respectively.

Distinctive Features of Sustained MLC_{20} Phosphorylation

The temporal shift from Ca^{2+}-dependent MLCK to Ca^{2+}-independent MLCK (ZIP kinase, ILK) is accompanied by a spatial shift of MLC_{20} phosphorylation from a central to a peripheral location in smooth muscle cells (83). Centrally located MLC_{20} phosphorylation monitored immunohistochemically with a Ser[19] antibody is susceptible to inhibition by ML-9. By contrast, the more sustained, peripherally located MLC_{20} phosphorylation is inhibited by Y27632 and upon expression of RhoA (T-19N) (83). As noted above, MLC_{20} is predominantly monophosphorylated at Ser[19] by MLCK but diphosphorylated with equal affinity at Ser[19] and Thr[18] by Ca^{2+}-independent MLCKs. In comparison with Ca^{2+}-dependent MLCK, both ZIP kinase and ILK are only moderately active, and their effectiveness in maintaining MLC_{20} phosphorylation appears to depend largely on RhoA-dependent mechanisms for concomitant inhibition of MLC phosphatase.

Distinctive Signaling in Esophageal Smooth Muscle

Although the focus of this review is on gastric and intestinal smooth muscle, signaling for contraction in esophageal smooth muscle deserves mention because of its distinctive aspects (2, 84–87). Contraction and MLC_{20} phosphorylation in muscle cells of the lower esophageal sphincter by acetylcholine are mediated by m3 receptors and involve the same mechanisms described above for gastric and intestinal circular smooth muscle cells. By contrast, contraction by acetylcholine of smooth muscle cells derived from the normally relaxed esophageal body is mediated by G_i-coupled m2 receptors, is dependent upon Ca^{2+} influx and activation of PKC-ε, and is not affected by calmodulin antagonists or MLCK inhibitors. Recent studies suggest that two MAP kinase pathways downstream of PKC-ε, involving ZIP kinase and ILK, mediate contraction in these muscle cells (87).

DESENSITIZATION OF RECEPTOR FUNCTION IN SMOOTH MUSCLE

G protein–coupled receptor signaling in smooth muscle of the gut is rapidly attenuated or terminated by mechanisms that target receptors, G proteins, or effector enzymes. The mechanisms include (*a*) receptor phosphorylation by G protein–coupled receptor kinases (GRKs) or second messenger–activated kinases (e.g., PKA or PKC), (*b*) inactivation of Gα.GTP via RGS proteins, and (*c*) phosphorylation of specific G proteins (i.e., Gα_{i1} and Gα_{i2}) and transient sequestration of activated Gα proteins in caveolae.

GRK-Mediated Receptor Desensitization

Homologous desensitization of agonist-occupied receptors is initiated by GRK-mediated phosphorylation of the receptors and binding of β-arrestin, which uncouples the receptors from the G protein as a prelude to receptor internalization (88–90). β-arrestin targets the phosphorylated receptor to clathrin-coated pits and acts as a scaffold for other proteins, including the terminal components of JNK/ERK/MAP kinase cascade and the cytosolic tyrosine kinase c-Src. The latter stimulates tyrosine phosphorylation of the large GTPase dynamin and promotes its ability to cleave the clathrin-coated vesicles from the cell surface. The vesicles fuse with endosomes, within which the receptors are dephosphorylated and either slowly or rapidly recycled to the surface (88). In smooth muscle of the gut, δ-opioid and m2 receptors are phosphorylated by G$\beta\gamma$-dependent GRK2, and cannabinoid CB_1 receptors by G$\beta\gamma$-independent GRK5 (91–92).

G Protein Deactivation by RGS Proteins

RGS proteins contain a conserved 120-amino acid RGS domain that binds exclusively to activated Gα subunits (Gα.GTP) and terminates G protein signaling by accelerating intrinsic Gα.GTPase activity ($\sim$1000-fold), thereby fostering reassociation of the G$\alpha\beta\gamma$ trimer (93). Several RGS categories have been identified on the basis of additional structures outside the RGS domain. Seven out of 25 known RGS proteins (RGS3, -4, -6, -8, -10, -12, and -16) have now been identified in smooth muscle cells of the gut (J. Huang & K.S. Murthy, unpublished observations). $G\alpha_q$ activated by m3 or motilin receptors is regulated by RGS4, whereas G_{i2} activated by CB_1 receptors is regulated by RGS6 (94, 95). RGS6, the only member of the R7 category of RGS proteins expressed in smooth muscle of the gut, contains a $G\gamma$-like domain that binds to $G\beta5$ and $G\alpha_{i2}$ to constitute an atypical G protein. Unlike classical $G\beta\gamma$, the $G\beta5$.RGS6 complex is incapable of activating $G\beta\gamma_i$-dependent downstream effectors, such as PLC-$\beta3$ (21, 22, 95).

G Protein–Dependent Desensitization: $G\alpha_{i1/2}$ Phosphorylation and Caveolar Sequestration

Phosphorylation of $G\alpha_{i1}$ and $G\alpha_{i2}$ (but not $G\alpha_{i3}$) by PKC limits the ability of these proteins to transduce signals. Phosphorylation appears to decrease the affinity of $G\alpha$ for $G\beta\gamma$, impeding reassociation of the subunits. In smooth muscle of the gut, activation of PKC with contractile agonists such as CCK induces phosphorylation of $G\alpha_{i1}$ and $G\alpha_{i2}$ and blocks inhibition of adenylyl cyclase V/VI (96). Blockade of PKC activity abolishes phosphorylation of $G\alpha_{i1/2}$ and restores their ability to inhibit adenylyl cyclase (96). Thus, phosphorylation of $G\alpha_{i1}$ and $G\alpha_{i2}$ by PKC derived from activation of $G\alpha_q$-coupled receptors contributes to heterologous desensitization.

$G\alpha$ proteins contain caveolin-binding motifs consisting of 10–15-amino acid sequences with characteristically spaced aromatic residues (97). These sequences bind to caveolin-3 expressed in smooth muscle of the gut (98). Receptor activation results in the transient binding of activated $G\alpha$ subunits by caveolin-3. This transient sequestration of $G\alpha$.GTP in caveolae hinders reassociation of $G\alpha$ and $G\beta\gamma$ and impedes the response mediated by a different receptor coupled to the same G protein. G protein sequestration peaks in 5–10 min and disappears within 30 min (98). The binding of $G\alpha$ subunits to caveolin-3 can be selectively blocked by caveolin-binding peptide fragments.

SIGNALING FOR RELAXATION IN SMOOTH MUSCLE OF THE GUT

Smooth muscle contraction is terminated spontaneously by mechanisms that target receptors or G proteins, or more directly by exposure to relaxant agonists. The mechanisms initiated by these agonists eventually converge to dephosphorylate MLC_{20} and thus may target one or more steps in the signaling pathways that mediate initial or sustained MLC_{20} phosphorylation. These targets are outlined below in more detail.

Relaxant Neurotransmitters and their Messengers in the Gut

Relaxation of smooth muscle in the gut reflects a unique interplay between neuronal NO synthase (nNOS) and the peptide neurotransmitters vasoactive intestinal peptide (VIP) and its homologue, pituitary adenylate cyclase–activating peptide (PACAP), which possesses identical properties in smooth muscle (only VIP is referred to below). NO formed in nerve terminals by nNOS regulates VIP release; in turn, VIP stimulates smooth muscle eNOS (nitric oxide synthase III) to generate NO. NO derived from both nerves and smooth muscle cells stimulates soluble guanylyl cyclase activity and cGMP formation (99–104). In addition, VIP interacts with $VPAC_2$ receptors to stimulate adenylyl cyclase activity and cAMP formation (103). Thus, concurrent generation of cAMP and cGMP and activation of PKA and PKG are the physiological norm. This fact has considerable bearing on the regulation of cyclic nucleotide levels, on cyclase and phosphodiesterase (PDE) activities, and on the activities of both PKA and PKG. PKA and/or PKG act upstream to induce feedback desensitization of the receptors, inhibition of adenylyl and guanylyl cyclase activities, and stimulation of cAMP-specific PDE3 and PDE4 and cGMP-specific PDE5 activities (105–107) (Figure 3). The kinases also act downstream to induce relaxation by inhibiting Ca^{2+} mobilization, MLC_{20} phosphorylation, and various targets that lead to inhibition of MLC phosphatase and sustained phosphorylation of MLC_{20} (Figure 4).

Receptors and G Proteins Mediating Relaxation

Three types of G protein–coupled receptors for VIP and PACAP have been identified: $VPAC_1$ and $VPAC_2$ receptors, which possess equally high affinity for VIP and PACAP, and PAC_1 receptors, which possess high affinity for PACAP and very low affinity for VIP (108). A recently cloned VIP-specific receptor has been identified in smooth muscle cells of guinea pig tenia coli; it differs from $VPAC_2$ receptors in the guinea pig stomach by only two amino acid residues (Phe^{40}/Phe^{41} in lieu of Leu^{40}/Leu^{41}) in the ligand-binding NH_2-terminal domain. Mutation of these two residues to Leu^{40}/Leu^{41} so as to mimic the gastric $VPAC_2$ receptor restores equal affinity for VIP and PACAP (109, 110).

Smooth muscle cells of the gut express predominantly $VPAC_2$ receptors (103). The receptors are coupled via G_s to activation of adenylyl cyclase V/VI. In addition, VIP interacts with a single-transmembrane natriuretic peptide receptor C (NPR-C) coupled via $G\alpha_{i1}/G\alpha_{i2}$, sequentially resulting in stimulation of Ca^{2+} influx, activation of Ca^{2+}/calmodulin-dependent eNOS, NO formation, and NO-mediated activation of soluble guanylyl cyclase and PKG (111). Radioligand binding studies using [^{125}I]ANP and [^{125}I]VIP confirmed that VIP interacts with both $VPAC_2$ and NPR-C (Figure 3). Heterologous expression of NPR-C in COS-1 cells provided decisive evidence that VIP interacts with NPR-C (112).

Unlike NPR-A and NPR-B, the intracellular domain of NPR-C is devoid of guanylyl cyclase or kinase activity and consists of 37 amino acids. A 17-amino acid sequence in the middle region of this domain is capable of selectively activating G_{i1} and G_{i2} (113). The G protein–activating region was identified using synthetic peptide sequences corresponding to various regions of the intracellular domain, and the results were confirmed by site-directed mutagenesis of the cloned receptor

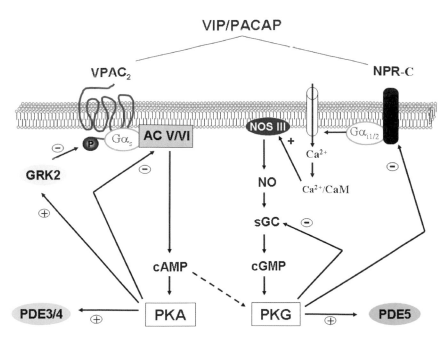

Figure 3   Dual signaling pathways initiated by the relaxant neuropeptides VIP and PACAP and feedback inhibition by PKA and PKG. VIP and its homologue PACAP activate cognate seven-transmembrane VPAC$_2$ receptors coupled via G$_s$ to adenylyl cyclase V/VI, leading to cAMP generation and PKA activation. They also activate a single-transmembrane natriuretic peptide receptor C (NPR-C) coupled via Gα_{i1} and Gα_{i2} to Ca^{2+} influx and Ca^{2+}/calmodulin activation of NOS III expressed in smooth muscle cells. This leads to NO-dependent activation of soluble guanylyl cyclase (sGC) and cGMP-dependent activation of PKG. Activation of G$\alpha_{i1/2}$ is mediated by a 17-amino acid sequence in the middle region of NPR-C. In the presence of cGMP, PKG also becomes highly sensitive to cAMP stimulation. Both PKA and PKG act separately upstream to desensitize VPAC$_2$ and NPR-C receptors, inhibiting adenylyl cyclase and soluble guanylyl cyclase, and activating cAMP-specific PDE3A and PDE4 and cGMP-specific PDE5. Stimulation, (+); inhibition, (−).

(114, 115). The active sequence possessed two NH$_2$-terminal Arg-Arg residues and the COOH-terminal motif His-Arg-Glu-Leu-Arg. Similar sequences have been identified in other single- and seven-transmembrane receptors (e.g., IGF-II and adrenoceptors α_{1b} and α_2) (116, 117).

VPAC$_2$ receptors in smooth muscle are internalized and desensitized via GRK2. PKA accelerates this process via feedback phosphorylation of GRK2 (118). A single Thr^{466} residue in the intracellular domain of NPR-C, located outside the G protein–activating sequence, is selectively phosphorylated by PKG, resulting in feedback desensitization of NPR-C (119).

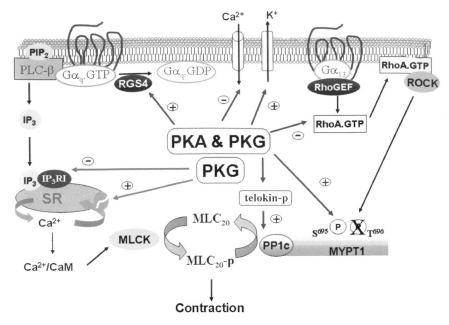

Contraction

Figure 4   Molecular targets of PKA and PKG during relaxation of smooth muscle. PKG or both PKG and PKA induce relaxation by targeting components of the contractile signaling pathways and inducing dephosphorylation of MLC_{20}. Both kinases phosphorylate RGS4 and stimulate deactivation of $G\alpha_q$, leading to inhibition of PLC-β1 activity and of IP_3 formation. In addition, PKG, but not PKA, phosphorylates IP_3R-I and the sarcoplasmic Ca^{2+}/ATPase pump, thereby inhibiting Ca^{2+} release and stimulating Ca^{2+} reuptake. Both kinases inhibit Ca^{2+} mobilization in longitudinal muscle by inhibiting $cPLA_2$ activity and ryanodine receptors (*not shown*). All of these mechanisms act to reduce $[Ca^{2+}]_i$ and thus inhibit MLCK activity and initial MLC_{20} phosphorylation and contraction. Both PKA and PKG inactivate RhoA by phosphorylating activated RhoA at Ser^{188} and translocating it to the cytosol. Further downstream, both kinases can phosphorylate Ser^{695} of MYPT1, thereby precluding phosphorylation of adjacent Thr^{696} by Rho kinase (ROCK). A PKG-specific mechanism reflecting interaction of leucine-zipper domains of PKG-Iα and MYPT1 has also been shown to stimulate dephosphorylation of MLC_{20}. A RhoA-independent mechanism involves stimulatory phosphorylation of telokin, an endogenous activator of MLC phosphatase, by both PKA and PKG. Stimulation, (+); inhibition, (−); red arrows, targets of both PKA and PKG; green arrows, targets of PKG only.

Regulation of cAMP and cGMP Levels

cAMP and cGMP depend on the rates of their synthesis by cyclases and degradation by PDEs (107, 120, 121). The activity of these enzymes is regulated via feedback by PKA and PKG. In smooth muscle of the gut, cAMP levels are regulated by PKA via stimulatory phosphorylation of PDE3A and PDE4D5, the predominant

PDE4 isozymes expressed in smooth muscle, and via inhibitory phosphorylation of adenylyl cyclase V/VI (105, 106). cGMP levels are regulated via stimulatory phosphorylation of PDE5 by PKG and PKA and via inhibitory phosphorylation of soluble guanylyl cyclase by PKG only (105). cAMP degradation by PDE3A is attenuated in the presence of cGMP, whereas cGMP degradation is augmented in the presence of cAMP as a result of PKA-mediated activation of PDE5.

Activation of Protein Kinases by cAMP and cGMP

cAMP is produced in $\sim$10-fold greater amounts than cGMP (122). cGMP selectively activates PKG, whereas cAMP selectively activates PKA at low concentrations but can cross-activate PKG at higher concentrations (122–125). Autophosphorylation of PKG in the presence of cGMP greatly augments the affinity of cAMP for PKG (126). Because both cyclic nucleotides are generated concurrently, the greater abundance of cAMP and its higher affinity for PKG in the presence of cGMP render cAMP an important activator of both PKA and PKG.

Molecular Targets of PKA and PKG

As noted above, PKA and PKG act on upstream targets to regulate cAMP and cGMP levels by desensitizing receptors, inhibiting cyclase activities, and stimulating PDE activities. In addition, both protein kinases act on several downstream targets in the cascades that mediate initial and sustained contraction, resulting in dephosphorylation of MLC_{20} and relaxation of smooth muscle (Figure 4). Thus, PKG and/or PKA inhibit Ca^{2+} mobilization by (a) inhibiting IP_3 formation in circular smooth muscle and arachidonic acid formation in longitudinal smooth muscle (127–129), (b) inhibiting Ca^{2+} release via PKG-Iα-specific phosphorylation of IP_3 receptor/Ca^{2+} channels in circular muscle and ryanodine receptor/Ca^{2+} channels in longitudinal muscle (36, 130), (c) stimulating Ca^{2+} uptake via PKG-specific phosphorylation of sarco-endoplasmic reticulum Ca^{2+}/ATPase (SERCA), and stimulating Ca^{2+} efflux via phosphorylation of plasmalemmal Ca^{2+}/ATPase (125, 131), and (d) inhibiting Ca^{2+} channel activity and stimulating K^+ channel activity, thereby reducing Ca^{2+} influx (132). These effects on plasma membrane Ca^{2+} and K^+ channels lead to membrane hyperpolarization and are relevant only to spontaneously depolarized smooth muscle (e.g., sphincter muscle) or to smooth muscle depolarized by electrical slow waves. All the mechanisms listed above influence Ca^{2+} mobilization and thus operate only during the initial phase of contraction. In addition, PKA and PKG influence sustained muscle contraction by acting directly on two major targets: RhoA and the regulatory subunit of MLC phosphatase (MYPT1) (66, 73, 74, 133, 134). A direct effect on CPI-17 has not been demonstrated.

Relaxation of Initial Ca^{2+}-Dependent Contraction

INHIBITION OF IP_3 FORMATION AND IP_3-DEPENDENT CA^{2+} RELEASE IN CIRCULAR SMOOTH MUSCLE   Both PKA and PKG inhibit Gα_q-dependent PLC-β1 activity and IP$_3$ formation. However, PKA and PKG do not directly phosphorylate Gα_q or PLC-β1. Rather, PLC-β1 activity indirectly is decreased via stimulatory phosphorylation of RGS4 by PKA and PKG (and of the RGS domain of GRK2 by PKA only) (129). This phosphorylation of RGS4 and GRK2 increases their binding to and rapidly inactivates Gα_q, leading to inhibition of both PLC-β1 activity and IP$_3$ formation.

In addition, PKG-Iα, one of two PKG isoforms expressed in smooth muscle of the gut—but not PKA—phosphorylates IP$_3$R-I, the only IP$_3$ receptor type mediating Ca^{2+} release in smooth muscle of the gut, resulting in inhibition of Ca^{2+} release (36). In circular smooth muscle, agonists that activate only PKA inhibit Ca^{2+} release by inhibiting IP$_3$ formation, whereas agonists that activate only PKG inhibit Ca^{2+} release by both inhibiting IP$_3$ formation and inactivating IP$_3$ receptors/Ca^{2+} channels. When both PKA and PKG are activated concurrently (see above), Ca^{2+} release is inhibited via both mechanisms.

INHIBITION OF ARACHIDONIC ACID FORMATION AND CA^{2+}/CADP RIBOSE–INDUCED CA^{2+} RELEASE IN LONGITUDINAL MUSCLE   In longitudinal smooth muscle, both PKA and PKG phosphorylate cPLA$_2$ and inhibit arachidonic acid formation, leading to a decrease in Ca^{2+} influx (128). PKG selectively inhibits arachidonic acid–mediated Ca^{2+} influx. This inhibition of Ca^{2+} influx leads to a decrease in both cyclic ADP ribose formation and cyclic ADP ribose– and Ca^{2+}-induced Ca^{2+} release via ryanodine receptors/Ca^{2+} channels.

Thus, during the initial phase of contraction, relaxation via PKA or PKG results from a decrease in [Ca^{2+}]$_i$, leading to inhibition of Ca^{2+}/calmodulin-dependent MLCK activity and MLC$_{20}$ phosphorylation. There is no evidence that either PKA or PKG inhibit MLCK activity directly in vivo.

Relaxation of Sustained Ca^{2+}-Independent Contraction

PKA and PKG can induce relaxation of sustained contraction by acting on one or more targets in the G$_{13}$/RhoA pathway, including G$_{13}$, p115 RhoGEF, and RhoA, as well as targets downstream of RhoA such as Rho kinase, PKC, MYPT1, and CPI-17. PKA inhibits RhoA activity in platelets by phosphorylating Thr^{203} in Gα_{13}, but this effect has not been determined in smooth muscle (135, 136). Both PKA and PKG can phosphorylate activated membrane-bound RhoA at Ser^{188} (73, 74). Phosphorylation of RhoA stimulates its translocation back to the cytosol and inhibits the activity of Rho-dependent, membrane-bound targets, such as Rho kinase and PLD (73). Inhibition of Rho kinase blocks phosphorylation of MYPT1 at Thr^{696}, whereas inhibition of PLD leads to decrease in DAG formation and in PKC activity and PKC-mediated CPI-17 phosphorylation at Thr^{38}. Thus, RhoA phosphorylation by PKA and PKG blocks the two Rho-dependent pathways that mediate inhibition of MLC phosphatase. The resultant increase in MLC phosphatase activity leads to dephosphorylation of MLC$_{20}$ and relaxation of sustained muscle contraction (Figure 4). Although several potential mechanisms may mediate relaxation of sustained muscle contraction (as described below), the location of RhoA at the inception of the signaling pathway suggests that inactivation of RhoA via PKA and/or PKG-mediated phosphorylation is the predominant mechanism for relaxation.

MYPT1 has been identified in recent studies as a target of PKA and/or PKG. Both kinases directly phosphorylate MYPT1 in vitro at a site (Ser^{695}) adjacent to the active site (Thr^{696}) (66). Phosphorylation of Ser^{695} by PKA or PKG prevents MYPT1 phosphorylation at Thr^{696} via Rho kinase, thereby blocking Rho kinase–dependent inhibition of MLC phosphatase activity (Figure 4). This mechanism has yet to be demonstrated in vivo. A distinct PKG-specific mechanism involves binding of the NH_2-terminal leucine-zipper domain of PKG-Iα to the COOH-terminal leucine-zipper domain of MYPT1. Blockade of this interaction in vascular smooth muscle prevents MLC_{20} dephosphorylation, implying that PKG can activate MLC phosphatase by binding to MYPT1 (133, 134). Relaxation of vascular smooth muscle via this mechanism appears to depend upon expression of MYPT1 isoforms with leucine-zipper motifs (137). Expression of these isoforms, which is tissue-specific, has not been determined in smooth muscle of the gut.

A distinct mechanism for muscle relaxation that targets MLC phosphatase but is not downstream of RhoA signaling involves telokin, an endogenous 17-kDa activator of MLC phosphatase (138–140) (Figure 4). The sequence of telokin is identical to the COOH-terminal 155-amino acid residues of MLCK but is expressed independently of it. Its expression is low in tonic smooth muscle; it is about five times higher in phasic smooth muscle (e.g., intestinal longitudinal muscle). The ability of telokin to stimulate MLC phosphatase is moderately enhanced upon phosphorylation of Ser^{13} by PKA or PKG (139). Telokin may be viewed as a counterpart to CPI-17, whose ability to inhibit MLC phosphatase activity is low but is greatly enhanced upon phosphorylation by PKC. Re-addition of telokin to permeabilized ileal longitudinal muscle strips inhibits Ca^{2+}-induced contraction and MLC_{20} phosphorylation; this effect is enhanced by 8-Br-cGMP (138). Addition of a phosphorylation site–deficient mutant of telokin has equivalent potency as a relaxant agent, but its effect is not enhanced by 8-Br-cGMP (140). The functional contribution of telokin to relaxation of sustained Ca^{2+}-independent contraction and MLC_{20} dephosphorylation has not been determined.

In summary, mechanisms of relaxation involving PKA and PKG converge on dephosphorylation of MLC_{20}. Relaxation of the initial Ca^{2+}-dependent contraction reflects inhibition of the Ca^{2+}-mobilizing messengers, IP_3 in circular muscle and arachidonic acid in longitudinal muscle, by PKA or PKG, accentuated by PKG-specific inactivation of sarcoplasmic IP_3 or ryanodine Ca^{2+} channels. Relaxation of sustained Ca^{2+}-independent contraction reflects inactivation of RhoA by PKA and PKG, which may be enhanced by mechanisms that target MLC phosphates, such as MYPT1 phosphorylation by PKA and PKG and PKG-specific MYPT1:PKG-Iα leucine-zipper interaction.

ACKNOWLEDGMENTS

The author's research is supported by grants DK15564 and DK28300 from the National Institute of Diabetes and Digestive and Kidney Diseases.

The *Annual Review of Physiology* is online at
http://physiol.annualreviews.org

LITERATURE CITED

1. Makhlouf GM. 2003. Smooth muscle of the gut. In *Textbook of Gastroenterology*, ed. T Yamada, pp. 92–116. Philadelphia: Lippincott Williams & Wilkins. 4th ed.

2. Harnett KM, Cao W, Biancani P. 2005. Signal transduction pathways that regulate smooth muscle function. I. Signal transduction in phasic (esophageal) and tonic (gastro-esophageal sphincter) smooth muscle. *Am. J. Physiol. Gastrointest. Liver Physiol.* 288:G407–16

3. Ward SM, Sanders KM, Hirst GD. 2004. Role of interstitial cells of Cajal in neural control of gastrointestinal smooth muscles. *Neurogastroenterol. Motil.* 16(Suppl. 1):112–17

4. Sanders KM, Koh SD, Ordog T, Ward SM. 2004. Ionic conductances involved in generation and propagation of electrical slow waves in phasic gastrointestinal muscles. *Neurogastroenterol. Motil.* 16(Suppl. 1):100–5

5. Szurszewski JH. 1987. Electrical basis of gastrointestinal motility. In *Physiology of the Gastrointestinal Tract*, ed. LR Johnson, pp. 383–422. New York: Raven. 2nd ed.

6. Furness JB, Young HM, Pompolo S, Bornstein JC, Kunze WAA, McConalogue K. 1995. Plurichemical transmission and chemical coding of neurons in the digestive tract. *Gastroenterology* 108:554–63

7. Bornstein JC, Costa M, Grider JR. 2004. Enteric motor and interneuronal circuits controlling motility. *Neurogastroenterol. Motil.* 16:34–38

8. Sarna SK. 1999. Tachykinins and in vivo gut motility. *Dig. Dis. Sci.* 44:S114–18

9. Spiegel S, Milstien S. 2003. Exogenous and intracellularly generated sphingosine-1-phosphate can regulate cellular processes by divergent pathways. *Biochem. Soc. Trans.* 31:1220–25

10. Izzo AA, Mascolo N, Capasso F. 2001. The gastrointestinal pharmacology of cannabinoids. *Curr. Opin. Pharmacol.* 1:597–603

11. Hellstrom PM, Murthy KS, Grider JR, Makhlouf GM. 1994. Coexistence of three tachykinins receptors coupled to Ca^{2+} signaling pathways in intestinal muscle cell. *J. Pharmacol. Exp. Ther.* 270:236–43

12. Hersh E, Huang J, Grider JR, Murthy KS. 2004. G_q/G_{13} signaling by ET-1 in smooth muscle: MYPT1 phosphorylation via ET_A and CPI-17 dephosphorylation via ET_B. *Am. J. Physiol. Cell Physiol.* 287:C1209–18

13. Huang J, Zhou H, Mahavadi S, Sriwai W, Lyall V, Murthy KS. 2005. Signaling pathways mediating gastrointestinal smooth muscle contraction and MLC_{20} phosphorylation by motilin receptors. *Am. J. Physiol. Gastrointest. Liver Physiol.* 288:G23–31

14. Morini G, Kuemmerle JF, Impicciatorre M, Grider JR, Makhlouf GM. 1993. Coexistence of histamine H_1 and H_2 receptors coupled to distinct signal transduction pathways in isolated intestinal muscle cells. *J. Pharmacol. Exp. Ther.* 264:598–603

15. Kuemmerle JF, Murthy KS, Grider JR, Martin DC, Makhlouf GM. 1995.

Co-expression of 5-HT$_{2A}$ and 5-IIT$_4$ receptors coupled to distinct signaling pathways in human intestinal muscle cells. *Gastroenterology* 109:1791–800

16. Murthy KS, Makhlouf GM. 1998. P2$_{X1}$ and P2$_{Y2}$ receptors coupled to distinct signaling pathways in smooth muscle. *J. Biol. Chem.* 273:4695–704

17. Zhou H, Murthy KS. 2004. Distinctive G protein-dependent signaling in smooth muscle by sphingosine 1-phosphate receptors, S1P$_1$ and S1P$_2$. *Am. J. Physiol. Cell Physiol.* 286:C1130–38

18. Zhou H, Huang J, Murthy KS. 2004. Lysophosphatidic acid (LPA) interacts with LPA$_3$ receptors to activate selectively Gα_q and induce initial and sustained MLC$_{20}$ phosphorylation and contraction. *Gastroenterology* 126:A278 (Abstr.)

19. Murthy KS, Makhlouf GM. 1997. Differential coupling of muscarinic m$_2$ and m$_3$ receptors to adenylyl cyclases V/VI in smooth muscle. Concurrent m$_2$-mediated inhibition via G$_{\alpha i3}$ and m$_3$-mediated stimulation via G$_{\beta \gamma q}$. *J. Biol. Chem.* 272:21317–24

20. Murthy KS, Zhou H, Grider JR, Brautigan DL, Eto M, Makhlouf GM. 2003. Differential signaling by m3 and m2 receptors in smooth muscle: m2-mediated inactivation of MLCK via G$_{i3}$, Cdc42/Rac1, and PAK1, and m3-mediated MLC$_{20}$ phosphorylation via Rho kinase/MYPT1 and PKC/CPI-17 pathways. *Biochem. J.* 374:145–55

21. Murthy KS, Coy DH, Makhlouf GM. 1996. Somatostatin-receptor mediated signaling in smooth muscle: activation of PLC-β3 by Gβ and inhibition of adenylyl cyclase by Gα_{i1} and Gα_o. *J. Biol. Chem.* 271:23458–63

22. Murthy KS, Makhlouf GM. 1996. Opioid μ, δ, κ receptor-induced activation of PLC-β3 and inhibition of adenylyl cyclase is mediated by G$_{i2}$ and G$_o$ in smooth muscle. *Mol. Pharmacol.* 50:870–77

23. Misra S, Murthy KS, Zhou H, Grider JR. 2004. Co-expression of Y$_1$, Y$_2$ and Y$_4$ receptors in smooth muscle coupled to distinct signaling pathways. *J. Pharmacol. Exp. Ther.* 311:1154–62

24. Murthy KS, Makhlouf GM. 1995. Adenosine A$_1$ receptor-mediated activation of PLC-β3 in intestinal muscle: dual requirement for α and $\beta\gamma$ subunits of G$_i$. *Mol. Pharmacol.* 47:1172–79

25. Rhee SG. 2001. Regulation of phosphoinositide-specific phospholipase C. *Annu. Rev. Biochem.* 70:281–312

26. Murthy KS, Makhlouf GM. 1991. Phosphoinositide metabolism in intestinal smooth muscle: preferential production of Ins(1,4,5)P$_3$ in circular muscle cells. *Am. J. Physiol. Gastrointest. Liver Physiol.* 261:G945–51

27. Murthy KS, Makhlouf GM. 1995. Functional characterization of phosphoinositide-specific phospholipase C-β1 and -β3 in intestinal smooth muscle. *Am. J. Physiol. Cell Physiol.* 269:C969–78

28. Murthy KS, Makhlouf GM. 1995. Agonist-mediated activation of phosphatidylcholine-specific phospholipase C and D in intestinal smooth muscle. *Mol. Pharmacol.* 48:293–304

29. Murthy KS, Zhou H, Grider JR, Makhlouf GM. 2001. Sequential activation of heterotrimeric and monomeric G proteins mediates PLD activity in smooth muscle. *Am. J. Physiol. Gastrointest. Liver Physiol.* 280:G381–88

30. Murthy KS, Zhou H, Huang J, Pentyala S. 2004. Activation of PLC-δ1 by G$_{i/o}$-coupled receptor agonists. *Am. J. Physiol. Cell Physiol.* 287:C1679–87

31. Kim YH, Park TJ, Lee YH, Baek KJ, Suh PG, et al. 1999. Phospholipase C-δ1 is activated capacitative calcium entry that follows phospholipase C-β activation upon bradykinin stimulation. *J. Biol. Chem.* 274:26127–34

32. Zhou H, Das S, Murthy KS. 2003. Erk1/2- and p38 MAP kinase-dependent phosphorylation and activation of cPLA$_2$ by m3 and m2 receptors. *Am. J. Physiol. Gastrointest. Liver Physiol.* 284:G472–80

33. Murthy KS, Grider JR, Makhlouf GM. 1991. InsP₃-dependent Ca²⁺ mobilization in circular but not longitudinal muscle cells of intestine. *Am. J. Physiol. Gastrointest. Liver Physiol.* 261:G937–44

34. Himpens B, Matthjis G, Somlyo AP. 1989. Desensitization to cytosolic Ca²⁺ and Ca²⁺ sensitivities of guinea pig ileum and rabbit pulmonary artery smooth muscle. *J. Physiol.* 413:489–503

35. Sun XP, Supplisson Mayer E. 1993. Chloride channels in myocytes from rabbit colon are regulated by a pertussis toxin-sensitive G protein. *Am. J. Physiol. Gastrointest. Liver Physiol.* 264:G774–85

36. Murthy KS, Zhou H. 2003. Selective phosphorylation of IP₃R-I in vivo by cGMP-dependent protein kinase in smooth muscle. *Am. J. Physiol. Gastrointest. Liver Physiol.* 284:G221–30

37. Kuemmerle JF, Murthy KS, Makhlouf GM. 1994. Agonist-mediated influx activates ryanodine-sensitive, IP₃-insensitive Ca²⁺ release channels in longitudinal muscle of intestine. *Am. J. Physiol. Cell Physiol.* 266:C1421–31

38. Kuemmerle JF, Makhlouf GM. 1995. Agonist-stimulated cADP ribose: endogenous modulation of Ca²⁺-induced Ca²⁺ release in intestinal longitudinal smooth muscle. *J. Biol. Chem.* 270:25488–94

39. Murthy KS, Kuemmerle JF, Makhlouf GM. 1995. Release of arachidonic acid by agonist-mediated activation of PLA₂ initiates Ca²⁺ mobilization in intestinal smooth muscle. *Am. J. Physiol. Gastrointest. Liver Physiol.* 269:G93–102

40. Wang X-B, Osugi T, Uchida S. 1993. Muscarinic receptors stimulate Ca²⁺ influx via phospholipase A₂ pathway in ileal smooth muscle. *Biophys. Biochem. Res. Commun.* 193:483–89

41. Somlyo AP, Somlyo AV. 2003. Ca²⁺ sensitivity of smooth muscle and non-muscle myosin II: modulated by G proteins, kinases and phosphatases. *Physiol. Rev.* 83:1325–58

42. Kamm KE, Stull JT. 2001. Dedicated myosin light chain kinase with diverse cellular functions. *J. Biol. Chem.* 276:4527–30

43. Tansey MG, Word RA, Hidaka H, Singer HA, Schworer CM, et al. 1992. Phosphorylation of myosin light chain kinase by the multifunctional calmodulin-dependent protein kinase II in smooth muscle cells. *J. Biol. Chem.* 267:12511–16

44. Wirth A, Schroeter M, Kock-Hauser C, Manser E, Chalovich JM, et al. 2003. Inhibition of contraction and myosin light chain phosphorylation in guinea-pig smooth muscle by p21-activated kinase 1. *J. Physiol.* 549:489–500

45. Kitazawa T, Kobayashi S, Horiuti K, Somlyo AV, Somlyo AP. 1989. Receptor-coupled, permeabilized smooth muscle. Role of the phosphoinositide cascade, G proteins, and modulation of the contractile response to Ca²⁺. *J. Biol. Chem.* 264:5339–42

46. Kubota Y, Nomura M, Kamm KE, Mumby MC, Stull JT. 1992. GTPγS-dependent regulation of smooth muscle contractile elements. *Am. J. Physiol. Cell Physiol.* 262:C405–10

47. Niiro N, Ikebe M. 2001. Zipper-interacting protein kinase induces Ca²⁺-free smooth muscle contraction via myosin light chain phosphorylation. *J. Biol. Chem.* 276:29567–74

48. Weber LP, Van Lierop JE, Walsh MP. 1999. Ca²⁺-indendent phosphorylation of myosin in rat caudal artery and chicken gizzard myofilaments. *J. Physiol.* 516:805–24

49. Murthy KS, Grider JR, Kuemmerle JF, Makhlouf GM. 2000. Sustained muscle contraction induced by agonists, growth factors, and Ca²⁺ mediated by distinc PKC isozymes. *Am. J. Physiol. Gastrointest. Liver Physiol.* 279:G201–10

50. Hartshorne DJ, Eto M, Erodi F. 1998 Myosin light chain phosphatase: subunit composition, interaction and regulation. *J. Muscle Res. Cell Motil.* 19:325–41

51. Fukata Y, Amano M, Kaibuchi K. 2001. Rho-kinase pathway in smooth muscle contraction and cytoskeletal reorganization of non-muscle cells. *Trends Pharmacol. Sci.* 22:32–39

52. Kitazawa T, Eto M, Woodsome TP, Khalequzzaman MD. 2003. Phosphorylation of the myosin phosphatase targeting subunit and CPI-17 during Ca^{2+} sensitization in rabbit smooth muscle. *J. Physiol.* 546:879–89

53. Kitazawa T, Eto M, Woodsome TP, Brautigan DL. 2000. Agonists trigger G protein-mediated activation of the CPI-17 inhibitor phosphoprotein of myosin light chain phosphatase to enhance vascular smooth muscle contractility. *J. Biol. Chem.* 275:9897–9900

54. Woodsome TP, Eto M, Everett A, Brautigan DL, Kitazawa T. 2001. Expression of CPI-17 and myosin phosphatase correlates with Ca^{2+} sensitivity of protein kinase C-induced contraction in rabbit smooth muscle. *J. Physiol.* 535:553–64

55. Endo A, Surks HK, Mochizuki S, Mochizuki N, Mendelsohn ME. 2004. Identification and characterization of zipper-interacting protein kinase as the unique vascular smooth muscle myosin phosphatase-associated kinase. *J. Biol. Chem.* 279:42055–61

56. Pfitzer G, Sonntag Bensch D, Brkic-Koric D. 2001. Thiophosphorylation-induced Ca^{2+} sensitization of guinea-pig ileum contractility is not mediated by Rho-associated kinase. *J. Physiol.* 533:651–64

57. MacDonald JA, Borman MA, Muranyi A, Somlyo AV, Hartshorne DJ, Haystead TAJ. 2001. Identification of the endogenous smooth muscle myosin phosphatase-associated kinase. *Proc. Natl. Acad. Sci. USA* 98:2419–24

58. Kureishi Y, Ito M, Feng J, Okinaka T, Isaka N, Nakano T. 1999. Regulation of Ca^{2+}-independent smooth muscle contraction by alternative staurosporine-sensitive kinase. *Eur. J. Pharmacol.* 376:315–20

59. Murthy KS, Huang J, Zhou H, Kuemmerle JF, Makhlouf GM. 2004. Receptors coupled to inhibitory G proteins induce MLC_{20} phosphorylation and muscle contraction via PI 3-kinase-dependent activation of integrin-linked kinase (ILK). *Gastroenterology* 126:A413

60. Kaibuchi K, Kuroda S, Amano M. 1999. Regulation of cytoskeleton and cell adhesion by Rho family GTPases in mammalian cells. *Annu. Rev. Biochem.* 68:459–86

61. Riento K, Ridley AJ. 2003. ROCKS: multifunctional kinases in cell behavior. *Nat. Rev.* 4:446–56

62. Exton JH. 2002. Regulation of phospholipase D. *FEBS Lett.* 531:58–61

63. Niiro N, Koga Y, Ikebe M. 2003. Agonist-induced changes in the phosphorylation of the myosin-binding subunit of myosin light chain phosphatase and CPI-17, two regulatory factors of myosin light chain phosphatase, in smooth muscle. *Biochem. J.* 369:117–28

64. MacDonald JA, Eto M, Borman MA, Brautigan DL, Haystead TAJ. 2001. Dual Ser and Thr phosphorylation of CPI-17, an inhibitor of myosin phosphatase, by MYPT-associated kinase. *FEBS Lett.* 493:91–94

65. Borman MA, MacDonald JA, Muranyi A, Hartshorne DJ, Haystead TAJ. 2002. Smooth muscle myosin phosphatase-associated kinase induces Ca^{2+}-sensitization via myosin phosphatase inhibition. *J. Biol. Chem.* 277:23441–46

66. Wooldridge AA, MacDonald JA, Erodi F, Ma C, Borman MA, et al. 2004. Smooth muscle phosphatase is regulated in vivo by exclusion of phosphorylation of threonine 696 of MYPT1 by phosphorylation of serine 695 in response to cyclic nucleotides. *J. Biol. Chem.* 279:34495–504

67. Eto M, Kitazawa T, Yazawa M, Mukai H, Ono Y, Brautigan DL. 2001. Histamine-induced vasoconstriction involves phosphorylation of a specific inhibitor protein for myosin phosphatase by protein

kinase C α and δ isoforms. *J. Biol. Chem.* 276:29072–78

68. Deng JT, Sutherland C, Brautigan DL, Eto M, Walsh MP. 2002. Phosphorylation of the myosin phosphatase inhibitors, CPI-17 and PHI, by integrin-linked kinase. *Biochem. J.* 367:517–24

69. Hamaguchi T, Ito M, Feng J, Seko T, Koyama M, et al. 2000. Phosphorylation of CPI-17, an inhibitor of myosin phosphatase, by protein kinase N. *Biochem. Biophys. Res. Commun.* 274:825–30

70. Koyama M, Ito M, Feng J, Seko T, Shiraki K, et al. 2000. Phosphorylation of CPI-17, an inhibitory phosphorylation of smooth muscle myosin phosphatase, by Rho kinase. *FEBS Lett.* 475:197–200

71. Pang H, Guo Z, Su W, Xie Z, Eto M, Gong MC. 2005. RhoA/Rho kinase mediates thrombin- and U-46619-induced phosphorylation of a myosin phosphatase inhibitor, CPI-17, in vascular smooth muscle cells. *Am. J. Physiol. Cell Physiol.* 289:C352–60

72. Etter EF, Eto M, Wardle RL, Brautigan DL, Murphy RA. 2001. Activation of myosin light chain phosphatase in intact arterial smooth muscle during nitric oxide-induced relaxation. *J. Bio. Chem.* 276:34681–85

73. Murthy KS, Zhou H, Grider JR, Makhlouf GM. 2003. Inhibition of sustained smooth muscle contraction by PKA and PKG preferentially meditated by phosphorylation of RhoA. *Am. J. Physiol. Gastrointest. Liver Physiol.* 284:G1006–16

74. Sauzeau V, Le Jeune H, Cario-Toumaniantz C, Smolenski A, Lohmann SM, et al. 2000. Cyclic GMP-dependent protein kinase signaling pathway inhibits RhoA-induced Ca^{2+} sensitization of contraction in vascular smooth muscle. *J. Biol. Chem.* 275:21722–29

75. Takizawa N, Niiro N, Ikebe M. 2002. Dephosphorylation of the two regulatory components of myosin phosphatase, MBS and CPI-17. *FEBS Lett.* 515:127–32

76. Liu Q, Hofmann PA. 2003. Modulation of protein phosphatase 2a by adenosine A_1 receptors in cardiomyocytes: role for p38 MAPK. *Am. J. Physiol. Heart Circ. Physiol.* 285:H97–103

77. Hirano K, Derkach DN, Hirano M, Nishimura J, Kanaide H. 2003. Protein kinase network in the regulation of phosphorylation and dephosphorylation of smooth muscle myosin light chain. *Mol. Cell. Biochem.* 248:105–14

78. Deng JT, Van Lierop JE, Sutherland C, Walsh MP. 2001. Ca^{2+}-independent smooth muscle contraction: a novel function of integrin-linked kinase. *J. Biol. Chem.* 276:16365–73

79. Komatsu S, Ikebe M. 2004. ZIP kinase is responsible for the phosphorylation of myosin II and necessary for cell motility in mammalian fibroblasts. *J. Cell Biol.* 165:243–54

80. Shani G, Marash L, Gozuacik D, Bialik S, Teitelbaum L, et al. 2004. Death-associated protein kinase phosphorylates ZIP kinase, forming a unique kinase hierarchy to activate its cell death functions. *Mol. Cell. Biol.* 24:8611–26

81. Graves PR, Winkfield KM, Haystead TAJ. 2005. Regulation of zipper-interacting protein kinase activity in vitro and in vivo by multisite phosphorylation. *J. Biol. Chem.* 280:9363–74

82. Kiss E, Muranyi A, Csortos C, Gergely P, Ito M, et al. 2002. Integrin-linked kinase phosphorylates the myosin phosphatase target subunit at the inhibitory subunit site in platelet cytoskeleton. *Biochem. J* 365:79–87

83. Miyazaki K, Yano T, Schmidt DJ, Tokui T, Shibata M, et al. 2002. Rho-dependent agonist-induced spatio-temporal changes in myosin phosphorylation in smooth muscle cells. *J. Biol. Chem.* 277:725–34

84. Sohn UD, Cao W, Tang DC, Stull JT, Haeberle JR, et al. 2001. Myosin light chain kinase- and PKC-dependent contraction of LES and esophageal smooth muscle. *Am. J. Physiol. Gastrointest. Liver Physiol.* 281:G467–78

85. Cao W, Chen Q, Sohn UD, Kim N, Kirber MT, et al. 2001. Ca^{2+}-induced contraction of cat esophageal circular smooth muscle cells. *Am. J. Physiol. Cell Physiol.* 280:C980–92

86. Sohn UD, Zoukhri D, Dartt D, Sergheraert C, Harnett KM, et al. 1997. Different protein kinase C isozymes mediate lower esophageal sphincter tone and phasic contraction of esophageal circular smooth muscle. *Mol. Pharmacol.* 51:462–70

87. Kim N, Cao W, Song IS, Kim CY, Harnett KM, et al. 2004. Distinct kinases are involved in contraction of cat esophageal and lower esophageal smooth muscles. *Am. J. Physiol. Cell Physiol.* 287:C384–94

88. Pierce KL, Premont RT, Lefkowitz RJ. 2002. Seven-transmembrane receptors. *Nat. Rev.* 3:639–50

89. Pitcher JA, Freedman NJ, Lefkowitz RJ. 1998. G protein-coupled receptor kinases. *Annu. Rev. Biochem.* 67:653–92

90. Grady EF, Bohm SK, Bunnett NW. 1997. Turning off the signal: mechanism that attenuate G protein-coupled receptors. *Am. J. Physiol. Gastrointest. Liver Physiol.* 273:G586–601

91. Huang J, Mahavadi S, Zhou H, Murthy KS. 2005. Agonist-induced internalization and desensitization of muscarinic m2 receptors in smooth muscle cells is mediated by two G protein-dependent pathways: a GRK2/Clathrin pathway and Src/Caveolar pathway. *Gastroenterology* 128:A190 (Abstr.)

92. Huang J, Zhou H, Mahavadi S, Murthy KS. 2005. Agonist-induced internalization of δ-opioid receptors (DOR) in smooth muscle cells is mediated exclusively by GRK2-dependent phosphorylation of Thr[358] and Ser[363] in the C-terminal domain. *Gastroenterology* 128: A359 (Abstr.)

93. Hollinger S, Hepler JR. 2002. Cellular regulation of RGS proteins: modulators and integrators of G protein signaling. *Pharmacol. Rev.* 54:527–59

94. Murthy KS, Huang J, Hu W, Makhlouf GM. 2005. Cross-regulation of m3-activated G proteins by m3 receptors is mediated via PI 3-kinase-dependent inhibition of RGS4 and stimulation of RGS16. *Gastroenterology* 128:A610

95. Mahavadi S, Zhou H, Murthy KS. 2004. Distinctive signaling by cannabinoid CB_1 receptors in smooth muscle cells: absence of $G\beta\gamma$-dependent activation of PLC-$\beta1$. *Gastroenterology* 126:A275 (Abstr.)

96. Murthy KS, Grider JR, Makhlouf GM. 2000. Heterologous desensitization of response mediated by selective PKC-dependent phosphorylation of G_{i1} and G_{i2}. *Am. J. Physiol. Cell Physiol.* 279:C925–34

97. Razani B, Woodman SE, Lisanti MP. 2002. Caveolae: from cell biology to animal physiology. *Pharmacol. Rev.* 54:431–67

98. Murthy KS, Makhlouf GM. 2000. Heterologous desensitization mediated by G protein-specific binding to caveolin. *J. Biol. Chem.* 275:30211–19

99. Grider JR, Murthy KS, Jin J-G, Makhlouf GM. 1992. Post-junctional stimulation of nitric oxide in muscle cells by the relaxant neurotransmitter VIP. *Am. J. Physiol. Gastrointest. Liver Physiol.* 262:G774–78

100. Jin J-G, Murthy KS, Grider JR, Makhlouf GM. 1996. Stoichiometry of VIP release, nitric oxide formation and relaxation induced by nerve stimulation of rabbit and rat gastric muscle. *Am. J. Physiol. Gastrointest. Liver Physiol.* 271:G357–69

101. Murthy KS, Makhlouf GM. 1994. VIP/PACAP-mediated activation of membrane-bound NO synthase in smooth muscle is mediated by pertussis toxin-sensitive G_{i1-2}. *J. Biol. Chem.* 269: 15977–80

102. Teng B-Q, Murthy KS, Kuemmerle JF, Grider JR, Michel T, Makhlouf GM. 1998. Constitutive endothelial nitric oxide

synthase: expression in human and rabbit gastrointestinal smooth muscle cells. *Am. J. Physiol. Gastrointest. Liver Physiol.* 275:G342–51

103. Murthy KS, Jin J-G, Grider JR, Makhlouf GM. 1997. Characterization of PACAP receptors and signaling pathways in rabbit gastric muscle cells. *Am. J. Physiol. Gastrointest. Liver Physiol.* 272:G1391–99

104. Teng B-Q, Murthy KS, Kuemmerle JF, Grider JR, Makhlouf GM. 1998. Selective expression of $VIP_2/PACAP_3$ receptors in rabbit and guinea pig gastric and tenia coli smooth muscle cells. *Reg. Pep.* 77:124–34

105. Murthy KS. 2001. Activation of PDE5 and inhibition of guanylyl cyclase by cGMP-dependent protein kinase in smooth muscle. *Biochem. J.* 360:199–208

106. Murthy KS, Zhou H, Makhlouf GM. 2002. Regulation of phosphodiesterase 3 (PDE3) and adenylyl cyclase by cAMP-dependent protein kinase. *Am. J. Physiol. Cell Physiol.* 282:C508–17

107. Francis SH, Turko IV, Corbin JD. 2001. Cyclic nucleotide phosphodiesterases: relating structure and function. *Prog. Nucleic Acids Res. Mol. Biol.* 65:1–52

108. Harmar AJ, Arimura A, Gozes I, Journot L, Laburthe M, et al. 1998. International union of pharmacology. XVIII. Nomenclature of receptors for vasoactive intestinal peptide and pituitary adenylate cyclase-activating polypeptide. *Pharmacol. Rev.* 50:265–70

109. Teng B-Q, Grider JR, Murthy KS. 2001. Identification of VIP-specific receptor in guinea pig tenia coli. *Am. J. Physiol. Gastrointest. Liver Physiol.* 281:G718–25

110. Zhou H, Huang J, Grider JR, Murthy KS. 2005. Molecular cloning of a VIP-specific receptor. *Gastroenterology* 128:A193 (Abstr.)

111. Murthy KS, Jin J-G, Teng B-Q, Makhlouf GM. 1998. G protein-dependent activation of smooth muscle eNOS mediated by the natriuretic peptide-C receptor. *Am.*

J. Physiol. Cell Physiol. 275:C1409–16

112. Murthy KS, Teng B-Q, Zhou H, Jin JG, Makhlouf GM. 2000. G_{i1}/G_{12}-dependent signaling by the single-transmembrane natriuretic peptide clearance receptor (NPR-C). *Am. J. Physiol. Gastrointest. Liver Physiol.* 278:G974–80

113. Anand-Srivastava MB, Sehl PD, Lowe DG. 1996. Cytoplasmic domain of natriuretic peptide receptor-C inhibits adenylyl cyclase: involvement of pertussis toxin-sensitive G protein. *J. Biol. Chem.* 271:19324–29

114. Murthy KS, Makhlouf GM. 1999. Identification of the G protein-activating domain of the natriuretic peptide clearance receptor (NPR-C). *J. Biol. Chem.* 274:17587–92

115. Zhou H, Murthy KS. 2003. Identification of the G protein-activating sequence of the single-transmembrane natriuretic peptide receptor C (NPR-C). *Am. J. Physiol. Cell Physiol.* 284:C1255–61

116. Okamoto T, Katada T, Murayama Y, Ui M, Ogata E, Nishimoto I. 1990. A simple structure encodes G protein-activating function of the IGF-II/mannose 6-phosphate receptor. *Cell* 67:723–30

117. Okamoto T, Nishimoto I. 1992. Detection of G protein-activator regions in m4 subtype muscarinic, cholinergic, and α_2 adrenergic receptors based on upon characteristics in primary structure. *J. Biol. Chem.* 267:8342–46

118. Zhou H, Grider JR, Murthy KS. 2002. Agonist-dependent, PKA-mediated phosphorylation of GRK2 augments homologous desensitization of VPAC2 receptors in smooth muscle. *Gastroenterology* 122:A136 (Abstr.)

119. Zhou H, Murthy KS. 2003. PKG-induced desensitization of the single transmembrane NPR-C receptor is mediated by phosphorylation of a single threonine residue (T505) in the intracellular domain. *Gastroenterology* 122:A377 (Abstr.)

120. Turko IV, Francis SH, Corbin JD. 1998

SIGNALING IN SMOOTH MUSCLE OF THE GUT **373**

Binding of cGMP to both allosteric sites of cGMP-binding cGMP-specific phosphodiesterase (PDE5) is required for its phosphorylation *Biochem. J.* 329:505–10

121. Rybalkin SD, Rybalkina IG, Feil R, Hofmann F, Beavo J. 2002. Regulation of cGMP-specific phosphodiesterase (PDE5) phosphorylation in smooth muscle cells. *J. Biol. Chem.* 277:3310–17

122. Francis SH, Corbin JD. 1999. Cyclic nucleotide-dependent protein kinases: intracellular receptors for cAMP and cGMP action. *Crit. Rev. Clin. Lab. Sci.* 36:275–328

123. Murthy KS, Makhlouf GM. 1995. Interaction of cA-kinase and cG-kinase in mediating relaxation of dispersed smooth muscle cell. *Am. J. Physiol. Cell Physiol.* 268:C171–80

124. Jiang H, Colbran JL, Francis SH, Corbin JD. 1992. Direct evidence for cross-activation of cGMP-dependent protein kinase by cAMP in pig coronary arteries. *J. Biol. Chem.* 267:1015–19

125. Lincoln TM, Dey N, Sellak H. 2001. cGMP-dependent protein kinase signaling mechanism in smooth muscle: from the regulation of tone to gene expression. *J. Appl. Physiol.* 91:1421–30

126. Landgraf W, Hulin R, Gobel C, Hofmann F. 1986. Phosphorylation of cGMP-dependent protein kinase increases the affinity for cAMP. *Eur. J. Biochem.* 154:113–17

127. Murthy KS, Severi C, Grider JR, Makhlouf GM. 1993. Inhibition of inositol 1,4,5-trisphosphate (IP_3) production and IP_3-dependent Ca^{2+} mobilization by cyclic nucleotides in isolated gastric muscle cells. *Am. J. Physiol. Gastrointest. Liver Physiol.* 264:G967–74

128. Murthy KS, Makhlouf GM. 1998. Differential regulation of phospholipase A_2 (PLA_2)-dependent Ca^{2+} signaling in smooth muscle by cAMP- and cGMP-dependent protein kinases. Inhibitory phosphorylation of PLA_2 by cyclic

nucleotide-dependent protein kinases. *J. Biol. Chem.* 273:34519–26

129. Zhou H, Murthy KS. 2003. Relative contribution of RGS4 and RGS domain of GRK2 to inhibition of PLC-β activity by PKA: direct evidence from site-directed mutagenesis of RGS4 and GRK2. *Gastroenterology* 124:A23 (Abstr.)

130. Komalavilas P, Lincoln TM. 1996. Phosphorylation of the inositol 1,4,5-trisphosphate receptor. Cyclic GMP-dependent protein kinase mediates cAMP and cGMP dependent phosphorylation in the intact rat aorta. *J. Biol. Chem.* 271:21933–38

131. Cornwell TL, Pryzwansky KB, Wyatt TA, Lincoln TM. 1991. Regulation of sarcoplasmic reticulum protein phosphorylation by localized cGMP-dependent protein kinase in vascular smooth muscle cells. *Mol. Pharmacol.* 40:923–31

132. Koh SD, Sanders KM, Carl A. 1996. Regulation of smooth muscle delayed rectifier K+ channels by protein kinase A. *Pflügers Arch.* 432:401–12

133. Surks HK, Mochizuki N, Kasai Y, Georgescu SP, Tang KM, et al. 1999. Regulation of myosin phosphatase by a specific interaction with cGMP-dependent protein kinase-Iα. *Science* 286:1583–87

134. Surks HK, Mendelsohn ME. 2003. Dimerization of cGMP-dependent protein kinase 1-α and the myosin-binding subunit of myosin phosphatase: role of leucine zipper domains. *Cell. Signal.* 15:937–44

135. Manganello JM, Djellas Y, Borg C, Atonakis K, Le Breton GC. 1999. Cyclic AMP-dependent phosphorylation of thromboxane A2 receptor-associated $G\alpha_{13}$. *J. Biol. Chem.* 274:28003–10

136. Manganello JM, Huang JS, Kozasa T, Voyno-Yasenetskaya TA, Le Breton GC. 2003. Protein kinase A-mediated phosphorylation of the $G\alpha_{13}$ switch I region alters the $G\alpha\beta\gamma_{13}$-G protein-coupled receptor complex and inhibits Rho activation. *J. Biol. Chem.* 278:124–30

137. Huang QQ, Fisher SA, Brozovich FV. 2004. Unzipping the role of myosin light chain phosphatase in smooth muscle cell relaxation. *J. Biol. Chem.* 279:597–603

138. Wu X, Haystead TAJ, Nakamoto RK, Somlyo AV, Somlyo AP. 1998. Acceleration of myosin light chain dephosphorylation and relaxation in smooth muscle by telokin. *J. Biol. Chem.* 273:11362–69

139. MacDonald JA, Walker LA, Nakamoto RK, Gorenne I, Somlyo AV, Somlyo AP. 2000. Phosphorylation of telokin by cyclic nucleotide kinases and the identification of *in vivo* phosphorylation sites in smooth muscle. *FEBS Lett.* 479:83–88

140. Walker LA, MacDonald JA, Liu X, Nakamoto RK, Haystead TAJ, et al. 2001. Site-specific phosphorylation and point mutations of telokin modulates its Ca^{2+}-desensitizing effect in smooth muscle. *J. Biol. Chem.* 276:24519–24

Annu. Rev. Physiol. 2006. 68:375–401
doi: 10.1146/annurev.physiol.68.040104.134728
First published online as a Review in Advance on September 28, 2005

CNG AND HCN CHANNELS: Two Peas, One Pod

Kimberley B. Craven[1] and William N. Zagotta[1,2]

[1]*Department of Physiology and Biophysics, University of Washington, Seattle, WA 98195; email: kcraven@u.washington.edu*
[2]*Howard Hughes Medical Institute, University of Washington, Seattle, WA 98195; email: zagotta@u.washington.edu*

Key Words cyclic nucleotide, activation, gating, CNBD

■ **Abstract** Cyclic nucleotide–activated ion channels play a fundamental role in a variety of physiological processes. By opening in response to intracellular cyclic nucleotides, they translate changes in concentrations of signaling molecules to changes in membrane potential. These channels belong to two families: the cyclic nucleotide–gated (CNG) channels and the hyperpolarization-activated cyclic nucleotide–modulated (HCN) channels. The two families exhibit high sequence similarity and belong to the superfamily of voltage-gated potassium channels. Whereas HCN channels are activated by voltage and CNG channels are virtually voltage independent, both channels are activated by cyclic nucleotide binding. Furthermore, the channels are thought to have similar channel structures, leading to similar mechanisms of activation by cyclic nucleotides. However, although these channels are structurally and behaviorally similar, they have evolved to perform distinct physiological functions. This review describes the physiological roles and biophysical behavior of CNG and HCN channels. We focus on how similarities in structure and activation mechanisms result in common biophysical models, allowing CNG and HCN channels to be viewed as a single genre.

INTRODUCTION

Ion channels are molecular machines that enable cells to communicate with the extracellular world (1). These allosteric proteins permit ions to flow into or out of cells through a conducting pore, thus allowing the cells to regulate their membrane potentials and intracellular Ca^{2+} concentrations in response to intracellular or extracellular events. The process of gating, the opening and closing of an ion-permeable pore, can be initiated by changes in membrane voltage, binding of ligands, or changes in membrane stretch due to mechanical stimuli, depending on the type of ion channel. This review focuses on similarities and differences between two families of ion channels that are gated by binding of cyclic nucleotide ligands: the cyclic nucleotide–gated (CNG) channels and the hyperpolarization-activated cyclic nucleotide–modulated (HCN) channels. CNG channels are nonselective cation channels that are activated very weakly by membrane depolarization. HCN channels are weakly K^+-selective cation channels that are activated by membrane hyperpolarization. Both channels, however, are activated by cyclic nucleotide binding, which produces a large increase in current in CNG channels and a depolarizing shift in the voltage dependence of activation for HCN channels. These electrophysiological properties tune each of these channels to different physiological roles.

PHYSIOLOGICAL ROLES

CNG Channels

Cyclic nucleotide–gated (CNG) channels generate the primary electrical signal in photoreceptors and olfactory sensory neurons. CNG channels are nonselective cation channels that are opened, or gated, by direct binding of intracellular cyclic nucleotides (2–4). In vertebrate rod photoreceptors, CNG channels open in response to guanosine 3′,5′-cyclic monophosphate (cGMP) binding. In the dark, CNG channels conduct a steady inward current (consisting primarily of Na^+ and Ca^{2+}) as a component of the dark current. Light striking the retina activates the phototransduction cascade on the disc membranes of rod photoreceptor outer segments (Figure 1) (5). Photoactivated rhodopsin activates phosphodiesterase via

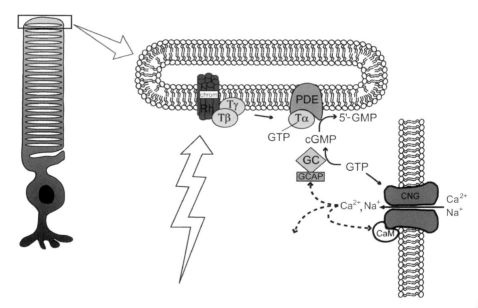

Figure 1 Phototransduction cascade in rod photoreceptor. Rod photoreceptor (*left*) with phototransduction cascade on rod disc membrane and rod outer segment (*right*). A photon activates rhodopsin (Rh), which in turn activates the heterotrimeric G protein transducin (T) whose α subunit activates a phosphodiesterase (PDE). A CNG channel is shown in the outer segment membrane, and the downstream effects of Ca^{2+} influx [calmodulin (CaM) binding, activation of guanylyl cyclase (GC) by guanylyl cyclase–activating protein (GCAP), and other effects] are shown with dotted arrows.

the GTP-binding protein transducin The phosphodiesterase hydrolyzes cGMP into 5′-GMP, leading to the closure of CNG channels and a decrease in inward current. This hyperpolarizes the outer segment and terminates the tonic release of neurotransmitter that occurs in the dark when cGMP levels are high and CNG channels are open. Therefore, it is the termination of a neurotransmitter signal to the downstream retinal cells and neurons that indicates detection of a photon. The decrease in Ca^{2+} influx also has an important downstream effect: It is detected by Ca^{2+}-sensing proteins such as guanylyl cyclase–activating protein (GCAP), which stimulates guanylyl cyclase, resulting in the increase of cGMP production from GTP (6–8).

CNG channels also play a crucial role in olfactory signal transduction. In olfactory sensory neurons, binding of an odorant to an olfactory receptor activates adenylate cyclase, which produces an increase in the intracellular concentration of adenosine 3′,5′-cyclic monophosphate (cAMP) (Figure 2). This increase in intracellular cAMP leads to opening of CNG channels, which depolarizes the olfactory sensory neurons. Depolarization results in the release of neurotransmitter, signaling to downstream neurons that an odorant has been detected (9). As in phototransduction, Ca^{2+} influx is likewise important in olfactory transduction. Ca^{2+} activates calmodulin, which can bind to and inhibit olfactory CNG channels, a process proposed to underlie olfactory adaptation (10–13). Additionally,

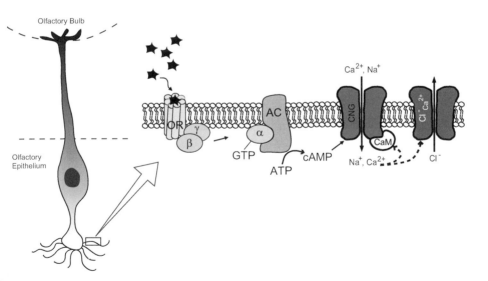

Figure 2   Olfactory transduction cascade in olfactory epithelium. Olfactory receptor cell (*left*) with olfactory transduction cascade in cilia (*right*). Odorants bind to and activate the olfactory receptor (OR), which activates a heterotrimeric G protein, whose α subunit activates adenylyl cyclase (AC). AC produces cAMP, which activates CNG channels. Downstream effects of Ca^{2+} [binding of calmodulin (CaM) and activating of a Ca^{2+}-sensitive Cl^- channel] are shown by dotted arrows.

Ca^{2+} activates Ca^{2+}-sensitive Cl^- channels to amplify this response. CNG channels are also located in other sensory tissues—such as taste receptors, and nonsensory tissues such as the hippocampus, heart, testis, kidney, pancreas, adrenal gland, and colon—in which their roles are not as well understood as in phototransduction and olfaction (3, 4).

There are six vertebrate CNG channel subunits: CNGA1, CNGA2, CNGA3, CNGA4, CNGB1, and CNGB3. These subunits can assemble in a variety of combinations to produce tetrameric channels. Three of these subunit types, CNGA1, CNGA2, and CNGA3, can form homomeric channels in heterologous expression systems. CNGA4, CNGB1, and CNGB3 do not form functional homomeric channels but can coassemble with other subunits to form functional heteromeric channels. In fact, most native channels are thought to be composed of more than one subunit type. For example, CNG channels of rod photoreceptors contain three CNGA1 subunits and one CNGB1 subunit (14–16). Those of cone photoreceptors are also heteromeric and are thought to form from two CNGA3 and two CNGB3 subunits (17). CNG channels of olfactory neurons have been proposed to be composed of two CNGA2 subunits, one CNGA4 subunit, and one alternatively sliced CNGB1 subunit, CNGB1b (18). The assorted combinations of these six subunits give each tissue-specific CNG channel unique properties to perform its physiological role.

HCN Channels

Hyperpolarization-activated cyclic nucleotide–modulated (HCN) channels constitute a related family of channels with a very different physiological role from CNG channels. HCN channels regulate neuronal and cardiac firing rates. HCN channels are cation channels that are modulated by binding of cyclic nucleotides, like CNG channels. Unlike CNG channels, however, they are activated by membrane hyperpolarization and are weakly selective for potassium. The current generated by HCN channels has been called the I_h (hyperpolarization), I_q (queer), or I_f (funny) current; one area in which it has been described extensively is the cardiac sinoatrial node (19–21). The sinoatrial node is the pacemaker region of the heart, and currents through several types of ion channels cooperate in order to generate spontaneous rhythmic firing of cardiac action potentials (Figure 3) (22, 23). HCN channels are activated by membrane hyperpolarization after a cardiac action potential. When activated, HCN channels conduct inward current and depolarize the cell toward the threshold of voltage-gated calcium channel activation, which in turn leads to firing of another cardiac action potential. In this manner, rhythmic firing is generated (21, 24). Neurotransmitters, such as norepinephrine, or pharmacological agents, such as a β-adrenergic agonist, can elevate cAMP levels and produce an accelerated heart rate (25, 26) This is due, in part, to the binding of cAMP to HCN channels, shifting the voltage dependence of activation to more depolarized potentials and increasing both the rate of channel opening and the maximal current level.

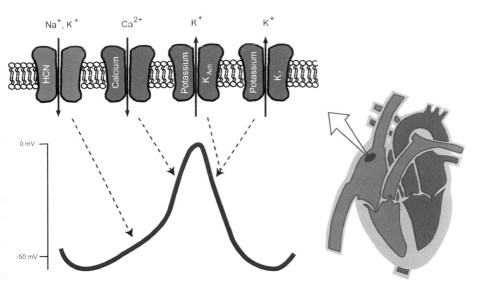

Figure 3   Heart sinoatrial node pacemaker current. Sinoatrial node (*black oval*) of the heart (*right*) and the ion channels responsible for pacemaker activity (*left*). Each channel contributes to a phase of the sinoatrial action potential, indicated by the dotted arrows.

HCN channels also mediate pacemaker activity in the nervous system by a similar mechanism as in the heart, although the gating kinetics in the nervous system must be faster than in the heart, as neuronal action potentials are faster than cardiac action potentials (27, 28). Neurons in the globus pallidus of the basal ganglia exhibit tonic activity whose rate and regularity are attributed to HCN channels (29). Moreover, as in the heart, cAMP speeds the activation kinetics and maximal current levels of HCN channels in the nervous system, and neurotransmitters can influence firing rates by either increasing or decreasing cAMP levels (30–33).

Besides acting as a pacemaker, HCN current also functions as a regulator of resting potential and membrane resistance. Moreover, HCN channels regulate synaptic transmission and contribute to nervous system development. In both neurons and muscle cells, HCN channels are activated at rest, and the inward sodium current leads to slightly depolarized resting potentials (34). HCN current stabilizes the resting membrane potential because small hyperpolarizations activate HCN channels, whose inward current depolarizes the cell. This depolarization deactivates HCN channels, preventing further departure from the resting potential. In retinal photoreceptors, hyperpolarization during bright light activates HCN channels, and the HCN current returns the membrane potential closer to depolarized levels as in the dark, allowing for adaptation of synaptic transmission during bright light (34). It has been suggested that I_h controls the spontaneous oscillatory activity during development that is required for neural network maturation. One example of this is in the cerebellum, in which HCN channels are expressed early in development in basket cells. Basket cells are inhibitory GABAergic cells that modulate Purkinje cell behavior (35). This HCN activity has also been shown to affect membrane resistance and dendritic integration (36, 37).

The HCN channel family, like the CNG channel family, comprises several subunit types. There are four known HCN subunit isoforms, HCN1–HCN4, which combine to form tetrameric channels in the heart and the nervous system. HCN1 is expressed in photoreceptors, dorsal root ganglia, cortex, cerebellum, and sino-atrial node. HCN2 is expressed in the sinoatrial node, dorsal root ganglia, and basal ganglia. HCN3 is widely expressed in the brain, but at low levels. HCN4 is expressed in the sinoatrial node and subcortical areas (38, 39). Although each subunit can form homomeric channels in heterologous expression systems, the subunits combine in diverse ways to form heteromeric channels in the heart and brain of different species. For example, HCN1 and HCN4 subunits have been proposed to form HCN channels in the rabbit sinoatrial node, whereas HCN2 and HCN4 subunits are thought to form HCN channels in the mouse sinoatrial node (23, 40). In the mouse brain, HCN channels of globus pallidus appear to be composed of the HCN1 and HCN2 subunits (29). Currently, the subunit stoichiometry of these native channels is unknown.

SIMILARITY IN STRUCTURES OF CNG AND HCN CHANNELS

Despite their different physiological roles, CNG and HCN channels are closely related in structure. Both channels belong to the superfamily of voltage-gated potassium (K^+) channels and are thought to be tetrameric proteins whose subunits are arranged around a central pore. Each subunit has six transmembrane domains and intracellular N and C termini (Figure 4). The fourth transmembrane domain (S4) contains several positively charged amino acids, as in K^+ channels. The HCN S4 functions as a voltage sensor, and the CNG S4 can function as a voltage sensor if placed in a permissive structural environment, such as the ether-a-go-go (Eag) K^+ channel (41). The pore region comprises the fifth transmembrane domain (S5), a reentrant pore loop that does not completely traverse the membrane, and the sixth transmembrane domain (S6). As with voltage-gated K^+ channels, the pore regions of CNG and HCN channels are thought to be similar to that of KcsA, the potassium channel from *Streptomyces lividans* whose crystal structure has been solved (42). The intracellular N-terminal regions are divergent in the various HCN and CNG subunits, and in CNG channels, the N terminus can be autoexcitatory and contribute to calmodulin modulation (43). Contained within the intracellular C termini of both HCN and CNG channels is a ligand-binding domain called the cyclic nucleotide–binding domain (CNBD). The CNBD is connected to the end of the S6 by the C-linker. Binding of ligand to the CNBD favors channel opening in both CNG and HCN channels, perhaps owing to a conformational change in the CNBD, which may be conferred to the pore through movement in the C-linker. The

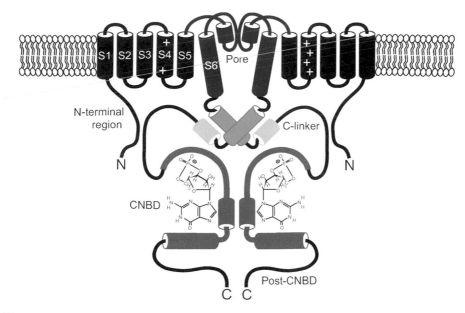

Figure 4   Membrane topology of CNG and HCN channels. Two of the four subunits of a CNG or HCN channel are shown. The transmembrane segments (S1–S6) are shown in black, except for the pore region (S6 and pore loop), which is shown in red. The voltage sensor (S4) is indicated by positive charges. The C-terminal region contains the cyclic nucleotide–binding domain (CNBD, *blue*, shown with cGMP bound) and the C-linker (*green*), which connects the CNBD to the pore.

region of C terminus past the CNBD, called the post-CNBD region, is proposed to be important in subunit assembly of CNG channels (16, 44).

The CNG and HCN channels exhibit high sequence similarity to two other channel families, Eag-like K^+ channels and plant K^+ channels, as illustrated by a phylogenetic tree (Figure 5). Like CNG and HCN channels, these channels also contain C-linkers and CNBDs (Figure 6), although the effects of cyclic nucleotides on these channels are weak or disputed (45–47). The difference in cyclic nucleotide modulation may be due to a lack of conservation in key residues involved in cyclic nucleotide coordination. This suggests that the fold of the CNBDs may serve other functions for these channels, such as binding other signaling molecules.

BIOPHYSICAL PROPERTIES OF CNG AND HCN CHANNELS

Permeability

Under physiological conditions, CNG channels carry an inward Na^+ and Ca^{2+} current. CNG channels are nonselective cation channels, allowing the monovalent

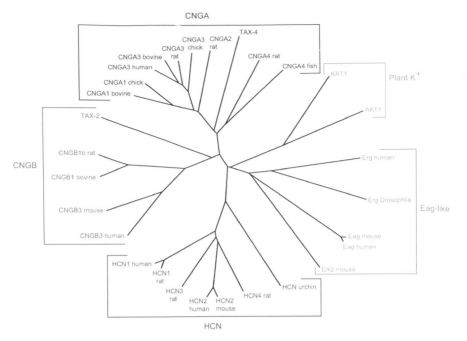

Figure 5   Unrooted phylogenetic tree showing relationships of the CNG, HCN, plant K$^+$, and ether-a-go-go (Eag)-like K$^+$ channel families. CNGA channel members (*red*): CNGA1 (formerly CNG1), CNGA2 (formerly CNG2), CNGA3 (formerly CNG3), CNGA4 (formerly CNG5), and TAX-4 (*Caenorhabditis elegans*). CNGB channel members (*purple*): CNGB1 (formerly CNG4), CNGB3 (formerly CNG6), and TAX-2 (*C. elegans*). HCN channel members (*blue*): HCN1 (formerly HAC2), HCN2 (formerly HAC1), HCN3 (formerly HAC3), HCN4 (formerly HAC4), and urchin HCN (also known as SpIH or spHCN). Eag-like channel members (*yellow*): Eag, Elk, ether-a-go-go-related gene (Erg), and human Erg, also known as HERG. Plant K$^+$ channel members (*green*): KAT1, AKT1. Sequences aligned with ClustalW, www.ebi.ac.uk/clustalw/, and tree drawn with PHYLIP software array, http://evolution.genetics.washington.edu/phylip.html.

cations K$^+$, Na$^+$, Li$^+$, Rb$^+$, and Cs$^+$ to conduct almost equally well (3). Although divalent cations can permeate the channel, at high concentrations they also block the channel. These permeability properties most closely resemble that of voltage-dependent Ca^{2+} channels and are due, in part, to an acidic residue in the outer mouth of CNG channels (48–52). This acidic residue replaces the YG of the GYG signature sequence in the pore loop of K$^+$ channels (53).

In contrast to the nonselective cation selectivity of CNG channels, HCN channels are more permeable to K$^+$ than Na$^+$ (with permeability ratios of about 4:1) and are blocked by millimolar concentrations of Cs$^+$ (54, 55). Despite this preference for K$^+$ conductance, perhaps conferred by the GYG motif in the pore loop,

HCN channels also carry an inward Na^+ current under physiological conditions (56). HCN channels can also conduct Ca^{2+}, but not as well as CNG channels. For example, with 2.5 mM external $[Ca^{2+}]$, the fractional Ca^{2+} current of HCN4 is 0.6%, whereas for CNGA3 it is 80% (57, 58). Unlike CNG channels, HCN channels are not blocked by divalent cations (59).

Voltage Dependence

An important difference between these two channel families is their degree of voltage dependence (Figure 7). As indicated above, the S4 of both CNG and HCN channels contains positively charged residues, as do the voltage-dependent K^+ channels that activate with depolarization. Surprisingly, however, HCN and CNG channels both behave differently from K^+ channels: HCN channels are activated by hyperpolarization and CNG channels are practically insensitive to voltage. S4s of both HCN and K^+ channels seem to respond similarly to changes in membrane potential, as they are accessible from the inside upon hyperpolarization and accessible from the outside upon depolarization (60, 61). The difference between HCN and K^+ channels is thought to be in the coupling to opening: HCN channels open with hyperpolarization whereas K^+ channels open with depolarization. It has also been proposed that, instead of merely having an opposite coupling mechanism, HCN channels couple voltage sensing to channel activation through a novel mechanism of transmembrane rearrangement (62). HCN channels are moderately voltage sensitive, on the average moving the equivalent of six electric charges through the membrane with activation, compared to thirteen electric charges for *Shaker* K^+ channels (63–65). The other subtypes of HCN channels have charge movements ranging from four to six electric charges (34, 54, 66–74). In addition to being activated by hyperpolarization instead of depolarization, HCN channels are also activated more slowly (0.2–2 sec) than most potassium channels. The voltage-activated gate has been shown to be on the intracellular side of the channel (75, 76).

Modulation of CNG and HCN Channels by Cyclic Nucleotides

A unifying biophysical property of CNG and HCN channels is their modulation by cyclic nucleotides. For both CNG and HCN channels, binding of cyclic nucleotides to the CNBD favors channel opening. The C-terminal regions of both channel families have high sequence similarity, suggesting perhaps that HCN and CNG channels sense and report ligand binding in similar ways (see Figure 6 for sequence alignment). This is supported by the fact that mutations in the CNBD regions of both CNG and HCN channels appear to have similar effects. For example, the same basic residue, Arg559 (CNGA1), Arg591 (HCN2), and Arg538 (HCN1), appears to interact with ligand in all three channels (66, 77, 78). If this Arg is mutated to an acidic amino acid, such as Glu, the channel retains normal kinetic and voltage behavior but binds cyclic nucleotides weakly. For HCN channels, the mutant channel no longer exhibits a shift in voltage dependence with cAMP, and for CNG channels, the mutant channel has a 1000-fold reduction in ligand affinity.

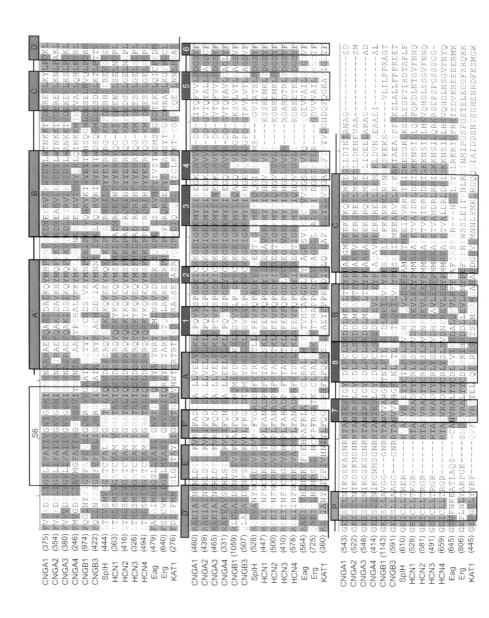

Another indication of the similarity in CNG and HCN C termini is that the same salt bridges exist in the C-linkers of both channels (67). These experiments suggest that CNG and HCN channels interact with ligand in the same way and then undergo similar conformational changes.

HCN2 C-TERMINAL CRYSTAL STRUCTURE   Recently, the crystal structure of the C-terminal region was solved for the HCN2 channel (73), shedding light on how cyclic nucleotides bind and how this binding may lead to channel opening. The crystal structure begins just after the end of S6 and extends through the C-linker and CNBD but does not include the post-CNBD region. The structure is a tetramer with a fourfold axis of symmetry and a hole down the central axis (Figure 8). The C-terminal region of HCN and CNG channels is thought to hang below the pore, like the "hanging gondola" of voltage-gated K^+ channels (79, 80). Although electronegative in character, the hole down the central axis has been shown not to be an obligatory part of the ion pathway (81). The top portion of the structure consists of the C-linkers, six α-helices (A′–F′), with the N terminus of each C-terminal fragment located close to each other on the very top of the structure (Figure 8). The bottom portion of the structure consists of the CNBDs, each of which demonstrates the same fold as the CNBD of other cyclic nucleotide–binding proteins such as catabolite gene activator protein (CAP) and cAMP-dependent protein kinase (PKA) (82, 83). Each CNBD contains an α-helix (A), followed by a β-roll formed by eight β-strands (1–8), followed by two α-helices (B and C). There is also an additional short α-helix, called the P-helix, between β-strands 6 and 7.

Most of the intersubunit contacts in the HCN2 C-terminal region crystal structure are in the C-linker region. The C-linker of each subunit contains six α-helices, and the first two helices (A′ and B′) form an antiparallel helix-turn-helix motif that interacts with the second two helices (C′ and D′) of the neighboring subunit (Figure 8). This interacting region has been likened to an "elbow on a shoulder," in which the "elbow" is the A′ and B′ helix-turn-helix motif that rests on the "shoulder," the C′- and D′-helices of the neighboring subunit. The interaction

---

Figure 6   Sequence alignment of CNG, HCN, and related families. Sequence alignment includes all members of the CNG channel family (bovine CNGA1, rat CNGA2, human CNGA3, rat CNGA4, bovine CNGB1, and human CNGB3) and HCN channel family (rat HCN1, mouse HCN2, rat HCN3, and rat HCN4) as well as SpIH (sea urchin HCN), two members of the Eag-like channels (mouse Eag and human Erg), and one plant K^+ channel (KAT1). The tertiary structure elements—the last transmembrane sequence (S6) as well as the α-helices of the C-linker (A′–F′) and the β-sheets (1–8) and α-helices (A–C, P) of the CNBD—are boxed (α-helices are in *orange boxes* and β-sheets in *indigo boxes*). The HCN2 C-terminal crystal structure extends from Asp443 to Leu643 (HCN2 amino acid numbers).

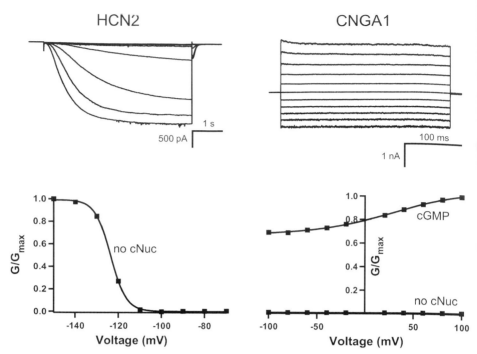

Figure 7 Voltage dependence of HCN2 and CNGA1. Behavior of HCN2 (*left*) and CNGA1 (*right*) channels in response to voltage. HCN2 currents (*left, top*) were recorded in response to voltage pulses from a holding potential of 0 mV to test potentials between −70 mV and −150 mV, returning to a tail potential of −40 mV. The conductance-voltage relation (*left, bottom*) was obtained from normalized tail currents. CNGA1 currents (*right, top*) were recorded in the presence of saturating cGMP and were recorded in response to voltage pulses from a holding potential of 0 mV to test potentials between −100 mV to +100 mV. The conductance-voltage relations with and without cGMP (*right, bottom*) were calculated from the currents and the driving force.

between the "elbow" and the "shoulder" involves hydrogen bonds, hydrophobic interactions, and salt bridges.

The C-terminal region of HCN2 was crystallized in the presence of two different ligands, cAMP and cGMP. The structures of the different ligand-bound channels were virtually identical, only differing in the configuration of the cyclic nucleotide itself. cAMP binds between the β-roll and the C-helix in the anti configuration, the configuration with the purine ring rotated away from the phosphoribose (Figure 9). Arg591 of the β-roll electrostatically interacts with the phosphate of cAMP. This has also been demonstrated in electrophysiology experiments, in which mutating this Arg to a Glu results in a large decrease in HCN2 affinity for cAMP (66). The β-roll and P-helix also interact with oxygens on the phosphoribose through hydrogen bonding. Additionally, the purine ring of cAMP interacts with Arg632 and Ile636

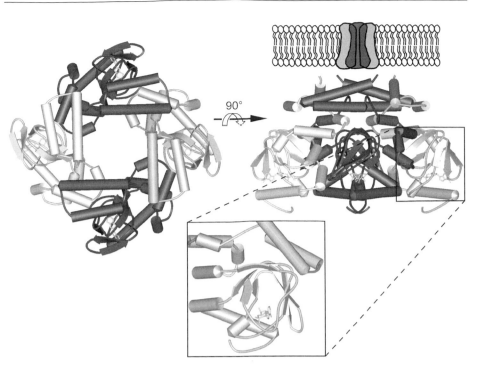

Figure 8   HCN2 C-terminal crystal structure. Structure of the HCN2 C-terminal region (73), viewed from the membrane (*left*) and from the side (*right*). The structure is positioned below the membrane-spanning portion of the channel, as it is thought to be in vivo. The structure contains four subunits, two in dark gray and two in light gray, with the C-linkers comprising the top half of the structure and the CNBDs the bottom half. cAMP (*yellow*) is bound in the CNBD of each subunit. The inset shows an enlargement of the β-roll of the CNBD, where cAMP binds.

of the C-helix. The phosphoribose of cGMP binds in the identical position, but cGMP binds in the syn configuration, the configuration with the purine ring folded back on top of the phosphoribose (Figure 9). In the syn configuration, the purine ring of cGMP can hydrogen bond with Thr592 of the β-roll. This interaction is also thought to occur in CNG channels, where Thr560 of rod channels and Thr537 of olfactory channels appear to confer cGMP selectivity. Mutation of these Thr residues decreases the channels' apparent affinity for cGMP while barely affecting cAMP affinity (84). The HCN2 C-terminal crystal structure not only provides insight into how the C-terminal regions of CNG and HCN channels are arranged and how ligands bind to the CNBD but also provides a framework for understanding biophysical data concerning cyclic nucleotide modulation.

EFFECTS OF CYCLIC NUCLEOTIDES ON HCN CHANNELS   HCN channels are activated by both membrane hyperpolarization and cyclic nucleotide binding. Voltage

cAMP   cGMP

anti configuration   syn configuration

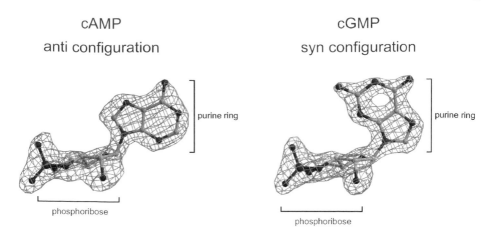

Figure 9   Cyclic nucleotide crystal structures. Density from X-ray crystallography (*webbing*) shown for cAMP (*left*) and cGMP (*right*), with the chemical structure of the molecules shown in ball-and-stick format. Elements are coded with color: carbon (*gray*), nitrogen (*blue*), phosphorous (*purple*), hydrogen (*red*).

and cyclic nucleotides, in fact, appear to activate the channel via the same gate (75, 76, 85). For homomeric HCN2 channels, saturating concentrations of both cAMP and cGMP produce the same effects. They both increase open probability at hyperpolarizing voltages, speed activation kinetics, and shift the voltage dependence of activation to more depolarized voltages (54). In response to hyperpolarization, HCN2 channels open with a predominantly exponential time course after an initial lag, with a midpoint of voltage activation (V_{half}) of approximately -130 mV (Figure 10). Saturating concentrations of cAMP shift the voltage dependence of activation approximately 15 mV in the depolarized direction (V_{half} of approximately -115 mV), speed the activation kinetics, and increase the maximal current at hyperpolarized voltages (67, 71, 73, 74). cAMP stabilizes the open state relative to the closed state, as measured by the change in free energy (ΔG) of channel opening in the presence of cAMP relative to the ΔG in the absence of ligand, $\Delta \Delta G$. The $\Delta \Delta G$ for HCN2 channels in the presence of cAMP is approximately -3 kcal mol^{-1} (73). Saturating cGMP produces a similar effect on HCN2 channels, shifting the activation curve approximately the same amount but not speeding the activation kinetics quite as much (54, 73, 85). However, these cGMP effects require a 10-fold higher concentration of agonist to produce the same shift as cAMP. This cyclic nucleotide specificity probably results from hydrogen bonding between Arg632 in the C-helix and the purine ring of cAMP as well as from hydrophobic interactions between the adenine ring and the binding pocket. Despite differences in affinity, both cyclic nucleotides stabilize the closed-to-open equilibrium for HCN2 channels, enabling activation with less hyperpolarization and more complete activation at hyperpolarized voltages. Other HCN isoforms are similarly modulated by cAMP, exhibiting depolarizing shifts in V_{half} and accelerated activation

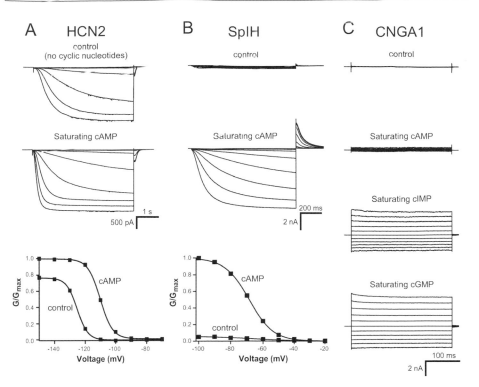

Figure 10 Cyclic nucleotide dependence of HCN2, SpIH, and CNGA1. Behavior of HCN2 (*A*), SpIH (*B*), and CNGA1 (*C*) channels in response to cyclic nucleotides. (*A*) HCN2 currents were recorded in the absence (*top*) or presence (*middle*) of saturating cAMP in response to voltage pulses from a holding potential of 0 mV to test potentials between –70 mV and –150 mV, returning to a tail potential of –40 mV. The conductance-voltage relations (*bottom*) were obtained from normalized tail currents. (*B*) SpIH currents were recorded in the absence (*top*) or presence (*middle*) of saturating cAMP in response to voltage pulses from a holding potential of 0 mV to test potentials between –20 mV and –100 mV, returning to a tail potential of +40 mV. The conductance-voltage relations (*bottom*) were obtained from normalized tail currents. (*C*) CNGA1 currents were recorded in the absence (*top*) and presence of saturating cAMP, cIMP, and cGMP in response to voltage pulses from a holding potential of 0 mV to test potentials between −100 mV to +100 mV.

kinetics, but they are not all modulated to the same degree. HCN1 channels display a shift in V_{half} of merely 2–5 mV in the presence of cAMP, whereas HCN4 channels shift approximately 15 mV (24, 71, 74).

A related HCN channel, SpIH (also known as spHCN) is found in sea urchin sperm. SpIH is a cation channel that, like other HCN channels, favors K$^+$ conductance and is responsive to both hyperpolarization and cAMP binding (86). The sequence of the SpIH subunit is most similar to that of the mammalian HCN channels (Figure 5). However, SpIH behavior differs from that of the rest of the HCN subunits in one noteworthy way: In the absence of cyclic nucleotide, hyperpolarization-activated currents inactivate (Figure 10). Binding of cAMP eliminates inactivation and dramatically increases the probability of opening with no shift in voltage dependence. SpIH channels are moderately voltage sensitive, moving approximately five equivalent charges through the membrane with activation, with a V_{half} of approximately -80 mV (76, 86). As in HCN channels, both voltage and cAMP are thought to open the same intracellular gate (75, 87). The gate must be coupled to both the voltage sensors and the CNBD for the channel to open in response to these stimuli. The mechanism of SpIH inactivation is thought to involve a "slip," or uncoupling, of the voltage sensors and the activation gate. Binding of cAMP must be able to prevent the channel from "slipping," as cAMP removes inactivation (75). SpIH can be thought of as functionally intermediate between HCN and CNG channels. SpIH channels are activated by hyperpolarization, as are HCN channels, but binding of cyclic nucleotides drastically increases the open probability of SpIH channels, as with CNG channels. These comparisons are especially apparent when looking at eletrophysiological data from all three types of channels, as in Figure 10.

EFFECTS OF CYCLIC NUCLEOTIDES ON CNG CHANNELS   In the absence of cyclic nucleotide, the open probability of CNG channels is extremely low (10^{-4}–10^{-6}) (88, 89). In CNGA1 channels, cGMP dramatically increases the open probability, whereas inosine $3',5'$-cyclic monophosphate (cIMP) and cAMP increase the open probability less (Figure 10). The change in free energy of CNG channel opening in the presence of cyclic nucleotides relative to unliganded channel opening, $\Delta\Delta G$, can be estimated for each ligand. If the free energy of opening for unliganded channels is 6.8 kcal mol^{-1} ($\Delta G = -RT\ln L$, where the equilibrium constant for channel opening $L = 1 \times 10^{-5}$), the $\Delta\Delta G$ values for each ligand are -8.8 kcal mol^{-1} for cGMP, -6.9 kcal mol^{-1} for cIMP, and -3.9 kcal mol^{-1} for cAMP (ΔG values from Reference 90). The $\Delta\Delta G$ values for cAMP are similar for CNGA1 and HCN2. cGMP, however, is a better agonist than cAMP for CNGA1 channels. This differential activation by agonists is observed in other CNG channels, although the relative effects of the different cyclic nucleotides are not always the same (74, 91–93). All of the cyclic nucleotides are thought to open the same activation gate in CNG channels, which is thought to be located in the pore loop, not intracellularly as it is in HCN and SpIH channels (94–97).

The greater agonist efficacy of cGMP in CNG channels is thought to be due, at least partially, to a residue in the C-helix that corresponds to Ile636 in HCN2. This residue interacts with cAMP in the HCN2 C-terminal crystal structure. If this residue, Asp604 of CNGA1, is mutated to a neutral residue, the channel's agonist selectivity switches from preferring cGMP to preferring cAMP (93, 98). These results suggest that Asp604 is hydrogen bonding to the N_1 and N_2 amines of the guanine ring of cGMP. Additionally, the high cGMP affinity of CNG channels is thought to arise from hydrogen bonding of Thr560 of CNGA1 with the N_2 amine of cGMP (84), as shown in the HCN2 C-terminal crystal structure (73). Therefore, it is possible that Thr560 holds cGMP in the syn configuration during initial binding and that Asp604 then interacts with cGMP during the allosteric conformational change of channel opening. Owing to the conservation of residues that interact with ligand in CNG and HCN channels, cyclic nucleotides probably bind in the same configuration in both channels.

MUTATIONS WITH SIMILAR EFFECTS ON CNG AND HCN CYCLIC NUCLEOTIDE MODULATION   If cyclic nucleotides activate CNG and HCN channels in a similar manner, then mutations that affect cyclic nucleotide modulation would be predicted to have similar effects on CNG and HCN channel behavior. As discussed above, the crystal structure of the HCN2 C-terminal region shows various intersubunit interactions, primarily in the C-linker regions, in which the A' and B' "elbow" of one subunit rests on the C' and D' "shoulder" of its neighbor (Figure 8). As these interactions are the points of subunit contact and are in the C-linker, they may assist in conferring ligand binding to the pore, resulting in channel opening. The crystal structure contains two salt bridges per subunit in the "elbow on the shoulder" region: one between the B'-helix and the D'-helix of the neighboring subunit (intersubunit salt bridge) and one between the same B'-helix residue and the β-roll of the same subunit (intrasubunit salt bridge). Mutation of these salt bridges revealed that they are present in intact HCN2 channels and in CNGA1 channels as well (67). Surprisingly, disruption of these salt bridges through mutation favors channel opening in both CNG and HCN channels. These results suggest that the C-linkers are in the closed configuration even though the CNBD is ligand bound in the HCN2 C-terminal crystal structure. Moreover, these findings highlight both the modular nature of gating as well as similarities between CNG and HCN channel structure and behavior.

Another recent result that suggests similar cyclic nucleotide modulation for CNG and HCN channels is the finding that three amino acids in the C-linker are crucial for normal ligand efficacy (74). This tripeptide is conserved among CNGA1, CNGA2, CNGA3, and HCN subunits, but not among CNGA4, CNGB1, or CNGB3 (channels that cannot produce functional homomers). If the C-linker of CNGA2 channels is substituted into CNGA4 channels, these chimeras are able to form functional homomers. When the conserved tripeptide is substituted with a tripeptide from CNGA4 in HCN channels, the efficacy of cAMP is greatly reduced. In fact, in this mutant, cAMP seems to inhibit rather than favor channel opening, as it does normally. Thus, it is suggested that this tripeptide confers the normal increase in open probability, with cyclic nucleotide binding in both CNG and HCN channels.

POSSIBLE MECHANISMS OF ACTIVATION BY CYCLIC NUCLEOTIDES

Understanding gating has long been a goal of ion channel research, and hence there have been numerous proposed mechanisms. Over time, and with advances in understanding of CNG and HCN channel behavior, quite a few mechanisms

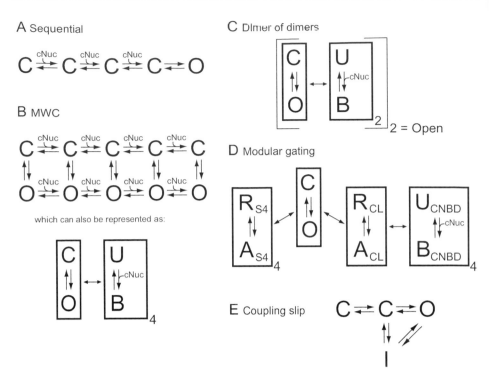

Figure 11 Proposed activation mechanisms (*A–E*) for CNG and HCN channels. States are shown as closed (C), open (O), resting (R), activated (A), unbound (U), and bound with ligand (B). Boxes around states indicate separable gating modules. Single-headed arrows indicate equilibrium between states, and double-headed arrows indicate coupling between gating modules.

have been suggested. The models that have been suggested to explain cyclic nucleotide activation of CNG and HCN channels are the sequential model; the Monod, Wyman, and Changeux (MWC) model; the dimer-of-dimers model; the modular gating model; and the coupling slip model (Figure 11).

The simplest model to explain cyclic nucleotide modulation is the sequential model. Originally formulated to describe retinal rod CNG channel response to increasing cyclic nucleotide concentration and voltage, the sequential model allows for sequential cyclic nucleotide–binding steps to closed channels, followed by a final closed-to-open transition when the channel is fully liganded (Figure 11*A*) (99). Variations of this model involving two to four ligand-binding steps have been proposed (99, 100). The two channel states, closed and open, are assumed to be the only conductance states. Additionally, the model assumes that the closed-to-open transition is very rapid and weakly voltage dependent. This model can reproduce the kinetic and steady-state behavior of CNG channels in response to various cyclic nucleotide concentrations and permits effects on ligand binding to be separated from effects on channel opening. However, this model does not completely reproduce channel behavior because it does not allow for channels to open when they are unliganded or partially liganded. Because CNG channels do show unliganded openings (89), and HCN channels open very well without ligand, this model is insufficient.

The next progression of modeling was the MWC model, a classic allosteric model that was originally formulated to describe hemoglobin behavior (101). The MWC model has been invoked to explain the dual activation by voltage and cyclic nucleotides for both CNG and HCN channels (89, 102–104). Generally, the MWC model is a gating scheme in which a concerted conformational transition occurs from the closed to the open state of the channel, and this transition is energetically stabilized by a constant amount for each ligand bound. This model therefore explicitly allows for opening to occur whether the channel is unliganded, partially liganded, or fully liganded (Figure 11B). Thus, at saturating concentrations of ligand, all four binding sites are assumed to be occupied and the model is reduced to one in which the channel is either closed but fully liganded or open and fully liganded. This is reminiscent of the final closed-to-open transition of the sequential model. The MWC model reproduces both the voltage dependence and the modulation of cyclic nucleotides for both HCN and CNG channels. However, the model is limited by the assumptions that there are four identical binding sites and that channel opening can be reduced to a single concerted allosteric transition of the entire protein.

In order to address these limitations of the MWC model, two different models have been proposed: a dimer-of-dimers model (Figure 11C) and modular gating model (Figure 11D). The dimer-of-dimers model arose from experiments designed to discover the energetic contribution of binding successive ligands to CNG and HCN channels (78, 105). In this model, each dimer acts as an MWC unit that undergoes a concerted allosteric transition from closed to open, but both dimers must be open for the channel to be open. This model suggests that in the absence of ligand, the C-terminal region assembles into a twofold symmetry and does not favor channel opening. Once the channel opens, it is thought to recover fourfold symmetry, as seen in the HCN2 crystal structure. Addition of cyclic nucleotide promotes channel opening and thus promotes fourfold symmetry. This model explains the energetic contribution of binding successive ligands in CNG and HCN channels.

The modular gating model arose from evidence that ion channels seem to undergo a series of coupled conformational changes in separate domains of the channel (106). This is in contrast to the single allosteric transition of the entire channel that is assumed by the MWC model. Cyclic nucleotide–dependent activation of CNG and HCN channels can be described as interactions between four domains: the pore, the S4 voltage sensor, the C-linker, and the CNBD (Figure 11D). Each domain, or module, is in equilibrium between two possible conformations. The pore can be closed or open, the voltage sensor and the C-linker can be resting or activated, and the CNBD can be unbound or bound with ligand. The modules are then coupled to each other, indicating that the conformation of one module affects the conformation of another module. This model can simulate the voltage dependence and cyclic nucleotide modulation for both HCN2 and CNGA1 channels (67).

In addition to reproducing voltage dependence and cyclic nucleotide modulation for both CNG and HCN channels, the modular gating model may also explain how, in the HCN2 C-terminal crystal structure, the CNBD can be bound by ligand while the C-linker can be in its resting conformation. If the resting state of the C-linker normally has an inhibitory effect on pore opening, then removing the pore (as for the HCN2 C-terminal crystal structure) should promote the resting state of the C-linker. Therefore, even in the presence of ligand, the "unloaded" (isolated from the pore) C-linker may return to its resting state. In fact, inhibition by the C-terminal regions has been proposed in a study with C-terminal deletion mutants of HCN1 and HCN2 channels (69). The CNBD exerts an inhibitory effect on channel activation by shifting the V_{half} of channel activation to more hyperpolarized potentials. Thus the modular gating model is a plausible model to explain HCN and CNG channel behavior.

A final model that has been proposed for HCN channels is the coupling slip model (Figure 11E). This model was proposed to explain the occurrence of inactivation in SpIH (and HCN) channels in the absence of cAMP (75). This model incorporates the idea that the voltage sensor and the activation gate are coupled and that, in the absence of cyclic nucleotide in SpIH, this coupling can "slip" such

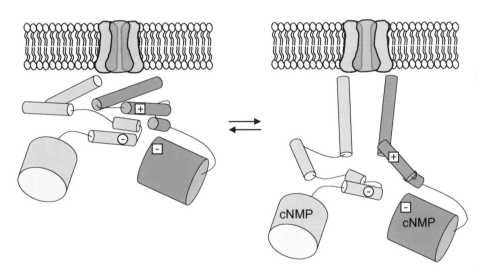

Figure 12   Gating model. Depiction of the HCN2 C-terminal regions from the side view, both in the resting configuration (*left*) and the active configuration (*right*). C-terminal regions of two subunits (*dark* and *light green*) are shown below the membrane-spanning portion of the channel. α-helices are shown with narrow cylinders, and the β-rolls of the CNBDs with wide cylinders. Salt bridge residues are shown: the positive B'-helix residue (*plus sign within square*) and the two negative residues on the D'-helix and the β-roll (*minus sign within circle* and *minus sign within square*).

that the channel will close even though the voltage sensor is activated, producing inactivation. In the presence of cyclic nucleotide, a coupling slip does not occur. A combination of the coupling slip model and the modular gating model may prove useful to predict behavior of CNG, HCN, and SpIH channels.

Using the modular gating scheme and conclusions derived from various CNG and HCN channel experiments, we can construct a cartoon demonstrating how these channels respond to cyclic nucleotide binding (Figure 12). In the absence of ligand, the C-linker is thought to be in a compact state, perhaps held in this position by "elbow on the shoulder" interactions, including salt bridges, between neighboring subunits and intrasubunit interactions between C-linkers and CNBDs. This compact state produces an inhibitory effect on channel opening. Then cyclic nucleotide binds to the CNBD, and there must be a quaternary rearrangement of the C-linkers, which removes the inhibitory effect of the C-terminal region. The C-terminal ends of the CNG S6 segments, which connect to the N-terminal ends of the A′-helices, are thought to be close together when the channel is closed and further apart when the channel is open (96, 107). Additionally, several His residues in the CNGA1 A′-helix, for example H420, are able to coordinate Ni^{2+} between the same residue on neighboring subunits in the open state of the channel (100, 108). In order for the A′-helices to be able to coordinate Ni^{2+}, the C-linkers must move such that the C-terminal ends of the A′-helices move closer together toward the central axis of the channel. This relaxed structure would disrupt the salt bridges that hold the C-terminal region in the tense structure, favoring channel opening.

CONCLUSIONS

Through structure-function analysis of CNG and HCN channels, a number of mechanisms have been proposed to explain gating and modulation by cyclic nucleotides. Initially these mechanisms only focused on certain features of channel behavior, such as liganded channel openings, and did not account for more complex behaviors, such as unliganded (spontaneous) openings and voltage-dependent openings. But as more was learned about these channels, the mechanisms became more elaborate. These latest mechanisms can apply to both CNG and HCN channels, two peas in a pod.

The *Annual Review of Physiology* is online at
http://physiol.annualreviews.org

LITERATURE CITED

1. Hille B. 2001. *Ion Channels of Excitable Membranes*. Sunderland, MA: Sinhauer Assoc., Inc. 814 pp.
2. Fesenko EE, Kolesnikov SS, Lyubarsky AL. 1985. Induction by cyclic GMP of cationic conductance in plasma membrane of retinal rod outer segment. *Nature* 313:310–13

3. Kaupp UB, Seifert R. 2002. Cyclic nucleotide-gated ion channels. *Physiol. Rev.* 82:769–824

4. Matulef K, Zagotta WN. 2003. Cyclic nucleotide-gated ion channels. *Annu. Rev. Cell Dev. Biol.* 19:23–44

5. Burns ME, Baylor DA. 2001. Activation, deactivation, and adaptation in vertebrate photoreceptor cells. *Annu. Rev. Neurosci.* 24:779–805

6. Detwiler P. 2002. Open the loop: Dissecting feedback regulation of a second messenger transduction cascade. *Neuron* 36:3–4

7. Palczewski K, Sokal I, Baehr W. 2004. Guanylate cyclase-activating proteins: Structure, function, and diversity. *Biochem. Biophys. Res. Commun.* 322:1123–30

8. Sharma RK. 2002. Evolution of the membrane guanylate cyclase transduction system. *Mol. Cell. Biochem.* 230:3–30

9. Frings S. 2001. Chemoelectrical signal transduction in olfactory sensory neurons of air-breathing vertebrates. *Cell. Mol. Life Sci.* 58:510–19

10. Bradley J, Reuter D, Frings S. 2001. Facilitation of calmodulin-mediated odor adaptation by cAMP-gated channel subunits. *Science* 294:2176–78

11. Chen TY, Yau KW. 1994. Direct modulation by Ca^{2+}-calmodulin of cyclic nucleotide-activated channel of rat olfactory receptor neurons. *Nature* 368:545–48

12. Kurahashi T, Menini A. 1997. Mechanism of odorant adaptation in the olfactory receptor cell. *Nature* 385:725–29

13. Munger SD, Lane AP, Zhong H, Leinders-Zufall T, Yau KW, et al. 2001. Central role of the CNGA4 channel subunit in Ca^{2+}-calmodulin-dependent odor adaptation. *Science* 294:2172–75

14. Weitz D, Ficek N, Kremmer E, Bauer PJ, Kaupp UB. 2002. Subunit stoichiometry of the CNG channel of rod photoreceptors. *Neuron* 36:881–89

15. Zheng J, Trudeau MC, Zagotta WN. 2002. Rod cyclic nucleotide-gated channels have a stoichiometry of three CNGA1 subunits and one CNGB1 subunit. *Neuron* 36:891–96

16. Zhong H, Molday LL, Molday RS, Yau KW. 2002. The heteromeric cyclic nucleotide-gated channel adopts a 3A:1B stoichiometry. *Nature* 420:193–98

17. Peng C, Rich ED, Varnum MD. 2004. Subunit configuration of heteromeric cone cyclic nucleotide-gated channels. *Neuron* 42:401–10

18. Zheng J, Zagotta WN. 2004. Stoichiometry and assembly of olfactory cyclic nucleotide-gated channels. *Neuron* 42:411–21

19. Accili EA, Redaelli G, DiFrancesco D. 1997. Differential control of the hyperpolarization-activated current (i(f)) by cAMP gating and phosphatase inhibition in rabbit sino-atrial node myocytes. *J. Physiol.* 500(Pt. 3):643–51

20. DiFrancesco D. 1986. Characterization of single pacemaker channels in cardiac sino-atrial node cells. *Nature* 324:470–73

21. DiFrancesco D. 1991. The contribution of the "pacemaker" current (if) to generation of spontaneous activity in rabbit sino-atrial node myocytes. *J. Physiol.* 434:23–40

22. Dokos S, Celler B, Lovell N. 1996. Ion currents underlying sinoatrial node pacemaker activity: A new single cell mathematical model. *J. Theor. Biol.* 181:245–72

23. Stieber J, Hofmann F, Ludwig A. 2004. Pacemaker channels and sinus node arrhythmia. *Trends Cardiovasc. Med.* 14:23–28

24. Altomare C, Terragni B, Brioschi C, Milanesi R, Pagliuca C, et al. 2003. Heteromeric HCN1-HCN4 channels: A comparison with native pacemaker channels from the rabbit sinoatrial node. *J. Physiol.* 549:347–59

25. Guth BD, Dietze T. 1995. I(f) current mediates beta-adrenergic enhancement of heart rate but not contractility in vivo. *Basic Res. Cardiol.* 90:192–202

26. Mangoni ME, Nargeot J. 2001. Properties of the hyperpolarization-activated current (I(f)) in isolated mouse sino-atrial cells. *Cardiovasc. Res.* 52:51–64

27. Ono K, Shibata S, Iijima T. 2003. Pacemaker mechanism of porcine sino-atrial node cells. *J. Smooth Muscle Res.* 39:195–204

28. Stein RB, Gossen ER, Jones KE. 2005. Neuronal variability: Noise or part of the signal? *Nat. Rev. Neurosci.* 6:389–97

29. Chan CS, Shigemoto R, Mercer JN, Surmeier DJ. 2004. HCN2 and HCN1 channels govern the regularity of autonomous pacemaking and synaptic resetting in globus pallidus neurons. *J. Neurosci.* 24:9921–32

30. Banks MI, Pearce RA, Smith PH. 1993. Hyperpolarization-activated cation current (Ih) in neurons of the medial nucleus of the trapezoid body: voltage-clamp analysis and enhancement by norepinephrine and cAMP suggest a modulatory mechanism in the auditory brain stem. *J. Neurophysiol.* 70:1420–32

31. Cuttle MF, Rusznak Z, Wong AY, Owens S, Forsythe ID. 2001. Modulation of a presynaptic hyperpolarization-activated cationic current (I(h)) at an excitatory synaptic terminal in the rat auditory brainstem. *J. Physiol.* 534:733–44

32. Ingram SL, Williams JT. 1996. Modulation of the hyperpolarization-activated current (Ih) by cyclic nucleotides in guinea-pig primary afferent neurons. *J. Physiol.* 492(Pt. 1):97–106

33. Saitow F, Konishi S. 2000. Excitability increase induced by beta-adrenergic receptor-mediated activation of hyperpolarization-activated cation channels in rat cerebellar basket cells. *J. Neurophysiol.* 84:2026–34

34. Moosmang S, Stieber J, Zong X, Biel M, Hofmann F, Ludwig A. 2001. Cellular expression and functional characterization of four hyperpolarization-activated pacemaker channels in cardiac and neuronal tissues. *Eur. J. Biochem.* 268:1646–52

35. Lujan R, Albasanz JL, Shigemoto R, Juiz JM. 2005. Preferential localization of the hyperpolarization-activated cyclic nucleotide-gated cation channel subunit HCN1 in basket cell terminals of the rat cerebellum. *Eur. J. Neurosci.* 21:2073–82

36. Magee JC. 1999. Dendritic Ih normalizes temporal summation in hippocampal CA1 neurons. *Nat. Neurosci.* 2:508–14

37. Williams SR, Stuart GJ. 2000. Site independence of EPSP time course is mediated by dendritic I(h) in neocortical pyramidal neurons. *J. Neurophysiol.* 83:3177–82

38. Moosmang S, Biel M, Hofmann F, Ludwig A. 1999. Differential distribution of four hyperpolarization-activated cation channels in mouse brain. *Biol. Chem.* 380:975–80

39. Santoro B, Chen S, Luthi A, Pavlidis P, Shumyatsky GP, et al. 2000. Molecular and functional heterogeneity of hyperpolarization-activated pacemaker channels in the mouse CNS. *J. Neurosci.* 20:5264–75

40. Baruscotti M, Difrancesco D. 2004. Pacemaker channels. *Ann. NY Acad. Sci.* 1015:111–21

41. Tang CY, Papazian DM. 1997. Transfer of voltage independence from a rat olfactory channel to the Drosophila ether-a-go-go K^+ channel. *J. Gen. Physiol.* 109:301–11

42. Doyle DA, Morais Cabral J, Pfuetzner RA, Kuo A, Gulbis JM, et al. 1998. The structure of the potassium channel: Molecular basis of K^+ conduction and selectivity. *Science* 280:69–77

43. Trudeau MC, Zagotta WN. 2003. Calcium/calmodulin modulation of olfactory and rod cyclic nucleotide-gated ion channels. *J. Biol. Chem.* 278:18705–8

44. Zhong H, Lai J, Yau KW. 2003. Selective heteromeric assembly of cyclic nucleotide-gated channels. *Proc. Natl. Acad. Sci. USA* 100:5509–13

45. Bruggemann A, Pardo LA, Stuhmer W, Pongs O. 1993. Ether-a-go-go encodes a voltage-gated channel permeable to K^+

and Ca^{2+} and modulated by cAMP. *Nature* 365:445–48

46. Cui J, Melman Y, Palma E, Fishman GI, McDonald TV. 2000. Cyclic AMP regulates the HERG K^+ channel by dual pathways. *Curr. Biol.* 10:671–74

47. Hoshi T. 1995. Regulation of voltage dependence of the KAT1 channel by intracellular factors. *J. Gen. Physiol.* 105:309–28

48. Eismann E, Muller F, Heinemann SH, Kaupp UB. 1994. A single negative charge within the pore region of a cGMP-gated channel controls rectification, Ca^{2+} blockage, and ionic selectivity. *Proc. Natl. Acad. Sci. USA* 91:1109–13

49. Heinemann SH, Terlau H, Stuhmer W, Imoto K, Numa S. 1992. Calcium channel characteristics conferred on the sodium channel by single mutations. *Nature* 356:441–43

50. Hess P, Lansman JB, Tsien RW. 1986. Calcium channel selectivity for divalent and monovalent cations. Voltage and concentration dependence of single channel current in ventricular heart cells. *J. Gen. Physiol.* 88:293–319

51. Sesti F, Eismann E, Kaupp UB, Nizzari M, Torre V. 1995. The multi-ion nature of the cGMP-gated channel from vertebrate rods. *J. Physiol.* 487(Pt. 1):17–36

52. Yang J, Ellinor PT, Sather WA, Zhang JF, Tsien RW. 1993. Molecular determinants of Ca^{2+} selectivity and ion permeation in L-type Ca^{2+} channels. *Nature* 366:158–61

53. Heginbotham L, Abramson T, MacKinnon R. 1992. A functional connection between the pores of distantly related ion channels as revealed by mutant K^+ channels. *Science* 258:1152–55

54. Ludwig A, Zong X, Jeglitsch M, Hofmann F, Biel M. 1998. A family of hyperpolarization-activated mammalian cation channels. *Nature* 393:587–91

55. Proenza C, Angoli D, Agranovich E, Macri V, Accili EA. 2002. Pacemaker channels produce an instantaneous current. *J. Biol. Chem.* 277:5101–9

56. DiFrancesco D. 1981. A new interpretation of the pace-maker current in calf Purkinje fibres. *J. Physiol.* 314:359–76

57. Dzeja C, Hagen V, Kaupp UB, Frings S. 1999. Ca^{2+} permeation in cyclic nucleotide-gated channels. *EMBO J.* 18:131–44

58. Yu X, Duan KL, Shang CF, Yu HG, Zhou Z. 2004. Calcium influx through hyperpolarization-activated cation channels (I(h) channels) contributes to activity-evoked neuronal secretion. *Proc. Natl. Acad. Sci. USA* 101:1051–56

59. Biel M, Schneider A, Wahl C. 2002. Cardiac HCN channels: Structure, function, and modulation. *Trends Cardiovasc. Med.* 12:206–12

60. Mannikko R, Elinder F, Larsson HP. 2002. Voltage-sensing mechanism is conserved among ion channels gated by opposite voltages. *Nature* 419:837–41

61. Vemana S, Pandey S, Larsson HP. 2004. S4 movement in a mammalian HCN channel. *J. Gen. Physiol.* 123:21–32

62. Bell DC, Yao H, Saenger RC, Riley JH, Siegelbaum SA. 2004. Changes in local S4 environment provide a voltage-sensing mechanism for mammalian hyperpolarization-activated HCN channels. *J. Gen. Physiol.* 123:5–19

63. Aggarwal SK, MacKinnon R. 1996. Contribution of the S4 segment to gating charge in the Shaker K^+ channel. *Neuron* 16:1169–77

64. Schoppa NE, McCormack K, Tanouye MA, Sigworth FJ. 1992. The size of gating charge in wild-type and mutant Shaker potassium channels. *Science* 255:1712–15

65. Zagotta WN, Hoshi T, Dittman J, Aldrich RW. 1994. Shaker potassium channel gating. II: Transitions in the activation pathway. *J. Gen. Physiol.* 103:279–319

66. Chen S, Wang J, Siegelbaum SA. 2001. Properties of hyperpolarization-activated pacemaker current defined by coassembly

of HCN1 and HCN2 subunits and basal modulation by cyclic nucleotide. *J. Gen. Physiol.* 117:491–504

67. Craven KB, Zagotta WN. 2004. Salt bridges and gating in the COOH-terminal region of HCN2 and CNGA1 channels. *J. Gen. Physiol.* 124:663–77

68. Ludwig A, Zong X, Stieber J, Hullin R, Hofmann F, Biel M. 1999. Two pacemaker channels from human heart with profoundly different activation kinetics. *EMBO J.* 18:2323–29

69. Wainger BJ, DeGennaro M, Santoro B, Siegelbaum SA, Tibbs GR. 2001. Molecular mechanism of cAMP modulation of HCN pacemaker channels. *Nature* 411: 805–10

70. Wang J, Chen S, Nolan MF, Siegelbaum SA. 2002. Activity-dependent regulation of HCN pacemaker channels by cyclic AMP: Signaling through dynamic allosteric coupling. *Neuron* 36:451–61

71. Wang J, Chen S, Siegelbaum SA. 2001. Regulation of hyperpolarization-activated HCN channel gating and cAMP modulation due to interactions of COOH terminus and core transmembrane regions. *J. Gen. Physiol.* 118:237–50

72. Yu HG, Lu Z, Pan Z, Cohen IS. 2004. Tyrosine kinase inhibition differentially regulates heterologously expressed HCN channels. *Pflugers Arch.* 447:392–400

73. Zagotta WN, Olivier NB, Black KD, Young EC, Olson R, Gouaux E. 2003. Structural basis for modulation and agonist specificity of HCN pacemaker channels. *Nature* 425:200–5

74. Zhou L, Olivier NB, Yao H, Young EC, Siegelbaum SA. 2004. A conserved tripeptide in CNG and HCN channels regulates ligand gating by controlling C-terminal oligomerization. *Neuron* 44:823–34

75. Shin KS, Maertens C, Proenza C, Rothberg BS, Yellen G. 2004. Inactivation in HCN channels results from reclosure of the activation gate: Desensitization to voltage. *Neuron* 41:737–44

76. Shin KS, Rothberg BS, Yellen G. 2001. Blocker state dependence and trapping in hyperpolarization-activated cation channels: Evidence for an intracellular activation gate. *J. Gen. Physiol.* 117:91–101

77. Tibbs GR, Liu DT, Leypold BG, Siegelbaum SA. 1998. A state-independent interaction between ligand and a conserved arginine residue in cyclic nucleotide-gated channels reveals a functional polarity of the cyclic nucleotide binding site. *J. Biol. Chem.* 273:4497–505

78. Ulens C, Siegelbaum SA. 2003. Regulation of hyperpolarization-activated HCN channels by cAMP through a gating switch in binding domain symmetry. *Neuron* 40:959–70

79. Kim LA, Furst J, Gutierrez D, Butler MH, Xu S, et al. 2004. Three-dimensional structure of I_to; Kv4.2-KChIP2 ion channels by electron microscopy at 21 Angstrom resolution. *Neuron* 41:513–19

80. Sokolova O, Kolmakova-Partensky L, Grigorieff N. 2001. Three-dimensional structure of a voltage-gated potassium channel at 2.5 nm resolution. *Structure* 9:215–20

81. Johnson JP Jr, Zagotta WN. 2005. The carboxyl-terminal region of cyclic nucleotide-modulated channels is a gating ring, not a permeation path. *Proc. Natl. Acad. Sci. USA* 102:2742–47

82. Su Y, Dostmann WR, Herberg FW, Durick K, Xuong NH, et al. 1995. Regulatory subunit of protein kinase A: Structure of deletion mutant with cAMP binding domains. *Science* 269:807–13

83. Weber IT, Gilliland GL, Harman JG, Peterkofsky A. 1987. Crystal structure of a cyclic AMP-independent mutant of catabolite gene activator protein. *J. Biol. Chem.* 262:5630–36

84. Altenhofen W, Ludwig J, Eismann E, Kraus W, Bonigk W, Kaupp UB. 1991. Control of ligand specificity in cyclic nucleotide-gated channels from rod photoreceptors and olfactory epithelium. *Proc. Natl. Acad. Sci. USA* 88:9868–72

85. DiFrancesco D, Tortora P. 1991. Direct activation of cardiac pacemaker channels by intracellular cyclic AMP. *Nature* 351:145–47

86. Gauss R, Seifert R, Kaupp UB. 1998. Molecular identification of a hyperpolarization-activated channel in sea urchin sperm. *Nature* 393:583–87

87. Rothberg BS, Shin KS, Phale PS, Yellen G. 2002. Voltage-controlled gating at the intracellular entrance to a hyperpolarization-activated cation channel. *J. Gen. Physiol.* 119:83–91

88. Ruiz ML, Karpen JW. 1997. Single cyclic nucleotide-gated channels locked in different ligand-bound states. *Nature* 389:389–92

89. Tibbs GR, Goulding EH, Siegelbaum SA. 1997. Allosteric activation and tuning of ligand efficacy in cyclic-nucleotide-gated channels. *Nature* 386:612–15

90. Sunderman ER, Zagotta WN. 1999. Mechanism of allosteric modulation of rod cyclic nucleotide-gated channels. *J. Gen. Physiol.* 113:601–20

91. Gordon SE, Zagotta WN. 1995. Localization of regions affecting an allosteric transition in cyclic nucleotide-activated channels. *Neuron* 14:857–64

92. Shapiro MS, Zagotta WN. 2000. Structural basis for ligand selectivity of heteromeric olfactory cyclic nucleotide-gated channels. *Biophys. J.* 78:2307–20

93. Varnum MD, Black KD, Zagotta WN. 1995. Molecular mechanism for ligand discrimination of cyclic nucleotide-gated channels. *Neuron* 15:619–25

94. Becchetti A, Gamel K, Torre V. 1999. Cyclic nucleotide-gated channels. Pore topology studied through the accessibility of reporter cysteines. *J. Gen. Physiol.* 114:377–92

95. Becchetti A, Roncaglia P. 2000. Cyclic nucleotide-gated channels: Intra- and extracellular accessibility to Cd^{2+} of substituted cysteine residues within the P-loop. *Pflugers Arch.* 440:556–65

96. Flynn GE, Zagotta WN. 2001. Conformational changes in S6 coupled to the opening of cyclic nucleotide-gated channels. *Neuron* 30:689–98

97. Liu J, Siegelbaum SA. 2000. Change of pore helix conformational state upon opening of cyclic nucleotide-gated channels. *Neuron* 28:899–909

98. Sunderman ER, Zagotta WN. 1999. Sequence of events underlying the allosteric transition of rod cyclic nucleotide-gated channels. *J. Gen. Physiol.* 113:621–40

99. Karpen JW, Zimmerman AL, Stryer L, Baylor DA. 1988. Gating kinetics of the cyclic-GMP-activated channel of retinal rods: Flash photolysis and voltage-jump studies. *Proc. Natl. Acad. Sci. USA* 85:1287–91

100. Gordon SE, Zagotta WN. 1995. Subunit interactions in coordination of Ni^{2+} in cyclic nucleotide-gated channels. *Proc. Natl. Acad. Sci. USA* 92:10222–26

101. Monod J, Wyman J, Changeux JP. 1965. On the nature of allosteric transitions: A plausible model. *J. Mol. Biol.* 12:88–118

102. DiFrancesco D. 1999. Dual allosteric modulation of pacemaker (f) channels by cAMP and voltage in rabbit SA node. *J. Physiol.* 515(Pt. 2):367–76

103. Goulding EH, Tibbs GR, Siegelbaum SA. 1994. Molecular mechanism of cyclic-nucleotide-gated channel activation. *Nature* 372:369–74

104. Varnum MD, Zagotta WN. 1996. Subunit interactions in the activation of cyclic nucleotide-gated ion channels. *Biophys. J* 70:2667–79

105. Liu DT, Tibbs GR, Paoletti P, Siegelbaum SA. 1998. Constraining ligand-binding site stoichiometry suggests that a cyclic nucleotide-gated channel is composed of two functional dimers. *Neuron* 21:235–48

106. Horrigan FT, Aldrich RW. 2002. Coupling between voltage sensor activation

Ca^{2+} binding and channel opening in large conductance (BK) potassium channels. *J. Gen. Physiol.* 120:267–305

107. Flynn GE, Zagotta WN. 2003. A cysteine scan of the inner vestibule of cyclic nucleotide-gated channels reveals architecture and rearrangement of the pore. *J. Gen. Physiol.* 121:563 82

108. Johnson JP Jr, Zagotta WN. 2001. Rotational movement during cyclic nucleotide-gated channel opening. *Nature* 412: 917–21

Annu. Rev. Physiol. 2006. 68:403–29
doi: 10.1146/annurev.physiol.68.040104.131404
Copyright © 2006 by Annual Reviews. All rights reserved
First published online as a Review in Advance on September 21, 2005

CLAUDINS AND EPITHELIAL PARACELLULAR TRANSPORT

Christina M. Van Itallie[1] and James M. Anderson[2]

[1]Department of Medicine, Division of Gastroenterology and Hepatology, University of North Carolina, Chapel Hill, North Carolina 27599-7545; email: vitallie@med.unc.edu

[2]Department of Cell and Molecular Physiology, University of North Carolina, Chapel Hill, North Carolina 27599-7545; email: jandersn@med.unc.edu

Key Word tight junction, permselectivity, occludin, transepithelial electrical resistance, flux

■ **Abstract** Tight junctions form continuous intercellular contacts controlling solute movement through the paracellular pathway across epithelia. Paracellular barriers vary among epithelia in electrical resistance and behave as if they are lined with pores that have charge and size selectivity. Recent evidence shows that claudins, a large family (at least 24 members) of intercellular adhesion molecules, form the seal and its variable pore-like properties. This evidence comes from the study of claudins expressed in cultured epithelial cell models, genetically altered mice, and human mutants. We review information on the structure, function, and transcriptional and posttranslational regulation of the claudin family as well as of their evolutionarily distant relatives called the PMP22/EMP/MP20/claudin, or pfam00822, superfamily.

TIGHT JUNCTIONS AND PARACELLULAR TRANSPORT

Tight junctions encircle the apical end of the lateral surface of adjacent epithelial cells, making a continuous paracellular seal between the apical/mucosal and basolateral/serosal fluid compartments. In transmission electron microscopic images, tight junctions appear as a series of very close membrane appositions, which in freeze-fracture replicas correspond to interconnected networks of transmembrane proteins (Figure 1). Protein strands adhere to matching strands on adjacent cells, forming a series of barriers across the paracellular space. The number of strands and the complexity of the network correlate approximately with barrier "tightness" (1a), but more recent evidence shows barrier properties are also defined by the expression profile of claudins. Various excellent reviews have described the tight junction's many (almost 40 integral and peripheral) protein components (2, 3) and their roles in epithelial cell polarity and signaling (4–6). In this review we focus on claudins and their role in controlling paracellular transport.

Transepithelial movement of ions and noncharged solutes results from the interdependent processes of transcellular and paracellular transport. Movement of

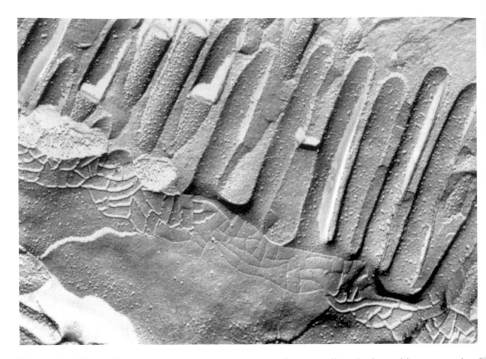

Figure 1   Freeze-fracture electron microscopic replica revealing the branching network of claudins, which form the tight junctions of epithelial cells in the mouse intestine. Apical microvilli are oriented upward. Adapted from Reference 1, figure 1, with permission from and courtesy of Dr. S. Bullivant.

material across cell membranes occurs in an energy-dependent fashion through a large number of cell type–specific pumps, channels, and transporters. In contrast, paracellular transport results from the passive movement of material down electrochemical gradients established either by the activity of the transcellular transporters or by externally imposed gradients such as those created by ingestion of solutes into the gastrointestinal tract. The paracellular pathway varies in its overall magnitude, typically estimated by the transepithelial electrical resistance (TER), and in its selectivity for ionic charge and solute size, or permselectivity (7–9). Resistance varies by $\sim 10^5$-fold along the spectrum from so-called tight to leaky epithelia. Discrimination between cations and anions varies perhaps 30-fold when measured in cultured cell monolayers (10), but less so in native tissues, possibly because the latter represent the average of different cell types, each with a different charge preference. Tight junctions show size discrimination, but controversy remains as to whether the size differs among different tissues.

Tight epithelia of high resistance can generate and maintain high transepithelial electrical potentials and ionic gradients and use these to form intralumenal fluids with compositions that deviate significantly from that of interstitial fluid

(7). Examples are found in the distal nephron and urinary bladder. At the opposite extreme, leaky epithelia of low resistance have low transepithclial potential and typically function to move larger volumes of iso-osomtic fluid. Examples of such epithelia include most epithelia of the gastrointestinal tract. General aspects of paracellular transport and pathophysiology have been reviewed elsewhere (7, 8, 11–14).

It is now clear that the tight junction contacts and their selective permeation properties are created by a large family of transmembrane proteins called claudins (15–17). A rapidly growing literature is beginning to explain how claudins control normal paracellular transport and how claudins are altered in, or contribute to, disease. This body of research includes evidence of several inherited human diseases resulting from mutations of claudins (18–20). The general field of tight junctions is also rapidly growing to reveal their role in epithelial cell polarity and oncogenesis (21, 22). In this review we limit our focus to claudins and their structure, function, regulation, and contribution to the formation of selective barriers.

THE CLAUDIN SUPERFAMILY: STRUCTURE AND DISTRIBUTION

The composition of tight junctions is quite complex and diverse, apparently much more so than the other epithelial junctions: gap and adherens junctions and desmosomes. Tight junctions are built from almost 40 different proteins, including members from multigene families (2). These proteins include three transmembrane proteins—claudins, occludin (23) and JAMs (24)—as well as cytoplasmic proteins fulfilling roles in scaffolding, cytoskeletal attachment, cell polarity, signaling, and vesicle trafficking. These proteins have been recently and comprehensively reviewed by Gonzalez-Mariscal and coworkers (25) and by Schneeberger & Lynch (2).

Many lines of evidence have shown that claudins, since their discovery in 1998 (15), are the principle barrier-forming proteins. Occludin, the first transmembrane protein described (23), was assumed to be the key component until occludin$^{-/-}$ mice were observed to have structurally and functionally normal junctions (26). This led the Tsukita group (15) to re-examine the same tight junction–enriched fraction they had used to isolate occludin, and from it identify claudin-1 and -2. In a series of seminal papers, this group showed that these claudins are incorporated into tight junction strands when expressed in cultured epithelial cells (15), can form strands de novo if expressed in fibroblasts (27), and are cell-cell adhesion molecules (28). There are at least 24 claudins in the annotated mammalian genomes (9), 56 in the puffer fish *Takifugu* (29), and 15 in zebrafish (in which some are required for proper organ morphogenesis) (30, 31). In retrospect, because occludin is encoded by a single gene and has no extracellular charges, it likely cannot explain the tissue-specific variations in resistance and charge selectivity. There is growing evidence

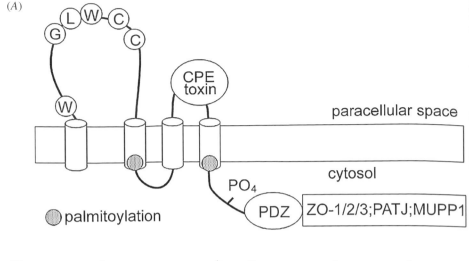

Figure 2 (A) Model depicting the conserved structural and functional features of claudins. The first extracellular loop contains the signature residues W-GLW-C-C [*shaded* in (B)]. See text for details. (B) Sequence alignment of the first extracellular loops of claudin-2, -14, and -16. When expressed in monolayers, claudin-2 confers higher permeability for Na^+ than for Cl^-, and vice versa for claudin-14. Coincidently, claudin-2 has more negative and claudin-14 more positive charges, supporting the notion that this loop is the charge-selectivity filter for the pore. The many negative charges in claudin-16 (also known as paracellin-1) suggest that it forms a cation pore.

that occludin is more important for cell signaling than for forming the paracellular barrier (32).

Claudins are predicted to have four transmembrane helices, with a very short internal N-terminal sequence (2–6 residues), two extracellular domains, and an internal C terminus (Figure 2A). Although this topology has not been directly proven, strong circumstantial evidence supports it. The first loop influences paracellular charge selectivity (33), the second loop is the receptor for a bacterial toxin (34), and the C terminus binds cytoplasmic proteins through a PDZ motif (35) (Figure 2A). Claudins are small, ranging from 20–27 kDa, and are recognized by a set of highly conserved amino acids in the first extracellular loop, with the residues

W-GLW C-C (Figure 2*B*). Although there has been no direct demonstration that the two extracellular cysteines form a disulfide bond, this seems likely given the oxidizing extracellular environment and the total conservation of these cysteines among claudins. The first extracellular loop ranges from 49–52 residues, followed by two transmembrane domains separated by a short (~15 amino acid) intracellular domain. The second extracellular loop is smaller, containing 16–33 amino acid residues. The cytoplasmic tails are the most diverse in sequence and vary in length from 21–63 residues. Claudins may form hexamers, as suggested by studies of claudin-4 (36) and a distant relative, MP20 (37).

All claudins (except claudin-12) end in PDZ-binding motifs, which interact with PDZ domains in the cytoplasmic scaffolding proteins ZO-1, -2, and -3 (35); MUPP1 (38, 39); and PATJ (40); and possibly other proteins. ZO-1 has 3 PDZ domains, MUPP1, MUPP13, and PATJ 10, suggesting that there exists a dense localized trap of PDZ interactions under the claudin strands. Claudins lacking the last three amino acids or those in which the PDZ-binding sites are blocked by addition of an epitope-tag (27) still localize to cell-cell contacts and form freeze-fracture strands, suggesting that they have an inherent ability to polymerize, independent of PDZ interactions. However, the strands formed by PDZ-blocked claudins are poorly organized and not restricted to the apical border (41). Perhaps the dense network of the PDZ domains in the ZOs, MUPP1, and PATJ functions to tether strands at the apical pole.

Not much more is currently known about the function of the cytoplasmic domains. Although the last three amino acids are critical for PDZ-dependent interactions, other regions of the cytoplasmic tail may be important in targeting, although this hypothesis is controversial. Furuse et al. (42) replaced the entire cytoplasmic domain of claudin-1 with a FLAG tag and found that this claudin still accumulates at sites of cell contact and forms strands in fibroblasts. However, a more recent study, using a similarly truncated claudin-1 tagged with GFP at the N terminus, found that this protein fails to localize to the plasma membrane and instead accumulates in intracellular vesicles (43). The reason for these different findings is unclear, but may lie in the location/type of tag. The membrane-proximal region of the C terminus is palmitoylated on conserved cysteines (44), and it is likely that most claudins can be phosphorylated on serines and/or threonines in the cytoplasmic tail (Figure 2*A*). A number of short motifs can be recognized as shared or different among subsets of tails, suggesting the existence of shared and distinct binding partners. One study that swapped tails between two different claudins showed that the tails controlled their unique protein half-lives. Presumably this results from interaction of the tails with different cytoplasmic binding partners (45).

The expression pattern of claudins is quite variable among different cell types and tissues. Some, e.g., claudin-1, are ubiquitously expressed, whereas others, e.g., claudin-16, are restricted to specific cell types (18, 46) or periods of development (47, 48). Some cells express a single claudin—e.g., Sertoli cells express only claudin-11 (46)—and others express many. Several groups have described segment-specific expression of different claudins along the nephron, whose paracellular properties are known to differ in each segment (49, 50). Presumably, these claudins confer gradients of resistance and charge selectivity, but the physiological implications for controlling paracellular transport remain unclear because the functional characteristics of most claudins are unknown. At the transcript level, claudins can differ by several thousandfold within a defined segment of the gastrointestinal tract (personal observations, J. Holmes, C. Van Itallie, & J. Anderson). The emerging data suggest that there may be "housekeeping" claudins expressed widely and at high levels as well as specialized claudins expressed at low levels or restricted to certain cell types.

Invertebrates also express claudins despite lacking tight junctions. Their epithelial barriers are formed by septate junctions with wide intercellular gaps, very unlike the near fusions at tight junctions. *Drosophila* have six claudin sequences (51–53), two of which, Megatrachea (51) and Sinuous (53), are located at septate junctions; mutations in Megatrachea disrupt the barrier. Mutations of either claudin also result in developmental defects in the size and shape of the tracheal epithelium. Five claudin-like sequences have been identified in *Caenorhabditis elegans*; RNAi-mediated ablation of claudin-like protein 1 disrupts the barrier between epithelial cells of the hypodermis (54). Although invertebrate claudins clearly contribute to the barrier, it is inconceivable that they can traverse the wide septate gap to form an occluding contact, as we envision occurs for vertebrate claudins. Perhaps invertebrate claudins share a signaling rather than pore-like function that is required to establish the barrier. For example, claudin-1 expression regulates epithelial-mesenchymal transformation through Wnt/beta-catenin signaling in human cancer cell lines (54b).

The claudins are members of the large PMP22/EMP/MP20/claudin mammalian superfamily also designated as pfam00822 by the NCBI. Nonclaudins range from approximately 20–47 kD and share the tetraspan topology and W-GLW-C-C signature motif residues; some but not all end in a PDZ-binding motif. Although historically these claudins were described in contexts other than that of the tight junction, it since has been confirmed that several share adhesive or barrier-forming properties (9).

The proteins most homologous to the orthodox claudins are eye lens–specific membrane protein (MP20) (55), the epithelial membrane proteins (EMP1, 2, 3) (56), and peripheral myelin protein 22 (PMP22) (57). MP20 is the second most prevalent membrane protein of the lens fiber cells after aquaporin 0; mutations in human MP20 cause cataracts (58). Although fiber cell junctions are ultrastructurally distinct from tight junctions, the location of MP20 coincides with the paracellular diffusion barrier for dyes (59). PMP22 is highly expressed in Schwann cells and required for myelin formation; it also influences cell proliferation and adhesion (60). Human gene duplications and mutations in PMP22 result in peripheral polyneuropathies (61). Although only 19% identical to human claudin-1, it is a component of tight junctions in liver, intestine (57), and the blood-brain barrier (62). When expressed in MDCK cells, PMP22 increases TER and paradoxically increases paracellular flux of tracer molecules, as well as inhibiting cell migration after wounding (63). Whereas PMP22 makes a good case for acceptance as an orthodox claudin, the relationship of EMPs to claudins remains unresolved. EMP2 interacts with β_1 integrin and influences cell adhesion to the extracellular matrix (64). EMP1 through 3, recently described as binding partners for the purineurgic $P2X_7$ ATP-gated family of ion channels, are required for ATP to induce apoptosis in a culture cell model (65).

CLP24 was originally cloned as a transcript unregulated by hypoxia. Although it is only 8% identical to claudin-1, when expressed in cultured MDCK cells it increases paracellular flux of tracer molecules and localizes to the apical junction complex (66). Interestingly, it localizes at the adherens junction and not the tight junction. Very distant members of the pfam00822 are the γ subunits of voltage-dependent calcium channels. These proteins are larger than the conventional claudins but share their topology and signature residues. The γ subunits bind pore-forming α subunits and are required for proper membrane delivery of α (67). One of these γ subunits, stargazin, which is an AMPA receptor regulator, was recently shown to mediate cell-cell adhesion in fibroblasts, suggesting even distantly related members of this family may have retained claudin adhesive function (68). A blast search of the NCBI database reveals more claudin-like proteins as uncharacterized open reading frames; some appear to be orthodox claudins, and others more distantly related. It will be interesting to see how large this family grows and to learn its members' common and unique functions.

CLAUDINS FORM SIZE- AND CHARGE-SELECTIVE PORES

Size

When determined by tracer flux, the apparent paracellular permeability of hydrophilic nonelectrolyte markers is inversely related to size down to an inflection point, which is interpreted as the pore size. The reported sizes vary widely, although most are in the range of 8–9 Å in diameter, which is much larger than are transmembrane pores and channels. Figure 3 shows illustrative evidence for existence of paracellular pores with a defined size cut-off; Watson et al. (69) used a continuous series of polyethylene glycol tracers to determine permeability properties in the T84 human intestinal cell line (69). In the older literature, when investigators tried to define the size of tight junction pores in native tissues, the values showed a wider range, from about 4–40 Å (14, 70). These latter values may be unreliable; most studies did not account for the possibility of transcytosis or passage of tracer through regions of damaged tissue. More importantly, some studies used charged tracers that may have been subject to tissue-specific charge discrimination. An early, excellent, and often-cited study from Salas et al. (71) employed a graded series of nitrogenous cations. Interpretation of this study was confounded by the

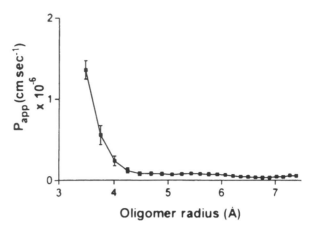

Figure 3   Apparent permeability, P_{app}, for polyethylene glycols of increasing sizes across T84 monolayers. A sharp inflection corresponds to molecules of 4.25 Å radius. Adapted from Reference 69, figure 1, with permission.

possibility of charge selectivity; nonetheless, these researchers observed an upper diameter cut-off of 8.8 Å in rabbit gallbladder and 16.2 Å in frog. Some endothelial tight junctions permit passage of larger solutes (40–60 Å) (72), although this may result from discontinuous cell-cell contacts or transcytosis.

The possibility that claudins create variable pore sizes was raised by vascular permeability studies in mice in which the claudin-5 gene was inactivated (73). Brain endothelial cells express claudin-5, -12, and possibly others. Deletion of claudin-5 leads to an incremental increase in the size of tracers allowed to exit from the vascular space into the brain. The authors of those studies interpreted this to mean that the remaining claudins form slightly larger pores. This issue might be tested in a more controlled fashion in a cultured cell model or when the molecular structure of the claudin pore is determined.

It remains surprisingly controversial whether tight junctions allow any significant passage of water. Despite the existence of the long-accepted solvent drag model describing how paracellular water carries along dissolved solutes, there are also elegant studies failing to document any paracellular water movement, even in very leaky epithelia (74). Perhaps through experimental manipulation of claudins we can learn more about the relative importance of transcellular (aquaporin) versus transjunctional water transport.

Resistance and Charge Discrimination

As noted above, native epithelia differ in resistance by $\sim 10^5$-fold. Most of this variation is accounted for by differences in resistance across the tight junction and not in plasma membrane resistance. Theoretically, variations in paracellular resistance may result from alterations in the physical dimensions of the pore (although

TABLE 1 Effects of expressing claudins through transfection in cultured epithelial monolayers

Protein	Cell type	R^a	P_{Na}	P_{Cl}	Flux[b]	Ref.
Cldn-1	MDCK	↑	—	—	↓	(123)
	MDCK	↑	—	—	No Δ	(41)
Cldn-2	LLC-PK$_1$	↓	↑	—	—	(10)
	MDCK-I	↓	—	-	—	(76)
	MDCK-C7[c]	↓	↑	No Δ	No Δ	(124)
	MDCK-II	No Δ	—	—	—	(10)
Cldn-4	MDCK-II	↑	↓	No Δ	No Δ	(125)
	LLC-PK$_1$	↑	-	—	—	(10)
Cldn-5	MDCK-II	↑	↓	No Δ	No Δ	(126)
Cldn-7	LLC-PK$_1$	↑	↑	↓		(127)
Cldn-8	MDCK	↑	↓	No Δ	No Δ	(128)
Cldn-11	MDCK-II	↑	—	—	—	(10)
	LLC-PK$_1$	↓	—	—	—	
Cldn-14	MDCK-II	↑	↓	No Δ	—	(83)
Cldn-15	LLC-PK$_1$	↓	↑	—	—	(10)
	MDCK-II	↑	—	—	—	(33)

[a] Transmonolayer electrical resistance compared with control.
[b] Transmonolayer flux of small tracer molecules, e.g., mannitol and dextrans.
[c] Like MDCK-I, the MDCK-C7 clone has high resistance and low P_{Na}.

this is unlikely, as discussed above), changes in the number of pores, or differences in the number of fixed charges on the pores. Since the charged-pore model was first proposed in the late 1960s (70), compelling evidence has supported the model and, more recently, the hypothesis that claudins' extracellular loops create the pores (Figure 2). Although there is insufficient experimental work to explain the basis of variations in resistance (e.g., density or location of charges), some evidence shows how charges on the extracellular loops discriminate between cations and anions (128).

When specific claudins are expressed in cultured epithelial monolayers, they consistently either increase or decrease electrical resistance (Table 1). In the limited examples available, behavior revealed in vitro appears to correlate nicely with the physiology of epithelia in which they are expressed in vivo. For example, claudin-4, -7, and -8 induce higher resistance in cultured models and are located only in distal nephron segments with high resistance (75). Likewise, claudin-2 induces lower resistance (76) and is found in vivo in leaky epithelia, such as the proximal renal tubule (25) and intestinal crypts (77).

Both tissues and culture cell lines show variable charge selectivity. The relative permeability for Na$^+$ to Cl$^-$ is easily measured, at least in leaky epithelia, using dilution potential methods (15). Most epithelia in the body are a mixture of cell types and are, on average, slightly cation selective. Cultured epithelial lines can show more extreme selectivity; the MDCK II line is cation selective (P_{Na+}/P_{Cl-} = 10), whereas the LLC-PK$_1$ line is anion selective (P_{Na+}/P_{Cl-} = 0.3). Expression of specific claudins on the background of MDCK or LLC-PK$_1$ monolayers either increases or decreases Na^+ permeability, with smaller effects on Cl^- permeability (Table 1). Charge-reversing mutagenesis of residues in the first extracellular loops of claudin-4 and -15 (10, 33) shows that permeability for solute ions is increased by opposite charges on the protein. Apparently some charged residues line the pore, creating a selectivity filter; however, not all the key residues have been resolved. Paracellular resistance and ionic charge selectivity likely are functions of the number and combinations of claudins expressed in individual tight junctions.

The Conductance-Flux Paradox

No discussion of flux would be complete without mentioning a long-standing puzzle in the field: Charged and noncharged solutes appear to have different barrier-regulating permeabilities. If paracellular transport were governed only by molecular dimensions of the pores, one would expect that reducing the pore size would simultaneously increase electrical resistance for charged solutes and decrease the flux of noncharged solutes. In fact, there is frequently a dissociation (78, 79). The traditional explanation for this is that the barrier must be dynamic. Resistance is an instantaneous (on the scale of milliseconds) measurement of the profile of pores in the barrier, whereas flux is measured over hours and includes additional pathways from breaks in the barrier or fluctuating pores (78). A recent study visualizing GFP-tagged claudin in live cells revealed the strands to be quite dynamic (80): They break, migrate, and reconnect in a manner that would allow saltatory movements of solutes over time (view Figure 1 for reference). In terms of claudins and instantaneous pore properties, the mutagenesis studies strongly suggest that charges on the claudin lining the pore create different effective pore dimensions depending on the solute charge or lack of it (Figure 2B; Table 1). This at least would explain why the electrical resistance can change without affecting flux of noncharged molecules. The structure of claudin pores and their possible dynamic nature deserve more investigation.

CLAUDIN GENE DELETION MODELS IN MICE

Various studies, some attempting to model human claudin mutants, have reported genetic deletion of several orthodox claudins in mice. The phenotypes of these mice confirm claudins' general role in creating barriers and in some cases reveal highly specialized roles in particular cell types and in permselectivity (Table 2). By contrast, when occludin was deleted, mice were viable and showed morphologically normal tight junctions, but also exhibited a constellation of phenotypes that are not easily interpreted as barrier defects (26). In fact, a recent detailed

TABLE 2   Tight junction gene deletions and transgenics in mouse

Gene	Phenotype	Ref.
Cldn-1 KO	Skin barrier defect	(85)
Cldn-5 KO	Size-selective BBB defect	(73)
Cldn-6 TG	Skin barrier defect	(86)
Cldn-11 KO	CNS myelin defect	(46)
Cldn-11 KO	Loss of endocochlear potential	(82)
Cldn-14 KO	Phenocopy of human deafness	(83)
Cldn-19 KO	Swann cell barrier defect	(84)
Occludin	a. Viable with complex phenotype	(85)
	b. No paracellular defects	(81)
	No gastric parietal cells	

electrophysiologic analysis of epithelia from the stomach, colon, and urinary bladder failed to reveal any change in electrical properties or [^{3}H] mannitol flux (81). This report also documented the absence of gastric parietal cells, which, if it is the result of a developmental signaling defect, is consistent with a report that occludin binds and augments the activity of the TGF-β Type 1 receptor (32). Thus, the absence of barrier defects in occludin$^{-/-}$ mice and evidence that occludin influences TGF-β signaling suggest that occludin may be more important in junctional signaling than in sealing.

Claudin-11, also known as oligodendrocyte-specific protein, was first thought specific to brain before the larger claudin family was recognized (46). This protein actually is expressed in many epithelia in the body. In oligodentrocytes it forms the continuous adhesive contacts around axons that are required for efficient salta-tory conduction between nodes. Homozygous null animals no longer show rows of freeze-fracture particles and have severely delayed CNS conduction velocities (46). Although these defects can be interpreted as simple adhesion defects, the additional loss of hearing suggests a more specific defect in permselectivity. Claudin-11 normally is expressed in the epithelium lining the endocochlear space, which generates a peculiar fluid composition: 150 mM K$^+$ and 90 mV positive relative to the adjacent tissue space. These characteristics are required for mechanotransduction by the hair cells. The endolymph of knockout (KO) animals maintains a high K$^+$ concentration but a reduced (approximately 30 mV) electrical potential, perhaps owing to a defect in the paracellular barrier not compensated for by transcellular transport (82). Claudin-14, like claudin-11, is expressed in the epithelia lining the endolymph; both mouse KOs and humans with mutations in the claudin-14 gene are deaf (83) (Tables 2 and 3). Here the mechanism by which claudin-14 acts is more obscure because the mice initially show a normal endocochlear potential but subsequently suffer progressive degeneration of the hair cells (83).

TABLE 3   Genetic diseases of tight junction proteins

Gene	Disease	Pathology/Mechanism	Ref.
Cldn-1	Ichthyosis and sclerosing cholangitis	Affects skin and bile ducts	(19)
Cldn-14	Nonsyndromic deafness, *DFNB29*	Cochlear hair cell degeneration	(83)
Cldn-16			
Human	HHNC[a]	Defective renal Mg^{2+} reabsorption	(18)
Bovine	Interstitial nephritis		(129)
PMP22	Peripheral polyneuropathies HNPP[b] Charcot-Marie-Tooth type 1A Dejerine-Sottas syndrome	Demyelination Gene deletion Gene duplication Mutations	(87)
ZO-2	Familial hypercholanemia	Defective PDZ-claudin binding	(88)

[a]Hypomagnesemia hypercalciuria with nephrocalcinosis.
[b]Hereditary neuropathy with liability to pressure palsies.

Claudin-19, normally found at the sealing contacts of the axon-insulating Schwann cells, appears to serve a role in the peripheral nervous system analogous to that of claudin-11 in the CNS. Mice with homozygous deletions lack freeze-fracture particles, although the histology of wrapping is not obviously affected (84). Nerve conduction velocity is severely impaired, consistent with loss of the ionic seal.

Claudin-1 KO mice studies revealed the surprising insight that tight junctions, or at least components of the junction, are critical for the epidermal barrier. Previously it was assumed that the skin barrier was formed by intracellular accumulation of hydrophobic material and the cumulative layers of dead squamous cells. Claudin-1 KO mice die shortly after birth from transdermal water loss (85). Wild-type mice show continuous junctional contacts formed by both claudin-1 and -4 between cells in the stratum granulosum. A similar loss of the skin barrier was observed in transgenic mice expressing claudin-6 with the skin-specific involucrin promoter (86).

HUMAN CLAUDIN MUTATIONS

Inherited human diseases can result from mutation in the orthodox claudin-1 (19), -14 (20), and -16 (18). Within the extended claudin family, mutations in lens MP20 lead to cataracts (58), and in Schwann cell PMP22 to peripheral neuropathies (87) (Table 3). A missense mutation in ZO-2, a cytoplasmic tight junction scaffold, is associated with defective bile secretion (88). Interestingly, the mutation is in a PDZ domain and reduces the affinity of claudin for this domain in vitro (89).

Mutations in the claudin-1 gene have been reported in a few children presenting with scaling of the skin, and progressive scarring and obstruction of the bile ducts, so-called neonatal sclerosing cholangitis with ichthyosis (19). Both these features can be nonspecific responses to injury of the ducts and skin, and the pathophysiology remains obscure. The clinical course is markedly variable. One child developed liver failure requiring transplantation; in another, symptoms resolved in the second decade. This variability in penetrance suggests the existence of modifying genes. Biliary disease in the mouse claudin-1 KO model was not detected, as the KO mice died at birth.

Two missense mutations, both leading to a form of recessive nonsyndromic deafness designated DFNB29, have been identified in claudin-14. The V85D mutation was first discovered in two consanguineous Pakistani families (20), and more recently, a G101R mutation was found in individuals from Greece and Spain (90). Both mutations are predicted to introduce a charged residue into the second transmembrane helix. Not unexpectedly, both mutant proteins fail to localize to cell contacts when expressed in cultured cells (90).

The phenotype of mutations in claudin-16 suggests a defect in paracellular ion selectivity. This claudin is highly restricted to junctions of the thick ascending limb of the loop of Henle in the kidney, a segment that is specialized to reabsorb cations (principally Na^+ and K^+) from the lumen, in part by generating an intralumenal positive potential, driving them back to the interstitial space. Magnesium and calcium also are resorbed here, largely through the paracellular route. Loss of claudin-16, also called paracellin-1, results in a rare Mg^{2+}-wasting disease called familial hypomagnesaemia with hypercalciuria and nephrocalcinosis (FHHN) (18). One interpretation of the FHHN phenotype is that claudin-16 is a magnesium-specific pore and that when it is missing, less magnesium is resorbed. However, none of the cultured cell studies show that claudins discriminate very much among similarly charged ions. Thus, another interpretation is that nonselective reduction in cation permeability reduces the intralumenal electrical gradient required to drive magnesium back to the blood. Inspection of the sequence of the first loop of claudin-16 reveals a large number of negative charges, consistent with cation selectivity (Figure 2B). Changes in intralumenal potential and paracellular selectivity for cations have not been measured in humans with FHHN, but perhaps the specific influence of claudin-16 on selectivity could be measured after expressing the protein in a cultured epithelial model. Alternatively, claudin-16 mutations have also been documented in cows, in which they lead to interstitial nephritis, and direct study of nephrons isolated from these animals should be feasible (91).

A distinct form of abnormal renal calcium loss, grouped under idiopathic hypercalcuria, was recently shown to result from novel mutations in the claudin-16 gene, which interrupts binding to the PDZ domain of ZO-1 (89). These patients display self-limited childhood hypercalciuria and not the full spectrum of FHHN symptoms described above. When the mutant protein was expressed in cultured cells, it no longer localized to tight junctions but accumulated in lysosomes, in contrast to other claudins in which removing PDZ-binding motifs had no effect (35, 43). Why null and targeting-defective mutants exhibit distinct clinical syndromes is not clear.

Mutations in other claudins likely will be discovered to contribute to human disease through barrier or signaling defects. Perhaps some of these mutations will manifest as more subtle changes in transport or drug absorption.

ACUTE REGULATION OF CLAUDIN FUNCTION

The paracellular barrier is affected by a wide range of physiologic inputs, including hormones, cytokines, Na^+-coupled solute transport (92), myosin activity (93), and many cell signaling pathways (11). In some cases a reduction in the barrier may result simply from disruption of cell-cell contacts and not molecularly discrete changes among tight junction proteins. We review below some of the few examples of acute regulation that seems to occur at the level of claudins. The field is nascent, and we predict more examples.

Phosphorylation

Although there is an increasing body of evidence implicating occludin as a site for acute physiologic regulation at the tight junction (reviewed in Reference 2), we know far less about the modulation of claudins as short-term control mechanisms. Two potential areas of regulation that have attracted recent interest are posttranslational modifications, in particular phosphorylation, and removal of claudins from the tight junction by endocytosis.

Most claudins contain potential serine and/or threonine phosphorylation sites in their cytoplasmic C-terminal domains. A number of studies have demonstrated phosphorylation of these sites (Figure 1). Increased phosphorylation is associated with either decreased or increased barrier function. Treatment of cultured endothelial cells with cAMP enhances barrier function in both a protein kinase A-dependent and -independent fashion. In the same experiments, claudin-5, but not claudin-1, was phosphorylated on threonine in response to cAMP (94). In similar experiments, expression of wild-type claudin-1 and a mutant version with threonine 203 replaced with alanine demonstrated that phosphorylation at this site contributes to the ability of claudin-1 to form a paracellular barrier to either inulin or mannitol (95). However, in both sets of experiments, the effects of phosphorylation were subtle, suggesting that an inability for claudin-1 to be phosphorylated may not be critically important to maintenance of the endothelial tight junction barrier.

Several other lines of evidence suggest that claudin phosphorylation may selectively decrease barrier function in epithelial cells. Yamauchi et al. (96) found that the threonine-serine kinase WNK4 can bind and phosphorylate claudins-1 through -4 and that a human disease–causing mutant of WNK4 hyperphosphorylates claudins and increases paracellular Cl^- permeability. Lifton and coworkers

(97) came to a similar conclusion, finding that expression of WNK4 in MDCK cells decreases TER through a selective increase in paracellular Cl⁻ permeability. D'Souza et al. (98) demonstrated that both claudin-3 and -4 are phosphorylated in ovarian cancer cells and that mimicking constitutive phosphorylation by mutating threonine 199 to aspartic acid in claudin-3 results in decreased TER and increased paracellular permeability. Forskolin, which stimulates phosphorylation of claudin-3, similarly results in a threefold drop in TER. These authors observed that both forsokolin treatment and expression of the mutant claudin results in more diffuse localization of the claudin when assessed by immunofluorescence, suggesting that phosphorylation may negatively regulate claudin integration into the tight junction. In summary, phosphorylation of the claudin tails does appear to influence the barrier, but the mechanisms remain to be defined.

Other Potential Forms of Regulation

Endocytic recycling is a potential mechanism for acute tight junction regulation, which may be modulated by phosphorylation or other signaling pathways. This is a well-established mechanism for regulating plasma membrane transporters, e.g., aquaporin and CFTR. Recent studies have demonstrated that claudins are continually removed and replaced at the apical junctional complex. This process, which can be visualized in normal cells (99) and is stimulated by calcium removal (100), may contribute to tight junction disruption in inflammatory bowel disease (100). Rab13 modulates continuous occludin recycling at the junction; regulated recycling of the integral membrane proteins may provide a mechanism for acute physiologic regulation.

Other posttranslational modifications, including palmitoylation (44) and glycosylation (101), of claudins and members of the larger PMP22 family have been reported. These modifications influence protein stability or membrane delivery, although neither of these has been confirmed to be a mechanism for acute regulation of claudin function.

REGULATION OF CLAUDIN EXPRESSION

There are numerous studies reporting changes in claudin expression level in response to a variety of physiological and pharmacological stimuli. Unfortunately, few of these studies were correlated with paracellular properties. Published reports can be divided into two partially overlapping categories. The first deals with differentiation-specific gene expression, a regulatory program likely to be important in oncogenesis as well, as there are many examples in which up- or downregulation of claudin gene expression has been reported in association with a wide variety of tumors (102, 103). The second category is hormones or other factors that result in acute changes in claudin levels. We describe selected examples of both categories below and offer further examples in Table 4.

TABLE 4 Effectors of claudin expression and function

Effector	Claudins	Function	Cell/Tissue	Mechanism	Ref.
EGF	Cldn-2↓	TER ↑	MDCKII	MAPK/ERK 1/2	(130)
	Cldn-1, -3, -4 ↑				
HGF	Cldn-2 ↓	TER ↑	MDCKII	MAPK/ERK 1/2	(109)
	Cldn-2 ↑	TER ↓	MDCKI		
Retinoic acid	Cldn-6, -7 ↑	Induce TJ	F9	(−)	(131)
GATA-4 CDX2, HNF1α	Cldn-2 ↑	Maintains expression	Caco-2 HIEC	Transcription	(107)
CDX2 HNF1α	Cldn-2 ↑	(−)	Caco-2	Transcription	(132)
HNF-4α	Cldn-6, -7 ↑	Induce TJ	F9	(−)	(133)
T/EBP/NKX2.1	Cldn-18 ↑	Splicing	Lung	Transcription	(108)
Snail	Cldn-2, -4, -7 ↓	TER ↓	MDCKII	(−)	(106)
Snail	Cldn-1 ↓	(−)	MDCK	Translation	(105)
Snail	Cldn-3, -4, -7 ↓	EMT	Eph4, CSG1	Transcription	(104)
Sp1	Cldn-19 ↑	(−)	Kidney cells	Transcription	(134)
Mg²⁺, vit D	Cldn-16 ↓	(−)	OK	Transcription	(135)
HIV-1 Tat	Cldn-1, -5 ↓	(−)	BMEC	VEGF2, MEK1	(136)
IL-1β	Cldn-2 ↑	(−)	Liver, hepat	MAP, PI3-kin	(137)
IL-17	Cldn-1, -2↑	TER↑	T84	MEK	(113)
Hypoxia	Cldn-3 ↑	(−)	HUVEC	Transcription	(138)
Isch/reperfusion	Cldn-1, -3, -7 ↑	(−)	Kidney	Transcription	(139)
IFN, TNF, IL1β	Cldn-1 ↑	TER ↓	Caco-2	(−)	(140)
Uropath E. coli	Cldn-4 ↑	(−)	Urothel cell		(141)

Transcriptional Regulation

Snail (104–106) is the best characterized of several transcription factors that bind directly to claudin promoters and/or influence claudin gene expression. Snail triggers an epithelial-to-mesenchymal transition (EMT) and downregulates other junction molecules, including cadherin and occludin. Tsukita and colleagues (104) demonstrated that overexpression of Snail also decreases expression of claudin-3, -4, and -7 and further showed that Snail binds directly to and regulates transcription from E-boxes in the promoter region of the claudin-7 gene. They concluded that the effect of Snail on EMT is likely mediated through coordinated repression of several different adhesive tight and adherens junction proteins. Ohkubo & Ozawa (105) described a second mechanism for Snail, finding that Snail expression decreases claudin-1 levels through an effect not on transcription but on translation. Carrozzino et al. (106) further explored the effect of Snail on claudin levels by expressing Snail in MDCK cells under control of an inducible promoter. Upon titrating Snail levels below the amount that triggered full EMT, Carrozzino et al. found a number of phenotypic alterations, including decreased TER with increases in both Na^+ and Cl^- permeability. They documented large decreases in claudin-4 and -7 levels but a smaller decrease in claudin-2 expression under these conditions. Thus, increased Snail expression has distinct effects on the expression of different claudins, resulting in selective changes in tight junction physiologic properties. The above authors speculate that alterations in Snail expression may be a physiologically relevant regulatory mechanism for modulating tight junction permeability.

Other transcription factors have been reported to regulate specific claudins. These factors include GATA-4 (107), which binds to the promoter region of claudin-2 and which, when expressed in conjunction with either CDX or HNF-1α, increases claudin-2 expression. Additionally, in Caco-2 cells, claudin-2 expression declines with increasing time in culture. This trend can be reversed by forced expression of GATA-4, suggesting this transcription factor is critical for the expression of this claudin. Expression of a lung-specific splice form of claudin-18 is regulated by T/EBP/NKX2.1, a homeodomain transcription factor that is expressed in lung, thyroid, and part of the brain (108). This tissue-specific transcript is generated by alternative splicing; the longer stomach-specific form of claudin-18 lacks the promoter regions that bind this transcription factor. This mechanism for tissue-specific claudin expression has not yet been described for any other claudin and suggests that the lung-specific form of claudin-18, which lacks the C-terminal domain, may serve a unique physiologic function.

Humoral Regulation

Much evidence exists for hormonal and cytokine effects on claudin gene regulation. Two lines of evidence discussed below demonstrate the effects of growth factors on claudin expression through regulation of extracellular signal–regulated kinases (ERKs) 1/2; other examples are listed in Table 4. Singh et al. (130) have demonstrated that epidermal growth factor receptor activation inhibits claudin-2 expression in MDCK II cells and increases expression of claudin-1, -3, and -4, with a concomitant threefold increase in TER. Pretreatment of the cells with an inhibitor of ERK 1/2 completely blocks this effect. Balkovetz and coworkers (109) demonstrated that either ERK 1/2 activation by hepatocyte growth factor or overexpression of ERK 1/2 by transfection inhibits claudin-2 expression in low-resistance MDCK II cells, which is correlated with a large transient increase in TER. Conversely, inhibition of ERK 1/2 in the high-resistance MDCK I cells results in induction of claudin-2 and a large fall in TER. By blocking the effect with actinomycin D and documenting increased claudin-2 RNA levels with quantitative real-time PCR, the Balkovetz group (109) demonstrated that this response is due to increased transcription. Both groups discussed above suggest that this signaling pathway likely represents an important physiologic regulatory mechanism of renal transport characteristics.

Several cytokines have general effects on epithelial tight junctions (11, 110); there are dozens of reports and we cite only a few that are related to claudins. For example, in cultured human intestinal epithelial monolayers, TNF-α (111) and IFN-γ (112) decrease the electrical resistance. By contrast, IL-17 increases resistance and gene transcription for claudin-1 and -2 (113). Coincidently, colonic tissue from patients with ulcerative colitis has elevated levels of TNF-α and decreased electrical impedance (114, 115). Extensive evidence supports the likelihood that the tight junction barrier is downregulated at transcriptional and posttranscriptional levels by cytokines associated with inflammation (112).

CLAUDINS AS THERAPEUTICS TARGETS

Manipulation of tight junctions has long been considered as a possible way in which to enhance transepithelial drug absorption. Attempts have already been made to enhance drug absorption through targeting occludin (116) or, less specifically, through the endogenous zonulin receptor system (117) or through manipulation of signaling pathways that control the barrier (118). The large number of claudins and their restricted locations suggest that they are selective targets for this effort.

The intriguing possibility of selectivity is supported by studies of the *Clostridia perfringins* enterotoxin, CPE, which uses a limited number of the claudins as its receptor. Normally, the toxin binds claudin-3 and -4 in the intestine, inducing cytolysis and diarrhea (119). Because certain human tumors, notably those of the pancreas and ovary, dramatically upregulate claudin-4, there is interest in using the toxin as a selective chemotherapeutic agent (120). In fact, CPE can eliminate human ovarian cancer cells grown in the peritoneal cavity of mice (121); these cells show a distribution typical of that seen in advanced human ovarian cancer. Interestingly, in a rat model, a fragment of the toxin enhances intestinal tracer flux without causing cell damage (122).

CONCLUSION

The discovery of claudins is beginning to provide a molecular-level understanding of the properties of paracellular transport. The claudins' large number and diverse expression patterns have revealed an unexpected degree of complexity and subtlety in the control of permeability. As we move beyond the phase of molecular bird-watching to that of molecular mechanisms, our remaining goals are to understand the general structure-function properties (e.g., atomic structure) and the specific contributions of individual claudins. Hopefully, we will learn to exploit their complexity to fashion specific therapies such as targeting of drug delivery to specific sites.

ACKNOWLEDGMENTS

Our program is supported by grants from the National Institutes of Health (DK 45134) and the State of North Carolina.

The *Annual Review of Physiology* is online at
http://physiol.annualreviews.org

LITERATURE CITED

1. Stevenson BR, Anderson JM, Bullivant S. 1988. The epithelial tight junction: Structure, function and preliminary biochemical characterization. *Mol. Cell Biochem.* 83(2):129–45

1a. Claude P. 1978. Morphological factors influencing transepithelial permeability: a model for the resistance of the zonula occludens. *J. Membr. Biol.* 39:219–32

2. Schneeberger EE, Lynch RD. 2004. The tight junction: A multifunctional complex. *Am. J. Physiol. Cell Physiol.* 286: C1213–28

3. Turksen K, Troy TC. 2004. Barriers built on claudins. *J. Cell Sci.* 117:2435–47

4. D'Atri F, Citi S. 2002. Molecular complexity of vertebrate tight junctions. *Mol. Membr. Biol.* 19:103–12

5. Cereijido M, Valdes J, Shoshani L, Contreras RG. 1998. Role of tight junctions in establishing and maintaining cell polarity. *Annu. Rev. Physiol.* 60:161–77

6. Cereijido M, Shoshani L, Contreras RG. 2000. Molecular physiology and pathophysiology of tight junctions. I. Biogen-

esis of tight junctions and epithelial polarity. *Am. J. Physiol. Gastrointest. Liver Physiol.* 279:G477–82

7. Powell DW. 1981. Barrier function of epithelia. *Am. J. Physiol.* 241:G275–88

8. Reuss L. 2001. Tight junction permeability to ions and water. In *Tight Junctions*, ed. M Cereijido, JM Anderson, pp. 61–88. Boca Raton: CRC

9. Van Itallie CM, Anderson JM. 2004. The molecular physiology of tight junction pores. *Physiology* 19:331–38

10. Van Itallie CM, Fanning AS, Anderson JM. 2003. Reversal of charge selectivity in cation or anion-selective epithelial lines by expression of different claudins. *Am. J. Physiol. Renal Physiol.* 285:F1078–84

11. Nusrat A, Turner JR, Madara JL. 2000. Molecular physiology and pathophysiology of tight junctions. IV. Regulation of tight junctions by extracellular stimuli: nutrients, cytokines, and immune cells. *Am. J. Physiol. Gastrointest. Liver Physiol.* 279:G851–57

12. Madara JL. 1998. Regulation of the movement of solutes across tight junctions. *Annu. Rev. Physiol.* 60:143–59

13. Stevenson BR, Siliciano JD, Mooseker MS, Goodenough DA. 1986. Identification of ZO-1: A high-molecular-weight polypeptide associated with the tight junction (zonula occludens) in a variety of epithelia. *J. Cell Biol.* 103:755–66

14. Tang VW, Goodenough DA. 2003. Paracellular ion channel at the tight junction. *Biophys. J.* 84:1660–73

15. Furuse M, Fujita K, Hiiragi T, Fujimoto K, Tsukita S. 1998. Claudin-1 and -2: Novel integral membrane proteins localizing at tight junctions with no sequence similarity to occludin. *J. Cell Biol.* 141: 1539–50

16. Morita K, Furuse M, Fujimoto K, Tsukita S. 1999. Claudin multigene family encoding four-transmembrane domain protein components of tight junction strands. *Proc. Natl. Acad. Sci. USA* 96:511–16

17. Tsukita S, Furuse M, Itoh M. 2001. Multifunctional strands in tight junctions. *Nat. Rev. Mol. Cell Biol.* 2:285–93

18. Simon DB, Lu Y, Choate KA, Velazquez H, Al Sabban E, et al. 1999. Paracellin-1, a renal tight junction protein required for paracellular Mg^{2+} resorption. *Science* 285:103–6

19. Hadj-Rabia S, Baala L, Vabres P, Hamel-Teillac D, Jacquemin E, et al. 2004. Claudin-1 gene mutations in neonatal sclerosing cholangitis associated with ichthyosis: a tight junction disease. *Gastroenterology* 127:1386–90

20. Wilcox ER, Burton QL, Naz S, Riazuddin S, Smith TN, et al. 2001. Mutations in the gene encoding tight junction claudin-14 cause autosomal recessive deafness DFNB29. *Cell* 104:165–72

21. Matter K, Balda MS. 2003. Signalling to and from tight junctions. *Nat. Rev. Mol. Cell Biol.* 4:225–36

22. Dhawan P, Singh AB, Deane NG, No Y, Shiou SR, et al. 2005. Claudin-1 regulates cellular transformation and metastatic behavior in colon cancer. *J. Clin. Invest.* 115:1765–76

23. Furuse M, Hirase T, Itoh M, Nagafuchi A, Yonemura S, et al. 1993. Occludin—a novel integral membrane-protein localizing at tight junctions. *J. Cell Biol.* 123:1777–88

24. Bazzoni G. 2003. The JAM family of junctional adhesion molecules. *Curr. Opin. Cell Biol.* 15:525–30

25. Gonzalez-Mariscal L, Betanzos A, Nava P, Jaramillo BE. 2003. Tight junction proteins. *Prog. Biophys. Mol. Biol.* 81:1–44

26. Saitou M, Furuse M, Sasaki H, Schulzke JD, Fromm M, et al. 2000. Complex phenotype of mice lacking occludin, a component of tight junction strands. *Mol. Biol. Cell* 11:4131–42

27. Furuse M, Sasaki H, Fujimoto K, Tsukita S. 1998. A single gene product, claudin-1 or -2, reconstitutes tight junction strands and recruits occludin in fibroblasts. *J. Cell Biol.* 143:391–401

28. Kubota K, Furuse M, Sasaki H, Sonoda N, Fujita K, et al. 1999. Ca^{2+}-independent cell-adhesion activity of claudins, a family of integral membrane proteins localized at tight junctions. *Curr. Biol.* 9:1035–38

29. Loh YH, Christoffels A, Brenner S, Hunziker W, Venkatesh B. 2004. Extensive expansion of the claudin gene family in the teleost fish, *Fugu rubripes*. *Genome Res.* 14:1248–57

30. Kollmar R, Nakamura SK, Kappler JA, Hudspeth AJ. 2001. Expression and phylogeny of claudins in vertebrate primordia. *Proc. Natl. Acad. Sci. USA* 98:10196–201

31. Hardison AL, Lichten L, Banerjee-Basu S, Becker TS, Burgess SM. 2005. The zebrafish gene claudinj is essential for normal ear function and important for the formation of the otoliths. *Mech. Dev.* 122:949–58

32. Barrios-Rodiles M, Brown KR, Ozdamar B, Bose R, Liu Z, et al. 2005. High-throughput mapping of a dynamic

signaling network in mammalian cells. *Science* 307:1621–25

33. Colegio OR, Van Itallie CM, Mccrea HJ, Rahner C, Anderson JM. 2002. Claudins create charge-selective channels in the paracellular pathway between epithelial cells. *Am. J. Physiol. Cell Physiol.* 283: C142–47

34. Fujita K, Katahira J, Horiguchi Y, Sonoda N, Furuse M, Tsukita S. 2000. *Clostridium perfringens* enterotoxin binds to the second extracellular loop of claudin-3, a tight junction integral membrane protein. *FEBS Lett.* 476:258–61

35. Itoh M, Furuse M, Morita K, Kubota K, Saitou M, Tsukita S. 1999. Direct binding of three tight junction–associated MAGUKs, ZO-1, ZO-2 and ZO-3, with the COOH termini of claudins. *J. Cell Biol.* 147:1351–63

36. Mitic LL, Unger VM, Anderson JM. 2003. Expression, solubilization, and biochemical characterization of the tight junction transmembrane protein claudin-4. *Protein Sci.* 12:218–27

37. Jarvis LJ, Louis CF. 1995. Purification and oligomeric state of the major lens fiber cell membrane proteins. *Curr. Eye Res.* 14:799–808

38. Hamazaki Y, Itoh M, Sasaki H, Furuse M, Tsukita S. 2002. Multi-PDZ domain protein 1 (MUPP1) is concentrated at tight junctions through its possible interaction with claudin-1 and junctional adhesion molecule. *J. Biol. Chem.* 277:455–61

39. Jeansonne B, Lu Q, Goodenough DA, Chen YH. 2003. Claudin-8 interacts with multi-PDZ domain protein 1 (MUPP1) and reduces paracellular conductance in epithelial cells. *Cell Mol. Biol.* 49:13–21

40. Roh MH, Liu CJ, Laurinec S, Margolis B. 2002. The carboxyl terminus of zona occludens-3 binds and recruits a mammalian homologue of discs lost to tight junctions. *J. Biol. Chem.* 277:27501–9

41. McCarthy KM, Francis SA, McCormack JM, Lai J, Rogers RA, et al. 2000. Inducible expression of claudin-1-myc but not occludin-VSV-G results in aberrant tight junction strand formation in MDCK cells. *J. Cell Sci.* 113:3387–98

42. Furuse M, Sasaki H, Tsukita S. 1999. Manner of interaction of heterogeneous claudin species within and between tight junction strands. *J. Cell Biol.* 147:891–903

43. Ruffer C, Gerke V. 2004. The C-terminal cytoplasmic tail of claudins 1 and 5 but not its PDZ-binding motif is required for apical localization at epithelial and endothelial tight junctions. *Eur. J. Cell Biol.* 83:135–44

44. Van Itallie CM, Gambling TM, Carson JL, Anderson JM. 2005. Palmitoylation of claudins is required for efficient tight-junction localization. *J. Cell Sci.* 118:1427–36

45. Van Itallie CM, Colegio OR, Anderson JM. 2004. The cytoplasmic tails of claudius can influence tight junction barrier properties through effects on protein stability. *J. Membr. Biol.* 199:29–38

46. Gow A, Southwood CM, Li JS, Pariali M, Riordan GP, et al. 1999. CNS myelin and sertoli cell tight junction strands are absent in Osp/claudin-11 null mice. *Cell* 99:649–59

47. Fukuhara A, Shimizu K, Kawakatsu T, Fukuhara T, Takai Y. 2003. Involvement of nectin-activated Cdc42 small G protein in organization of adherens and tight junctions in Madin-Darby canine kidney cells. *J. Biol. Chem.* 278:51885–93

48. Turksen K, Troy TC. 2001. Claudin-6: A novel tight junction molecule is developmentally regulated in mouse embryonic epithelium. *Dev. Dyn.* 222:292–300

49. Reyes JL, Lamas M, Martin D, Namorado MD, Islas S, et al. 2002. The renal segmental distribution of claudins changes with development. *Kidney Int.* 62:476–87

50. Kiuchi-Saishin Y, Gotoh S, Furuse M, Takasuga A, Tano Y, Tsukita S. 2002. Differential expression patterns of claudins, tight junction membrane proteins, in

mouse nephron segments. *J. Am. Soc. Nephrol.* 13:875–86

51. Behr M, Riedel D, Schuh R. 2003. The claudin-like megatrachea is essential in septate junctions for the epithelial barrier function in Drosophila. *Dev. Cell* 5:611–20

52. Tepass U, Tanentzapf G, Ward R, Fehon R. 2001. Epithelial cell polarity and cell junctions in Drosophila. *Annu. Rev. Genet.* 35:747–84

53. Wu VM, Schulte J, Hirschi A, Tepass U, Beitel GJ. 2004. Sinuous is a Drosophila claudin required for septate junction organization and epithelial tube size control. *J. Cell Biol.* 164:313–23

54. Asano A, Asano K, Sasaki H, Furuse M, Tsukita S. 2003. Claudins in *Caenorhabditis elegans*: their distribution and barrier function in the epithelium. *Curr. Biol.* 13:1042–46

54a. Dhawan P, Singh AB, Deane NG, No Y, Shiou SR, et al. 2005. Claudin-1 regulates cellular transformation and metastatic behavior in colon cancer. *J. Clin. Invest.* 115:1765–76

55. Steele EC, Lyon MF, Favor J, Guillot PV, Boyd Y, Church RL. 1998. A mutation in the connexin 50 (Cx50) gene is a candidate for the No2 mouse cataract. *Curr. Eye Res.* 17:883–89

56. Jetten AM, Suter U. 2000. The peripheral myelin protein 22 and epithelial membrane protein family. *Prog. Nucleic Acid. Res. Mol. Biol.* 64:97–129

57. Notterpek L, Roux KJ, Amici SA, Yazdanpour A, Rahner C, Fletcher BS. 2001. Peripheral myelin protein 22 is a constituent of intercellular junctions in epithelia. *Proc. Natl. Acad. Sci. USA* 98:14404–9

58. Steele EC Jr, Kerscher S, Lyon MF, Glenister PH, Favor J, et al. 1997. Identification of a mutation in the MP19 gene, Lim2, in the cataractous mouse mutant To3. *Mol. Vis.* 3:5

59. Grey AC, Jacobs MD, Gonen T, Kistler J, Donaldson PJ. 2003. Insertion of MP20 into lens fibre cell plasma membranes correlates with the formation of an extracellular diffusion barrier. *Exp. Eye Res.* 77:567–74

60. Bronstein JM. 2000. Function of tetraspan proteins in the myelin sheath. *Curr. Opin. Neurobiol.* 10:552–57

61. Brancolini C, Edomi P, Marzinotto S, Schneider C. 2000. Exposure at the cell surface is required for Gas3/PMP22 to regulate both cell death and cell spreading: Implication for the Charcot-Marie-Tooth type 1A and Dejerine-Sottas diseases. *Mol. Biol. Cell* 11:2901–14

62. Roux KJ, Amici SA, Notterpek L. 2004. The temporospatial expression of peripheral myelin protein 22 at the developing blood-nerve and blood-brain barriers. *J. Comp. Neurol.* 474:578–88

63. Roux KJ, Amici SA, Fletcher BS, Notterpek L. 2005. Modulation of epithelial morphology, monolayer permeability, and cell migration by growth arrest specific 3/peripheral myelin protein 22. *Mol. Biol. Cell* 16:1142–51

64. Wadehra M, Iyer R, Goodglick L, Braun J. 2002. The tetraspan protein epithelial membrane protein-2 interacts with beta1 integrins and regulates adhesion. *J. Biol. Chem.* 277:41094–100

65. Wilson HL, Wilson SA, Surprenant A, North RA. 2002. Epithelial membrane proteins induce membrane blebbing and interact with the P2X(7) receptor C terminus. *J. Biol. Chem.* 277:34017–23

66. Kearsey J, Petit S, De Oliveira C, Schweighoffer F. 2004. A novel four transmembrane spanning protein, CLP24. *Eur. J. Biochem.* 271:2584–92

67. Tomita S, Fukata M, Nicoll RA, Bredt DS. 2004. Dynamic interaction of stargazin-like TARPs with cycling AMPA receptors at synapses. *Science* 303:1508–11

68. Price MG, Davis CF, Deng F, Burgess DL. 2005. The alpha-amino-3-hydroxyl-5-methyl-4-isoxazolepropionate receptor trafficking regulator "stargazin" is related

to the claudin family of proteins by its ability to mediate cell-cell adhesion. *J. Biol. Chem.* 280:19711–20

69. Watson CJ, Rowland M, Warhurst G. 2001. Functional modeling of tight junctions in intestinal cell monolayers using polyethylene glycol oligomers. *Am. J. Physiol. Cell Physiol.* 281:C388–97

70. Diamond JM. 1978. Channels in epithelial cell membranes and junctions. *Fed. Proc.* 37:2639–44

71. Salas PJI, Moreno JH. 1982. Single-file diffusion multi-ion mechanism of permeation in paracellular epithelial channels. *J. Membr. Biol.* 64:103–12

72. Firth JA. 2002. Endothelial barriers: from hypothetical pores to membrane proteins. *J. Anat.* 200:541–48

73. Nitta T, Hata M, Gotoh S, Seo Y, Sasaki H, et al. 2003. Size-selective loosening of the blood-brain barrier in claudin-5-deficient mice. *J. Cell Biol.* 161:653–60

74. Kovbasnjuk O, Leader JP, Weinstein AM, Spring KR. 1998. Water does not flow across the tight junctions of MDCK cell epithelium. *Proc. Natl. Acad. Sci. USA* 95:6526–30

75. Li WY, Huey CL, Yu ASL. 2004. Expression of claudin-7 and -8 along the mouse nephron. *Am. J. Physiol. Renal Physiol.* 286:F1063–71

76. Furuse M, Furuse K, Sasaki H, Tsukita S. 2001. Conversion of Zonulae occludentes from tight to leaky strand type by introducing claudin-2 into Madin-Darby canine kidney I cells. *J. Cell Biol.* 153:263–72

77. Rahner C, Mitic LL, Anderson JM. 2001. Heterogeneity in expression and subcellular localization of claudins 2, 3, 4, and 5 in the rat liver, pancreas, and gut. *Gastroenterology* 120:411–22

78. Matter K, Balda MS. 2003. Functional analysis of tight junctions. *Methods* 30:228–34

79. Yap AS, Mullin JM, Stevenson BR. 1998. Molecular analyses of tight junction phys-iology: insights and paradoxes. *J. Membr. Biol.* 163:159–67

80. Sasaki H, Matsui C, Furuse K, Mimori-Kiyosue Y, Furuse M, Tsukita S. 2003. Dynamic behavior of paired claudin strands within apposing plasma membranes. *Proc. Natl. Acad. Sci. USA* 100: 3971–76

81. Schulzke JD, Gitter AH, Mankertz J, Spiegel S, Seidler U, et al. 2005. Epithelial transport and barrier function in occludin-deficient mice. *Biochim. Biophys. Acta* 1669:34–42

82. Kitajiri SI, Miyamoto T, Mineharu A, Sonoda N, Furuse K, et al. 2004. Compartmentalization established by claudin-11-based tight junctions in stria vascularis is required for hearing through generation of endocochlear potential. *J. Cell Sci.* 117:5087–96

83. Ben Yosef T, Belyantseva IA, Saunders TL, Hughes ED, Kawamoto K, et al. 2003. Claudin 14 knockout mice, a model for autosomal recessive deafness DFNB29, are deaf due to cochlear hair cell degeneration. *Hum. Mol. Genet.* 12:2049–61

84. Miyamoto T, Morita K, Takemoto D, Takeuchi K, Kitano Y, et al. 2005. Tight junctions in Schwann cells of peripheral myelinated axons: a lesson from claudin-19-deficient mice. *J. Cell Biol.* 169:527–38

85. Furuse M, Hata M, Furuse K, Yoshida Y, Haratake A, et al. 2002. Claudin-based tight junctions are crucial for the mammalian epidermal barrier: a lesson from claudin-1-deficient mice. *J. Cell Biol.* 156:1099–111

86. Turksen K, Troy TC. 2002. Permeability barrier dysfunction in transgenic mice overexpressing claudin 6. *Development* 129:1775–84

87. Gabreels-Festen A, Wetering RV. 1999. Human nerve pathology caused by different mutational mechanisms of the PMP22 gene. *Ann. NY Acad. Sci.* 883:336–43

88. Carlton VE, Harris BZ, Puffenberger EG, Batta AK, Knisely AS, et al. 2003.

Complex inheritance of familial hyper-cholanemia with associated mutations in TJP2 and BAAT. *Nat. Genet.* 34:91–96

89. Muller D, Kausalya PJ, Claverie-Martin F, Meij IC, Eggert P, et al. 2003. A novel claudin 16 mutation associated with child-hood hypercalciuria abolishes binding to ZO-1 and results in lysosomal mistarget-ing. *Am. J. Hum. Genet.* 73:1293–301

90. Wattenhofer M, Reymond A, Falciola V, Charollais A, Caille D, et al. 2005. Different mechanisms preclude mutant CLDN14 proteins from forming tight junctions in vitro. *Hum. Mutat.* 25:543–49

91. Hirano T, Hirotsune S, Sasaki S, Kikuchi T, Sugimoto Y. 2002. A new deletion mutation in bovine Claudin-16 (CL-16) deficiency and diagnosis. *Animal Genet.* 33:118–22

92. Madara JL. 1994. Sodium-glucose co-transport and epithelial permeability. *Gastroenterology* 107:319–320

93. Berglund JJ, Riegler M, Zolotarevsky Y, Wenzl E, Turner JR. 2001. Regulation of human jejunal transmucosal re-sistance and MLC phosphorylation by Na$^+$-glucose cotransport. *Am. J. Physiol. Gastrointest. Liver Physiol.* 281:G1487–93

94. Ishizaki T, Chiba H, Kojima T, Fujibe M, Soma T, et al. 2003. Cyclic AMP induces phosphorylation of claudin-5 immuno-precipitates and expression of claudin-5 gene in blood-brain-barrier endothelial cells via protein kinase A-dependent and -independent pathways. *Exp. Cell Res.* 290:275–88

95. Fujibe M, Chiba H, Kojima T, Soma T, Wada T, et al. 2004. Thr203 of claudin-1, a putative phosphorylation site for MAP kinase, is required to promote the barrier function of tight junctions. *Exp. Cell Res.* 295:36–47

96. Yamauchi K, Rai T, Kobayashi K, Sohara E, Suzuki T, et al. 2004. Disease-causing mutant WNK4 increases paracel-lular chloride permeability and phospho-rylates claudins. *Proc. Natl. Acad. Sci. USA* 101:4690–94

97. Kahle KT, Macgregor GG, Wilson FH, Van Hoek AN, Brown D, et al. 2004. Para-cellular Cl$^-$ permeability is regulated by WNK4 kinase: insight into normal physi-ology and hypertension. *Proc. Natl. Acad. Sci. USA* 101:14877–82

98. D'Souza T, Agarwal R, Morin PJ. 2005. Phosphorylation of claudin-3 at threonine 192 by PKA regulates tight junction bar-rier function in ovarian cancer cells. *J. Biol. Chem.* 280:26233–40

99. Matsuda M, Kubo A, Furuse M, Tsukita S. 2004. A peculiar internalization of claudins, tight junction–specific adhesion molecules, during the intercellular move-ment of epithelial cells. *J. Cell Sci.* 117:1247–57

100. Ivanov AI, Nusrat A, Parkos CA. 2004. The epithelium in inflammatory bowel disease: potential role of endocytosis of junctional proteins in barrier disruption. *Novartis. Found. Symp.* 263:115–24

101. Ryan MC, Notterpek L, Tobler AR, Liu N, Shooter EM. 2000. Role of the pe-ripheral myelin protein 22 N-linked gly-can in oligomer stability. *J. Neurochem.* 75:1465–74

102. Al Moustafa AE, Alaoui-Jamali MA, Batist G, Hernandez-Perez M, Serruya C, et al. 2002. Identification of genes asso-ciated with head and neck carcinogene-sis by cDNA microarray comparison be-tween matched primary normal epithelial and squamous carcinoma cells. *Oncogene* 21:2634–40

103. Swisshelm K, Macek R, Kubbies M. 2005. Role of claudins in tumorigenesis. *Adv. Drug Deliv. Rev.* 57:919–28

104. Ikenouchi J, Matsuda M, Furuse M, Tsukita S. 2003. Regulation of tight junc-tions during the epithelium-mesenchyme transition: direct repression of the gene expression of claudins/occludin by Snail. *J. Cell Sci.* 116:1959–67

105. Ohkubo T, Ozawa M. 2004. The tran-scription factor Snail downregulates the

tight junction components independently of E-cadherin downregulation. *J. Cell Sci.* 117:16/5–85

106. Carrozzino F, Soulie P, Huber D, Mensi N, Orci L, et al. 2005. Inducible expression of Snail selectively increases paracellular ion permeability and differentially modulates tight junction proteins. *Am. J. Physiol. Cell Physiol.* 289:1002–14

107. Escaffit F, Boudreau F, Beaulieu JF. 2005. Differential expression of claudin-2 along the human intestine: Implication of GATA-4 in the maintenance of claudin-2 in differentiating cells. *J. Cell. Physiol.* 203:15–26

108. Niimi T, Nagashima K, Ward JM, Minoo P, Zimonjic DB, et al. 2001. Claudin-18, a novel downstream target gene for the T/EBP/NKX2.1 homeodomain transcription factor, encodes lung- and stomach-specific isoforms through alternative splicing. *Mol. Cell Biol.* 21:7380–90

109. Lipschutz JH, Li S, Arisco A, Balkovetz DF. 2005. Extracellular signal-regulated kinases 1/2 control claudin-2 expression in Madin-Darby canine kidney strain I and II cells. *J. Biol. Chem.* 280:3780–88

110. Clayburgh DR, Shen L, Turner JR. 2004. A porous defense: The leaky epithelial barrier in intestinal disease. *Lab. Invest.* 84:282–91

111. Schmitz H, Barmeyer C, Gitter AH, Wullstein F, Bentzel CJ, Fromm M, Riecken EO, Schulzke JD. 2000. Epithelial barrier and transport function of the colon in ulcerative colitis. *Ann. NY Acad. Sci.* 915:312–26

112. Walsh SV, Hopkins AM, Nusrat A. 2000. Modulation of tight junction structure and function by cytokines. *Adv. Drug Deliv. Rev.* 41:303–13

113. Kinugasa T, Sakaguchi T, Gu XB, Reinecker HC. 2000. Claudins regulate the intestinal barrier in response to immune mediators. *Gastroenterology* 118:1001–11

114. May GR, Sutherland LR, Meddings JB. 1993. Is small intestinal permeability really increased in relatives of patients with Crohn's disease? *Gastroenterology* 104:1627–32

115. Schmitz H, Barmeyer C, Fromm M, Runkel N, Foss HD, et al. 1999. Altered tight junction structure contributes to the impaired epithelial barrier function in ulcerative colitis. *Gastroenterology* 116:301–9

116. Wong V, Gumbiner BM. 1997. A synthetic peptide corresponding to the extracellular domain of occludin perturbs the tight junction permeability barrier. *J. Cell Biol.* 136:399–409

117. Wang W, Uzzau S, Goldblum SE, Fasano A. 2000. Human zonulin, a potential modulator of intestinal tight junctions. *J. Cell Sci.* 113(Pt. 24):4435–40

118. Clarke H, Marano CW, Peralta SA, Mullin JM. 2000. Modification of tight junction function by protein kinase C isoforms. *Adv. Drug Deliv. Rev.* 41:283–301

119. McClane BA. 2001. The complex interactions between *Clostridium perfringens* enterotoxin and epithelial tight junctions. *Toxicon* 39:1781–91

120. Michl P, Buchholz M, Rolke M, Kunsch S, Lohr M, et al. 2001. Claudin-4: A new target for pancreatic cancer treatment using *Clostridium perfringens* enterotoxin. *Gastroenterology* 121:678–84

121. Santin AD, Cane S, Bellone S, Palmieri M, Siegel ER, et al. 2005. Treatment of chemotherapy-resistant human ovarian cancer xenografts in C.B-17/SCID mice by intraperitoneal administration of *Clostridium perfringens* enterotoxin. *Cancer Res.* 65:4334–42

122. Masuyama A, Kondoh M, Seguchi H, Takahashi A, Harada M, et al. 2005. Role of N-terminal amino acids in the absorption-enhancing effects of the C-terminal fragment of *Clostridium perfringens* enterotoxin. *J. Pharmacol. Exp. Ther.* 314:789–95

123. Inai T, Kobayashi J, Shibata Y. 1999.

Claudin-1 contributes to the epithelial barrier function in MDCK cells. *Eur. J. Cell Biol.* 78:849–55

124. Amasheh S, Meiri N, Gitter AH, Schoneberg T, Mankertz J, et al. 2002. Claudin-2 expression induces cation-selective channels in tight junctions of epithelial cells. *J. Cell Sci.* 115:4969–76

125. Van Itallie C, Rahner C, Anderson JM. 2001. Regulated expression of claudin-4 decreases paracellular conductance through a selective decrease in sodium permeability. *J. Clin. Invest.* 107:1319–27

126. Wen HJ, Watry DD, Marcondes MCG, Fox HS. 2004. Selective decrease in paracellular conductance of tight junctions: role of the first extracellular domain of claudin-5. *Mol. Cell. Biol.* 24:8408–17

127. Alexandre MD, Lu Q, Chen YH. 2005. Overexpression of claudin-7 decreases the paracellular Cl⁻ conductance and increases the paracellular Na⁺ conductance in LLC-PK1 cells. *J. Cell Sci.* 118:2683–93

128. Yu AS, Enck AH, Lencer WI, Schneeberger EE. 2003. Claudin-8 expression in Madin-Darby canine kidney cells augments the paracellular barrier to cation permeation. *J. Biol. Chem.* 278:17350–59

129. Hirano T, Kobayashi N, Itoh T, Takasuga A, Nakamaru T, et al. 2000. Null mutation of PCLN-1/claudin-16 results in bovine chronic interstitial nephritis. *Genome Res.* 10:659–63

130. Singh AB, Tsukada T, Zent R, Harris RC. 2004. Membrane-associated HB-EGF modulates HGF-induced cellular responses in MDCK cells. *J. Cell Sci.* 117:1365–79

131. Kubota H, Chiba H, Takakuwa Y, Osanai M, Tobioka H, et al. 2001. Retinoid X receptor α and retinoic acid receptor γ mediate expression of genes encoding tight-junction proteins and barrier function in F9 cells during visceral endodermal differentiation. *Exp. Cell Res.* 263:163–72

132. Sakaguchi T, Gu XB, Golden HM, Suh ER, Rhoads DB, Reinecker HC.

2002. Cloning of the human claudin-2 5′-flanking region revealed a TATA-less promoter with conserved binding sites in mouse and human for caudal-related homeodomain proteins and hepatocyte nuclear factor-1 α. *J. Biol. Chem.* 277:21361–70

133. Chiba H, Gotoh T, Kojima T, Satohisa S, Kikuchi K, et al. 2003. Hepatocyte nuclear factor (HNF)-4α triggers formation of functional tight junctions and establishment of polarized epithelial morphology in F9 embryonal carcinoma cells. *Exp. Cell Res.* 286:288–97

134. Luk JM, Tong MK, Mok BW, Tam PC, Yeung WS, Lee KF. 2004. Sp1 site is crucial for the mouse claudin-19 gene expression in the kidney cells. *FEBS Lett.* 578:251–56

135. Efrati E, Arsentiev-Rozenfeld J, Zelikovic I. 2005. The human paracellin-1 gene (hPCLN-1): renal epithelial cell-specific expression and regulation. *Am. J. Physiol. Renal Physiol.* 288:F272–83

136. Andras IE, Pu H, Tian J, Deli MA, Nath A, et al. 2005. Signaling mechanisms of HIV-1 Tat-induced alterations of claudin-5 expression in brain endothelial cells. *J. Cereb. Blood Flow Metab.* 25:1159–70

137. Yamamoto T, Kojima T, Murata M, Takano K, Go M, et al. 2004. IL-1β regulates expression of Cx32, occludin, and claudin-2 of rat hepatocytes via distinct signal transduction pathways. *Exp. Cell Res.* 299:427–41

138. Scheurer SB, Rybak JN, Rosli C, Neri D, Elia G. 2004. Modulation of gene expression by hypoxia in human umbilical cord vein endothelial cells: A transcriptomic and proteomic study. *Proteomics* 4:1737–60

139. Kieran NE, Doran PP, Connolly SB, Greenan MC, Higgins DF, et al. 2003. Modification of the transcriptomic response to renal ischemia/reperfusion injury by lipoxin analog. *Kidney Int.* 64:480–92

140. Han X, Fink MP, Delude RL. 2003. Proinflammatory cytokines cause NO*-dependent and -independent changes in expression and localization of tight junction proteins in intestinal epithelial cells. *Shock* 19:229–37

141. Mysorekar IU, Mulvey MA, Hultgren SJ, Gordon JI. 2002. Molecular regulation of urothelial renewal and host defenses during infection with uropathogenic *Escherichia coli. J. Biol. Chem.* 277:7412–19

Annu. Rev. Physiol. 2006. 68:431–59
doi: 10.1146/annurev.physiol.68.040104.131852
Copyright © 2006 by Annual Reviews. All rights reserved
First published online as a Review in Advance on August 25, 2005

ROLE OF FXYD PROTEINS IN ION TRANSPORT

Haim Garty and Steven J.D. Karlish

*Department of Biological Chemistry, Weizmann Institute of Science, Rehovot 76100,
Israel; email: h.garty@weizmann.ac.il; steven.karlish@weizmann.ac.il*

Key Words   Na,K-ATPase, phospholemman, γ subunit, Mat-8, CHIF, RIC

■ **Abstract**   The FXYD proteins are a family of seven homologous single transmembrane segment proteins (FXYD1–7), expressed in a tissue-specific fashion. The
FXYD proteins modulate the function of Na,K-ATPase, thus adapting kinetic properties of active Na^+ and K^+ transport to the specific needs of different cells. Six FXYD
proteins (1–5, 7) are known to interact with Na,K-ATPase and affect its kinetic properties in specific ways. Although effects of FXYD proteins on parameters such as
$K_{1/2}Na^+$, $K_{1/2}K^+$, $K_m ATP$, and V_{max} are modest, usually twofold, these effects may
have important long-term consequences for homeostasis of cation balance. In this review we summarize basic features of FXYD proteins and present recent evidence for
functional effects, structure-function relations and structural interactions with Na,K-
ATPase. We then discuss possible physiological roles, based on in vitro observations
and newly available knockout mice models. Finally, we also consider evidence that
FXYD proteins affect functioning of other ion transport systems.

INTRODUCTION

Work in several laboratories has led to the identification of a new family of seven
single-span transmembrane proteins named after the invariant extracellular motif
FXYD (1). Early studies indicated a variety of unrelated functions for these proteins, summarized in Table 1. It is, however, clear today that one central role of
FXYD proteins is to interact with the Na,K-ATPase and modulate its properties
(14, 15). Specific association with the Na,K-ATPase and effects on the pump function have been reported for FXYD1 (also known as phospholemman, or PLM) (16,
17), FXYD2 (also known as the γ subunit of Na,K-ATPase) (18–22), FXYD3 (also
known as Mat-8) (23), FXYD4 (also known as corticosteroid hormone-induced
factor, or CHIF) (24–26), FXYD7 (27), and PLM-like protein from shark rectal
gland (PLMS) (28–30). Because each of these proteins has a different tissue distribution and functional effects, the current hypothesis is that FXYD proteins act
as tissue-specific modulators of Na,K-ATPase that adjust or fine-tune its kinetic
properties to the specific needs of a given tissue, cell type, or physiological state,
without affecting it elsewhere (14, 15). In general, FXYD proteins behave as modulatory subunits. The exceptions are PLM and PLMS, the effects of which may be

TABLE 1   The FXYD protein family

	Common name(s)	Original observations	Reference(s)
FXYD1	Phospholemman	Major membrane phosphoprotein in heart and muscle. Evokes Cl⁻ conductance in *Xenopus* oocytes. Conducts other ions and osmolytes	(2–5)
FXYD2	The γ subunit of Na,K-ATPase	Proteolipid associated with the Na,K-ATPase	(6, 7)
FXYD3	Mat-8	*neu-* and *ras-*induced gene. Evokes Cl⁻ conductance in *Xenopus* oocytes	(8, 9)
FXYD4	CHIF	Aldosterone-induced gene evokes K⁺ conductance in *Xenopus* oocytes	(10)
FXYD5	RIC, dysadherin	Cancer-associated E2a-Pbx1 induce gene that downregulates E-cadherin	(11, 12)
FXYD6	Phosphohippolin	EST clones homologous to other FXYD proteins	(1)
FXYD7		EST clones homologous to other FXYD proteins	(1)

altered by PKA- and PKC-dependent phosphorylation (28, 31–34). These FXYD proteins may therefore be classed as regulatory subunits, which alter the pump properties in response to external signals.

This review summarizes existing data on the structural and functional interactions between the Na,K-ATPase and the FXYD proteins. It also discusses these proteins' putative physiological role(s) as well as possible interactions with other ion transport systems.

PRIMARY STRUCTURE, SEQUENCE HOMOLOGY, AND MEMBRANE TOPOLOGY OF FXYD PROTEINS

FXYD proteins are short polypeptides (>100 amino acids) with a single transmembrane segment and with or without a signal peptide (Figure 1). The only exception is FXYD5 (related to ion channel, or RIC, also termed dysadherin), which has an atypically long N-terminal sequence. Phylogenetic analysis indicates that RIC was the first member to diverge from a common ancestor gene. Marked homology among family members is observed in a stretch of 35 amino acids around the transmembrane domain, but is not observed outside this region. The primary structure of at least some of the FXYD proteins is not very well conserved through evolution. Thus, more than 11% of the amino acids of PLM and CHIF vary between the rat and mouse sequences. Only six residues, marked in Figure 1 by asterisks

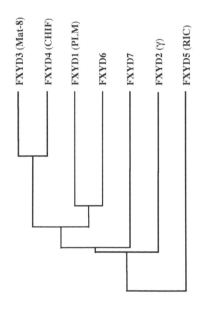

Figure 1 Sequence alignment of FXYD family members. (*Upper*) Sequence alignment of the seven family members (mouse sequences). In the case of RIC (FXYD5), the sequence depicted starts at amino acid 106. The six invariant amino acids are marked by asterisks. (*Lower*) Phylogenetic tree of FXYD proteins.

are fully conserved in all family members and species. These are the extracellular phenylalanine; tyrosine and aspartic acid, which define the FXYD motif; two transmembrane glycines; and a serine residue at the membrane-cytoplasm interface. It is likely that these residues are involved in an essential function common to all FXYD proteins, such as the structural interaction with the Na,K-ATPase. One study has shown that the FXYD motif is required for this structural interaction (24).

The genomic organization of the seven FXYD members is unusual in that each transcript is made up of at least six exons and that the 35 amino acid conserved region is formed by three separate exons (35). This may serve to enable high structural diversity, although structural diversity is not well documented experimentally. Multiple transcripts have been reported for γ and PLM only (36, 37). At the protein level, however, only γ exists, as two splice variants termed γa and γb with different N termini (38). Two additional γ species, termed $\gamma a'$ and $\gamma b'$, have been seen in cells transfected with γ cDNA coding for a single open-reading frame (19, 21) and represent posttranslationally modified proteins. The as yet unidentified posttranslational modifications are located in the extracellular N terminus (25).

Analysis of the membrane topology of PLM, γ, CHIF, and FXYD7 established that they are type I single-span membrane proteins, with an extracellular N terminus and cytoplasmic C terminus. PLM, CHIF, and RIC have a signal peptide that is cleaved in the mature protein (2, 18, 20, 24, 27, 39). In Mat-8, on the other hand, signal peptide cleavage is not observed, and this FXYD protein may have a second transmembrane-like domain (23). Helical wheel projection of the transmembrane segment demonstrates that most of the residues conserved among the FXYD proteins are preferentially located on one face of the helix (1).

TISSUE DISTRIBUTION OF FXYD PROTEINS

The available information on organ and tissue distributions of FXYD mRNA and protein is summarized in Table 2. At least some of the FXYD family members exhibit a high degree of tissue specificity. Thus, γ is detected only in kidney, and CHIF in kidney and colon, but neither is detected in any other tissue (22, 26, 42, 46). PLM is detected mainly in heart and skeletal muscle, whereas FXYD7 is brain specific (17, 27, 40, 41).

The detailed expression patterns of γ (γa and γb), CHIF, and PLM in kidney provide insights into their possible physiological roles. Immunohistochemistry of rat kidney sections shows that γ is expressed at high levels in the thick ascending limb of the loop of Henle (TAL), in the distal convoluted tubule (DCT), at lower levels in the proximal convoluted tubule, and at low but detectable levels in the inner medullary collecting duct (IMCD) (22, 42, 43). CHIF is expressed only along the cortical collecting duct (CCD) principal cells, and in the outer medullary collecting duct and IMCD segments (26, 43, 46). In general, γ and CHIF are not expressed in the same cells, an exception being the middle and final segment of the IMCD, in which both γa and CHIF are expressed (43). Also, high magnification confocal imaging and electron microscopy have demonstrated exclusive basolateral localization of both CHIF and γ (43, 46). In different studies, use of splice variant–specific antibodies of γ have revealed overlapping but distinct expression patterns of γa and γb in different nephron segments (22, 49). In the proximal convoluted tubule, both splice variants are detected at low to moderate levels, but γa predominates over γb (49). In medullary TAL, both γa and γb are expressed at high levels, whereas in cortical TAL only γb is detected (22, 49). γa, but not γb, was observed in macula densa epithelial cells (22). γb was found at a high level in the DCT and CNT (49). γa, but not γb, was found in the intercalated cells (but not in the principal cells) of the initial quarter of the IMCD, as well as in the principal cells of the middle and final portions of the collecting duct, IMCD2 and IMCD3 (43; Pihakaski-Maunsbach K, Vorum H, Honoré B, Tokonabe S, Frøkiær J, et al., submitted for publication). Recently, expression in the kidney of a third FXYD protein, RIC, was reported (39). RIC was found in the basolateral membrane of the collecting tubule, in the connecting tubule, and in the intercalated cells of the collecting duct. This expression pattern is complementary to those of CHIF and γ but also predicts some overlap that will give rise to a colocalization of RIC and γ in the same cell.

TABLE 2   Major organs and tissues expressing FXYD mRNA and proteins

FXYD	mRNA	Reference(s)	Protein	Reference(s)
PLM	Heart, liver, skeletal muscle	(2, 37)	Heart (myocytes), brain (cerebellum and choroids plexus), kidney (extraglomerular mesangium)	(17, 40, 41)
γ	Kidney, heart, stomach	(18)	Kidney (TAL, DCT, IMCD, and macula densa)	(22, 42, 43)
Mat-8	Colon, stomach, uterus	(9)	Stomach (mucous cells)	(23)
CHIF	Kidney collecting duct, distal colon	(10, 45)	Kidney collecting duct, distal colon	(26, 43, 46)
RIC	Heart, brain, spleen, lung, liver, skeletal muscle, kidney, testis	(11)	Kidney, intestine, lung, heart, and spleen	(39)
FXYD6 (phosphohippolin)	Brain and kidney	(47)	CNS	(48)
FXYD7	Brain (cerebellum, cerebrum, hippocampus, and stem)	(27)	Brain cerebellum, cerebrum, hippocampus, and stem	(27)
PLMS	Brain, heart, kidney, intestine, and rectal gland	(30)	Rectal gland	(30)

PLM has been found in selected structures of the heart, brain, and kidney. In cardiac myocytes, it was detected in all regions exposed to the extracellular space: sarcolemma, intercalated disks, and transverse tubules (34, 40). In the brain, it was visualized in the cerebellum and enriched in choroids plexus, colocalized with Na,K-ATPase, implying a possible role in secretion of CSF (17). In the kidney, PLM has been found in extraglomerular mesangial cells and afferent arterioles but not in nephrons, suggesting a role in renal blood flow and tubuloglomerular feedback (41). The location of PLM in this region contrasts with the expression of γa in the adjacent macula densa epithelial cells (22).

Only limited information is available on the cell and tissue specificity of the other FXYD proteins. FXYD7 was detected in brain (both neurons and glia) but not in any of the other tissues expressing PLM, γ, or CHIF (27). Mat-8 is expressed in gastric epithelia, mainly in the mucosal cells forming the upper part of the gland (23). Finally, FXYD6 (phosphohippolin) exhibits a unique distribution in the central nervous system (48). It is as yet unknown whether Mat-8 and FXYD6 expression is specific to stomach and brain, respectively.

FUNCTIONAL INTERACTIONS OF FXYD PROTEINS AND Na,K-ATPase

For many years, the renal Na,K-ATPase was known to contain a proteolipid, the γ subunit (50–52). Association of this short polypeptide with the Na,K-ATPase was first demonstrated convincingly by covalent labeling with photoreactive derivatives of ouabain (6), and subsequently by co-immunoprecipitation (7). However, neither its functional effects nor physiological role were known. The functional effects of the other FXYD proteins (e.g., PLM and CHIF) were not considered, as they were not known to interact with Na,K-ATPase. In the past seven years, the situation has changed completely. As shown in the various expression systems, association with the FXYD proteins is not required for Na,K-ATPase activity, but there is now compelling evidence for specific associations with the α/β complex, demonstrated mainly by co-immunoprecipitation, as well as for significant functional effects of six FXYD proteins: PLM, γ, Mat-8, CHIF, FXYD7, and PLMS (14, 15, 23, 53, 54). Expression of FXYD proteins in *Xenopus* oocytes and mammalian cells with endogenous α/β subunits or coexpressing exogenous α/β subunits has provided the most extensive functional information. Neutralizing interactions of antibodies specific for accessible epitopes, or proteolytic digestion of FXYD proteins, have also been useful. Most recently, functional effects in membranes obtained from PLM and γ knockout (KO) mice have been reported. Table 3 summarizes the data on modulatory effects of the different FXYD proteins on apparent affinities for cytoplasmic Na^+ and extracellular K^+ as well as ATP and maximal rates of Na,K-ATPase. Although the effects of FXYD proteins on the kinetic parameters are small, usually twofold or less, they are likely to be physiologically significant (see below, "Possible Physiological Roles of the FXYD Proteins"). However, some functional effects have been seen more consistently than others. Some differences are attributable to different conditions, posttranslational modifications (e.g., membrane potentials in *Xenopus* oocytes or phosphorylation of PLM), or variability in the molar ratio of expressed FXYD proteins to the α subunit, leading to variability of the magnitude and detection of functional effects.

One of the initial indications for a functional effect of γ came from a demonstration that an antibody raised against a C-terminal sequence (anti-γC) partially inhibited activity of renal Na,K-ATPase (20). Subsequently, expression of γa in a kidney cell line (HEK293) was found to reduce the K_m for activation by ATP in isolated membranes, an effect abrogated by anti-γC (21). A number of observations indicated that the rise in apparent affinity for ATP is secondary to a γ-induced shift of the E_1-E_2 conformational equilibrium toward E_1 (reduced vanadate sensitivity and inhibition by K^+ at low ATP, and the pH dependence of the effect) (20, 22). Expression of γa in NRK-52E rat renal cells provided the first evidence in mammalian cells for modulation of both Na^+ and K^+ affinities, involving a reduction in apparent affinities for both cytoplasmic Na^+ ions and extracellular K^+ ions (19, 56). Significant effects of γ on K_m for ATP have not been observed in membranes isolated from these cells (56). Expression of γa or γb in HeLa cells has confirmed the effect of γ on the $K_{1/2}$ for Na^+ and shown that it is mainly the result of increased K:Na antagonism at the cytoplasmic surface, as the magnitude of the effect depends on the concentration of K^+ ions; little or no effect on Na-ATPase activity is seen in the absence of K^+ ions (22). This effect was not abrogated by anti-γC. Recent work using intact HeLa cells showed an decrease in apparent affinity for cytoplasmic Na^+, an increase in affinity for intracellular ATP—consistent with prior observations of isolated membranes—and a modest increase in apparent affinity for extracellular K^+, which was not seen with isolated membranes (57). Experiments with *Xenopus* oocytes provided an initial indication of a specific $\gamma - \alpha / \beta$ interaction (18) and, subsequently, important functional information on several FXYD proteins due to the fact that Na, K-pump currents can be measured with controlled cytoplasmic Na^+ and extracellular K^+ concentrations and transmembrane voltages (14, 18). In this system, expression of γ has also been found to raise the $K_{1/2}$ for cytoplasmic Na^+ (24). γ somewhat reduces $K_{1/2}$ for extracellular K^+ in the presence of extracellular Na^+ and a hyperpolarizing membrane voltage (-150 mV), at 0 mV it slightly raises $K_{1/2}$ for extracellular K^+, and at -50 mV—a normal cell membrane voltage—there is little effect (18, 24). In the absence of extracellular Na^+ ions, there are also no detectable effects. γ does not affect the apparent binding affinity of ouabain (24).

In HeLa and HEK293 cells, no significant differences in functional effects were found between γb and γa or between γb and γb' (expressed in HeLa cells) and γa and γa' (expressed in HEK293 cells) (22). Similarly, in *Xenopus* oocytes, significant functional differences between γa and γb have not been detected (24). By contrast, it was reported that in NRK-52E cells, γa, γb, and γb', but not γa', reduce cytoplasmic Na^+ affinity, whereas γa, γa', and γb', but not γb, reduce extracellular K^+ affinity (19, 56). However, whether splice variants, and in particular posttranslationally modified variants, have specific functional effects is uncertain. Na:K antagonism is still found in γa lacking the first seven N-terminal residues (60).

TABLE 3 Effects of FXYD proteins on Na,K-ATPase activity

FXYD protein	Experimental system	Kinetic parameter	Fold-effect of the FXYD protein as compared to control	Reference(s)
PLM	Xenopus oocytes[a]	$K_{1/2}$ Na^+	1.8 (α1), 1.51 (α2)	(16)
		$K_{1/2}$ K^+	$\leq$1.36 (α1), 1.42 (α2)	
		V_{max}	No change	
	Choroid plexus, neutralizing antibody	$K_{1/2}$ Na^+	No change	(17)
		V_{max}	1.4	
	KO mouse-cardiac SLM	$K_{1/2}$ Na^+	No change	(55)
		V_{max}	2.14	
	Cardiac SLM	V_{max}	2.8 (PLM phosphorylated)	(33)
	Cardiac myocytes	V_{max}	1.36 (α1) (PLM phosphorylated)	(34)
PLMS	Proteolytic digestion	V_{max}	1.4	(30)
γ	Xenopus oocytes	$K_{1/2}$ K^+ (at -150 mV)	$\leq$0.7–0.8 (γa, or γb)	(24)
		$K_{1/2}$ K^+ (at -50 mV)	No change	
		$K_{1/2}$ Na^+	1.29 (γa) 1.18 (γb)	
	Renal Na,K-ATPase neutralizing antibody	K'_{ATP}	0.5	(20)
	Mammalian cell membranes	K'_{ATP}	0.5 (γa, or γb)	(21, 22)
		K'_{ATP}	No change	(19, 56)
		$K_{1/2}$ Na^+	1.48 (γa)[b] 1.56 (γa) 1.58 (γb)	(19) (22)
		$K_{1/2}$ K^+	1.61 (γa) No change (γa, or γb)	(19) (22)
	Mammalian cells	K'_{ATP}	0.78 (γb)	(57)
		$K_{1/2}$ Na^+	1.71–1.98 (γa, or γb)	
		$K_{1/2}$ K^+	0.64	
	Reconstitution with γ peptide	$K_{1/2}$ Na^+	2–3	(58)
	KO mouse. Kidney membranes	$K_{1/2}$ Na^+	1.33	(59)
		$K_{1/2}$ K^+	No change	
		$K_{1/2}$ ATP	1.33	
CHIF	Xenopus oocytes	$K_{1/2}$ Na^+	0.64	(24)
		$K_{1/2}$ K^+ (at -50 mV)	1.55	
	Mammalian cells	$K_{1/2}$ Na^+	0.3	(26)
		$K_{1/2}$ K^+	No change	
FXYD7	Xenopus oocytes	$K_{1/2}$ Na^+	No change	(27)
		$K_{1/2}$ K^+	1.3–1.9	
Mat-8	Xenopus oocytes	$K_{1/2}$ Na^+	1.2	(23)
		$K_{1/2}$ K^+	1.15–1.4	

$K_{1/2}K^+$ refers to extracellular K^+ and $K_{1/2}Na^+$ to intracellular Na^+. Note that in each case the effect of the FXYD protein is quoted. In practice, with transfected cells one measures a gain of effect by the FXYD protein, whereas for inhibitory antibodies or measurements on membranes from KO mice one measures loss of effects of the FXYD protein.

[a] Figures quoted for $K_{1/2}$ K^+ measured in Xenopus oocytes are in the presence of extracellular Na^+. The effects of PLM and FXYD7 also depend on the α isoform used.

[b] In this study the γ subunit was expressed as a doublet in some experiments, which showed effects on both $K_{1/2}$ Na^+ and $K_{1/2}$ K^+, or as a single band in other experiments, which showed only the $K_{1/2}$ K^+ effect.

This segment includes T2 and S5, which are proposed to be posttranslationally modified (56). In addition, the modified forms $\gamma a'$(HEK) and $\gamma b'$ (HeLa) are expressed in a cell-specific fashion (22, 38), arguing against any invariant functional role of the modifications. Furthermore, in intact rat kidney outer medulla membranes, no $\gamma a'$ and very little $\gamma b'$ have been detected (38). Overall, no convincing evidence for specific roles of the two splice variants has been produced.

A recent report (59) on properties of γ KO mice documented the following effects on activity of Na,K-ATPase partially purified from membranes prepared from renal cortex plus medulla. The $K_{1/2}$ for cytoplasmic Na$^+$ ions was reduced 1.33-fold whereas the K_m for ATP was also reduced 1.33-fold (i.e., both the apparent affinities were raised). The effect on $K_{1/2}$ for Na$^+$ was expected, but the effect on K_m for ATP was the opposite of that expected from prior experiments with the expression systems. Thus, the investigators concluded that "regulation of Na$^+$ affinity is the major functional role for this protein while regulation of ATP affinity may be context-specific." In addition, γ stabilized the renal Na,K-ATPase against thermal inactivation, suggesting that this subunit may also have a structural role (59).

To summarize, γ raises the $K_{1/2}$ for cytoplasmic Na$^+$ ions in all systems examined (i.e., reduces apparent affinity), whereas the different observations on $K_{1/2}$ for extracellular K$^+$ ions in *Xenopus* oocytes and mammalian cell membranes are attributable primarily to the presence or absence of the membrane voltage. There is no obvious explanation why the effect on K_m for ATP observed in HeLa cells and renal microsomal Na,K-ATPase is not observed in membranes prepared from kidneys of γ KO mice or NRK-E52 cells.

The functional effects of CHIF in *Xenopus* oocytes (24) and mammalian cells (26) are similar, quite distinct from and opposite to those of γ. In *Xenopus* oocytes, CHIF induces an almost twofold decrease in $K_{1/2}$ for cell Na$^+$, with no change in maximal pump current. An increased $K_{1/2}$ for extracellular K$^+$ was also detected at strongly hyperpolarized membrane potentials (-150 mV), but not at 0 mV, and only in the presence of 90 mM external Na$^+$ (24). Ouabain binding is unaffected. In whole HeLa cells, CHIF expression increases the apparent affinity for cytoplasmic Na$^+$ by two- to threefold, but has no significant effect either on the affinity for external K$^+$ or on the maximal rate of active K$^+$(Rb$^+$) uptake (26). Similar effects were observed for isolated membranes, and no effect on K_m for ATP was seen. The lack of an observed effect of CHIF on $K_{1/2}$ for extracellular K$^+$ in whole HeLa cells does not contradict the findings in *Xenopus* oocytes because the medium in the former experiments contained only 30 mM Na$^+$ and the cell membrane potential was unlikely to exceed -50 mV. Enzyme kinetic considerations indicate that an effect of CHIF on apparent $K_{1/2}$ for cytoplasmic Na$^+$ or extracellular K$^+$ affinities at high extracellular Na$^+$ (in *Xenopus* oocytes), without parallel effects on the maximal rate of active transport, is consistent with modulation of the intrinsic Na$^+$-binding affinities at cytoplasmic or extracellular sites.

When expressed in *Xenopus* oocytes, PLM interacts with both the $\alpha 1 \beta 1$ and $\alpha 2 \beta 1$ isoforms. This is associated with a roughly twofold reduction in the apparent affinity for cytoplasmic Na$^+$ ions and a small decrease in affinity for extracellular

K^+ ions, but there is no effect on the maximal Na, K-pump current (16). In cardiac sarcolemma membranes, co-immunoprecipitation of PLM with $\alpha 1\beta 1$ and $\alpha 2\beta 1$ was also detected, albeit more weakly for the $\alpha 2\beta 1$ complex. PLM does not affec ouabain binding. PLM has also been detected in bovine choroid plexus membrane: (17), and it has been observed that an anti-PLM antibody reduces V_{max} withou affecting the Na^+ affinity (17). This latter result does not itself contradict the Xenopus oocytes findings because the antibody can only neutralize interaction: at the C terminus of PLM, whereas the Na^+ affinity effect may be mediated by the transmembrane segments of PLM (see below). However, a recent study (55 using sarcolemma membrane obtained from hearts of PLM KO mice showed a lower V_{max} of Na,K-ATPase, even after taking into account a $\sim 20\%$ lowe expression of the α subunit, but again, no difference in apparent Na^+ affinity wa detected. The implication is that PLM stimulates Na,K-ATPase activity, with no change in Na^+ affinity; this does not fit well with the observations in Xenopu oocytes. Purification of recombinant α/β/PLM complexes expressed in Pichia pastoris, by methods described in Reference 61, have been achieved recentl (62). In this system, PLM has been observed to raise the apparent affinity for Na^+ ions, again in contrast to the findings regarding Xenopus oocytes (16). On possible explanation of these discrepancies is that the phosphorylation states o PLM differ in the above tissues and expression systems. As reported recently, Na,K ATPase in sarcolemma membranes isolated from ischemic rat hearts is strongl stimulated compared to the controls, and this is accompanied by activation o PKA and PKC and phosphorylation of PLM (but not of the α subunit) (33 In addition, treatment of isolated guinea pig cardiac myocytes with forskolin which activates PKA, has been shown to significantly stimulate the Na, K-pum current (34). Only the $\alpha 1\beta 1$ (Ip$\alpha 1$) isoform—and not the $\alpha 2\beta 1$(Ip$\alpha 2$) isoforn distinguishable by its sensitivity to dihydroouabain—was affected by the forskoli treatment. The forskolin treatment was accompanied by phosphorylation of S6 on PLM, but not of S63, which is a substrate for PKC. Based on these finding: one can hypothesize that the effects of PLM observed in Xenopus oocytes (16 reflect the unphosphorylated state of PLM, whereas those observed in the wild type sarcolemma membranes reflect the phosphorylated state of PLM and ar absent in membranes obtained from KO mice (55). Evidently, there is good reaso to investigate the effects of PLM phosphorylation in greater detail.

The FXYD protein in shark rectal gland membranes, which copurifies wit the α and β subunits, is termed PLMS owing to its PLM-like phosphorylatio sites, although the closest mammalian sequence homolog is in fact Mat-8 (28 30). Phosphorylation of PLMS by PKC or selective proteolysis of the C terminu: which removes the phosphorylated serine, increases Na,K-ATPase activity (28, 30 Na,K-ATPase activity is partially inhibited when the enzyme interacts with intac or unphosphorylated PLMS, and inhibition is relieved upon proteolysis or PK phosphorylation. Phosphorylation of PLMS is proposed to impair the interactio of the C-terminal sequence of PLMS with the α/β complex. These effects ma appear opposite to those reported for PLM in cardiac sarcolemma membrane (3.

55) and choroid plexus (17), which suggests that PLM stimulates Na,K-ATPase activity. However, as discussed above, it may be phosphorylated PLM that has the effects seen in References 33 and 34. Thus, the mechanism of action of PLM may be similar to that of PLMS, justifying the latter's name.

FXYD7 is the fourth mammalian family member for which a functional inter-action with the Na,K-ATPase has been demonstrated. In *Xenopus* oocytes, FXYD7 decreases the apparent affinity for extracellular K^+ ions when expressed with $\alpha 1 \beta 1$ or $\alpha 2 \beta 1$, but not when expressed with $\alpha 3 \beta 1$ (63). This effect was observed with-out or with extracellular Na^+, implying an effect of FXYD7 on intrinsic binding affinity for extracellular K^+ sites. No effect on cytoplasmic Na^+ activation was detected. In brain, this protein was found to interact with $\alpha 1 \beta$ isoforms only. Thus, FXYD7 appears to exhibit an isoform-specific interaction with the pump. FXYD7 undergoes posttranslational modifications at the N termini (T3, T5, and T7), but these modifications do not affect the functional effects. Finally, recent work shows that Mat-8, when expressed in *Xenopus* oocytes, decreases the apparent affinities for both Na^+ and K^+ of Na,K-ATPase (23).

Whereas the functional effects of six FXYD proteins on Na,K-ATPase activity or Na, K-pump currents are well documented, detailed mechanistic information is lacking. Such information would provide a better understanding of these proteins' functional roles as well as complement studies of structure-function relations. Functional modulators are likely to affect Na,K-ATPase activity by altering rate-limiting steps, particularly the E_1P-E_2P and $E_2(K)$-E_1 conformational transitions, or binding of the transported cations, particularly cytoplasmic Na^+ ions (or com-peting K^+ ions), which limit the rate of active Na^+ pumping in vivo. Although some inferences on the effects of FXYD proteins on cation binding and confor-mational transitions have already been drawn, direct observations using purified α/β/FXYD complexes would be more conclusive. Availability of purified com-plexes of α/β/FXYD expressed in *P. pastoris* (61, 62) should soon make it feasible to measure directly cation occlusion and E_1-E_2 conformational transitions.

STRUCTURE-FUNCTION RELATIONS AND STRUCTURAL INTERACTIONS OF FXYD PROTEINS AND Na,K-ATPase

How do FXYD proteins and the α/β complex interact, and how do these in-teractions account for FXYD proteins' functional effects? There is now good evidence for multiple sites of interaction between FXYD and the α/β subunits. An initial indication came from the fact that the anti-γC terminus neutralizes the effect of γ on the apparent ATP affinity in renal Na,K-ATPase or HeLa cells transfected with γ, but not that on the K:Na antagonism (20–22). Also, in *Xenopus* oocytes, both the extracytoplasmic FXYD motif and cytoplasmic, posi-tively charged residues RKK are necessary for efficient association of γ and CHIF (24). Expression in HeLa cells of γ with either 4 or 10 C-terminal residues or 7 deleted N-terminal residues removes the effect of γ on ATP affinity but does not affect the K:Na antagonism. Replacement of the deleted 7 N-terminal residues with 7 alanines restores the effect on ATP affinity (60). Thus, different regions mediate different functional effects, and these effects can be long range.

In a systematic study of roles of the different segments, a series of γ-CHIF chimeric molecules was prepared in which extracellular, transmembrane, and cytoplasmic sequences were interchanged (25). The chimera were expressed in HeLa cells together with rat $\alpha1\beta1$, and the stability of the α/β/FXYD chimera was determined in a co-immunoprecipitation assay, taking advantage of the fact that the α/β/CHIF complex is more sensitive to excess detergent than is $\alpha/\beta/\gamma$. The stability of the α/β/FXYD chimera was found to depend on the origin of the transmembrane segment. Chimerae with a transmembrane segment derived from CHIF are less stable (like CHIF itself), whereas chimerae with a transmembrane segment derived from γ behave like γ. Upon the exchange of residues [55]MA in the transmembrane segment of CHIF with the corresponding IL residues in γ, the stability reverts to that of $\alpha/\beta/\gamma$. A similar effect was obtained by mutating [45]G in the transmembrane segment of CHIF to the corresponding γ residue (A) (25). Helical wheel projection shows that the three CHIF residues identified as important for the stability of the CHIF-α/β complex face two different planes of the membrane, i.e., they are likely to interact with at least two different helixes on α and/or β.

The opposite functional effect of γ and CHIF on apparent Na^+ affinity is also determined by the origin of the transmembrane segments. Thus, a CHIF construct with a transmembrane segment replaced by that of γ decreases the Na^+ affinity similar to the effect seen for γ, and vice versa (25). On the other hand, exchanging the extracellular or C-terminal domains has no effect on the FXYD-dependent change in Na^+ affinity. However, different residues determine the functional effect and stability. Mutating [45]G into A and [55]MA into IL does not alter the effect of CHIF on the pump kinetics. The functional role of the transmembrane segments has been confirmed in a study showing that peptides corresponding to the transmembrane segment of γ reduce apparent Na^+ affinity of the α/β complex in HeLa cell membranes, as found previously for full-length transfected γ (58). Peptides with G41R or G41L substitutions do not mediate the effect. The G41R mutation in γ is associated with familial renal Mg^{2+} wasting disease (64). The result in Reference 58 shows that [41]G is an important residue, probably involved in the functional effect. [35]G on the opposite face of the helix is not important.

Before we address the question of where in the α/β complex FXYD proteins interact, we must consider how many of the FXYD proteins can interact with each α/β complex. Information on possible stoichiometries of α/β to γ or CHIF subunits in renal Na,K-ATPase, and HeLa cells expressing γ and CHIF, has been obtained by co-immunoprecipitation in conditions shown to stabilize the protein and to optimize efficiency of co-immunoprecipitation (solubilization with $C_{12}E_1$ in the presence of Rb^+/ouabain) (26). Previously, purified renal Na,K-ATPase was shown to contain $\alpha:\beta:\gamma$ (both γa and γb) at a ratio close to 1:1:1 (7, 38), suggestive of an equal mixture of the $\alpha/\beta/\gamma$a and $\alpha/\beta/\gamma$b complexes, but compatible also with equal proportions of mixed $\alpha/\beta/\gamma$a/γb complexes and α/β complexes without γ. However, co-immunoprecipitation assays have effectively excluded mixed $\alpha/\beta/\gamma a/\gamma b$ or $\alpha/\beta/\gamma/CHIF$ complexes as well as heterodimeric complexes such as $\alpha/\beta/\gamma a - \alpha/\beta/\gamma b$ and $\alpha/\beta/\gamma - \alpha/\beta/CHIF$, at least in the optimal conditions of detergent solubilization (26; M. Lindzen, K. Gottschalk, M. Füzesi, H. Garty, S. Karlish, manuscript in preparation). For purified renal Na,K-ATPase, a specific anti-γa antibody precipitates γa (as well as α and β) but not γb, whereas an anti-γC terminus, which does not discriminate between γa and γb, precipitates both splice variants (as well as α and β). Thus, neither $\alpha/\beta/\gamma a/\gamma b$ nor $\alpha/\beta/\gamma a - \alpha/\beta/\gamma b$ is detected in these conditions. In renal papilla membranes, anti-γC immunoprecipitates γ (as well as α plus β) but not CHIF, and anti-CHIF precipitates CHIF but not γ (26). Although this experiment is suggestive, it does not exclude the possibility of $\alpha/\beta/\gamma/CHIF$ or a heterodimer $\alpha/\beta/\gamma - \alpha/\beta/CHIF$ because γ and CHIF are largely expressed in different nephron segments. However, the same result, which has now been obtained for γ and CHIF coexpressed together in HeLa cells, does exclude both of these complexes (M. Lindzen, K. Gottschalk, M. Füzesi, H. Garty, S. Karlish, manuscript in preparation). Coexpression of CHIF and of an HA-tagged CHIF in HeLa cells also clearly excludes $\alpha/\beta/CHIF/HA$-CHIF and the heterodimeric $\alpha/\beta/CHIF - \alpha/\beta/HA$-CHIF complex (M. Lindzen, K. Gottschalk, M. Füzesi, H. Garty, S. Karlish, manuscript in preparation). Taken together, these results suggest that either the $\alpha/\beta/\gamma$ or $\alpha/\beta/CHIF$ is the major complex in the detergent solution. A report inferring the possibility of mixed complexes of α/β with γa and γb employs different detergent conditions ($C_{12}E_8$ without added pump ligands) that may stabilize $\alpha/\beta/\gamma a - \alpha/\beta/\gamma b$ or higher macromolecular complexes, but it is not known if these complexes represent native structures (49).

Direct structural information on the Na,K-ATPase-FXYD interaction is as meager as structural information on the Na,K-ATPase itself. Cryoelectron microscopy of renal Na,K-ATPase at 9–10 Å resolution identified electron densities assumed to correspond to transmembrane helices of the α, β, and γ subunits (65). By analogy with 10 α subunit helices to Ca-ATPase helices, additional electron densities were assigned to the transmembrane helices of the β and γ subunits. The γ subunit helix was proposed to lie in a groove bounded by M2, M6, and M9 of the α subunit. This is analogous to the proposed location of phospholamban (PLN) in the Ca-ATPase (66).

A denaturation study suggested that γ may interact in the M8-M10 region (67). Recently, a role for M9 of the α subunit has been inferred from studies on *Xenopus* oocytes investigating the effects of mutants in M9 on the stability of the α/β-γ, CHIF, and FXYD 7 complexes and on their functional consequences (68). L964 and F967 are important for stability of the complexes, whereas F956 and E960 are required for mediation of the FXYD protein's effects on K^+ affinity. Thus, the structural and functional interactions are separable, as mutational studies in CHIF have also found (69). Interestingly, the F956 and E960 mutations do not alter the effects of γ and CHIF on Na^+ affinity, implying that still other interactions mediate these effects. Modeling of the FXYD7 transmembrane helix is consistent with docking in the M2, M6, and M9 groove.

Direct evidence for structural interactions has now been obtained by covalent cross-linking (70–72). Specific intramolecular covalent cross-links of the γ to α and γ to β subunits of pig kidney Na,K-ATPase, as well as those of rat γ to α coexpressed in HeLa cells, have been detected and analyzed (70). An α-γ cross-link has been located to residues K55 or K56 in the cytoplasmic tail of γ and to K347 in the cytoplasmic stalk segment S4 of the α subunit. On the basis of the cross-linking and other data on α-γ proximities, a model has been constructed to account for interactions of the transmembrane γ-helix and an unstructured cytoplasmic segment SKRLRCGGKKHR of γ with the α subunit (70). According to the model, the transmembrane segment fits in the groove between M2, M6, and M9, and the cytoplasmic segment interacts with loops L6/7 and L8/9 and stalk S5. An overview and details of the transmembrane segments and cytoplasmic sequence interactions are shown in Figure 2a and 2b, respectively. The model explains available evidence on proximities and also makes testable predictions, which should lead to its modification and further refinement. For example, the residues I54, I55, and A45, which were found to be important for the structural interaction (25) and were suggested to be in contact with more than one helix of the α subunit, interact with both M2 and M9. Moreover, new experiments showing cross-linking between C140 in M2 and C49 of CHIF or an F36C mutant of γ, expressed in HeLa cells, fit well with predicted proximity of the transmembrane

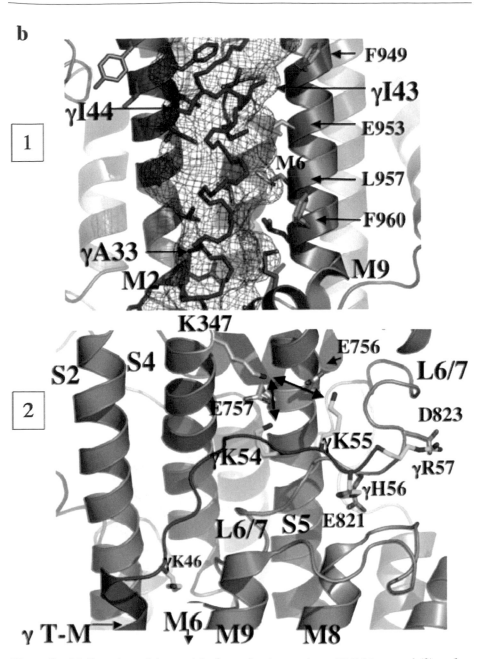

Figure 2   (*a*) Overview of the model of γ and α interactions. (1) Ribbons and (2) surface view of the α subunit with docked γ(*red*). K347 and K352 on the α subunit, which are candidates for cross-linking to γ, are shown in blue. (*b*) Detail of proposed interactions of (1) the transmembrane helix and (2) the cytoplasmic segment of the γ subunit with the α subunit. Numbering is for the pig α1 and pig γα subunits.

segment to M2 of α (71). New work on shark rectal gland PLMS shows a cross-link between the C-terminal cysteine and a cysteine in the A domain of the α subunit (72). This is different from the predicted location of the cytoplasmic domain of γ (Figure 2), and raises the question of whether the cytoplasmic domains of the different FXYD proteins interact with different domains of the α subunit. In Reference 70, two β-γ cross-links were detected in renal Na,K-ATPase, a major one at the extracellular surface within the segment G143-S302 of the β subunit and another within G1-R142. In colonic membranes, both α-CHIF and β-CHIF cross-links can also be detected, implying a similar disposition of γ and CHIF with respect to the α/β complex (M. Lindzen, K. Gottschalk, M. Füzesi, H. Garty, S. Karlish, manuscript in preparation). Selectivity for association of FXYD7 with β1 but not with β2 isoforms has been reported; this also suggests interactions between the FXYD protein and the β1 subunit (63).

Structural modeling of FXYD proteins with homology models of the α subunit may be improved by use of structures of FXYD proteins determined by NMR methods, which appears to be within sight (73). It is hoped that molecular structure of Na,K-ATPase with FXYD proteins eventually will become elucidated.

POSSIBLE PHYSIOLOGICAL ROLES OF FXYD PROTEINS

The physiological roles of FXYD proteins can be proposed on the basis of their observed effects in vitro and, especially, on the basis of phenotypic analysis of KO mice models and genetic disorders associated with mutations. KO mice models for CHIF have been described (74, 75), and initial analyses of γ (59) and PLM (55) KO mice models have also been reported recently. In addition, a single nucleotide mutation in human γ (G41R) has been linked to familial dominant renal hypomagnesemia (64). It is important to be aware of the fact that deletion of FXYD proteins may have complex physiological consequences, in particular owing to turning on of compensatory homeostatic mechanisms, which obscure phenotypic changes in the absence of additional physiological stresses (e.g., osmotic stress, salt loading, etc.). For example, γ KO mice are viable, without observable pathology, and the urinary secretion of Na^+, K^+, and Mg^{2+} is unaffected (59), despite the connection of the G41R mutation with renal hypomagnesemia (64). In addition, γ was reported to be expressed in the embryo and to be required for the cavitation process to the blastocyst stage (76), but in the KO mouse no delay in blastocoel formation or embryonic development was detected (59). Another aspect of the physiological complexity of FXYD protein function is that maintenance of ionic balance in vivo must involve both short-term and long-term regulatory mechanisms. Finally, it cannot be excluded that FXYD proteins modulate ion transport proteins other than Na,K-ATPase (see next section, below).

The distinct and opposite functional effects of γ and CHIF on the Na, K-pump in the kidney are consistent with their different patterns of expression along the nephron (Table 2), and presumably relate to the different physiological needs of the cells in which they are expressed. In the case of γ, the red outer medulla TAL segments contain high levels of γ and active Na,K-ATPase ($\alpha/\beta/\gamma$ complexes) and are characterized by a very high rate of Na^+ pumping and transepithelial Na^+ reabsorption, which generate the renal salt gradients. The effect of γ in lowering cellular Na^+ affinity, deduced from experiments on membranes of the γ KO mice (59) and transfected cells (19, 22, 24), implies a major homeostatic function of this feature in the physiological role of the γ subunit. In whole cells, the rate of active Na^+ and K^+ pumping is limited by the cytoplasmic Na^+ concentration, and also determines levels of intracellular Na^+ by altering the balance of active Na^+ extrusion and passive Na^+ leak. All other things being equal, a reduced Na^+ affinity should be associated with a higher steady-state level of intracellular Na^+. A decreased Na^+ affinity can then allow the pump to respond sensitively to increases of Na^+ concentration, at higher set-point levels of cytoplasmic Na^+, so as to restore intracellular Na^+ concentration. Cytoplasmic Na^+ concentrations in renal cells, measured by electron microprobe analysis, appear to be correlated with expression of γ or CHIF. In rat outer medulla outer stripe TAL cells, with the highest expression level of γ, the measured Na^+ content is significantly higher than in rat or mouse CCD principal cells, which contains CHIF but no γ. In rat and mice proximal tubule cells (PTCs) with moderate levels of γ, an intermediate Na^+ content was found [rat outer stripe medullary TAL 17.1 ± 1.4 (77), rat or mouse principal cell of CCD, 9.0–9.8 (78, 79), and rat or mouse proximal tubule 14.3–16.9 (78, 80) mmol Na^+ per kg wet weight]. In intact HeLa cells, expression of γb has also been reported to produce a significant but modestly raised steady-state Na^+ (14.5 ± 0.5 mM compared to 10.4 ± 0.6 mM), whereas γa raises intracellular Na^+ slightly (57). Whether or not the small changes in observed γ-induced changes in extracellular K^+ affinity play any physiological role is uncertain. The most pronounced change, an increase in K^+ affinity seen in cells with membrane potentials more negative than -50 mV (18, 24), may help to maintain pumping in energy-compromised conditions.

The γ-induced increase in ATP affinity revealed in in vitro studies (20–22) suggests that, in vivo, this may maintain the Na, K-pump rate in the face of transient decreases in ATP concentration in regions such as the renal outer medulla, which has a tendency to become hypoxic owing to low blood flow, low oxygen tension, and the high energy requirements of the medullary TAL cells (81). In fact, the Na, K-pumping rate in intact nephrons (82) and HeLa cells (57, 83) is linearly dependent, up to 2–3 mM, on the ATP concentration. This is in contrast to the saturation kinetics observed in membrane preparation (K_mATP of 400 μM) (57, 82), and indicates that the pumping rate in these cells is limited by the ATP concentration. On the other hand, a major physiological role of the effect on ATP affinity is not supported by recent observations on renal membranes of the γ KO mice model (59). Although there is no obvious explanation why the predicted effect of γ on ATP affinity is absent in these membranes, it is hypothetically possible that the context dependence, suggested in Reference 59, is related to the fact that the membranes were derived largely from the renal cortex. Renal cortex and proximal tubules are well supplied with blood, and unlike renal medulla, they are not highly sensitive to hypoxia (81).

Thus, tissue ATP levels should not fluctuate widely and modulation of ATP affinity by γ may be superfluous. Conceivably, the null mutation triggers slow adaptive processes, which may annul this kinetic effect in the specific cellular context.

In contrast to γ, CHIF is expressed in the collecting duct and distal colon and is induced by mineralocorticoids and K^+ loading (10, 43, 45, 46, 84, 85). Thus, it may be expected to function in the specific and crucial role of the principal cells of CCD in K^+ homeostasis and its regulation by mineralocorticoids (for reviews see References 86–88). A CHIF-induced increased Na^+ affinity should adapt the pumping rate to a lower intracellular Na^+ level resulting from the low level of luminal Na^+, and should respond effectively to increases in cytosolic Na^+ associated with mineralocorticoid-induced Na^+ permeability at the luminal surface. A lowered intracellular Na^+ concentration also maintains the driving force for Na^+ entry across the apical membranes and the transepithelial potential, which drive Na^+ reabsorption from the luminal fluid. A higher affinity of the Na, K-pump for cytosolic Na^+ observed in the cortical collecting ducts (CCDs) as compared to that of other nephron segments (89) may be the result of the modulatory interaction of CHIF, and may contribute to the lower cell Na^+ content of the principal cells.

CHIF KO mice are viable and fertile, and under normal conditions cannot be distinguished from wild-type litter mates (74). Only following K^+ loading or Na^+ deprivation do they manifest mild kidney phenotypes such as increased glomerular filtration rate and urine volume, but even then, there is no difference in the fractional excretion of Na^+ and K^+. Short-circuit current measurements in distal colon show a two-fold decrease in amiloride-blockable Na^+ absorption and forskolin-induced Cl^- secretion (75). These observations are consistent with a specific decrease in the Na^+, K^+-pumping rate in the collecting duct and distal colon under electrolyte stress (K^+-rich or Na^+-depleted diets). It is assumed that this primary deficit in kidney Na^+ and K^+ transport is well compensated in the more proximal nephron segments, whereas the secondary volume imbalance is only partly compensated by regulation of the glomerular filtration rate.

The single nucleotide mutation in human γ linked to familial dominant renal hypomagnesemia causes a G41R replacement in the transmembrane segment (64). As a result, γ fails to be incorporated into the plasma membrane and accumulates in the Golgi, although routing of the α/β subunits is not affected. The distal convoluted tubule (DCT) plays an important role in determining the final urinary excretion of Mg^{2+} even though it reabsorbs only approximately 10% of the filtered Mg^{2+} (90). The loss of Mg^{2+} may be an indirect effect of the absence of γ. For example, reduced ATP affinity and Na,K-ATPase activity may lead to elevated cell Na^+ and secondary reduction of Mg^{2+} entry into the DCT (64). Again, the underlying assumptions are that the primary effect of the impairment of Na^+ and K^+ transport is fully compensated by nephron segments that do not express γ and that only secondary effects that are specific to γ-enriched segments are manifested in the phenotype.

Long-term adaptation of cultured mouse inner medullary collecting duct cells (IMCD3) to hypertonic salt conditions (from 300 mOsm to 600 and 900 mOsm

NaCl) is accompanied by amplified synthesis of organic osmolytes and many proteins, including the Na,K ATPase α and β subunits (91). Although IMCD3 cells cultured in isotonic media express neither γa nor γb, both the γa and γb subunits are expressed upon long-term adaptation to hypertonicity and also, but weakly, after acute exposure (92). The γ subunit is incorporated into the basolateral cell membranes of cultured IMCD3 cells in response to hypertonicity (93). The process involves the C-terminal Jun kinase (JunK2 kinase) and phosphatidylinositol 3-kinase (PI3 kinase) signaling pathways. By contrast, the adaptive expression of the α subunit does not involve either of these signaling pathways. The JunK2 pathway involves transcriptional regulation of γ mRNA, whereas the PI3 kinase pathway involves translational regulation and appears to be important in cell-specific expression or suppression of the protein (94). Both mRNA and protein are expressed in IMCD3 cells, but only mRNA is found in the CCD cell line M1. Significantly, CCD principal cells in intact kidney express CHIF but not γ whereas IMCD expresses both CHIF and γ (43). Also interesting is that expression of the γ subunit depends on Cl^- and not Na^+ ions (95), whereas adaptation of α and β subunit expression is dependent on Na^+ ions. In addition, water hydration of mice partially suppresses expression of inner medulla γ (92). These findings all suggest that observations on cultured mouse IMCD3 cells are relevant and that γ plays an important role in the long-term adaptive response of inner medulla cells to hypertonicity. As a result of increased passive Na^+ and Cl^- influx, the cells adapt by increasing their Na^+ pumping capacity (α/β units) and γ, respectively. It appears, however, that expression of γ, and not α/β alone, is critical for long-term adaptation to hypertonic salt solutions (92, 95). Indeed, in hypertonic but not isotonic conditions, inhibition of the JunK pathway, the presence of a Cl^- channel blocker, or replacement of Cl^- ions with acetate is associated with loss of cell viability (92, 95). It is striking that different signaling pathways are used to control expression of α/β and γ and that regulation of such a small protein occurs at both the transcriptional and translational levels and is critical for cell survival. These observations raise the question of whether fine-tuning of the Na, K-pump kinetics is the sole role and mechanism of action of the γ subunit.

A number of other observations also suggest the possibility of additional, as yet unknown mechanisms and functions of γ and CHIF. As discussed above, γ exists as two splice variants, γa and γb, which have no apparent functional differences (22, 24). Yet the two splice variants have somewhat different tissue distribution, implying separate roles in a specific cellular setting (43). Recently it was reported that γa is detected without γb in preparations of rat renal caveolar membrane (96). Because caveolae are the site of many signaling pathways and a ouabain-dependent signaling pathway is known to exist (97), γa may participate in a signaling cascade. Also, γ and CHIF were recently detected within the same cells of native IMCD (43). Because γ and CHIF have opposite effects on Na^+ affinity and are not found in the mixed complexes $\alpha/\beta/\gamma$/CHIF (26), this finding raises the issue of whether there are additional roles for γ or CHIF. Alternatively, this may suggest a regulatory mechanism by which some posttranslational modification determines the interaction of α with either γ or CHIF.

The PLM KO mice described in Reference 55 are characterized by mild cardiac hypertrophy; increased ejection fraction; and 23% or 60% reduced expression, respectively, of the α1 and α2 isoforms of the Na,K-ATPase. The V_{max} of the Na,K-ATPase activity is reduced roughly 50%, even after accounting for the lower expression levels, i.e., reflecting a reduced turnover rate of the expressed pumps. The cardiac hypertrophy is not secondary to hypertension, as blood pressures were normal. The increased ejection fraction can be predicted on the basis of reduced Na,K-ATPase activity in the KO mice, by analogy to the mechanism of the positive inotropic effect of ouabain. Thus, these initial observations are compatible with PLM-induced activation of Na,K-ATPase in intact cardiac cells (34) and isolated membranes (17, 33). In principle, an increase in V_{max} or Na^+ affinity of the Na, K-pump may serve as a homeostatic mechanism that, accompanying or following transiently increased cardiac activity, acts to moderate or restore changes of intracellular Na^+ (increased), Ca^{2+} (increased), and K^+ (decreased) ion concentrations. As one example, β-adrenergic stimulation of cardiac activity is accompanied by stimulation of the Na, K-pump (see Reference 98) via activation of PKA. It appears from recent work that this is associated with phosphorylation of PLM at S68, and not of the α subunit (33, 34). Because PLM is a substrate for several protein kinases (31, 99, 100), including PKA, PKC, and NIMA, which phosphorylate different residues, there may be complex combinations of effects, and PLM may provide a means to integrate the effects of hormonal and other influences on the pump. A reduced affinity for Na^+ ions induced by PLM, observed upon expression in *Xenopus* oocytes (16), may allow the Na, K-pump to respond to an increase in Na^+ concentration at a relatively elevated set-point Na^+ concentration. This "reserve capacity" would not exist for pumps with a $K_{1/2}$ too low compared to the cytoplasmic Na^+ concentration, for the rate already would be close to saturation. In cardiac cells, such a situation can be envisaged for the basal electrical activity and cardiac contractility and may serve to efficiently control intracellular Na^+ concentrations. This mechanism is similar to that proposed for γ in kidney, and may apply to cells with high Na^+ influx rates. As hypothesized above, the PLM would not be phosphorylated, or the state of phosphorylation would be different from that in cells activated with β-adrenergic agents. Also relevant to its physiological role, PLM interacts strongly with α1 (16, 17, 34), whereas its interaction with the α2 isoform is less uniform—varying from unselective between isoforms in the brain (17) to less efficient in cardiac and skeletal muscle membranes (16) to undetectable in cardiac sarcolemma and guinea pig cardiac myocytes (33, 34). By contrast with the uniform distribution of the α1 isoform in neurons, vascular smooth muscle (101), and the cardiac myocyte membranes (34), and a similar uniform distribution of PLM (34), the α2 isoform is expressed in patches (101), similar to the distribution of the Na^+/Ca^{2+} exchanger (NCX1) (102). These patches, which have been termed "plasmerosomes," are adjacent to the endoplasmic or sarcoplasmic reticulum (SR). α2 has been proposed to function in tandem with NCX1 to regulate locally Na^+

and Ca^{2+} concentrations in the submembrane space, and thus Ca^{2+} accumulation in the SR (103). For example, $\alpha 2$ is involved in the mediation of inotropic effects of ouabain(104, 105). The relationship between NCX1 and PLM is discussed further below, but in any event regulation of Na^+ and Ca^{2+} concentrations by PLM in the cardiac cell must reflect its strong association with $\alpha 1$ as well as any effects mediated via $\alpha 2$, with which its association seems to be more variable.

DO FXYD PROTEINS REGULATE ION TRANSPORT PROTEINS OTHER THAN Na,K-ATPase?

For most of the FXYD proteins, the initial publications reported findings that are unrelated to regulation of the Na,K-ATPase (cf. Table 1). Thus, PLM and Mat-8 were shown to evoke a Cl^- channel activity in *Xenopus* oocytes (3, 9). Other studies suggested permeation, mediated by PLM, of various other ions as well as large molecules (4, 5, 106–110). CHIF and γ also were reported to manifest ionic channel activity in expression systems (10, 111–113). γ evokes a large non-specific conductance that allows permeation of inulin, whereas the CHIF-induced conductance is K^+-specific and may reflect activation of KCNQ1 channels. RIC (dysadherin) was reported to downregulate E-cadherin and promote metastasis (12). Finally, recent findings suggest direct regulation of NCX1 by PLM (40, 114–117). These observations raise the possibility that FXYD proteins perform other cellular functions in addition to the modulation of the pump kinetics and, in particular, that they modulate other ion transporters.

The ionic conductance recorded in *Xenopus* oocytes expressing PLM, γ, Mat-8, or CHIF may suggest that the FXYD proteins form ionic channels (possibly by self-oligomerization) or can modulate ionic channels present in the oocyte membrane. This suggests the hypothesis that FXYD proteins contribute to intracellular ionic homeostasis by simultaneously regulating the Na,K-ATPase and cation conductances in the same cell. However, because the conductances induced by PLM, γ, Mat-8, and CHIF are not readily seen in native cells expressing these FXYD proteins, other possibilities need to be considered. One option is that the channel activity recorded in expression systems is an artifact caused by overexpression of the relevant FXYD protein. Such an artifact may be either activation of an endogenous channel, as part of a stress response to this overexpression, or channel-like activity caused by oligomerization of excess FXYD protein in the cell membrane. Indeed, the slowly activating hyperpolarization-induced Cl^- conductance characteristic of PLM and Mat-8 is seen in *Xenopus* oocytes expressing many other unrelated proteins as well as in noninjected oocytes (118–120).

There is evidence that FXYD proteins can undergo self-oligomerization. Oligomers of both native γ (28) and peptides corresponding to the transmembrane segment are detected in the presence of perfluoroctanoate (121). Oligomerization is not observed for peptides with the G41R mutation but is retained in G35L mutants, suggesting that this process depends on a specific interaction. The γ sequence

LAFVVGLLILLS is similar to the sequence LAXXVGXXIGXXI, which is known to mediate homo-oligomerization (122). The functional role of oligomerization is unknown. One possibility is that it is associated with the channel-like properties, similar to the channel-like property of the pentameric form of phospholamban (PLN) (123). Another possibility is that oligomers represent an uncomplexed store of FXYD proteins.

A fairly large body of evidence indicates involvement of PLM in the regulation of NCX1 (40, 114–117). Overexpression of PLM in adult rat myocytes increases their contraction amplitudes and internal Ca^{2+} transients, inhibits NCX1-mediated currents, and slows down relaxation from caffeine-induced contractions (40, 116). One possibility is that the observed responses are secondary to Na,K-ATPase-dependent changes in intracellular Na^+. However, because PLM inhibits both influx and efflux of Ca^{2+} through NCX1, a decrease in Na^+ gradient cannot account for these findings (40, 117). Other experiments have shown colocalization and co-immunoprecipitation of PLM and NCX1 (40, 114, 115). Mutating putative phosphorylation sites, in particular S68, in the carboxy tail of PLM abolishes the effects of PLM overexpression on contractility, Ca^{2+} transients, and the NCX1-mediated current (114). The above effects were also reported for HEK293 cells cotransfected with PLM and NCX1 (117). The combined effects of PLM on Na,K-ATPase ($\alpha 1$ and $\alpha 2$) and NCX1 in the heart may, together, promote Ca^{2+} accumulation and cardiac contractility as well as Ca^{2+} extrusion and cardiac relaxation. For example, inhibition of NCX1 following β-adrenergic stimulation and phosphorylation of PLM, causing a local rise of Ca^{2+} concentration, may combine with activation of the sarcoplasmic reticulum (SR) Ca^{2+} pump isoform (SERCA2a) that accompanies phosphorylation of its regulator PLN, thereby amplifying Ca^{2+} accumulation into the SR (124). Amplified Ca^{2+} release upon electrical excitation would then increase cardiac muscle contractility. Activation of the Na,K-ATPase $\alpha 1$ by PLM during β-adrenergic stimulation would help to restore Na^+ and Ca^{2+} gradients and cardiac muscle relaxation.

CONCLUSIONS

Much data obtained in recent years have shown clearly that members of the FXYD family interact specifically with the Na,K-ATPase and alter its apparent affinities for Na^+, K^+, ATP, and V_{max}. These kinetic effects, although modest, may be of profound physiological importance. It is uncertain, however, whether the kinetic effects detected in vitro are the only outcome of the FXYD/Na,K-ATPase interactions. Also, mechanistic detail on these functional effects is lacking, as is information on the structural interaction between FXYD proteins and the α and β subunits. New work on phenotypic and biochemical analyses of KO mice models promises to provide important information on the physiological roles of FXYD proteins in addition to those inferred from in vitro studies. Finally, accumulating data indicate that FXYD proteins may control activity of other transporters in addition to modulating the Na,K-ATPase.

ACKNOWLEDGMENTS

Work in the authors' laboratories was funded by research grants from the Israel Science foundation, the Minerva foundation, and the Weizmann Institute Renal Research Fund. H. Garty and S. Karlish are the incumbents of the Hella & Derrick Klecnian and William Smithburg Chairs of Biochemistry, respectively.

The *Annual Review of Physiology* is online at
http://physiol.annualreviews.org

LITERATURE CITED

1. Sweadner KJ, Rael E. 2000. The FXYD gene family of small ion transport regulators or channels: cDNA sequence, protein signature sequence, and expression. *Genomics* 68:41–56

2. Palmer CJ, Scott BT, Jones LR. 1991. Purification and complete sequence determination of the major plasma membrane substrate for cAMP-dependent protein kinase and protein kinase C in myocardium. *J. Biol. Chem.* 266:11126–30

3. Moorman JR, Palmer CJ, John JE 3rd, Durieux ME, Jones LR. 1992. Phospholemman expression induces a hyperpolarization-activated chloride current in Xenopus oocytes. *J. Biol. Chem.* 267:14551–54

4. Kowdley GC, Ackerman SJ, Chen Z, Szabo G, Jones LR, Moorman JR. 1997. Anion, cation, and zwitterion selectivity of phospholemman channel molecules. *Biophys. J.* 72:141–45

5. Moorman JR, Jones LR. 1998. Phospholemman: a cardiac taurine channel involved in regulation of cell volume. *Adv. Exp. Med. Biol.* 442:219–28

6. Forbush B III, Kaplan JH, Hoffman JF. 1978. Characterization of a new photoaffinity derivative of ouabain: labeling of the large polypeptide and of a proteolipid component of the Na, K-ATPase. *Biochemistry* 17:3667–76

7. Mercer RW, Biemesderfer D, Bliss DP Jr., Collins JH, Forbush B III. 1993. Molecular cloning and immunological characterization of the gamma polypeptide, a small protein associated with the Na,K-ATPase. *J. Cell Biol.* 121:579–86

8. Morrison BW, Leder P. 1994. neu and ras initiate murine mammary tumors that share genetic markers generally absent in c-myc and int-2-initiated tumors. *Oncogene* 9:3417–26

9. Morrison BW, Moorman JR, Kowdley GC, Kobayashi YM, Jones LR, Leder P. 1995. Mat-8, a novel phospholemman-like protein expressed in human breast tumors, induces a chloride conductance in Xenopus oocytes. *J. Biol. Chem.* 270:2176–82

10. Attali B, Latter H, Rachamim N, Garty H. 1995. A corticosteroid-induced gene expressing an "IsK-like" K^+ channel activity in Xenopus oocytes. *Proc. Natl. Acad. Sci. USA* 92:6092–96

11. Fu X, Kamps MP. 1997. E2a-Pbx1 induces aberrant expression of tissue-specific and developmentally regulated genes when expressed in NIH 3T3 fibroblasts. *Mol. Cell Biol.* 17:1503–12

12. Ino Y, Gotoh M, Sakamoto M, Tsukagoshi K, Hirohashi S. 2002. Dysadherin, a cancer-associated cell membrane glycoprotein, down-regulates E-cadherin and promotes metastasis. *Proc. Natl. Acad. Sci. USA* 99:365–70

13. Deleted in press

14. Crambert G, Geering K. 2003. FXYD proteins: new tissue-specific regulators of

the ubiquitous Na,K-ATPase. *Sci STKE* 2003:RE1

15. Garty H, Karlish SJD. 2005. FXYD proteins: tissue specific regulators of the Na, K ATPase. *Semin. Nephrol.* 25:304–11

16. Crambert G, Fuzesi M, Garty H, Karlish S, Geering K. 2002. Phospholemman (FXYD1) associates with Na,K-ATPase and regulates its transport properties. *Proc. Natl. Acad. Sci. USA* 99:11476–81

17. Feschenko MS, Donnet C, Wetzel RK, Asinovski NK, Jones LR, Sweadner KJ. 2003. Phospholemman, a single-span membrane protein, is an accessory protein of Na,K-ATPase in cerebellum and choroid plexus. *J. Neurosci.* 23:2161–69

18. Beguin P, Wang X, Firsov D, Puoti A, Claeys D, et al. 1997. The γ subunit is a specific component of the Na,K-ATPase and modulates its transport function. *EMBO J.* 16:4250–60

19. Arystarkhova E, Wetzel RK, Asinovski NK, Sweadner KJ. 1999. The γ subunit modulates Na$^+$ and K$^+$ affinity of the renal Na,K-ATPase. *J. Biol. Chem.* 274: 33183–85

20. Therien AG, Goldshleger R, Karlish SJ, Blostein R. 1997. Tissue-specific distribution and modulatory role of the γ subunit of the Na,K-ATPase. *J. Biol. Chem.* 272:32628–34

21. Therien AG, Karlish SJ, Blostein R. 1999. Expression and functional role of the γ subunit of the Na, K-ATPase in mammalian cells. *J. Biol. Chem.* 274:12252–56

22. Pu HX, Cluzeaud F, Goldshleger R, Karlish SJ, Farman N, Blostein R. 2001. Functional role and immunocytochemical localization of the γ a and γ b forms of the Na,K-ATPase γ subunit. *J. Biol. Chem.* 276:20370–78

23. Crambert G, Li C, Claeys D, Geering K. 2005. FXYD3 (Mat-8), a new regulator of Na,K-ATPase. *Mol. Biol. Cell.* 16(5):2363–71

24. Beguin P, Crambert G, Guennoun S, Garty H, Horisberger JD, Geering K.

25. Lindzen M, Aizman R, Lifshitz Y, Lubarski I, Karlish SJ, Garty H. 2003. Structure-function relations of interactions between Na,K-ATPase, the γ subunit, and corticosteroid hormone-induced factor. *J. Biol. Chem.* 278:18738–43

26. Garty H, Lindzen M, Scanzano R, Aizman R, Fuzesi M, et al. 2002. A functional interaction between CHIF and Na-K-ATPase: implication for regulation by FXYD proteins. *Am. J. Physiol. Renal Physiol.* 283:F607–15

27. Beguin P, Crambert G, Monnet-Tschudi F, Uldry M, Horisberger JD, et al. 2002. FXYD7 is a brain-specific regulator of Na,K-ATPase α 1-β isozymes. *EMBO J.* 21:3264–73

28. Mahmmoud YA, Vorum H, Cornelius F. 2000. Identification of a phospholemman-like protein from shark rectal glands. Evidence for indirect regulation of Na,K-ATPase by protein kinase c via a novel member of the FXYDY family. *J. Biol. Chem.* 275:35969–77

29. Mahmmoud YA, Cornelius F. 2002. Protein kinase C phosphorylation of purified Na,K-ATPase: C-terminal phosphorylation sites at the α- and γ-subunits close to the inner face of the plasma membrane *Biophys. J.* 82:1907–19

30. Mahmmoud YA, Cramb G, Maunsbach AB, Cutler CP, Meischke L, Cornelius F. 2003. Regulation of Na,K-ATPase by PLMS, the phospholemman-like protein from shark: molecular cloning, sequence expression, cellular distribution, and functional effects of PLMS. *J. Biol. Chem* 278:37427–38

31. Walaas SI, Czernik AJ, Olstad OK, Sletten K, Walaas O. 1994. Protein kinase C and cyclic AMP-dependent protein kinase phosphorylate phospholemman, an insulin and adrenaline-regulated membrane phosphoprotein, at specific sites in

the carboxy terminal domain. *Biochem. J.* 304:635–40

32. Walaas O, Horn RS, Walaas SI. 1999. Inhibition of insulin-stimulated phosphorylation of the intracellular domain of phospholemman decreases insulin-dependent GLUT4 translocation in streptolysin-O-permeabilized adipocytes. *Biochem. J.* 343:151–57

33. Fuller W, Eaton P, Bell JR, Shattock MJ. 2004. Ischemia-induced phosphorylation of phospholemman directly activates rat cardiac Na/K-ATPase. *FASEB J.* 18:197–99

34. Silverman BZ, Fuller W, Eaton P, Deng J, Moorman JR, et al. 2005. Serine 68 phosphorylation of phospholemman: acute isoform-specific activation of cardiac Na/K ATPase. *Cardiovasc. Res.* 65:93–103

35. Sweadner KJ, Arystarkhova E, Donnet C, Wetzel RK. 2003. FXYD proteins as regulators of the Na,K-ATPase in the kidney. *Ann. NY Acad. Sci.* 986:382–87

36. Jones DH, Golding MC, Barr KJ, Fong GH, Kidder GM. 2001. The mouse Na$^+$-K$^+$-ATPase γ-subunit gene (Fxyd2) encodes three developmentally regulated transcripts. *Physiol. Genomics* 6:129–35

37. Bogaev RC, Jia LG, Kobayashi YM, Palmer CJ, Mounsey JP, et al. 2001. Gene structure and expression of phospholemman in mouse. *Gene* 271:69–79

38. Kuster B, Shainskaya A, Pu HX, Goldshleger R, Blostein R, et al. 2000. A new variant of the γ subunit of renal Na,K-ATPase. Identification by mass spectrometry, antibody binding, and expression in cultured cells. *J. Biol. Chem.* 275:18441–46

39. Lubarski I, Pihakaski-Maunsbach K, Karlish SJ, Maunsbach AB, Garty H. 2005. Interaction with the Na, K ATPase and tissue distribution of FXYD5 (RIC). *J. Biol. Chem.* 280:37717–24

40. Zhang XQ, Qureshi A, Song J, Carl LL, Tian Q, et al. 2003. Phospholemman mod-

ulates Na$^+$/Ca^{2+} exchange in adult rat cardiac myocytes. *Am. J. Physiol. Heart Circ. Physiol.* 284:H225–33

41. Wetzel RK, Sweadner KJ. 2003. Phospholemman expression in extraglomerular mesangium and afferent arteriole of the juxtaglomerular apparatus. *Am. J. Physiol. Renal Physiol.* 285:F121–29

42. Wetzel RK, Sweadner KJ. 2001. Immunocytochemical localization of Na-K-ATPase α- and γ-subunits in rat kidney. *Am. J. Physiol. Renal Physiol.* 281:F531–45

43. Pihakaski-Maunsbach K, Vorum H, Locke EM, Garty H, Karlish SJ, Maunsbach AB. 2003. Immunocytochemical localization of Na,K-ATPase γ subunit and CHIF in inner medulla of rat kidney. *Ann. NY Acad. Sci.* 986:401–9

44. Deleted in proof

45. Capurro C, Coutry N, Bonvalet JP, Escoubet B, Garty H, Farman N. 1996. Cellular localization and regulation of CHIF in kidney and colon. *Am. J. Physiol.* 271:C753–62

46. Shi H, Levy-Holzman R, Cluzeaud F, Farman N, Garty H. 2001. Membrane topology and immunolocalization of CHIF in kidney and intestine. *Am. J. Physiol. Renal Physiol.* 280:F505–12

47. Yamaguchi F, Yamaguchi K, Tai Y, Sugimoto K, Tokuda M. 2001. Molecular cloning and characterization of a novel phospholemman-like protein from rat hippocampus. *Brain Res. Mol. Brain Res.* 86:189–92

48. Kadowaki K, Sugimoto K, Yamaguchi F, Song T, Watanabe Y, et al. 2004. Phosphohippolin expression in the rat central nervous system. *Brain Res. Mol. Brain Res.* 125:105–12

49. Arystarkhova E, Wetzel RK, Sweadner KJ. 2002. Distribution and oligomeric association of splice forms of Na$^+$-K$^+$-ATPase regulatory γ-subunit in rat kidney. *Am. J. Physiol. Renal Physiol.* 282:F393–407

50. Rivas E, Lew V, De Robertis E. 1972.

(3 H) Ouabain binding to a hydrophobic protein from electroplax membranes. *Biochim. Biophys. Acta* 290:419–23

51. Reeves AS, Collins JH, Schwartz A. 1980. Isolation and characterization of (Na,K)-ATPase proteolipid. *Biochem. Biophys. Res. Commun.* 95:1591–98

52. Collins JH, Leszyk J. 1987. The "γ subunit" of Na,K-ATPase: a small, amphiphilic protein with a unique amino acid sequence. *Biochemistry* 26:8665–68

53. Therien AG, Pu HX, Karlish SJ, Blostein R. 2001. Molecular and functional studies of the γ subunit of the sodium pump. *J. Bioenerg. Biomembr.* 33:407–14

54. Cornelius F, Mahmmoud YA. 2003. Functional modulation of the sodium pump: the regulatory proteins "Fixit." *News Physiol. Sci.* 18:119–24

55. Jia LG, Donnet C, Bogaev RC, Blatt RJ, McKinney CE, et al. 2005. Hypertrophy, increased ejection fraction, and reduced Na-K-ATPase activity in phospholemman-deficient mice. *Am. J. Physiol. Heart Circ. Physiol.* 288:H1982–88

56. Arystarkhova E, Donnet C, Asinovski NK, Sweadner KJ. 2002. Differential regulation of renal Na,K-ATPase by splice variants of the γ subunit. *J. Biol. Chem.* 277:10162–72

57. Zouzoulas A, Dunham PB, Blostein R. 2005. The effect of the gamma modulator on Na/K pump activity of intact HeLa cells. *J. Membr. Biol.* 204:49–56

58. Zouzoulas A, Therien AG, Scanzano R, Deber CM, Blostein R. 2003. Modulation of Na,K-ATPase by the γ subunit: studies with transfected cells and transmembrane mimetic peptides. *J. Biol. Chem.* 278:40437–41

59. Jones DH, Li TY, Arystarkhova E, Barr KJ, Wetzel RK, et al. 2005. Na,K-ATPase from mice lacking the γ subunit (FXYD2) exhibits altered Na^+ affinity and decreased thermal stability. *J. Biol. Chem.* 280:19003–11

60. Pu HX, Scanzano R, Blostein R. 2002. Distinct regulatory effects of the Na,K-ATPase gamma subunit. *J. Biol. Chem.* 277:20270–76

61. Cohen E, Goldshleger R, Shainskaya A, Tal DM, Ebel C et al. 2005. Purification of Na^+,K^+-ATPase expressed in *P. pastoris* reveals an essential role of phospholipid-protein interactions. *J. Biol. Chem.* 280:16610–18

62. Lifshitz Y, Garty H, Karlish SJD. 2005. Purification of $\alpha/\beta/PLM$ (FXYD1) complexes of Na,K-ATPase expressed in *Pichia pastoris*. *J. Gen. Physiol.* 126:50a (Abstr.)

63. Crambert G, Beguin P, Uldry M, Monnet-Tschudi F, Horisberger JD, et al. 2003. FXYD7, the first brain- and isoform-specific regulator of Na,K-ATPase: biosynthesis and function of its posttranslational modifications. *Ann. NY Acad. Sci.* 986:444–48

64. Meij IC, Koenderink JB, van Bokhoven H, Assink KF, Tiel Groenestege W, et al. 2000. Dominant isolated renal magnesium loss is caused by misrouting of the Na^+,K^+-ATPase γ-subunit. *Nat. Genet.* 26:265–66

65. Hebert H, Purhonen P, Vorum H, Thomsen K, Maunsbach AB. 2001. Three-dimensional structure of renal Na,K-ATPase from cryo-electron microscopy of two-dimensional crystals. *J. Mol. Biol.* 314:479–94

66. Toyoshima C, Asahi M, Sugita Y, Khanna R, Tsuda T, MacLennan DH. 2003. Modeling of the inhibitory interaction of phospholamban with the Ca^{2+} ATPase. *Proc. Natl. Acad. Sci. USA* 100:467–72

67. Donnet C, Arystarkhova E, Sweadner KJ. 2001. Thermal denaturation of the Na,K-ATPase provides evidence for $\alpha-\alpha$ oligomeric interaction and γ subunit association with the C-terminal domain. *J. Biol. Chem.* 276:7357–65

68. Li C, Grosdidier A, Crambert G, Horisberger JD, Michielin O, Geering K. 2004. Structural and functional interaction sites between Na,K-ATPase and FXYD

proteins. *J. Biol. Chem.* 279:38895–902

69. Lindzen M, Aizman R, Lifshitz Y, Fuzesi M, Karlish SJ, Garty H. 2003. Domains involved in the interactions between FXYD and Na,K-ATPase. *Ann. NY Acad. Sci.* 986:530–31

70. Fuzesi M, Gottschalk KE, Lindzen M, Shainskaya A, Kuster B, et al. 2005. Covalent cross-links between the γ subunit(FXYD2) and α and β subunits of Na,K-ATPase. Modeling the α-γ interaction. *J. Biol. Chem.* 280:18291–19301

71. Lindzen M, Gottschalk K-E, Fuzesi M, Garty H, Karlish SJD. 2006. Structural interactions between FXYD proteins and Na⁺,K⁺-ATPase: α/β/FXYD sub-unit stoichiometry and cross-linking. *J. Biol. Chem.* In press

72. Mahmmoud YA, Vorum H, Cornelius F. 2005. Interaction of FXYD10 (PLMS) with Na,K-ATPase from shark rectal glands: Close proximity of Cys74 of FXYD10 to Cys254 in the a domain of the α-subunit revealed by intermolecular thiol cross-linking. *J Biol. Chem.* 280:27776–82

73. Marassi FM, Crowell KJ. 2003. Hydration-optimized oriented phospholipid bilayer samples for solid-state NMR structural studies of membrane proteins. *J. Magn. Reson.* 161:64–69

74. Aizman R, Asher C, Fuzesi M, Latter H, Lonai P, et al. 2002. Generation and phenotypic analysis of CHIF knockout mice. *Am. J. Physiol. Renal Physiol.* 283:F569–77

75. Goldschimdt I, Grahammer F, Warth R, Schulz-Baldes A, Garty H, et al. 2004. Kidney and colon electrolyte transport in CHIF knockout mice. *Cell Physiol. Biochem.* 14:113–20

76. Jones DH, Davies TC, Kidder GM. 1997. Embryonic expression of the putative γ subunit of the sodium pump is required for acquisition of fluid transport capacity during mouse blastocyst development. *J. Cell Biol.* 139:1545–52

77. Beck FX, Sone M, Dorge A, Thurau K. 1992. Effect of increased distal sodium delivery on organic osmolytes and cell electrolytes in the renal outer medulla. *Pflugers Arch.* 422:233–38

78. Beck FX, Dorge A, Rick R, Schramm M, Thurau K. 1988. The distribution of potassium, sodium and chloride across the apical membrane of renal tubular cells: effect of acute metabolic alkalosis. *Pflugers Arch.* 411:259–67

79. Beck FX, Neuhofer W, Dorge A, Giebisch G, Wang T. 2003. Intracellular Na concentration and Rb uptake in proximal convoluted tubule cells and abundance of Na/K-ATPase α1-subunit in NHE3-/- mice. *Pflugers Arch.* 446:100–5

80. Matsuda O, Beck FX, Dorge A, Thurau K. 1988. Electrolyte composition of renal tubular cells in gentamicin nephrotoxicity. *Kidney Int.* 33:1107–12

81. Brezis M, Epstein FH. 1993. Cellular mechanisms of acute ischemic injury in the kidney. *Annu. Rev. Med.* 44:27–37

82. Soltoff SP, Mandel LJ. 1984. Active ion transport in the renal proximal tubule. III. The ATP dependence of the Na pump. *J. Gen. Physiol.* 84:643–62

83. Ikehara T, Yamaguchi H, Hosokawa K, Sakai T, Miyamoto H. 1984. Rb⁺ influx in response to changes in energy generation: effect of the regulation of the ATP content of HeLa cells. *J. Cell. Physiol.* 119:273–82

84. Wald H, Popovtzer MM, Garty H. 1997. Differential regulation of CHIF mRNA by potassium intake and aldosterone. *Am. J. Physiol.* 272:F617–23

85. Wald H, Goldstein O, Asher C, Yagil Y, Garty H. 1996. Aldosterone induction and epithelial distribution of CHIF. *Am. J. Physiol.* 271:F322–29

86. Feraille E, Doucet A. 2001. Sodium-potassium-adenosinetriphosphatase-dependent sodium transport in the kidney: hormonal control. *Physiol. Rev.* 81:345–418

87. Palmer LG. 1999. Potassium secretion

and the regulation of distal nephron K channels. *Am. J. Physiol.* 277:F821–25

88. Giebisch G. 2001. Renal potassium channels: function, regulation, and structure. *Kidney Int.* 60:436–45

89. Barlet-Bas C, Cheval L, Khadouri C, Marsy S, Doucet A. 1990. Difference in the Na affinity of Na^+-K^+-ATPase along the rabbit nephron: modulation by K. *Am. J. Physiol.* 259:F246–50

90. Dai LJ, Ritchie G, Kerstan D, Kang HS, Cole DE, Quamme GA. 2001. Magnesium transport in the renal distal convoluted tubule. *Physiol. Rev.* 81:51–84

91. Capasso JM, Rivard CJ, Berl T. 2001. Long-term adaptation of renal cells to hypertonicity: role of MAP kinases and Na-K-ATPase. *Am. J. Physiol. Renal Physiol.* 280:F768–76

92. Capasso JM, Rivard C, Berl T. 2001. The expression of the γ subunit of Na-K-ATPase is regulated by osmolality via C-terminal Jun kinase and phosphatidylinositol 3-kinase-dependent mechanisms. *Proc. Natl. Acad. Sci. USA* 98:13414–19

93. Pihakaski-Maunsbach K, Tokonabe S, Vorum H, Rivard CJ, Capasso JM, et al. 2005. The γ-subunit of Na-K-ATPase is incorporated into plasma membranes of mouse IMCD3 cells in response to hypertonicity. *Am. J. Physiol. Renal Physiol.* 288:F650–57

94. Capasso JM, Rivard CJ, Berl T. 2005. Synthesis of the Na-K-ATPase γ-subunit is regulated at both the transcriptional and translational levels in IMCD3 cells. *Am. J. Physiol. Renal Physiol.* 288:F76–81

95. Capasso JM, Rivard CJ, Enomoto LM, Berl T. 2003. Chloride, not sodium, stimulates expression of the γ subunit of Na/K-ATPase and activates JNK in response to hypertonicity in mouse IMCD3 cells. *Proc. Natl. Acad. Sci. USA* 100:6428–33

96. Ferrandi M, Molinari I, Barassi P, Minotti E, Bianchi G, Ferrari P. 2004. Organ hypertrophic signaling within caveolae membrane subdomains triggered by ouabain and antagonized by PST 2238. *J. Biol. Chem.* 279:33306–14

97. Xie Z, Askari A. 2002. Na^+/K^+-ATPase as a signal transducer. *Eur. J. Biochem.* 269:2434–39

98. Glitsch HG. 2001. Electrophysiology of the sodium-potassium-ATPase in cardiac cells. *Physiol. Rev.* 81:1791–826

99. Lu KP, Kemp BE, Means AR. 1994. Identification of substrate specificity determinants for the cell cycle–regulated NIMA protein kinase. *J. Biol. Chem.* 269:6603–7

100. Mounsey JP, John JE 3rd, Helmke SM, Bush EW, Gilbert J, et al. 2000. Phospholemman is a substrate for myotonic dystrophy protein kinase. *J. Biol. Chem.* 275:23362–67

101. Juhaszova M, Blaustein MP. 1997. Na^+ pump low and high ouabain affinity α subunit isoforms are differently distributed in cells. *Proc. Natl. Acad. Sci. USA* 94:1800–5

102. Juhaszova M, Shimizu H, Borin ML, Yip RK, Santiago EM, et al. 1996. Localization of the Na^+-Ca^{2+} exchanger in vascular smooth muscle, and in neurons and astrocytes. *Ann. NY Acad. Sci.* 779:318–35

103. Juhaszova M, Blaustein MP. 1997. Distinct distribution of different Na^+ pump α subunit isoforms in plasmalemma. Physiological implications. *Ann. NY Acad. Sci.* 834:524–36

104. Blaustein MP, Juhaszova M, Golovina VA. 1998. The cellular mechanism of action of cardiotonic steroids: a new hypothesis. *Clin. Exp. Hypertens.* 20:691–703

105. James PF, Grupp IL, Grupp G, Woo AL, Askew GR, et al. 1999. Identification of a specific role for the Na,K-ATPase α 2 isoform as a regulator of calcium in the heart. *Mol. Cell.* 3:555–63

106. Davis CE, Patel MK, Miller JR, John JE 3rd, Jones LR, et al. 2004. Effects of phospholemman expression on swelling-activated ion currents and volume

regulation in embryonic kidney cells. *Neurochem. Res.* 29:177–87

107. Chen ZH, Jones LR, Moorman JR. 1999. Ion currents through mutant phospholemman channel molecules. *Receptors Channels* 6:435–47

108. Moorman JR, Ackerman SJ, Kowdley GC, Griffin MP, Mounsey JP, et al. 1995. Unitary anion currents through phospholemman channel molecules. *Nature* 377:737–40

109. Morales-Mulia M, Pasantes-Morales H, Moran J. 2000. Volume sensitive efflux of taurine in HEK293 cells overexpressing phospholemman. *Biochim. Biophys. Acta* 1496:252–60

110. Moran J, Morales-Mulia M, Pasantes-Morales H. 2001. Reduction of phospholemman expression decreases osmosensitive taurine efflux in astrocytes. *Biochim. Biophys. Acta* 1538:313–20

111. Minor NT, Sha Q, Nichols CG, Mercer RW. 1998. The γ subunit of the Na,K-ATPase induces cation channel activity. *Proc. Natl. Acad. Sci. USA* 95:6521–25

112. Sha Q, Lansbery KL, Distefano D, Mercer RW, Nichols CG. 2001. Heterologous expression of the Na$^+$,K$^+$-ATPase γ subunit in Xenopus oocytes induces an endogenous, voltage-gated large diameter pore. *J. Physiol.* 535:407–17

113. Jespersen T, Grunnet M, Rasmussen HB, Jørgensen NB, Angelo K, et al. 2003. The corticosteroid hormone induced factor (CHIF): a new modulator of KCNQ1 channels? *Biophys. J.* 84:16a

114. Song J, Zhang XQ, Ahlers BA, Carl LL, Wang J, et al. 2005. Serine 68 of phospholemman is critical in modulation of contractility, [Ca^{2+}]i transients and Na$^+$/Ca^{2+} exchange in adult rat cardiac myocytes. *Am. J. Physiol. Heart Circ. Physiol.* 288: H2342–54

115. Mirza MA, Zhang XQ, Ahlers BA, Qureshi A, Carl LL, et al. 2004. Effects of phospholemman downregulation on contractility and [Ca^{2+}]i transients in adult rat

cardiac myocytes. *Am. J. Physiol. Heart Circ. Physiol.* 286:H1322–30

116. Song J, Zhang XQ, Carl LL, Qureshi A, Rothblum LI, Cheung JY. 2002. Overexpression of phospholemman alters contractility and [Ca^{2+}](i) transients in adult rat myocytes. *Am. J. Physiol. Heart Circ. Physiol.* 283:H576–83

117. Ahlers BA, Zhang XQ, Moorman JR, Rothblum LI, Carl LL, et al. 2005. Identification of an endogenous inhibitor of the cardiac Na$^+$/Ca^{2+} exchanger: Phospholemman. *J. Biol. Chem.* 280:19875–82

118. Kowdley GC, Ackerman SJ, John JE 3rd, Jones LR, Moorman JR. 1994. Hyperpolarization-activated chloride currents in Xenopus oocytes. *J. Gen. Physiol.* 103:217–30

119. Shimbo K, Brassard DL, Lamb RA, Pinto LH. 1995. Viral and cellular small integral membrane proteins can modify ion channels endogenous to Xenopus oocytes. *Biophys. J.* 69:1819–29

120. Tzounopoulos T, Maylie J, Adelman JP. 1995. Induction of endogenous channels by high levels of heterologous membrane proteins in Xenopus oocytes. *Biophys. J.* 69:904–8

121. Therien AG, Deber CM. 2002. Oligomerization of a peptide derived from the transmembrane region of the sodium pump γ subunit: effect of the pathological mutation G41R. *J. Mol. Biol.* 322:583–90

122. Russ WP, Engelman DM. 2000. The GxxxG motif: a framework for transmembrane helix-helix association. *J. Mol. Biol.* 296:911–19

123. Arkin IT, Adams PD, Brunger AT, Smith SO, Engelman DM. 1997. Structural perspectives of phospholamban, a helical transmembrane pentamer. *Annu. Rev. Biophys. Biomol. Struct.* 26:157–79

124. MacLennan DH, Kranias EG. 2003. Phospholamban: a crucial regulator of cardiac contractility. *Nat. Rev. Mol. Cell. Biol.* 4: 566–77

Annu. Rev. Physiol. 2006. 68:461–90
doi: 10.1146/annurev.physiol.68.040104.131654
Copyright © 2006 by Annual Reviews. All rights reserved
First published online as a Review in Advance on September 28, 2005

Sgk Kinases and Their Role in Epithelial Transport

Johannes Loffing,[1] Sandra Y. Flores,[2] and Olivier Staub[2]

[1] Department of Medicine: Unit of Anatomy, University of Fribourg,
CH-1700 Fribourg, Switzerland; email: johannes.loffing@unifr.ch
[2] Department of Pharmacology and Toxicology, University of Lausanne, CH-1005
Lausanne, Switzerland; email: syflores@gmail.com, olivier.staub@unil.ch

Key Words phosphorylation, cell survival, gene expression, ubiquitin, ion
transport, kidney

■ **Abstract** The serum/glucocorticoid-induced kinase Sgk1 plays an important role
in the regulation of epithelial ion transport. This kinase is very rapidly regulated at the
transcriptional level as well as via posttranslational modifications involving phosphorylation by the MAP or PI-3 kinase pathways and/or ubiquitylation. Although Sgk1 is
a cell survival kinase, its primary role likely concerns the regulation of epithelial ion
transport, as suggested by the phenotype of Sgk1-null mice, which display a defect in
Na^+ homeostasis owing to disturbed renal tubular Na^+ handling. In this review we first
discuss the molecular, cellular, and regulatory aspects of Sgk1 and its paralogs. We
then discuss its roles in the physiology and pathophysiology of epithelial ion transport.

INTRODUCTION

Serum- and glucocorticoid-induced kinase (Sgk[1]) was originally isolated in a differential screen searching for glucocorticoid-induced transcripts in a mammary
tumor cell line (1). It was found to be induced within 30 minutes, either by glucocorticoids or serum and in both mammary epithelial cells and fibroblasts (1, 2).
This induction persisted in mammary cells (1), whereas in fibroblasts it was transient and returned to normal after 4 hours of induction. Sgk mRNA is extremely
unstable in fibroblasts (2), a finding that subsequently has been confirmed in other
cells (3). Sgk (specifically Sgk1, as there are two closely related paralogs referred
to as Sgk2 and Sgk3) is remarkable in that it is tightly regulated by numerous signaling molecules via its expression levels, intrinsic kinase activity, and subcellular
localization. These properties allow it to integrate numerous signaling pathways. It
is best known as a kinase that acts as a cell survival molecule in different situations
and that is able to regulate numerous membrane transport and channel proteins.

---

[1] For a list of abbreviations used in this chapter, please see the Appendix.

Sgk Is an Inducible Ser/Thr Kinase

Inducible Ser/Thr kinases include Polo-like kinases (Plks) Plk1 (4–6), Plk2 (formerly referred to as Snk) (7), and Plk3 (Fnk/Prk) (8–10); Pim-1; Pim-3 (11); and Sgk1. Various mitogenic or homonal signals induce the expression of these kinases. The transcriptional regulation of the Sgk1 gene [on chromosome 6q23 (12)] has been extensively studied, and this gene was found to contain several consensus sequences for transcription factors (13). The Sgk1 promoter contains a functional glucocorticoid response element (GRE) located at −1000 bp from the transcription start site that is highly homologous to the consensus GRE (1, 14, 15). Other putative GREs have been found further upstream (between −2.5 kbp and −3.2 kbp), but they have not been characterized yet. Additional regulatory transcription sites include p53-binding sites (−1380 bp to −1345 bp and −1550 bp to −1532 bp); a Sp1-responsive DNA element and an as-yet untermed hyperosmotic stress-regulated element, both between −50 bp to −40 bp; and cAMP-responsive element–binding (CREB) protein sites at approximately −2.2 kb, −2.7 kb, and −3.6 kb. A reporter gene driven by the Sgk1 5′-regulatory region is stimulated by the mineralocorticoid receptor (MR) (16). For more details on elements in the Sgk1 promoter region, we refer the reader to References 13 and 17.

Sgk1 is induced by a very large spectrum of stimuli distinct from glucocorticoids and serum (Table 1). These include aldosterone (18, 19), cell shrinkage (20, 21), cell swelling (22), TGF-β (23–26), ischemic injury of the brain (27, 28), neuronal excitotoxicity (29), memory consolidation (30), chronic viral hepatitis (25), DNA-damaging agents (31), vitamin D3 (32), psychophysiological stress (33), iron (34), glucose (26), endothelin-1 (26), granulocyte-macrophage colony–stimulating factor (GM-CSF) (35), fibroblast growth factor (FGF) (36), platelet-derived growth factor (PDGF) (36), phorbolesters (36), follicle-stimulating hormone (FSH) (37), sorbitol (31, 38), heat shock (31, 38), oxidative stress (31, 38), UV irradiation (31, 38), and p53 (13, 15, 31, 39). This list is likely not exhaustive. Many of these stimuli are highly cell-specific, as is the case, for example, for aldosterone, which has been found to stimulate Sgk1 expression only in cells derived from aldosterone-responsive epithelia (see below).

Sgk Belongs to the Family of AGC Kinases

With respect to its primary structure, Sgk1 is part of the family of AGC kinases, which include protein kinase A (PKA), protein kinase G (PKG), protein kinase C (PKC), and protein kinase B/Akt/rac (PKB/Akt) (Figure 1). Its catalytic domain shares 54% identity with those of Akt/PKB/rac kinases, 48% with that of PKC-β, 50% with that of rat p70^{S6K} kinase, and 45% with that of PKA (1). PKB/Akt kinases are the closest relatives. However, as mentioned above, there are two additional Sgk paralogs, referred to as Sgk2 and Sgk3 [the latter of which is also known as cytokine-independent survival kinase (CISK)] (40–42); these are discussed further below. Sgk kinases appear to be highly conserved in eukaryotic cells from human to yeast. In *Caenorhabditis elegans*, Sgk has been found to be part of a PKB

TABLE 1   Stimuli that cause induction of Sgk1 expression

Stimuli	Cell or tissue type	Reference(s)
Serum	Ubiquitous	(1)
Glucocorticoids	Ubiquitous	(1, 38)
Aldosterone	ASDN	(18, 19)
Cell shrinkage	HepG2, MDCK, neuroblastoma	(20, 21)
Cell swelling	A6 cells	(22)
TGF-β	U937, HepG2, intestine, fibroblasts, endothelial cells	(23–26)
Chronic viral hepatitis	Liver	(25)
Ischemic injury of brain	Brain	(27, 28)
Neuronal excitotoxicity	Brain, glial cells	(29)
Memory consolidation	Brain, hypocampus	(30)
DNA damage	Fibroblasts	(31)
1a,25-dihydroxyvitamin D3	Squamous cell carcinoma	(32)
Psychophysiological stress	Brain, heart, kidney	(33)
Iron	Intestine	(34)
Glucose	Endothelial cells	(26)
Endothelin-1	Endothelial cells	(26)
GM-CSF	Granulocytes	(35)
FGF	NIH 3T3 cells	(36)
PDGF	NIH 3T3 cells	(36)
Phorbolesters	NIH 3T3 cells	(36)
FSH	Rat ovarian granulosa cells	(37)
Sorbitol	NMuMg mammary epithelial cells	(38)
UV radiation	NMuMg mammary epithelial cells	(38)
Heat shock	NMuMg mammary epithelial cells	(38)
Oxidative stress/H_2O_2	NMuMg mammary epithelial cells	(38)
PPARγ	ASDN	(184)
p53	Mammary epithelial cells	(13, 15, 31, 39)

complex and to control stress resistance and life span (43). In yeast there are two orthologs, referred to as Ypk1 and Ypk2, which share 55% identity in the catalytic domain with that of Sgk1 (44). The inactivation of these two genes is lethal for the yeast cell, but they can be rescued by expression of mammalian Sgk1 (44). Recently it has been shown that Ypk1 and Ypk2 are essential for internalization of Ste2, a G protein–coupled receptor in yeast (45). Notably, Sgk kinases are missing in *Drosophila*.

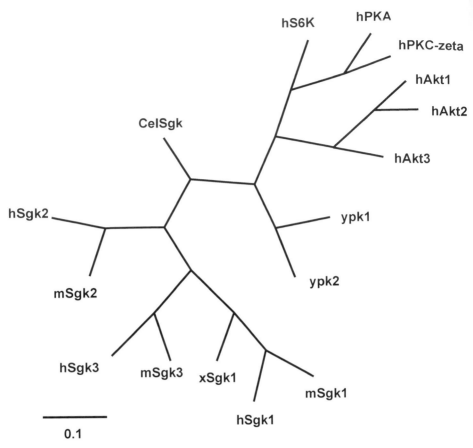

Figure 1 Schematic presentation and phylogenetic analysis of serum- and glucocorticoid induced kinase (Sgk) and more distantly related kinases. The amino acids of the catalytic domain of the indicated kinases were analyzed using Clustal W and viewed by Treetop. The scalebar represents 10 estimated changes per 100 sites.

The Sgk1 Isoforms Sgk2 and Sgk3

As mentioned above, there are two paralogs of Sgk1, Sgk2 and Sgk3/CISK, which share 80% amino acid identity with Sgk1 and with each other in their catalytic domains (40, 42, 46). The three enzymes differ in the region N-terminal of the C-terminal catalytic domain: Sgk2 contains a relatively short N terminus (98 amino acids), with no discernable domain, whereas Sgk3 has a longer N terminus (16 amino acids) comprising a phox homology (PX) domain (41). PX domains were originally found as conserved domains in the p40[phox] and p47[phox] subunits of the neutrophil NADPH oxidase (phox) superoxide-generating complex (47). These domains are part of many proteins involved in intracellular protein trafficking, such as the sorting nexins (48). PX domains are phosphoinositide-binding domains that appear to be important for localization of these proteins to membranes (especially endosomes) enriched in phosphoinositides. In this respect, these domains resemble other domains such as the pleckstrin homology (PH), FYFE, FERM, and ENTH domains. PKB/Akt contains a PH domain in its N terminus, which is important for PKB/Akt activation by phosphoinositide-3 kinases (PI-3Ks). This domain enables the colocalization with the 3-phosphoinositide-dependent protein kinase-1 (PDK-1), which is known to phosphorylate and activate PKB/Akt. Similarly, Sgk3's PX domain is involved in Sgk3 localization and activity: It is necessary for phosphoinositide binding, endosomal localization, and proper kinase activity (41, 42). Moreover, structural studies indicate that it may play a role in dimerization of the kinase (49).

With respect to their physiological role(s), it has been shown in vitro that both the Sgk2 and Sgk3 enzymes have the same phosphorylation consensus as Sgk1 (and PKB/Akt), namely R-X-R-X-X-(S/T) (40). It is likely, however, that other factors, such as surrounding amino acids, subcellular localization, or cofactors are important for the specificity of and functional differences between the enzymes. For example, in *Xenopus* A6 cells, only Sgk1 and not the coexpressed PKB modulates the activity of the epithelial Na^+ channel (ENaC) (50). The role of Sgk2 has mainly been studied in heterologous expression systems such as *Xenopus laevis* oocytes or Hek293 cells and with respect to numerous transport and channel proteins. These studies revealed that Sgk2 can stimulate the activity of K^+ channels such as the voltage-gated K^+ channel Kv1.3 (51, 52), Na^+,K^+-ATPase (53), KCNE1 (54), ENaC (55), the glutamate transporter EEAT4 (56), and the glutamate receptors GluR6 (57) and GluR1 (58). All of these transport proteins are also stimulated in the same cellular systems by Sgk1, Sgk3, and/or PKB; hence, the physiological relevance of these findings has to be considered with caution. To define more precisely the role of Sgk2, it will be necessary to carry out additional studies, using more relevant cell or animal systems and knocking down Sgk2 by either RNA interference protocols or by gene inactivation. Sgk3/CISK, which is better characterized than Sgk2, was identified in a screen for antiapoptotic genes (42) and found to act downstream of the PI-3K pathway and in parallel with PKB/Akt. Moreover, it was demonstrated to phosphorylate and inhibit Bad (a proapoptotic protein) and FKHRL1, a proapoptotic transcription factor (42). Knockout (KO) mice have been generated; these mice are viable and fertile and have normal Na^+ handling and glucose tolerance, as opposed to the KO mice of Sgk1 or PKB/Akt2 (59–62). However, they display after birth a defect in hair follicle development, a defect preceded by disturbances in the β-catenein/Lef1 gene regulation (59). Like Sgk2, Sgk3 has been implicated in the regulation of numerous transporters and channels, including K^+ channels (51, 52, 54), Na^+,K^+-ATPase (53), the glutamate transporter EEAT1 (63), the cardiac voltage-gated Na^+ channel SCN5A (64), ENaC (55), Na^+-dicarboxylate cotransporter 1 (65), the chloride channel ClCa/barttin (66), the epithelial Ca^{2+} channel TRPV5 (67), the Na^+-phosphate cotransporter NaPi1b (68), the amino acid transporter ASCT2 (69), GluR1, and GluR6 (57, 58).

For the same reasons mentioned above for Sgk2, additional studies on Sgk3 will be necessary to evaluate the physiological relevance of these findings.

Regulation of Sgk1 by Phosphorylation

Many AGC kinases contain in their catalytic domain and in the C-terminal region conserved phosphorylation motifs. The motif in the catalytic domain, situated on the so-called activation loop (A-loop), was first identified in PKB/Akt and was shown to be a target of PDK-1 (70, 71). Hence, PKB/Akt is a kinase regulated by an upstream kinase, which depends on the PI-3K system. As outlined above, both PKB/Akt and PDK-1 contain a PH domain, enabling them to interact with phosphoinositides [phosphatidylinositol-(3,4)-biphosphate and phosphatidylinositol-(3,4,5)-triphosphate, or PtdIns(3,4)P2 and PtdIns(3,4,5)P3, respectively] generated by PI-3K. The PH domain recruits both kinases to membranes and allows PDK-1 to phosphorylate PKB/Akt. This activation is very fast (approximately 2 minutes after stimulation with insulin or growth factor) (72). PDK-1-dependent phosphorylation sites are also found in PKA (73), PKCζ (74, 75), p70^{S6K} (76, 77), and the Sgk paralogs (40, 78, 79). In Sgk1 it is Thr256 in the A-loop of the catalytic domain. However, although Sgk1 is also dependent on the PI-3K pathway and on PDK-1 activity, the mode of activation of Sgk1 is different from that of PKB/Akt, as Sgk1 does not contain a PH or any similar phosphoinositide-binding domain. Sgk1 becomes phosphorylated on Ser422 in the C-terminal region termed the hydrophobic motif (H-motif) by a so-far unknown kinase that is PI-3K-dependent. This kinase activity has been referred to as PDK-2 or H-motif kinase activity; its activity can be indirectly inhibited by PI-3K inhibitors such as Wortmannin or LY-29,4002 (78, 79) or mimicked by mutation of Ser422 to aspartate, thereby generating a constitutively active kinase. Phosphorylation of Ser422 or mutation to aspartate transforms Sgk1 into a substrate for PDK-1, which binds via its PDK-1-interacting fragment (PIF)–binding pocket to the H-motif on Sgk1, promoting phosphorylation of the A-loop motif and consequently activation of Sgk1 (80). Hence, in this context, and in contrast to PKB/Akt activation, phosphorylation of Sgk1 by PDK-1 is independent of phosphoinositides or the PH domain in PDK-1. This renders the activation of Sgk1 much slower (with maximal activation occurring typically after 10 to 40 minutes) than that of PKB/Akt. The physiological significance of this difference is unclear but suggests a bigger need for rapid activation of PKB/Akt than of Sgk1. Changes in Na$^+$ balance (whereby Sgk1 is activated via aldosterone) usually occur less rapidly than changes in blood glucose levels (whereby PKB/Akt is activated).

cAMP also may activate Sgk1 (81). In this context, a consensus sequence for PKA-dependent phosphorylation surrounding Thr369 (K-K-I-T-P), which is necessary for cAMP-dependent phosphorylation, was identified. However, this consensus for phosphorylation is imperfect, and other groups have been unable to detect activation of Sgk1 by cAMP under conditions in which cAMP stimulated the phosphorylation of the transcription factor CREB (82). Sgk1 is also phosphorylated and activated by various MAP kinases, namely by bone marrow kinase (BMK) [also known as extracellular signal–regulated kinase 5 (ERK5)] or by p38α (83, 84). Both these kinases phosphorylate Sgk1 on Ser78, which lies outside of the catalytic domain. Such phosphorylation leads to an increase in Sgk1 activity, independent of the phosphorylation state of Thr256 (in the A-loop). It is not known how this phosphorylation activates Sgk1, but other kinases are also activated by phosphorylation outside of the catalytic domain (85). Sgk1 has also been reported to interact with WNK1 (_w_ith _n_o lysine _k_inase _1_), a kinase playing a role in Na$^+$ homeostasis and blood pressure control. In this process WNK1 is able to activate Sgk1. The mechanism of activation is not completely clear, but it may involve phosphorylation of Ser422 or mutation of Ser422 to aspartate in addition to phosphorylation of Thr256, suggesting a similar mechanism as that involved in the activation via PDK-1 (see Figure 2, below). However, direct phosphorylation of Sgk1 by WNK1 has not been demonstrated (86).

Regulation of Sgk1 by Ubiquitylation

As can be expected from a protein that is regulated rapidly via transcription of its RNA, Sgk1 is a protein that is unstable and has a rapid turnover, with a half-life of approximately 30 minutes (87). Rapid degradation of Sgk1 involves its ubiquitylation followed by degradation by the proteasome. Moreover, it appears that ubiquitylated Sgk1 is preferentially associated with intracellular membranes (87). The first 60 N-terminal amino acids likely control the degradation and thus the stability of the kinase; removal of this region causes near-complete loss of ubiquitin modification, predominant cytosolic location, and significant stabilization of the enzyme. Currently, it is not clear how the N terminus controls this stability. It contains a partial PX-domain (88), which usually localizes proteins to endosomes, but endosomal localization of Sgk1 has not been described. There are six lysines, i.e., potential ubiquitylation sites in the N terminus, but when these are mutated to arginines, the kinase still becomes ubiquitylated, suggesting that the ubiquitylation sites are outside of this region. With respect to the ubiquitin-protein ligase involved in Sgk1 ubiquitylation, Snyder and collaborators (89) showed that the ubiquitin-protein ligase Nedd4-2 (_n_euronal precursor cells _e_xpressed _d_evelopmentally _d_ownregulated 2), itself a target of Sgk1 (see below), does increase Sgk1 ubiquitylation and degradation. This effect depends on the presence of the N terminus described above and, intriguingly, requires phosphorylation of Nedd4-2 by Sgk1. This implies a feedback regulation in which Sgk1 phosphorylates Nedd4-2, which in turn suppresses Sgk1 by ubiquitylating and degrading it (89).

Sgk1 Phosphorylation Sites and Target Proteins

The consensus phosphorylation motifs for Sgk1 have been determined using in vitro peptide phosphorylation assays (78, 79). It was found that arginines at positions 5 and 3 were important for efficient phosphorylation. Moreover, basic residues C terminal of S/T appear to be inhibitory, and there is no requirement for bulky

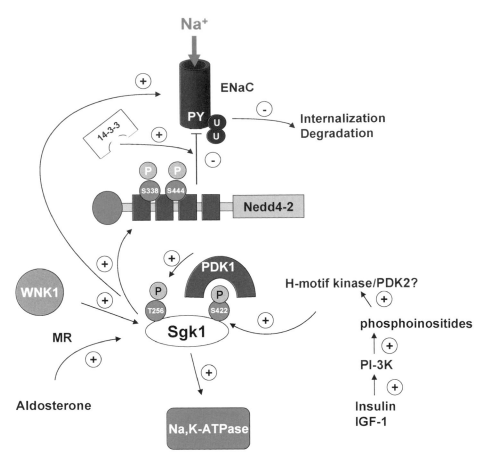

Figure 2   Regulation of epithelial Na$^+$ transport via the aldosterone-induced serum- and glucocorticoid-induced kinase (Sgk). Aldosterone induces the expression of Sgk1 by binding to the mineralocorticoid receptor (MR) and translocating into the nucleus. Sgk1 becomes phosphorylated on Ser422 by an unknown kinase referred to tentatively as 3-phosphoinositide-dependent protein kinase (PDK)-2 or hydrophobic motif (H-motif) kinase; this latter kinase depends on the PI-3K pathway and is stimulated by insulin or IGF-1. Phosphorylated Ser422 then serves as a binding site for PDK-1 via PIF-pocket, leading to the phosphorylation of Thr256 on the activation loop and activation of Sgk1. Alternatively, Sgk1 is also activated by WNK1 (_w_ith _n_o lysine _k_inase _1_) via an unknown mechanism. Once activated, Sgk1 phosphorylates Nedd4-2 (_n_euronal precursor cell _e_xpressed _d_evelopmentally _d_ownregulated 2) on Ser444 and Ser338, which allows binding of 14-3-3 on Ser444. This interferes with Nedd4-2/ENaC interaction, causing reduced ubiquitylation of and accumulation of epithelial Na$^+$ channel (ENaC) at the plasma membrane. Alternatively, Sgk1 phosphorylates and activates ENaC directly. ENaC may also be stimulated by a mechanism involving cAMP (_not depicted_), stimulating either Sgk1 or phosphorylation of the Nedd4-2 Ser444.

TABLE 2 Proteins phosphorylated by Sgk1

Substrate protein	Tissue	Phosphorylation site(s)	Reference(s)
Glycogen synthase kinase 3 (GSK3)	Heterologous expression in Hek293 cells	RARTSS	(78)
FKHRL1	Cerebellar granule neurons	RPRSCT RSRTNS	(93)
Raf	Heterologous expression in Hek293 cells	RDRSSS	(94)
MEKK3	Heterologous expression in COS cells	RSRHLS RGRLPS	(95)
Nedd4-2	Oocytes, mpkCCD$_{cl4}$	RLRSCS	(68, 99, 100, 160)
NDRG-2	Skeletal muscle	RLSRSRT RSRTAS RSRTLS	(98)
NDRG-1	Skeletal muscle	RMSRSRT RSRTAS RSRTLS	(98)
NHE3	In vitro phosphorylation assay, heterologous expression in IEC6 cells	RKRLESF	(96)
ROMK	Heterologous expression in Xenopus oocytes	RARLVS	(182)
α-ENaC	Heterologus expression in Xenopus oocytes	RSRYWS	(101)

hydrophobic residues. Thus, this translates into a consensus of R-X-R-X-X-(S/T), which is similar to the consensus for phosphorylation by PKB/Akt or by Sgk2 and/or Sgk3 (see above).

Sgk1 was shown to phophorylate a variety of proteins (Table 2). The first demonstrated substrate for Sgk1 was glycogen synthase kinase 3 (GSK3), a kinase that is involved in the regulation of glycogen and protein synthesis by insulin and that is also a substrate of PKB/Akt (78, 90). Phosphorylation of GSK3 by both Sgk1 and PKB/Akt leads to an increase in the synthesis of glycogen (91). The respective contributions of each of the two kinases is not yet clearly established. Some light has been shed by data from Collins and collaborators (92), who generated knock-in embryonic stem cells that express a PIF-pocket mutant of PDK-1. As described above, such PKD-1 mutants would not be expected to activate Sgk1. Nevertheless, such cells display normal phosphorylation of GSK3, likely via PKB/Akt (or Sgk3), suggesting that Sgk1 is not essential for proper GSK3 phosphorylation (92). Another described substrate is the forkhead transcription factor FKHRL1 (*forkh*ead in *r*habdomyosarcoma-*l*ike 1) (also known as FOXO3), a transcription factor that promotes cell cycle arrest and apoptosis. FKHRL1 becomes phosphorylated by Sgk1 on Thr32 and Ser315 (93), promoting exit of FKRHL1 from the nucleus and interference with FKRHL1-dependent transcription. Consistent with Sgk1's role as a cell survival kinase, this phosphorylation leads to reduction of cell cycle arrest and apoptosis (93). Again, in the embryonic stem cells expressing the PIF-binding pocket mutant of PDK-1, IGF-1 (insulin growth factor 1) induces phosphorylation of the above-mentioned sites, suggesting that Sgk1 is not required for phosphorylation in this process (92). Another substrate of Sgk1 is the Raf kinase, which plays a fundamental role in the transmission of signals from Ras to the MAP kinase pathway. Phosphorylation of Raf leads to its inactivation (94). Another member of the MAP kinase pathway is MEKK3 (*m*itogen-activated protein kinase/*E*RK *k*inase *k*inase 3), which also gets phosphorylated and inhibited by Sgk1 (95). The identification of Raf and MEKK3 as Sgk1 substrates is especially interesting because the MAP kinases p38 and BMK can phosphorylate and activate Sgk1. These kinases are downstream of Raf and MEKK3, indicating that there is some feedback control regulating these pathways. As discussed above, all of the membrane transport and channel proteins shown to be regulated in heterologous expression systems by Sgk2 and/or Sgk3 are also controlled by Sgk1. Other recently identified Sgk1 substrates are the Na^+,H^+ exchanger 3 (NHE3) (96), N-myc downstream-regulated gene 2 (NDRG-2) (97, 98), Nedd4-2 (66, 68, 99, 100), and ENaC (101). The roles of ENaC, Nedd4-2, and NDRG-2 are discussed below.

Tissue Distribution of SGK Isoforms

Given the large spectrum of signals regulating the Sgk isoforms, it is not surprising that these isoforms are expressed in numerous tissues and cell lines. Among the three kinases, Sgk1 (1, 20, 40) and Sgk3 (40) show the broadest distribution, with expression in many tissues, including the brain, placenta, lung, liver, pancreas, kidney, heart, and skeletal muscle. In situ hybridization studies localized Sgk1 mRNA in several epithelial and/or nonepithelial cells within the brain (21, 28, 30, 102, 103), eye (104, 105), lung (106), liver (25), ovary (107), pancreas (108), intestine (23), and kidney (18, 109, 110). Sgk1 mRNA expression is established very early in embryonic development, as indicated by in situ hybridizations on whole-mount preparations of mouse embryo (111). By embryonic day (E) 8.5, Sgk1 is already highly expressed in the decidua and yolk sac. By days E9.5–E12.5 it is found in the developing heart, eye, and lung, and it becomes highly expressed by days E13.5–E16.5 in the brain choroid plexus, kidney distal tubules, bronchi/brochioli, adrenal glands, liver, thymus, and intestine (111). In contrast to Sgk1 and Sgk3, Sgk2 reveals a more restricted distribution and is highly abundant only in the liver, kidney, and pancreas, where it is found in two different Sgk2 species, referred to as Sgk2α and Sgk2β (40).

Sgk isoform expression varies also between cell lines cultured in vitro. Similar to its expression pattern in vivo, Sgk1 is broadly expressed in cultured cells and is readily detectable in, for example, hepatoma cells, fibroblasts, and mammary tumor cells (1, 40). By contrast, Sgk2 mRNA is expressed in hepatoma cells but not in fibroblasts, whereas Sgk3 is found in fibroblasts but not in hepatoma cells. Remarkably, all three Sgk isoforms are expressed in cells derived from the renal cortical collecting duct (112).

Only little is known so far about the distribution of the Sgk isoforms at the protein level. Moreover, reported data on Sgk1 in the kidney differ with respect to its cellular and subcellular distribution, which may be rooted at least in part in the fact that the available antibodies are not isoform-specific and thus may recognize different Sgk isoforms. Canessa and coworkers (113, 114) developed a Sgk antibody that recognizes heterologously expressed Sgk1–3 but not closely related Akt1. In tissue homogenates (e.g., those from kidney), the antibody recognizes only a single band that migrates at the same molecular size as does heterologously expressed Sgk1 (114). This band is only found in rat kidneys and intestine, but not in brain, liver, or skeletal muscle (113), although these latter organs reveal high levels of Sgk1 mRNA expression (see above). One possible explanation is that Sgk1 mRNA expression does not necessarily translate into a corresponding protein expression. A dissociation of Sgk1 mRNA and protein expression also has been proposed to explain why the antibody does not detect Sgk1 in the renal papilla despite high expression levels of Sgk1 mRNA at this site (114). Immunohistochemistry with this particular Sgk1 antibody revealed a predominant basolateral immunostaining in all renal distal tubules, (i.e., the thick ascending limbs, distal convoluted tubules, connecting tubules, and collecting ducts). Buse and coworkers (115) generated another Sgk antibody. In cell lysates from mammary tumor cells, this antibody recognizes a protein of approximately 51 kDa that is strongly induced by serum and glucocorticoids (115). In vivo in the kidney, the antibody reveals a strong aldosterone-induced labeling in the cytoplasm of aldosterone-sensitive distal nephron (ASDN) cells (116). The subcellular localization is consistent with the reported intracellular localization of Sgk1 in stably transfected CCD cells (112), in mammary tumor cells expressing exogenous or endogenous Sgk1 (115), and in *C. elegans*–expressing GFP-tagged Sgk1 (43), but is in contrast with the predominant basolateral membrane localization seen in the antibody generated by Canessa and coworkers (113, 114). Huber et al. (110) raised an additional anti-Sgk1 antibody that recognizes an endogenous protein of approximately 50 kDa in HepG2 and HEK293 cells. In immunofluorescent studies on embryonic mouse kidneys, the antibody specifically stained developing collecting ducts (110). To what extent these two latter antibodies may also recognize other Sgk isoforms or members of the AGC kinase family is unclear. The use of the different Sgk antibodies on tissues from the now-available Sgk1 (60), Sgk3 (59), Akt1 (117), and Akt2 (61, 62) KO mice may help to determine the specificity of the respective antibodies.

It is important to note that the subcellular distribution of Sgk1 may change according to the stimulus received. Sgk1 can be localized to the nucleus, to the cytoplasm, and/or at the plasma membrane (107, 113, 115, 118, 119). For example, IL6-dependent stimulation leads to p38-dependent phosphorylation and subsequent localization of Sgk1 in the nucleus, where it may activate transcription factors, such as forkhead transcription factor FKHRL1 (84). The nuclear import receptor importin-α is an Sgk1-interacting protein that recognizes a nuclear localization signal sequence, KKAILKKKEEK, within the central domain of Sgk1 between amino acids 131–141 (119), allowing Sgk1 to shuttle actively between the nucleus and the cytoplasm in synchrony with the cell cycle (115). Consistent with this, Sgk1/importin-α interaction may control the nuclear localization of Sgk1 during the S and G_2/M phases of the cell cycle and the cytoplasmic localization of the kinase during the G_1 phase (119). Leong et al. (38) have reported that in mammary epithelial cells, oxidative stress, heat shock, and UV-radiation exposure transiently induces Sgk1 localization throughout the cell, whereas dexamethasone or sorbitol directs Sgk1 to the nucleus. Additionally, Sgk1 contains a putative mitochondrial import signal at the N terminus necessary to localize cytoplasmic Sgk1 to this organelle in a stimulus-dependent fashion (13). Compartmentalization of Sgk1 may control the stimulus-regulated accessibility of the kinase to its targets and its nonsubstrate-interacting proteins.

SGK AND EPITHELIAL ION TRANSPORT

Role of Sgk1 in Aldosterone-Dependent Na$^+$ Reabsorption

Although Sgk isoforms are expressed in various tissues and cell types, the role of Sgk1 in aldosterone-dependent regulation of Na$^+$ homeostasis is the best-studied function of these kinases with respect to epithelial ion transport. The kidneys play a pivotal role in the maintenance of Na$^+$ homeostasis. Urinary Na$^+$ excretion must be tightly regulated to maintain a constant extracellular volume during varying dietary Na$^+$ intake and extrarenal Na$^+$ losses. The final control of renal Na$^+$ excretion is achieved by the ASDN, i.e., the late distal convoluted tubule the connecting tubule, and the cortical as well as the medullary collecting ducts (CCD and MCD, respectively) (120). Transepithelial Na$^+$ transport in these segments is accomplished by Na$^+$ entry into the epithelial cells via the epithelial Na$^+$ channel (ENaC) in the luminal membrane and by exit of Na$^+$ through the Na$^+$,K$^+$-ATPase in the basolateral plasma membrane. ENaC represents the rate limiting step in this process and is highly regulated (121). It is composed of three subunits (α, β, and γ) (122–125) with a stoichiometry of $2\alpha 1\beta 1\gamma$ (126), although other stoichiometries have also been proposed (octa- or nonamers) (127, 128). It subunits have a similar topology, with two transmembrane domains, one extracellular loop, and two cytoplasmic ends (129–131). Each subunit also contains at its C-terminal end, a PY-motif (P-P-X-Y, where P is a proline, Y a tyrosine and X any amino acid), which is known as protein:protein interaction motifs that can interact with tryptophan (W)-rich WW domains (132, 133). The importance of these PY-motifs for ENaC regulation has been recognized by the findings that most cases of Liddle's syndrome (134), an inherited form of salt-sensitive hypertension are caused by mutations in the genes encoding β- and γ-ENaC (135, 136). These mutations invariably cause either the deletion or the mutation of the PY-motifs on these subunits. When such Liddle channels are expressed in heterologous systems, increases in both the density at the cell surface and the open probability of ENaC are observed (137–140). Our research, and subsequently the research of many other laboratories, has demonstrated that these PY-motifs are the binding sites for ubiquitin-protein ligases of the Nedd4/Nedd4-like family (Figure 2) (141–147), and particularly of Nedd4-2 (145–149). It is thought that Nedd4-2 binds via its WW domains with the PY-motifs of ENaC and ubiquitylates ENaC on its α and γ subunits, consequently leading to the internalization and degradation of ENaC in the endosomal/lysosomal system (148, 150). In Liddle's syndrome, this mechanism is impaired owing to the inactivation of a PY-motif, causing the accumulation of ENaC at the plasma membrane (for a review, see also References 151, 152).

The activity of ENaC and the Na^+,K^+-ATPase is tightly regulated by aldosterone and by Sgk1, as recently outlined in References 121 and 153–155. Experiments in heterologous expression systems (i.e., *X. laevis* oocytes) revealed that coexpression of either ENaC (18, 19) or Na^+,K^+-ATPase (156, 157) with Sgk1 profoundly increases the activity of both Na^+-transporting proteins. Likewise, Sgk2 and Sgk3 stimulate ENaC (55) and Na^+,K^+-ATPase (53). The stimulatory effect of Sgk1 on ENaC is related both to an increased number of channels in the plasma membrane (24, 99, 120, 158) and an activation of channels already present in the membrane (101, 159). The first effect likely involves the action of Nedd4-2, as there are several consensus phosphorylation motifs (2–3 depending on the splice variant) on Nedd4-2 and a PY-motif on Sgk1 that may serve as a binding site for Nedd4-2 (Figure 2). In *Xenopus* oocytes, Sgk1 induces Nedd4-2 phosphorylation on two of these phosphorylation sites (primarily Ser444, but also Ser338) (66, 68, 99, 100, 160), which decreases the interaction of Nedd4-2 with ENaC and finally leads to an enhanced expression and activity of ENaC at the cell surface (99, 100). This inhibitory effect of Sgk1 on Nedd4-2 likely involves 14-3-3 proteins as phosphorylation of Ser444 in Nedd4-2 creates a possible binding site for such proteins (consensus: R-S-X-pS-X-P). Indeed, in *X. laevis* oocytes, Sgk1 increases the binding of 14-3-3 to Nedd4-2 in a phosphorylation-dependent manner, a dominant-negative 14-3-3 mutant profoundly attenuates Sgk1-dependent stimulation of ENaC, and overexpression of the 14-3-3 protein impairs Nedd4-2-dependent ubiquitylation of ENaC (161, 161a).

In addition to this indirect action of Sgk1 on ENaC cell surface abundance, it was proposed that Sgk1 can directly interact with ENaC (162) and increase ENaC channel activity by phosphorylating the α-ENaC subunits (101). Diakov & Korbmacher (101) used outside-out membrane patches of *X. laevis* oocytes expressing rat ENaC to demonstrate that addition of recombinant, constitutively active Sgk1 directly stimulates ENaC currents two- to threefold. An alanine mutation of the serine residue in the Sgk1 consensus R-X-R-X-X-S phosphorylation motif abolishes the stimulatory effect on ENaC in this experimental setting. Experiments in native *Xenopus* A6 cells expressing endogenous Sgk1 and ENaC further confirmed that the action of Sgk1 on ENaC is complex and likely involves (*a*) increases in the subunit abundance in the plasma membrane and (*b*) activation of channels already in the plasma membrane combined with an increase in ENaC open probability (163). However, in this model the stimulatory effect on ENaC channel activity cannot be explained by a direct Sgk1-dependent phosphorylation of α-ENaC because *Xenopus* α-ENaC does not contain the Sgk1 consensus phosphorylation motif. That direct phosphorylation of ENaC at the Sgk1 consensus site is not essential for ENaC activation is also supported by data from Lang and coworkers (24, 109) that showed that channels with a serine-to-alanine mutation within the consensus site of α-ENaC are still rigorously upregulated by coexpression of Sgk1 in *Xenopus* oocytes. NDRG-2, which is an aldosterone-induced protein in the ASDN, is another target of Sgk1 (97, 98). Although the functional role of NDRG-2 in the ASDN is not known, this protein may also have some function in the Sgk1-dependent signaling cascade related to Na$^+$ transport.

As an aldosterone-induced protein, Sgk1 is thought to mediate at least some of the physiological effects of aldosterone on ENaC and Na$^+$,K$^+$-ATPase. The stimulatory effect of aldosterone (or of dexamethasone) on Sgk1 expression has now been firmly documented in several studies on various in vitro and in vivo systems, including *Xenopus* A6 cells (154), primary rabbit CCD cells (19), mouse inner MCD cells (164), mouse mpkCCD$_{cl4}$ (160), mouse M1 cells (88), and mouse and rat kidneys (18, 120, 165–169). Corticosteroids rapidly (within 30 minutes) induce Sgk1 at the mRNA and/or protein levels. This induction precedes or at least coincides with enhanced phosphorylation of Nedd4-2 (160), the activation of transepithelial Na$^+$ transport in cultured renal epithelia (19, 154, 160), and reduced renal Na$^+$ secretion in intact animals (165). At least part of the stimulatory effect of aldosterone on Sgk1 appears to be mediated by activation of the MR, as indicated by findings in primary rabbit collecting duct cells in vitro (19) and kidneys in vivo (165). Consistently, physiologically relevant concentrations of aldosterone are sufficient to significantly induce Sgk1 mRNA in the renal cortex and outer medulla (167, 168). The physiological importance of aldosterone in Sgk induction is also supported by the fact that dietary Na$^+$ restriction, which physiologically increases plasma aldosterone, induces Sgk1 mRNA in the renal cortex (170). The aldosterone-dependent induction of Sgk1 occurs specifically in the ENaC-positive cells of the ASDN, whereas Sgk1 expression in other nephron portions such as the thick ascending limb or the proximal tubule is not increased by aldosterone. Likewise, the high level of expression of Sgk1 in the renal papilla is not further stimulated by aldosterone, suggesting that Sgk1 expression at this site is controlled by factors other than aldosterone. The renal papilla plays an important role for the urinary concentration mechanism, and the cells in the renal papilla can be exposed to a large variation in extracellular osmolarity depending on the requirements for diuresis to antidiuresis. Sgk1 expression is strongly modulated by osmotic cell shrinkage and swelling (20, 22, 171), and it is therefore conceivable that Sgk1 participates in the functional adaptation of the renal papilla cells to fluctuation of extracellular osmolarity. Consistent with this notion, recent data suggest that Sgk1 mediates the osmotic induction of the type A natriuretic peptide receptor (NPR-A) in rat inner MCD cells (172).

Aldosterone also controls Sgk1 expression in the distal colon (113, 165, 166, 169). Aldosterone-dependent Na^+ reabsorption at this site may help to limit extrarenal Na^+ losses during conditions of dietary Na^+ restriction. Transepithelial Na^+ transport is achieved mainly by epithelial cells that are situated at the tips of colonic crypts and that express high levels of ENaC (113, 173) and Sgk1 (23, 113). In spite of these data pointing to aldosterone-dependent regulation of ENaC via Sgk1, recent Western blot and immunohistochemical studies on rat kidney and colon, which reported no or rather modest aldosterone-dependent induction of Sgk1 at the protein level, were interpreted to question the significance of aldosterone-dependent induction of Sgk1 for ENaC-mediated Na^+ transport regulation (113).

Support for a functional significance of Sgk1 in regulation of transepithelial Na^+ transport comes from experiments in X. laevis A6 cells and in mouse M1 CCD cells. Transfection of A6 or M1 cells with Sgk1 leads to an increase in transepithelial Na^+ transport, whereas transfection of a dominant-negative "kinase-dead" Sgk1 mutant or an antisense Sgk1 transcript abolishes dexamethasone- and/or insulin-dependent regulation of transepithelial Na^+ transport (88, 174, 175). Likewise, the use of interfering RNA to knockdown Sgk1 expression in A6 cells results in a significant reduction in Sgk1 protein levels and a ~50% reduction in dexamethasone-induced short-circuit currents (154). Consistent with these in vitro findings, experiments in Sgk1 KO ($Sgk1^{-/-}$) mice supported the importance of Sgk1 for aldosterone-dependent regulation of renal Na^+ transport (60). Under a standard diet, the KO mice have unaltered Na^+ excretion as compared to their wild-type littermates. However, plasma aldosterone levels are significantly increased in $Sgk1^{-/-}$ mice, suggesting extracellular volume contraction. Under dietary Na^+ restriction, activated compensatory mechanisms are no longer sufficient to keep the mice in Na^+ balance, and mice disclosed significant loss in renal NaCl and in body weight. Experiments on collecting ducts perfused ex vivo revealed significantly lower transepithelial amiloride-sensitive potential differences, consistent with a reduced Na^+ transport activity in the CCD. Although apical localization of ENaC was seen in both Na^+-restricted $Sgk^{+/+}$ and $Sgk^{-/-}$ mice, the apical localization of ENaC is inappropriately low in the $Sgk^{-/-}$ mice given the severalfold higher plasma aldosterone levels in the KO mice. Nevertheless, these data, together with the rather mild phenotype of $Sgk1^{-/-}$ mice, as compared to the much more severe and life-threatening phenotypes of MR or ENaC KO mice, suggest that (a) aldosterone-dependent control of ENaC function does not solely rely on the induction and activation of Sgk1 and (b) some redundancy exists in the signal transduction pathway that controls ENaC activity. Consistent with these ideas, we recently found significant phosphorylation of the Sgk1 target Nedd4-2 in mouse $mpkCCD_{cl4}$ cells in vitro and in rat collecting ducts in vivo in the absence of any aldosterone and detectable Sgk1 protein expression (160).

In addition to aldosterone-dependent regulation of renal Na^+ reabsorption, Sgk1 appears to be involved also in the regulation of aldosterone-induced salt appetite. $Sgk1^{+/+}$ and $Sgk^{-/-}$ mice show a similar salt intake under standard conditions. Treatment with the synthetic aldosterone analogue deoxycorticosterone-acetate (DOCA) increases Na^+ intake much more in $Sgk1^{+/+}$ mice than in $Sgk1^{-/-}$ mice. The underlying mechanism for the reduced mineralocorticoid-induced salt intake is unclear (176).

Role of Sgk1 in Renal K^+ Secretion

Aside from its stimulatory effect on renal Na^+ reabsorption, aldosterone has strong kaliuretic action. Renal K^+ secretion also takes place in the ASDN and is likely mediated by the renal outer medullary K^+ channel ROMK. ROMK is coexpressed with ENaC in the ASDN cells, and Na^+ reabsorption via ENaC provides the necessary driving force for K^+ secretion. Consistently, pharmacological inhibition (i.e. by amiloride) or genetic loss of function [i.e., pseudohypoaldosteronism (PHA type 1] of ENaC lower renal K^+ secretion and predispose one to hyperkalemia. It remains unresolved whether the kaliuretic effect of aldosterone is entirely secondary to the activation of ENaC-mediated Na^+ reabsorption or whether aldosterone directly regulates ROMK function. Patch-clamp studies on rat CCDs found no measurable effect of acute aldosterone administration on K^+ channel number, open probability, or conductance (177, 178). However, some data suggested that aldosterone induces renal K^+ secretion already at aldosterone concentrations that do not exhibit any measurable effect on urinary Na^+ excretion (165). Moreover, high K^+ intake increases ROMK activity more efficiently in intact rats than in adrenalectomized animals, suggesting that aldosterone may have at least a permissive effect on ROMK activation (177). Consistent with a possible role of aldosterone in ROMK regulation, recent studies in heterologous expression systems advocate a regulatory action of aldosterone-induced Sgk1 on ROMK cell surface activity and abundance (179–182). The regulatory role of Sgk1 with regard to ROMK may be indirect via increased interaction with the Na^+,H^+ exchanger–regulating factor 2 (NHERF2) (179–181) or direct via increased phosphorylation of ROMK at serine residue within the canonical Sgk1 consensus phosphorylation motif (182). The in vivo significance of Sgk1 in regulation of renal K^+ transport was recently analyzed in Sgk1 KO mice. These mice indeed show a disturbed adaptation to an acute and chronic K^+ load, but, as indicated by electrophysiological and immunohistochemical data obtained from these mice after a chronic potassium load, this maladaptation likely is related to altered ENaC (or Na^+,K^+-ATPase) activity in the ASDN cells rather than to inhibition of ROMK cell surface targeting or activity (183).

Pathophysiological Role of Sgk1

In addition to playing a role in the physiological regulation of renal ion transport, Sgk1 may also be involved in the initiation and/or maintenance of deranged transepithelial ion transport under various pathophysiological conditions. For example, enhanced renal expression of Sgk1 was found in kidneys of diabetic rats (3) and humans (24). In situ hybridization studies localized the upregulation of Sgk1 predominantly to the thick ascending limbs (TALs) of Henle's loop (24). Na^+ transport in the TALs depends on the activity of the furosemide-sensitive Na-K-2Cl cotransporter (NKCC2), which is a putative target for Sgk1, as indicated by experiments in heterologous expression systems (24). Sgk1-dependent stimulation of NKCC2 may contribute to disturbed renal Na^+ handling in diabetes mellitus. Induction of Sgk1 may also explain extracellular volume expansion and edema formation in diabetic patients treated with activators of peroxisome proliferator–activated receptor gamma (PPARγ). PPARγ can bind to specific response elements in the Sgk1 promoter and PPARγ agonists augment Sgk1 mRNA expression and α-ENaC cell surface abundance in human collecting duct cells (184). Likewise, increased phosphorylation of Sgk1 may contribute to the renal Na^+ and fluid retention associated with nephrotic syndrome (185). Whether the enhanced activation of Sgk1 can explain the inappropriately high activity of ENaC in this disease is still an open question. Enhanced renal Na^+ reabsorption and ascites formation in liver cirrhosis apparently occurs without any detectable Sgk1 upregulation (186), although the therapeutic efficiency of MR inhibition by spironolactone suggests the involvement of MR-dependent signaling in cirrhosis-associated renal Na^+ retention.

Dysregulated Sgk1 expression also may be involved in the development of salt-sensitive hypertension. Indeed, salt-sensitive Dahl rats show a paradoxical increase in renal Sgk1 expression in response to high dietary Na^+ intake, whereas salt-resistant Sprague Dawley rats reveal the expected decrease of renal Sgk1 expression that is likely related to reduced plasma aldosterone levels and that likely reduces apical ENaC localization to prevent the development of hypertension (170). However, whether salt sensitivity in Dahl rats is caused by ENaC dysfunction and whether this relates to Sgk1 upregulation awaits confirmation. Support for a linkage between arterial hypertension and Sgk1 comes also from genetic analysis of twins. Busjahn and coworkers (187) described an association between arterial blood pressure and two single-nucleotide polymorphisms in Sgk1 exon 8 and intron 6. Studies on unrelated subjects, however, failed to show a significant interdependence between Sgk1 single-nucleotide polymorphisms and blood pressure and estimated the risk for homozygosity for nonsynonymous, nonconservative mutations in the coding region of the Sgk1 gene as rather low (below 1/300,000 in the white Caucasian population) (188). Recent studies in heterologous expression systems linked Sgk1 to arterial hypertension in patients with PHA type 2 caused by mutation in WNK1. In transfected HEK and CHO cells, WNK1 activates Sgk1 and increases ENaC activity at the cell surface. On the basis of these in vitro findings, it was suggested that mutated WNK1 may also enhance ENaC activity in the kidney in vivo, which may contribute to the hypertensive phenotype of patients with PHA type 2 (86). However, this proposed pathogenetic mechanism is difficult to reconcile with the disease-characteristic hyperkalemia and acidosis that are usually seen as side effects of lowered ENaC activity.

CONCLUSION AND PERSPECTIVE

The identification of Sgk1 as an aldosterone-induced transcript and protein in renal epithelia disclosed the importance of this kinase in the regulation of epithelial ion transport. During the past six years, several in vitro and in vivo experiments have provided conclusive evidence that aldosterone-induced Sgk1 promotes Na^+ reabsorption in the ASDN by activating ENaC in the luminal membrane and Na^+,K^+-ATPase in the basolateral plasma membrane. Putative upstream activators (e.g., PI3K, PDK1) and downstream effectors (e.g., Nedd4-2) have been identified, placing Sgk1 into a complex signaling network that likely integrates many incoming hormonal and nonhormonal signals. Dysfunctions of Sgk1 in the kidneys have been linked to altered renal Na^+ transport and may also affect other ion transport processes such as renal potassium secretion.

However, many open questions remain. Sgk1 has a broad tissue and cellular distribution and apparently phosphorylates and activates a great variety of different substrates, at least in in vitro settings. In vivo experiments, however, suggest a much more narrow and specific action of the kinase. Proper identification of the substrates specifically activated in vivo as well as the cellular and molecular mechanisms that determine the specificity of Sgk1 for these substrates will be crucial to further understand the relevance of this kinase in the physiology and pathophysiology of epithelial ion transport.

That Sgk1-deficient mice exhibit a rather mild phenotype points to a certain degree of redundancy in Sgk1-dependent signal transduction cascades. In this context, the putative roles of other members of the AGC kinase family in the regulation of epithelial ion transport warrant further attention. Cross-breeding of mouse lines with targeted inactivation for different types of AGC kinases will likely provide informative insights into the redundancy of AGC kinase–dependent signaling. Moreover, the development of specific activators and inhibitors of Sgk1 and the other members of the AGC kinase family are needed to dissect further the functions of these kinases under various experimental conditions. These compounds also may eventually represent interesting drug candidates for the treatment of patients with disturbed epithelial ion transport.

APPENDIX

Abbreviations used in this chapter: ASDN, aldosterone-sensitive distal nephron; BMK, bone marrow kinase; CHO, Chinese hamster ovary; CISK, cytokine-independent survival kinase; CREB, cAMP-responsive element–binding; ENaC, epithelial Na^+ channel; ERK, extracellular signal–regulated kinase; FGF, fibroblast growth factor; FKHRL, forkhead in rhabdomyosarcoma-like 1; Fnk, FGF-inducible kinase; FSH, follicle-stimulating hormone; GluR, glutamate receptor; GM-CSF, granulocyte-macrophage colony–stimulating factor; GRE, glucocorticoid response element; GSK3, glycogen synthase kinase 3; HEK, human embryonic kidney; H-motif, hydrophobic motif; IGF-1, insulin growth factor 1

MAPK, mitogen-activated protein kinase; MEKK3, mitogen-activated protein kinase/ERK kinase kinase 3; NDRG-2, N-myc downstream-regulated gene 2; Nedd4-2, neuronal precursor cell expressed developmentally downregulated 2; NHE3, Na^+,H^+ exchanger 3; NHERF2, Na^+,H^+ exchanger–regulating factor 2; NKCC2, $Na^+,K^+,2Cl^-$-cotransporter 2; PDGF, platelet-derived growth factor; PDK, 3-phosphoinositide-dependent kinase; PH domain, pleckstrin homology domain; PHA, pseudohypoaldosteronism; phox, phagocyte oxidase; PI-3K, phosphoinositide-3 kinase; Pif, PDK-1-interacting fragment; Pim, provirus integration site for Moloney murine leukemia virus; PKB, protein kinase B; PKC, protein kinase C; Plk, polo-like kinase; PPARγ, peroxisome proliferator–activated receptor gamma; Prk, phosphoribulokinase; PtdIns, phosphatidylinositol; PtdIns(3,4)P2, phosphatidylinositol-(3,4)-biphosphate; PtdIns(3,4,5)P3, phosphatidylinositol-(3,4,5)-triphosphate; PX domain, phox domain; ROMK, renal outer medulla K^+ channel; Sgk, serum/glucocorticoid-induced kinase; Snk, serum-inducible kinase; TAL, thick ascending limb; TGF-β, transforming growth factor β; WNK, with no lysine kinase.

ACKNOWLEDGMENTS

Cited work by the author's laboratories was supported by the Swiss National Science Foundation (to O.S., grant number 3100A0-10,3779/1, and to J.L., grant number 3200B0-10,5769/1), the Roche Research Foundation in Basel (to O.S.), the Cloëtta Foundation (to J.L.), the EMDO Foundation (to J.L.), and the Swiss Diabetes Foundation (to J.L.).

The *Annual Review of Physiology* is online at
http://physiol.annualreviews.org

LITERATURE CITED

1. Webster MK, Goya L, Ge Y, Maiyar AC, Firestone GL. 1993. Characterization of sgk, a novel member of the serine/threonine protein kinase gene family which is transcriptionally induced by glucocorticoids and serum. *Mol. Cell. Biol.* 13:2031–40

2. Webster MK, Goya L, Firestone GL. 1993. Immediate-early transcriptional regulation and rapid mRNA turnover of a putative serine/threonine protein kinase. *J. Biol. Chem.* 268:11482–85

3. Kumar JM, Brooks DP, Olson BA, Laping NJ. 1999. Sgk, a putative serine/threonine kinase, is differentially expressed in the kidney of diabetic mice

and humans. *J. Am. Soc. Nephrol.* 10: 2488–94

4. Golsteyn RM, Schultz SJ, Bartek J, Ziemiecki A, Ried T, Nigg EA. 1994. Cell cycle analysis and chromosomal localization of human Plk1, a putative homologue of the mitotic kinases *Drosophila* polo and *Saccharomyces cerevisiae* Cdc5. *J. Cell Sci.* 107(Pt. 6): 1509–17

5. Hamanaka R, Maloid S, Smith MR, O'Connell CD, Longo DL, Ferris DK. 1994. Cloning and characterization of human and murine homologues of the *Drosophila* polo serine-threonine kinase. *Cell Growth Differ.* 5:249–57

6. Holtrich U, Wolf G, Brauninger A, Karn T, Bohme B, et al. 1994. Induction and down-regulation of PLK, a human serine/threonine kinase expressed in proliferating cells and tumors. *Proc. Natl. Acad. Sci. USA* 91:1736–40

7. Simmons DL, Neel BG, Stevens R, Evett G, Erikson RL. 1992. Identification of an early-growth-response gene encoding a novel putative protein kinase. *Mol. Cell. Biol.* 12:4164–69

8. Donohue PJ, Alberts GF, Guo Y, Winkles JA. 1995. Identification by targeted differential display of an immediate early gene encoding a putative serine/threonine kinase. *J. Biol. Chem.* 270:10351–57

9. Kauselmann G, Weiler M, Wulff P, Jessberger S, Konietzko U, et al. 1999. The polo-like protein kinases Fnk and Snk associate with a Ca^{2+}- and integrin-binding protein and are regulated dynamically with synaptic plasticity. *EMBO J.* 18:5528–39

10. Clay FJ, McEwen SJ, Bertoncello I, Wilks AF, Dunn AR. 1993. Identification and cloning of a protein kinase-encoding mouse gene, Plk, related to the polo gene of Drosophila. *Proc. Natl. Acad. Sci. USA* 90:4882–86

11. Konietzko U, Kauselmann G, Scafidi J, Staubli U, Mikkers H, et al. 1999. Pim kinase expression is induced by LTP stimulation and required for the consolidation of enduring LTP. *EMBO J.* 18:3359–69

12. Waldegger S, Erdel M, Nagl UO, Barth P, Raber G, et al. 1998. Genomic organization and chromosomal localization of the human SGK protein kinase gene. *Genomics* 51:299–302

13. Firestone GL, Giampaolo JR, O'Keeffe BA. 2003. Stimulus-dependent regulation of serum and glucocorticoid inducible protein kinase (SGK) transcription, subcellular localization and enzymatic activity. *Cell Physiol. Biochem.* 13:1–12

14. Itani OA, Liu KZ, Cornish KL, Campbell JR, Thomas CP. 2002. Glucocorticoids stimulate human sgk1 gene expression by activation of a GRE in its 5′-flanking region. *Am. J. Physiol. Endocrinol. Metab.* 283:E971–79

15. Maiyar AC, Phu PT, Huang AJ, Firestone GL. 1997. Repression of glucocorticoid receptor transactivation and DNA binding of a glucocorticoid response element within the serum/glucocorticoid-inducible protein kinase (sgk) gene promoter by the p53 tumor suppressor protein. *Mol. Endocrinol.* 11:312–29

16. Pearce D, Bhargava A, Cole TJ. 2003. Aldosterone: Its receptor, target genes, and actions. *Vitam. Horm.* 66:29–76

17. Lang F, Cohen P. 2001. Regulation and physiological roles of serum- and glucocorticoid-induced protein kinase isoforms. *Sci. STKE* 2001:RE17

18. Chen S, Bhargava A, Mastroberardino L, Meijer OC, Wang J, et al. 1999. Epithelial sodium channel regulated by aldosterone-induced protein sgk. *Proc. Natl. Acad. Sci. USA* 96:2514–19

19. Naray-Fejes-Toth A, Canessa CM, Cleaveland ES, Aldrich G, Fejes-Toth G. 1999. sgk is an aldosterone-induced kinase in the renal collecting duct. *J. Biol. Chem.* 274:16973–78

20. Waldegger S, Barth P, Raber G, Lang F. 1997. Cloning and characterization of a putative human serine/threonine protein kinase transcriptionally modified during anisotonic and isotonic alterations of cell volume. *Proc. Natl. Acad. Sci. USA* 94:4440–45

21. Warntges S, Friedrich B, Henke G, Duranton C, Lang PA, et al. 2002. Cerebral localization and regulation of the cell volume-sensitive serum- and glucocorticoid-dependent kinase SGK1. *Pflugers Arch.* 443:617–24

22. Rozansky DJ, Wang J, Doan N, Purdy T, Faulk T, et al. 2002. Hypotonic induction of SGK1 and Na^+ transport in A6 cells. *Am. J. Physiol. Renal Physiol.* 283:F105–13

23. Waldegger S, Klingel K, Barth P, Sauter M, Rfer ML, et al. 1999. h-sgk serine-threonine protein kinase gene as transcriptional target of transforming growth factor beta in human intestine. *Gastroenterology* 116:1081–88

24. Lang F, Klingel K, Wagner CA, Stegen C, Warntges S, et al. 2000. Deranged transcriptional regulation of cell-volume-sensitive kinase hSGK in diabetic nephropathy. *Proc. Natl. Acad. Sci. USA* 97:8157–62

25. Fillon S, Klingel K, Warntges S, Sauter M, Gabrysch S, et al. 2002. Expression of the serine/threonine kinase hSGK1 in chronic viral hepatitis. *Cell Physiol. Biochem.* 12:47–54

26. Khan ZA, Barbin YP, Farhangkhoee H, Beier N, Scholz W, Chakrabarti S. 2005. Glucose-induced serum- and glucocorticoid-regulated kinase activation in oncofetal fibronectin expression. *Biochem. Biophys. Res. Commun.* 329:275–80

27. Imaizumi K, Tsuda M, Wanaka A, Tohyama M, Takagi T. 1994. Differential expression of sgk mRNA, a member of the Ser/Thr protein kinase gene family, in rat brain after CNS injury. *Brain Res. Mol. Brain Res.* 26:189–96

28. Nishida Y, Nagata T, Takahashi Y, Sugahara-Kobayashi M, Murata A, Asai S. 2004. Alteration of serum/glucocorticoid regulated kinase-1 (sgk-1) gene expression in rat hippocampus after transient global ischemia. *Brain Res. Mol. Brain Res.* 123:121–25

29. Hollister RD, Page KJ, Hyman BT. 1997. Distribution of the messenger RNA for the extracellularly regulated kinases 1, 2 and 3 in rat brain: Effects of excitotoxic hippocampal lesions. *Neuroscience* 79:1111–19

30. Tsai KJ, Chen SK, Ma YL, Hsu WL, Lee EH. 2002. sgk, a primary glucocorticoid-induced gene, facilitates memory consolidation of spatial learning in rats. *Proc. Natl. Acad. Sci. USA* 99:3990–95

31. You H, Jang Y, You-Ten AI, Okada H, Liepa J, et al. 2004. p53-dependent inhibition of FKHRL1 in response to DNA damage through protein kinase SGK1. *Proc. Natl. Acad. Sci. USA* 101:14057–62

32. Akutsu N, Lin R, Bastien Y, Bestawros A, Enepekides DJ, et al. 2001. Regulation of gene expression by $1\alpha,25$-dihydroxyvitamin D3 and its analog EB1089 under growth-inhibitory conditions in squamous carcinoma cells. *Mol. Endocrinol.* 15:1127–39

33. Murata S, Yoshiara T, Lim CR, Sugino M, Kogure M, et al. 2005. Psychophysiological stress-regulated gene expression in mice. *FEBS Lett.* 579:2137–42

34. Marzullo L, Tosco A, Capone R, Andersen HS, Capasso A, Leone A. 2004. Identification of dietary copper- and iron-regulated genes in rat intestine. *Gene* 338:225–33

35. Cowling RT, Birnboim HC. 2000. Expression of serum- and glucocorticoid-regulated kinase (sgk) mRNA is up-regulated by GM-CSF and other proinflammatory mediators in human granulocytes. *J. Leukoc. Biol.* 67:240–48

36. Mizuno H, Nishida E. 2001. The ERK MAP kinase pathway mediates induction of SGK (serum- and glucocorticoid-inducible kinase) by growth factors. *Genes Cells* 6:261–68

37. Alliston TN, Maiyar AC, Buse P, Firestone GL, Richards JS. 1997. Follicle stimulating hormone-regulated expression of serum/glucocorticoid-inducible kinase in rat ovarian granulosa cells: A functional role for the Sp1 family in promoter activity. *Mol. Endocrinol.* 11:1934–49

38. Leong ML, Maiyar AC, Kim B, O'Keeffe BA, Firestone GL. 2003. Expression of the serum- and glucocorticoid-inducible protein kinase, Sgk, is a cell survival response to multiple types of environmental stress stimuli in mammary

epithelial cells. *J. Biol. Chem.* 278:5871–82

39. Maiyar AC, Huang AJ, Phu PT, Cha HH, Firestone GL. 1996. p53 stimulates promoter activity of the sgk serum/glucocorticoid-inducible serine/threonine protein kinase gene in rodent mammary epithelial cells. *J. Biol. Chem.* 271:12414–22

40. Kobayashi T, Deak M, Morrice N, Cohen P. 1999. Characterization of the structure and regulation of two novel isoforms of serum- and glucocorticoid-induced protein kinase. *Biochem. J.* 344:189–97

41. Xu J, Liu D, Gill G, Songyang Z. 2001. Regulation of cytokine-independent survival kinase (CISK) by the Phox homology domain and phosphoinositides. *J. Cell Biol.* 154:699–705

42. Liu D, Yang X, Songyang Z. 2000. Identification of CISK, a new member of the SGK kinase family that promotes IL-3-dependent survival. *Curr. Biol.* 10:1233–36

43. Hertweck M, Gobel C, Baumeister R. 2004. *C. elegans* SGK-1 is the critical component in the Akt/PKB kinase complex to control stress response and life span. *Dev Cell* 6:577–88

44. Casamayor A, Torrance PD, Kobayashi T, Thorner J, Alessi DR. 1999. Functional counterparts of mammalian protein kinases PDK1 and SGK in budding yeast. *Curr. Biol.* 9:186–97

45. de Hart AK, Schnell JD, Allen DA, Hicke L. 2002. The conserved Pkh-Ypk kinase cascade is required for endocytosis in yeast. *J. Cell Biol.* 156:241–48

46. Dai F, Yu L, He H, Zhao Y, Yang J, et al. 1999. Cloning and mapping of a novel human serum/glucocorticoid regulated kinase-like gene, SGKL, to chromosome 8q12.3–q13.1. *Genomics* 62:95–97

47. Ponting CP. 1996. Novel domains in NADPH oxidase subunits, sorting nexins, and PtdIns 3-kinases: Binding partners of SH3 domains? *Protein Sci.* 5:2353–57

48. Worby CA, Dixon JE. 2002. Sorting out the cellular functions of sorting nexins. *Nat. Rev. Mol. Cell Biol.* 3:919–31

49. Xing Y, Liu D, Zhang R, Joachimiak A, Songyang Z, Xu W. 2004. Structural basis of membrane targeting by the Phox homology domain of cytokine-independent survival kinase (CISK-PX). *J. Biol. Chem.* 279:30662–69

50. Arteaga MF, Canessa CM. 2005. Functional specificity of Sgk1 and Akt1 on ENaC activity. *Am. J. Physiol. Renal Physiol.* 289:F90–96

51. Gamper N, Fillon S, Feng Y, Friedrich B, Lang PA, et al. 2002. K$^+$ channel activation by all three isoforms of serum- and glucocorticoid-dependent protein kinase SGK. *Pflugers Arch.* 445:60–66

52. Henke G, Maier G, Wallisch S, Boehmer C, Lang F. 2004. Regulation of the voltage gated K$^+$ channel Kv1.3 by the ubiquitin ligase Nedd4-2 and the serum and glucocorticoid inducible kinase SGK1. *J. Cell. Physiol.* 199:194–99

53. Henke G, Setiawan I, Bohmer C, Lang F. 2002. Activation of Na$^+$/K$^+$-ATPase by the serum and glucocorticoid-dependent kinase isoforms. *Kidney Blood Press. Res.* 25:370–74

54. Embark HM, Bohmer C, Vallon V, Luft F, Lang F. 2003. Regulation of KCNE1-dependent K$^+$ current by the serum and glucocorticoid-inducible kinase (SGK) isoforms. *Pflugers Arch.* 445:601–6

55. Friedrich B, Feng Y, Cohen P, Risler T, Vandewalle A, et al. 2003. The serine/threonine kinases SGK2 and SGK3 are potent stimulators of the epithelial Na$^+$ channel α,β,γ-ENaC. *Pflugers Arch.* 445:693–96

56. Bohmer C, Philippin M, Rajamanickam J, Mack A, Broer S, et al. 2004. Stimulation of the EAAT4 glutamate transporter by SGK protein kinase isoforms and PKB. *Biochem. Biophys. Res. Commun.* 324:1242–48

57. Strutz-Seebohm N, Seebohm G, Shu-milina E, Mack AF, Wagner HJ, et al. 2005. Glucocorticoid adrenal steroids and glucocorticoid-inducible kinase iso-forms in the regulation of GluR6 expres-sion. *J. Physiol.* 565:391–401

58. Strutz-Seebohm N, Seebohm G, Mack AF, Wagner HJ, Just L, et al. 2005. Reg-ulation of GluR1 abundance in murine hippocampal neurones by serum- and glucocorticoid-inducible kinase 3. *J. Physiol.* 565:381–90

59. McCormick JA, Feng Y, Dawson K, Behne MJ, Yu B, et al. 2004. Tar-geted disruption of the protein kinase SGK3/CISK impairs postnatal hair fol-licle development. *Mol. Biol. Cell* 15:4278–88

60. Wulff P, Vallon V, Huang DY, Volkl H, Yu F, et al. 2002. Impaired renal Na$^+$ re-tention in the sgk1-knockout mouse. *J. Clin. Invest.* 110:1263–68

61. Garofalo RS, Orena SJ, Rafidi K, Torchia AJ, Stock JL, et al. 2003. Severe dia-betes, age-dependent loss of adipose tis-sue, and mild growth deficiency in mice lacking Akt2/PKB beta. *J. Clin. Invest.* 112:197–208

62. Cho H, Mu J, Kim JK, Thorvaldsen JL, Chu Q, et al. 2001. Insulin resistance and a diabetes mellitus-like syndrome in mice lacking the protein kinase Akt2 (PKB β). *Science* 292:1728–31

63. Boehmer C, Henke G, Schniepp R, Pal-mada M, Rothstein JD, et al. 2003. Regulation of the glutamate transporter EAAT1 by the ubiquitin ligase Nedd4-2 and the serum and glucocorticoid-inducible kinase isoforms SGK1/3 and protein kinase B. *J. Neurochem.* 86:1181–88

64. Boehmer C, Wilhelm V, Palmada M, Wallisch S, Henke G, et al. 2003. Serum and glucocorticoid inducible ki-nases in the regulation of the cardiac sodium channel SCN5A. *Cardiovasc. Res.* 57:1079–84

65. Boehmer C, Embark HM, Bauer A, Pal-mada M, Yun CH, et al. 2004. Stimula-tion of renal Na$^+$ dicarboxylate cotrans-porter 1 by Na$^+$/H$^+$ exchanger regulat-ing factor 2, serum and glucocorticoid inducible kinase isoforms, and protein kinase B. *Biochem. Biophys. Res. Com-mun.* 313:998–1003

66. Embark HM, Bohmer C, Palmada M, Rajamanickam J, Wyatt AW, et al. 2004. Regulation of CLC-Ka/barttin by the ubiquitin ligase Nedd-4-2 and the serum-and glucocorticoid-dependent kinases. *Kidney Int.* 66:1918–25

67. Embark HM, Setiawan I, Poppendieck S, van de Graaf SF, Boehmer C, et al. 2004. Regulation of the epithelial Ca^{2+} channel TRPV5 by the NHE regulat-ing factor NHERF2 and the serum and glucocorticoid inducible kinase isoforms SGK1 and SGK3 expressed in *Xenopus* oocytes. *Cell Physiol. Biochem.* 14:203–12

68. Palmada M, Dieter M, Speil A, Bohmer C, Mack AF, et al. 2004. Regulation of intestinal phosphate cotransporter NaPi IIb by ubiquitin ligase Nedd-4-2 and by serum- and glucocorticoid-dependent kinase 1. *Am. J. Physiol. Gastrointest. Liver Physiol.* 287:G143–50

69. Palmada M, Speil A, Jeyaraj S, Bohmer C, Lang F. 2005. The serine/threonine kinases SGK1, 3 and PKB stimulate the amino acid transporter ASCT2. *Biochem. Biophys. Res. Commun.* 331:272–77

70. Alessi DR, Deak M, Casamayor A, Caudwell FB, Morrice N, et al. 1997. 3-phosphoinositide-dependent protein kinase-1 (PDK1): Structural and func-tional homology with the *Drosophila* DSTPK61 kinase. *Curr. Biol.* 7:776–89

71. Stokoe D, Stephens LR, Copeland T, Gaffney PR, Reese CB, et al. 1997. Dual role of phosphatidylinositol-3,4,5-trisphosphate in the activation of protein kinase B. *Science* 277:567–70

72. Vanhaesebroeck B, Alessi DR. 2000. The PI3K-PDK1 connection: More than

just a road to PKB. *Biochem. J.* 346(Pt. 3):561–76

73. Cheng X, Ma Y, Moore M, Hemmings BA, Taylor SS. 1998. Phosphorylation and activation of cAMP-dependent protein kinase by phosphoinositide-dependent protein kinase. *Proc. Natl. Acad. Sci. USA* 95:9849–54

74. Chou MM, Hou W, Johnson J, Graham LK, Lee MH, et al. 1998. Regulation of protein kinase C ζ by PI 3-kinase and PDK-1. *Curr. Biol.* 8:1069–77

75. Le Good JA, Ziegler WH, Parekh DB, Alessi DR, Cohen P, Parker PJ. 1998. Protein kinase C isotypes controlled by phosphoinositide 3-kinase through the protein kinase PDK1. *Science* 281:2042–45

76. Alessi DR, Kozlowski MT, Weng QP, Morrice N, Avruch J. 1997. 3-phosphoinositide-dependent protein kinase 1 (PDK1) phosphorylates and activates the p70 S6 kinase in vivo and in vitro. *Curr. Biol.* 8:69–81

77. Pullen N, Dennis PB, Andjelkovic M, Dufner A, Kozma SC, et al. 1998. Phosphorylation and activation of p70s6k by PDK1. *Science* 279:707–10

78. Kobayashi T, Cohen P. 1999. Activation of serum- and glucocorticoid-regulated protein kinase by agonists that activate phosphatidylinositide 3-kinase is mediated by 3- phosphoinositide-dependent protein kinase-1 (PDK1) and PDK2. *Biochem. J.* 339:319–28

79. Park J, Leon MLL, Buse P, Maiyar AC, Firestone GL, Hemmings BA. 1999. Serum and glucocorticoid-inducible kinase (SGK) is a target of the PI 3-kinase-stimulated signaling pathway. *EMBO J.* 18:3024–33

80. Biondi RM, Kieloch A, Currie RA, Deak M, Alessi DR. 2001. The PIF-binding pocket in PDK1 is essential for activation of S6K and SGK, but not PKB. *EMBO J.* 20:4380–90

81. Perrotti N, He RA, Phillips SA, Haft CR, Taylor SI. 2001. Activation of serum- and glucocorticoid-induced protein kinase (Sgk) by cyclic AMP and insulin. *J. Biol. Chem.* 276:9406–12

82. Shelly C, Herrera R. 2002. Activation of SGK1 by HGF, Rac1 and integrin-mediated cell adhesion in MDCK cells: PI-3K-dependent and -independent pathways. *J. Cell Sci.* 115:1985–93

83. Hayashi M, Tapping RI, Chao TH, Lo JF, King CC, et al. 2001. BMK1 mediates growth factor-induced cell proliferation through direct cellular activation of serum and glucocorticoid-inducible kinase. *J. Biol. Chem.* 276:8631–34

84. Meng F, Yamagiwa Y, Taffetani S, Han J, Patel T. 2005. Interleukin-6 activates serum and glucocorticoid kinase via a p38α mitogen activated protein kinase pathway. *Am. J. Physiol. Cell Physiol.* In press

85. Stokoe D, Campbell DG, Nakielny S, Hidaka H, Leevers SJ, et al. 1992. MAP-KAP kinase-2: A novel protein kinase activated by mitogen-activated protein kinase. *EMBO J.* 11:3985–94

86. Xu BE, Stippec S, Chu PY, Lazrak A, Li XJ, et al. 2005. WNK1 activates SGK1 to regulate the epithelial sodium channel. *Proc. Natl. Acad. Sci. USA* 102:10315–20

87. Brickley DR, Mikosz CA, Hagan CR, Conzen SD. 2002. Ubiquitin modification of serum and glucocorticoid-induced protein kinase-1 (SGK-1). *J. Biol. Chem.* 277:43064–70

88. Helms MN, Fejes-Toth G, Naray-Fejes-Toth A. 2003. Hormone-regulated transepithelial Na⁺ transport in mammalian CCD cells requires SGK1 expression. *Am. J. Physiol.* 284:F480–87

89. Zhou R, Snyder PM. 2005. Nedd4-2 phosphorylation induces serum and glucocorticoid-regulated kinase (SGK) ubiquitination and degradation. *J. Biol. Chem.* 280:4518–23

90. Cross DA, Alessi DR, Cohen P, Andjelkovich M, Hemmings BA. 1995. Inhibition of glycogen synthase kinase-3

by insulin mediated by protein kinase B. *Nature* 378:785–89

91. Sakoda H, Gotoh Y, Katagiri H, Kurokawa M, Ono H, et al. 2003. Differing roles of Akt and serum- and glucocorticoid-regulated kinase in glucose metabolism, DNA synthesis, and oncogenic activity. *J. Biol. Chem.* 278:25802–7

92. Collins BJ, Deak M, Arthur JS, Armit LJ, Alessi DR. 2003. In vivo role of the PIF-binding docking site of PDK1 defined by knock-in mutation. *EMBO J.* 22:4202–11

93. Brunet A, Park J, Tran H, Hu LS, Hemmings BA, Greenberg ME. 2001. Protein kinase SGK mediates survival signals by phosphorylating the forkhead transcription factor FKHRL1 (FOXO3a). *Mol. Cell. Biol.* 21:952–65

94. Zhang BH, Tang ED, Zhu T, Greenberg ME, Vojtek AB, Guan KL. 2001. Serum- and glucocorticoid-inducible kinase SGK phosphorylates and negatively regulates B-Raf. *J. Biol. Chem.* 276:31620–26

95. Chun J, Kwon T, Kim DJ, Park I, Chung G, et al. 2003. Inhibition of mitogen-activated kinase kinase kinase 3 activity through phosphorylation by the serum- and glucocorticoid-induced kinase 1. *J. Biochem.* 133:103–8

96. Wang D, Sun H, Lang F, Yun CC. 2005. Activation of NHE3 by dexamethasone requires phosphorylation of NHE3 at Ser663 by SGK1. *Am. J. Physiol. Cell Physiol.* In press

97. Boulkroun S, Fay M, Zennaro MC, Escoubet B, Jaisser F, et al. 2002. Characterization of rat NDRG2 (N-Myc downstream regulated gene 2), a novel early mineralocorticoid-specific induced gene. *J. Biol. Chem.* 277:31506–15

98. Murray JT, Campbell DG, Morrice N, Auld GC, Shpiro N, et al. 2004. Exploitation of KESTREL to identify NDRG family members as physiological substrates for SGK1 and GSK3. *Biochem. J.* 384:477–88

99. Debonneville C, Flores SY, Kamynina E, Plant PJ, Tauxe C, et al. 2001. Phosphorylation of Nedd4-2 by Sgk1 regulates epithelial Na$^+$ channel cell surface expression. *EMBO J.* 20:7052–59

100. Snyder PM, Olson DR, Thomas BC. 2002. Serum and glucocorticoid-regulated kinase modulates Nedd-4-2-mediated inhibition of the epithelial Na$^+$ channel. *J. Biol. Chem.* 277:5–8

101. Diakov A, Korbmacher C. 2004. A novel pathway of epithelial sodium channel activation involves a serum- and glucocorticoid-inducible kinase consensus motif in the C terminus of the channel's α-subunit. *J. Biol. Chem.* 279:38134–42

102. Stichel CC, Schoenebeck B, Foguet M, Siebertz B, Bader V, et al. 2005. sgk1, a member of an RNA cluster associated with cell death in a model of Parkinson's disease. *Eur. J. Neurosci.* 21:301–16

103. Gonzalez-Nicolini V, McGinty JF. 2002. Gene expression profile from the striatum of amphetamine-treated rats: A cDNA array and in situ hybridization histochemical study. *Brain Res Gene Expr Patterns* 1:193–98

104. Rauz S, Walker EA, Hughes SV, Coca-Prados M, Hewison M, et al. 2003. Serum- and glucocorticoid-regulated kinase isoform-1 and epithelial sodium channel subunits in human ocular ciliary epithelium. *Invest. Ophthalmol. Vis. Sci.* 44:1643–51

105. Rauz S, Walker EA, Murray PI, Stewart PM. 2003. Expression and distribution of the serum and glucocorticoid regulated kinase and the epithelial sodium channel subunits in the human cornea. *Exp. Eye Res.* 77:101–8

106. Waerntges S, Klingel K, Weigert C, Fillon S, Buck M, et al. 2002. Excessive transcription of the human serum and glucocorticoid dependent kinase hSGK1 in lung fibrosis. *Cell Physiol. Biochem.* 12:135–42

107. Alliston TN, Gonzalez-Robayna IJ,

Buse P, Firestone GL, Richards JS. 2000. Expression and localization of serum/glucocorticoid-induced kinase in the rat ovary: Relation to follicular growth and differentiation. *Endocrinology* 141:385–95

108. Klingel K, Warntges S, Bock J, Wagner CA, Sauter M, et al. 2000. Expression of cell volume-regulated kinase h-sgk in pancreatic tissue. *Am. J. Physiol. Gastrointest. Liver Physiol.* 279:G998–1002

109. Friedrich B, Warntges S, Klingel K, Sauter M, Kandolf R, et al. 2002. Upregulation of the human serum and glucocorticoid-dependent kinase 1 in glomerulonephritis. *Kidney Blood Press. Res.* 25:303–7

110. Huber SM, Friedrich B, Klingel K, Lenka N, Hescheler J, Lang F. 2001. Protein and mRNA expression of serum and glucocorticoid-dependent kinase 1 in metanephrogenesis. *Dev. Dyn.* 221:464–69

111. Lee E, Lein ES, Firestone GL. 2001. Tissue-specific expression of the transcriptionally regulated serum and glucocorticoid-inducible protein kinase (Sgk) during mouse embryogenesis. *Mech. Dev.* 103:177–81

112. Naray-Fejes-Toth A, Helms MN, Stokes JB, Fejes-Toth G. 2004. Regulation of sodium transport in mammalian collecting duct cells by aldosterone-induced kinase, SGK1: Structure/function studies. *Mol. Cell Endocrinol.* 217:197–202

113. Coric T, Hernandez N, Alvarez de la Rosa D, Shao D, Wang T, Canessa CM. 2004. Expression of ENaC and serum- and glucocorticoid-induced kinase 1 in the rat intestinal epithelium. *Am. J. Physiol. Gastrointest. Liver Physiol.* 286:G663–70

114. Alvarez de la Rosa D, Coric T, Todorovic N, Shao D, Wang T, Canessa CM. 2003. Distribution and regulation of expression of serum- and glucocorticoid-induced kinase-1 in the rat kidney. *J. Physiol.* 551:455–66

115. Buse P, Tran SH, Luther E, Phu PT, Aponte GW, Firestone GL. 1999. Cell cycle and hormonal control of nuclear-cytoplasmic localization of the serum- and glucocorticoid-inducible protein kinase, Sgk, in mammary tumor cells. A novel convergence point of anti-proliferative and proliferative cell signaling pathways. *J. Biol. Chem.* 274:7253–63

116. Loffing J, Pietri L, Aregger F, Bloch-Faure M, Ziegler U, et al. 2000. Differential subcellular localization of ENaC subunits in mouse kidney in response to high- and low-Na diets. *Am. J. Physiol.* 279:F252–58

117. Chen WS, Xu PZ, Gottlob K, Chen ML, Sokol K, et al. 2001. Growth retardation and increased apoptosis in mice with homozygous disruption of the Akt1 gene. *Genes Dev.* 15:2203–8

118. Pearce D. 2001. The role of SGK1 in hormone-regulated sodium transport. *Trends Endocrinol. Metab.* 12:341–47

119. Maiyar AC, Leong ML, Firestone GL. 2003. Importin-α mediates the regulated nuclear targeting of serum- and glucocorticoid-inducible protein kinase (Sgk) by recognition of a nuclear localization signal in the kinase central domain. *Mol. Biol. Cell* 14:1221–39

120. Loffing J, Zecevic M, Féraille E, Kaissling B, Asher C, et al. 2001. Aldosterone induces rapid apical translocation of ENaC in early portion of renal collecting system: Possible role of SGK. *Am. J. Physiol.* 280:F675–82

121. Kellenberger S, Schild L. 2002. Epithelial sodium channel/degenerin family of ion channels: A variety of functions for a shared structure. *Physiol. Rev.* 82:735–67

122. Canessa CM, Schild L, Buell G, Thorens B, Gautschi I, et al. 1994. Amiloride-sensitive epithelial Na+ channel is made of three homologous subunits. *Nature* 367:463–67

123. Canessa CM, Horisberger J-D, Rossier

BC. 1993. Epithelial sodium channel related to proteins involved in neurodegeneration. *Nature* 361:467–70

124. Lingueglia E, Voilley N, Waldmann R, Lazdunski M, Barbry P. 1993. Expression cloning of an epithelial amiloride-sensitive Na+ channel. A new channel type with homologies to *Caenorhabditis elegans* degenerins. *FEBS Lett.* 318:95–99

125. Lingueglia E, Renard S, Waldmann R, Voilley N, Champigny G, et al. 1994. Different homologous subunits of the amiloride-sensitive Na+ channel are differently regulated by aldosterone. *J. Biol. Chem.* 269:13736–39

126. Firsov D, Gautschi I, Mérillat A-M, Rossier BC, Schild L. 1998. The heterotetrameric architecture of the epithelial sodium channel (ENaC). *EMBO J.* 17:344–52

127. Eskandari S, Snyder PM, Kreman M, Zampighi GA, Welsh MJ, Wright EM. 1999. Number of subunits comprising the epithelial sodium channel. *J. Biol. Chem.* 274:27281–86

128. Snyder PM, Cheng C, Prince LS, Rogers JC, Welsh MJ. 1998. Electrophysiological and biochemical evidence that DEG/ENaC cation channels are composed of nine subunits. *J. Biol. Chem.* 273:681–84

129. Renard S, Lingueglia E, Voilley N, Lazdunski M, Barbry P. 1994. Biochemical analysis of the membrane topology of the amiloride-sensitive Na+ channel. *J. Biol. Chem.* 269:12981–86

130. Canessa CM, Mérillat A-M, Rossier BC. 1994. Membrane topology of the epithelial sodium channel in intact cells. *Am. J. Physiol.* 267:C1682–90

131. Snyder PM, McDonald FJ, Stokes JB, Welsh MJ. 1994. Membrane topology of the amiloride-sensitive epithelial sodium channel. *J. Biol. Chem.* 269:24379–83

132. Chen HI, Sudol M. 1995. The WW domain of Yes-associated protein binds a novel proline-rich ligand that differs from the consensus established for SH3-binding modules. *Proc. Natl. Acad. Sci. USA* 92:7819–23

133. Staub O, Rotin D. 1996. WW domains. *Structure* 4:495–99

134. Liddle GW, Bledsoe T, Coppage WS Jr. 1963. A familial renal disorder simulating primary aldosteronism but with negligible aldosterone secretion. *Trans. Assoc. Am. Physicians* 76:199–213

135. Hansson JH, Nelson-Williams C, Suzuki H, Schild L, Shimkets RA, et al. 1995. Hypertension caused by a truncated epithelial sodium channel γ subunit: Genetic heterogeneity of Liddle syndrome. *Nat. Genet.* 11:76–82

136. Shimkets RA, Warnock DG, Bositis CM, Nelson-Williams C, Hansson JH, et al. 1994. Liddle's syndrome: Heritable human hypertension caused by mutations in the β subunit of the epithelial sodium channel. *Cell* 79:407–14

137. Firsov D, Schild L, Gautschi I, Mérillat A-M, Schneeberger E, Rossier BC. 1996. Cell surface expression of the epithelial Na+ channel and a mutant causing Liddle syndrome: A quantitative approach. *Proc. Natl. Acad. Sci. USA* 93:15370–75

138. Snyder PM, Price MP, McDonald FJ, Adams CM, Volk KA, et al. 1995. Mechanism by which Liddle's syndrome mutations increase activity of a human epithelial Na+ channel. *Cell* 83:969–78

139. Schild L, Canessa CM, Shimkets RA, Warnock DG, Lifton RP, Rossier BC. 1995. A mutation in the epithelial sodium channel causing Liddle's disease increases channel activity in the *Xenopus laevis* oocyte expression system. *Proc. Natl. Acad. Sci. USA* 92:5699–703

140. Schild L, Lu Y, Gautschi I, Schneeberger E, Lifton RP, Rossier BC. 1996. Identification of a PY motif in the epithelial Na+ channel subunits as a target sequence for mutations causing channel activation found in Liddle syndrome. *EMBO J.* 15:2381–87

141. Staub O, Dhu S, Henry PC, Correa J, Ishikawa T, et al. 1996. WW domains of Nedd4 bind to the proline-rich PY motifs in the epithelial Na⁺ channel deleted in Liddle's syndrome. *EMBO J.* 15:2371–80

142. Goulet CC, Volk KA, Adams CM, Prince LS, Stokes JB, Snyder PM. 1998. Inhibition of the epithelial Na⁺ channel by interaction of Nedd4 with a PY motif deleted in Liddle's syndrome. *J. Biol. Chem.* 273:30012–17

143. Dinudom A, Harvey BJ, Komwatana P, Young JA, Kumar S, Cook DI. 1998. Nedd4 mediates control of an epithelial Na⁺ channel in salivary duct cells by cytosolic Na⁺. *Proc. Natl. Acad. Sci. USA* 95:7169–73

144. Kanelis V, Rotin D, Forman-Kay JD. 2001. Solution structure of a Nedd4 WW domain-ENaC peptide complex. *Nat. Struct. Biol.* 8:1–6

145. Abriel H, Loffing J, Rebhun JF, Pratt JH, Horisberger J-D, et al. 1999. Defective regulation of the epithelial Na⁺ channel (ENaC) by Nedd4 in Liddle's syndrome. *J. Clin. Invest.* 103:667–73

146. Kamynina E, Debonneville C, Bens M, Vandewalle A, Staub O. 2001. A novel mouse Nedd4 protein suppresses the activity of the epithelial Na⁺ channel. *FASEB J.* 15:204–14

147. Kamynina E, Tauxe C, Staub O. 2001. Differential characteristics of two human Nedd4 proteins with respect to epithelial Na⁺ channel regulation. *Am. J. Physiol.* 281:F469–77

148. Snyder PM, Steines JC, Olson DR. 2004. Relative contribution of Nedd4 and Nedd-4-2 to ENaC regulation in epithelia determined by RNA interference. *J. Biol. Chem.* 279:5042–46

149. Harvey KF, Dinudom A, Cook DI, Kumar S. 2001. The Nedd4-like protein KIAA0439 is a potential regulator of the epithelial sodium channel. *J. Biol. Chem.* 276:8597–601

150. Staub O, Gautschi I, Ishikawa T, Bre-itschopf K, Ciechanover A, et al. 1997. Regulation of stability and function of the epithelial Na⁺ channel (ENaC) by ubiquitination. *EMBO J.* 16:6325–36

151. Kamynina E, Staub O. 2002. Concerted action of ENaC, Nedd-4-2, and Sgk1 in transepithelial Na⁺ transport. *Am. J. Physiol.* 283:F377–87

152. Snyder PM. 2002. The epithelial Na⁺ channel: Cell surface insertion and retrieval in Na⁺ homeostasis and hypertension. *Endocr. Rev.* 23:258–75

153. Vallon V, Wulff P, Huang DY, Loffing J, Volkl H, et al. 2005. Role of Sgk1 in salt and potassium homeostasis. *Am. J. Physiol. Regul. Integr. Comp. Physiol.* 288:R4–10

154. Bhargava A, Wang J, Pearce D. 2004. Regulation of epithelial ion transport by aldosterone through changes in gene expression. *Mol. Cell. Endocrinol.* 217:189–96

155. Verrey F, Loffing J, Zecevic M, Heitzmann D, Staub O. 2003. SGK1: Aldosterone-induced relay of Na+ transport regulation in distal kidney nephron cells. *Cell Physiol. Biochem.* 13:21–28

156. Zecevic M, Heitzmann D, Camargo SM, Verrey F. 2004. SGK1 increases Na,K-ATP cell-surface expression and function in *Xenopus laevis* oocytes. *Pflugers Arch.* 448:29–35

157. Setiawan I, Henke G, Feng Y, Bohmer C, Vasilets LA, et al. 2002. Stimulation of Xenopus oocyte Na⁺,K⁺ATPase by the serum and glucocorticoid-dependent kinase sgk1. *Pflugers Arch.* 444:426–31

158. Alvarez de la Rosa D, Zhang P, Naray-Fejes-Toth A, Fejes-Toth G, Canessa CM. 1999. The serum and glucocorticoid kinase sgk increases the abundance of epithelial sodium channels in the plasma membrane of *Xenopus* oocytes. *J. Biol. Chem.* 274:37834–39

159. Vuagniaux G, Vallet V, Jaeger NF, Hummler E, Rossier BC. 2002. Synergistic activation of ENaC by three

membrane-bound channel-activating serine proteases (mCAP1, mCAP2, and mCAP3) and serum and glucocorticoid-regulated kinase (Sgk1) in *Xenopus* oocytes. *J. Gen. Physiol.* 120:191–201

160. Flores SY, Loffing-Cueni D, Kamynina E, Daidie D, Gerbex C, et al. 2005. Aldosterone induced Sgk1 expression is accompanied by Nedd-4-2 phosphorylation and increased Na^+ transport in cortical collecting duct cells. *J. Am. Soc. Nephrol.* 16:2279–87

161. Ichimura T, Yamamura H, Sasamoto K, Tominaga Y, Taoka M, et al. 2005. 14-3-3 proteins modulate the expression of epithelial Na^+ channels by phosphorylation-dependent interaction with Nedd4-2 ubiquitin ligase. *J. Biol. Chem.* 280:13187–94

161a. Bhalla V, Daidie D, Li H, Pao AC, Lagrange LP, et al. 2005. SGK1 regulates ubiquitin ligase Nedd4-2 by inducing interaction with 14-3-3. *Mol. Endocrinol.* In press

162. Wang J, Barbry P, Maiyar AC, Rozansky DJ, Bhargava A, et al. 2001. SGK integrates insulin and mineralocorticoid regulation of epithelial sodium transport. *Am. J. Physiol.* 280:F303–13

163. Alvarez de la Rosa D, Paunescu TG, Els WJ, Helman SI, Canessa CM. 2004. Mechanisms of regulation of epithelial sodium channel by SGK1 in A6 cells. *J. Gen. Physiol.* 124:395–407

164. Gumz ML, Popp MP, Wingo CS, Cain BD. 2003. Early transcriptional effects of aldosterone in a mouse inner medullary collecting duct cell line. *Am. J. Physiol. Renal Physiol.* 285:F664–73

165. Bhargava A, Fullerton MJ, Myles K, Purdy TM, Funder JW, et al. 2001. The serum- and glucocorticoid-induced kinase is a physiological mediator of aldosterone action. *Endocrinology* 142:1587–94

166. Brennan FE, Fuller PJ. 2000. Rapid up-regulation of serum and glucocorticoid-regulated kinase (sgk) gene expression by corticosteroids in vivo. *Mol.Cell. Endocrinol.* 166:129–36

167. Hou J, Speirs HJ, Seckl JR, Brown RW. 2002. Sgk1 gene expression in kidney and its regulation by aldosterone: Spatiotemporal heterogeneity and quantitative analysis. *J. Am. Soc. Nephrol.* 13:1190–98

168. Muller OG, Parnova RG, Centeno G, Rossier BC, Firsov D, Horisberger JD. 2003. Mineralocorticoid effects in the kidney: Correlation between αENaC, GILZ, and Sgk-1 mRNA expression and urinary excretion of Na^+ and K^+. *J. Am. Soc. Nephrol.* 14:1107–15

169. Shigaev A, Asher C, Latter H, Garty H, Reuveny E. 2000. Regulation of sgk by aldosterone and its effects on the epithelial Na^+ channel. *Am. J. Physiol.* 278:F613–19

170. Farjah M, Roxas BP, Geenen DL, Danziger RS. 2003. Dietary salt regulates renal SGK1 abundance: Relevance to salt sensitivity in the Dahl rat. *Hypertension* 41:874–78

171. Bell LM, Leong ML, Kim B, Wang E, Park J, et al. 2000. Hyperosmotic stress stimulates promoter activity and regulates cellular utilization of the serum- and glucocorticoid-inducible protein kinase (Sgk) by a p38 MAPK-dependent pathway. *J. Biol. Chem.* 275:25262–72

172. Chen S, McCormick JA, Prabaker K, Wang J, Pearce D, Gardner DG. 2004. Sgk1 mediates osmotic induction of NPR-A gene in rat inner medullary collecting duct cells. *Hypertension* 43:866–71

173. Duc C, Farman N, Canessa CM, Bonvalet J-P, Rossier BC. 1994. Cell-specific expression of epithelial sodium channel α, β, and γ subunits in aldosterone-responsive epithelia from the rat: Localization by in situ hybridization and immunocytochemistry. *J. Cell Biol.* 127:1907–21

174. Alvarez De La Rosa D, Canessa CM. 2003. Role of SGK in hormonal

regulation of epithelial sodium channel in A6 cells. *Am. J. Physiol.* 284:C404–14

175. Faletti CJ, Perrotti N, Taylor SI, Blazer-Yost BL. 2002. sgk: An essential convergence point for peptide and steroid hormone regulation of ENaC-mediated Na⁺ transport. *Am. J. Physiol.* 282:C494–500

176. Vallon V, Huang DY, Grahammer F, Wyatt AW, Osswald H, et al. 2005. SGK1 as a determinant of kidney function and salt intake in response to mineralocorticoid excess. *Am. J. Physiol. Regul. Integr. Comp. Physiol.* 289: R395–401

177. Palmer LG, Antonian L, Frindt G. 1994. Regulation of apical K and Na channels and Na/K pumps in rat cortical collecting tubule by dietary K. *J. Gen. Physiol.* 104:693–710

178. Palmer LG, Frindt G. 1999. Regulation of apical K channels in rat cortical collecting tubule during changes in dietary K intake. *Am. J. Physiol.* 277:F805–12

179. Palmada M, Embark HM, Yun C, Bohmer C, Lang F. 2003. Molecular requirements for the regulation of the renal outer medullary K⁺ channel ROMK1 by the serum- and glucocorticoid-inducible kinase SGK1. *Biochem. Biophys. Res. Commun.* 311:629–34

180. Palmada M, Embark HM, Wyatt AW, Bohmer C, Lang F. 2003. Negative charge at the consensus sequence for the serum- and glucocorticoid-inducible kinase, SGK1, determines pH sensitivity of the renal outer medullary K⁺ channel, ROMK1. *Biochem. Biophys. Res. Commun.* 307:967–72

181. Yun CC, Palmada M, Embark HM, Fedorenko O, Feng Y, et al. 2002. The serum and glucocorticoid-inducible kinase SGK1 and the Na⁺/H⁺ exchange regulating factor NHERF2 synergize to

stimulate the renal outer medullary K⁺ channel ROMK1. *J. Am. Soc. Nephrol.* 13:2823–30

182. Yoo D, Kim BY, Campo C, Nance L, King A, et al. 2003. Cell surface expression of the ROMK (Kir 1.1) channel is regulated by the aldosterone-induced kinase, SGK-1, and protein kinase A. *J. Biol. Chem.* 278:23066–75

183. Huang DY, Wulff P, Völkl H, Loffing J, Richter K, et al. 2004. Impaired regulation of renal K⁺ elimination in the sgk1-knockout mouse. *J. Am. Soc. Nephrol.* 15:885–91

184. Hong G, Lockhart A, Davis B, Rahmoune H, Baker S, et al. 2003. PPARγ activation enhances cell surface ENaCα via up-regulation of SGK1 in human collecting duct cells. *FASEB J.* 17:1966–68

185. Bistrup C, Thiesson HC, Jensen BL, Skott O. 2005. Reduced activity of 11β-hydroxysteroid dehydrogenase type 2 is not responsible for sodium retention in nephrotic rats. *Acta Physiol. Scand.* 184:161–69

186. Yu Z, Serra A, Sauter D, Loffing J, Ackermann D, et al. 2005. Sodium retention in rats with liver cirrhosis is associated with increased renal abundance of NaCl cotransporter (NCC). *Nephrol. Dial. Transplant.* 20:1833–41

187. Busjahn A, Aydin A, Uhlmann R, Krasko C, Bahring S, et al. 2002. Serum- and glucocorticoid-regulated kinase (SGK1) gene and blood pressure. *Hypertension* 40:256–60

188. Trochen N, Ganapathipillai S, Ferrari P, Frey BM, Frey FJ. 2004. Low prevalence of nonconservative mutations of serum and glucocorticoid-regulated kinase (SGK1) gene in hypertensive and renal patients. *Nephrol. Dial. Transplant.* 19:2499–504

Annu. Rev. Physiol. 2006. 68:491–505
doi: 10.1146/annurev.physiol.68.040104.131050
First published online as a Review in Advance on August 18, 2005

THE ASSOCIATION OF NHERF ADAPTOR PROTEINS WITH G PROTEIN–COUPLED RECEPTORS AND RECEPTOR TYROSINE KINASES[*]

Edward J. Weinman,[1] Randy A. Hall,[2] Peter A. Friedman,[3] Lee-Yuan Liu-Chen,[4] and Shirish Shenolikar[5]

[1]Departments of Medicine and Physiology, University of Maryland School of Medicine, and Medical Service, Department of Veterans Affairs Medical Center, Baltimore, Maryland 21201; email: eweinman@medicine.umaryland.edu

[2]Department of Pharmacology, Rollins Research Center, Emory University School of Medicine, Atlanta, Georgia 30322; email: rhall@pharm.emory.edu

[3]Departments of Pharmacology and Medicine, University of Pittsburgh School of Medicine, Pittsburgh, Pennsylvania 15261; email: friedman@server.pharm.pitt.edu

[4]Department of Pharmacology and Center for Substance Abuse Research, Temple University School of Medicine, Philadelphia, Pennsylvania 19140; email: lliuche@astro.temple.edu

[5]Department Pharmacology and Cancer Biology, Duke University Medical Center, Durham, North Carolina 27710; email: sheno001@mc.duke.edu

Key Words   PDZ adaptor proteins, multiple protein complexes, hormone receptors, growth factor receptors

■ **Abstract**   The sodium-hydrogen exchanger regulatory factors (NHERF-1 and NHERF-2) are a family of adaptor proteins characterized by the presence of two tandem PDZ protein interaction domains and a C-terminal domain that binds the cytoskeleton proteins ezrin, radixin, moesin, and merlin. The NHERF proteins are highly expressed in the kidney, small intestine, and other organs, where they associate with a number of transporters and ion channels, signaling proteins, and transcription factors. Recent evidence has revealed important associations between the NHERF proteins and several G protein–coupled receptors such as the β_2-adrenergic receptor, the κ-opioid receptor, and the parathyroid hormone receptor, as well as growth factor tyrosine kinase receptors such as the platelet-derived growth factor receptor and the epidermal growth factor receptor. This review summarizes the emerging data on the biochemical mechanisms, physiologic outcomes, and potential clinical implications of the assembly and disassembly of receptor/NHERF complexes.

---

[*]The U.S. Government has the right to retain a nonexclusive, royalty-free license in and to any copyright covering this paper.

0066-4278/06/0315-0491$20.00
491

INTRODUCTION

The sodium-hydrogen exchanger regulatory factors NHERF-1 (also known as EBP50) and NHERF-2 (also known as E3KARP) represent a family of adaptor proteins characterized by two tandem PSD-95/Drosophila discs large/ZO-1 (PDZ) protein interaction domains and a C-terminal domain that binds the ezrin-radixin-moesin-merlin (ERM) family of cytoskeletal proteins (1–4). The NHERF proteins were the first PDZ proteins found to be localized to the apical membranes of renal epithelial cells. These proteins were initially characterized as facilitating the formation of a multiprotein complex that mediated protein kinase A phosphorylation of the renal sodium-hydrogen exchanger 3 (NHE3) and the downregulation of its activity (5–9). NHERF-1 and NHERF-2 form homo- and heterodimers and also bind to other PDZ proteins to form an extended membrane/submembrane scaffold that in turn binds an array of transmembrane and soluble proteins, including other transporters and ion channels, signaling proteins, transcription factors, and cellular structural proteins (10–14).

This review focuses on a new class of NHERF targets, namely G protein–coupled receptors (GPCRs) and growth factor receptor tyrosine kinases. A growing body of evidence indicates that the NHERF proteins regulate the localization and function of several GPCRs, including the β_2-adrenergic receptor, the κ-opioid receptor, and the parathyroid hormone (PTH) receptor (15–18). NHERF proteins also interact with selected growth factor receptors such as the platelet-derived growth factor receptor (PDGFR) and the epidermal growth factor receptor (EGFR) to modulate mitogenic signaling by these receptor tyrosine kinases (19–20). Table 1 lists the membrane receptors currently known to associate with the NHERF proteins.

TABLE 1 Hormone receptor targets of NHERF*

G protein–coupled receptors

β_2-adrenergic receptor (15, 21)

κ-opioid receptor (18)

Purinergic P2Y1 receptor (21)

Parathyroid hormone 1 receptor (16)

Adenosine 2b receptor (52)

Lysophosphatidic acid receptor A2 (53)

Luteinizing hormone receptor (54)

RAMP3 of the adrenomedullin receptor (55)

Tyrosine receptor kinases

Platelet-derived growth factor receptor (19)

Epidermal growth factor receptor (20)

*References are indicated in parentheses.

This review summarizes data pointing to the role of the NHERF proteins in regulating selected hormone and growth factor receptors. It also outlines the underlying biochemical mechanisms, the physiologic outcomes, and potential clinical implications of the dynamic assembly and disassembly of known receptor/NHERF complexes. By highlighting common themes, we hope this review provides new insights in hormone signaling as well as direction for future studies of cell signaling by membrane receptor complexes.

G PROTEIN–COUPLED RECEPTORS AS TARGETS OF NHERF

NHERF Associates with the β_2-Adrenergic Receptor

The first receptor found to associate with the NHERF proteins was the β_2-adrenergic receptor (β_2-AR) (15). This interaction was discovered using a biochemical purification approach in which the β_2-AR C terminus was expressed as a fusion protein, linked to an agarose matrix, and used to purify interacting proteins from crude tissue extracts. A single 50 kDa protein was purified from kidney lysates, and amino sequencing revealed this protein to be NHERF-1. Mutagenesis studies demonstrated that the last four amino acids (D-S-L-L) of the β_2-AR C terminus are critical for the association of β_2-AR with NHERF-1, with the serine at the -2 position and the terminal leucine residue being of particular importance (15, 21). NHE3, the first identified NHERF target, had been shown to bind to PDZ-II of NHERF-1 (22). β_2-AR, by contrast, was the first of now a large number of polypeptides found to bind to PDZ-I of NHERF-1 (15). β_2-AR also binds to the first PDZ domain of NHERF-2 (21).

NHERF-1 Regulates β_2-AR Trafficking

The NHERF proteins play an important role in the regulation of β_2-AR trafficking. The association between β_2-AR and NHERF-1 occurs in an agonist-promoted manner. Agonist-bound wild-type β_2-AR normally recycles to the plasma membrane; this process requires NHERF-1. Introduction of mutations in the NHERF-binding motif of β_2-AR resulted in agonist-promoted internalization of the receptor and shunting of β_2-AR to lysosomes, a pathway leading to degradation of the receptor (23). Overexpression of the first PDZ domain of NHERF-1, which disrupts the association of β_2-AR with endogenous NHERF, altered β_2-AR trafficking in a fashion similar to the effect of mutation of the receptor's NHERF-binding motif (23). Furthermore, fusion of the NHERF-binding motif to other GPCRs, such as the δ-opioid receptor, enhances receptor recycling to the membrane following agonist-induced endocytosis (24, 25). These findings provide strong evidence that the NHERF proteins play a pivotal role in directing β_2-AR endocytic sorting. A regulated mechanism controlling the interaction between β_2-AR and NHERF-1 was suggested by the observation that G protein–coupled receptor kinase 5 (GRK 5) phosphorylates residues in the C terminus of the β_2-AR, impairing binding of the receptor to NHERF-1 (23). Said in another way, NHERF-1 readily engages the unphosphorylated receptor, but not β_2-AR phosphorylated by GRK5. This observation provided a critical insight into the dynamic association of the processes that direct the β_2-AR receptor to recycling or degradatory pathways.

The Effect of the β_2-AR/NHERF-1 Interaction on NHE3 and CFTR Activity

Prior studies had demonstrated that β_2-AR stimulation in the kidney proximal tubule raised cyclic AMP levels but that, despite the activation of protein kinase A, β_2-AR agonists increased rather than decreased sodium and water transport (26, 27). Following the characterization of the β_2-AR/NHERF-1 interactions, additional studies were performed to attempt to address this paradox. When renal brush border membrane proteins containing NHE3 were reconstituted into soybean lipid vesicles, the addition of a polypeptide representing C-terminal β_2-AR residues blocked the inhibitory effect of protein kinase A on Na^+-H^+ exchange activity (15). In studies utilizing a fibroblast cell line expressing rat NHE3 and wild-type β_2-AR, treatment with a β_2-adrenergic agonist resulted in stimulation of NHE3 activity, whereas forskolin (which increased cAMP) inhibited the transporter. By contrast, coexpression of a mutated form of β_2-AR that was incapable of binding NHERF-1 resulted in inhibition of NHE3 activity in response to both the β_2-adrenergic agonist and forskolin (15). These experiments were interpreted to indicate that the presence of β_2-AR displaced NHERF-1 binding to NHE3. It is postulated that the NHERF-1 protein cannot accommodate simultaneous occupancy by both β_2-AR and NHE3, and that the agonist-promoted binding of NHERF-1 to β_2-AR represents a mechanism by which β_2-AR stimulation can modulate NHE3 activity.

Following the elucidation of the association between β_2-AR and NHERF-1, three independent groups simultaneously reported binding of the cystic fibrosis transmembrane conductance regulator (CFTR) to NHERF-1 (21, 28, 29). The CFTR C terminus ends in D-T-R-L, a sequence nearly identical to the NHERF-1 binding motif found at the β_2-AR C terminus. Mutation of this motif in the CFTR polypeptide abrogated its association with NHERF-1 (21, 28, 29). Recent studies have indicated that β_2-AR, CFTR, and NHERF-1 form a triple protein complex in airway epithelial cells and that the formation of this complex plays a key role in β_2-AR-mediated regulation of CFTR function (30, 31). These findings suggest another physiologically relevant role for the association between the β_2-AR and the NHERF proteins.

NHERF INTERACTIONS WITH THE κ-OPIOID RECEPTOR

NHERF Binds to the κ-Opioid Receptor

Given the C-terminal amino acid sequences of receptors known to bind to PDZ domain–containing proteins, studies were undertaken to determine whether the κ-opioid receptor, which has a C-terminal valine, interacts with NHERF-1 (18, 32). It was initially demonstrated, using CHO cells stably expressing the human κ-opioid receptor (hKOR), that NHERF-1 co-immunoprecipitated with hKOR. When the cells were treated with the selective κ-opioid agonist U50,488H, there was a significant increase in the amount of co-immunoprecipitated NHERF-1, indicating that the agonist-occupied receptor had greater binding affinity for NHERF-1 than did the unbound receptor. A polypeptide representing the PDZ-I domain of NHERF-1 (amino acids 1–151) co-immunoprecipitated with hKOR, but a fragment representing PDZ-II[152–358] did not. Thus, like β_2-AR and PTH1R, hKOR preferentially binds to PDZ-I of NHERF-1. Mutant hKOR containing an additional C-terminal alanine (hKOR-A) or glutamic acid residues (hKOR-EE) did not co-immunoprecipitate with NHERF-1, suggesting that the interaction between hKOR and NHERF-1 involved a PDZ domain interaction. Additional studies indicated that purified glutathione-S-transferase (GST)-hKOR C-terminal domain (hKOR 334–380) interacted with purified NHERF-1, whereas GST or GST-C terminus of the μ- or δ-opioid receptors did not (32). The C terminus of hKOR-bound NHERF-1 in extracts of CHO cells transfected with the construct. The hKOR C terminus also associated with the PDZ-I but not the PDZ-II domain of NHERF-1.

Although the C-terminal sequence of hKOR, N-K-P-V, is not a canonical PDZ domain–binding motif, these results suggest a PDZ domain interaction between hKOR and NHERF-1. There are, however, differences between β_2-AR and hKOR in their interactions with NHERF-1. Substitution of alanine for serine at the −2 position of β_2-AR or mutation of the C-terminal leucine to valine reduced the interaction with NHERF-1, indicating a class I PDZ interaction (15, 21, 33). Because the −2 position of hKOR is lysine and the C-terminal residue of hKOR is valine, the interaction of hKOR with NHERF-1 appears not to represent a prototypic class I interaction. This suggestion is supported by pull-down studies that indicate that the binding of NHERF-1 to the C terminus of hKOR is much weaker than to the C terminus of β_2-AR (32). Recently, Heydorn et al. (34) screened the C termini of 59 GPCRs for interaction with NHERF-1 and found that β_2-AR was the only one that yielded a positive result, whereas hKOR did not interact with NHERF-1 in this assay. As these investigators used relative signals and β_2-AR yielded a very strong signal, weaker interactions may have been overlooked. In pull-down assays, the hKOR C terminus–bound NHERF-1 expressed in CHO cells or endogenous NHERF-1 in the brain with apparently higher affinities than did purified NHERF-1 (P. Huang & L.-Y. Liu-Chen, unpublished observations). Such findings may suggest that there are cell-specific posttranslational modifications of NHERF-1 or that accessory proteins affect the binding characteristics. Whatever the explanation, the results with hKOR suggest that NHERF-1, and particularly PDZ-I, has broad binding specificity by virtue of the conformation of the binding groove when associated with different PDZ-I targets. It is, in fact, the flexibility and diversity of PDZ- I of NHERF-1 that may allow interaction with a subset of GPCRs.

NHERF-1 Affects the Trafficking of the κ-Opioid Receptor

Treatment of CHO cells expressing hKOR with U50,488H resulted in a 30% reduction in the number of hKOR receptors (18, 35). Of interest, expression of NHERF-1 in CHO cells coexpressing hKOR abolished U50,488H-induced downregulation of the receptor (18). By contrast, U50,488H-induced downregulation of the mutant receptors hKOR-A and hKOR-EE, which do not co-immunoprecipitate with NHERF-1, was not affected by expression of NHERF-1. Other studies have indicated that interactions between NHERF and the ERM proteins are important in the regulation of hKOR (36). Expression of NHERF-1 (1-298) containing the PDZ domains but lacking the ERM-binding domain did not block downregulation of hKOR, indicating that binding of NHERF-1 to the cortical actin cytoskeleton is critical for its inhibitory effect on downregulation. As KOR internalization is required for downregulation, a reduction in downregulation may be due to a decrease in KOR internalization (35). Alternatively, because internalized receptors either are routed for degradation in lysosomes and/or proteasomes or recycled back to plasma membranes, attenuation in downregulation may be attributed to enhanced recycling. Additional studies have now indicated that coexpression of NHERF-1 accelerated the recycling of internalized hKOR without affecting U50,488H-induced internalization (18). By contrast, expression of NHERF-1 did not affect U50,488H binding affinity, U50,488H-stimulated $[^{35}S]GTP\gamma S$ binding, activation of p42/p44 MAP kinase, or U50,488H-induced desensitization of hKOR. These results indicate that NHERF-1 binds to the C terminus of the hKOR; this association appears to serve as a signal for internalized hKOR to be sorted to the recycling pathway.

Physiologic Consequences of NHERF-1/κ-Opioid Receptor Association

The physiologic consequences of NHERF-1 interaction with the κ-opioid receptor were determined by measuring Na^+-H^+ exchange activity in OK proximal tubule epithelial cells. OK cells demonstrate the presence of κ-opioid binding sites and, using pull-down techniques, the hKOR C terminus interacted directly with endogenous OK NHERF-1 (18, 37). Incubation of OK cells with U50,488H significantly enhanced Na^+-H^+ exchange (32). U50,488H-stimulated Na^+-H^+ exchange was blocked by naloxone but not by pertussis toxin pretreatment, indicating that it is mediated by KOR but independent of G_i/G_o proteins. In OKH cells, a subclone of OK cells expressing a much lower level of NHERF-1, U50,488H had no effect on Na^+-H^+ exchange. By contrast, in OKH cells, as in OK cells, U50,488H stimulated p44/p42 MAP kinase phosphorylation via κ-opioid receptors and pertussis toxin-sensitive G proteins. Stable transfection of NHERF-1 into OKH cells restored the stimulatory effect of U50,488H on Na^+-H^+ exchange. Thus, NHERF-1 binds directly to KOR, and this association plays an important role in accelerating Na^+-H^+ exchange. How the κ-opioid receptor interacts with NHERF-1 to stimulate NHE3 in OK cells is not currently well understood. The hKOR C terminus facilitates oligomerization of NHERF-1, similar to the C-terminal domain of the platelet-derived growth factor receptor (PDGFR) and β_2-AR (13, 19, 32). NHERF-1 oligomers may represent a pool of inactive NHERF-1 that is incapable of inhibiting NHE3 (14, 38). A working hypothesis is that there is a basal level of PKA activity in OK cells that mediates NHERF-1-associated inhibition of NHE3. Activation of the KOR enhances its association with and oligomerization of NHERF-1. This, in turn, eliminates the inhibition of NHE3 by NHERF-1, unmasking the observed stimulatory effect.

NHERF INTERACTIONS WITH THE PARATHYROID HORMONE RECEPTOR

Structural Determinants of PTH1R Binding to NHERF

Parathyroid hormone (PTH) receptors transduce the effects of PTH and of PTH-related peptide. The type 1 PTH receptor (PTH1R) is a Class II GPCR and is widely expressed in tissues responsible for mineral ion homeostasis. PTH1R stimulates adenylyl cyclase and/or phospholipase C (PLC), and it has been established that multiple G proteins mediate the bifurcating signaling pathways activated by PTH (39, 40). The PTH1R binds to PDZ-I of NHERF-1, but to PDZ-II of NHERF-2 (41). The C-terminal amino acid sequence of PTH1R, E-T-V-M, is consistent with a class 1 PDZ-binding domain motif. Mutation of E, T, or M of the PTH1R PDZ recognition domain prevents binding to full-length NHERF-1 or NHERF-2 (16, 17). The -1 position is permissive, and mutation to alanine does not interfere with the interaction with NHERF-2. Mahon and Segre (16) identified an 18-amino acid C-terminal PTH1R fragment (residues 573–591) that interacted with NHERF-1 in pull-down assays. A smaller receptor fragment, residues 583–591, however, did not, implying that the interaction between NHERF-1 and the PTH1R extends upstream to include some residues of the C-terminal 18 amino acids. The residues in this region were identified as E585 and E586. This finding suggests that these acidic residues contribute to stabilizing the interaction of the PTH1R with NHERF-1. From studies using a truncated form of the PTH1R lacking virtually all of the C-terminal intracellular tail, Sneddon et al. (42) found that although interaction of PTH1R, through its PDZ-binding domain with NHERF-1, modified the response to activating and nonactivating ligands, it was not required for receptor internalization. NHERF-1 lacking the ERM domain was incapable of suppressing PTH1R endocytosis evoked by PTH[7–34], a PTH fragment missing the N terminus required for adenylyl cyclase activation. Preliminary experiments have indicated that mutations in the PDZ-I-binding groove reduced the effect of NHERF-1 on PTH[7–34]-induced internalization (41). Actin colocalizes with NHERF-1 and the PTH1R in apical domains of OK proximal tubule–like opossum kidney cells (29). In OKH cells expressing little or no NHERF-1, PTH1Rs were diffusely distributed throughout the cytoplasm with some punctate localization in apical cell membranes and along the basolateral membrane. Treatment of the cells with cytochalasin D, a membrane-permeant inhibitor of actin polymerization, promoted actin aggregation and disrupted apical membrane PTH1R localization, attended by redistribution to the cytoplasm. Colchicine, a microtubule inhibitor, had no effect on PTH1R endocytosis (17, 43). In contrast to cells expressing NHERF-1, cytochalasin D had no effect on PTH1R endocytosis in response to PTH[1–34] in cells deficient in NHERF-1. When considered together, these findings suggest that the interaction between NHERF-1 and cortical actin plays a role in internalization of PTH1R and/or stabilizing PTH1R in the plasma membrane.

NHERF and the Trafficking of PTH1R

PTH1R, like other G protein–coupled receptors, undergoes cyclical receptor activation, desensitization, and internalization. Following endocytosis, PTH1R either can be recycled to the cell membrane, resulting in receptor resensitization, or targeted for degradation, leading to receptor downregulation (41, 42, 44). In distal tubule cells lacking endogenous NHERF-1, PTH[1–31], a PTH fragment that lacks the 32–34 sequence required for PTH1R activation of protein kinase C, had no discernible effect on PTH1R internalization, despite the fact that it stimulated adenylyl cyclase and phospholipase C (PLC) (44). PTH[1–34], a polypeptide that recapitulates all the classic effects of full-length PTH[1–84], promoted receptor endocytosis. PTH[7–34] lacking the N terminus required for adenylyl cyclase activation induced prompt PTH1R internalization that was faster and greater in magnitude than that observed with PTH[1–34]. Similar results were obtained with PTH[1–84] and PTH[7–84] (17). These observations indicate a striking dissociation between receptor activation and inactivation by native peptides. By contrast, the profile of responses was markedly different in proximal tubule cells that constitutively express NHERF-1. In proximal tubule cells, PTH[7–34] and PTH[1–31] had little or no effect on PTH1R endocytosis, whereas PTH[1–34] caused prompt internalization. Distal tubule cells and rat osteoblast-like cells, which also display little constitutive NHERF-1 expression, responded to PTH[7–34] and PTH[1–34] in a manner similar to proximal tubule cells when NHERF-1 was stably expressed. Conversely, proximal tubule cells transfected with a dominant negative form of NHERF-1 assumed a distal tubule response profile when treated with PTH[7–34]. These results are consistent with the view that NHERF-1 conditionally modulates the response to PTH fragments that are full agonists or antagonists. The ability of peptides with only weak affinity for the PTH1R to promote conformational changes is enhanced in the absence of NHERF-1. This suggests that NHERFs stabilize the receptor in the membrane such that only full agonists are able to elicit biological responses. These observations also suggest that PTH1R activation and endocytosis are mediated through distinct structural states that depend on the nature of the specific interactions between ligand and receptor.

NHERF Binding to PTH1R Functions as a Molecular Switch in the Generation of Downstream Signals

When expressed in PS120 fibroblasts that contain little or no NHERF-1 or NHERF-2, PTH1R signaled robustly and exclusively through adenylyl cyclase (16). When NHERF-2 was coexpressed, cAMP formation was decreased, whereas PLC signaling and inositol phosphate formation were dramatically increased. These important and novel findings suggest a unique role for NHERF-2 as a bimolecular signaling switch. Pertussis toxin treatment of PS120 cells coexpressing PTH1R and NHERF-2 inhibited PTH activation of PLC and partially restored activation of adenylyl cyclase. This indicates that PTH stimulates G_i/G_o proteins when the PTH1R is bound to NHERF-2. These findings are consistent with the idea that, upon activation of G_i/G_o, dissociation of the $\beta\gamma$- subunits activates PLC-β, whereas the α_1-subunits inhibit adenylyl cyclase (40).

NHERF, PTH1R Signaling, and Sodium-Dependent Phosphate Transport

By contrast to wild-type OK cells, PTH stimulated adenylyl cyclase but had no effect on phosphate transport in OKH cells (45). When OKH cells were transfected with NHERF-1, inhibition of phosphate transport by PTH was restored. Although introducing NHERF-1 in OKH cells reduced cAMP signaling and promoted inositol phosphate formation, there was little correlation between the changes in second-messenger formation and phosphate transport. Moreover, inhibiting PKA or PKC pathways failed to curtail PTH-sensitive phosphate transport. Thus, the precise mechanism by which NHERF-1 transduces PTH signals to inhibit phosphate transport in OK cells remains unresolved.

NHERF INTERACTIONS WITH THE RECEPTOR TYROSINE KINASES

In addition to associating with G protein–coupled receptors, the NHERF proteins have been found to interact with various receptor tyrosine kinases. These kinases include the platelet-derived growth factor receptor (PDGFR) and epidermal growth factor receptor (EGFR).

NHERF Binds to the Platelet-Derived Growth Factor Receptor

The possibility of an association between NHERF-1 and PDGFR was first examined because of the striking similarity between the C termini of PDGFR (D-S-F-L) and β_2-AR (D-S-L-L). NHERF-1 was found to associate with PDGFR and potentiate the receptor's signaling activity (19). It had been established previously that minimal truncations to the distal C terminus of PDGFR strongly impaired

PDGFR signaling, but the relevant protein-protein interactions disrupted by such truncations were completely unknown (46). Characterization of the effects of the association between PDGFR and NHERF on the receptor's activity provided a specific molecular mechanism to help explain the importance of the distal PDGFR C terminus in controlling the receptor's signaling.

The interaction between PDGFR and NHERF is of special importance in mediating PDGFR reorganization of the actin cytoskeleton (47). Fibroblasts expressing mutant PDGFRs incapable of binding NHERF exhibit markedly reduced migration as well as alterations to their actin cytoskeleton (48). Recent studies have indicated that the interaction between PDGFR and NHERF-1 can be disrupted by phosphorylation of PDGFR by GRK2. GRK2 phosphorylates the serine residue at the -2 position on the PDGFR C terminus and, by blocking binding to NHERF proteins, alters PDGFR activity (49). These findings parallel earlier experiments that demonstrated that the β_2-AR/NHERF interaction can be disrupted by phosphorylation of the β_2-AR C terminus by GRK5 (23). Taken together, these studies suggest that GRK-mediated phosphorylation of receptor C termini may be a general mechanism by which interactions with the NHERF proteins are regulated.

NHERF Binds the C Terminus of the Epidermal Growth Factor Receptor

Despite the fact that the C terminus of the epidermal growth factor receptor (EGFR) does not terminate in a canonical NHERF PDZ-binding motif, EGFR does associate with NHERF-1. A yeast two-hybrid screen of a mouse embryo cDNA library for potential EGFR-interacting proteins, using the distal EGFR C terminus as bait, revealed four positive clones, all of which corresponded to NHERF-1 (20). The interaction between EGFR and NHERF-1 is likely mediated by an internal peptide motif within the EGFR C terminus. The association between EGFR and NHERF-1 appears to stabilize EGFR at the cell surface and prevent its agonist-induced receptor downregulation, which parallels the aforementioned effects of NHERF-1 in preventing agonist-induced downregulation of β_2-AR (20, 23).

CONCLUSIONS

Study of the association between specific hormone and growth factor receptors and the two NHERF isoforms, NHERF-1 and NHERF-2, has indicated a role for these adaptor proteins beyond the regulation of ion transporters and channels. In fact, receptors currently represent the largest group of NHERF targets identified and suggest a critical role of these adaptor proteins in signal transduction. Emerging evidence suggests that NHERFs use a variety of modes to elicit hormonal control of physiological processes, ranging from the assembly of signaling complexes containing protein kinase A and its substrates, such as NHE3 and CFTR to the trafficking and tethering of membrane proteins to promote signaling, as discussed for PDGFR; to the modulation of second-messenger generation, as seen for PTH1R. Recent studies suggest additional targets of NHERF-1 such as $G\alpha q$ and coatomer proteins may also contribute to hormone signaling as well as receptor internalization and recycling (50, 51).

As study of the role of the NHERF proteins in regulating components of intracellular signaling pathways and other cell machinery expands, the remarkable diversity of these PDZ adaptor proteins is emphasized. Studies of EGFR and other targets whose C termini bind one or both NHERF PDZ domains suggest that the structural determinants that mediate their association with NHERF may be quite different from those of other targets studied to date. This may further expand the array of NHERF targets present in mammalian tissues. Even among those proteins that bind NHERF via their C termini, such as β_2-AR and hKOR, there appear to be significant differences in their affinities for the NHERF PDZ domains. This suggests that in diseases, such as aggressive forms of human breast cancer, that have been associated with elevated NHERF-1 expression, the disease state may arise from the increased association of NHERF-1 with a wide range of cellular targets. Mutations in some NHERF targets, such as the ERM protein merlin (also known as NF2) or transporters such as CFTR and Npt2a, have been linked with their altered association with NHERF, and it has been suggested that disruption of cellular NHERF complexes may contribute to human disease. Finally, recent studies have identified mutations in the NHERF proteins in specific cancers and in individuals with renal disease. These studies also suggest that defects in NHERF-regulated hormone signaling and other cellular functions may be critical in the development of these diseases.

ACKNOWLEDGMENTS

All authors contributed equally to this review. We thank the many students, fellows, and collaborators who have contributed to the efforts of their laboratories and whose work is cited. Our studies were supported by grants from the Department of Veterans Affairs (E.J.W.), the W.M. Kech Foundation (R.A.H.), and the National Institutes of Health (E.J.W., S.S., R.A.H., P.A.F., and L.-Y. L.-C.).

The *Annual Review of Physiology* is online at
http://physiol.annualreviews.org

LITERATURE CITED

1. Voltz JW, Weinman EJ, Shenolikar S. 2001. Expanding the role of NHERF, a PDZ domain containing adaptor to growth regulation. *Oncogene* 20:6309–14

2. Shenolikar S, Voltz JW, Cunningham R, Weinman EJ. 2004. Regulation of ion trans-

port by the NHERF family of PDZ proteins. *Physiology* 19:48–54

3. Shenolikar S, Weinman EJ. 2001. NHERF: Targeting and trafficking membrane proteins. *Am. J. Physiol. Renal Physiol.* 280:F389–95

4. Weinman EJ, Shenolikar S. 1993. Regulation of the renal brush border membrane Na^+-H^+ exchanger. *Annu. Rev. Physiol.* 55: 289–304

5. Zizak M, Lamprecht G, Steplock D, Tariq N, Shenolikar S, et al. 1999. cAMP-induced phosphorylation and inhibition of Na^+/H^+ exchanger 3 (NHE3) is dependent on the presence but not the phosphorylation of NHERF. *J. Biol. Chem.* 274:24753–58

6. Lamprecht G, Weinman EJ, Yun C-HC. 1998. The role of NHERF and E3KARP in the cAMP-mediated inhibition of NHE3. *J. Biol. Chem.* 273:29972–78

7. Yun CH, Oh S, Zizak M, Steplock D, Tsao S, et al. 1997. cAMP-mediated inhibition of the epithelial brush border Na^+/H^+ exchanger, NHE3, requires an associated regulatory protein. *Proc. Natl. Acad. Sci. USA* 94:3010–15

8. Weinman EJ, Steplock D, Wang Y, Shenolikar S. 1995. Characterization of a protein co-factor that mediates protein kinase A regulation of the renal brush border membrane Na^+-H^+ exchanger. *J. Clin. Invest.* 95:2143–49

9. Weinman EJ, Steplock D, Shenolikar S. 1993. cAMP-mediated inhibition of the renal BBM Na^+-H^+ exchanger requires a dissociable phosphoprotein co-factor. *J. Clin. Invest.* 92:1781–86

10. Custer M, Spindler B, Verrey F, Murer H, Biber J. 1997. Identification of a new gene product (diphor-1) regulated by dietary phosphate. *Am. J. Physiol. Renal Physiol.* 273:F801–6

11. Fouassier L, Yun CC, Fitz JG, Doctor RB. 2000. Evidence for ezrin-radixin-moesin-binding phosphoprotein 50 (EBP50) self-association through PDZ-PDZ interactions. *J. Biol. Chem.* 275:25039–45

12. Gisler SM, Pribanic S, Bacic D, Forrer P, Gantenbein A, et al. 2003. PDZK1: I. A major scaffolder in brush borders of proximal tubular cells. *Kidney Int.* 64:1733–45

13. Lau AG, Hall RA. 2001. Oligomerization of NHERF-1 and NHERF-2 PDZ domains: differential regulation by association with receptor carboxyl-termini and by phosphorylation. *Biochemistry* 40:8572–80

14. Shenolikar S, Minkoff CM, Steplock D, Chuckeree C, Liu M-Z, Weinman EJ. 2001. N-terminal PDZ domain is required for NHERF dimerization. *FEBS Lett.* 489:233–36

15. Hall RA, Premont RT, Chow CW, Blitzer JT, Pitcher JA, et al. 1998. The β2-adrenergic receptor interacts with the Na^+/H^+-exchanger regulatory factor to control Na^+/H^+ exchange. *Nature* 392:626–30

16. Mahon MJ, Donowitz M, Yun CC, Segre GV. 2002. Na^+/H^+ exchanger regulatory factor 2 directs parathyroid hormone 1 receptor signalling. *Nature* 417:858–61

17. Sneddon WB, Syme CA, Bisello A, Magyar CE, Rochi MD, et al. 2003. Activation-independent parathyroid hormone receptor internalization is regulated by NHERF1 (EBP50). *J. Biol. Chem.* 278:43787–96

18. Li J-G, Chen C, Liu-Chen L-Y. 2002. Ezrin-radixin-moesin-binding phosphoprotein-50/Na^+/H^+ exchanger regulatory factor (EBP50/NHERF) blocks U50,488H-induced down-regulation of the human kappa opioid receptor by enhancing its recycling rate. *J. Biol. Chem.* 277: 27545–52

19. Maudsley S, Zamah AM, Rahman N, Blitzer JT, Luttrell LM, et al. 2000. Platelet-derived growth factor receptor association with Na^+/H^+ exchanger regulatory factor potentiates receptor activity. *Mol. Cell Biol.* 20:8352–63

20. Lazar CS, Cresson CM, Lauffenburger DA, Gill GN. 2004. The Na^+/H^+ exchanger regulatory factor stabilizes epidermal growth factor receptors at the cell surface. *Mol. Biol. Cell* 15:5470–80

21. Hall RA, Ostedgaard LS, Premont RT, Blitzer JT, Rahman N, et al. 1998. A C-terminal motif found in the β2-adrenergic receptor, P2Y1 receptor and cystic fibrosis transmembrane conductance regulator determines binding to the Na^+/H^+ exchanger regulatory factor family of PDZ proteins. *Proc. Natl. Acad. Sci. USA* 95:8496–501

22. Weinman EJ, Steplock D, Tate K, Hall RA, Spurney RF, Shenolikar S. 1998. Structure-function of the Na/H exchanger regulatory factor (NHE-RF). *J. Clin. Invest.* 101: 2199–206

23. Cao TT, Deacon HW, Reczek D, Bretscher A, von Zastrow M. 1999. A kinase-regulated PDZ-domain interaction controls endocytic sorting of the β2-adrenergic receptor. *Nature* 401:286–90

24. Gage RM, Kim KA, Cao TT, von Zastrow M. 2001. A transplantable sorting signal that is sufficient to mediate rapid recycling of G protein–coupled receptors. *J. Biol. Chem.* 276:44712–20

25. Gage RM, Matveeva EA, Whiteheart SW, von Zastrow M. 2005. Type I PDZ ligands are sufficient to promote rapid recycling of G protein–coupled receptors independent of binding to N-ethylmaleimide-sensitive factor. *J. Biol. Chem.* 280:3305–13

26. Bello-Reuss E. 1980. Effect of catecholamines on fluid reabsorption by the isolated proximal convoluted tubule. *Am. J. Physiol. Renal Physiol.* 238:F347–52

27. Weinman EJ, Sansom SC, Knight TF, Senekjian HO. 1982. Alpha and beta adrenergic agonists stimulate water absorption in the rat proximal tubule. *J. Membr. Biol.* 69:107–11

28. Short DB, Trotter KW, Reczek D, Kreda SM, Bretscher A, et al. 1998. An apical PDZ protein anchors the cystic fibrosis transmembrane conductance regulator to the cytoskeleton. *J. Biol. Chem.* 273: 19797–801

29. Wang S, Raab RW, Schatz PJ, Guggino WB, Li M. 1998. Peptide binding consensus of the NHE-RF-PDZ1 domain matches the C-terminal sequence of cystic fibrosis transmembrane conductance regulator (CFTR). *FEBS Lett.* 427:103–8

30. Naren AP, Cobb B, Li CY, Roy K, Nelson D, et al. 2003. A macromolecular complex of β2 adrenergic receptor, CFTR, and ezrin/radixin/moesin-binding phosphoprotein 50 is regulated by PKA. *Proc. Natl. Acad. Sci. USA* 100:342–46

31. Taouil K, Hinnrasky J, Hologne C, Corlieu P, Klossek JM, Puchelle E. 2003. Stimulation of β2-adrenergic receptor increases cystic fibrosis transmembrane conductance regulator expression in human airway epithelial cells through a cAMP/protein kinase A-independent pathway. *J. Biol. Chem.* 278:17320–27

32. Huang P, Steplock D, Weinman EJ, Hall RA, Ding Z, et al. 2004. κ opioid receptor interacts with Na+/H+-exchanger regulatory factor-1/ezrin-radixin-moesin-binding phosphoprotein-50 (NHERF-1/EBP50) to stimulate Na+/H+ exchange independent of G_i/G_o proteins. *J. Biol. Chem.* 279: 25002–9

33. Sheng M, Sala C. 2001. PDZ domains and the organization of supramolecular complexes. *Annu. Rev. Neurosci.* 24:1–29

34. Heydorn A, Sondergaard BP, Ersboll B, Holst B, Nielsen FC, et al. 2004. A library of 7TM receptor C-terminal tails. Interactions with the proposed post-endocytic sorting proteins ERM-binding phosphoprotein 50 (EBP5 0), N-ethylmaleimide-sensitive factor (NSF), sorting nexin 1 (SNX1), and G protein–coupled receptor-associated sorting protein (GASP). *J. Biol. Chem.* 279:54291–303

35. Li J-G, Benovic JL, Liu-Chen L-Y. 2000. Mechanisms of agonist-induced down-regulation of the human κ-opioid receptor: internalization is required for down-regulation. *Mol. Pharmacol.* 58:795–801

36. Reczek D, Berryman M, Bretscher A. 1997. Identification of EBP50: A PDZ-containing phosphoprotein that associates with members of the ezrin-radixin-moesin family. *J. Cell Biol.* 139:169–79

37. Hatzoglou A, Bakogeorgou E, Papakonstanti E, Stournaras C, Emmanouel DS, Castanas E. 1996. Identification and characterization of opioid and somatostatin binding sites in the opossum kidney (OK) cell line and their effect on growth. *J. Cell Biochem.* 63:410–21

38. Weinman EJ, Steplock D, Shenolikar S. 2001. Acute regulation of NHE3 by protein kinase A requires a multiprotein signal complex. *Kidney Int.* 60:450–54

39. Brown EM, Segre GV, Goldring SR. 1996. Serpentine receptors for parathyroid hormone, calcitonin and extracellular calcium ions. *Baillieres Clin. Endocrinol. Metab.* 10:123–61

40. Birnbaumer L. 1992. Receptor-to-effector signaling through G proteins: Roles for $\beta\gamma$ dimers as well as α subunits. *Cell* 71:1069–72

41. Sneddon WB, Syme CA, Bisello A, Magyar CE, Weinman EJ, Friedman PA. 2003. PTH receptor internalization by inactive PTH fragments. Regulation by NHERF1 (EBP50). *J. Bone Miner. Res.* 18:S37

42. Sneddon WB, Bisello A, Magyar CE, Willick GE, Syme CA, et al. 2004. Ligand-selective dissociation of activation and internalization of the parathyroid hormone receptor. Conditional efficacy of PTH peptide fragments. *Endocrinology* 145:2815–23

43. Bacskai BJ, Friedman PA. 1990. Activation of latent Ca^{2+} channels in renal epithelial cells by parathyroid hormone. *Nature* 347:388–91

44. Sneddon WB, Bisello A, Friedman PA. 2002. PTH receptor desensitization accompanies receptor activation and not internalization. *J. Bone Miner. Res.* 17(Suppl. 1):S216

45. Mahon MJ, Cole JA, Lederer ED, Segre GV. 2003. Na^+/H^+ exchanger-regulatory factor 1 mediates inhibition of phosphate transport by parathyroid hormone and second messengers by acting at multiple sites in opossum kidney cells. *Mol. Endocrinol.* 11:2355–64

46. Seedorf K, Millauer B, Kostka G, Schlessinger J, Ullrich A. 1992. Differential effects of carboxy-terminal sequence deletions on platelet-derived growth factor receptor signaling activities and interactions with cellular substrates. *Mol. Cell Biol.* 12:4347–56

47. Demoulin JB, Seo JK, Ekman S, Grapengiesser E, Hellman U, et al. 2003. Ligand-induced recruitment of Na^+/H^+ exchanger regulatory factor to the PDGF (platelet-derived growth factor) receptor regulates actin cytoskeleton reorganization by PDGF. *Biochem. J.* 376:505–10

48. James MF, Beauchamp RL, Manchanda N, Kazlauskas A, Ramesh V. 2004. A NHERF binding site links the βPDGFR to the cytoskeleton and regulates cell spreading and migration. *J. Cell Sci.* 117:2951–61

49. Hildreth KL, Wu JH, Barak LS, Exum ST, Kim LK, et al. 2004. Phosphorylation of the platelet-derived growth factor receptor β by G protein-coupled receptor kinase-2 reduces receptor signaling and interaction with the Na^+/H^+ exchanger regulatory factor. *J. Biol. Chem.* 279:41775–82

50. Rochdi MD, Watier V, La Madeleine C, Nakata H, Kozasa T, Parent JL. 2002. Regulation of GTP-binding protein α_q ($G\alpha_q$) signaling by the ezrin-radixin-moesin binding phosphoprotein-50 (EBP50). *J. Biol. Chem.* 277:40751–59

51. Kim JH, Lee-Kwon W, Park JB, Ryu SH, Yun CH, Donowitz M. 2002. Ca(2+)-dependent inhibition of Na^+/H^+ exchange 3 (NHE3) requires an NHE3-E3KARP-α-actinin-4 complex for oligomerization and endocytosis. *J. Biol. Chem.* 277:23714–24

52. Sitaraman SV, Wang L, Wong M, Bruewer M, Hobert M, et al. 2002. The adenosine 2b receptor is recruited to the plasma membrane and associates with E3KARP and ezrin upon agonist stimulation. *J. Biol. Chem.* 277:33188–95

53. Oh YS, Jo NW, Choi JW, Kim HS, Seo SW, et al. 2004. NHERF2 specifically interacts with LPA(2) receptor and defines the specificity and efficiency of receptor-mediated phospholipase C-β3 activation. *Mol. Cell Biol.* 11:5069–79

54. Hirakawa T, Galet C, Kishi M, Ascoli M. 2003. GIPC binds to the human lutropin receptor (hLHR) through an unusual PD

domain binding motif, and it regulates the sorting of the internalized human chorio-gonadotropin and the density of cell surface hLHR. *J. Biol. Chem.* 278:49348–57

55. Bomberger JM, Spielman WS, Hall CS, Weinman EJ, Parameswaran N. 2005. RAMP isoform-specific regulation of adrenomedullin receptor trafficking by NHERF-1. *J. Biol. Chem.* 280:23926–35

Annu. Rev. Physiol. 2006. 68:507–41
doi: 10.1146/annurev.physiol.68.072304.114110
Copyright © 2006 by Annual Reviews. All rights reserved
First published online as a Review in Advance on October 31, 2005

Stress Transmission in the Lung:
Pathways from Organ to Molecule

Jeffrey J. Fredberg[1] and Roger D. Kamm[2]

[1]*Department of Environmental Health, Harvard School of Public Health, Boston, Massachusetts 02115; email: jfredber@hsph.harvard.edu*
[2]*Department of Mechanical Engineering, Massachusetts Institute of Technology, Cambridge, Massachusetts 02139; email: rdkamm@mit.edu*

Key Words   force, pressure, remodeling, stretch, tension, cytoskeleton, focal adhesion, airway, alveolus

■ **Abstract**   Gas exchange, the primary function of the lung, can come about only with the application of physical forces on the macroscale and their transmission to the scale of small airway, small blood vessel, and alveolus, where they serve to distend and stabilize structures that would otherwise collapse. The pathway for force transmission then continues down to the level of cell, nucleus, and molecule; moreover, to lesser or greater degrees most cell types that are resident in the lung have the ability to generate contractile forces. At these smallest scales, physical forces serve to distend the cytoskeleton, drive cytoskeletal remodeling, expose cryptic binding domains, and ultimately modulate reaction rates and gene expression. Importantly, evidence has now accumulated suggesting that multiscale phenomena span these scales and govern integrative lung behavior.

INTRODUCTION

This review deals with mechanical stress in the lung at the levels of tissue, cell, and molecule. Our goal is to elucidate biophysical mechanisms that underlie mechanical function at each of these levels, the connections between levels, and their alterability in selective but biologically important departures from the normal state. There are many excellent reviews of lung mechanics on the macroscale (1, 2) and the microscale (3), but for the pulmonary cell biologist that literature may be to some extent unfamiliar, and many of its underlying concepts may not be readily accessible. More importantly, we know of no review that emphasizes the pathway for transmission of mechanical stresses from the pulmonary macroscale to the cell microenvironment, across the cell wall, through the cytoskeleton (CSK), and ultimately to the nucleus. This review is designed to fill that gap.

It is important to remember at the outset that the main function of the lung is gas exchange. For a long time biologists have sought to understand the pathway for oxygen delivery from the atmosphere, through the lung, across the air-blood

0066-4278/06/0315-0507$20.00

barrier, through the circulation, and, ultimately, to the mitochondria, where the oxygen is consumed. To achieve this feat the lung provides branching networks of conducting airways and blood vessels for air and blood to convect to the alveolar septum, and there they come into intimate contact. At that level the surface area is large, the distances are small, and therefore, the main mechanism of gas exchange is passive diffusion. "The pathway for oxygen" and its structure and function have come to be well understood and are summarized best in the text of the same name by Weibel (4).

The primary function of the lung may center on gas exchange, but there are other demands that must be simultaneously met. Some of these demands come into conflict with gas exchange and therefore lead to interesting questions of biological design. Coming to the forefront in that regard are mechanical considerations. More than anything else, the lung is a mechanical organ; the act of breathing is a mechanical process that entails the cyclic application of physical stresses at the pleural surface as well as the transmission of those stresses throughout the lung tissue and its complement of adherent cells. If strain is defined as the resulting length change of a structure per unit initial length, then in the course of a lifetime, the lung, as well as the cells within it, must withstand 10^9 strain cycles with amplitudes that approach 4% during quiet tidal breathing and 10^7 strain cycles with amplitudes that approach 25% during sighs, deep inspirations, or heavy exercise (5). By the standards of common engineering materials, these strains are extreme and would appear to call for tissue structures that are rather substantial. But at the same time, gas exchange is favored when the tissue diffusional barrier is minimal. To explain lung structure in the context of such competing demands, Taylor, Weibel & Hoppeler (6, 7) introduced the principle of symmorphosis. Symmorphosis holds that at each level along the pathway for oxygen, there is enough structure, but just barely enough, to support the required oxygen flux. That is to say, in the biological design of the pathway for oxygen, there is little wasted structure and no overdesign. The closely related issues of the scaling of oxygen demand and oxygen transport with body size are treated in the excellent volumes by Schmidt-Nielson (8) and McMahon (9). In addition to supporting oxygen flux, the lung must provide the appropriate mechanical microenvironment for epithelial cells, endothelial cells and the other 60 resident cell types in between, many of which sense and respond to mechanical stress or strain.

THE LUNG MICROENVIRONMENT AND ITS STATE OF STRESS

The lung is like a balloon that requires an externally generated distending stress to maintain its state of inflation; unlike most other organs, the lung is a pressure-supported structure (10–14). That distending stress is provided by the chest wall and associated muscles of respiration (15–19). At functional residual capacity, the

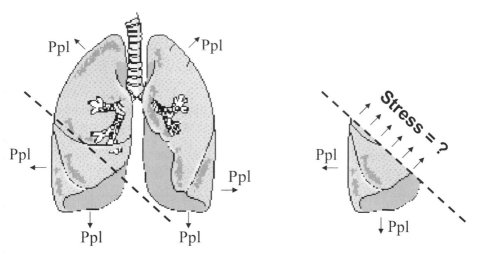

Figure 1   To remain in mechanical equilibrium, the normal (perpendicular) stress acting upon an imaginary cut surface must equal the pleural pressure (Ppl). Because pleural pressure is usually negative, arrows depicting that stress are shown as exerting a distending stress upon the visceral pleura.

lung is distended by an inflating pressure[1] of approximately 2–5 cm H_2O. That is, if the pressure at the airway opening, Pao, were atmospheric and the pressure in the pleural space, Ppl, were 5 cm H_2O below atmospheric, then the inflating pressure— also called the transpulmonary pressure (P_L)—would be 5 cm H_2O; at total lung capacity (TLC) the inflating pressure is approximately 30 cm H_2O. Because P_L = Pao – Ppl, lung-inflating pressure depends upon this pressure difference rather than either Pao or Ppl individually; as far as lung inflation and tissue distension are concerned, it makes no difference whether Pao is increased or Ppl is decreased. For example, during a Valsalva maneuver, Pao, alveolar pressure (P_A), and Ppl can approach 200 cm H_2O, whereas P_L changes hardly at all, merely decreasing slightly secondary to the small decrease of lung volume attributable to alveolar gas compression.

What is the state of distending stress within lung parenchymal tissues? The answer to this question is complicated, but we can make extremely good estimates of the average value of that stress. If we ignore for the moment the effects of gravity (20, 21), the lung is subjected on all of its outside surfaces to a pleural pressure that is roughly uniform (22). We imagine a plane cutting through the lung at an arbitrary position and angle, and then ask, what is the state of stress that must be acting on that imaginary cut face for the cut lung segment to remain in mechanical equilibrium (Figure 1)? Below, we deal with local variations of

---

[1]A note on units: 1 Pa $= 1$ N m^{-2} $= 10$ dynes cm^{-2} $= 1.033 \times 10^{-2}$ cm H_2O.

stress at a finer level of resolution, but here we seek only a coarse-grained average that blurs all microstructural detail. Newton's third law requires that if the lung segment in question is not accelerating, then the sum of all forces acting upon it must be zero. As a result, the average stress acting on the cut face must be exactly the pleural pressure, although over that cut face there may be systematic regional variations of stress near large intrapulmonary airways and blood vessels as well as the visceral pleura (23–31). In this thought experiment, the position and orientation of the cutting plane are arbitrary, and as such, the average state of stress everywhere within the lung must be precisely the pleural pressure. Moreover, that the pleural pressure is usually negative implies a state of tension in the tissues. The parenchymal tissue network transmits this tension from the pleura into the lung tissue.

We next consider this lung parenchyma tissue network at the level of microstructure (Figure 2). It is now well established that the parenchymal microstructure

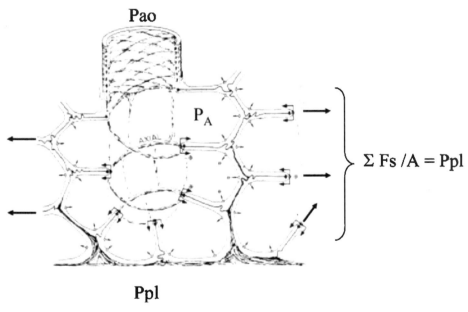

$$\sum Fs /A = Ppl$$

Figure 2   At the level of lung microstucture, alveolar geometry and elastic recoil are set by a balance of interfacial and tissue forces. In the absence of gas flow, gas pressures are everywhere equal, in which case we can set $Pao = P_A = 0$. Although distending stresses are concentrated in and transmitted by discrete alveolar septa, these concentrated septal forces (indicated by *arrows*) can be summed over a parenchymal region encompassing many alveoli and divided by the total cross-sectional area, yielding a distending stress. From Figure 1, this stress must be equal to Ppl. If the pressure on the inside of an airway or an alveolus is Pao and that on the outside is Ppl, then the transmural pressure for either structure is Pao-Ppl or simply P_L, the transpulmonary pressure. Adapted from Reference 13.

functions as a tension-supported lattice (10–14). As described below, the key distending stress—the transpulmonary pressure—is transmitted throughout parenchymal tissues and distends all intrapulmonary structures. This distending stress originates largely in the mechanics of the alveolar surface film and confers most of the parenchymal shear stiffness and the associated forces that are critical in maintaining the shape and function of the organ, including each vessel, airway, and alveolus. At each of these levels this distending stress is essential to lung function; it is a primary determinant of lung stability, the distributions of perfusion and ventilation, ventilation-perfusion matching, airway obstruction, and expiratory flow limitation (32–43).

The central importance of the surface film and the distending stress in these processes are not in dispute, and as described below, the roles of physical forces at the cellular and subcellular level are now becoming well established. But before moving to the cellular level, we ask, what are the stresses that act to distend an alveolus, intrapulmonary airway, or intrapulmonary blood vessel?

The Special Role of Transpulmonary Pressure

Although there are systematic departures, as described below (23–29), the effective stress acting on the outside of each of these structures is the same and approximately equal to the pleural pressure. As Ppl becomes more negative during the course of inspiration, the lung, the airway, the alveolus, and the vessel all experience an increase in distending stress. The effective stress acting on the inside of these structures is quite different, however. In the case of the blood vessel, the internal pressure is the local vascular pressure, Pv, and therefore the pressure difference distending the vessel wall, or the transmural pressure, is Pv–Ppl. During a normal lung inflation, pleural pressure becomes more negative, causing the lung to inflate, vascular transmural pressure to increase, and blood to accumulate in the thoracic cavity. Below we come to the issue of shear stresses acting on the endothelium owing to the flow of blood.

The transmural distending pressures of airways or alveoli are, respectively, Paw – Ppl and P_A – Ppl, where Paw is the local gas pressure in the airway and P_A the local gas pressure in the alveolus. If we limit attention to the case in which gas flow velocity is zero, such as at end-inspiration, end-expiration, or during a breath-hold, all gas in the lung will be at hydrostatic equilibrium and therefore the gas pressure will be everywhere equal and the same as the pressure at the airway opening, such that P_A = Paw = Pao. In that case, the transmural pressure distending an airway or an alveolus will become approximately Pao – Ppl, which we recognize immediately as the transpulmonary pressure. Thus, the transpulmonary pressure is the pressure difference that acts to distend not only the organ but also the alveolus and the airway.

An important departure from this simple description arises because intraparenchymal vessels and airways have mechanical properties different from those of the parenchymal tissues in which they are embedded. In the vicinity of an airway or blood vessel, therefore, parenchymal distortions arise during the act of breathing, and the resulting stresses that act to distend these structures depart systematically from Ppl. This phenomenon is called elastic interdependence (10), and its effects on the distensibility of airways and vessels have been investigated extensively (23–29, 44–46).

STRESS INVARIANTS, LENGTH SCALES, AND TIMESCALES

Several characteristic scales for stress and time can be used to guide our thinking. Indeed, some of these seem to be invariants or regulated set points (Table 1).

Transpulmonary Pressure

At the macro level, the total lung capacity (TLC) in the adult human is achieved when P_L approaches 30 cm H_2O. Interestingly, 0–30 cm H_2O is the working range of transpulmonary pressures independent of lung size, body mass, or even lung development from neonate to adult (47–49). Even more remarkably, in species spanning a wide range of body masses, the working range of transpulmonary pressure is always the same, from 0 to 30 cm H_2O (50–52).

Interfacial and Tissue Elastic Stresses

Moving to a smaller length scale, we encounter the gas-liquid interface and interfacial forces. Most of the transpulmonary pressure is supported by alveolar surface tension; the balance is supported by lung tissue stresses (13, 30, 31, 53, 54). Although the surface tension can be higher during film expansion and approach zero during film compression, across all species the equilibrium surface tension in the alveolus is close to 25 dynes cm^{-1} (54). The law of Laplace states that the pressure difference acting across the gas-liquid interface is proportional to the surface tension and inversely proportional to the radius of curvature of the interface, and across species spanning three decades of body mass, M_b, from bat

TABLE 1 Characteristic levels of stress or force in lung structures

Equilibrium surface tension	25 dynes cm^{-1}
Transpulmonary pressure	0–30 cm H_2O (0–3000 Pa)
Shear stress at vessel wall	$\sim$1 Pa
Tensile stress in CSK	$\sim$$10^{2}$–$10^{5}$ Pa
Stress in focal adhesion	$\sim$5000 Pa
Stress-generating capacity of contractile machinery	$\sim$$10^{5}$ Pa
Forces generated by molecular motors	$\sim$1–10 pN

to dog, alveolar radius increases only weakly, with M_b, going as $M_b^{1/8}$ (9, 55). Therefore, if transpulmonary pressure and equilibrium surface tension are both invariant with body mass, then interfacial forces must play a relatively larger role in smaller species, and conversely (51).

Tissue Frictional Stresses and the Structural Damping Law

When they are deformed, lung tissues generate both elastic and frictional stresses. If tissue frictional stress is expressed as a tissue resistance, then lung resistance can be given as the sum of airway resistance and tissue resistance (56, 57), and at frequencies corresponding to quiet tidal breathing, lung tissue resistance typically exceeds airway resistance (58–60). This description is convenient and widely used, but to describe tissue friction as a flow resistance is to imply that there arises within lung tissues a stress analogous to that which results from the flow of a viscous fluid. The concept of a viscous stress as applied to lung tissue friction is now understood to be erroneous, however (61). Bayliss & Robertson (62) and Hildebrandt (63) demonstrated that frictional stress in lung tissue is dependent upon the amount of expansion but not the rate of expansion, findings that are fundamentally incompatible with the notion that friction is caused by a tissue viscous stress. If not in a viscous stress, how then does tissue friction arise, and how is it properly described?

With different lung tissue compositions and various physiological circumstances, tissue elastic and tissue frictional stresses vary greatly, but the relationship between them turns out to be very nearly invariant; the frictional stress in lung tissues is almost invariably between 0.1 and 0.2 times the elastic stress, where this fraction is called the structural damping coefficient (61). It is a simple phenomenological fact, therefore, that for each unit of peak elastic strain energy that is stored in lung tissue during a cyclic deformation, 10% to 20% of that elastic energy is taxed and lost irreversibly to friction. This fixed relationship holds at the level of the whole lung (60, 64), isolated lung parenchymal tissue strips (65), isolated smooth muscle strips (61, 66), and even isolated living cells (67–69). This close relationship between tissue frictional and elastic stresses is called the structural damping law (61, 63, 70, 71) or sometimes the constant phase model (64). The structural damping law implies that frictional energy loss and elastic energy storage are tightly coupled. As described below (see The Molecular Basis for Lung Mechanics, below), the sawtooth length-extension curve of some proteins offers a clue, but the precise molecular origin of this phenomenon remains unknown (72).

Endothelial Shear Stress and Murray's Law

The flow of blood in the pulmonary vascular tree exerts a shear stress upon the endothelial cell. The shear stress is proportional to the fluid viscosity and the shear rate, where the shear rate increases in rough proportion with blood velocity and in inverse proportion with vessel radius. At each bifurcation within a branching vascular network, the vessel radius tends to decrease slightly; this tendency might thus seem to favor an increased shear stress, but because blood flow is divided into two daughter vessels, the flow velocity deceases as well and thus would seem to favor a decreased shear stress. As described by Murray in 1926 (73), as one proceeds along the vascular tree from larger to smaller vessels, which of these competing factors wins out and how wall shear stress varies with branching generation depend upon details of vessel geometry. Murray considered that issue as a question of biological design; he reasoned that increased vessel radius would favor decreased vascular resistance and thus reduced cardiac work required to pump blood, but increased vessel radius would also require more blood, more tissue mass, and associated metabolic energy costs for maintenance of these tissues. With those two energetic factors in mind, he then calculated the branching configuration that would minimize the rate of net energy utilization needed to support a given flow of blood. He showed that the rate of energy utilization is minimized when the cube of the parent vessel radius is equal to the sum of the cubes of the daughter radii. This result has come to be called Murray's law, and in his result Murray seems to have captured an integrative principle of biology, for not only the vascular tree but also water transport in plants seem to conform to Murray's law (74).

How could this state of affairs come about? It is easy to show that Murray's law implies that the shear stress at the walls of the daughter vessels is equal to that in the parent. The shear stress would be roughly invariant throughout the vascular tree, therefore, because the effects of decreasing radius are just offset by those of decreasing velocity. Although systematic departures have been noted, it is usually a good first approximation that the wall shear stress anywhere in the vascular tree is the nearly the same and assumes a value on the order of 10 dynes cm^{-2}. It is now well established that shear flow over endothelial cells modulates release and/or activation of many mediators, growth factors, cytokines, and nuclear transcription factors (75–77). The endothelium exhibits mechanosensation, therefore, capable of local regulation of a multitude of cellular functions including vascular remodeling. Later in this review we address mechanisms for mechanosensing.

Stresses in the Focal Adhesion Complex

On a still-smaller-length scale, another important stress invariant arises at the level of the focal adhesion (FA). Force transmission between the adherent cell and the extracellular matrix (ECM) is not distributed continuously across the cell membrane but rather is localized to the spot welds called focal adhesions (FAs). The focal adhesion complex is composed of integrins that are attached in the extracellular domain to the ECM and in the cytoplasmic domain to vinculin, talin, α-actinin, paxillin, tensin, and the microfilament terminus (78–80). The focal adhesion complex of an adherent cell may play a role analogous to the peg of a pup-tent; the external tensile distending stress is transmitted through the peg and confers stiffness and shape stability to the tent (Figure 3). Similarly, the adherent cell may be dependent upon external distending stresses to maintain its shape stability (81–84).

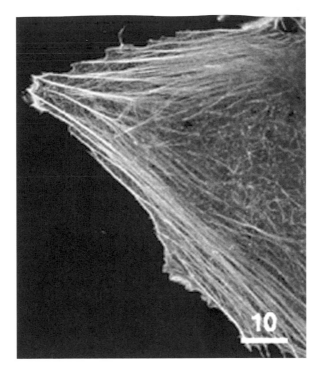

Figure 3 The CSK of the adherent cell appears to be under tension, with tensile stresses supported at the cell boundary by focal adhesions. If deformed, does the cell generate restoring forces mostly by resistance to elongation of line elements, or by reorienting them? Rat fibroblast is stained by antiactin antibody; scale bar = 10 μm.

Cell stiffness or contractile forces must be considered in the context of the ECM in which they reside. In this connection, there are marked differences between cells grown on two-dimensional (2D) versus three-dimensional (3D) substrates (85, 86). When grown on a flat 2D surface, cells initially attach to matrix molecules (e.g., collagen, fibronectin) laid down on the substrate but over time develop a matrix of their own that more closely mimics their in vivo environment. During this process, cells first adhere to the substrate through focal complexes that contain a relatively small number of FA proteins [vinculin, paxillin, focal adhesion kinase (FAK)] and that are roughly 1 μm in size (87, 88). Focal complexes grow into FAs, which serve as a point of attachment between the ECM and the cellular CSK, typically using the $\alpha_v\beta_3$ integrin. These, in turn, evolve into fibrillar adhesions that exert a tension on the matrix laid down by the cell, giving rise to extracellular remodeling through the exposure of cryptic binding sites in fibronectin, causing it to bundle and strengthen. Given ample time, the substrate takes on the nature of a 3D matrix, at which point the adhesions come to resemble those found in vivo, with a long, thin morphology in which paxillin, fibronectin, and α_5 integrin are colocalized (85). This takes time, however typically days—even in a dense cell culture (85). By comparison, cells grown either on or inside a 3D matrix in which the fibrous architecture—filament size, density, matrix compliance—from the on-set resembles the cell environment in vivo immediately form 3D adhesions. Most prominently, these differences are reflected by cells that adhere more strongly to and migrate more rapidly through 3D matrices than to/on 2D substrates and that take on a different, more elongated morphology. Some of these differences are likely attributable to differences in integrin-mediated signaling. For example, autophosphorylation of FAK is downregulated in 3D relative to 2D systems, whereas phosphorylation of two other FA proteins, paxillin and mitogen-activated protein kinase (MAPK), is unaffected (85). Throughout this maturation process, various integrins are used, each with its own unique signaling characteristics, and both a compliant substrate and the $\alpha_5\beta_1$ integrin appear to be critical for the formation of typical 3D adhesions (85, 86).

Contractile Machinery and Cytoskeleton

Airways and blood vessels in the lung are each surrounded by a sheath of smooth muscle (89–91). To within an order of magnitude, virtually all muscles, whether striated or smooth, have the same intrinsic maximal stress-generating capacity, which is on the order of 10^5 Pa (9). The individual myosin motor generates forces that are in the pN range.

Virtually all cells attached to a substrate tend to contract. Although this is an obvious function of skeletal, cardiac, or smooth muscle cells, the role of contraction in nonmuscle cells is less clear. Contractility is, however, an important element in cell migration and therefore in the inflammatory response, wound healing, and metastatic disease. It may also play a fundamental role in maintaining cell adhesions in a state of tension, which may be important for signaling via receptors such as integrins, and is also an important determinant of cytoskeletal tension, as discussed further below.

Much of what we know about the forces that cells exert on the surrounding matrix comes from measurements by a method known as traction microscopy. Dembo & Wang (92) were the first to show that the spatial distribution of these tractions can be mapped from measurements of the displacement field created in a flexible substrate on which the cell is adherent. Measurement of the displacement field is accomplished by tracking small beads, typically 0.2 µm in diameter, embedded near the surface of the substrate gel. To mitigate the extreme computational requirements associated with the method of Dembo & Wang, Butler et al. (93) developed Fourier Transform Traction Microscopy (FTTM) (Figure 4). Although there remains some controversy concerning competing computational approaches (94), FTTM is equally rigorous and has become widely used (84, 95–97). In an alternative approach, elastomer microposts and micropatterned elastic substrates have been used to measure cell tractions (87, 98). It is important to recognize, however, that all three approaches are performed with the cells spread on a flat surface.

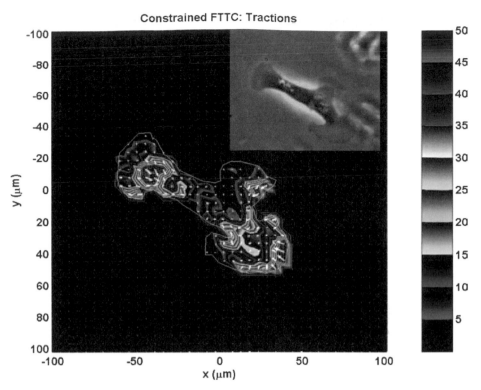

Figure 4   Traction microscopy creates an image of the physical stresses (tractions) applied by an endothelial cell to its substrate, shown here from a study of responses to hypoxia via p38 MAPK and Rho kinase–dependent pathways. Colors show the magnitude of the tractions in Pa; see color scale. Arrows show the directions and relative magnitudes of the tractions. From Reference 97.

Measurements have been made on a variety of cell types, but most often contractile cells are used, including fibroblasts, myocytes (87), or smooth muscle cells (96). In general, cells exert tractions on the substrate that are consistent with intracellular contraction, that tend to be concentrated at FAs, and that tend to be greatest near the outer perimeter of the cell. Total contractile forces have been measured in the range of ~20 nN in fibroblasts but as high as ~70 nN in contracting myocytes (87). Interestingly, the stress acting at individual FAs is found to be relatively constant, in the range of approximately 5.5 kPa (55 cm H_2O), and similar for both cell types (87). This value appears to be invariant, so that as the contractile force exerted by the cell varies, so does the contact area of the FAs and, presumably, the number of integrin receptors involved. The mechanism remains unclear but may include local recruitment of filamin, actin, and talin 1 (99, 100).

Perhaps of greater relevance to the lung, Stamenovic and coworkers (84, 96) measured the contractile stresses generated within airway smooth muscle cells cultured on a 2D surface and found values ranging up to approximately 2 kPa. Note that this is the computed mean stress acting over the cross section of the cell normal to its primary axis of contraction and is not directly comparable to the values for FA stress given above. Other measures in more realistic external environments or with isolated muscle fibers demonstrate that contractile stress is almost invariant, having a value on the order of 10^5 Pa in all contractile tissues (9). Differences between this value and the one for smooth muscle cells grown in culture may be attributable to two facts: (*a*) The CSK in cultured cells is organized more for adhesion than contraction (Figure 3), and (*b*) in 2D culture the level of expression of contractile proteins is reduced (101).

Elastic Similarity

McMahon (9) has suggested a unifying principle of biological design with far-reaching implications and remarkable explanatory power. This principle, called elastic similarity, holds that across species spanning a wide range of body mass, M_b, the bones and muscles are designed so as to bend by approximately the same amount. Although deceptively simple, from elastic similarity it follows more or less directly that energy metabolism scales as $M_b^{3/4}$, respiratory frequency and heart rate as $M_b^{-1/4}$, lung volume and tidal volume as M_b^{1}, alveolar radius as $M_b^{1/8}$, transpulmonary pressure and arterial pressure as M_b^{0}, airway resistance as $M_b^{-3/4}$, lung and chest wall compliance as M_b^{1}, mechanical time constants as $M_b^{1/4}$, and clearance rates and reaction rates as $M_b^{1/4}$; in each instance the indicated relationship falls close to observations. McMahon has pointed out that if time constants go as $M_b^{1/4}$, respiratory frequency goes as $M_b^{-1/4}$, and lung recoil goes as M_b^{0}, this guarantees that lungs of all sizes are given the same number of time constants to empty during a passive expiration. The questions of how bending might be sensed at cellular and molecular levels, and how mechano-sensing cues might trigger remodeling, are addressed below.

CYTOSKELETAL STRUCTURE, STIFFNESS, AND STRESS TRANSMISSION

Although the lung parenchyma (Figure 2) and the CSK (Figure 3) are superficially distinct in structure and widely disparate in scale, they share a deep similarity in the mechanism that confers stability of shape. Ordinary solid materials retain a defined shape in the absence of a distending stress, but not so for pressure-supported structures such as the lung, which would collapse, or the adherent cell, which would round up. Shape stability of both the lung and the adherent cell is determined mainly by their respective distending stresses.

Understanding of the distending stress as the main source of shape stability began approximately 30 years ago, when one of the major conceptual problems in respiratory mechanics was parenchyma resistance to distortion of shape. This problem is of great physiological and pathophsyiological importance because, besides the transpulmonary pressure itself, it is the resistance of the lung parenchyma to shear distortion that is the preeminent factor acting to maintain the patency of intrapulmonary airways, intrapulmonary blood vessels, and individual alveoli. Each of these structures has a natural tendency to collapse, but shear distortion of the surrounding parenchyma creates restoring stresses that act to oppose that collapse (10–14, 102). Importantly, this elastic interdependence creates restoring stresses that are proportional to P_L and comparable in magnitude. The question was, why?

Shape Stability of Stress-Supported Structures

The measure of resistance to shape distortion of a material is the shear modulus (shear stiffness), but early predictions of the shear modulus for lung parenchyma far exceeded observed values (3, 103, 104); given the amounts of its constituents and their material properties, the lung seemed to be far floppier than it reasonably ought to have been. Perhaps more importantly, these initial theories also failed to explain the central organizing feature of lung tissue mechanical behavior, namely that its shear stiffness increases in direct proportion to inflating pressure; the shear modulus of the lung is simply $0.7\ P_L$ (14). On a much smaller scale, an entirely parallel relationship was subsequently shown to hold between the shear stiffness of the CSK and its distending stress (82–84, 96, 105–109).

Kimmell, Kamm, & Shapiro (11) first identified, and Kimmel & Budiansky (12, 110) and Stamenovic (3) later elaborated upon and generalized, the physical basis of such behavior. As applied either to the lung parenchyma or to the CSK, earlier theories had made the traditional assumption that as one proceeds to smaller and smaller scales, the microscale strain of structural members faithfully follows the local macroscale strain of the matrix or network; this condition is called affine deformation. The approach introduced by Kimmel et al. relaxed that assumption and instead considered the discrete nature of stress-bearing members at the microscale and the nature of their connectedness. It is the limited nature of this connectedness that turned out to be the radical feature. Kimmel and coworkers showed that depending on that connectedness, the structural matrix composed of discrete stress-bearing members resists shape distortion not so much by resistance to elongation in those microstructural members, as it must in an affine continuum, but more so by reorientation of those members. When placed under load, the structural elements move relative to one another in a nonaffine manner, changing their orientation and spacing until a new equilibrium configuration is attained. Importantly, the bigger the initial tension carried by those elements, i.e., the bigger the prestress[2] (the inflating pressure in the case of the lung and tractions in the case of the adherent cell, as described below), the smaller the deformation that the structure

---

[2]The terms prestress and distending stress are often used interchangeably. Prestress, in particular, refers to the value of the distending stress before the system is deformed incrementally or before the distending stress is varied incrementally.

must undergo before attaining a new equilibrium configuration. As the prestress increases, therefore, the resistance to a change of shape increases in proportion. Indeed, in such structures the macroscopic shear stiffness can remain finite and small even as the stiffness of microstructural members becomes infinitely large.

This line of thinking explained why the material shear stiffness of lung parenchyma is much smaller than earlier theories had predicted and at the same time revealed why the shear stiffness must increase in direct proportion to the distending stress. The same physics applies at the level of the CSK, where this idea subsumed the controversial tensegrity hypothesis of cell mechanics (111–113), generalized it, and put it on firm theoretical and experimental footing (82–84, 96, 105, 106, 108, 114–117).

A complementary perspective to CSK mechanics derives from the mechanics of polymer solutions or gels (118–122). Each filament in the matrix is treated as a semiflexible biopolymer and is characterized by its bending stiffness as influenced by thermal fluctuations. As with the earlier models, when no initial stress is imposed, as that by external tractions, the theory cannot account for the observed stiffness of the CSK. When an initial stress or strain is imposed, however, these gels progressively stiffen as the fluctuations in individual filaments are pulled out and the polymer chains lose configurational entropy. In both the tensegrity and biopolymer theories, two features are critical: the finite initial prestress imposed by the combined effects of internal contraction and external tethering, and the nonaffine nature of the deformations at the microscale.

Once one understands behavior at the microscale and derives values for macroscopic parameters such as the shear modulus, a continuum approach can once again be adopted to analyze cellular stress and deformation, but all information about strain in individual microstructural elements is then lost (82, 84, 106, 120, 123).

Stress Transmission Pathways Within the Cytoskeleton

What is the pathway of stress transmission within the CSK? To address this question, Hu et al. (124) developed intracellular stress tomography. They coupled a ferromagnetic microbead bound tightly to the CSK via integrins and FA plaques (Figure 5, *pink arrow*). Next they applied a sinusoidal magnetic field, causing a cyclic mechanical torque on the bead and resulting oscillatory distortions of the CSK (in the range of 5–300 nm). They then created images of the synchronous displacements of specific cytoskeletal structures. If elastic moduli of the CSK were homogeneous and isotropic, then the corresponding map of displacement gradients would represent intracellular stress. However, because that assumption is unlikely to be accurate, these maps can only be thought of as a rough approximation to the intracellular stress distribution, although they are no less interesting. If anything, they probably underestimate the degree of stress heterogeneity because the CSK is likely to be both denser and have a higher concentration of stress fibers in the vicinity of FAs. These maps show that in the living adherent cell, the distribution of mechanical strain is punctate and transmitted over long distances through the cytoplasm. They also show convergence of strain fields at the intracellular domain of the FA plaque.

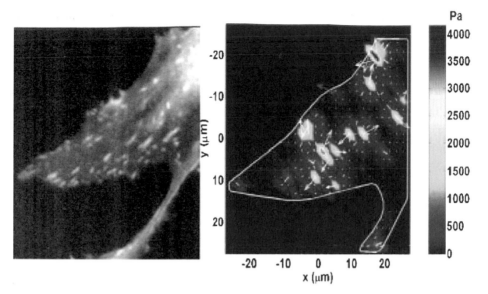

Figure 5 Intracellular stress tomography. (*Left*) Fluorescently labeled paxillin in focal contacts imaged 1 μm above the cell base. (*Right*) Stresses induced in those structures by the cyclic torque of a bead on the cell apex (*pink arrow*). From Reference 124.

Stress Transmission To and Through the Nucleus

Maniotis et al. (125) showed that when integrins are pulled by micromanipulating bound microbeads, cytoskeletal filaments transmit stresses to the cell nucleus and cause nucleoli redistribution. Using the stress tomography method described above, Hu et al. (126) mapped the displacements of the nucleolus, an intranuclear organelle crucial for ribosomal RNA synthesis. Using micropipette aspiration and atomic force microscopy indentation of isolated nuclei, Dahl et al. (127) showed that rheology of the nucleus exhibits power law rheology and implies an infinite spectrum of timescales for structural reorganization; they speculated that this behavior may have fundamental implications for regulating genome expression kinetics. As described below, power law behaviors seem to be a pervasive feature of cytoskeletal dynamics.

SCALE-FREE BEHAVIOR, CYTOSKELETON REMODELING, AND GLASSY DYNAMICS

Recently there has been established a striking analogy in which the CSK of the adherent cell is seen to adjust its mechanical properties and modulate its malleability in much the same way that a glassblower fashions a work of glass (67, 68, 128, 129). However, instead of changing temperature, the cell changes a temperature-like property that has much the same effect.

Fabry et al. (67, 68) first suggested that the CSK may behave as a glassy material. They measured stiffness and friction in a variety of cell types in culture, including the human airway smooth muscle cell. Fabry et al.'s reports contained two surprising results. First, they had set out to identify distinct internal timescales (i.e., molecular relaxation times or time constants) that might reflect molecular dynamics of proteins integrated within the CSK lattice, and in particular they had expected to find relaxation timescales corresponding to myosin-actin cycling rates. Surprisingly, over a spectrum spanning five decades of frequency, f, and in all five of the cell types investigated, they found no characteristic timescales. Rather, stiffness increased as f^{x-1}, implying that relaxation times were distributed as a power law, with a great many relaxation processes contributing when the frequency of the imposed deformation was small but fewer as the frequency was increased and slower processes became progressively frozen out of the response. Thus, no distinct internal timescale could typify protein-protein interactions; all timescales were present simultaneously but distributed very broadly. Because no distinct timescales characterize the response, behavior of this kind is called scale-free. Using other techniques and differing cell types, others have confirmed that CSK dynamics are scale-free (130–134).

Fabry et al.'s second surprise was the discovery of an unexpected relationship among changes in cell mechanics that arise when CSK constituents are modulated. In the living cell, rheology of the CSK matrix is influenced by many structural proteins and their interactions. Despite this complexity, if stiffness or friction data are appropriately scaled and then plotted versus x (which is readily determined from the power law exponent of stiffness versus frequency), all data collapse onto the very same relationship (Figure 6) regardless of the cell type studied, the signal transduction pathway activated, or the particular molecule manipulated (67, 135).

Purely as a matter of phenomenology, the observations described above firmly establish that the parameter x determines where the CSK sits along a continuous spectrum of solid-like versus fluid-like states (68); in the limit that x approaches 1, the behavior approaches that of a Hookean elastic solid, and in the limit that x approaches 2, the behavior approaches that of a Newtonian viscous fluid. Values of x for adherent living cells fall in the range of 1.1 to 1.3, placing them closer to the solid than the fluid state (Figure 6). Moreover, the parameter x was subsequently found to bring together into one phenomenological picture not only cytoskeletal stiffness and friction but also remodeling (128, 129). Moreover, there is a direct relationship between the effective temperature x and the structural damping coefficient η; $\eta = \tan((x-1)\pi/2)$, and when x is close to 1, x is approximately $1 + \eta$.

With regard to mechanism, classical theories of viscoelasticity and equilibrium systems fail to account for these observations (128), and the physical significance of the parameter x remains elusive. The working hypothesis holds that the CSK behaves as glassy system and that the parameter x, which is easily measured, corresponds to the effective temperature of the CSK lattice (67, 128, 129,

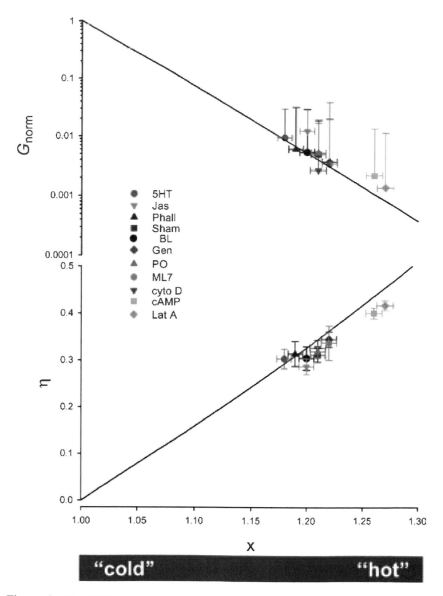

Figure 6   The CSK sits along a continuous spectrum of solid-like versus fluid-like states, where the limit $x = 1$ corresponds to a perfect Hookean elastic solid and the limit $x = 2$ corresponds to a perfect Newtonian viscous fluid. Master curves: (*upper panel*) normalized stiffness $G_{norm} = g'$ (0.75 Hz)$/g_0$ and (*lower panel*) hysteresivity η (0.75 Hz) versus x. The solid lines are as predicted from the equations below: ln $G_{norm} = (x - 1) \ln(\omega/\omega_0)$ (*upper panel*), $\eta = \tan((x - 1)\pi/2)$ (*lower panel*). g_0 and f_0 are determined. Symbols denote baseline, latrunculin A, cytochalasin D, phallacidin, phalloidin oleate, jasplakinolide, genistein, 5-HT, db cAMP, and ML-7. Adapted from Reference 193.

135–137). In that case the system is regarded as being frozen when x is close to 1 and melted when x is close to 2. Although the underlying physics remains poorly understood, strong experimental support for x as an effective temperature of the cytoskeletal lattice has recently been reported (128, 129). In a recent publication, this analogy has been supported by a broader range of biophysical observations (128, 129).

To explain these findings, Bursac et al. (128) have suggested the following physical picture. They imagine trapping of a CSK structural protein or protein complex in a deep energy well, i.e., a well so deep that thermal collisions are insufficient to drive it out of the well (67, 68, 128, 129, 135, 138). If thermal forces are insufficient, then hopping out of the well is imagined to be driven by nonthermal energies that are much larger than k_BT (where k_B is Boltzmann's constant and T is thermodynamic temperature) but can be expressed nonetheless as an "effective temperature" of the CSK matrix. Formally, the term "temperature" means molecular motion; temperature carries with it the connotation of molecular collisions and resulting agitation caused by ongoing molecular bombardments of thermal origin. Here the term "effective temperature" is meant to carry with it a similar connotation of molecular agitation caused, however, by processes that may be of nonthermal origin. A protein conformational change fueled by ATP hydrolysis, for example, releases energy that is greater than thermal energy by 25-fold (139) and does so at the rate of more than 10^4 events per second per cubic micron of cytoplasm (138). Accordingly, such events would have the potential to jostle a neighboring structure rather substantially and dislodge it from a relatively deep energy well.

Physical interactions that lead to trapping may include molecular crowding, hydrogen bonding, or weak cross-linking (140–144). With regard to molecular crowding, the volume fraction of macromolecules within the cell approaches the maximum packing fraction that is possible; the mean distance between macromolecules is only 2 nm (145). The molecular space within a cell is so crowded, in fact, that crowding can change binding affinities between specific proteins by orders of magnitude (142, 144, 146, 147). If trapping is due to nonspecific mechanisms such as molecular crowding or hydrophobic interactions, then the specific properties of cross-linking molecules may be of only secondary importance. But if trapping has its origins in cross-linking of CSK filaments, then the density and the specific type of cross-links and CSK filaments, as well as the dynamics of their specific interactions, may be crucial (120). Many of the cytoskeletal modulations used to date alter both macromolecular packing and the dynamics of cross-linked filament networks (148), and thus they cannot determine which of the two interactions may play a dominant role. Regardless of the particular interaction, trapping is hypothesized to arise owing to the insufficiency of thermal energy to drive CSK structural rearrangements (140). The collapse of data onto master curves (Figure 6) suggests that specific molecular interactions influence CSK stiffness and friction, but only to the extent that these interactions can modify the effective temperature.

Is the Effective Temperature Set by the Distending Stress?

Recently reported data from mechanical measurements of cultured cells show that stiffness of the cytoskeletal matrix is determined by the extent of distending stress borne by the CSK, where distending stress can be changed either by modulating the contractile state of the cell or by mechanical stretch (82–84, 96, 105, 106, 108, 114–117, 149). At the same time, rheological measurements show that cytoskeletal stiffness changes with frequency of imposed mechanical loading according to a power law (150). Stamenovic et al. (150) and Trepat et al. (149) examined the possibility that these two empirical observations might be interrelated. Their findings revealed that the transition between solid-like and liquid-like states of the CSK, and the effective temperature that describes that transition, seemed to be controlled by cytoskeletal distending stress. They took this result to imply that the depth of the energy wells described above must increase as cytoskeletal tension increases. A similar behavior was subsequently reported in tensed cross-linked actin gels (120), and an alternate explanation derived from the theory of semiflexible biopolymers has been proposed (123).

SENSATION OF FORCE: MECHANOSENSING

That cells sense and respond to externally applied or internally generated stresses is now widely recognized as an essential feature of cell function. Endothelial cells, for example, respond to hemodynamic shear stresses as low as 1 Pa through activation of Src, FAK, and MAPKs. Signals are often mediated by the $\alpha_v \beta_3$ and $\alpha_5 \beta_1$ integrins (151) and involve the small Rho GTPases. Consequences of stimulation are many and varied but include intracellular reorganization, changes in protein expression, and the regulation of proteins involved in extracellular remodeling. Many of these same effects can be observed in endothelial cells subjected to mechanical strain, so the response to stress may be relatively independent of the manner in which force is applied.

Candidates for the transducer that converts mechanical force into a biochemical signal include mechosensitive ion channels whose conductivity changes in response to stresses in the cell membrane (152, 153), as well as other force-induced changes in protein conformation that may alter either binding affinity or kinase activity of proteins subjected to force. Mechanosensitive ion channels presumably change their conductance either as a result of stresses within the cell membrane or as a consequence of forces transmitted directly to the channel protein(s) via membrane-associated proteins. An example of the latter is found in hair cells of the inner ear, in which forces are transmitted from the tip of one stereocilium to a channel on the side of a neighboring one via a thin connecting filament, the tip link. As the collection of stereocilia is displaced by vibrations in the cochlear fluid, forces in the tip link are transmitted to the channel, causing it to change its conductance. In other mechanosensitive channels, such as the mechanosensitive channel of large conductance, it is thought that membrane tension itself directly mediates channel opening. Support for this mechanism comes from molecular dynamic simulation, although the simulated changes in pore dimension still appear insufficient to explain the entire process of channel activation (154, 155).

In the lung, current evidence points to a role for ion pumps in the activation of alveolar epithelial cells by stretch (153). Through the use of epithelial cells grown on a flexible substrate to simulate the cyclic stretch that these cells experience during normal or assisted breathing, Na^+ pump activity was found to increase in proportion to the magnitude of stretching (153). This may be a beneficial response to increases in epithelial layer permeability associated with damage to the alveolar wall from overdistention (156, 157). Interestingly, cells cultured for five days, sufficient time to lay down a matrix of their own, responded only when the levels of substrate stretch were very high, corresponding to an approximately 37% increase in surface area (153). Although these levels of strain may seem high, direct measurement of changes in epithelial basement membrane area can be as high as 40% (corresponding to a linear strain of 18%) when lung volume increases from 24% TLC to 100% TLC (158). Cells subjected to noncyclic, steady stretch showed no increase in pump activity, suggesting that the cells can accommodate to static changes in area, but not to cyclic ones.

Other studies have examined the role of increased production of reactive oxygen species in ventilator-induced lung injury (159), which in turn can lead to increased production of inflammatory cytokines (160) and increased neutrophil sequestration (161). Again, cyclic stretch, this time a 15–30% biaxial strain (32–69% increase in surface area) was used, and such large strains may contribute to either cell detachment or loss of cell viability. Although the underlying mechanism causing the change in the production of reactive oxygen species remains unclear, it may involve the possible role of changes in mitochondrial shape or size (159), as may occur in endothelial cells (162, 163).

Much of the stress-induced signaling originates in FAs, and these are also the sites at which forces tend to be focused (77, 124). Thus, many have speculated that FAs represent also the location of mechanosensation, the biological process that transduces a mechanical force into a biochemical signal. It is hardly surprising that some of the same conformational changes that occur as a result of ligand-receptor binding, for example, may be brought about by forces transmitted by these same proteins, especially those in the FA complex. That such conformational changes typically lead to nanometer-scale domain movements under forces on the order of 10 pN lends indirect support for this concept. Corresponding energies for such deformations are then in the range of 10 pN·nm, or a few multiples of the thermal energy, $k_B T$—large enough to avoid frequent, unintended thermal activation but low enough so that levels of force typical of those transmitted through an FA are sufficient to cause activation. Compare this to estimates that each integrin in a mature FA under normal cell tension experiences a force of at least 1 pN (87), given the assumption that the integrins are close-packed. In separate experiments, it has been demonstrated that forces as small as 2 pN can break bonds in an initial contact (164), thought to be a bond between talin-1 and actin filaments, whereas subsequent strengthening under force, presumably by the recruitment of vinculin, can increase bond strengths to more than 60 pN. Proteins such as titin unfold or denature with forces in the range of tens to several hunded pN, depending on the rate of increase in force, and many important molecular bonds (e.g., biotin-streptavidin) rupture at forces in the range of 100 pN (reviewed in Reference 165). Therefore, the range of force between 1–100 pN and energies of 10–100 pN·nm are likely crucial in the control of cellular processes by mechanical force.

Experimental evidence for the role of conformational change in mechanosensing is gradually accumulating. In one set of experiments, Sawada & Sheetz (166) grew cells on a flexible substrate and then incubated them in a detergent to remove the cell membranes and thereby eliminate any potential influence of membrane ion channels. Then, comparing the proteins that bound to these exposed CSKs with the substrate stretched or relaxed, they found significant differences, notably that FAK, paxillin, and p130Cas (166) preferentially bound to the stretched substrate. In a separate set of experiments, Src activation was imaged directly and locally as force was applied to a cell through a tethered bead with an optical trap (167). The response was immediate and generally localized to the FAs experiencing force. Interestingly, the researchers observed a wave of Src activation emanating from the site of force application, propagating at a speed of approximately 18 nm s^{-1}. Although these experiments are not capable of identifying the mechanism that initiates the signal, they do demonstrate that the signaling is immediate, is initiated in the regions of high force, and can be propagated to other regions within the cell.

More direct experimental and computational evidence, through the use of molecular dynamics, demonstrates that force application can expose cryptic binding sites, for example in fibronectin (168), leading to bundling of this ECM protein. Similar cryptic binding sites have been identified in several intracellular proteins—vinculin (169), α-actinin (170), and potentially talin (171, 172)—all present in a FA. Although not compelling, this at least suggests that similar phenomena may occur intracellularly, leading either to mechanical reinforcement or the initiation of a signaling cascade.

Cytoskeletal changes can also occur rapidly and may be locally mediated. Experiments on human neutrophils subjected to elongation in narrow channels comparable in size to the pulmonary capillaries exhibit a rapid (within seconds) and significant (by a factor of two) reduction in shear modulus (173). This drop in modulus can be reversed on a timescale of less than one minute. Such deformations give rise to numerous changes, such as increased calcium ion concentration and increased expression of two adhesion receptors, CD18 and CD11b (174), but the change in modulus appears to be linked to a transient reduction in polymerized actin (175). These experiments highlight the dynamic nature of force-induced changes in cell structure and also point to the limitations of assuming that the cell can be treated, even on relatively short timescales, as an inert material with constant properties.

Changes in protein conformation or modulation of channel conductivity are not the only mechanisms by which a cell transduces stress. Recent studies demonstrate that as the transepithelial pressure rises—as during airway constriction, when epithelial folds push the opposing epithelial surfaces against one another—the resulting stresses cause the spaces between neighboring cells to shrink, squeezing out some of the intracellular fluid and bringing the lateral cell membranes into closer proximity. One consequence of this reduction in the lateral intercellular space is that the concentration of ligands shed from the lateral membrane, specifically heparin-binding epidermal growth factor, increases owing to the diminished width of the pathway for diffusion out of the lateral intercellular space, leading to enhanced activation of the EGF receptor (176). This then initiates a signaling cascade characterized by increased phosphorylation of the MAPK ERK1/2 within 15 minutes, and longer-term upregulation of early growth response-1 and transforming growth factor-β (177). These upregulated proteins also stimulate fibroblasts in close proximity to increase their collagen production (178), contributing to airway wall remodeling, one of the hallmarks of asthma.

Remodeling also occurs in the pulmonary vasculature. Vascular endothelial cells respond to stress in two stages. In the early stage, elevated shear stress due to increased blood flow, potentially in combination with other factors such as hypoxia, is thought to initiate a process that leads to vasoconstriction and possibly proliferation of vascular wall cells (179). This consequently leads to elevated resistance to blood flow and increased pulmonary arterial pressure, initiating the second stage. Increased hoop stress associated with higher arterial pressure (*a*) stimulates the endothelium to initiate a process that leads to increased levels of collagen synthesis (180) and (*b*) influences release, from the endothelium, of factors that ultimately produce smooth muscle hypertrophy and adventitial thickening. These processes cause the observed arterial wall remodeling (181). Thus, the process of arterial wall remodeling appears to be influenced strongly by the forces experienced by the pulmonary arteries, an effect that is predominantly mediated by the arterial endothelium.

THE MOLECULAR BASIS FOR LUNG MECHANICS

Whether we consider the extracellular matrix, whose structural integrity is primarily conferred by the collagen and elastin fibers, or the matrix of resident cells, whose stiffness is derived from the CSK of actin microfilaments, intermediate filaments, and microtubules, a full understanding of mechanical stiffness requires a look at the individual molecular constituents. In addition, as we discuss above, most of the fundamental mechanisms by which cells sense and respond to mechanical stimulus occur at the level of single molecules, such as those that comprise the FA complex.

In principle, all macroscopic elastic and frictional properties of cells or tissues can be derived from knowledge of single-molecule mechanics and the manner in which the constituent molecules assemble into a 3D matrix. In practice, we are far from achieving that goal but have begun to take some important steps. Here we use, as an example, the microfilament matrix of the CSK to show how, once we attain a molecularly based appreciation of the structures, we can use this foundation to expand our understanding of critical mechanical properties.

Two evolving technologies form the basis for this approach: molecular dynamics simulation and experimental measurements on single molecules. Nearly a decade ago, the first measurements of the elasticity of a single protein were made using an atomic force microscope (AFM) tip (182) or a bead in an optical trap (183) functionalized so that it bound to the molecule of interest. One of the earliest proteins studied was titin, which was thought to be responsible for much of the elastic character of passive or relaxed skeletal and cardiac muscle. Through these initial and subsequent investigations, investigators revealed that titin consists of three elastic regions called the PEVK (rich in proline, glutamate, valine, lysine) domains, fibronectin type III domains, and the immunoglobulin (Ig)-like β-barrel domains. The first two of these behave as entropic springs (their resistance to stretch is due to a loss of configurational entropy), and the third, which is initially entropic in nature, exhibits a sawtooth pattern of force plotted against extension if stretched excessively (Figure 7A) (182). Each peak in the sawtooth pattern corresponds to the point at which one of the Ig domains reaches its limit, at approximately 250 pN with a pulling rate of 1 μm s^{-1}, and unfolds to an extended configuration, as depicted for a fibronectin domain in Figure 7B (194). In this manner, and owing to the stochastic nature of domain unfolding, a collection of titin (or fibronectin) molecules arranged in parallel, as in a sarcomere of skeletal muscle, would exhibit a constant force resisting pulling, while lengthening many times the initial length of the individual folded domains. As it turns out, domain unfolding is not thought to occur frequently in muscle under normal physiological conditions (185), but the same behavior has been discovered in several other structural proteins, notably spectrin (186), known for its role in the cortex of a red blood cell, and filamin A (187), one of the primary actin cross-linking agents in the CSK. Molecular dynamics simulations of a single Ig domain subsequently revealed the rich detail of behavior that occurs during unfolding and refolding as well as subtle differences between variations of the domain (188). One drawback of steered molecular dynamics (SMD), however, is that it is currently possible to simulate only a very short time period, on the order of 1 ns, so the forces needed to cause unfolding are orders of magnitude higher ($\sim$2000 pN) than those observed in the corresponding experiments.

In the case of filamin A (previously known as ABP-280), the molecular structure is particularly interesting and may be an essential element in cytoskeletal stiffness. Filamin A is a homodimer that is joined at its subunits' C-terminal domains and that binds with actin, at the end of an extended rod region comprised largely of Ig domains similar to those in titin, to an N-terminal domain. Because of the conformation of the protein and the nature of its binding to actin, filamin A tends to form an isotropic matrix, especially in fibroblasts and macrophages, with actin filaments crossing at nearly right angles. Another important feature of filamin A

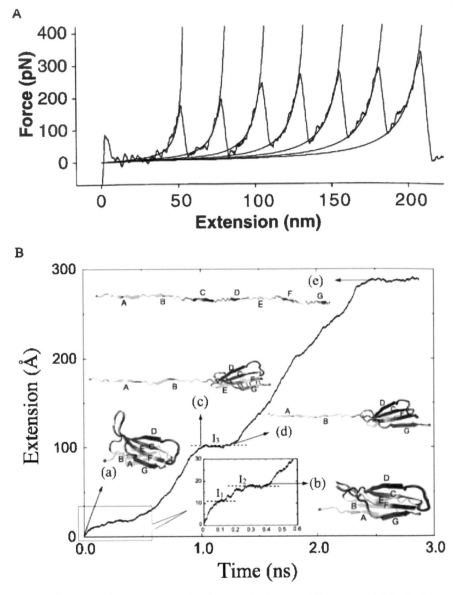

Figure 7   Stretching a single molecule. (*A*) The "sawtooth" pattern exhibited when a single fragment of titin (Ig8) is pulled using an AFM probe and plotting applied force against protein extension. Smooth lines show agreement of each segment of the curve to the worm-like chain model equation. From Reference 182. (*B*) A plot of extension against time obtained by steered molecular dynamics simulation of a single fibronectin domain when pulled at a constant force of 500 pN. Sketches show the conformations that are observed at different stages of domain unfolding from its native state (*a*) to the completely extended state (*e*). From Reference 194.

is a flexible hinge domain located about halfway along each of the two "legs" of the dimer.

Experimental measurements on single filamin A proteins exhibit the same sawtooth force-displacement character found initially with titin (Figure 7A), with domain unfolding occurring at a force in the range of 50–220 pN depending on the rate of pulling (189). Because these domains can reversibly refold when the force is relaxed, this unfolding/refolding process constitutes a means by which two cross-linked actin filaments can slide some distance relative to each other, yet remain bound so that they can spring back to their original position when the force is released. The role of the hinged domain has not yet been elucidated, but it, too, likely contributes to the unique characteristics of CSK elasticity.

Coupled with knowledge of the protein structure, these experimental results can provide enormous insight into the molecular basis for elastic behavior. Although a high-resolution structure of the entire filamin A (which can be obtained either by X-ray crystallography or NMR spectroscopy) has not yet been published, we do know the structure of some of the individual Ig domains (190) and can use the method of SMD to simulate the unfolding process. This has been done for titin (184), fibronectin (194), and spectrin (191), but not yet for filamin, thus allowing for the identification of the individual molecular bonds that essentially provide the glue holding the protein together and that ultimately rupture, allowing the domain to unfold. SMD has also been used to demonstrate that forces can expose cryptic, or inaccessible, binding sites, leading, in the case of fibronectin for example, to filament bundling and strengthening of the fibronectin matrix as a direct consequence of applied stress. These experimental results also provide insight into the molecular basis for frictional behavior and the structural damping law; a sawtooth force extension curve (Figure 7A) implies that energy dissipated upon rupture is tightly coupled to the elastic strain energy stored prior to rupture (61).

CONCLUDING PERSPECTIVE

With knowledge of the elastic character of each of the constituent proteins, one can now envision constructing a model that incorporates both the elastic properties of the individual proteins (e.g., actin and its cross-linker, filamin A) and their geometrical arrangements. In cross-linked actin gels, for example, actin filaments behave as a collection of entropic springs at low cross-link densities, but their enthalpic or mechanical stiffness (e.g., bending and extensional stiffness) becomes increasingly important with increased cross-linking and filament bundling into stress fibers. As with filamin A, actin filament elasticity is determined by the protein conformation and intermolecular bonds, giving rise to an effective elastic modulus of approximately 1.8 GPa (192). Microstructural models that incorporate these characteristics, not simply those of the CSK but also those of the extracellular matrix, are just now beginning to be developed (120), but this field will certainly grow along with our knowledge of molecular structure and our computational capacity for molecular dynamics simulation. Similarly, molecular-level models for protein conformational change and the consequent changes in binding affinity or kinase activity will soon be developed.

In this review we have emphasized the pathway for force transmission from organ to molecule and shown evidence to suggest that even at the smallest scales, physical forces serve to distend the CSK, drive cytoskeletal remodeling, expose cryptic binding domains, and ultimately modulate reaction rates and gene expression. Evidence has now accumulated suggesting that multiscale phenomena span these scales and govern integrative lung behavior, but it is important to recognize that a sequence of reductionist models, each at a different scale and piled one upon the next, does not necessarily add up to a multiscale model or a multiscale phenomenon. Multiscale phenomena are those that intrinsically span scales; prime examples are symmorphosis, elastic similarity, structural damping, Murray's law, and glassy dynamics. Each of these represents an integrative biological phenomenon whose molecular basis remains largely unknown.

ACKNOWLEDGMENTS

We gratefully acknowledge support from grants HL33009, HL65960, HL59682, and HL64858 from the National Heart, Lung, and Blood Institute.

The *Annual Review of Physiology* is online at
http://physiol.annualreviews.org

LITERATURE CITED

1. Mead J. 1973. Respiration: pulmonary mechanics. *Annu. Rev. Physiol.* 35:169–92

2. Otis AB. 1983. A perspective of respiratory mechanics. *J. Appl. Physiol.* 54:1183–87

3. Stamenovic D. 1990. Micromechanical foundations of pulmonary elasticity. *Physiol. Rev.* 70:1117–34

4. Weibel E. 1984. *The Pathway for Oxygen.* Cambridge, MA: Harvard University Press. 425 pp.

5. Gump A, Haughney L, Fredberg J. 2001. Relaxation of activated airway smooth muscle: relative potency of isoproterenol vs. tidal stretch. *J. Appl. Physiol.* 90:2306–10

6. Weibel ER, Taylor CR, Hoppeler H. 1991. The concept of symmorphosis: a testable hypothesis of structure-function relationship. *Proc. Natl. Acad. Sci. USA* 88:10357–61

7. Weibel ER, Taylor CR, Hoppeler H. 1992. Variations in function and design: testing symmorphosis in the respiratory system. *Respir. Physiol.* 87:325–48

8. Schmidt-Nielson K. 1984. *Scaling: Why Is Animal Size So Important?* New York: Cambridge Univ. Press. 241 pp.

9. McMahon TA. 1984. *Muscles, Reflexes, and Locomotion.* Princeton, NJ: Princeton Univ. Press. 331 pp.

10. Mead J, Takishima T, Leith D. 1970. Stress distribution in lungs: a model of pulmonary elasticity. *J. Appl. Physiol.* 28:596–608

11. Kimmel E, Kamm RD, Shapiro AH. 1987. A cellular model of lung elasticity. *J. Biomech. Eng.* 109:126–31

12. Kimmel E, Budiansky B. 1990. Surface

tension and the dodecahedron model for lung elasticity. *J. Biomech. Eng.* 112:160–67

13. Wilson TA, Bachofen H. 1982. A model for mechanical structure of the alveolar duct. *J. Appl. Physiol.* 52:1064–70

14. Lai-Fook SJ, Wilson TA, Hyatt RE, Rodarte JR. 1976. Elastic constants of inflated lobes of dog lungs. *J. Appl. Physiol.* 40:508–13

15. Boriek AM, Wilson TA, Rodarte JR. 1994. Displacements and strains in the costal diaphragm of the dog. *J. Appl. Physiol.* 76:223–29

16. Boriek AM, Liu S, Rodarte JR. 1993. Costal diaphragm curvature in the dog. *J. Appl. Physiol.* 75:527–33

17. Macklem PT, Macklem DM, De Troyer A. 1983. A model of inspiratory muscle mechanics. *J. Appl. Physiol.* 55:547–57

18. Loring SH, Mead J. 1982. Action of the diaphragm on the rib cage inferred from a force-balance analysis. *J. Appl. Physiol.* 53:756–60

19. Sprung J, Deschamps C, Hubmayr RD, Walters BJ, Rodarte JR. 1989. In vivo regional diaphragm function in dogs. *J. Appl. Physiol.* 67:655–62

20. Faridy EE, Kidd R, Milic-Emili J. 1967. Topographical distribution of inspired gas in excised lobes of dogs. *J. Appl. Physiol.* 22:760–66

21. Milic-Emili J, Henderson JA, Dolovich MB, Trop D, Kaneko K. 1966. Regional distribution of inspired gas in the lung. *J. Appl. Physiol.* 21:749–59

22. Hoppin FG Jr, Green ID, Mead J. 1969. Distribution of pleural surface pressure in dogs. *J. Appl. Physiol.* 27:863–73

23. Hajji MA, Wilson TA, Lai-Fook SJ. 1979. Improved measurements of shear modulus and pleural membrane tension of the lung. *J. Appl. Physiol.* 47:175–81

24. Kallok MJ, Lai-Fook SJ, Hajji MA, Wilson TA. 1983. Axial distortion of airways in the lung. *J. Appl. Physiol.* 54:185–90

25. Lai-Fook SJ. 1979. A continuum mechanics analysis of pulmonary vascular inter-dependence in isolated dog lobes. *J. Appl. Physiol.* 46:419–29

26. Lai-Fook SJ. 1981. Elasticity analysis of lung deformation problems. *Ann. Biomed. Eng.* 9:451–62

27. Lai-Fook SJ, Hyatt RE. 1979. Effect of parenchyma and length changes on vessel pressure-diameter behavior in pig lungs. *J. Appl. Physiol.* 47:666–69

28. Lai-Fook SJ, Hyatt RE, Rodarte JR. 1978. Effect of parenchymal shear modulus and lung volume on bronchial pressure-diameter behavior. *J. Appl. Physiol.* 44:859–68

29. Lai-Fook SJ, Kallok MJ. 1982. Bronchial-arterial interdependence in isolated dog lung. *J. Appl. Physiol.* 52:1000–7

30. Oldmixon EH, Hoppin FG Jr. 1984. Comparison of amounts of collagen and elastin in pleura and parenchyma of dog lung. *J. Appl. Physiol.* 56:1383–88

31. Oldmixon EH, Carlsson K, Kuhn C 3rd, Butler JP, Hoppin FG Jr. 2001. α-actin: disposition, quantities, and estimated effects on lung recoil and compliance. *J. Appl. Physiol.* 91:459–73

32. Otis AB, McKerrow CB, Bartlett RA, Mead J, McIlroy MB, et al. 1956. Mechanical factors in distribution of pulmonary ventilation. *J. Appl. Physiol.* 8:427–43

33. Engel LA, Landau L, Taussig L, Martin RR, Sybrecht G. 1976. Influence of bronchomotor tone on regional ventilation distribution at residual volume. *J. Appl. Physiol.* 40:411–16

34. Crawford AB, Cotton DJ, Paiva M, Engel LA. 1989. Effect of lung volume on ventilation distribution. *J. Appl. Physiol.* 66:2502–10

35. Wagner PD, Saltzman HA, West JB. 1974. Measurement of continuous distributions of ventilation-perfusion ratios: theory. *J. Appl. Physiol.* 36:588–99

36. Wagner PD, Laravuso RB, Uhl RR, West JB. 1974. Continuous distributions of ventilation-perfusion ratios in normal subjects breathing air and 100 per cent O2. *J. Clin. Invest.* 54:54–68

37. Mead J, Turner JM, Macklem PT, Little JB. 1967. Significance of the relationship between lung recoil and maximum expiratory flow. *J. Appl. Physiol.* 22:95–108

38. Hubmayr RD, Walters BJ, Chevalier PA, Rodarte JR, Olson LE. 1983. Topographical distribution of regional lung volume in anesthetized dogs. *J. Appl. Physiol.* 54:1048–56

39. West JB, Dollery CT. 1960. Distribution of blood flow and ventilation-perfusion ratio in the lung, measured with radioactive carbon dioxide. *J. Appl. Physiol.* 15:405–10

40. West JB, Dollery CT, Naimark A. 1964. Distribution of blood flow in isolated lung; relation to vascular and alveolar pressures. *J. Appl. Physiol.* 19:713–24

41. Hughes JM, Glazier JB, Maloney JE, West JB. 1968. Effect of lung volume on the distribution of pulmonary blood flow in man. *Respir. Physiol.* 4:58–72

42. Permutt S, Bromberger-Barnea B, Bane HN. 1962. Alveolar pressure, pulmonary venous pressure, and the vascular waterfall. *Med. Thorac.* 19:239–60

43. Skloot G, Permutt S, Togias A. 1995. Airway hyperresponsiveness in asthma: a problem of limited smooth muscle relaxation with inspiration. *J. Clin. Invest.* 96:2393–403

44. Hoppin FG Jr, Lee GC, Dawson SV. 1975. Properties of lung parenchyma in distortion. *J. Appl. Physiol.* 39:742–51

45. Sasaki H, Hoppin FG Jr, Takishima T. 1978. Peribronchial pressure in excised dog lungs. *J. Appl. Physiol.* 45:858–69

46. Elad D, Kamm RD, Shapiro AH. 1988. Tube law for the intrapulmonary airway. *J. Appl. Physiol.* 65:7–13

47. Mortola JP, Rossi A, Zocchi L. 1984. Pressure-volume curve of lung and lobes in kittens. *J. Appl. Physiol.* 56:948–53

48. Stocks J. 1999. Respiratory physiology during early life. *Monaldi Arch. Chest Dis.* 54:358–64

49. Shen X, Ramchandani R, Dunn B, Lambert R, Gunst SJ, Tepper RS. 2000. Effect of transpulmonary pressure on airway diameter and responsiveness of immature and mature rabbits. *J. Appl. Physiol.* 89:1584–90

50. Schroter RC. 1980. Quantitative comparisons of mammalian lung pressure volume curves. *Respir. Physiol.* 42:101–7

51. Mercer RR, Russell ML, Crapo JD. 1994. Alveolar septal structure in different species. *J. Appl. Physiol.* 77:1060–66

52. Agostoni E, D'Angelo E. 1970. Comparative features of the transpulmonary pressure. *Respir. Physiol.* 11:76–83

53. Bachofen H, Hildebrandt J, Bachofen M. 1970. Pressure-volume curves of air- and liquid-filled excised lungs-surface tension in situ. *J. Appl. Physiol.* 29:422–31

54. Bachofen H, Schurch S, Urbinelli M, Weibel ER. 1987. Relations among alveolar surface tension, surface area, volume, and recoil pressure. *J. Appl. Physiol.* 62:1878–87

55. Tenney SM, Remmers JE. 1963. Comparative quantitative morphology of the mammalian lung: diffusing area. *Nature* 197:54–56

56. Ludwig MS, Dreshaj I, Solway J, Munoz A, Ingram RH Jr. 1987. Partitioning of pulmonary resistance during constriction in the dog: effects of volume history. *J. Appl. Physiol.* 62:807–15

57. Peslin R, Fredberg J. 1986. Oscillation mechanics of the respiratory system. In *Handbook of Physiology: The Respiratory System III, Respiration*, PT Macklem, J Mead, eds, pp. 145–77. Bethesda, MD: Am. Physiol. Soc.

58. Ludwig MS, Robatto FM, Simard S, Stamenovic D, Fredberg JJ. 1992. Lung tissue resistance during contractile stimulation: structural damping decomposition. *J. Appl. Physiol.* 72:1332–37

59. Nagase T, Dallaire MJ, Ludwig MS. 1996. Airway and tissue behavior during early response in sensitized rats: role of 5-HT and LTD4. *J. Appl. Physiol.* 80:583–90

60. Jensen A, Atileh H, Suki B, Ingenito EP, Lutchen KR. 2001. Airway caliber in

healthy and asthmatic subjects: effects of bronchial challenge and deep inspirations. *J. Appl. Physiol.* 91:506–15

61. Fredberg JJ, Stamenovic D. 1989. On the imperfect elasticity of lung tissue. *J. Appl. Physiol.* 67:2408–19

62. Bayliss L, Robertson G. 1939. The viscoelastic properties of the lungs. *Q. J. Exp. Physiol.* 29:27–47

63. Hildebrandt J. 1969. Comparison of mathematical models for cat lung and viscoelastic balloon derived by Laplace transform methods from pressure-volume data. *Bull. Math Biophys.* 31:651–67

64. Hantos Z, Daroczy B, Suki B, Nagy S, Fredberg JJ. 1992. Input impedance and peripheral inhomogeneity of dog lungs. *J. Appl. Physiol.* 72:168–78

65. Fredberg JJ, Bunk D, Ingenito E, Shore SA. 1993. Tissue resistance and the contractile state of lung parenchyma. *J. Appl. Physiol.* 74:1387–97

66. Fredberg JJ, Jones KA, Nathan M, Raboudi S, Prakash YS, et al. 1996. Friction in airway smooth muscle: mechanism, latch, and implications in asthma. *J. Appl. Physiol.* 81:2703–12

67. Fabry B, Maksym GN, Butler JP, Glogauer M, Navajas D, Fredberg JJ. 2001. Scaling the microrheology of living cells. *Phys. Rev. Lett.* 87:148102

68. Fabry B, Maksym GN, Butler JP, Glogauer M, Navajas D, et al. 2003. Time scale and other invariants of integrative mechanical behavior in living cells. *Phys. Rev. E Stat. Nonlin. Soft Matter Phys.* 68:041914

69. Fabry B, Maksym GN, Shore SA, Moore PE, Panettieri RA Jr, et al. 2001. Time course and heterogeneity of contractile responses in cultured human airway smooth muscle cells. *J. Appl. Physiol.* 91:986–94

70. Crandal S. 1970. The role of damping in vibration theory. *J. Sound Vibrat.* 11:3–18

71. Fung Y. 1988. *Biomechanics: Mechanical Properties of Living Tissues.* New York: Springer-Verlag. 433 pp.

72. Hubmayr RD. 2000. Biology lessons from oscillatory cell mechanics. *J. Appl. Physiol.* 89:1617–18

73. Murray C. 1926. The physiological principle of minimum work. I. The vascular system and the cost of blood volume. *Proc. Natl. Acad. Sci. USA* 12:207–14

74. McCulloh KA, Sperry JS, Adler FR. 2003. Water transport in plants obeys Murray's law. *Nature* 421:939–42

75. Chien S, Shyy JY. 1998. Effects of hemodynamic forces on gene expression and signal transduction in endothelial cells. *Biol. Bull.* 194:390–91

76. Nollert MU, Panaro NJ, McIntire LV. 1992. Regulation of genetic expression in shear stress-stimulated endothelial cells. *Ann. NY Acad. Sci.* 665:94–104

77. Bershadsky AD, Balaban NQ, Geiger B. 2003. Adhesion-dependent cell mechanosensitivity. *Annu. Rev. Cell. Dev. Biol.* 19:677–95

78. Geiger B, Bershadsky A, Pankov R, Yamada KM. 2001. Transmembrane crosstalk between the extracellular matrix–cytoskeleton crosstalk. *Nat. Rev. Mol. Cell Biol.* 2:793–805

79. Saez AO, Zhang W, Wu Y, Turner CE, Tang DD, Gunst SJ. 2004. Tension development during contractile stimulation of smooth muscle requires recruitment of paxillin and vinculin to the membrane. *Am. J. Physiol. Cell Physiol.* 286:C433–47

80. Tang DD, Wu MF, Opazo Saez AM, Gunst SJ. 2002. The focal adhesion protein paxillin regulates contraction in canine tracheal smooth muscle. *J. Physiol.* 542:501–13

81. Davies PF, Barbee KA, Volin MV, Robotewskyj A, Chen J, et al. 1997. Spatial relationships in early signaling events of flow-mediated endothelial mechanotransduction. *Annu. Rev. Physiol.* 59:527–49

82. Stamenovic D, Ingber DE. 2002. Models of cytoskeletal mechanics of adherent cells. *Biomech. Model. Mechanobiol.* 1:95–108

83. Stamenovic D, Fredberg JJ, Wang N, Butler JP, Ingber DE. 1996. A microstructural approach to cytoskeletal mechanics based on tensegrity. *J. Theor. Biol.* 181:125–36

84. Stamenovic D, Mijailovich SM, Tolic-Norrelykke IM, Chen J, Wang N. 2002. Cell prestress. II. Contribution of microtubules. *Am. J. Physiol. Cell Physiol.* 282:C617–24

85. Cukierman E, Pankov R, Stevens DR, Yamada KM. 2001. Taking cell-matrix adhesions to the third dimension. *Science* 294:1708–12

86. Geiger B. 2001. Cell biology. Encounters in space. *Science* 294:1661–63

87. Balaban NQ, Schwarz US, Riveline D, Goichberg P, Tzur G, et al. 2001. Force and focal adhesion assembly: a close relationship studied using elastic micropatterned substrates. *Nat. Cell Biol.* 3:466–72.

88. Zamir E, Katz M, Posen Y, Erez N, Yamada KM, et al. 2000. Dynamics and segregation of cell-matrix adhesions in cultured fibroblasts. *Nat. Cell Biol.* 2:191–96

89. Weichselbaum M, Sparrow MP. 1999. A confocal microscopic study of the formation of ganglia in the airways of fetal pig lung. *Am. J. Respir. Cell. Mol. Biol.* 21:607–20

90. Sparrow MP, Weichselbaum M, McCray PB. 1999. Development of the innervation and airway smooth muscle in human fetal lung. *Am. J. Respir. Cell. Mol. Biol.* 20:550–60

91. Seow CY, Fredberg JJ. 2001. Historical perspective on airway smooth muscle: the saga of a frustrated cell. *J. Appl. Physiol.* 91:938–52

92. Dembo M, Wang YL. 1999. Stresses at the cell-to-substrate interface during locomotion of fibroblasts. *Biophys. J.* 76:2307–16

93. Butler JP, Tolic-Norrelykke IM, Fabry B, Fredberg JJ. 2002. Traction fields, moments, and strain energy that cells exert on their surroundings. *Am. J. Physiol. Cell Physiol.* 282:C595–605

94. Schwarz US, Balaban NQ, Riveline D, Bershadsky A, Geiger B, Safran SA. 2002. Calculation of forces at focal adhesions from elastic substrate data: the effect of localized force and the need for regularization. *Biophys. J.* 83:1380–94

95. Tolic-Norrelykke IM, Butler JP, Chen J, Wang N. 2002. Spatial and temporal traction response in human airway smooth muscle cells. *Am. J. Physiol. Cell Physiol.* 283:C1254–66

96. Wang N, Tolic-Norrelykke IM, Chen J, Mijailovich SM, Butler JP, et al. 2002. Cell prestress. I. Stiffness and prestress are closely associated in adherent contractile cells. *Am. J. Physiol. Cell Physiol.* 282:C606–16

97. An SS, Pennella CM, Gonnabathula A, Chen J, Wang N, et al. 2005. Hypoxia alters biophysical properties of endothelial cells via p38 MAPK-and Rho kinase-dependent pathways. *Am. J. Physiol. Cell Physiol.* 289:C521–30

98. du Roure O, Saez A, Buguin A, Austin RH, Chavrier P, et al. 2005. Force mapping in epithelial cell migration. *Proc. Natl. Acad. Sci. USA* 102:2390–95

99. Glogauer M, Arora P, Chou D, Janmey PA, Downey GP, McCulloch CA. 1998. The role of actin-binding protein 280 in integrin-dependent mechanoprotection. *J. Biol. Chem.* 273:1689–98

100. Giannone G, Jiang G, Sutton DH, Critchley DR, Sheetz MP. 2003. Talin1 is critical for force-dependent reinforcement of initial integrin-cytoskeleton bonds but not tyrosine kinase activation. *J. Cell Biol.* 163:409–19

101. Panettieri RA, Murray RK, DePalo LR, Yadvish PA, Kotlikoff MI. 1989. A human airway smooth muscle cell line that retains physiological responsiveness. *Am. J. Physiol.* 256:C329–35

102. Stamenovic D, Wilson TA. 1992. Parenchymal stability. *J. Appl. Physiol.* 73:596–602

103. Lambert RK, Wilson TA. 1973. A model for the elastic properties of the lung and

their effect of expiratory flow. *J. Appl. Physiol.* 34:34–48

104. Stamenovic D, Wilson TA. 1985. A strain energy function for lung parenchyma. *J. Biomech. Eng.* 107:81–86

105. Stamenovic D, Coughlin MF. 1999. The role of prestress and architecture of the cytoskeleton and deformability of cytoskeletal filaments in mechanics of adherent cells: a quantitative analysis. *J. Theor. Biol.* 201:63–74

106. Stamenovic D, Liang Z, Chen J, Wang N. 2002. Effect of the cytoskeletal prestress on the mechanical impedance of cultured airway smooth muscle cells. *J. Appl. Physiol.* 92:1443–50

107. Pourati J, Maniotis A, Spiegel D, Schaffer JL, Butler JP, et al. 1998. Is cytoskeletal tension a major determinant of cell deformability in adherent endothelial cells? *Am. J. Physiol.* 274:C1283–89

108. Wang N, Naruse K, Stamenovic D, Fredberg JJ, Mijailovich SM, et al. 2001. Mechanical behavior in living cells consistent with the tensegrity model. *Proc. Natl. Acad. Sci. USA* 98:7765–70

109. Griffin MA, Engler AJ, Barber TA, Healy KE, Sweeney HL, Discher DE. 2004. Patterning, prestress, and peeling dynamics of myocytes. *Biophys. J.* 86:1209–22

110. Budiansky B, Kimmel E. 1987. Elastic moduli of lungs. *J. Appl. Mech.* 54S:351–58

111. Heidemann SR, Lamoureaux P, Buxbaum RE. 2000. Opposing views on tensegrity as a structural framework for understanding cell mechanics. *J. Appl. Physiol.* 89:1670–78

112. Ingber DE. 2000. Opposing views on tensegrity as a structural framework for understanding cell mechanics. *J. Appl. Physiol.* 89:1663–70

113. Ingber DE. 1998. The architecture of life. *Sci. Am.* 278:48–57

114. Wang N, Butler JP, Ingber DE. 1993. Mechanotransduction across the cell surface and through the cytoskeleton. *Science* 260:1124–27

115. Wang N, Ingber DE. 1994. Control of cytoskeletal mechanics by extracellular matrix, cell shape, and mechanical tension. *Biophys. J.* 66:2181–89

116. Wang N, Ostuni E, Whitesides GM, Ingber DE. 2002. Micropatterning tractional forces in living cells. *Cell Motil. Cytoskeleton* 52:97–106

117. Parker KK, Brock AL, Brangwynne C, Mannix RJ, Wang N, et al. 2002. Directional control of lamellipodia extension by constraining cell shape and orienting cell tractional forces. *FASEB J.* 16:1195–204

118. MacKintosh FC, Käs J, Janmey PA. 1995. Elasticity of semiflexible biopolymer networks. *Phys. Rev. Lett.* 75:4425–28

119. Gardel ML, Shin JH, MacKintosh FC, Mahadevan L, Matsudaira PA, Weitz DA. 2004. Scaling of F-actin network rheology to probe single filament elasticity and dynamics. *Phys. Rev. Lett.* 93:188102

120. Gardel ML, Shin JH, MacKintosh FC, Mahadevan L, Matsudaira P, Weitz DA. 2004. Elastic behavior of cross-linked and bundled actin networks. *Science* 304:1301–5

121. Mahaffy RE, Shih CK, MacKintosh FC, Kas J. 2000. Scanning probe-based frequency-dependent microrheology of polymer gels and biological cells. *Phys. Rev. Lett.* 85:880–83

122. Gittes F, MacKintosh FC. 1998. Dynamic shear modulus of a semiflexible polymer network. *Phys. Rev. E* 58:R1241–44

123. Head DA, Levine AJ, MacKintosh FC. 2003. Deformation of cross-linked semiflexible polymer networks. *Phys. Rev. Lett.* 91:108102

124. Hu S, Chen J, Fabry B, Numaguchi Y, Gouldstone A, et al. 2003. Intracellular stress tomography reveals stress focusing and structural anisotropy in cytoskeleton of living cells. *Am. J. Physiol. Cell Physiol.* 285:C1082–90

125. Maniotis AJ, Chen CS, Ingber DE. 1997. Demonstration of mechanical connections between integrins, cytoskeletal filaments, and nucleoplasm that stabilize

nuclear structure. *Proc. Natl. Acad. Sci. USA* 94:849–54

126. Hu S, Eberhard L, Chen J, Love JC, Butler JP, et al. 2004. Mechanical anisotropy of adherent cells probed by a 3D magnetic twisting device. *Am. J. Physiol. Cell Physiol.* 287:1184–91

127. Dahl KN, Engler AJ, Pajerowski JD, Discher DE. 2005. Power-law rheology of isolated nuclei with deformation mapping of nuclear sub-structures. *Biophys. J.* 89(4):2855–64, 2005

128. Bursac P, Lenormand G, Fabry B, Oliver M, Weitz DA, et al. 2005. Cytoskeletal remodelling and slow dynamics in the living cell. *Nat. Mat.* 4:557–71

129. Seow C. 2005. Fashionable glass. *Nature* 435:1172–73

130. Puig-De-Morales M, Millet E, Fabry B, Navajas D, Wang N, et al. 2004. Cytoskeletal mechanics in the adherent human airway smooth muscle cell: probe specificity and scaling of protein-protein dynamics. *Am. J. Physiol. Cell Physiol.* 287:C643–54

131. Lenormand G, Millet E, Fabry B, Butler J, Fredberg J. 2004. Linearity and time-scale invariance of the creep function in living cells. *J. Roy. Soc. Interface* 1:91–97

132. Alcaraz J, Buscemi L, Grabulosa M, Trepat X, Fabry B, et al. 2003. Microrheology of human lung epithelial cells measured by atomic force microscopy. *Biophys. J.* 84:2071–79

133. Balland M, Richert A, Gallet F. 2005. The dissipative contribution of myosin II in the cytoskeleton dynamics of myoblasts. *Eur. Biophys. J.* 34:255–61

134. Desprat N, Richert A, Simeon J, Asnacios A. 2005. Creep function of a single living cell. *Biophys. J.* 88:2224–33

135. Fabry B, Fredberg JJ. 2003. Remodeling of the airway smooth muscle cell: are we built of glass? *Respir. Physiol. Neurobiol.* 137:109–24

136. Sollich P. 1998. Rheological constitutive equation for a model of soft glassy materials. *Physiol. Rev.* 58:738–59

137. Sollich P, Lequeux F, Hebraud P, Cates ME. 1997. Rheology of soft glassy mateirals. *Phys. Rev. Lett* 78:2020–23

138. Gunst SJ, Fredberg JJ. 2003. The first three minutes: smooth muscle contraction, cytoskeletal events, and soft glasses. *J. Appl. Physiol.* 95:413–25

139. Howard J. 2001. *Mechanics of Motor Proteins and the Cytoskeleton.* Sunderland, MA: Sinauer. 367 pp.

140. Trappe V, Prasad V, Cipelletti L, Segre PN, Weitz DA. 2001. Jamming phase diagram for attractive particles. *Nature* 411:772–75

141. Segre PN, Prasad V, Schofield AB, Weitz DA. 2001. Glasslike kinetic arrest at the colloidal-gelation transition. *Phys. Rev. Lett.* 86:6042–45

142. Ellis RJ, Minton AP. 2003. Cell biology: join the crowd. *Nature* 425:27–28

143. Ellis RJ. 2001. Macromolecular crowding: obvious but underappreciated. *Trends Biochem. Sci.* 26:597–604

144. Ellis RJ. 2001. Macromolecular crowding: an important but neglected aspect of the intracellular environment. *Curr. Opin. Struct. Biol.* 11:114–19

145. Pollack GH. 2001. *Cells, Gels and the Engines of Life.* Seattle, WA: Ebner. 305 pp.

146. Goodsell D. 2000. Biomolecules and nanotechnology. *Am. Sci.* 88:230–37

147. Dobson CM. 2004. Chemical space and biology. *Nature* 432:824–28

148. Mehta D, Gunst SJ. 1999. Actin polymerization stimulated by contractile activation regulates force development in canine tracheal smooth muscle. *J. Physiol.* 519(Pt. 3):829–40

149. Trepat X, Grabulosa M, Puig F, Maksym GN, Navajas D, Farre R. 2004. Viscoelasticity of human alveolar epithelial cells subjected to stretch. *Am. J. Physiol. Lung Cell Mol. Physiol.* 287:L1025–34

150. Stamenovic D, Suki B, Fabry B, Wang N, Fredberg JJ. 2004. Rheology of airway smooth muscle cells is associated with cytoskeletal contractile stress. *J. Appl. Physiol.* 96:1600–5

151. Davies PF, Robotewskyj A, Griem ML. 1994. Quantitative studies of endothelial cell adhesion. Directional remodeling of focal adhesion sites in response to flow forces. *J. Clin. Invest.* 93:2031–38

152. Hamill OP, Martinac B. 2001. Molecular basis of mechanotransduction in living cells. *Physiol. Rev.* 81:685–740

153. Fisher JL, Margulies SS. 2002. Na$^+$-K$^+$-ATPase activity in alveolar epithelial cells increases with cyclic stretch. *Am. J. Physiol. Lung Cell Mol. Physiol.* 283:L737–46

154. Gullingsrud J, Kosztin D, Schulten K. 2001. Structural determinants of MscL gating studied by molecular dynamics simulations. *Biophys. J.* 80:2074–81

155. Gullingsrud J, Schulten K. 2003. Gating of MscL studied by steered molecular dynamics. *Biophys. J.* 85:2087–99

156. Dreyfuss D, Martin-Lefevre L, Saumon G. 1999. Hyperinflation-induced lung injury during alveolar flooding in rats: effect of perfluorocarbon instillation. *Am. J. Respir. Crit. Care Med.* 159:1752–57

157. Dreyfuss D, Saumon G. 1998. Ventilator-induced lung injury: lessons from experimental studies. *Am. J. Respir. Crit. Care Med.* 157:294–323

158. Tschumperlin DJ, Margulies SS. 1999. Alveolar epithelial surface area-volume relationship in isolated rat lungs. *J. Appl. Physiol.* 86:2026–33

159. Chapman KE, Sinclair SE, Zhuang D, Hassid A, Desai L, Waters CM. 2005. Cyclic mechanical strain increases reactive oxygen species production in pulmonary epithelial cells. *Am. J. Physiol. Lung Cell Mol. Physiol.* 289:834–41

160. Wilson MR, Choudhury S, Goddard ME, O'Dea KP, Nicholson AG, Takata M. 2003. High tidal volume upregulates intrapulmonary cytokines in an in vivo mouse model of ventilator-induced lung injury. *J. Appl. Physiol.* 95:1385–93

161. Choudhury S, Wilson MR, Goddard ME, O'Dea KP, Takata M. 2004. Mechanisms of early pulmonary neutrophil sequestration in ventilator-induced lung injury in mice. *Am. J. Physiol. Lung Cell Mol. Physiol.* 287:L902–10

162. Ali MH, Pearlstein DP, Mathieu CE, Schumacker PT. 2004. Mitochondrial requirement for endothelial responses to cyclic strain: implications for mechanotransduction. *Am. J. Physiol. Lung Cell Mol. Physiol.* 287:L486–96

163. Ichimura H, Parthasarathi K, Quadri S, Issekutz AC, Bhattacharya J. 2003. Mechano-oxidative coupling by mitochondria induces proinflammatory responses in lung venular capillaries. *J. Clin. Invest.* 111:691–99

164. Jiang G, Giannone G, Critchley DR, Fukumoto E, Sheetz MP. 2003. Two-piconewton slip bond between fibronectin and the cytoskeleton depends on talin. *Nature* 424:334–37

165. Evans E. 2001. Probing the relation between force—lifetime–and chemistry in single molecular bonds. *Annu. Rev. Biophys. Biomol. Struct.* 30:105–28

166. Sawada Y, Sheetz MP. 2002. Force transduction by Triton cytoskeletons. *J. Cell Biol.* 156:609–15

167. Wang Y, Botvinick EL, Zhao Y, Berns MW, Usami S, et al. 2005. Visualizing the mechanical activation of Src. *Nature* 434:1040–45

168. Zhong C, Chrzanowska-Wodnicka M, Brown J, Shaub A, Belkin AM, Burridge K. 1998. Rho-mediated contractility exposes a cryptic site in fibronectin and induces fibronectin matrix assembly. *J. Cell Biol.* 141:539–51

169. Gilmore AP, Burridge K. 1995. Cell adhesion. Cryptic sites in vinculin. *Nature* 373:197

170. Sampath R, Gallagher PJ, Pavalko FM. 1998. Cytoskeletal interactions with the leukocyte integrin beta2 cytoplasmic tail. Activation-dependent regulation of associations with talin and alpha-actinin. *J. Biol. Chem.* 273:33588–94

171. Papagrigoriou E, Gingras AR, Barsukov IL, Bate N, Fillingham IJ, et al. 2004. Activation of a vinculin-binding site in the

talin rod involves rearrangement of a five-helix bundle. *EMBO J.* 23:2942–51

172. Fillingham I, Gingras AR, Papagrigoriou E, Patel B, Emsley J, et al. 2005. A vinculin binding domain from the talin rod unfolds to form a complex with the vinculin head. *Structure* 13:65–74

173. Yap B, Kamm RD. 2005. Mechanical deformation of neutrophils into narrow channels induces pseudopod projection and changes in biomechanical properties. *J. Appl. Physiol.* 98:1930–39

174. Kitagawa Y, Van Eeden SF, Redenbach DM, Daya M, Walker BA, et al. 1997. Effect of mechanical deformation on structure and function of polymorphonuclear leukocytes. *J. Appl. Physiol.* 82:1397–405

175. Yap B, Kamm RD. 2005. Cytoskeletal remodeling and cellular activation during deformation of neutrophils into narrow channels. *J. Appl. Physiol.*:doi: 10.1152/japplphysiol.00503.2005. In press

176. Tschumperlin DJ, Dai G, Maly IV, Kikuchi T, Laiho LH, et al. 2004. Mechanotransduction through growth-factor shedding into the extracellular space. *Nature* 429:83–86

177. Ressler B, Lee RT, Randell SH, Drazen JM, Kamm RD. 2000. Molecular responses of rat tracheal epithelial cells to transmembrane pressure. *Am. J. Physiol. Lung Cell Mol. Physiol.* 278:L1264–72

178. Swartz MA, Tschumperlin DJ, Kamm RD, Drazen JM. 2001. Mechanical stress is communicated between different cell types to elicit matrix remodeling. *Proc. Natl. Acad. Sci. USA* 98:6180–85

179. Budhiraja R, Tuder RM, Hassoun PM. 2004. Endothelial dysfunction in pulmonary hypertension. *Circulation* 109:159–65

180. Tozzi CA, Poiani GJ, Harangozo AM, Boyd CD, Riley DJ. 1989. Pressure-induced connective tissue synthesis in pulmonary artery segments is dependent on intact endothelium. *J. Clin. Invest.* 84:1005–12

181. Strauss BH, Rabinovitch M. 2000. Adventitial fibroblasts: defining a role in vessel wall remodeling. *Am. J. Respir. Cell. Mol. Biol.* 22:1–3

182. Rief M, Gautel M, Oesterhelt F, Fernandez JM, Gaub HE. 1997. Reversible unfolding of individual titin immunoglobulin domains by AFM. *Science* 276:1109–12

183. Kellermayer MS, Smith SB, Granzier HL, Bustamante C. 1997. Folding-unfolding transitions in single titin molecules characterized with laser tweezers. *Science* 276:1112–16

184. Gao M, Lu H, Schulten K. 2002. Unfolding of titin domains studied by molecular dynamics simulations. *J. Muscle Res. Cell Motil.* 23:513–21

185. Granzier H, Labeit S. 2002. Cardiac titin: an adjustable multi-functional spring. *J. Physiol.* 541:335–42

186. Discher DE, Carl P. 2001. New insights into red cell network structure, elasticity, and spectrin unfolding–a current review. *Cell Mol. Biol. Lett.* 6:593–606

187. Furuike S, Ito T, Yamazaki M. 2001. Mechanical unfolding of single filamin A (ABP-280) molecules detected by atomic force microscopy. *FEBS Lett.* 498:72–75

188. Gao M, Lu H, Schulten K. 2001. Simulated refolding of stretched titin immunoglobulin domains. *Biophys. J.* 81:2268–77

189. Yamazaki M, Furuike S, Ito T. 2002. Mechanical response of single filamin A (ABP-280) molecules and its role in the actin cytoskeleton. *J. Muscle Res. Cell Motil.* 23:525–34

190. Schwaiger I, Kardinal A, Schleicher M, Noegel AA, Rief M. 2004. A mechanical unfolding intermediate in an actin-crosslinking protein. *Nat. Struct. Mol. Biol.* 11:81–85

191. Ortiz V, Nielsen SO, Klein ML, Discher DE. 2005. Unfolding a linker between helical repeats. *J. Mol. Biol.* 349:638–47

192. Gittes F, Mickey B, Nettleton J, Howard

J. 1993. Flexural rigidity of microtubules and actin filaments measured from thermal fluctuations in shape. *J. Cell Biol.* 120:923–34

193. Laudadio R, Millet E, Fabry F, An S, Butler J, Fredberg J. The rat airway smooth muscle cell during actin modula- tion: rheology and glassy dynamics. *Am. J. Physiol. Cell Physiol.*; doi:10.1152/ ajpcell.00060.2005. In press

194. Gao M, Craig D, Vogel V, Schulten K. 2002. Identifying unfolding intermediates of FN-III10 by steered molecular dynam- ics. *J. Mol. Biol.* 323:939–50

Annu. Rev. Physiol. 2006. 68:543–61
doi: 10.1146/annurev.physiol.68.072304.112754
First published online as a Review in Advance on October 31, 2005

REGULATION OF NORMAL AND CYSTIC FIBROSIS AIRWAY SURFACE LIQUID VOLUME BY PHASIC SHEAR STRESS

Robert Tarran, Brian Button, and Richard C. Boucher

Cystic Fibrosis/Pulmonary Research & Treatment Center, University of North Carolina, Chapel Hill, North Carolina 27599-7248; email: robert_tarran@med.unc.edu, brian_button@med.unc.edu, r_boucher@med.unc.edu

Key Words CFTR, chloride, ENaC, mucus clearance, nucleotides

■ **Abstract** The physical removal of viruses and bacteria on the mucociliary escalator is an important aspect of the mammalian lung's innate defense mechanism. The volume of airway surface liquid (ASL) present in the respiratory tract is a critical determinant of both mucus hydration and the rate of mucus clearance from the lung. ASL volume is maintained by the predominantly ciliated epithelium via coordinated regulation of (*a*) absorption, by the epithelial Na^+ channel, and (*b*) secretion, by the Ca^{2+}-activated Cl^- channel (CaCC) and CFTR. This review provides an update on our current understanding of how shear stress regulates ASL volume height in normal and cystic fibrosis (CF) airway epithelia through extracellular ATP- and adenosine (ADO)-mediated pathways that modulate ion transport and ASL volume homeostasis. We also discuss (*a*) how derangement of the ADO-CFTR pathway renders CF airways vulnerable to viral infections that deplete ASL volume and produce mucus stasis, and (*b*) potential shear stress–dependent therapies for CF.

INTRODUCTION

Airway surface liquid (ASL) is separated into two layers: a mucus layer containing secreted mucins, which traps inhaled particles, and a periciliary liquid layer (PCL), which keeps these mucins at a sufficient distance from the underlying epithelium to optimize mucus clearance (1, 2). In normal airways, the PCL height is defined by the length of the outstretched cilia (~7 μm), whereas the mucus layer varies in height [7 μm to 70 μm in vivo; (3, 4)]. In cystic fibrosis (CF), lack of cystic fibrosis transmembrane conductance regulator (CFTR) Cl^- channel function and inhibition of the epithelial Na^+ channel (ENaC) are thought to cause depletion of the PCL, resulting in mucus accumulation. However, whereas in vitro data from standard (static) culture systems describe rapid depletion of ASL height and a complete failure of mucus transport (5, 6), young CF patients exhibit reduced but measurable rates of mucus clearance in vivo (7). This inconsistency suggests that mechanisms for ASL volume regulation operating in vivo are absent in standard static culture systems.

Tidal volume expansion and airflow impart shear stress to airway surfaces (8), a stimulus that releases nucleotides from many cell types into the extracellular environment (9, 10). In mammalian airways, extracellular nucleotides interact with $P2Y_2$ purinoceptors ($P2Y_2$-R) to regulate airway ion transport (11). We therefore hypothesized that the failure to recreate in vitro the phasic motion generated by tidal breathing in vivo has hindered the identification of key ASL volume autoregulatory systems. Accordingly, we designed devices to mimic in vivo shear stress in vitro. Furthermore, the established body of literature regarding airway ion transport is largely based on experiments performed with large volumes of Ringer/saline bathing epithelial mucosal surfaces. Under these conditions, endogenously produced ASL is grossly diluted and may even be removed entirely. Thus, potential signaling molecules that could accumulate in the ASL would be lost, which would likely alter epithelial ion transport rates. Hence, we interfaced a culture system that allowed the epithelium to generate and maintain an endogenous "thin film" of ASL on airway surfaces with the phasic motion devices for studies of interactions of airway shear stress and ASL volume homeostasis. We have used this system to investigate the regulation of CFTR, ENaC, and a second, alternate Ca^{2+}-activated Cl^- channel (CaCC) that is present in CF lungs, via extracellular release of ATP and its metabolic product, adenosine (ADO). This system is reviewed below, along with evidence that viral infections can episodically abrogate residual ion transport functions in CF lungs by depleting the ASL of ATP, leading to mucus stasis.

SALT AND WATER TRANSPORT

Airway epithelia effect vectoral ion transport to maintain ASL volume via the coordinated action of pumps, transporters, and channels in both the apical and basolateral membranes (12). This review focuses on the regulation of apical ion channels (CFTR, CaCC, and ENaC). However, brief descriptions of some of the underlying assumptions are useful in interpreting these data.

Airway epithelia have a low transepithelial resistance (12) and are highly water permeable (13–15). These features suggest that large, osmotic gradients cannot be maintained across airway epithelia. This property initially appears disadvantageous for vectoral ion/water transport because extra energy is expended to move ions/water against a continual backflux. However, leaky epithelia are capable of both absorbing Na^+ and secreting Cl^- (16), which may permit fine-tuning of the ASL volume to produce efficient mucus clearance (17). Thus, because ASL NaCl concentrations are generally isotonic with plasma, one should consider the mass of NaCl in the ASL rather than its concentration as being the determining factor in influencing volume/composition. For example, if the epithelium secretes more salt into the ASL, water follows passively, and the ASL volume rapidly increases to maintain isotonicity. Conversely, if NaCl is absorbed, water also follows passively but in the other direction, and ASL volume decreases. Thus, the mass of salt in the ASL, and hence the volume, is set by the actions of CFTR, CaCC, and ENaC.

Despite previous controversy, a consensus seems to have been reached that the ASL is not only isotonic with plasma but is also not different in normal (NL) versus CF patients (1, 18–20). The mucus layer possesses a similar ion concentration to that of the PCL (1, 21), and water and ions may move between the mucus and PCL without interrupting the discreteness of these layers. This "reservoir" effect serves to buffer PCL volume, keeping cilia at a suitable distance from the mucus layer to mediate efficient mucus transport (21). This reservoir effect likely reflects the properties of mucins, which, despite their low concentration (2–3%), give the mucus layer its characteristic visco-elastic and gel-forming properties (22).

ALTERED ION TRANSPORT AND CYSTIC FIBROSIS AIRWAY DISEASE

One of the hallmarks of CF is the accumulation of viscous, sticky mucus on airway surfaces, leading to occlusion of the small airways and bacterial infection (23, 24). However, the primary genetic defect in CF is the dysfunction of an apical membrane Cl^- channel, the CFTR (25), and a long-term challenge has been to reconcile knowledge of this genetic defect with clinical observations. Researchers recently have proposed that the initiating event in CF lung disease is the rapid depletion of the ASL due to Na^+ hyperabsorption and Cl^- hyposecretion (Figure 1). Overexpression of the β subunit of ENaC in mice in vivo also causes ASL depletion (26). In time, this ASL depletion is predicted to cause mucus adherence to airway surfaces, preventing mucus clearance (1). Indeed, the β ENaC mouse exhibited such severe mucus plugging as to cause death. The failure of mechanical clearance in CF airways allows dehydrated mucus plaques to adhere to airway surfaces and increase in height until the airways become occluded. Ultimately, bacteria colonize mucus adherent to airways and form biofilms. White blood cells (WBCs) (macrophages and neutrophils) are attracted to the CF airway lumen and try to infiltrate the mucus plaques to attack the bacteria, but they are excluded as a consequence of the reduced porosity (mesh size) of the concentrated mucus (27). Consequently, a persistent but ineffective inflammatory response is triggered, and WBCs remain at the periphery of the mucus plugs (28), where they release contents such as neutrophil elastase that cause airway damage (23, 24). Interestingly, the β ENaC mouse exhibits inflammation that appears to be associated with the low ASL volume/mucus plugging, even in the absence of bacterial infections, providing further evidence that a low ASL volume may be deleterious for airways health (29).

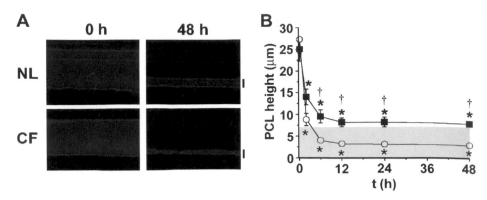

Figure 1 Abnormal regulation of perciliary liquid layer (PCL) height by CF airway epithelia. (*A*) XZ confocal images of PCL at 0 and 48 h after mucosal addition of 20 µl PBS containing Texas-red dextran to normal (NL) and CF bronchial epithelial cultures. Scale bars, 7 µm. (*B*) Mean PCL height with time taken from NL (*squares*; n = 9) and CF (*circles*; n = 8) cultures. Blue-shaded region denotes normal height of outstretched cilia (i.e., normal PCL height). Data shown as mean ± SEM. * = significantly different from t = 0. † = significantly different between NL and CF cultures.

NORMAL TIDAL BREATHING AND PHASIC SHEAR STRESS

Airflow due to normal tidal breathing imparts shear stress on airway surfaces (8). Cyclic compressive (transmural) shear and stretch also contribute to the overall magnitude of cellular shear stress. However, calculations of total shear stress on airway surfaces have not been performed owing to difficulties in making accurate measurements of these parameters in vivo. We focused for our studies of ASL volume homeostasis on the best-characterized component of breathing-induced shear, i.e., the shear stress due to tidal oscillations in airflow. Estimates of the magnitude of this shear stress under resting tidal breathing range between ∼0.4– 2 dynes cm^{-2} (Figure 2*A*) (29a), and airway wall shear stress varies little with airway order in the lung (29a, 30). This constancy of shear stress throughout the branching airway tree occurs because the dramatic decreases in linear velocities as the airways branch distally are proportionately matched by a decrease in airways diameter. Thus, the shear rate, which is roughly the velocity divided by the diameter, varies little, and wall shear stress is nearly conserved across all diameters of airways. This finding is analogous to that of Cecil Murray (31), who examined branching morphology of the vasculature and showed likewise that wall shear stress in that system is also conserved.

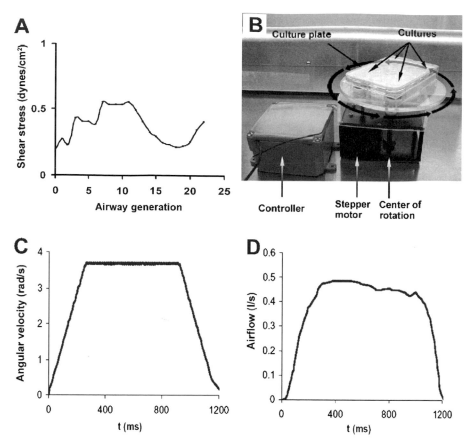

Figure 2   Airflow-induced shear stress. (*A*) Shear stress is constant throughout the respiratory tract. Generation 0 is the trachea. (*B*). Phasic motion device. To generate shear stress, four cultures were rotated in a stop-go fashion (*black arrows*) for up to 48 h inside a highly humidified incubator. The device was powered by a computer-controlled stepper motor and, by varying the rate of acceleration, could produce from 0.06–6 dynes cm^{-2} shear stress. (*C, D*) Velocity profiles associated with phasic shear stress. Angular velocity over time during one start/stop of the phasic motion device (*C*). Change in airflow through one inspiration during normal tidal breathing (*D*).

METHODS OF GENERATING PHASIC SHEAR ON AIRWAY SURFACES

We built a series of devices to control the magnitude and frequency of the phasic motion/shear stress exerted on human bronchial airway epithelial cultures in a quantitative and controllable manner. These devices accelerated and decelerated cultures inside a highly humidified incubator to generate liquid flow over airway epithelial surfaces while avoiding airflow-induced dehydration of the ASL (29a).

To calculate shear stress with this system, we assumed that the ASL/mucus compartment had a density equal to that of water (ASL is ~97% water) (6, 21), and consequently, the shear stress induced by each rotation was:

$$\tau shear = \frac{\rho h \theta r}{\Delta t}$$

where ρ is the density of water (~1 g/cm^3), h the initial liquid layer height (0.0027 cm) as measured by confocal microscopy, θ the angular velocity (maximum 3.7 radians s^{-1}), r the distance from the center of rotation (4.5 cm), and Δt the time of acceleration assuming a constant rate of acceleration (the plate is accelerated smoothly by a stepper motor over 100 ms). Thus, the shear stress under these conditions is estimated to be on the order of 0.5 dynes cm^{-2}. This rate of shear would decline in a linear fashion as excess ASL was absorbed, falling to 0.13 dynes cm^{-2} when ASL height was 8 µm at 48 h. More detailed calculations are provided elsewhere (29a).

PHASIC MOTION REBALANCES ASL HEIGHT

Under static culture conditions, NL cultures regulate ASL height to 7 µm, a height suitable for cilia to effect mucus transport (21) (Figure 1A,B). By contrast, CF cultures rapidly deplete ASL height to ~3 µm (Figure 1A,B). Under these static conditions, NL cultures rebalance ion transport as ASL height decreases, with a shift from Na$^+$ absorption to Cl$^-$ secretion. By contrast, CF cultures maintain inappropriately high levels of Na$^+$ absorption and do not initiate Cl$^-$ secretion regardless of ASL height (29a).

As we discussed above, phasic motion is missing from standard static culture systems, which may omit an important variable that regulates ASL height. With phasic motion, NL cultures approximately double ASL height to 14 µm (Figure 3A,B). Importantly, CF cultures increase ASL height to 7 µm (Figure 3A,B) and are capable of maintaining mucus transport for protracted intervals (Figure 3C). This increase in CF ASL height is accompanied by inhibition of Na$^+$ absorption (amiloride-sensitive V$_t$) and the appearance of Cl$^-$ secretion (bumetanide-sensitive V$_t$) (29a), suggesting that shear stress has activated pathways in CF cultures that are absent under static conditions.

HOW IS SHEAR STRESS TRANSDUCED BY AIRWAY EPITHELIA INTO ION TRANSPORT REGULATION?

Shear stress has multiple effects on cell function. One almost-universal mechanism to couple shear stress to autocrine, and indeed paracrine, effects is through the regulated release of ATP into the extracellular environment. Similar to other cell

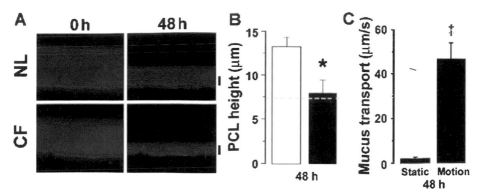

Figure 3 Phasic motion–induced changes in PCL volume in NL versus CF cultures. (*A*) XZ confocal images of PCL immediately (0 h) and 48 h after mucosal PBS addition to NL and CF epithelia cultured under phasic motion. (*B*) Mean PCL heights after 48 h of phasic motion culture for NL (*open bars*; n = 7) and CF (*closed bars*; n = 8) cultures. (*C*) Rotational mucus transport rates in static CF cultures (*closed bars*; n = 12) and phasic motion cultures (*gray bars*; n = 14) 48 h after volume addition. Data shown as mean ± SEM. * = data significantly different between NL and CF cultures. ‡ = data significantly different between static and phasic motions.

types [e.g., endothelia; (9)], cultured airway epithelia continually release purines and increase their rate of ATP release whenever the mucosal media is disturbed (32), suggesting that culturing airway epithelia under phasic motion may alter extracellular purine concentrations. It appears that ATP is released from NL and CF airway epithelia at equal rates (33) and that NL and CF ATP concentrations in vivo do not differ (34).

Guyot & Hanrahan (10) grew CALU3 cells (immortalized cells of glandular origin) in hollow biofibers and observed that increasing the perfusion rate through the lumen of the fibers causes a rapid, reversible increase in perfusate ATP concentrations. Under thin film conditions, we found that phasic shear stress causes a dose-dependent increase in ASL ATP levels in primary cultures of bronchial surface epithelial cells (Figure 4) (29a). Interestingly, the epithelium appears geared to secrete graded amounts of ATP over a wide range of shear stress values. For example, when static ASL ATP levels are $\sim$0.1 nM (29a), imposition of small increments of phasic shear (0.01 dynes cm^{-2}) causes an $\sim$100-fold increase in ATP levels to $\sim$10 nM (Figure 4), into the range in which apical membrane ATP-sensing purinoceptors ($P2Y_2$-R) are activated. Increasing phasic shear stress from 0.01–6 dynes cm^{-2} causes only an additional sixfold increase in PCL ATP levels to $\sim$70 nM (Figure 4A).

Perhaps consistent with the nature of the shear stresses experienced by superficial airway epithelia (i.e., phasic shear stress due to phasic airflow and expansions/contractions of the lung), it is likely the change in shear stress, rather than

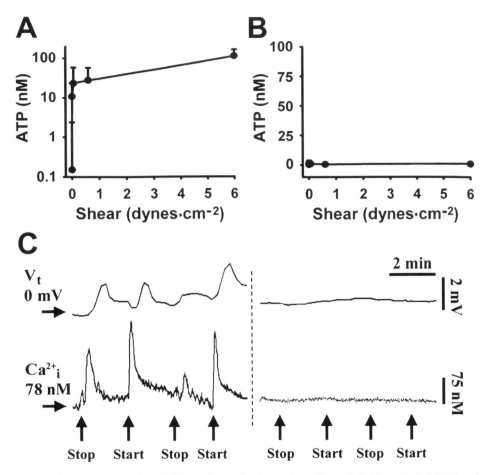

Figure 4   ATP release into PCL mediates phasic motion effects. (*A*) Mean PCL [ATP] and (*B*) mean serosal bath [ATP] obtained from CF cultures 1 h after 50 μl PBS addition under variable phasic motion (both n = 4). NL PCL [ATP] was not significantly different to CF PCL [ATP] (1.9 ± 0.6, 12 ± 4, 95 ± 13, and 131 ± 41 nM at 0, 0.006, 0.6, and 6 dynes cm^{-2}, respectively; all n = 4 and p < 0.05). (*C*) Simultaneous measurements of V_t and intracellular calcium (Ca^{2+}_i) in NL cultures perfused bilaterally with Ringer solution. (*Left*) Changes in V_t/Ca^{2+}_i induced by stopping and then restarting mucosal perfusion (denoted by *arrows*). (*Right*) Altered perfusion rates in the presence of mucosal apyrase (5 U ml^{-1}). The mean changes in V_t and Ca^{2+}_i responses to phasic perfusion with KBR are –7.8 ± 0.3 mV and 198 ± 12 nM, respectively (p < 0.05; n = 7); the changes in each parameter are significantly reduced in the presence of apyrase: ΔV_t = –0.2 ± 0.1 mV, ΔCa^{2+}_i = 3 ± 2 nM, respectively (p < 0.05; n = 6). Data shown as mean ± SEM. * = data significantly different from t = 0. † = data significantly different between NL and CF cultures.

absolute shear stress, that signals the airway epithelia to release ATP. Evidence for this notion emanated from studies demonstrating that airway epithelial intra cellular Ca^{2+} and transepithelial PD (V_t) spikes are initiated by starting/stopping of perfusion (Figure 4C) (29a). The addition of apyrase to the mucosal solution, which metabolizes ATP and abolishes the effects of shear stress, abolished these shear stress–induced effects and established that extracellular ATP mediated these cellular events (Figure 4C) (29a). Similarly, Balcells et al. (35) found that the frequency of shear is more important than the magnitude of shear for a variety of tissues, including rat lung epithelia. These data suggest that Cl^- (and ASL) secretion in vivo are pulsatile rather than constant. Because each "pulse" of secretion is much smaller than the ambient volume of liquid on airway surfaces, this mechanism likely results in a constant overall increase in ASL volume, broadly analogous to a dripping tap filling up a bath tub. In addition, the increase in ASL volume caused by Cl^-/ASL secretion is likely to be offset by Na^+ absorption in proximal airways and/or expectoration of ASL containing mucus, so that ASL volume remains constant overall.

Although it is known that phasic shear stress stimulates ATP release, the underlying mechanism whereby mechanical shear stresses are recognized and transduced into ATP release by airway epithelia is not known. Primary cilia act as flow sensors in renal epithelia (36), and direct mechanical force applied to these cilia results in Ca^{2+} influx, which may control ATP release via vesicular release. However, although we cannot exclude the possibility that primary cilia may transduce shear in the airways, primary cilia in airway epithelia are surrounded by continuously beating motile cilia, which may obfuscate the shear signal. Similarly, rat endothelial cells are reliant on an intact glycocalyx for cytoskeletal reorganization (37). The glycocalyx on the cell surface and/or on the cilia in airway epithelia may also sense (be strained by) shear stress.

Ion channels, including ENaC and TRP channels, are shear-sensitive (37a, 38). Despite the regulation of these channels by shear stress, there are no reports that these channels can conduct ATP itself. Thus, if these channels are important in transducing ATP release, they presumably interact with vesicle-release mechanisms and/or ATP-conducting channels, e.g., VDAC (38a). Other proteins known to be mechanosensitive include integrins, which are inserted in the plasma membrane and also attached to the extracellular matrix (39). This class of proteins is thought of as being actual force sensors that may be "pulled" during shear stress, leading to downstream activation of RhoA and other GTPases that may be involved in regulating ATP release (39, 40).

Force transduction in endothelia has been extensively investigated, and it is possible to speculate about the nature of shear stress sensing in airways on the basis of this previous research. For example, it is widely known that endothelial cells reorganize their cytoskeletons in response to shear stress (41), which likely occurs because shear stress transiently causes cellular deformation. This deformation has been linked with activation of small GTPases, such as RhoA, which feed into downstream cell signaling pathways (42). Interestingly, changes in the activity of another GTPase (Rac1) also occur. These changes in protein activity are orientated in the direction of the applied shear stress and are not accompanied by changes in levels of Rac1 gene expression. This protein is known, however, to control gene expression of other proteins, including NFκB (43).

ATP STIMULATES CA^{2+}-MEDIATED CL$^-$ SECRETION AND INHIBITS NA$^+$ ABSORPTION

CaCC can be activated by ATP- (and UTP-) mediated stimulation of apical P2Y$_2$ receptors (44) via G protein–mediated activation of PLC and increases in inositol triphosphate (IP$_3$) and Ca^{2+}$_i$ (45, 46). By contrast, ENaC is inactivated by purines (47), likely via depletion of PIP2 rather than by direct actions of IP$_3$/Ca^{2+} (48, 49). ATP and UTP are equipotent in stimulating P2Y$_2$ receptors, and EC$_{50}$ estimates have ranged from 0.2–1.0 μM (50–52). We have recently established that mucosal nucleotide addition to airway epithelia induces ASL secretion in both NL and CF airways (21, 53). However, the effects are relatively short lived, and ASL height returns to baseline values within 1 h (6, 53), consistent with transient activation of Ca^{2+}-mediated anion secretion in Ussing chambers (54, 55). This short duration may in part be due to the rapid ecto-metabolism (~30 s) of even very large administered doses of ATP [>200 μM; (6, 56)] and perhaps also to internalization of the P2Y$_2$-R and/or transience of the IP$_3$/Ca^{2+} signal (45). However, despite the rapid metabolism of ATP by ecto-enzymes, sufficient ATP is released to increase ASL ATP concentrations at steady state under shear stress. Intracellular ATP levels are ~5–10 mM, so relatively little cellular ATP must be released during shear stress to maintain ASL ATP concentrations in these ranges (0.2–1 μM) to activate P2Y$_2$-R tonically (Figure 4A).

Although there are no good antagonists of P2Y$_2$-R, extracellular apyrase addition depletes ASL ATP levels and abolishes the effects of phasic shear stress in CF cultures (29a). Similarly, while Cl$^-$ channel antagonists are relatively nonspecific, DIDS, which is known to inhibit CaCC, abolishes the CF shear response and partially attenuates the NL response (29a). Collectively, these data suggest that shear stress–induced ATP release activates the CaCC channel.

ROLE OF ADENOSINE IN MODULATING AIRWAY ION TRANSPORT VIA CFTR AND ENAC

On the basis of their distinct molecular structures and pharmacological profiles, adenosine (or P1) purinoceptors have been divided into four subtypes: A1, A2a, A2b, and A3 (57). All ADO receptors are associated with G proteins and coupled to adenylyl cyclase either positively (A2a, A2b) or negatively (A1, A3) to increase or decrease cAMP, respectively (57, 58). Clancy et al. (59) noted that global increases in cAMP mediated by ADO constituted less than 20% of the measured increase in cAMP induced by forskolin, despite equal increases in CFTR-mediated Cl^- secretion. Huang et al. (60) provided a possible explanation for this incongruity. Based on patch clamp data, they proposed that luminal A2b adenosine receptors and CFTR are in close proximity and are compartmentalized to the apical membrane. They also noted that stimulation with ADO elicited changes in cAMP close to the apical membrane that fully activated CFTR Cl^- channels without altering global cAMP levels. Similarly, other investigators have proposed that asymmetric barriers exist within lateral planes along the interior of the plasma membrane that restrict cAMP movement (61).

CFTR appears to be activated in an autocrine/paracrine fashion by ADO formed in the ASL from the metabolism of ATP, ADP, and AMP by ecto-nucleotidases and ecto-apyrases. These ecto-enzymes are typically located on the apical membrane of the superficial epithelia (32, 62). Under static culture conditions, the ability of ASL height to be maintained at 7 μm was ablated following the addition of the A2b-R antagonist 8-SPT (63) or of adenosine deaminase (63), suggesting that the A2b system and CFTR are primarily responsible for regulating ASL under these conditions. Similarly, in normal human airways under basal conditions in vivo, CFTR is ~30% active (64), which may reflect basal stimulation of CFTR via adenosine receptors (65).

Under phasic motion conditions, 8-SPT-addition to normal culture surfaces reduced ASL height by approximately two- to threefold, implying that the ADO and CFTR are the primary regulators of ASL volume. However, apyrase addition, in the presence of 8-SPT, further reduced ASL height from ~7 μm to 3.5 μm, suggesting that CaCC was also active under shear stress, but less so than CFTR (29a). In keeping with these observations, addition of the CFTR antagonist $CFTR_{inh172}$ greatly reduced ASL height under shear stress conditions, and the CaCC antagonist DIDS inhibited any residual ASL height regulation (29a).

A complex regulatory relationship appears to exist between CFTR and the epithelial Na^+ channel (ENaC). In NL airways, ADO/cAMP regulates CFTR positively and ENaC negatively (66, 67). This reciprocal regulation is required to facilitate Cl^- secretion against its chemical gradient because the concentration of Cl^- is three- to fourfold higher in ASL covering normal and CF epithelia, as compared to intracellular Cl^- levels (12, 21). Consequently, the absence of a chemical gradient for Cl^- secretion can be circumvented by the inactivation of ENaC, which hyperpolarizes the apical membrane, providing the necessary electrical driving force for Cl^- to exit the cells into the ASL (12).

In CF airways, cAMP positively regulates ENaC, with disastrous consequences. ENaC is active under basal conditions, and following cAMP stimulation, ENaC activity is further increased [rather than being inactivated; (68)], resulting in Na^+/ASL hyperabsorption (66, 67). Despite an increase in ADO formation consequent to shear stress, ENaC was inactivated in CF airways (29a), suggesting that ATP-mediated inactivation of ENaC (see above) dominated over ADO/cAMP activation of ENaC. Thus, there may be a hierarchy of purinergic receptors. However, we predict that β agonists, which are typically used to prevent bronchoconstriction, may independently increase cAMP levels to stimulate ENaC in CF airways despite shear stress–induced (ATP-induced) inactivation of ENaC (68).

PHASIC MOTION AND LUNG DEFENSE

Phasic motion is an important component of airway physiology. Normal tidal breathing, interspersed with sighs, is important for maintaining alveolar integrity, likely by ATP stimulation of surfactant secretion (69). However, our data provide a mechanism that links the phasic motion of pulmonary ventilation to mucus clearance, indicating that phasic motion is also important for airway defense. For example, the phasic motion associated with respiration tonically releases sufficient ATP to regulate ion transport to optimize local normal mucus transport rates and sustain CF mucus transport.

Shear stress has been reported to increase bacterial adherence and promote bacterial infections in other organs such as the gastrointestinal tract (70). In the airways, however, the primary site of bacterial adhesion is to the mucus layer (28). Consequently, we speculate that the net effect of shear stress in the lung is to benefit the host, because shear stress increases the rate of mucus transport and hence promotes the clearance of mucus-containing adherent pathogens.

The importance of mucus hydration is reflected in the redundancy of Cl^- secretory mechanisms in NL airways, which are mediated via both CFTR and CaCC. CF airways, however, exhibit only one pathway for Cl^- secretion (i.e., CaCC), potentially leaving them unable to regulate ASL volume under adverse conditions. For example, although young CF patients can clear mucus (7) and hence appear to have physiologically adequate ASL volume, these patients appear to be more vulnerable to "catastrophic events." Specifically, clinical studies have suggested that the onset of acute exacerbations of bacterial infection in CF often follows a viral infection such as respiratory syncytial virus (RSV) (70a–70c). The importance of ASL regulation in mucus clearance (1) led us to investigate whether RSV-infected CF airway epithelial cells also lose the capacity to preserve an effective ASL volume. Indeed, RSV infection reduced ASL height under phasic motion conditions without inducing detectable epithelial damage (29a). Our data suggest this effect was mediated by a direct effect of RSV on the epithelium that led to upregulation of a class of extracellular ATPases that metabolize ATP in the ranges relevant for activation of $P2Y_2$-R under phasic motion conditions. This observation suggests that in CF patients reduction of ASL ATP concentrations by increased metabolism may contribute to bacterial infection by reducing ASL volume, leading to mucus stasis/adherence to airway surfaces. These data may explain why RSV is such a relatively dangerous pathogen in CF patients.

This effect may be specific to RSV, however, and parainfluenza virus appears to activate Cl^- secretion and inhibit Na^+ absorption (71), which is predicted to increase mucus clearance and help prevent further pathogenic infections. To date, this phenomenon has been investigated only in NL airway epithelia, and neither the effect of parainfluenza virus on CF airways, nor the role of nucleotides in this phenomenon, is known.

IMPLICATIONS OF PHASIC MOTION FOR CYSTIC FIBROSIS AIRWAY DISEASE

There are available a number of therapeutic maneuvers that are designed to increase CF airways hydration to enhance mucus clearance (71a). For example, INS37217 activates apical membrane $P2Y_2$ receptors to stimulate Cl^- secretion via CaCC and to inhibit ENaC (47, 49). Another type of therapeutic maneuver that has shown promise in enhancing mucus clearance in CF patients has been the use of aerosolized hypertonic saline. Administration of hypertonic saline (7% NaCl) increases mucociliary clearance in the lungs of CF patients (72). The mechanism for this effect likely involves increasing the volume of liquid on the airway surface by osmotically drawing water from the serosal (interstitial) spaces. In CF, hypertonic saline is uniquely active, as diffusion of the deposited hypertonic saline (NaCl) from airway surfaces is slow, reflecting the absence of the CFTR Cl^- conductance. The net effect is to increase the mass of NaCl on airway surfaces, hence increasing the osmotic driving force favoring water flow to the lumen.

Exercise is important in promoting mucus clearance and slowing the progression of CF lung disease (73, 74). Recent studies have suggested that postexercise, the Na^+ transport path in CF airway epithelia is inhibited (75, 76). Our data, showing that phasic motion induces ATP release and that ATP inhibits Na^+ transport and initiates Cl^- secretion (29a), provide a mechanism to account for amelioration of the severity of CF lung disease by exercise. Thus, we would stress the importance of regular exercise programs in CF to promote a favorable balance of ion transport and facilitate mucus clearance. We speculate that the application of additional therapies that increase shear stress to airway surfaces, e.g., exercise or physical devices, may also maintain mucus clearance and prevent the mucus stasis that produces exacerbations of the underlying bacterial airway infection.

Finally, there has always been the perplexing question in the natural history of CF lung disease whereby the upper lobes develop disease prior to the lower lobes (77). Owing to gravitational effects, the upper lobes are ventilated less (by ~50%) than the lower lobes (78). The reduced upper lobe tidal volume reduces airflow-induced shear stress on airway epithelia, which is predicted to result in less local nucleotide release. We speculate that the reduction in shear-induced ATP release in the upper lobes lowers the ASL concentration of the sole regulator of CF mucus clearance (ATP) and contributes to the vulnerability of this region to bacterial infection.

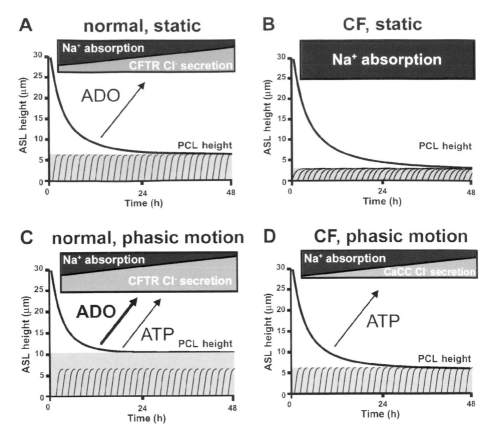

Figure 5 Schema describing PCL height regulation by active ion transport. (*A*) NL airway epithelia under static conditions coordinately regulate the rates of Na^+ absorption and Cl^- secretion to adjust PCL volume from an "excessive" PCL height (25 µm) to the physiologic PCL height with time. The blue color depicts PCL height as referenced to extended cilia. (*B*) In CF epithelia, the higher basal rate of Na^+ absorption, the failure to inhibit Na^+ transport rates, and the failure to initiate Cl^- secretion under static conditions lead to PCL depletion (note "flattened" cilia). (*C*) NL airway epithelia under phasic motion respond to increased nucleotide/nucleoside release into the PCL by shifting the balance toward Cl^- secretion via CFTR (and CaCC; *not shown*) and toward a higher PCL. (*D*) CF cultures under phasic motion conditions release sufficient ATP into the PCL to inhibit Na^+ absorption and initiate CaCC-mediated Cl^- secretion to restore PCL to a physiologically adequate height.

CONCLUSIONS

In normal airways, the concentration of nucleotides/nucleosides within the ASL modulates the relative rates of Na^+ absorption and Cl^- secretion to maintain ASL volume at levels adequate for mucus transport under basal and phasic motion conditions (Figure 5*A*,*B*). In a CF airway, ASL volume regulation is crippled by the absence of CFTR-mediated Cl^- secretion and regulation of Na^+ transport (Figure 5*C*). Importantly, our data demonstrate that CF airway epithelia can maintain an adequate ASL and mucus transport under conditions of phasic motion that promote ATP release (Figure 5*D*). The CF lung is vulnerable because it relies solely on this motion-induced ATP release to slow Na^+ transport and initiate CaCC-dependent Cl^- secretion to maintain ASL homeostasis and mucus transport (i.e., the ADO path is ineffective owing to absent CFTR function). Thus, our studies point to the central role of shear stress in ASL regulation in health and disease. Perhaps our laboratory model will be of general use to investigators who wish to investigate interactions of shear stress with epithelial function in vitro.

ACKNOWLEDGMENTS

We gratefully acknowledge the assistance of the University of North Carolina Cystic Fibrosis Center Tissue and Histology Cores, and we thank the "virtual lung group" for interesting discussions on shear stress. Our work was funded by the National Institutes of Health and the Cystic Fibrosis Foundation.

The *Annual Review of Physiology* is online at
http://physiol.annualreviews.org

LITERATURE CITED

1. Knowles MR, Boucher RC. 2002. Mucus clearance as a primary innate defense mechanism for mammalian airways ("Perspective"). *J. Clin. Invest.* 109:571–77

2. Grubb BR, Pickles RJ, Ye H, Yankaskas JR, Vick RN, et al. 1994. Inefficient gene transfer by adenovirus vector to cystic fibrosis airway epithelia of mice and humans. *Nature* 371:802–6

3. Rahmoune H, Shephard KL. 1995. State of airway surface liquid on guinea pig trachea. *J. Appl. Physiol.* 78:2020–24

4. Sims DE, Horne MM. 1997. Heterogeneity of the composition and thickness of tracheal mucus in rats. *Am. J. Physiol.* 273:L1036–41

5. Matsui H, Grubb BR, Tarran R, Randell SH, Gatzy JT, et al. 1998. Evidence for periciliary liquid layer depletion, not abnormal ion composition, in the pathogenesis of cystic fibrosis airways disease. *Cell* 95:1005–15

6. Tarran R, Grubb BR, Parsons D, Picher M, Hirsh AJ, et al. 2001. The CF salt controversy: *in vivo* observations and therapeutic approaches. *Mol. Cell* 8:149–58

7. Robinson M, Bye PTB. 2002. Mucociliary clearance in cystic fibrosis. *Pediatr. Pulmonol.* 33:293–306

8. Basser PJ, McMahon TA, Griffith P. 1989. The mechanism of mucus clearance in cough. *J. Biomech. Eng.* 111:288–97

9. Burnstock G. 2002. Potential therapeutic targets in the rapidly expanding field of purinergic signalling. *Clin. Med.* 2:45–53

10. Guyot A, Hanrahan JW. 2002. ATP release from human airway epithelial cells studied using a capillary cell culture system. *J. Physiol.* 545:199–206

11. Mason SJ, Paradiso AM, Boucher RC.

1991. Regulation of transepithelial ion transport and intracellular calcium by extracellular adenosine triphosphate in human normal and cystic fibrosis airway epithelium. *Br. J. Pharmacol.* 103:1649–56

12. Boucher RC. 1994. Human airway ion transport (Part 1). *Am. J. Respir. Crit. Care Med.* 150:271–81

13. Crews A, Taylor AE, Ballard ST. 2001. Liquid transport properties of porcine tracheal epithelium. *J. Appl. Physiol.* 91: 797–802

14. Farinas J, Kneen M, Moore M, Verkman AS. 1997. Plasma membrane water permeability of cultured cells and epithelia measured by light microscopy with spatial filtering. *J. Gen. Physiol.* 110:283–96

15. Matsui H, Davis CW, Tarran R, Boucher RC. 2000. Osmotic water permeabilities of cultured, well-differentiated normal and cystic fibrosis airway epithelia. *J. Clin. Invest.* 105:1419–27

16. Spring KR. 1999. Epithelial fluid transport: a century of investigation. *News Physiol. Sci.* 14:92–98

17. Kilburn KH. 1968. A hypothesis for pulmonary clearance and its implications. *Am. Rev. Respir. Dis.* 98:449–63

18. Verkman AS. 2001. Lung disease in cystic fibrosis: Is airway surface liquid composition abnormal? *Am. J. Physiol.* 281:L306–8

19. Kotaru C, Hejal RB, Finigan JH, Coreno AJ, Skowronski ME, et al. 2003. Desiccation and hypertonicity of the airway surface fluid and thermally induced asthma. *J. Appl. Physiol.* 94:227–33

20. Effros RM, Peterson B, Casaburi R, Su J, Dunning M, et al. 2005. Epithelial lining fluid solute concentrations in chronic obstructive lung disease patients and normal subjects. *J. Appl. Physiol.* 99:1286–92

21. Tarran R, Grubb BR, Gatzy JT, Davis CW, Boucher RC. 2001. The relative roles of passive surface forces and active ion transport in the modulation of airway surface liquid volume and composition. *J. Gen. Physiol.* 118:223–36

22. Bansil R, Stanley E, LaMont JT. 1995. Mucin biophysics. *Annu. Rev. Physiol.* 57:635–57

23. Chmiel JF, Davis PB. 2003. State of the art: why do the lungs of patients with cystic fibrosis become infected and why can't they clear the infection? *Respir. Res.* 4:8

24. Gibson RL, Burns JL, Ramsey BW. 2003. Pathophysiology and management of pulmonary infections in cystic fibrosis. *Am. J. Respir. Crit. Care Med.* 168:918–51

25. Welsh MJ, Smith AE. 1993. Molecular mechanisms of CFTR chloride channel dysfunction in cystic fibrosis. *Cell* 73:1251–54

26. Mall M, Grubb BR, Harkema JR, O'Neal WK, Boucher RC. 2004. Increased airway epithelial Na$^+$ absorption produces cystic fibrosis-like lung disease in mice. *Nat. Med.* 10:487–93

27. Matsui H, Verghese MW, Kesimer M, Schwab UE, Randell SH, et al. 2005. Reduced 3-dimensional motility in dehydrated airway mucus prevents neutrophil capture and killing bacteria on airway epithelial surfaces. *J. Immunol.* 175:1090–99

28. Worlitzsch D, Tarran R, Ulrich M, Schwab U, Cekici A, et al. 2002. Effects of reduced mucus oxygen concentration in airway *Pseudomonas* infections of cystic fibrosis patients. *J. Clin. Invest.* 109:317–25

29. Frizzell RA, Pilewski JM. 2004. Finally, mice with CF lung disease. *Nat. Med.* 10:452–54

29a. Tarran R, Button B, Picher M, Paradiso AM, Ribeiro CM, et al. 2005. Normal and cystic fibrosis airway surface liquid homeostasis: the effects of phasic shear stress and viral infections. *J. Biol. Chem.* 280:35751–59

30. Fredberg JJ, Hoenig A. 1978. Mechanical response of lungs at high frequencies. *J. Biochem. Eng.* 100:57–66

31. Murray CD. 1926. The physiological principle of minimum work. I. The vascular

system and the cost of blood. *Proc. Natl. Acad. Sci. USA* 12:201–14

32. Lazarowski ER, Boucher RC, Harden TK. 2000. Constitutive release of ATP and evidence for major contribution of ecto-nucleotide pyrophosphatase and nucleoside diphosphokinase to extracellular nucleotide concentrations. *J. Biol. Chem.* 275:31061–68

33. Watt WC, Lazarowski ER, Boucher RC. 1998. Cystic fibrosis transmembrane regulator-independent release of ATP. Its implications for the regulation of P2Y$_2$ receptors in airway epithelia. *J. Biol. Chem.* 273:14053–58

34. Donaldson SH, Picher M, Boucher RC. 2002. Secreted and cell-associated adenylate kinase and nucleoside diphosphokinase contribute to extracellular nucleotide metabolism on human airway surfaces. *Am. J. Respir. Cell Mol. Biol.* 26:209–15

35. Balcells M, Fernandez Suarez M, Vazquez M, Edelman ER. 2005. Cells in fluidic environments are sensitive to flow frequency. *J. Cell. Physiol.* 204:329–35

36. Praetorius HA, Spring KR. 2005. A physiological view of the primary cilium. *Annu. Rev. Physiol.* 67:515–29

37. Thi MM, Tarbell JM, Weinbaum S, Spray DC. 2004. The role of the glycocalyx in reorganization of the actin cytoskeleton under fluid shear stress: a "bumper-car" model. *Proc. Natl. Acad. Sci. USA* 101:16483–88

37a. Nilius B, Droogmans G, Wondergem R. 2003. Transient receptor potential channels in endothelium: solving the calcium entry puzzle? *Endothelium* 10:5–15

38. Carattino MD, Sheng S, Kleyman TR. 2004. Epithelial Na$^+$ channels are activated by laminar shear stress. *J. Biol. Chem.* 279:4120–26

38a. Okada SF, O'Neal WK, Huang P, Nicholas RA, Ostrowski LE, et al. 2004. Voltage-dependent anion channel-1 (VDAC-1) contributes to ATP release and cell volume regulation in murine cells. *J. Gen. Physiol.* 124:513–26

39. Shyy JY, Chien S. 2002. Role of integrins in endothelial mechanosensing of shear stress. *Circ. Res.* 91:769–75

40. Resnick N, Yahav H, Shay-Salit A, Shushy M, Schubert S, et al. 2003. Fluid shear stress and the vascular endothelium: for better and for worse. *Prog. Biophys. Mol. Biol.* 81:177–99

41. Chrzanowska-Wodnicka M, Burridge K. 1992. Rho, rac and the actin cytoskeleton. *Bioessays* 14:777–78

42. Dudek SM, Garcia JG. 2001. Cytoskeletal regulation of pulmonary vascular permeability. *J. Appl. Physiol.* 91:1487–500

43. Tzima E, del Pozo MA, Kiosses WB, Mohamed SA, Li S, et al. 2002. Activation of Rac1 by shear stress in endothelial cells mediates both cytoskeletal reorganization and effects on gene expression. *EMBO J.* 21:6791–800

44. Cressman VL, Lazarowski E, Homolya L, Boucher RC, Koller BH, Grubb BR. 1999. Effect of loss of P2Y$_2$ receptor gene expression on nucleotide regulation of murine epithelial Cl$^-$ transport. *J. Biol. Chem.* 274:26461–68

45. Lazarowski ER, Boucher RC. 2000. UTP as an extracellular signaling molecule. *News Physiol. Sci.* 16:1–5

46. Schwiebert EM. 2001. ATP release mechanisms, ATP receptors and purinergic signalling along the nephron. *Clin. Exp. Pharmacol. Physiol.* 28:340–50

47. Devor DC, Pilewski JM. 1999. UTP inhibits Na$^+$ absorption in wild-type and DeltaF508 CFTR-expressing human bronchial epithelia. *Am. J. Physiol.* 276:C827–37

48. Yue G, Malik B, Yue G, Eaton DC. 2002. Phosphatidylinositol 4,5-bisphosphate (PIP2) stimulates epithelial sodium channel activity in A6 cells. *J. Biol. Chem.* 277:11965–69

49. Kunzelmann K, Bachhuber T, Regeer R, Markovich D, Sun J, Schreiber R. 2005. Purinergic inhibition of the epithelial Na$^+$ transport via hydrolysis of PIP2. *FASEB J.* 19:142–43

50. Clarke LL, Harline MC, Otero MA, Glover GG, Garrad RC, et al. 1999. Desensitization of P2Y2 receptor-activated transepithelial anion secretion. *Am. J. Physiol.* 276:C777–87

51. Lazarowski ER, Paradiso AM, Watt WC, Harden TK, Boucher RC. 1997. UDP activates a mucosal-restricted receptor on human nasal epithelial cells that is distinct from the $P2Y_2$ receptor. *Proc. Natl. Acad. Sci. USA* 94:2599–603

52. Otero M, Garrad RC, Velazquez B, Hernandez-Perez MG, Camden JM, et al. 2000. Mechanisms of agonist-dependent and -independent desensitization of a recombinant P2Y2 nucleotide receptor. *Mol. Cell. Biochem.* 205:115–23

53. Tarran R, Loewen ME, Paradiso AM, Olsen JC, Gray MA, et al. 2002. Regulation of murine airway surface liquid volume by CFTR and Ca^{2+}-activated Cl^- conductances. *J. Gen. Physiol.* 120:407–18

54. Clarke LL, Grubb BR, Yankaskas JR, Cotton CU, McKenzie A, Boucher RC. 1994. Relationship of a non-CFTR mediated chloride conductance to organ-level disease in *cftr*($-/-$) mice. *Proc. Natl. Acad. Sci. USA* 91:479–83

55. Grubb BR, Vick RN, Boucher RC. 1994. Hyperabsorption of Na^+ and raised Ca^{2+}-mediated Cl^- secretion in nasal epithelia of CF mice. *Am. J. Physiol.* 266:C1478–83

56. Picher M, Burch LH, Boucher RC. 2004. Metabolism of P2 receptor agonists in human airways: implications for mucociliary clearance and cystic fibrosis. *J. Biol. Chem.* 279:20234–41

57. Klinger M, Freissmuth M, Nanoff C. 2002. Adenosine receptors: G protein-mediated signalling and the role of accessory proteins. *Cell. Signal.* 14:99–108

58. Feoktistov I, Biaggioni I. 1997. Adenosine A_{2B} receptors. *Pharmacol. Rev.* 49:381–402

59. Clancy JP, Ruiz FE, Sorscher EJ. 1999. Adenosine and its nucleotides activate wild-type and R117H CFTR through an A2B receptor-coupled pathway. *Am. J. Physiol.* 276:C361–69

60. Huang P, Lazarowski ER, Tarran R, Milgram SL, Boucher RC, Stutts MJ. 2001. Compartmentalized autocrine signaling to cystic fibrosis transmembrane conductance regulator at the apical membrane of airway epithelial cells. *Proc. Natl. Acad. Sci. USA* 98:14120–25

61. Schwartz JH. 2001. The many dimensions of cAMP signaling. *Proc. Natl. Acad. Sci. USA* 98:13482–84

62. Picher M, Boucher RC. 2003. Human airway ecto-adenylate kinase. A mechanism to propagate ATP signaling on airway surfaces. *J. Biol. Chem.* 278:11256–64

63. Lazarowski ER, Tarran R, Grubb BR, van Heusden CA, Okada S, Boucher RC. 2004. Nucleotide release provides a mechanism for airway surface liquid homeostasis. *J. Biol. Chem.* 279:36855–64

64. Knowles MR, Stutts MJ, Spock A, Fischer N, Gatzy JT, Boucher RC. 1983. Abnormal ion permeation through cystic fibrosis respiratory epithelium. *Science* 221:1067–70

65. Hentchel-Franks K, Lozano D, Eubanks-Tarn V, Cobb B, Fan L, et al. 2004. Activation of airway Cl^- secretion in human subjects by adenosine. *Am. J. Respir. Cell Mol. Biol.* 31:140–46

66. Stutts MJ, Canessa CM, Olsen JC, Hamrick M, Cohn JA, et al. 1995. CFTR as a cAMP-dependent regulator of sodium channels. *Science* 269:847–50

67. Kunzelmann K, Schreiber R, Boucherot A. 2001. Mechanisms of the inhibition of epithelial Na^+ channels by CFTR and purinergic stimulation. *Kidney Int.* 60:455–61

68. Boucher RC, Stutts MJ, Knowles MR, Cantley L, Gatzy JT. 1986. Na^+ transport in cystic fibrosis respiratory epithelia. Abnormal basal rate and response to adenylate cyclase activation. *J. Clin. Invest.* 78:1245–52

69. Dietl P, Haller T, Mair N, Frick M. 2001. Mechanisms of surfactant exocytosis in alveolar type II cells *in vitro* and *in vivo*. *News Physiol. Sci.* 16:239–43

70. Thomas WE, Trintchina E, Forero M, Vogel V, Sokurenko EV. 2002. Bacterial adhesion to target cells enhanced by shear force. *Cell* 109:913–23

70a. Armstrong D, Grimwood K, Carlin JB, Carzino R, Hull J, et al. 1998. Severe viral respiratory infections in infants with cystic fibrosis. *Pediatr. Pulmonol.* 26:371–79

70b. Hiatt PW, Grace SC, Kozinetz CA, Raboudi SH, Treece DG, et al. 1999. Effects of viral lower respiratory tract infection on lung function in infants with cystic fibrosis. *Pediatrics* 103:619–26

70c. Abman SH, Ogle JW, Butler-Simon N, Rumack CM, Accurso FJ. 1988. Role of respiratory syncytial virus in early hospitalizations for respiratory distress of young infants with cystic fibrosis. *J. Pediatr.* 113:826–30

71. Kunzelmann K, Konig J, Sun J, Markovich D, King NJ, et al. 2004. Acute effects of parainfluenza virus on epithelial electrolyte transport. *J. Biol. Chem.* 279:48760–66

71a. Boucher RC. 2004. New concepts of the pathogenesis of cystic fibrosis lung disease. *Eur. Respir. J.* 23:146–58

72. Robinson M, Regnis JA, Bailey DL, King M, Bautovich GJ, Bye PT. 1996. Effect of hypertonic saline, amiloride, and cough on mucociliary clearance in patients with cystic fibrosis. *Am. J. Respir. Crit. Care Med.* 153:1503–9

73. Schneiderman-Walker J, Pollock SL, Corey M, Wilkes DD, Canny GJ, et al. 2000. A randomized controlled trial of a 3-year home exercise program in cystic fibrosis. *J. Pediatr.* 136:304–10

74. Bradley J, Moran F. 2002. Physical training for cystic fibrosis. *Cochrane Database Syst. Rev.* CD002768

75. Alsuwaidan S, Li Wan Po A, Morrison G, Redmond A, Dodge JA, et al. 1994. Effect of exercise on the nasal transmucosal potential difference in patients with cystic fibrosis and normal subjects. *Thorax* 49:1249–50

76. Hebestreit A, Kersting U, Basler B, Jeschke R, Hebestreit H. 2001. Exercise inhibits epithelial sodium channels in patients with cystic fibrosis. *Am. J. Respir. Crit. Care Med.* 164:443–46

77. Santis G, Hodson ME, Strickland B. 1991. High resolution computed tomography in adult cystic fibrosis patients with mild lung disease. *Clin. Radiol.* 44:20–22

78. Milic-Emili J, Henderson JA, Dolovich MB, Trop D, Kaneko K. 1966. Regional distribution of inspired gas in the lung. *J. Appl. Physiol.* 21:749–59

Annu. Rev. Physiol. 2006. 68:563–83
doi: 10.1146/annurev.physiol.68.072304.113102
First published online as a Review in Advance on September 30, 2005

CHRONIC EFFECTS OF MECHANICAL FORCE ON AIRWAYS

Daniel J. Tschumperlin and Jeffrey M. Drazen

Physiology Program, Department of Environmental Health, Harvard School of Public Health, Boston, Massachusetts 02115; email: dtschump@hsph.harvard.edu, jdrazen@nejm.org

Key Words mechanotransduction, airway remodeling, morphogenesis, epithelium, smooth muscle

■ **Abstract**   Airways are embedded in the mechanically dynamic environment of the lung. In utero, this mechanical environment is defined largely by fluid secretion into the developing airway lumen. Clinical, whole lung, and cellular studies demonstrate pivotal roles for mechanical distention in airway morphogenesis and cellular behavior during lung development. In the adult lung, the mechanical environment is defined by a dynamic balance of surface, tissue, and muscle forces. Diseases of the airways modulate both the mechanical stresses to which the airways are exposed as well as the structure and mechanical behavior of the airways. For instance, in asthma, activation of airway smooth muscle abruptly changes the airway size and stress state within the airway wall; asthma also results in profound remodeling of the airway wall. Data now demonstrate that airway epithelial cells, smooth muscle cells, and fibroblasts respond to their mechanical environment. A prominent role has been identified for the epithelium in transducing mechanical stresses, and in both the fetal and mature airways, epithelial cells interact with mesenchymal cells to coordinate remodeling of tissue architecture in response to the mechanical environment.

INTRODUCTION

The human airway tree forms in the pseudoglandular stage of lung development through a series of lateral and dichotomous branchings (1, 2). This developmental process gives rise to a conducting network of tubes that facilitates convective gas transport to and from the alveolar gas exchange region of the lung. Although there is no airflow in utero, the branching airways experience significant transpulmonary pressure as a result of epithelial fluid secretion into the airway lumen (3). The epithelial and mesodermal cells of the developing airways, influenced in part by this mechanical environment, undergo rapid proliferation (4, 5) and orchestrate continuous remodeling of the extracellular matrix (5). After birth, the mechanical environment of the airways continues to be shaped by growth (although this shaping slows with age), but in the presence of both surface tension at the air-liquid interface of the luminal surface and continuously varying transpulmonary pressures that fluctuate with breathing (6). Before birth, the primitive mesoderm differentiates into a smooth muscle layer wrapped around the epithelium (7, 8) and thus provides a means by which the mechanical environment (and lumen size) of the mature airway can be abruptly and profoundly changed. Activation and shortening of airway smooth muscle can result in buckling of the airway and substantial compressive mechanical forces in the airway wall (9).

The major structural cells of the airways (epithelium, fibroblast, smooth muscle cells) all sense and respond to this highly dynamic mechanical environment (10–12). In turn, these cells cooperate to establish, maintain, and remodel the architecture and mechanical properties of the airway wall (1, 13). Diseases of the airways and lungs modify the mechanical stresses to which the airways are exposed and alter the structure and mechanical behavior of the airways (14). This review focuses on a growing and diverse experimental record demonstrating the marked influence that mechanical force, and more broadly the mechanical environment, exert on the airways in development, health, and disease.

Specifically, this review focuses on the chronic and large-scale mechanical perturbations that occur in airways, as distinct from those in alveolar airspaces. The chapter by Fredberg in this volume defines the source, magnitude, and direction of mechanical forces that are extant in the lung during development and in health and disease (14a). Boucher's chapter details the effects of acute and small-scale forces on airway function (14b), whereas dos Santos & Slutsky's chapter sets forth the consequences of mechanical perturbation on the alveolar zone (14c).

MECHANICAL FORCE AND THE DEVELOPING AIRWAYS IN UTERO

Developing organisms and tissues undergo rapid cellular growth, proliferation, and migration during morphogenesis; these processes are broadly influenced by the mechanical environment (15–17). Recent evidence from Drosophila oogenesis and embryogenesis studies have demonstrated key roles for mechanical forces in developmental processes (18, 19). In both developing and mature organisms, cell proliferation and apoptosis are intrinsically coupled to coordinate tissue growth and maintenance (20). Although substantial attention has been focused on the role played by biochemical signaling in this coupling process (21), a fundamental contribution from mechanical feedback has been proposed (22) and seems likely.

In the developing lung, the cellular processes of proliferation and migration, which themselves generate substantial forces (17, 22), are superimposed onto mechanical stresses that result from fluid inflation of the fetal lung and intermittent fetal breathing movements that move fluid along the developing airway (reviewed in Reference 23). During development the lumen of the airways are filled predominantly by fluid secreted from the epithelium (3); in fetal sheep, secretion rates can reach 3.5–5.5 ml·h^{-1}·kg^{-1} (24, 25). Flow of luminal fluid out of the lung is resisted by the larynx and nasopharynx, with the vocal cords functioning as a one-way valve (26, 27). Secretion is thus sufficient to increase luminal pressure 2–3 torr above amniotic fluid pressure (28). This transpulmonary pressure gradient actively distends the lung and passively distends the chest wall to a volume roughly the equivalent of functional residual capacity (29).

The important developmental role of the airway distention (volume change) induced by fluid secretion is amply illustrated by clinical scenarios in which the capacity for distention fails to develop adequately. In oligohydramnios, there is a decrease in fluid secretion, whereas in congenital diaphragmatic hernia, skeletal dysplasia, and neuromuscular paralysis of the diaphragm, the modified mechanical environment of the forming thorax results in reduced lung distention (30). These clinical scenarios are accompanied by lung hypoplasia and a reduced number of airway generations (30–32). Experiments in fetal sheep have demonstrated the direct linkage between reduced distention and lung hypoplasia (33). These experiments have also highlighted the counterprocess of lung hyperplasia in the presence of increased lung distention, which is accomplished via occlusion of the developing trachea (33). It is important to note that the bulk of experimental evidence suggests that these developmental effects are driven by change in volume, not pressure, suggesting a critical link between tissue deformation and cellular signaling processes in the developing lung (34). The acceleration of fetal-lung growth and development caused by increased distention in animal models (35, 36) has led to a potential clinical treatment, in human fetuses, of congenital diaphragmatic hernia, a condition in which the lung is hypoplastic. This treatment involves a surgery in which a device is introduced to achieve occlusion of the fetal trachea. Although this approach is theoretically sound, the disappointing clinical results obtained thus far (37) suggest that there are as yet unappreciated additional factors that modulate formation of the lung in utero.

In addition to manifesting lung hypoplasia, human fetal lungs that develop in the presence of reduced distention exhibit a paucity of cells staining positively for smooth muscle α-actin in the space around the developing airways (38). In freshly dissociated embryonic mesenchymal cells, static uniaxial stretch enhances smooth muscle α-actin expression, and in embryonic day 11 mouse lung explants, airway distention, caused by luminal osmotic stress, enhances periluminal smooth muscle α-actin expression (38). This latter response is accompanied by elongation of the periluminal mesenchymal cells, supporting a fundamental linkage between mechanical force, cell shape, and cell differentiation that appears to span a variety of cell types (39–42).

Experiments with fetal mouse lung explants have demonstrated that the internal forces generated by cells during morphogenesis contribute to the pattern of branching in the developing lung. Incubation of embryonic lung buds from the twelfth day of gestation with cytotoxic necrotizing factor 1, an activator of the small GTPase Rho, revealed a concentration-dependent effect on the number of terminal airway buds formed, presumably mediated through the effects of Rho on generation of cytoskeletal tension (43). More detailed experiments have demonstrated that inhibition of cytoskeletal tension with inhibitors of Rho-associated kinase, myosin light chain kinase, myosin ATPase, and microfilament integrity decreases the number of terminal buds formed (44). The authors of these reports demonstrated that these soluble mediators also influence gradients in basement membrane thickness, leading them to suggest that cellular force generation plays an important role in defining the spatial differentials of cell growth and matrix remodeling that drive embryonic lung development (44). Indeed, it is well known that there are differences in cell proliferation among geographic regions of the developing lung (4, 5); these are thought to occur through the local effects of diffusible factors (1, 45). In addition to mechanical stresses generated by cytoskeletal tension, mechanical stresses also arise out of local gradients in cell proliferation (22, 46). These data suggest that further experiments with embryonic lung buds may provide continuing insight into the influences that the local mechanical environment exerts on branching morphogenesis and lung development.

FETAL AIRWAY CELL RESPONSES TO MECHANICAL STRESS

Limitations in human tissue availability and the technical challenges inherent to fetal-lung studies have made in vitro approaches an attractive alternative with which to study mechanical effects on cellular behavior. These in vitro approaches have followed two basic strategies: culture of cells on matrix-coated two-dimensional elastic substrates that can be uni- or biaxially stretched, or culture of cells within a three-dimensional sponge (approximating some features of alveolar morphology) that can be uniaxially elongated (Table 1).

Two-dimensional stretching experiments have focused primarily on fibroblasts derived from fetal lungs. IMR-90 human fetal fibroblasts grown on collagen-coated membranes and stretched (10% change in surface area) at 1 Hz for 2–4 days exhibited increased cell numbers when compared to cells grown without stretch. Conditioned media from stretched fibroblasts, when compared to a conditioned medium from fibroblasts that were not stretched, exhibited increased mitogenic activity (47). The same fibroblast line exposed to stretch (maximum 20% elongation) for 48 h on laminin-, fibronectin-, and elastin-coated substrates exhibited no increase in proliferation on any of the three coatings (48). Rather than exhibiting enhanced proliferation, these cells increased their expression of procollagen $\alpha 1(I)$ mRNA and synthesis of new collagen protein, but only when stretched on elastin or laminin and not when stretched on fibronectin (48). In contrast to both of the previous findings, primary fibroblasts isolated from embryonic day 18–20 mouse lungs (canalicular stage) exposed to 5% biaxial strain at 1 Hz for 15 minutes out of every hour (intermittent) showed no increase in proliferative index. Rather, the day 19–derived fibroblasts exhibited an increased apoptotic index and increased

TABLE 1 Fetal cell stretch responses: dependence on culture and mechanical parameters

Cell(s)	Mechanical deformation	Magnitude	Frequency	Duration	Matrix attachment	Outcome	Ref.
Human IMR-90 fibroblast	Nonuniform biaxial stretch	$10\% \leftrightarrows SA$	1 Hz	2–4 days	Col I	↑ Cell numbers	(47)
Human IMR-90 fibroblast	Nonuniform biaxial stretch	Maximum 20% elongation	1 Hz	2 days	Elastin Laminin Fibronectin	↑ Collagen synthesis ↑ Collagen synthesis No ⇆	(48)
Embryonic mouse fibroblast	Uniform biaxial stretch	5%	1 Hz	15 min h^{-1} up to 24 h	Col I	Day 19 cells: ↑ apoptosis Days 18 & 20: no ⇆	(49)
Mixed embryonic mouse epithelium and fibroblast	Sponge uniaxial elongation	5%	1 Hz	15 min h^{-1} various durations	Gelatin (processed collagen)	↑ Cell numbers ↑ DNA synthesis ↓ Cell doubling time ↑ GAGs and PGs ↑ Col I, IV, biglycan ↑ FN synthesis	(52) (56) (57) (58)

Abbreviations: SA, surface area; Col I, type I collagen; Col IV, type IV collagen; ⇆, change; ↑, increase; ↓, decrease; GAGs, glycosaminoglycans; PGs, proteoglycans; FN, fibronectin.

cell cycle arrest (49). The cells derived from embryonic days 18 and 20 did not respond differentially to mechanical stress in terms of proliferation and apoptosis (49).

These results emphasize the complexity in mechanical responses among fetal-lung fibroblasts. The cellular responses clearly depend greatly on the mechanical environment (frequency, magnitude, uniformity, duration), the substrate interface to which the cells are adherent, and the source of the cells studied (primary or cell line, embryonic stage, human or mouse; summarized in Table 1). Careful control and examination of these variables are needed to dissect precisely the mechanical influences on fetal-lung fibroblast behavior. A necessary step in defining the relevant responses will be comparison of in vitro outcomes to responses in the whole animal or isolated lung, as exemplified in only a limited number of studies thus far (38, 50).

Another level of complexity has been added to in vitro investigations with the use of an organotypic cell-stretching system in which purified or mixed cell populations isolated from embryonic day 18–21 mouse lungs have been grown on a substrate resembling the alveolar microstructure of the lung. These sponge-like substrates are subject to mechanical stress using a stretching protocol that mimics the intermittent fetal breathing movements that occur in utero (51). The deformations are therefore small (5% linear elongation of the sponge) and intermittent (1 Hz, 15 minutes per hour). Nevertheless, when compared to cells on substrates not subjected to stretch, cells subject to these deformations show a significant increase in cell number, reduction in cell doubling time, and increase in DNA synthesis (52). The DNA synthetic response was replicated when purified fetal epithelial cells and fibroblasts were stretched alone (53, 54) but not when the purified cells or cocultures were exposed to similar deformations in a two-dimensional stretch system, demonstrating the importance of tissue architecture in the response to stretch (53).

Mechanical stretch of these organotypic three-dimensional cultures causes the elaboration of growth factors that influenced epithelial, but not fibroblast, proliferation (55). This observation indicates that the response of the mixed cultures likely represents both paracrine (fibroblast-to-epithelium) stimulation induced by stress as well as direct and/or autocrine responses to mechanical stress within each cell type. Cells grown in mixed epithelial and fibroblast cultures also respond to intermittent mechanical stretching with increased synthesis of glycosaminoglycans and proteoglycans (56), enhanced expression of collagen types I and IV and biglycan (57), and fibronectin synthesis and secretion (58), demonstrating that mechanical stresses likely provide a prominent regulatory cue for deposition of the extracellular matrix during lung development (Table 1).

Taken together, the observations in clinical scenarios, fetal-lung bud cultures and isolated cell stretching experiments all emphasize the pivotal roles the underlying mechanical environment plays in the morphogenesis and maturation of the developing airway. Continuing investigations that span these approaches are likely to reveal the integrated mechanisms that link mechanics and morphogenesis.

MECHANICAL FORCE AND THE AIR-FILLED AIRWAYS

At the time of birth, the mechanical environment of the airway abruptly changes: As soon as the air-liquid interface is formed, the fluid pressure that distended the lung in utero is replaced by a complex and dynamic balance of surface, tissue, and muscle forces. The rapid transition of the epithelium from net secretion to net absorption of fluid creates an air-filled lumen for gas transport and exchange (3). Formation of the air-liquid interface in the airway lumen results in surface tension, which exerts a net negative pressure on the airway (if we adopt the convention that pressures acting to collapse a tube are defined as negative and those tending to inflate a tube are defined as positive). Under static (no-flow) conditions, the pressure within the airway lumen is equivalent to that in the alveoli surrounding the airway, and thus the net transmural pressure across the airway wall is zero. What then distends the airway? It is the tension transmitted from the pleural surface to the airway through the solid structures of the lung (59). The size of the airway lumen is set by the interaction of these forces with the intrinsic mechanical properties of the airway wall tissue.

The mechanical properties of the airway wall are complex, defined by active and passive contributions from multiple cell types embedded in a heterogeneous, nonlinear, three-dimensional matrix architecture (14). Passive airway tissue embedded in an inflated lung (in which smooth muscle tone is minimized) appears to contribute to geometric stability of the airway such that the airway resists both excessive expansion (60) and collapse (9, 61, 62). When active, the contractile force–generating capacity of airway smooth muscle can profoundly change the dimensions (60), geometry (62), and mechanical behavior of the airways (63, 64), along with the stress state within the airway wall (9). Asthma is a common disease process in which there is substantial airway smooth muscle constriction with an attendant mechanical impact on the airway lumen and wall (65). To further complicate matters, the force-generating capacity of smooth muscle is extraordinarily plastic and dynamically determined by its dimensional and loading history (66–68). The mechanical environment of the airway wall is thus dynamically defined by the volume of the lung, the surface tension in the airways, the mechanical properties of the airway wall, and the state of activation of the airway smooth muscle. Each of these properties is interrelated and subject to change with diseases of the lung; thus, alterations in the mature airway mechanical environment are gaining increasing attention as potential modifiers of airway structure and function.

MECHANICAL FORCES AND AIRWAY RESPONSES

Unlike the case for the developing lung in utero, for the airways in the mature lung, no clinical scenarios arise that allow separation of the effects of altered mechanical environment from other environmental, inflammatory, or disease effects. This has led to the development of a variety of whole organism and tissue models to begin elucidating the effects of the mechanical environment on mature airways.

The simplest way to experimentally perturb the mechanical environment of the airways is to change lung volume. Although the effects of this intervention on alveolar function are covered in detail in the chapter by dos Santos & Slutsky (14c), it is important to consider also lung volume effects on airways. In an open-chest mechanical ventilation preparation in rabbits, four hours ventilation with high end-expiratory pressures (9 cm H_2O) increased expression of mRNA for procollagen III and IV, fibronectin, bFGF, and TGF-β1 (69). Similarly, four hours of high-pressure ventilation (peak airway pressure 35 cm H_2O) in isolated rat lungs increased mRNA levels for laminin and type I and type IV procollagen relative to those seen in baseline pressure ventilation (70). Direct mechanical stretch of isolated mouse lobes led to increased activation of NF-κB, AP-1, and mitogen-activated protein kinases within 5 to 15 minutes (71).

In the above studies it was presumed that the chief effect of lung volume was in the alveolus, but recent studies demonstrate that the airways are also a target of changes in lung volume and pressure. In a closed-chest mechanical ventilation preparation in rats, two hours of high tidal volume ventilation (30 ml kg^{-1}) increased versican, biglycan, and heparin sulfate proteoglycan expression around the airways (as measured by immunohistochemistry), which was accompanied by increased tissue elastance and damping (72). Through the use of cDNA microarrays, high tidal volume ventilation in rats for 30 min was shown to significantly increase expression of 10 genes and suppress expression of 12 genes (73). Somewhat surprisingly, when four genes were selected for further analysis, the changes in expression of all four (Egr-1, c-Jun, heat shock protein 70, and IL-1β) were found to be localized primarily in the bronchial epithelium (73).

In a related approach aimed at elucidating the effect of chronic strain on airway behavior, isolated bronchial segments were exposed to a continuous positive transmural pressure of 7 cm H_2O, and tissue explants were subjected to a 50% change in lumen diameter (74). These studies complement the well-known effects of both static (75–77) and dynamic (78) strain on airway smooth muscle contractile behavior by demonstrating that chronic strain can increase the size and specific compliance of passive rabbit airways (74). Together, the ventilation and airway strain models show that increased distention of the mature airway lumen can lead to signaling, structural, and mechanical responses, much as in the developing lung.

Another method to modulate the mechanical environment of the airways is to activate smooth muscle and study the effects on the cells of the airway wall (79). In a method previously developed in guinea pig lungs (80), broncho-active agents were administered via tracheal superfusion to isolated mouse lungs. The superfusate was administered via the trachea at a constant flow rate, draining through multiple holes punctured in the pleura. The pressure at the airway opening was monitored in real time, reflecting the degree of airway constriction. Perfusion-fixation of the lungs at the apex of the constriction event preserved both the geometric and biochemical effects of constriction and allowed signaling responses to be investigated by immunohistochemistry. Methacholine-constricted airways exhibited an increase in immunostaining with an antibody to the phosphorylated, active form of the epidermal growth factor receptor localized in the airway epithelium (80). Co-perfusion of airways with methacholine and isoproterenol prevented airway constriction, confirmed by a lack of increase in perfusion pressure and reduced morphological signs of constriction. This co perfusion significantly attenuated the epidermal growth factor receptor phosphorylation, suggesting a direct response of the airway epithelium to the mechanical forces of airway constriction.

The results discussed above, although limited in scope, demonstrate that alterations in the airway mechanical environment can elicit both acute and sustained biological responses in the cells and tissue of the airway wall. Further development of these models will refine our understanding of airway responses to the mechanical environment, and future studies can be designed to superimpose this understanding onto relevant disease models (71).

MECHANICAL FORCES AND RESPONSES OF AIRWAY CELLS IN CULTURE

As in the fetal lung, tissue-level results have prompted the development of cell culture models to complement and extend our molecular and mechanistic understanding of responses to the airway mechanical environment.

Airway Smooth Muscle

The acute effects of mechanical loading and length oscillation on smooth muscle biophysics have been reviewed in detail elsewhere (68, 81, 82). Additional biological effects of stretch and the mechanical environment on smooth muscle have been extensively studied, using strips of intact muscle and isolated cells grown on elastic substrates. Sinusoidal stretching of intact bovine tracheal smooth muscle decreases expression of smooth muscle α-actin in inactive samples, but not in histamine- or carbachol-contracted samples, suggesting the existence of antagonistic interactions between mechanical and biochemical signaling (83). In a two-dimensional cell-stretching device, cyclic stretching of canine airway smooth muscle cells led to increased proliferation (84) and contractile protein expression and activity (85–87), accompanied by increases in the magnitude and velocity of contraction (88). Stretch also led to stiffening of smooth muscle cells via rearrangement of the cytoskeletal lattice (89) and recruitment of actin to local areas of stress application (90). Transduction of stretch was associated with rapid phosphorylation of focal adhesion proteins (91) and with increased calcium sensitivity (92). These in vitro responses are similar to the mechanosensitive responses identified in intact tracheal muscle strips (93, 94). The responses to stretch of smooth muscle cells isolated from mature organisms appear to counteract partially the loss of differentiated cell function that occurs in culture (95). These responses to stretch may be related to the previously discussed specification of smooth muscle phenotype by the mechanical environment in the developing lung (38, 41), although contrary evidence points to the complexity of this issue (96). In addition to affecting smooth muscle cytoskeleton and differentiation, stretching of isolated smooth muscle cells increases synthesis of the inflammatory cytokine IL-8, hinting at additional biological responses important to disease (97). The IL-8 response to stretch is driven by increased kinase activity of ERK1/2, JNK1, and p38, and the DNA-binding activity of the AP-1 and C/EBP transcription factors (97).

The Airway Epithelium

The micromechanical environment of the airway epithelium in situ remains somewhat mysterious. Until recently it had been widely assumed that the basement membrane underlying the epithelium behaved as an inextensible membrane (laterally flexible, but rigid in compression and tension—similar to a bicycle chain) (98); the necessary corollary to this assumption is that changes in airway lumen dimensions (60) must be accommodated by folding and unfolding of the epithelial lining, not by lateral stretching or compression of the epithelium. Although the folding of the airway wall is a well-recognized phenomenon (99), recent evidence has called into question the assumption that the epithelial basement membrane is inextensible (100). Even in the presence of a minimally extensible basement membrane, bending of the epithelium and its substrate gives rise to local shear deformations and pressure gradients (9).

The poorly defined mechanical environment, along with the difficulty in combining differentiated primary bronchial epithelial cultures with mechanically relevant experimental systems, has limited the in vitro study of this cell type. Transient mechanical perturbations of the epithelial surface have been shown to induce a number of signaling pathways, including those involving calcium, inositol 1,4,5-trisphosphate, protein kinase C and ATP/UTP (101–105), helping to coordinate ciliary beating activity (106, 107). Airway epithelial cells also possess mechanosensitive cell membrane channels that can regulate ion fluxes gated to membrane stress (108). Together these results suggest that, at a minimum, airway epithelial cells possess sensitive molecular means with which to sample their mechanical environment; these changes are covered in detail in the Boucher chapter (14b).

Via the use of fluid-submerged cultures of cat and human airway epithelial cells, cyclic application of a nonuniform stretch field has been shown to downregulate prostaglandin E2, I2, and thromboxane A2 synthesis in a frequency-dependent and oxidant-mediated fashion (109). Both cyclic stretch and compression in the same nonuniform deformation device attenuate closure of epithelial wounds in cat and human airway epithelium (110). Bronchial epithelial cell line BEAS-2B responds to 20% biaxial cyclic stretch with increased production of IL-8 via activation of the p38 pathway, and mechanical stretch broadly increases signaling via mitogen-activated protein kinase and cAMP-responsive pathways (111).

Over the past 30 years, the culturing of primary airway epithelial cells at air-liquid interface on microporous surfaces has developed into the gold-standard technique for study of this cell type; under these conditions cells recapitulate the mixed mucus-secretory and ciliated surface phenotype expressed in vivo (112). The difficulty inherent in mechanically manipulating cultures on microporous substrates stems from the relatively inextensible nature of the substrates (113). One alternative is to grow the cells on gels of collagen type I (114–121). Intriguingly, in addition to having effects on receptor ligation, the soft collagen gels influence the function and phenotype of cells (see below for further discussion of the possible mechanism), as opposed to rigid substrates or substrates coated with only a thin layer of collagen (116). Further changes are induced when the gels are cultivated in a manner that allows them to deform in response to the cells cultured on their surface (118, 120, 121); these changes suggest interesting parallels with the mechanical and developmental context of the rapidly proliferating epithelium and remodeling matrix present in the fetal lung (5, 44).

An alternative method to mechanically manipulate differentiated airway epithelial cells is to expose them to a transcellular pressure gradient (Figure 1A). Primary human bronchial epithelial cells and rat tracheal epithelial cells grown on microporous substrates respond to an apical-to-basal pressure gradient with magnitude-dependent changes in gene expression and signaling through ERK phosphorylation (113, 122). The mechanism for this response has been revealed by a combination of imaging, analytical modeling, and biochemical inhibitor studies (79). Three-dimensional microscopic imaging studies reveal that under a pressure gradient, the cells undergo predominantly shear deformation (little change in volume), which results in slightly shorter and wider cells. The cells are separated on their lateral surfaces by a lateral intercellular space (LIS) which shrinks (expels fluid) as the cells encroach (Figure 1B).

The observed loss of the lateral intercellular space prompted development of an analytical model to predict changes in concentration of ligands that are spontaneously released into this space (79). The model predicts that ligand concentration is roughly inversely proportional to the change in LIS dimension: As the LIS shrinks, the ligand concentration within the LIS increases (Figure 1C). Biochemical inhibitor studies reveal that the ERK signaling response to the pressure gradient is a key intermediate in downstream biological effects (122) and that activation of ERK depends on autocrine activation of the epidermal growth factor receptor (79). The ERK phosphorylation response was found to match closely that stimulated by exogenously administered growth factors, in a manner proportional to that predicted by the analytical model. This mechanotransduction mechanism appears to be at work in the epithelium in situ, as constricted airways also exhibit increased phosphorylation of the epidermal growth factor receptor in the epithelium (79). Despite the modest and transient signaling response activated by the imposition of a pressure gradient, this mechanical stimulus has been shown to produce prominent

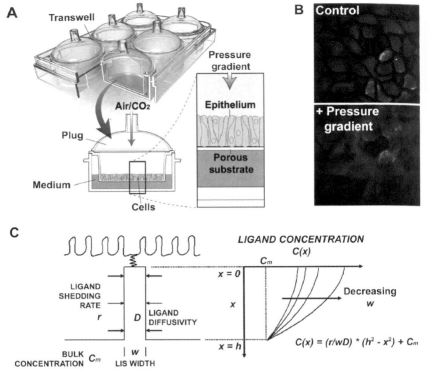

Figure 1 Combined use of imaging and analytical modeling to reveal mechanotransduction occurring through an extracellular autocrine signaling loop. (*A*) Schematic of device used to apply an apical-to-basal transcellular pressure gradient across bronchial epithelial cells grown at air-liquid interface on a microporous substrate. (*B*) Two-photon imaging of cells under control conditions reveals the lateral intercellular space (LIS) separating bronchial epithelial cells; application of a pressure gradient dramatically shrinks this compartment (similar to Reference 79). (*C*) The ligand concentration in an idealized LIS depends on the shedding rate and diffusivity of the ligand, and the height and width of the LIS. Thus, mechanical force–induced changes in LIS width can alter ligand concentration in the cellular microenvironment.

and lasting changes in growth factor expression (123, 124). Together these results suggest potent and sensitive responses of the airway epithelium to alterations in its mechanical environment.

Multicellular Models

Mechanical responses in the airway and the cellular behaviors that shape and remodel airway structure likely represent the collective actions and simultaneous influences of many cell types and stimuli (125–127). This has led to the development of a new generation of coculture methods in which cellular interactions and collective responses to stimuli can be studied in vitro (128–133). In these systems the individual cell type responses reflect a multicellular environment that begins to approach that of the native cellular context. Moreover, these multicellular systems lend themselves to the study of integrated behaviors (e.g., matrix remodeling) that are likely influenced by multiple cell types.

Employing the apical-to-basal pressure gradient model of airway epithelial compression with unstressed, cocultured reporter fibroblasts reveals an integrated profibrotic response to mechanical stress (127) similar to that observed in other types of epithelial "injury" (131, 133). The soluble mediators communicating the response from the epithelium to the fibroblasts have been partially defined and include endothelin and TGF-β2 (124). Taking one step further, a coculture model has been developed in which epithelial cells grown atop matrix-embedded fibroblasts are simultaneously exposed to lateral compression (130). Combining this multicellular lateral compression with eosinophils induces epithelial remodeling in this system (130).

The results from these multicellular systems, although preliminary, highlight the presence, complexity, and importance of bidirectional communication between epithelial and mesenchymal cells in mechanical responses, and reemphasize the fundamental nature of the multicellular unit (1, 13). The capacity of these systems to recapitulate many features of airway remodeling sets the stage for systematic investigations into the cellular biology and cell-cell communication underlying formation and remodeling of airway structure.

NONLINEAR ELASTICITY AND CELLULAR RESPONSES

Cells respond not only to external forces (134, 135) but to the stiffness of the matrix within which they reside (136, 137). Cells sample their mechanical microenvironment by exerting internally generated tensile forces on the surrounding matrix and adjust their phenotype in a cell- and tissue-specific manner in response to mechanical cues (138–140). Because the lung, like many biological tissues, exhibits nonlinear elastic properties (141), tissue stiffness changes with strain (lung volume). Conditions that alter the volume range of ventilation, or the relative distention of the airways, may thus change the stiffness regime in which resident cells in the airways grow and function. Similarly, diseases that remodel and alter the mechanical stiffness of the airway wall may influence the biology of resident cells through mechanosensing of these microenvironmental changes.

CONCLUSION

Airways in the lung are embedded in a mechanically active structure; from in utero development through adulthood they are subject to a dynamic and complex mechanical environment. Data now indicate that cells of the airways sense and respond to the mechanical environment and that these changes represent both normal physiology and the response to disease. For example, in the growing lung in utero, strains resulting from the secretion of liquid into the airways modify airway morphogenesis and cellular differentiation. In asthma, a condition characterized by smooth muscle constriction, airway epithelial cells sense the modified mechanical environment, adapt their phenotype, and signal to nearby mesenchymal cells to coordinate airway remodeling. These responses to mechanical perturbations in health and disease shape the structure and modulate the functional properties of the airways.

The *Annual Review of Physiology* is online at
http://physiol.annualreviews.org

LITERATURE CITED

1. Shannon JM, Hyatt BA. 2004. Epithelial-mesenchymal interactions in the developing lung. *Annu. Rev. Physiol.* 66:625–45
2. Cardoso WV. 2001. Molecular regulation of lung development. *Annu. Rev. Physiol.* 63:471–94
3. Olver RE, Walters DV, Wilson SM. 2004. Developmental regulation of lung liquid transport. *Annu. Rev. Physiol.* 66:77–101
4. Nogawa H, Morita K, Cardoso WV. 1998. Bud formation precedes the appearance of differential cell proliferation during branching morphogenesis of mouse lung epithelium in vitro. *Dev. Dyn.* 213:228–35
5. Mollard R, Dziadek M. 1998. A correlation between epithelial proliferation rates, basement membrane component localization patterns, and morphogenetic potential in the embryonic mouse lung. *Am. J. Respir. Cell Mol. Biol.* 19:71–82
6. Mead J. 1961. Mechanical properties of lungs. *Physiol. Rev.* 41:281–330
7. McHugh KM. 1995. Molecular analysis of smooth muscle development in the mouse. *Dev. Dyn.* 204:278–90
8. Sparrow MP, Lamb JP. 2003. Ontogeny of airway smooth muscle: structure, innervation, myogenesis and function in the fetal lung. *Respir. Physiol. Neurobiol.* 137:361–72
9. Wiggs BR, Hrousis CA, Drazen JM, Kamm RD. 1997. On the mechanism of

mucosal folding in normal and asthmatic airways. *J. Appl. Physiol.* 83:1814–21
10. Liu M, Post M. 2000. Invited review: mechanochemical signal transduction in the fetal lung. *J. Appl. Physiol.* 89:2078–84
11. Liu M, Tanswell AK, Post M. 1999. Mechanical force–induced signal transduction in lung cells. *Am. J. Physiol.* 277:L667–83
12. Wirtz HR, Dobbs LG. 2000. The effects of mechanical forces on lung functions. *Respir. Physiol.* 119:1–17
13. Holgate ST, Davies DE, Lackie PM, Wilson SJ, Puddicombe SM, Lordan JL. 2000. Epithelial-mesenchymal interactions in the pathogenesis of asthma. *J. Allergy Clin. Immunol.* 105:193–204
14. Kamm RD. 1999. Airway wall mechanics. *Annu. Rev. Biomed. Eng.* 1:47–72
14a. Fredberg J, Kamm RD. 2006. Stress transmission in the lung: pathways from organ to molecule. *Annu. Rev. Physiol.* 68:507–41
14b. Tarran R, Button B, Boucher RC. 2006. Regulation of normal and cystic fibrosis airway surface liquid volume by phasic shear stress. *Annu. Rev. Physiol.* 68:543–61
14c. dos Santos CC, Slutsky AS. 2006. The contribution of biophysical lung injury to the development of biotrauma. *Annu. Rev. Physiol.* 68:585–618

15. Keller R, Davidson LA, Shook DR. 2003. How we are shaped: the biomechanics of gastrulation. *Differentiation* 71:171–205

16. Swartz MA. 2003. Signaling in morphogenesis: transport cues in morphogenesis. *Curr. Opin. Biotechnol.* 14:547–50

17. Lauffenburger DA, Horwitz AF. 1996. Cell migration: a physically integrated molecular process. *Cell* 84:359–69

18. Farge E. 2003. Mechanical induction of twist in the Drosophila foregut/stomodeal primordium. *Curr. Biol.* 13:1365–77

19. Somogyi K, Rorth P. 2004. Evidence for tension-based regulation of Drosophila MAL and SRF during invasive cell migration. *Dev. Cell* 7:85–93

20. Abrams JM, White MA. 2004. Coordination of cell death and the cell cycle: linking proliferation to death through private and communal couplers. *Curr. Opin. Cell Biol.* 16:634–38

21. Day SJ, Lawrence PA. 2000. Measuring dimensions: the regulation of size and shape. *Development* 127:2977–87

22. Shraiman BI. 2005. Mechanical feedback as a possible regulator of tissue growth. *Proc. Natl. Acad. Sci. USA* 102:3318–23

23. Kitterman JA. 1996. The effects of mechanical forces on fetal lung growth. *Clin. Perinatol.* 23:727–40

24. Strang LB. 1991. Fetal lung liquid: secretion and reabsorption. *Physiol. Rev.* 71:991–1016

25. Harding R, Hooper SB. 1996. Regulation of lung expansion and lung growth before birth. *J. Appl. Physiol.* 81:209–24

26. Brown MJ, Olver RE, Ramsden CA, Strang LB, Walters DV. 1983. Effects of adrenaline and of spontaneous labour on the secretion and absorption of lung liquid in the fetal lamb. *J. Physiol.* 344:137–52

27. Harding R, Bocking AD, Sigger JN. 1986. Influence of upper respiratory tract on liquid flow to and from fetal lungs. *J. Appl. Physiol.* 61:68–74

28. Vilos GA, Liggins GC. 1982. Intrathoracic pressures in fetal sheep. *J. Dev. Physiol.* 4:247–56

29. Klaus M, Tooley WH, Weaver KH, Clements JA. 1962. Lung volume in the newborn infant. *Pediatrics* 30:111–16

30. Wigglesworth JS. 1988. Lung development in the second trimester. *Br. Med. Bull.* 44:894–908

31. Hislop A, Hey E, Reid L. 1979. The lungs in congenital bilateral renal agenesis and dysplasia. *Arch. Dis. Child.* 54:32–38

32. Nobuhara KK, Wilson JM. 1996. The effect of mechanical forces on in utero lung growth in congenital diaphragmatic hernia. *Clin. Perinatol.* 23:741–52

33. Alcorn D, Adamson TM, Lambert TF, Maloney JE, Ritchie BC, Robinson PM. 1977. Morphological effects of chronic tracheal ligation and drainage in the fetal lamb lung. *J. Anat.* 123:649–60

34. Nardo L, Hooper SB, Harding R. 1998. Stimulation of lung growth by tracheal obstruction in fetal sheep: relation to luminal pressure and lung liquid volume. *Pediatr. Res.* 43:184–90

35. Joe P, Wallen LD, Chapin CJ, Lee CH, Allen L, et al. 1997. Effects of mechanical factors on growth and maturation of the lung in fetal sheep. *Am. J. Physiol.* 272:L95–105

36. Maltais F, Seaborn T, Guay S, Piedboeuf B. 2003. In vivo tracheal occlusion in fetal mice induces rapid lung development without affecting surfactant protein C expression. *Am. J. Physiol. Lung Cell. Mol. Physiol.* 284:L622–32

37. Harrison MR, Keller RL, Hawgood SB, Kitterman JA, Sandberg PL, et al. 2003. A randomized trial of fetal endoscopic tracheal occlusion for severe fetal congenital diaphragmatic hernia. *N. Engl. J. Med.* 349:1916–24

38. Yang Y, Beqaj S, Kemp P, Ariel I, Schuger L. 2000. Stretch-induced alternative splicing of serum response factor promotes bronchial myogenesis and is defective in lung hypoplasia. *J. Clin. Invest.* 106:1321–30

39. Yang Y, Relan NK, Przywara DA, Schuger L. 1999. Embryonic mesenchymal

cells share the potential for smooth muscle differentiation: myogenesis is controlled by the cell's shape. *Development* 126:3027–33

40. McBeath R, Pirone DM, Nelson CM, Bhadriraju K, Chen CS. 2004. Cell shape, cytoskeletal tension, and RhoA regulate stem cell lineage commitment. *Dev. Cell* 6:483–95

41. Jakkaraju S, Zhe X, Schuger L. 2003. Role of stretch in activation of smooth muscle cell lineage. *Trends Cardiovasc. Med.* 13:330–35

42. Engler AJ, Griffin MA, Sen S, Bonnemann CG, Sweeney HL, Discher DE. 2004. Myotubes differentiate optimally on substrates with tissue-like stiffness: pathological implications for soft or stiff microenvironments. *J. Cell Biol.* 166:877–87

43. Moore KA, Huang S, Kong Y, Sunday ME, Ingber DE. 2002. Control of embryonic lung branching morphogenesis by the Rho activator, cytotoxic necrotizing factor 1. *J. Surg. Res.* 104:95–100

44. Moore KA, Polte T, Huang S, Shi B, Alsberg E, et al. 2005. Control of basement membrane remodeling and epithelial branching morphogenesis in embryonic lung by Rho and cytoskeletal tension. *Dev. Dyn.* 232:268–81

45. Freeman M, Gurdon JB. 2002. Regulatory principles of developmental signaling. *Annu. Rev. Cell Dev. Biol.* 18:515–39

46. Padera TP, Stoll BR, Tooredman JB, Capen D, di Tomaso E, Jain RK. 2004. Pathology: cancer cells compress intratumour vessels. *Nature* 427:695

47. Bishop JE, Mitchell JJ, Absher PM, Baldor L, Geller HA, et al. 1993. Cyclic mechanical deformation stimulates human lung fibroblast proliferation and autocrine growth factor activity. *Am. J. Respir. Cell Mol. Biol.* 9:126–33

48. Breen EC. 2000. Mechanical strain increases type I collagen expression in pulmonary fibroblasts in vitro. *J. Appl. Physiol.* 88:203–9

49. Sanchez-Esteban J, Wang Y, Cicchiello LA, Rubin LP. 2002. Cyclic mechanical stretch inhibits cell proliferation and induces apoptosis in fetal rat lung fibroblasts. *Am. J. Physiol. Lung Cell. Mol. Physiol.* 282: L448–56

50. Breen EC, Fu Z, Normand H. 1999. Calcyclin gene expression is increased by mechanical strain in fibroblasts and lung. *Am. J. Respir. Cell Mol. Biol.* 21:746–52

51. Dawes GS, Fox HE, Leduc BM, Liggins GC, Richards RT. 1972. Respiratory movements and rapid eye movement sleep in the foetal lamb. *J. Physiol.* 220:119–43

52. Liu M, Skinner SJ, Xu J, Han RN, Tanswell AK, Post M. 1992. Stimulation of fetal rat lung cell proliferation in vitro by mechanical stretch. *Am. J. Physiol.* 263:L376–83

53. Liu M, Xu J, Souza P, Tanswell B, Tanswell AK, Post M. 1995. The effect of mechanical strain on fetal rat lung cell proliferation: comparison of two- and three-dimensional culture systems. *In Vitro Cell. Dev. Biol. Anim.* 31:858–66

54. Xu J, Liu M, Tanswell AK, Post M. 1998. Mesenchymal determination of mechanical strain–induced fetal lung cell proliferation. *Am. J. Physiol.* 275:L545–50

55. Liu M, Xu J, Tanswell AK, Post M. 1993. Stretch-induced growth-promoting activities stimulate fetal rat lung epithelial cell proliferation. *Exp. Lung Res.* 19:505–17

56. Xu J, Liu M, Liu J, Caniggia I, Post M. 1996. Mechanical strain induces constitutive and regulated secretion of glycosaminoglycans and proteoglycans in fetal lung cells. *J. Cell Sci.* 109(Pt. 6):1605–13

57. Xu J, Liu M, Post M. 1999. Differential regulation of extracellular matrix molecules by mechanical strain of fetal lung cells. *Am. J. Physiol.* 276:L728–35

58. Mourgeon E, Xu J, Tanswell AK, Liu M, Post M. 1999. Mechanical strain–induced posttranscriptional regulation of fibronectin production in fetal lung cells. *Am. J. Physiol.* 277:L142–49

59. Mead J, Takishima T, Leith D. 1970. Stress distribution in lungs: a model of pulmonary elasticity. *J. Appl. Physiol.* 28: 596–608

60. Brown RH, Mitzner W. 2003. Understanding airway pathophysiology with computed tomography. *J. Appl. Physiol.* 95:854–62

61. Seow CY, Wang L, Pare PD. 2000. Airway narrowing and internal structural constraints. *J. Appl. Physiol.* 88:527–33

62. Lambert RK, Codd SL, Alley MR, Pack RJ. 1994. Physical determinants of bronchial mucosal folding. *J. Appl. Physiol.* 77:1206–16

63. Brown RH, Scichilone N, Mudge B, Diemer FB, Permutt S, Togias A. 2001. High-resolution computed tomographic evaluation of airway distensibility and the effects of lung inflation on airway caliber in healthy subjects and individuals with asthma. *Am. J. Respir. Crit. Care Med.* 163:994–1001

64. Brown RH, Mitzner W. 1996. Effect of lung inflation and airway muscle tone on airway diameter in vivo. *J. Appl. Physiol.* 80:1581–88

65. Bousquet J, Jeffery PK, Busse WW, Johnson M, Vignola AM. 2000. Asthma. From bronchoconstriction to airways inflammation and remodeling. *Am. J. Respir. Crit. Care Med.* 161:1720–45

66. Fredberg JJ, Inouye D, Miller B, Nathan M, Jafari S, et al. 1997. Airway smooth muscle, tidal stretches, and dynamically determined contractile states. *Am. J. Respir. Crit. Care Med.* 156:1752–59

67. Gunst SJ, Meiss RA, Wu MF, Rowe M. 1995. Mechanisms for the mechanical plasticity of tracheal smooth muscle. *Am. J. Physiol.* 268:C1267–76

68. Fredberg JJ. 2004. Bronchospasm and its biophysical basis in airway smooth muscle. *Respir. Res.* 5:2

69. Berg JT, Fu Z, Breen EC, Tran HC, Mathieu-Costello O, West JB. 1997. High lung inflation increases mRNA levels of ECM components and growth factors in lung parenchyma. *J. Appl. Physiol.* 83:120–28

70. Parker JC, Breen EC, West JB. 1997. High vascular and airway pressures increase interstitial protein mRNA expression in isolated rat lungs. *J. Appl. Physiol.* 83:1697–705

71. Kumar A, Lnu S, Malya R, Barron D, Moore J, et al. 2003. Mechanical stretch activates nuclear factor-κB, activator protein-1, and mitogen-activated protein kinases in lung parenchyma: implications in asthma. *FASEB J.* 17:1800–11

72. Al-Jamal R, Ludwig MS. 2001. Changes in proteoglycans and lung tissue mechanics during excessive mechanical ventilation in rats. *Am. J. Physiol. Lung Cell. Mol. Physiol.* 281:L1078–87

73. Copland IB, Kavanagh BP, Engelberts D, McKerlie C, Belik J, Post M. 2003. Early changes in lung gene expression due to high tidal volume. *Am. J. Respir. Crit. Care Med.* 168:1051–59

74. Tepper RS, Ramchandani R, Argay E, Zhang L, Xue Z, et al. 2005. Chronic strain alters the passive and contractile properties of rabbit airways. *J. Appl. Physiol.* 98:1949–54

75. Naghshin J, Wang L, Pare PD, Seow CY. 2003. Adaptation to chronic length change in explanted airway smooth muscle. *J. Appl. Physiol.* 95:448–53

76. Wang L, Pare PD, Seow CY. 2001. Selected contribution: effect of chronic passive length change on airway smooth muscle length-tension relationship. *J. Appl. Physiol.* 90:734–40

77. Pratusevich VR, Seow CY, Ford LE. 1995. Plasticity in canine airway smooth muscle. *J. Gen. Physiol.* 105:73–94

78. Latourelle J, Fabry B, Fredberg JJ. 2002. Dynamic equilibration of airway smooth muscle contraction during physiological loading. *J. Appl. Physiol.* 92:771–79

79. Tschumperlin DJ, Dai G, Maly IV, Kikuchi T, Laiho LH, et al. 2004. Mechanotransduction through growth-factor

shedding into the extracellular space. *Nature* 429:83–86

80. Martins MA, Shore SA, Gerard NP, Gerard C, Drazen JM. 1990. Peptidase modulation of the pulmonary effects of tachykinins in tracheal superfused guinea pig lungs. *J. Clin. Invest.* 85:170–76

81. Gunst SJ, Tang DD, Opazo Saez A. 2003. Cytoskeletal remodeling of the airway smooth muscle cell: a mechanism for adaptation to mechanical forces in the lung. *Respir. Physiol. Neurobiol.* 137:151–68

82. Stephens NL, Li W, Jiang H, Unruh H, Ma X. 2003. The biophysics of asthmatic airway smooth muscle. *Respir. Physiol. Neurobiol.* 137:125–40

83. Wahl M, Eddinger TJ, Hai CM. 2004. Sinusoidal length oscillation- and receptor-mediated mRNA expression of myosin isoforms and α-SM actin in airway smooth muscle. *Am. J. Physiol. Cell Physiol.* 287:C1697–708

84. Smith PG, Janiga KE, Bruce MC. 1994. Strain increases airway smooth muscle cell proliferation. *Am. J. Respir. Cell Mol. Biol.* 10:85–90

85. Smith PG, Tokui T, Ikebe M. 1995. Mechanical strain increases contractile enzyme activity in cultured airway smooth muscle cells. *Am. J. Physiol.* 268:L999–1005

86. Smith PG, Moreno R, Ikebe M. 1997. Strain increases airway smooth muscle contractile and cytoskeletal proteins in vitro. *Am. J. Physiol.* 272:L20–27

87. Smith PG, Roy C, Zhang YN, Chauduri S. 2003. Mechanical stress increases RhoA activation in airway smooth muscle cells. *Am. J. Respir. Cell Mol. Biol.* 28:436–42

88. Smith PG, Roy C, Dreger J, Brozovich F. 1999. Mechanical strain increases velocity and extent of shortening in cultured airway smooth muscle cells. *Am. J. Physiol.* 277:L343–48

89. Smith PG, Deng L, Fredberg JJ, Maksym GN. 2003. Mechanical strain increases cell stiffness through cytoskeletal filament reorganization. *Am. J. Physiol. Lung Cell. Mol. Physiol.* 285:L456–63

90. Deng L, Fairbank NJ, Fabry B, Smith PG, Maksym GN. 2004. Localized mechanical stress induces time-dependent actin cytoskeletal remodeling and stiffening in cultured airway smooth muscle cells. *Am. J. Physiol. Cell Physiol.* 287:C440–48

91. Smith PG, Garcia R, Kogerman L. 1998. Mechanical strain increases protein tyrosine phosphorylation in airway smooth muscle cells. *Exp. Cell Res.* 239:353–60

92. Smith PG, Roy C, Fisher S, Huang QQ, Brozovich F. 2000. Selected contribution: mechanical strain increases force production and calcium sensitivity in cultured airway smooth muscle cells. *J. Appl. Physiol.* 89:2092–98

93. Tang D, Mehta D, Gunst SJ. 1999. Mechanosensitive tyrosine phosphorylation of paxillin and focal adhesion kinase in tracheal smooth muscle. *Am. J. Physiol.* 276:C250–58

94. Tang DD, Gunst SJ. 2001. Selected contribution: roles of focal adhesion kinase and paxillin in the mechanosensitive regulation of myosin phosphorylation in smooth muscle. *J. Appl. Physiol.* 91:1452–59

95. Halayko AJ, Camoretti-Mercado B, Forsythe SM, Vieira JE, Mitchell RW, et al. 1999. Divergent differentiation paths in airway smooth muscle culture: induction of functionally contractile myocytes. *Am. J. Physiol.* 276:L197–206

96. Wang L, Liu HW, McNeill KD, Stelmack G, Scott JE, Halayko AJ. 2004. Mechanical strain inhibits airway smooth muscle gene transcription via protein kinase C signaling. *Am. J. Respir. Cell Mol. Biol.* 31:54–61

97. Kumar A, Knox AJ, Boriek AM. 2003. CCAAT/enhancer-binding protein and activator protein-1 transcription factors regulate the expression of interleukin-8 through the mitogen-activated protein kinase pathways in response to mechanical stretch of human airway smooth muscle cells. *J. Biol. Chem.* 278:18868–76

98. James AL, Hogg JC, Dunn LA, Pare PD. 1988. The use of the internal perimeter to compare airway size and to calculate smooth muscle shortening. *Am. Rev. Respir. Dis.* 138:136–39

99. James AL, Pare PD, Moreno RH, Hogg JC. 1987. Quantitative measurement of smooth muscle shortening in isolated pig trachea. *J. Appl. Physiol.* 63:1360–65

100. McParland BE, Pare PD, Johnson PR, Armour CL, Black JL. 2004. Airway basement membrane perimeter in human airways is not a constant; potential implications for airway remodeling in asthma. *J. Appl. Physiol.* 97:556–63

101. Felix JA, Chaban VV, Woodruff ML, Dirksen ER. 1998. Mechanical stimulation initiates intercellular Ca^{2+} signaling in intact tracheal epithelium maintained under normal gravity and simulated microgravity. *Am. J. Respir. Cell Mol. Biol.* 18:602–10

102. Felix JA, Woodruff ML, Dirksen ER. 1996. Stretch increases inositol 1,4,5-trisphosphate concentration in airway epithelial cells. *Am. J. Respir. Cell Mol. Biol.* 14:296–301

103. Homolya L, Steinberg TH, Boucher RC. 2000. Cell to cell communication in response to mechanical stress via bilateral release of ATP and UTP in polarized epithelia. *J. Cell Biol.* 150:1349–60

104. Sanderson MJ, Charles AC, Dirksen ER. 1990. Mechanical stimulation and intercellular communication increases intracellular Ca^{2+} in epithelial cells. *Cell Regul.* 1:585–96

105. Woodruff ML, Chaban VV, Worley CM, Dirksen ER. 1999. PKC role in mechanically induced Ca^{2+} waves and ATP-induced Ca^{2+} oscillations in airway epithelial cells. *Am. J. Physiol.* 276:L669–78

106. Sanderson MJ, Dirksen ER. 1986. Mechanosensitivity of cultured ciliated cells from the mammalian respiratory tract: implications for the regulation of mucociliary transport. *Proc. Natl. Acad. Sci. USA* 83:7302–6

107. Sanderson MJ, Dirksen ER. 1989. Mechanosensitive and β-adrenergic control of the ciliary beat frequency of mammalian respiratory tract cells in culture. *Am. Rev. Respir. Dis.* 139:432–40

108. Kim YK, Dirksen ER, Sanderson MJ. 1993. Stretch-activated channels in airway epithelial cells. *Am. J. Physiol.* 265:C1306–18

109. Savla U, Sporn PH, Waters CM. 1997. Cyclic stretch of airway epithelium inhibits prostanoid synthesis. *Am. J. Physiol.* 273:L1013–19

110. Savla U, Waters CM. 1998. Mechanical strain inhibits repair of airway epithelium in vitro. *Am. J. Physiol.* 274:L883–92

111. Oudin S, Pugin J. 2002. Role of MAP kinase activation in interleukin-8 production by human BEAS-2B bronchial epithelial cells submitted to cyclic stretch. *Am. J. Respir. Cell Mol. Biol.* 27:107–14

112. Perez-Vilar J, Sheehan JK, Randell SH. 2003. Making more MUCS. *Am. J. Respir. Cell Mol. Biol.* 28:267–70

113. Ressler B, Lee RT, Randell SH, Drazen JM, Kamm RD. 2000. Molecular responses of rat tracheal epithelial cells to transmembrane pressure. *Am. J. Physiol. Lung Cell. Mol. Physiol.* 278:L1264–72

114. Jetten AM, Brody AR, Deas MA, Hook GE, Rearick JI, Thacher SM. 1987. Retinoic acid and substratum regulate the differentiation of rabbit tracheal epithelial cells into squamous and secretory phenotype. Morphological and biochemical characterization. *Lab. Invest.* 56:654–64

115. Rearick JI, Deas M, Jetten AM. 1987. Synthesis of mucous glycoproteins by rabbit tracheal cells in vitro. Modulation by substratum, retinoids and cyclic AMP. *Biochem. J.* 242:19–25

116. Wu R, Martin WR, Robinson CB, St. George JA, Plopper CG, et al. 1990. Expression of mucin synthesis and secretion in human tracheobronchial epithelial cells grown in culture. *Am. J. Respir. Cell Mol. Biol.* 3:467–78

117. Wu R, Nolan E, Turner C. 1985. Expression of tracheal differentiated functions in serum-free hormone-supplemented medium. *J. Cell. Physiol.* 125:167–81

118. Kim KC, Zheng QX, Brody JS. 1993. Effect of floating a gel matrix on mucin release in cultured airway epithelial cells. *J. Cell Physiol.* 156:480–86

119. Kim KC. 1985. Possible requirement of collagen gel substratum for production of mucin-like glycoproteins by primary rabbit tracheal epithelial cells in culture. *In Vitro Cell Dev. Biol.* 21:617–21

120. Chevillard M, Hinnrasky J, Pierrot D, Zahm JM, Klossek JM, Puchelle E. 1993. Differentiation of human surface upper airway epithelial cells in primary culture on a floating collagen gel. *Epithelial Cell Biol.* 2:17–25

121. Benali R, Chevillard M, Zahm JM, Hinnrasky J, Klossek JM, Puchelle E. 1992. Tubule formation and functional differentiation by human epithelial respiratory cells cultured in a three-dimensional collagen matrix. *Chest* 101:7S–9S

122. Tschumperlin DJ, Shively JD, Swartz MA, Silverman ES, Haley KJ, et al. 2002. Bronchial epithelial compression regulates MAP kinase signaling and HB-EGF-like growth factor expression. *Am. J. Physiol. Lung Cell. Mol. Physiol.* 282:L904–11

123. Chu EK, Foley JS, Cheng J, Patel AS, Drazen JM, Tschumperlin DJ. 2005. Bronchial epithelial compression regulates epidermal growth factor receptor family ligand expression in an autocrine manner. *Am. J. Respir. Cell Mol. Biol.* 32:373–80

124. Tschumperlin DJ, Shively JD, Kikuchi T, Drazen JM. 2003. Mechanical stress triggers selective release of fibrotic mediators from bronchial epithelium. *Am. J. Respir. Cell Mol. Biol.* 28:142–49

125. Holgate ST, Davies DE, Puddicombe S, Richter A, Lackie P, et al. 2003. Mechanisms of airway epithelial damage: epithelial-mesenchymal interactions in the pathogenesis of asthma. *Eur. Respir. J. Suppl.* 44:24s–9s

126. Tschumperlin DJ, Drazen JM. 2001. Mechanical stimuli to airway remodeling. *Am. J. Respir. Crit. Care Med.* 164:S90–94

127. Swartz MA, Tschumperlin DJ, Kamm RD, Drazen JM. 2001. Mechanical stress is communicated between different cell types to elicit matrix remodeling. *Proc. Natl. Acad. Sci. USA* 98:6180–85

128. Agarwal A, Coleno ML, Wallace VP, Wu WY, Sun CH, et al. 2001. Two-photon laser scanning microscopy of epithelial cell–modulated collagen density in engineered human lung tissue. *Tissue Eng.* 7:191–202

129. Chakir J, Page N, Hamid Q, Laviolette M, Boulet LP, Rouabhia M. 2001. Bronchial mucosa produced by tissue engineering: a new tool to study cellular interactions in asthma. *J. Allergy Clin. Immunol.* 107:36–40

130. Choe MM, Sporn PH, Swartz MA. 2003. An in vitro airway wall model of remodeling. *Am. J. Physiol. Lung Cell. Mol. Physiol.* 285:L427–33

131. Morishima Y, Nomura A, Uchida Y, Noguchi Y, Sakamoto T, et al. 2001. Triggering the induction of myofibroblast and fibrogenesis by airway epithelial shedding. *Am. J. Respir. Cell Mol. Biol.* 24:1–11

132. Paquette JS, Tremblay P, Bernier V, Auger FA, Laviolette M, et al. 2003. Production of tissue-engineered three-dimensional human bronchial models. *In Vitro Cell. Dev. Biol. Anim.* 39:213–20

133. Zhang S, Smartt H, Holgate ST, Roche WR. 1999. Growth factors secreted by bronchial epithelial cells control myofibroblast proliferation: an in vitro co-culture model of airway remodeling in asthma. *Lab. Invest.* 79:395–405

134. Janmey PA, Weitz DA. 2004. Dealing with mechanics: mechanisms of force transduction in cells. *Trends Biochem. Sci.* 29:364–70

135. Silver FH, Siperko LM. 2003. Mechanosensing and mechanochemical transduction: how is mechanical energy sensed
and converted into chemical energy in an
extracellular matrix? *Crit. Rev. Biomed.
Eng.* 31:255–331

136. Yeung T, Georges PC, Flanagan LA, Marg
B, Ortiz M, et al. 2005. Effects of substrate
stiffness on cell morphology, cytoskeletal
structure, and adhesion. *Cell Motil. Cytoskeleton* 60:24–34

137. Engler A, Bacakova L, Newman C, Hate-
gan A, Griffin M, Discher D. 2004. Substrate compliance versus ligand density in
cell on gel responses. *Biophys. J.* 86:617–
28

138. Georges PC, Janmey PA. 2005. Cell type–
specific response to growth on soft materials. *J. Appl. Physiol.* 98:1547–53

139. Lo CM, Wang HB, Dembo M, Wang
YL. 2000. Cell movement is guided by
the rigidity of the substrate. *Biophys. J.*
79:144–52

140. Wang HB, Dembo M, Wang YL. 2000.
Substrate flexibility regulates growth and
apoptosis of normal but not transformed
cells. *Am. J. Physiol. Cell Physiol.*
279:C1345–50

141. Storm C, Pastore JJ, MacKintosh FC,
Lubensky TC, Janmey PA. 2005. Nonlinear elasticity in biological gels. *Nature*
435:191–94

Annu. Rev. Physiol. 2006. 68:585–618
doi: 10.1146/annurev.physiol.68.072304.113443
First published online as a Review in Advance on October 20, 2005

THE CONTRIBUTION OF BIOPHYSICAL LUNG INJURY TO THE DEVELOPMENT OF BIOTRAUMA

Claudia C. dos Santos and Arthur S. Slutsky

Departments of Medicine and Critical Care Medicine, St. Michael's Hospital, and Inter-Departmental Division of Critical Care, University of Toronto, Toronto, Ontario M5B 1W8, Canada; email: claudia.santos@utoronto.ca, arthur.slutsky@utoronto.ca

■ **Abstract** Patients with severe acute respiratory distress syndrome who die usually succumb to multiorgan failure as opposed to hypoxia. Despite appropriate resuscitation, some patients' symptoms persist on a downward spiral, apparently propagated by an uncontained systemic inflammatory response. This phenomenon is not well understood. However, a novel hypothesis to explain this observation proposes that it is related to the life-saving ventilatory support used to treat the respiratory failure. According to this hypothesis, mechanical ventilation per se, by altering both the magnitude and the pattern of lung stretch, can cause changes in gene expression and/or cellular metabolism that ultimately can lead to the development of an overwhelming inflammatory response—even in the absence of overt structural damage. This mechanism of injury has been termed biotrauma. In this review we explore the biotrauma hypothesis, the causal relationship between biophysical injury and organ failure, and its implications for the future therapy and management of critically ill patients.

INTRODUCTION

Acute respiratory distress syndrome (ARDS), the most severe manifestation of acute lung injury (ALI), represents a significant burden of illness in the intensive care setting. The incidence of ARDS has been estimated to be 5–15 per 100,000 cases of ALI per year (1–3), if not higher (4). International utilization reviews have reported that acute respiratory failure accounts for 33–66% of the indications for mechanical ventilation in major intensive care units (5, 6). In the absence of positive-pressure ventilation, the severe oxygen deprivation resulting from ARDS causes death in more than 90% of patients (7). Despite advances in life support, an unacceptably high mortality persists—on the order of 25–50%, depending on a number of factors, including age and the predisposing insult (8–11). In the absence of heart failure, ARDS is characterized by severe dysfunction of gas exchange and by chest radiographic abnormalities following a predisposing injury (12–14). Perplexingly, mortality in the vast majority of patients who die from ARDS is not related to the underlying pulmonary disease (hypoxia) but rather to the development of multiple organ dysfunction syndrome (MODS).

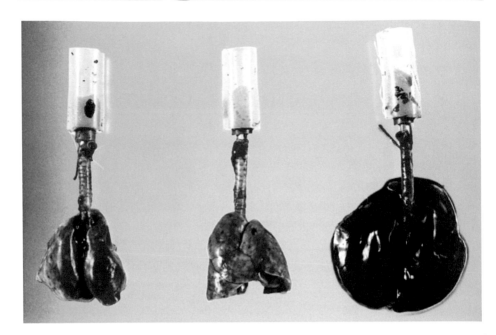

Figure 1   Photographs of rat lungs after mechanical ventilation at a peak airway pressure of 45 cm H_2O. (*Left*) Normal lungs. (*Middle*) After 5 min of high airway pressure mechanical ventilation. Note the focal zones of atelectasis (in particular at the left lung apex). (*Right*) After 20 min, the lungs are markedly enlarged and congestive; edema fluid fills the tracheal cannula. From Reference 21 with permission.

A number of animal and clinical studies have shown that mechanical ventilation per se can worsen preexisting lung injury and produce ventilator-induced lung injury (VILI) (Figure 1). The spectrum of VILI includes not only air leaks and increases in endothelial and epithelial permeability but also increases in pulmonary and systemic inflammatory mediators. These physiological consequences are encompassed in a process termed biotrauma (15–17). A recent hypothesis proposes that biophysical injury stemming from the forces generated by mechanical ventilation may be harmful to the lung, resulting in the propagation of the underlying lung injury and/or generation of a new injury, with translocation of mediators that may lead to systemic manifestations of inflammation, culminating in multiorgan failure (MOF) and ultimately death. The clinical significance of VILI on patient outcome became apparent when the National Institutes of Health–sponsored Acute Respiratory Distress Syndrome Network study of 861 patients reported a reduction in mortality from 40% to 31% when plateau pressures were limited in ventilated patients to 30 cm H_2O and tidal volumes were reduced from a conventional value of 12 ml per kg (predicted body weight) to 6 ml per kg. These results suggest that VILI has an attributable mortality of at least 9% (Figure 2). In this paper we

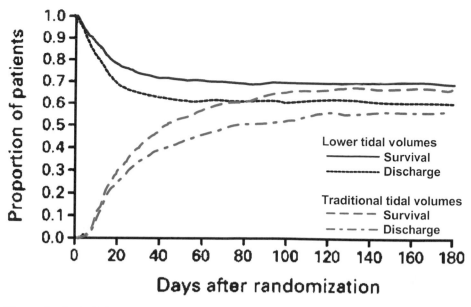

Figure 2 Probability of survival (*red* and *green curves*) and of being discharged home and breathing without assistance (*blue* and *orange curves*) during the first 180 days after randomization in patients with acute lung injury and acute respiratory distress syndrome. The traditional tidal volume strategy used a tidal volume of 12 ml per kg predicted body weight, and the lower tidal volume group used a tidal volume of 6 ml per kg predicted body weight. (From Reference 18 with permission.)

review the literature in support for the biotrauma hypothesis and its implication for clinical practice.

PULMONARY SUSCEPTIBILITY TO MECHANICAL INJURY

The ARDS Lung and Mechanical Forces

Patients with acute lung injury and on life support do not have normal lungs. Their injured lungs are strategically positioned to be the primary receptor/conductor of the effects of mechanical force from the outside world to the body. The alveolar surface is covered by epithelial cells that are in contact with inspired air. Alterations in the structure and function of alveoli have a direct and immediate impact on inflammatory homeostasis. Gas exchange takes 0.25 s, and the entire blood volume (5 liters) passes through the lungs each minute in the resting state, providing ample opportunity for mechanical forces generated by the machine at the bedside to be transmitted by different mechanisms (*vide infra*) to the patient.

For years, barotrauma (injury from high peak pressure) was synonymous with air leaks (extra-alveolar air). The development of a pressure gradient between the alveolus and its adjacent bronchovascular sheath can allow gas to extravasate into the interstitium and track along the bronchovascular sheaths into the mediastinum, retroperitoneum, or subcutaneous tissues, or break free into the pleural, pericardial, or peritoneal spaces (19, 20). High transpulmonary pressure has also been demonstrated to lead to much more subtle injury. Ultrastructural abnormalities such as endothelial cell detachment, intracapillary blebs, and disrupted or damaged type I pneumocytes with areas of denuded basement membrane have been described as evidence of volutrauma secondary to alveolar overdistension (21, 22).

Atelectrauma and Overdistension

Diffuse alveolar damage, whether caused by direct (e.g., pneumonia or acid aspiration) or indirect pulmonary injury (e.g., sepsis or severe trauma), is a hallmark of patients with ALI/ARDS. Structural changes in the alveolar-capillary unit lead to loss of integrity of the alveolar-capillary membrane. Protein-rich pulmonary edema fluid exudes into the alveolar spaces, followed by neutrophil influx, hyaline membrane deposition, and surfactant destruction (23). As a result, there is a dramatic reduction in the number of functional alveolar units, a decrease in pulmonary compliance, and a worsening of alveolar gas exchange. In the absence of positive-pressure ventilation, death ensues. However, the pulmonary abnormalities are not homogeneous. Along with the marked overall reduction in lung volumes ["baby lung" (24)], the patchy infiltrate distribution of lung injury results in exposure of the lung parenchyma's uninjured regions to mechanical forces not encountered normally, while alveoli in injured parts remain collapsed (Figure 3). These potentially pathogenic forces include repetitive (cyclic) tensile strain (stretch) from overdistension and interdependence as well as shear stress to the epithelial cells as lung units collapse and reopen (atelectrauma).

Much of the change in lung volume during positive-pressure ventilation is accommodated by either the elastic properties of the respiratory bronchioles or by the recruitment of new populations of acini. By contrast, in injured lungs, alveoli can be thought of as existing in one of three conditions: (1) fluid-filled or collapsed, and thus closed and not inflated; (2) collapsed or filled with fluid at end-exhalation, but opened and air-filled at end-inspiration; or (3) expanded and aerated throughout the respiratory cycle (25). Healthy alveoli in this last condition are the ones most vulnerable to overdistension due to uneven distribution of an inflated breath and interdependence. Interdependence refers to the force exerted on an alveolar region by adjacent regions (Figure 4). Collapse of an alveolus causes shear stress not only on its own walls but also on those of adjacent alveoli. In normal lungs, the alveolar distending force is equal to the transpulmonary pressure. Conversely, in heterogeneously inflated lungs, the alveolar distending force can be up to 140 cm H_2O even though the transpulmonary pressure is only 30 cm H_2O (26). Injured lungs (and indeed normal lungs) are susceptible to cyclic damage

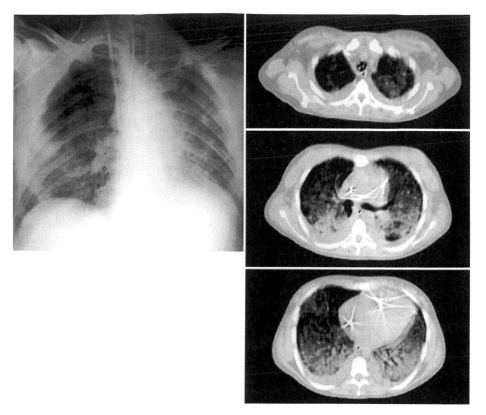

Figure 3   Anteroposterior chest radiography (*left*) and CT—apex, hilum, and base—(*right*) in a patient with sepsis who developed ARDS; taken at 5 cm H_2O end-expiratory pressure. Chest radiography shows diffuse ground glass opacification, sparing the right upper lung. CT shows nonhomogeneous disease and both the craniocaudal and sternovertebral gradients. (From Reference 24 with permission.)

by shear stress resulting from repetitive alveolar collapse and overdistension (or recruitment-derecruitment).

VENTILATOR-INDUCED LUNG INJURY AND BIOTRAUMA

In contrast to the three well-described mechanisms of VILI mentioned above (barotrauma, volutrauma, and atelectrauma), biotrauma describes an injury process that may occur in the lung in the absence of gross ultrastructural damage (17). The biotrauma hypothesis proposes that biophysical forces alter normal cellular physiology in the lung, leading to increase in local proinflammatory mediator levels (15, 16) and changes in pulmonary repair, remodeling, and apoptotic mechanisms.

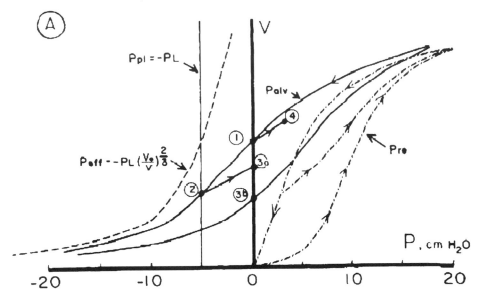

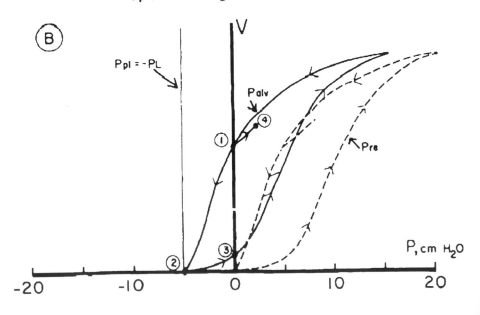

For the purpose of this review, we focus primarily on the effects of mechanical force on the inflammatory response. For further reading in the field of ventilator-induced changes in repair, remodeling, and apoptosis, we direct the reader to various excellent reviews in these respective fields (27–32). Moreover, although we focus primarily on pulmonary parenchymal cells, it must be noted that all pulmonary cell types are subject to the effects of mechanical forces and may contribute to VILI (33, 34).

Mechanical Forces Induce Changes in Gene Expression

During mechanical ventilation, bronchial, alveolar, and other parenchymal cells, as well as fibroblasts and macrophages, may be subjected to nonphysiologic forces and deformation. These forces include changes in transpulmonary pressure or stress (force per unit area), stretching or strain (fractional length change across an axis), and shear (the stress component parallel to a given surface) (28, 35, 36). Cellular strain and stretch occur primarily in lung regions in which overdistension or repeated atelectrauma occurs. One study of alveolar epithelial cell cultures exposed to mechanical strain in vitro suggests that a 25% increase in cell surface area corresponding to 8–12% linear distension likely correlates with physiological

---

Figure 4 *A* shows the relationship between alveolar pressure and volume for a region of lung exhibiting mechanical interdependence, with the surrounding regions in a lung exposed to a fixed transpulmonary pressure. This figure shows that individual air spaces do not expand indefinitely. Any influence tending to increase regional volumes results in opposing reduction in local expanding forces. $P_{pl} (V_0/V)^{2/3}$ denotes the expanding stress applied to a specific region of the lung by the recoil of surrounding tissue. This stress is called the effective pressure (P_{eff}), which is analogous to pleural pressure when the lung is homogeneous (shown as *broken line* on the *left*). The solid lines show how regional alveolar pressure changes under the condition of mechanical interdependence; for comparison, regional independence is shown in *B*. P_{alv} denotes the pressure in the alveolus, P_{pl} the pleural pressure, V_0 the volume at time 0, V the volume, PL the transpulmonary pressure, and P_{re} the recoil stress developed in a specific region of the lung. Immediately before fixation of PL, the lungs are fully expanded, and regional volume is at $P_{alv} = 0$ (this is the same in both *A* and *B*). As regional alveolar pressure is reduced below atmospheric, the mechanically independent regions reach nearly complete collapse (*point 2* in *B*) when $P_{alv} = P_{pl}$. By contrast, the interdependent lung retains more than half of its original volume (compare *point 1* with *point 2* in *A*). As alveolar pressure is returned to atmospheric, the independent regions remain nearly collapsed, while the interdependent region expands to below 20% of the original volume. Consequently, to achieve substantial collapse in the interdependent regions, a reduction in regional alveolar pressure far below atmospheric is required. Conversely, with the return of pressure to atmospheric, the region can expand to more than 50% of its original volume (*point 3b* in *A*), increasing the susceptibility to overdistension. From Reference 26 with permission.

levels of mechanical strain experienced by the alveolar epithelium (37). By contrast, cyclic stretch resulting in a 37–50% increase in cell surface area, corresponding to 17–22% linear distension, is relevant to pathophysiological conditions produced by mechanical ventilation and may cause progressive cell death (37). Shear stress has been postulated to exist, owing to the movement of pathological exudates in the alveolus over alveolar epithelial cells. In addition, heterogeneous constriction may also increase airflow-related shear stress (38).

How alveolar cells perceive alterations in mechanical forces (i.e., how they mechanosense) remains to be elucidated. Furthermore, because of the complexity of the lung, the variety of cell types, and the variety of physical forces to which cells are exposed, the mechanisms by which mechanical stimulation is perceived and the cellular response that is induced may vary widely (39, 40). Mechanotransduction is the conversion of mechanical stimuli, such as cellular deformation, into biochemical and biomolecular alterations (a detailed description of mechanosensing and mechanotransduction is beyond the scope of this review; the reader is directed to References 36 and 41–45). Three primary mechanisms have been thought to play a role in the mechanotransduction associated with VILI: (1) stretch-sensitive channels, (2) changes in plasma membrane integrity, and/or (3) a direct conformational change in membrane-associated molecules (28, 41). Although exactly how mechanically induced biochemical signals are initiated remains to be elucidated, force-induced directional and long-range activation of Src has been implicated in directing signals via the cytoskeleton to spatial destinations (45a). Recent data have also established a novel mechanism of mechanotransduction whereby compressive forces lead to changes in gene expression by shrinking the lateral intercellular spaces surrounding epithelial cells. This triggers cellular signals via autocrine binding of the epidermal growth factor receptor (46). Shikata et al. (47) have recently demonstrated that shear stress and cyclic stretch induce distinct patterns of gene expression, suggesting that understanding the role of particular forces in the generation of individual molecular responses may lead to the development of specific cell/stimulus-directed therapy in VILI. Central to the biotrauma hypothesis, irrespective of its specific mechanisms, is the concept that mechanical signals are converted to biomolecular responses.

Tremblay and colleagues (48) demonstrated in a model of VILI that mechanical ventilation can alter cellular gene expression and lead to release of cytokines (Figure 5). Moreover, levels of expression of c-*fos* and of various clinically relevant proinflammatory mediators increased according to the degree of lung injury in an ex vivo–perfused rat lung injury model (Figure 6) (48). Confirmation that repetitive cyclic stretch can lead to changes in gene expression was subsequently shown by Pugin and coworkers (49, 51) in alveolar macrophages and Vlahakis et al. (51) in A549 cells (transformed human alveolar epithelial cell line). Pressure-cycled ventilation caused human alveolar macrophages to release cytokines and proteases in vitro, and the effect was amplified by bacterial LPS (51). In both alveolar and bronchial epithelial cells, cyclic stretch leads to increased expression of interleukin (IL)-8. Augmentation of this response was seen with costimulation with tumor

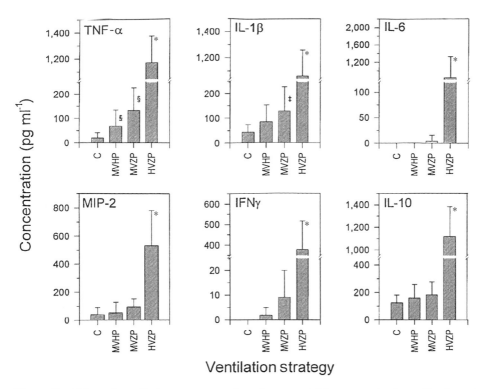

Figure 5 Effect of ventilation strategy on lung lavage cytokine concentrations. Ex vivo rat lungs ventilated for 2 h with ventilatory strategies thought to be injurious [MVHP, medium volume high positive end-expiratory pressure (PEEP); MVZP, medium volume zero PEEP; and HVZP, high volume zero PEEP)] led to an increase in broncoalveolar lavage (BAL) cytokines. A similar trend was seen for all cytokines, with the lowest levels in the control group C and the highest levels in the HVZP group. Although patients in the MVHP group exhibited similar end-expiratory distention, they had significantly lower BAL cytokine concentrations than did those in the HVZP group. $^*p < 0.05$ vs. C, MVHP, MVZP; $p < 0.05$ vs. C, MVHP; $\S p < 0.05$ vs. C (from Reference 48 with permission).

necrosis factor-α (TNF-α) (50, 52). These observations marked a turning point in the studies of VILI because it then became possible to propose that mechanical forces were independently modulating the expression of inflammatory genes in pulmonary-relevant cell types. Much attention was initially focused on the effects of cyclic stretch in directing the upregulation of proinflammatory genes such as TNF-α and IL-8. The latter has been shown to contain a stretch promoter sequence in its *cis*-regulatory region, suggesting a physiological role for cyclic stretch in the independent upregulation of the proinflammatory response (52).

To demonstrate that lung parenchymal cells, by altering their transcriptional program, can effectively sense and respond to the biophysical injury generated by

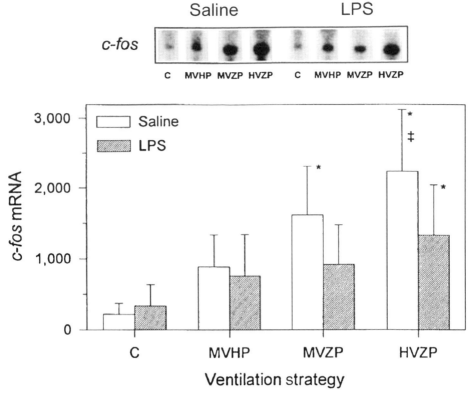

Figure 6   Northern blot analysis of lung homogenate *c-fos* mRNA for the various ventilation strategies (see Figure 5 legend for definition of abbreviations). Densitometric values for *c-fos* were standardized to 28S ribosomal RNA. A similar trend to that observed for the BAL cytokine concentrations was seen in both saline- and lipopolysaccharide (LPS)-treated animals. Whether animals received or did not receive LPS did not make a significant difference in *c-fos* mRNA levels. $^*p < 0.05$ vs. C; $p < 0.05$ vs. MVHP (from Reference 48 with permission).

the mechanical ventilator, Tremblay et al. (53) used in situ hybridization to demonstrate diffuse increase in expression of TNF-α and IL-6 in bronchial, bronchiolar, and alveolar epithelial cells from ex vivo rat lungs randomized to either a high (V_T 40 ml kg^{-1}) or a low tidal volume (V_T 7 ml kg^{-1}) ventilation strategy. Copland et al. (54) used gene expression profiling to dissect the pulmonary response to high tidal volume (V_T 25 ml kg^{-1}) ventilation in vivo. Although high tidal volume ventilation for 30 min did not lead to discernable levels of lung injury, it resulted in the upregulation of 10 genes and the downregulation of another 12 genes. In situ hybridization demonstrated that the majority of the overexpressed genes (Egr-1, c-Jun, and IL-1β) localized to the bronchiolar epithelium, whereas heat shock protein 70 was detected in the alveolar epithelium (Figure 7). This agrees with data

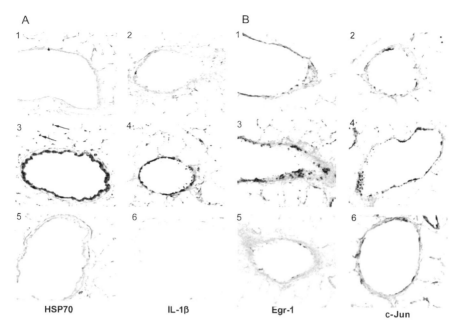

Figure 7   (*A*) Cellular localization of heat shock protein 70 (HSP70) and interleukin (IL)-1β mitochondrial RNA in lungs subjected to high tidal volume (HV) ventilation. Through the use of nonradioactive in situ hybridization, the cellular expression of HSP70 (*1, 3, 5*) and IL-1β (*2, 4, 6*) was compared between control (*1, 2*) and HV animals (*3, 4*). At 90 min, strong positive staining (*purple*) for both HSP70 (*arrows, 3*) and IL-1β (*4*) was localized to the bronchiolar epithelium of HV animals, in contrast to control animals (*1, 2*). In both control and HV animals, there was no appreciable staining noted for IL-1β in the alveolar and smooth muscle regions or in the vasculature. Positive staining was apparent in the alveolar region for HSP70 (*arrows, 3*), but not in the smooth muscle or vasculature. Hybridization with sense probes resulted in minimal nonspecific staining (*5, 6*) (from Reference 54 with permission). (*B*) Cellular localization of Egr-1 and c-Jun mitochondrial RNA in lungs subjected to HV ventilation. Through the use of nonradioactive in situ hybridization, the cellular expression of Egr-1 (*1, 3, 5*) and c-Jun (*2, 4, 6*) was compared between control (*1, 2*) and HV animals (*3, 4*). At 90 min, strong positive staining (*purple*) for both Egr-1 (*3*) and c-Jun (*4*) was localized to the bronchiolar epithelium of HV animals, in contrast to control animals (*1, 2*). In both control and HV animals, there was no appreciable staining in the alveolar and smooth muscle regions or in the vasculature. Hybridization with sense probes resulted in minimal nonspecific staining (*5, 6*) (from Reference 54 with permission).

from other groups suggesting that whereas gene expression at the distal airway epithelium is exquisitely sensitive to mechanical stretch, the alveolar epithelium for unknown reasons may be less responsive to direct mechanical stimulation (28). Few studies have expanded these observations to determine protein levels in the affected lung tissues or/and cells. Immunohistochemistry has been used to show

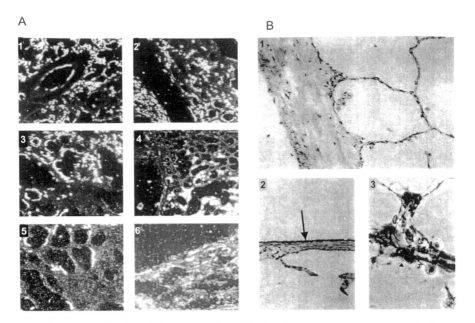

Figure 8 (*A*) In situ hybridization (dark field illumination) for tumor necrosis factor-α (TNF-α) and interleukin (IL)-6 after 120 min of ventilation. There was a diffuse increase in signal for TNF-α messenger RNA in lungs subjected to either high positive end-expiratory pressure (PEEP):tidal volume (V_T) 15 ml kg^{-1} or no PEEP:V_T 15 ml kg^{-1} ventilation for 120 min (*1* and *3*, respectively), in contrast to control ventilation (*5*). The changes in IL-6 expression mirrored the changes in TNF-α for all groups (panel *2* shows a representative section of in situ hybridization for IL-6 after high PEEP:V_T 15 ml kg^{-1} ventilation). Although increased expression of TNF-α and IL-6 was seen within 30 min of ventilation in the no PEEP:V_T 40 ml kg^{-1} group, by 2 h (*4*) the number of cells positive for signal was similar to controls. Simultaneous in situ hybridization (TNF-α; *white grains*) and immunocytochemistry (keratin; *dark gray stain*) demonstrated that the majority of TNF-α-positive cells were epithelial (*6*) (from Reference 53 with permission). (*B*) Immunocytochemistry for TNF-α in lungs subjected to injurious mechanical ventilation. (*1*) A representative section from the no PEEP:V_T 15 ml kg^{-1} group. Diffuse alveolar epithelial staining for TNF-α was found. Positive cells stain red (alkaline phosphatase antialkaline phosphatase technique). (*2*) Diffuse staining for TNF-α also was found in airway epithelium (*arrow*). (*3*) High-power magnification of dual immunocytochemistry for TNF-α (developed by peroxidase, *dark gray*) and keratin (developed by fast red), with the arrow illustrating an alveolar epithelial cell staining positive for both (from Reference 53 with permission).

that changes in TNF-α protein levels occur in pulmonary parenchymal cells as predicted by messenger levels: in a dose-dependent fashion, according to both the duration of ventilation and ventilation strategy used (Figure 8) (53).

Recently, two studies postulated that global genome analysis of dog, mouse, and rat models of VILI could be exploited to identify VILI-related genes and evolutionary conserved bioprocesses after exposure to injurious mechanical ventilation. Through the identification of orthologous genes that behaved in a similar fashion when animals were exposed to high V_T ventilation, more than 60 genes were identified as significantly affected by mechanical ventilation across different species. A number of these genes were represented by two major biological processes, inflammation and coagulation. Inflammation-related genes whose expression was also modulated by mechanical ventilation included IL-1β, IL-6, CC chemokine 2, and cyclo-oxigenase 2. Genes involved in the coagulation pathway that are also regulated by ventilation included plasminogen activator type I, tissue factor, and plasminogen activator urokinase (53a,b).

Further evidence of the consequence of mechanical force on gene expression comes from recent studies demonstrating that tidal volumes as low as 10 ml kg^{-1} can function as a co-factor in the initiation of ALI by modulating the transcriptional response to bacterial products (53c). In the future, understanding the pathways by which mechanical ventilation leads to amplification of inflammation and coagulation may be crucial in developing strategies to interrupt these pathways.

BIOTRAUMA AND MULTIORGAN FAILURE

The concept that biotrauma can lead to the development of multiorgan failure (MOF) presupposes that three steps must occur: (*a*) mechanical forces lead to a pulmonary proinflammatory response, which (*b*) escapes the confines of the lung and (*c*) directly or indirectly causes the development of MOF.

Support for the biotrauma hypothesis comes from experimental in vitro cell-stretch systems; ex vivo lung models (see above); and in vivo models of mechanical ventilation, either alone or following lung lavage, aspiration, and administration of drugs or endotoxin (48, 50, 53, 55, 56). Damaging of intact (noninjured) lungs, by the application of very high tidal volumes (30–40 ml kg^{-1}) or very high inspiratory pressures, has been utilized to document the proof of concept: High tidal volume alone (lung overdistension) can lead to increases in the levels of pulmonary proinflammatory cytokines in vivo. After 2 h of mechanical ventilation, the expression of proteins encoding for the TNF-α and IL-6 genes is increased in the lung (Figure 8) (48). These data are in keeping with work from various investigators (55, 57). However, the concept of biotrauma is relatively new and a source of debate (58, 59). In contrast to the studies mentioned above, Ricard et al. (60) could not detect a rise in TNF-α or macrophage inflammatory protein-2 (one of the putative murine homologues of the human chemokine IL-8) in either broncoalveolar lavage fluid (BAL fluid) or serum from intact or isolated nonperfused lungs from animals ventilated with 42 ml kg^{-1} for 2 h. They did, however, observe a fourfold increase in IL-1β with high tidal volume ventilation. Other studies have also failed to detect a rise in proinflammatory cytokines after high tidal volume ventilation (61, 62).

Despite controversies regarding the role of cyclic overstretch independently leading to a rise in pulmonary proinflammatory mediators, "injurious" mechanical ventilations strategies—e.g., large tidal volume (usually greater than 12 ml kg^{-1})

and zero PEEP—of previously injured lungs can promote the release of inflammatory mediators in the lungs and worsen lung injury (63–66). Conversely, protective ventilation strategies—approaches that favor small tidal volumes and higher PEEPs—have been associated with a decrease in pulmonary as well as systemic inflammatory mediator levels (67–69). The potential importance of proinflammatory mediators in the development of VILI is further supported by data from experimental studies assessing the effects of an anti-TNF-α antibody and IL-1 receptor agonist on lung injury following surfactant depletion. Imai et al. (70) reported that pretreatment of surfactant-depleted rabbits with anti-TNF-α antibody prior to the initiation of mechanical ventilation resulted in an attenuation of the histologic lung injury score and markedly improved oxygenation. In a similar model, IL-1 receptor agonist pretreatment reduced endothelial albumin permeability and neutrophil infiltration (71).

Clinical studies have further provided convincing evidence that high tidal volume ventilation can lead to an increase in production of pulmonary as well as systemic inflammatory mediators in humans. Ranieri et al. (72, 73) demonstrated that a ventilatory strategy using a mean tidal volume of 11 ml kg^{-1} leads to an increase in both local as well as systemic inflammatory mediators compared to a group treated with a more protective ventilatory strategy (Figure 9). They also showed that an increase in plasma IL-6 levels correlates with the development of MODS (Figure 10) (72, 73). In addition, a large trial recently demonstrated that the TNF-α receptor (74), IL-6, IL-8, and IL-10 are increased in the plasma of patients with ALI/ARDS and that each of these cytokines predicts adverse clinical outcomes, suggesting that the degree of activation of the inflammatory cascade can be significantly altered by mechanical ventilation (75). Furthermore, the fact that soluble TNF receptor (sTNFR) I and not sTNFRII is released from alveolar epithelial cells has catalyzed the hypothesis that the reduction in sTNFRI levels seen in the patient group that received the low tidal volume ventilation strategy may reflect in part a reduction in epithelial lung injury and the consequent parenchymal cell–driven inflammatory response (74, 76).

In a dynamic study of cytokine release following different modes of mechanical ventilation, Stuber et al. (77) demonstrated that the time course of cytokine release can be extremely rapid. This group ventilated patients with ARDS with a

Figure 9 Levels of inflammatory mediators in bronchoalveolar lavage fluid and plasma. Individual trends of TNF-α, interleukin (IL)-8, and IL-6 in plasma and bronchoalveolar lavage fluid in two patient groups. The control group was ventilated with a standard ventilatory strategy, whereas the lung-protective group was ventilated with a lower tidal volume and higher positive end-expiratory pressure levels on the basis of the patients' pressure-volume curves. Time 1 indicates 24–30 h after study entry, and time 2 36–40 h after study entry. Horizontal bars indicate mean values. p values are for repeated measures analysis of variance for time 2 vs. entry (from Reference 72 with permission).

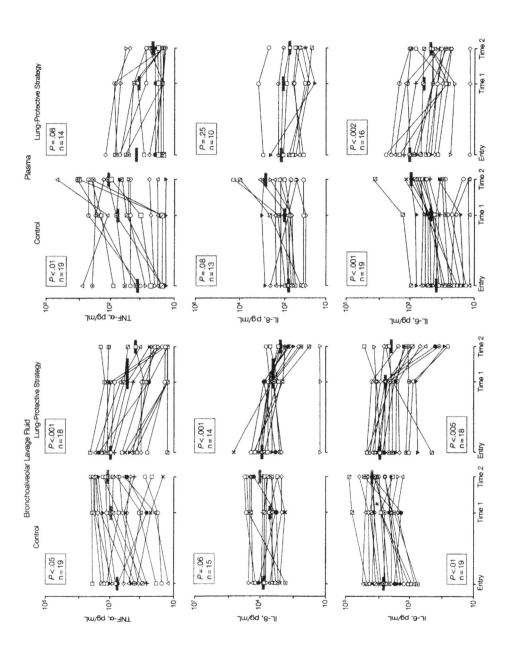

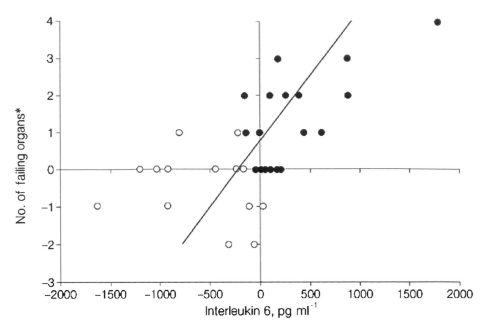

Figure 10 Changes in the total number of failing organs and changes in plasma concentration of interleukin 6 (IL-6). The patients in the two groups (control and lung protective) were treated as described in Figure 9 legend. There was a significant correlation between changes in IL-6 levels and the number of failing organs ($R^2 = 0.37$ and $p < .001$). The black dots denote a conventional ventilatory strategy ($n = 22$), and the hollow dots a lung-protective ventilatory strategy ($n = 22$). The asterisk indicates the number of failing organ systems at 72 h minus the number on entry. Changes in IL-6 were derived from the value at 36 h minus the entry value (from Reference 73 with permission).

lung-protective strategy for a period of time and then changed the strategy to a less protective approach, using high tidal volumes and zero PEEP for a few hours. An increase in measurable levels of inflammatory cytokines was noted within an hour of changing the ventilatory strategy (Figure 11). A decrease in cytokine levels was detected within an hour of reversal of the ventilatory strategy back to a protective mode using low tidal volumes. In large part, these studies attempt to explain the change in mortality associated with the decrease in tidal volume noted in the Acute Respiratory Distress Syndrome Network trial discussed above (18). The mechanism underlying the improved survival in patients ventilated with the tidal volume reduction strategy cannot be explained by changes in barotrauma or oxygenation. The incidence of barotrauma was not significantly different for the two groups, and the low tidal volume group, which had a better survival rate, had a lower PaO_2/fraction of inspired oxygen ratio for the first couple of days than did the higher tidal volume group. The decrease in plasma IL-6 levels in patients who were ventilated with a protective strategy (18) can be inferred to indicate that

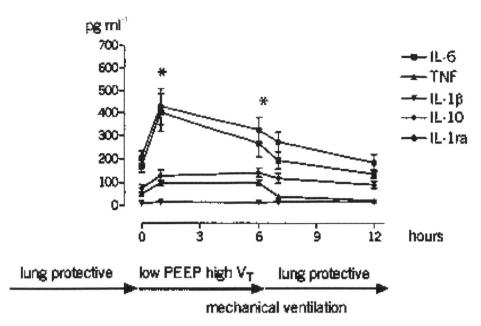

Figure 11   Plasma interleukin (IL)-1, IL-1ra, IL-6, IL-10, and TNF-α measured in 12 patients with ALI ventilated with a lung-protective strategy [positive end-expiratory pressure (PEEP) 15 cm H_2O; V_T 5 ml per kg predicted body weight] that was transiently switched to low PEEP and higher V_T mechanical ventilation (PEEP 5 cm H_2O; V_T 12 ml per kg predicted body weight). *$p < 0.05$ for all cytokines except IL-1 when compared to baseline cytokine levels at 0 h (Friedmann test and post hoc analysis). No statistically significant difference in cytokine levels was found between patients with and without mini-broncoalveolar lavage (Kruskal-Wallis test) (from Reference 77 with permission).

the lower IL-6 level in these patients reflects a reduction in the proinflammatory response secondary to decreased biotrauma to the lung.

Immunomodulatory Effect of Mechanical Ventilation

Studies focusing on the role of VILI have highlighted two important points about the effects of mechanical stretch on lung inflammation and the development of MODS. First, these studies have led to the development of the "two-hit" hypothesis of multiorgan failure (MOF) in VILI. Second, it has become apparent that mechanical stretch is immunomodulatory. Because patients with ARDS require life-saving intervention in the form of mechanical ventilation, the clinical phenotype can be modified through exposure to mechanical forces. Little is known about the pathogenetic development of ARDS and the modulatory effects of cyclic stretch on the molecular and gene expression profile of acutely injured lungs. However, the evidence available suggests that two hits are required to produce the multiorgan dysfunction associated with ARDS. The first hit (inflammatory insult) primes the lung, whereas the second hit (mechanical injury) enables the local inflammatory response to escape the confines of the lung. In this model of MOF, as pulmonary containment of the local inflammatory response is lost, systemic release of inflammatory mediators promotes a systemic proinflammatory response. This proinflammatory response is rapidly followed by the generation of an equally dramatic compensatory anti-inflammatory reaction that is designed to downregulate and attenuate the proinflammatory injury. Loss of appropriate immunomodulation and/or persistent inflammatory injury appears to contribute to the inability of organisms to bring about resolution of the proinflammatory response and ultimately leads to intractable organ dysfunction. Experiments in animal models of acute lung injury followed by randomization to either the protective or injurious ventilatory strategies have demonstrated that injurious ventilatory strategies can significantly alter pulmonary compartmentalization, permitting leakage of inflammatory mediators (e.g., cytokines, bacteria, and endotoxin) into the circulation (67, 68, 72–75, 78–80). Whether this step is the cause of organ failure in ARDS is an active area of debate and further research.

In addition to modulating cytokine and chemokine expression in the lung (and perhaps in the circulation), cyclic stretch has also been shown to interact with effector cells of the innate immune system. Overstretching of normal rabbit lungs with large tidal volumes (V_T 20 ml kg^{-1}) produces neutrophil influx and increases in IL-8 in bronchoalveolar lavage (BAL) fluid (81). Neutrophil depletion (vinblastine injection) attenuates the pulmonary increase in IL-8. Increasing the tidal volume in this model does not lead to further increases in IL-8 levels or in TNF-α in plasma or BAL fluid in neutropenic animals. No expression of P-selectin

---

$\longrightarrow$

Figure 12   (*Top*) In vivo neutralization of murine CXCR2 attenuates VILI. (*A*) VILI scores of H&E-stained histopathologic sections from mice placed on high peak pressure/stretch ventilation protocol treated with neutralizing antibodies to CXCR2 or with control antibodies. A cumulative score was based on leukocyte infiltration, exudative edema, hemorrhage, and alveolar wall thickness ($n = 15$; three random sections per lung and five lungs per group). $*p < 0.05$. (*B*) Evans blue permeability index ($n = 10$ mice per group). $*p < 0.05$. (*C*) VILI scores of H&E-stained histopathologic sections from CXCR2$^{-/-}$ mice placed on high peak pressure/stretch ventilation protocol, as compared with CXCR2$^{+/+}$ mice ($n = 15$; three random sections per lung and five lungs per group). $*p < 0.05$. (*D*) Wet-to-dry weight ratio of lungs from CXCR2$^{-/-}$ and CXCR2$^{+/+}$ mice placed on high peak pressure/stretch ventilation protocol ($n = 8$ mice per group). $*p < 0.05$ (from Reference 83 with permission). (*Bottom*) Histopathology of lungs is associated with mechanical ventilation. Representative photomicrographs ($\times400$) with H&E staining of the lung histopathology. Mice placed on high peak pressure/stretch mechanical ventilation treated with anti–murine CXCR2 (*E*) or with control antibodies (*F*). CXCR2$^{-/-}$ mice (*G*) placed on high peak pressure/stretch mechanical ventilation, as compared with CXCR2$^{+/+}$ mice (*H*) (from Reference 83 with permission).

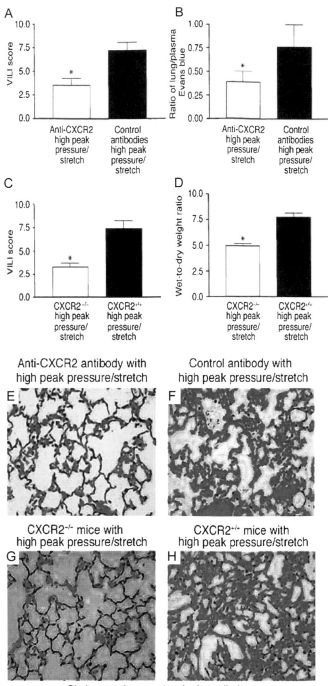

Anti-CXCR2 antibody with high peak pressure/stretch

Control antibody with high peak pressure/stretch

CXCR2⁻/⁻ mice with high peak pressure/stretch

CXCR2⁺/⁺ mice with high peak pressure/stretch

Six hours after mechanical ventilation

or intercellular adhesion molecule-1 (key cell membrane proteins involved in endothelial cell activation) was observed in animals depleted of their neutrophils. These results suggest that production of pulmonary IL-8 by lung overstretch may require interaction between resident lung cells and migrated neutrophils.

Owing to the consistent association between polymorphonucleocytes (PMNs) and lung injury in humans and experimental models, PMNs have been implicated as causative agents of both ALI and VILI. In animal models of VILI, neutrophil migration into the alveoli appears to be in large part dependent on stretch-induced production of macrophage inflamatory protein-2 (82). In a recent study, Belperio et al. (83) provided compelling evidence that injurious mechanical ventilation can lead not only to PMN accumulation but also to consequent changes in microvascular permeability in the lung. Importantly, the ability of neutrophils to cause lung damage is mediated by increased expression of the CXCR2 (chemokine receptor) ligands in lung tissues interacting with the CXCR2 receptors on PMNs after mechanical injury. Blocking CXCR2 or producing CXCR2 deficiency confers protection against the deleterious effects of VILI (Figure 12). More recently, Steinberg et al. (84) employed in vivo video microscopy to directly assess alveolar stability in normal and surfactant-deactivated lungs and showed that alveolar instability (atelectrauma) causes mechanical injury, initiating an inflammatory response that results in a secondary neutrophil-mediated proteolytic injury. These data suggest that a key inciting event in VILI—the event that leads to PMN activation—may indeed be related to cyclic stretch–induced lung injury.

That proinflammatory mediator levels can increase in response to VILI does not preclude moderate degrees of lung stretch from activating specific mechanotransduction pathways (41). In fact, Plotz et al. (85) noted that in 12 infants without prior lung injury, BAL levels of cytokines increased after 2 h of mechanical ventilation with conservative tidal volumes (in noninjured lungs) of 10 ml kg^{-1}. The clinical consequences of the activation of these pathways remain to be elucidated, but recent evidence suggests that this process may play an active role in predisposing the lung to further injury. In a rat model of cecal ligation perforation, Herrera et al. (86) showed that lung damage, cytokine synthesis and release, and mortality rates were significantly affected by the method of mechanical ventilation in the presence of sepsis. Moreover, stabilizing alveoli in septic animals with PEEP (thus presumably reducing atelectrauma) resulted in attenuation of lung injury, reduced systemic and local inflammatory responses as measured by levels of inflammatory mediators, and prevented death. Altemeier et al. (87) postulated that, in a rabbit model of ALI, mechanical ventilation with a moderately high tidal volume (15 ml kg^{-1}) can augment—independently of biotrauma—the inflammatory response of uninjured lungs to systemic LPS treatment. This group found that mechanical ventilation alone resulted in minimal cytokine expression in the lung but significantly enhanced LPS-induced expression of TNF-α, IL-8, and monocyte chemotactic protein-1. The tidal volume (15 ml kg^{-1}) used in this study did not lead to disruption of the epithelial cell membrane, as demonstrated by the preservation of barrier function and an absence of histological changes consistent with structural disruption. Based on these findings, the authors postulated that cyclic stretch interacts with innate immune components, allowing leakage of bacterial products and resulting in an enhanced inflammatory response (87). One potential interaction is with endotoxin; another potential mechanism is through activation of effector cells via the effects of mechanotransduction. Animal experimental data suggest that in fact ventilation of normal lungs with conventional tidal volumes (10 ml kg^{-1}) may potentiate the pulmonary injurious effects that occur during transient nonlethal endotoxemia (88), findings that may become increasingly relevant for patients with no lung injury undergoing major surgery.

BIOTRAUMA AND MULTIORGAN FAILURE

Although there is no direct evidence demonstrating that mediators generated in the lung can cause MODS, injurious ventilatory strategies can lead to release of a number of factors that theoretically can impact MODS. The cumulative evidence that implicates VILI as a direct causative agent in MOF has been recently reviewed (89–92). There are several principal mechanisms by which mediator release may occur in VILI-induced MOF (see above): (*a*) stress failure of the alveolar epithelial-endothelial barrier (decompartmentalization), (*b*) stress failure of the plasma membrane (necrosis/apoptosis), (*c*) alterations in cytoskeletal structure without ultrastructural damage (mechanotransduction), (*d*) effects on vasculature independent of stretch or rupture, and as most recently proposed, (*e*) cyclic stretch, which has been implicated in failure of the epithelial regenerative process (impeding repair of the alveolar barrier function) (93). Irrespective of the precise mechanism(s) of mediator release, the clinical consequences may be devastating.

Mechanical Force and Barrier Disruption

In the space limitations of this review, it is not possible to do justice to the complex and intriguing field of membrane stress failure and barrier disruption: the reader is referred to excellent reviews on the topic (28, 40, 94–97). As mentioned above, exposure to mechanical force produces alveolar epithelial pathology ranging from inter- and intracellular gap formations, denudation of the basement membrane, and cell destruction (21, 98, 99). In these studies, alveolar type II pneumocytes appeared relatively spared—possibly because they had experienced a smaller deformation force owing to their localization in the corner of the alveoli. The current hypothesis is that disruption of the integrity of the endothelial cellular barrier occurs via cytoskeletal rearrangement and generation of tensile forces within the cell. These changes result in cellular contraction; the interruption of intercellular adhesion complexes; and the creation of gaps between neighboring cells, or through cells, that allow the exudation of fluid, macromolecules, and leukocytes into the interstitium and ultimately into alveolar spaces (100–103). These gaps close rapidly upon removal of the deforming stress, restoring normal vascular permeability. In contrast to alveolar epithelial cells, pulmonary endothelial cells are exquisitely sensitive to the effects of cyclic stretch (104). Cyclic stretch, greatly accentuated by mechanical ventilation, may invoke cytoskeletal rearrangements, resulting in contraction of neighboring endothelial cells and disruption of intercellular adhesion, thus worsening endothelial barrier function and producing increased fluid flux across capillaries (104, 105).

There are a number of lines of evidence that VILI leads to disruption of the capillary membrane: (*a*) the presence of alveolar hemorrhage in animal models of VILI (99, 106–108); (*b*) ultrastructural damage to the alveolar-capillary membrane after high volume ventilation (98, 105, 106, 109–111); and (*c*) translocation (movement of intra-alveolar molecules from the alveolar compartment to the systemic compartment) of bacteria (78, 79, 112, 113), endotoxin (80), cytokines (56, 67, 114), and surfactant-associated proteins (115, 116). Surfactant-associated proteins are normal constituents of surfactant, a complex mixture of phospholipid and associated proteins. Although surfactant proteins (SPs) normally are found only in the lung (only trace amounts in the serum), leakage of SP-A (117, 118), SP-B, and SP-D (119) into the circulation has been reported in a number of respiratory disorders. The route by which these proteins enter the circulation is unknown; however, there is strong evidence that bidirectional plasma protein flux, the magnitude of which depends on disease severity, occurs in the lungs (120). Secreted primarily by alveolar epithelial type II pneumocytes, plasma SP-A and SP-D apparently exhibit an increase in plasma levels early in the clinical course of ARDS (121–123) as well as in animal models of ALI (124). Mechanical ventilation can also modulate SP expression (76). The progressive increase in the levels of these proteins early in ALI/ARDS may reflect pulmonary epithelial injury secondary to mechanical ventilation and consequent loss of barrier function. Elevated plasma levels of SP-D early in the course of ARDS may be associated with a worse clinical outcome, suggesting that SP-D may be used in the future as a biomarker of ARDS (115).

The effects of mechanical ventilation on barrier function may not be confined to the lungs. Guery et al. (125) evaluated gut permeability as a marker of a potential consequence of the activation of the systemic inflammation triggered by high tidal volume ventilation. In this study, VILI was associated with increased gut permeability and elevated levels of serum-TNF concentrations. Administration of an anti-TNF antibody prevented both increased lung and gut permeability.

Apoptosis and Necrosis

In general cells die in either an organized or semiorganized fashion (apoptosis), or they succumb randomly to endogenous or exogenous signals (necrosis). The extent to which apoptosis involves the integral mechanisms of inflammation, tissue remodeling, and repair in ARDS/VILI has been recently reviewed (31, 32, 126, 127). Type II–cell apoptosis has been described in patients with resolving ARDS, whereas cellular hyperplasia and necrosis dominates the acute phase (128). In brief, little is known about the mechanisms of alveolar epithelial repair in ARDS/ALI and how it is modulated by VILI. The alveolar epithelium exhibits extensive necrosis of alveolar type I cells, leaving a denuded, but mostly intact, basement membrane with overlying hyaline membranes. As discussed above, type I cells are highly susceptible to injury, whereas type II cells are more resistant and therefore can function as progenitor cells for regeneration of the alveolar epithelium after injury.

The balance between apoptosis and necrosis may be relevant not only in reestablishment of alveolar-capillary membrane barrier function and repair but also in resorption of pulmonary fibrinous exudate processes, resulting in restoration of normal lung function and architecture.

The apoptotic pathway most studied in VILI, the Fas/FasL system, is mediated by the TNF death membrane receptor Fas (CD95) and its natural ligand (FasL). FasL exists in membrane-bound and soluble forms (FasL and sFasL, respectively), both of which are capable of inducing apoptosis (129, 130). Both alveolar and airway epithelial cells express Fas/CD95 on their surfaces (131, 132); Fas expression in epithelial cells increases in response to inflammatory mediators such as bacterial endotoxin (133). Soluble FasL expression is increased in ARDS patients and can induce apoptosis of normal lung epithelial cells. Increased FasL concentrations in BAL from ARDS patients is associated with an increase in mortality (129). Incubation of ARDS BAL fluids with specific inhibitors of FasL abolishes the apoptotic effect of ARDS BAL on human lung epithelial cells. Accumulating evidence indicates that stimulation of Fas-dependent pathways in animal lungs causes alveolar epithelial apoptosis and lung injury. Instillation of human sFasL into the lungs of rabbits also causes alveolar hemorrhage and the production of proinflammatory cytokines, such as IL-8, by alveolar macrophages (134).

Apoptosis has been implicated in multiorgan failure (135, 136). sFasL has also been implicated in apoptosis in distal organs. This may represent one of the mechanisms by which mechanical ventilation may exert its detrimental effect (129).

Imai et al. (137) used a rabbit model of acid aspiration–induced acute lung injury and VILI to determine whether necrosis and apoptosis were coexisting phenomena that could explain the incidence of distal organ dysfunction in the setting of ARDS/VILI. Apoptosis was noted in lungs of animals treated with the noninjurious ventilation strategy. By contrast, in animals ventilated with an injurious strategy, necrosis predominated in lung tissues, and apoptosis was increased in distal organs (kidney, intestines, and liver) (Figure 13). The increase in distal organ apoptosis may occur because of the increase of FasL in the lungs; the ligand would extravasate into the peripheral circulation and bind to receptors on distal organs, thus mediating the onset of programmed cell death. Further research is necessary to determine if distal organ apoptosis is indeed Fas-mediated. In addition, the possible mechanisms by which FasL actually escapes the confines of the lung and leads to remote apoptosis remain to be elucidated.

CLINICAL RELEVANCE OF BIOTRAUMA AND FUTURE IMPLICATIONS FOR THERAPY

The primary clinical consequence of biotrauma pertains to how biotrauma affects the choice of mechanical ventilation strategy for patients with ARDS. Biotrauma offers a rationale for the use of low tidal volume ventilation. Moreover, increasing

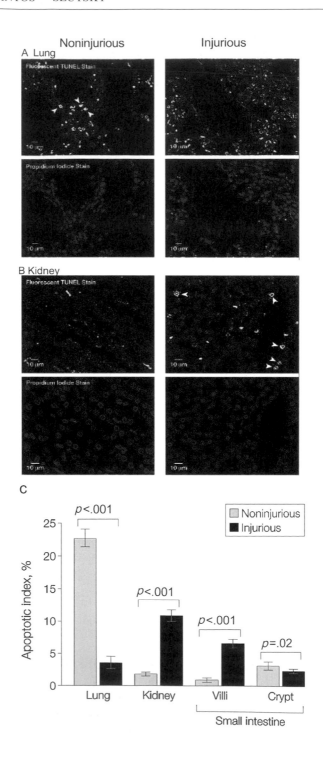

A Lung

B Kidney

C

Figure showing Noninjurious vs Injurious conditions. Panel A: Lung with Fluorescent TUNEL Stain and Propidium Iodide Stain. Panel B: Kidney with Fluorescent TUNEL Stain and Propidium Iodide Stain. Panel C: Bar graph of Apoptotic index (%) for Lung ($p<.001$), Kidney ($p<.001$), Villi ($p<.001$), and Crypt ($p=.02$) in Small intestine, comparing Noninjurious and Injurious conditions.

data suggest that in addition to reducing tidal volume, stabilization of alveoli in the form of positive end-expiratory pressure to reduce interdependence (138) may reduce further the proinflammatory effects of mechanical ventilation and ultimately affect outcome. Data from human studies examining serum cytokine levels in patients ventilated with a conventional strategy versus a protective strategy suggest that a ventilation strategy that decreases VILI is associated with decreased levels of serum proinflammatory mediators (18, 72, 73). Furthermore, this correlation between serum levels of proinflammatory cytokines and ventilation strategies suggests a possible optimal ventilatory strategy: titration based on levels of systemic mediators (74, 75, 77). Moreover, other markers that may correlate specifically with the degree of capillary-alveolar membrane disruption and loss of pulmonary compartmentalization may be used to monitor and anticipate the advent of both the development of a systemic inflammatory syndrome and the need for reduction of the iatrogenic lung injury. The hope is to modulate the mechanical ventilation strategy so as to avoid MOF (115, 116).

Of potential clinical importance is the fact that, unlike other proinflammatory events such as sepsis, we know exactly when VILI begins: at the time that mechanical ventilation is initiated. This knowledge provides clinicians with the opportunity to institute early and perhaps preventive therapy. In addition, current research has enabled the identification of genes that predispose individuals to an increased susceptibility not only to ARDS but also to VILI (139–141). Genetic susceptibility to the biophysical injury component of VILI has been termed "ventilogenomics." In the future, anticipating that patients with key mutations will be more susceptible to the biophysical component of the lung injury will enable early treatment to be initiated. To date, anti-TNF-α and anti-IL-1β have been used as potential therapies

---

Figure 13   (*A,B*) Representative photomicrographs of the lung and kidney in noninjurious and injurious groups. Rabbits were given intratracheal hydrochloric acid and then ventilated with either a strategy thought to be relatively noninjurious (V_T 6 ml kg^{-1}, positive end-expiratory pressure 11 cm H_2O) or one thought to be injurious (V_T 17 ml kg^{-1}, zero PEEP). Terminal deoxynucleotidyltransferase–mediated dUTP nick end-labeling (TUNEL) staining was used to identify apoptotic-positive cells; propidium iodide staining identified nuclei and helped ensure that noncellular background staining was not inadvertently identified as a TUNEL-positive cell. (*A*) Lung (original magnification $\times$ 320). TUNEL-positive nuclei (*arrowheads*) were markedly increased in the noninjurious group. (*B*) Kidney (original magnification $\times$ 400). There were numerous TUNEL-positive apoptotic tubular epithelial cells (*arrowheads*) in the injurious group, with fewer numbers in the noninjurious group (from Reference 137 with permission). (*C*) Apoptotic index percentages in the lung, kidney, and small intestine. Values are mean (SE). The injurious ventilatory strategy led to increased epithelial cell apoptosis in the kidney [injurious, 10.9% (0.88%); noninjurious, 1.86% (0.17%); $p <$ .001] and small intestine villi [injurious, 6.7% (0.66%); noninjurious, 0.97% (0.14%); $p <$ .001]. From Reference 137 with permission.

for VILI and VILI-related loss of membrane permeability (70, 71, 125). More recently, McVerry et al. (142) explored the use of the lipid growth factor sphingosine 1-phosphate to reduce the profound vascular leak and the intense inflammatory response resulting from the detrimental effects of excessive mechanical stress in an animal model of endotoxin-induced lung injury. In this study, sphingosine 1-phosphate significantly attenuated both alveolar and vascular barrier dysfunction while significantly reducing shunt formation associated with lung injury (142). The key concept is that the current therapy-related studies are proof-of-concept studies that have begun to address the question of whether the biotrauma hypothesis identifies an important biological mechanism (i.e., a causal relationship between biophysical injury and organ failure) rather than a coincidental association.

SUMMARY AND CONCLUSIONS

The impact of our increased understanding of VILI on the care of patients with acute lung injury has been great. The importance of this iatrogenic complication is reflected in the overwhelming efforts of the research community to comprehend the role of mechanical forces in modulating the clinical phenotype in this prototype acute inflammatory disorder. Before 1970 there were no published reports on VILI; in the past five years more than 230 papers have been published on the topic and are available on Medline. In less than half a century, life support due to mechanical ventilation has moved from obscurity to become one of the most advanced medical resources available. Our knowledge of the effects of positive pressure has advanced from a view of gross barotrauma injury to an understanding that mechanical forces can modulate the inflammatory and immune responses of the lung and determine outcome in ARDS/ALI. The future holds novel challenges, specifically in the area of therapeutics. We hope that we will be able to develop novel nonventilatory therapies based on the mediators released owing to biotrauma and hence treat patients with acute lung injury in whom a noninjurious ventilatory strategy is not possible. Moreover, identifying individuals who are more susceptible to injury and providing therapy in a site-directed, tailored fashion may further benefit not only patients with ARDS/ALI but all critically ill patients requiring mechanical ventilation.

The *Annual Review of Physiology* is online at
http://physiol.annualreviews.org

LITERATURE CITED

1. Goss CH, Brower RG, Hudson LD, Rubenfeld GD. 2003. Incidence of acute lung injury in the United States. *Crit. Care Med.* 31:1607–11

2. Luhr OR, Antonsen K, Karlsson M, Aardal S, Thorsteinsson A, et al. 1999. Incidence and mortality after acute respiratory failure and acute respiratory

distress syndrome in Sweden, Denmark, and Iceland. The ARF Study Group. *Am. J. Respir. Crit. Care Med.* 159:1849–61

3. Hudson LD, Steinberg KP. 1999. Epidemiology of acute lung injury and ARDS. *Chest* 116:74S–82S

4. Rubenfeld GD, Angus DC, Pinsky MR, Curtis JR, Connors AF Jr, Bernard GR. 1999. Outcomes research in critical care: results of the American Thoracic Society Critical Care Assembly Workshop on Outcomes Research. The Members of the Outcomes Research Workshop. *Am. J. Respir. Crit. Care Med.* 160:358–67

5. Esteban A, Anzueto A, Alia I, Gordo F, Apezteguia C, et al. 2000. How is mechanical ventilation employed in the intensive care unit? An international utilization review. *Am. J. Respir. Crit. Care Med.* 161:1450–58

6. Esteban A, Anzueto A, Frutos F, Alia I, Brochard L, et al. 2002. Characteristics and outcomes in adult patients receiving mechanical ventilation: a 28-day international study. *JAMA* 287:345–55

7. Petty TL, Ashbaugh DG. 1971. The adult respiratory distress syndrome. Clinical features, factors influencing prognosis and principles of management. *Chest* 60:233–39

8. Brun-Buisson C, Minelli C, Bertolini G, Brazzi L, Pimentel J, et al. 2004. Epidemiology and outcome of acute lung injury in European intensive care units. Results from the ALIVE study. *Intensive Care Med.* 30:51–61

9. Rubenfeld GD. 2003. Epidemiology of acute lung injury. *Crit. Care Med.* 31: S276–S284

10. Bersten AD, Edibam C, Hunt T, Moran J. 2002. Incidence and mortality of acute lung injury and the acute respiratory distress syndrome in three Australian states. *Am. J. Respir. Crit. Care Med.* 165: 443–48

11. Abel SJ, Finney SJ, Brett SJ, Keogh BF, Morgan CJ, Evans TW. 1998. Reduced mortality in association with the acute respiratory distress syndrome (ARDS). *Thorax* 53:292–94

12. Bernard GR, Artigas A, Brigham KL, Carlet J, Falke K, et al. 1994. The American-European consensus conference on ARDS. Definitions, mechanisms, relevant outcomes, and clinical trial coordination. *Am. J. Respir. Crit. Care Med.* 149:818–24

13. Bernard GR, Artigas A, Brigham KL, Carlet J, Falke K, et al. 1994. Report of the American-European consensus conference on ARDS: definitions, mechanisms, relevant outcomes and clinical trial coordination. The Consensus Committee. *Intensive Care Med.* 20:225–32

14. Bernard GR, Artigas A, Brigham KL, Carlet J, Falke K, et al. 1994. Report of the American-European Consensus conference on acute respiratory distress syndrome: definitions, mechanisms, relevant outcomes, and clinical trial coordination. The Consensus Committee. *J. Crit. Care* 9:72–81

15. Tremblay LN, Slutsky AS. 1998. Ventilator-induced injury: from barotrauma to biotrauma. *Proc. Assoc. Am. Physicians* 110:482–88

16. Slutsky AS, Tremblay LN. 1998. Multiple system organ failure. Is mechanical ventilation a contributing factor? *Am. J. Respir. Crit. Care Med.* 157:1721–25

17. Slutsky AS. 1999. Lung injury caused by mechanical ventilation. *Chest* 116:9S–15S

18. The Acute Respiratory Distress Syndrome Network. 2000. Ventilation with lower tidal volumes as compared with traditional tidal volumes for acute lung injury and the acute respiratory distress syndrome. *N. Engl. J. Med.* 342:1301–8

19. Sosin VV. 1959. On clinical manifestations of barotrauma of the lungs. *Vojnosanit. Pregl.* 8:46–48 (In Russian)

20. Denney MK, Glas WW. 1964. Experimental studies in barotrauma. *J. Trauma* 27:791–96

21. Dreyfuss D, Saumon G. 1998. Ventilator-induced lung injury: lessons from experimental studies. *Am. J. Respir. Crit. Care Med.* 157:294–323

22. American Thoracic Society. 1999. International consensus conferences in intensive care medicine: Ventilator-associated lung injury in ARDS. *Am. J. Respir. Crit. Care Med.* 160:2118–24

23. Barth PJ, Holtermann W, Muller B. 1998. The spatial distribution of pulmonary lesions in severe ARDS. An autopsy study of 35 cases. *Pathol. Res. Pract.* 194:465–71

24. Gattinoni L, Pesenti A, Avalli L, Rossi F, Bombino M. 1987. Pressure-volume curve of total respiratory system in acute respiratory failure. Computed tomographic scan study. *Am. Rev. Respir. Dis.* 136:730–36

25. Schiller HJ, McCann UG, Carney DE, Gatto LA, Steinberg JM, Nieman GF. 2001. Altered alveolar mechanics in the acutely injured lung. *Crit. Care Med.* 29:1049–55

26. Mead J, Takishima T, Leith D. 1970. Stress distribution in lungs: a model of pulmonary elasticity. *J. Appl. Physiol.* 28:596–608

27. Geiser T. 2003. Mechanisms of alveolar epithelial repair in acute lung injury—a translational approach. *Swiss. Med. Wkly.* 133:586–90

28. Vlahakis NE, Hubmayr RD. 2005. Cellular stress failure in ventilator injured lungs. *Am. J. Respir. Crit. Care Med.* 171:1328–42

29. Matthay MA, Bhattacharya S, Gaver D, Ware LB, Lim LH, et al. 2002. Ventilator-induced lung injury: in vivo and in vitro mechanisms. *Am. J. Physiol. Lung Cell Mol. Physiol.* 283:L678–82

30. Mendez JL, Hubmayr RD. 2005. New insights into the pathology of acute respiratory failure. *Curr. Opin. Crit. Care* 11:29–36

31. Matute-Bello G, Martin TR. 2003. Science review: apoptosis in acute lung injury. *Crit. Care* 7:355–58

32. Martin TR, Nakamura M, Matute-Bello G. 2003. The role of apoptosis in acute lung injury. *Crit. Care Med.* 31:S184–88

33. Marini JJ, Hotchkiss JR, Broccard AF. 2003. Bench-to-bedside review: microvascular and airspace linkage in ventilator-induced lung injury. *Crit. Care* 7:435–44

34. Marini JJ. 2004. Microvasculature in ventilator-induced lung injury: target or cause? *Minerva Anestesiol.* 70:167–73

35. Pugin J. 2002. Is the ventilator responsible for lung and systemic inflammation? *Intensive Care Med.* 28:817–19

36. Pugin J. 2003. Molecular mechanisms of lung cell activation induced by cyclic stretch. *Crit. Care Med.* 31:S200–6

37. Tschumperlin DJ, Oswari J, Margulies AS. 2000. Deformation-induced injury of alveolar epithelial cells. Effect of frequency, duration, and amplitude. *Am. J. Respir. Crit. Care Med.* 162:357–62

38. Nucci G, Suki B, Lutchen K. 2003. Modeling airflow-related shear stress during heterogeneous constriction and mechanical ventilation. *J. Appl. Physiol.* 95:348–56

39. Liu M, Post M. 2000. Invited review: mechanochemical signal transduction in the fetal lung. *J. Appl. Physiol.* 89:2078–84

40. Liu M, Tanswell AK, Post M. 1999. Mechanical force-induced signal transduction in lung cells. *Am. J. Physiol.* 277:L667–83

41. dos Santos CC, Slutsky AS. 2000. Invited review: mechanisms of ventilator-induced lung injury: a perspective. *J. Appl. Physiol.* 89:1645–55

42. Uhlig S. 2002. Ventilation-induced lung injury and mechanotransduction: stretching it too far? *Am. J. Physiol. Lung Cell Mol. Physiol.* 282:L892–96

43. Waters CM, Ridge KM, Sunio G, Venetsanou K, Sznajder JI. 1999. Mechanical stretching of alveolar epithelial cells

increases Na^+-K^+-ATPase activity. *J. Appl. Physiol.* 87:715–21

44. Waters CM, Chang JY, Glucksberg MR, DePaola N, Grotberg JB. 1997. Mechanical forces alter growth factor release by pleural mesothelial cells. *Am. J. Physiol.* 272:L552–57

45. Lionetti V, Recchia FA, Ranieri VM. 2005. Overview of ventilator-induced lung injury mechanisms. *Curr. Opin. Crit. Care* 11:82–86

45a. Wang Y, Botvinick EL, Zhao Y, Berns MW, Usami S, et al. 2005. Visualizing the mechanical activation of Src. *Nature* 434:1040–45

46. Tschumperlin DJ, Dai G, Maly IV, Kikuchi T, Laiho LH, et al. 2004. Mechanotransduction through growth-factor shedding into the extracellular space. *Nature* 429:83–86

47. Shikata Y, Rios A, Kawkitinarong K, DePaola N, Garcia JG, Birukov KG. 2005. Differential effects of shear stress and cyclic stretch on focal adhesion remodeling, site-specific FAK phosphorylation, and small GTPases in human lung endothelial cells. *Exp. Cell Res.* 304:40–49

48. Tremblay L, Valenza F, Ribeiro SP, Li J, Slutsky AS. 1997. Injurious ventilatory strategies increase cytokines and c-fos mRNA expression in an isolated rat lung model. *J. Clin. Invest.* 99:944–52

49. Pugin J, Dunn I, Jolliet P, Tassaux D, Magnenat JL, et al. 1998. Activation of human macrophages by mechanical ventilation in vitro. *Am. J. Physiol.* 275:L1040–50

50. Vlahakis NE, Schroeder MA, Limper AH, Hubmayr RD. 1999. Stretch induces cytokine release by alveolar epithelial cells in vitro. *Am. J. Physiol.* 277:L167–73

51. Dunn I, Pugin J. 1999. Mechanical ventilation of various human lung cells in vitro: identification of the macrophage as the main producer of inflammatory mediators. *Chest* 116:95S–97S

52. Oudin S, Pugin J. 2002. Role of MAP kinase activation in interleukin-8 production by human BEAS-2B bronchial epithelial cells submitted to cyclic stretch. *Am. J. Respir. Cell Mol. Biol.* 27:107–14

53. Tremblay LN, Miatto D, Hamid Q, Govindarajan A, Slutsky AS. 2002. Injurious ventilation induces widespread pulmonary epithelial expression of tumor necrosis factor-α and interleukin-6 messenger RNA. *Crit. Care Med.* 30:1693–700

53a. Ma SF, Grigoryev DN, Taylor AD, Nonas S, Sammani S, et al. 2005. Bioinformatic identification of novel early stress response genes in rodent models of lung injury. *Am. J. Physiol Lung Cell Mol. Physiol.* 289:L468–77

53b. Grigoryev DN, Ma SF, Irizarry RA, Ye SQ, Quackenbush J, Garcia JG. 2004. Orthologous gene-expression profiling in multi-species models: search for candidate genes. *Genome Biol.* 5:R34

53c. Altemeier WA, Matute-Bello G, Gharib SA, Glenny RW, Martin TR, Liles WC. 2005. Modulation of lipopolysaccharide-induced gene transcription and promotion of lung injury by mechanical ventilation. *J. Immunol.* 175:3369–76

54. Copland IB, Kavanagh BP, Engelberts D, McKerlie C, Belik J, Post M. 2003. Early changes in lung gene expression due to high tidal volume. *Am. J. Respir. Crit. Care Med.* 168:1051–59

55. Wilson MR, Choudhury S, Goddard ME, O'Dea KP, Nicholson AG, Takata M. 2003. High tidal volume upregulates intrapulmonary cytokines in an in vivo mouse model of ventilator-induced lung injury. *J. Appl. Physiol.* 95:1385–93

56. von Bethmann AN, Brasch F, Nusing R, Vogt K, Volk HD, et al. 1998. Hyperventilation induces release of cytokines from perfused mouse lung. *Am. J. Respir. Crit. Care Med.* 157:263–72

57. Choudhury S, Wilson MR, Goddard ME, O'Dea KP, Takata M. 2004. Mechanisms of early pulmonary neutrophil sequestration in ventilator-induced lung injury in

mice. *Am. J. Physiol. Lung Cell Mol. Physiol.* 287:L902–10

58. Dreyfuss D, Ricard JD, Saumon G. 2003. On the physiologic and clinical relevance of lung-borne cytokines during ventilator-induced lung injury. *Am. J. Respir. Crit. Care Med.* 167:1467–71

59. Uhlig S, Ranieri M, Slutsky AS. 2004. Biotrauma hypothesis of ventilator-induced lung injury. *Am. J. Respir. Crit. Care Med.* 169:314–15

60. Ricard JD, Dreyfuss D, Saumon G. 2001. Production of inflammatory cytokines in ventilator-induced lung injury: a reappraisal. *Am. J. Respir. Crit. Care Med.* 163:1176–80

61. Verbrugge SJ, Uhlig S, Neggers SJ, Martin C, Held HD, et al. 1999. Different ventilation strategies affect lung function but do not increase tumor necrosis factor-α and prostacyclin production in lavaged rat lungs in vivo. *Anesthesiology* 91:1834–43

62. Wrigge H, Uhlig U, Zinserling J, Behrends-Callsen E, Ottersbach G, et al. 2004. The effects of different ventilatory settings on pulmonary and systemic inflammatory responses during major surgery. *Anesth. Analg.* 98:775–81

63. Bregeon F, Delpierre S, Chetaille B, Kajikawa O, Martin TR, et al. 2005. Mechanical ventilation affects lung function and cytokine production in an experimental model of endotoxemia. *Anesthesiology* 102:331–39

64. Bregeon F, Roch A, Delpierre S, Ghigo E, Autillo-Touati A, et al. 2002. Conventional mechanical ventilation of healthy lungs induced pro-inflammatory cytokine gene transcription. *Respir. Physiol. Neurobiol.* 132:191–203

65. Kurahashi K, Ota S, Nakamura K, Nagashima Y, Yazawa T, et al. 2004. Effect of lung-protective ventilation on severe *Pseudomonas aeruginosa* pneumonia and sepsis in rats. *Am. J. Physiol. Lung Cell Mol. Physiol.* 287:L402–10

66. Moriyama K, Ishizaka A, Nakamura M, Kubo H, Kotani T, et al. 2004. Enhance-

ment of the endotoxin recognition pathway by ventilation with a large tidal volume in rabbits. *Am. J. Physiol. Lung Cell Mol. Physiol.* 286:L1114–21

67. Chiumello D, Pristine G, Slutsky AS. 1999. Mechanical ventilation affects local and systemic cytokines in an animal model of acute respiratory distress syndrome. *Am. J. Respir. Crit. Care Med.* 160:109–16

68. Haitsma JJ, Uhlig S, Goggel R, Verbrugge SJ, Lachmann U, Lachmann B. 2000. Ventilator-induced lung injury leads to loss of alveolar and systemic compartmentalization of tumor necrosis factor-α. *Intensive Care Med.* 26:1515–22

69. Imai Y, Kawano T, Miyasaka K, Takata M, Imai T, Okuyama K. 1994. Inflammatory chemical mediators during conventional ventilation and during high frequency oscillatory ventilation. *Am. J. Respir. Crit. Care Med.* 150:1550–54

70. Imai Y, Kawano T, Iwamoto S, Nakagawa S, Takata M, Miyasaka K. 1999. Intratracheal anti-tumor necrosis factor-α antibody attenuates ventilator-induced lung injury in rabbits. *J. Appl. Physiol.* 87:510–15

71. Narimanbekov IO, Rozycki HJ. 1995. Effect of IL-1 blockade on inflammatory manifestations of acute ventilator-induced lung injury in a rabbit model. *Exp. Lung Res.* 21:239–54

72. Ranieri VM, Suter PM, Tortorella C, De Tullio R, Dayer JM, et al. 1999. Effect of mechanical ventilation on inflammatory mediators in patients with acute respiratory distress syndrome: a randomized controlled trial. *JAMA* 282:54–61

73. Ranieri VM, Giunta F, Suter PM, Slutsky AS. 2000. Mechanical ventilation as a mediator of multisystem organ failure in acute respiratory distress syndrome. *JAMA* 284:43–44

74. Parsons PE, Matthay MA, Ware LB, Eisner MD. 2005. Elevated plasma levels of soluble TNF receptors are associated with morbidity and mortality in patients with

acute lung injury. *Am. J. Physiol. Lung Cell Mol. Physiol.* 288:L426–31

75. Parsons PE, Eisner MD, Thompson BT, Matthay MA, Ancukiewicz M, et al. 2005. Lower tidal volume ventilation and plasma cytokine markers of inflammation in patients with acute lung injury. *Crit. Care Med.* 33:1–6

76. Frank JA, Gutierrez JA, Jones KD, Allen L, Dobbs L, Matthay MA. 2002. Low tidal volume reduces epithelial and endothelial injury in acid-injured rat lungs. *Am. J. Respir. Crit. Care Med.* 165:242–49

77. Stuber F, Wrigge H, Schroeder S, Wetegrove S, Zinserling J, et al. 2002. Kinetic and reversibility of mechanical ventilation-associated pulmonary and systemic inflammatory response in patients with acute lung injury. *Intensive Care Med.* 28:834–41

78. Verbrugge SJ, Sorm V, van't Veen A, Mouton JW, Gommers D, Lachmann B. 1998. Lung overinflation without positive end-expiratory pressure promotes bacteremia after experimental *Klebsiella pneumoniae* inoculation. *Intensive Care Med.* 24:172–77

79. Nahum A, Hoyt J, Schmitz L, Moody J, Shapiro R, Marini JJ. 1997. Effect of mechanical ventilation strategy on dissemination of intratracheally instilled *Escherichia coli* in dogs. *Crit. Care Med.* 25:1733–43

80. Murphy DB, Cregg N, Tremblay L, Engelberts D, Laffey JG, et al. 2000. Adverse ventilatory strategy causes pulmonary-to-systemic translocation of endotoxin. *Am. J. Respir. Crit. Care Med.* 162:27–33

81. Kotani M, Kotani T, Ishizaka A, Fujishima S, Koh H, et al. 2004. Neutrophil depletion attenuates interleukin-8 production in mild-overstretch ventilated normal rabbit lung. *Crit. Care Med.* 32:514–19

82. Li LF, Yu L, Quinn DA. 2004. Ventilation-induced neutrophil infiltration depends on c-Jun N-terminal kinase. *Am. J. Respir. Crit. Care Med.* 169:518–24

83. Belperio JA, Keane MP, Burdick MD, Londhe V, Xue YY, et al. 2002. Critical role for CXCR2 and CXCR2 ligands during the pathogenesis of ventilator-induced lung injury. *J. Clin. Invest.* 110:1703–16

84. Steinberg JM, Schiller HJ, Halter JM, Gatto LA, Lee HM, et al. 2004. Alveolar instability causes early ventilator-induced lung injury independent of neutrophils. *Am. J. Respir. Crit. Care Med.* 169:57–63

85. Plotz FB, Vreugdenhil HA, Slutsky AS, Zijlstra J, Heijnen CJ, van Vught H. 2002. Mechanical ventilation alters the immune response in children without lung pathology. *Intensive Care Med.* 28:486–92

86. Herrera MT, Toledo C, Valladares F, Muros M, Diaz-Flores L, et al. 2003. Positive end-expiratory pressure modulates local and systemic inflammatory responses in a sepsis-induced lung injury model. *Intensive Care Med.* 29:1345–53

87. Altemeier WA, Matute-Bello G, Frevert CW, Kawata Y, Kajikawa O, et al. 2004. Mechanical ventilation with moderate tidal volumes synergistically increases lung cytokine response to systemic endotoxin. *Am. J. Physiol. Lung Cell Mol. Physiol.* 287:L533–42

88. Bregeon FDS, Chetaille B, Kajikawa O, Martin TR, Autillo-Touati A, et al. 2005. Mechanical ventilation affects lung function and cytokine production in an experimental model of endotoxemia. *Anesthesiology* 102(2):331–39

89. Plotz FB, Slutsky AS, van Vught AJ, Heijnen CJ. 2004. Ventilator-induced lung injury and multiple system organ failure: a critical review of facts and hypotheses. *Intensive Care Med.* 30:1865–72

90. Frank JA, Matthay MA. 2003. Science review: mechanisms of ventilator-induced injury. *Crit. Care* 7:233–41

91. Whitehead T, Slutsky AS. 2002. The pulmonary physician in critical care. 7. Ventilator induced lung injury. *Thorax* 57:635–42

92. Pinhu L, Whitehead T, Evans T, Griffiths

M. 2003. Ventilator-associated lung injury. *Lancet* 361:332–40

93. Reynolds SD, Giangreco A, Hong KU, McGrath KE, Ortiz LA, Stripp BR. 2004. Airway injury in lung disease pathophysiology: selective depletion of airway stem and progenitor cell pools potentiates lung inflammation and alveolar dysfunction. *Am. J. Physiol. Lung Cell Mol. Physiol.* 287:L1256–65

94. Vlahakis NE, Hubmayr RD. 2003. Response of alveolar cells to mechanical stress. *Curr. Opin. Crit. Care* 9:2–8

95. Vlahakis NE, Hubmayr RD. 2000. Invited review: plasma membrane stress failure in alveolar epithelial cells. *J. Appl. Physiol.* 89:2490–96

96. Wirtz HR, Dobbs LG. 2000. The effects of mechanical forces on lung functions. *Respir. Physiol.* 119:1–17

97. Mendez JL, Rickman OB, Hubmayr RD. 2004. Plasma membrane stress failure in ventilator-injured lungs. A hypothesis about osmoregulation and the pharmacologic protection of the lungs against deformation injury. *Biol. Neonate* 85:290–92

98. John E, McDevitt M, Wilborn W, Cassady G. 1982. Ultrastructure of the lung after ventilation. *Br. J. Exp. Pathol.* 63:401–7

99. Dreyfuss D, Soler P, Saumon G. 1995. Mechanical ventilation-induced pulmonary edema. Interaction with previous lung alterations. *Am. J. Respir. Crit. Care Med.* 151:1568–75

100. Neal CR, Michel CC. 1997. Transcellular openings through frog microvascular endothelium. *Exp. Physiol.* 82:419–22

101. Michel CC, Neal CR. 1999. Openings through endothelial cells associated with increased microvascular permeability. *Microcirculation* 6:45–54

102. Feng D, Nagy JA, Hipp J, Pyne K, Dvorak HF, Dvorak AM. 1997. Reinterpretation of endothelial cell gaps induced by vasoactive mediators in guinea-pig, mouse and rat: many are transcellular pores. *J. Physiol.* 504(Pt. 3):747–61

103. Elliott AR, Fu Z, Tsukimoto K, Prediletto R, Mathieu-Costello O, West JB. 1992. Short-term reversibility of ultrastructural changes in pulmonary capillaries caused by stress failure. *J. Appl. Physiol.* 73:1150–58

104. Birukov KG, Birukova AA, Dudek SM, Verin AD, Crow MT, et al. 2002. Shear stress-mediated cytoskeletal remodeling and cortactin translocation in pulmonary endothelial cells. *Am. J. Respir. Cell Mol. Biol.* 26:453–64

105. Parker JC, Hernandez LA, Peevy KJ. 1993. Mechanisms of ventilator-induced lung injury. *Crit. Care Med.* 21:131–43

106. Dreyfuss D, Basset G, Soler P, Saumon G. 1985. Intermittent positive-pressure hyperventilation with high inflation pressures produces pulmonary microvascular injury in rats. *Am. Rev. Respir. Dis.* 132:880–84

107. Webb HH, Tierney DF. 1974. Experimental pulmonary edema due to intermittent positive pressure ventilation with high inflation pressures. Protection by positive end-expiratory pressure. *Am. Rev. Respir. Dis.* 110:556–65

108. Broccard A, Shapiro RS, Schmitz LL, Adams AB, Nahum A, Marini JJ. 2000. Prone positioning attenuates and redistributes ventilator-induced lung injury in dogs. *Crit. Care Med.* 28:295–303

109. Fu Z, Costello ML, Tsukimoto K, Prediletto R, Elliott AR, et al. 1992. High lung volume increases stress failure in pulmonary capillaries. *J. Appl. Physiol.* 73:123–33

110. Costello ML, Mathieu-Costello O, West JB. 1992. Stress failure of alveolar epithelial cells studied by scanning electron microscopy. *Am. Rev. Respir. Dis.* 145:1446–55

111. Parker JC, Townsley MI. 2004. Evaluation of lung injury in rats and mice. *Am. J. Physiol. Lung Cell Mol. Physiol.* 286:L231–46

112. Cakar N, Akinci O, Tugrul S, Ozcan PE,

Esen F, et al. 2002. Recruitment maneuver: does it promote bacterial translocation? *Crit. Care Med.* 30:2103–6

113. Savel RH, Yao EC, Gropper MA. 2001. Protective effects of low tidal volume ventilation in a rabbit model of *Pseudomonas aeruginosa*-induced acute lung injury. *Crit. Care Med.* 29:392–98

114. Douzinas EE, Tsidemiadou PD, Pitaridis MT, Andrianakis I, Bobota-Chloraki A, et al. 1997. The regional production of cytokines and lactate in sepsis-related multiple organ failure. *Am. J. Respir. Crit. Care Med.* 155:53–59

115. Eisner MD, Parsons P, Matthay MA, Ware L, Greene K. 2003. Plasma surfactant protein levels and clinical outcomes in patients with acute lung injury. *Thorax* 58:983–88

116. Cheng IW, Ware LB, Greene KE, Nuckton TJ, Eisner MD, Matthay MA. 2003. Prognostic value of surfactant proteins A and D in patients with acute lung injury. *Crit. Care Med.* 31:20–27

117. De Pasquale CG, Arnolda LF, Doyle IR, Grant RL, Aylward PE, Bersten AD. 2003. Prolonged alveolocapillary barrier damage after acute cardiogenic pulmonary edema. *Crit. Care Med.* 31:1060–67

118. De Pasquale CG, Arnolda LF, Doyle IR, Aylward PE, Chew DP, Bersten AD. 2004. Plasma surfactant protein-B: a novel biomarker in chronic heart failure. *Circulation* 110:1091–96

119. Kendall M, Brown L, Trought K. 2004. Molecular adsorption at particle surfaces: a PM toxicity mediation mechanism. *Inhal. Toxicol.* 16(Suppl. 1):99–105

120. De Pasquale CG, Bersten AD, Doyle IR, Aylward PE, Arnolda LF. 2003. Infarct-induced chronic heart failure increases bidirectional protein movement across the alveolocapillary barrier. *Am. J. Physiol. Heart Circ. Physiol.* 284:H2136–45

121. Greene R. 1999. Acute lobar collapse: adults and infants differ in important ways. *Crit. Care Med.* 27:1677–79

122. Doyle IR, Nicholas TE, Bersten AD. 1995. Serum surfactant protein-A levels in patients with acute cardiogenic pulmonary edema and adult respiratory distress syndrome. *Am. J. Respir. Crit. Care Med.* 152:307–17

123. Doyle IR, Bersten AD, Nicholas TE. 1997. Surfactant proteins-A and -B are elevated in plasma of patients with acute respiratory failure. *Am. J. Respir. Crit Care Med.* 156:1217–29

124. Pan T, Nielsen LD, Allen MJ, Shannon KM, Shannon JM, et al. 2002. Serum SP-D is a marker of lung injury in rats. *Am. J. Physiol. Lung Cell Mol. Physiol.* 282:L824–32

125. Guery BP, Welsh DA, Viget NB, Robriquet L, Fialdes P, et al. 2003. Ventilation-induced lung injury is associated with an increase in gut permeability. *Shock* 19:559–63

126. Edwards YS. 2001. Stretch stimulation: its effects on alveolar type II cell function in the lung. *Comp. Biochem. Physiol. A Mol. Integr. Physiol.* 129:245–60

127. Del RV, van Tuyl M, Post M. 2004. Apoptosis in lung development and neonatal lung injury. *Pediatr. Res.* 55:183–89

128. Bardales RH, Xie SS, Schaefer RF, Hsu SM. 1996. Apoptosis is a major pathway responsible for the resolution of type II pneumocytes in acute lung injury. *Am. J. Pathol.* 149:845–52

129. Matute-Bello G, Liles WC, Steinberg KP, Kiener PA, Mongovin S, et al. 1999. Soluble Fas ligand induces epithelial cell apoptosis in humans with acute lung injury (ARDS). *J. Immunol.* 163:2217–25

130. Matute-Bello G, Winn RK, Jonas M, Chi EY, Martin TR, Liles WC. 2001. Fas (CD95) induces alveolar epithelial cell apoptosis in vivo: implications for acute pulmonary inflammation. *Am. J. Pathol.* 158:153–61

131. Wen LP, Madani K, Fahrni JA, Duncan SR, Rosen GD. 1997. Dexamethasone inhibits lung epithelial cell apoptosis induced by IFN-gamma and Fas. *Am. J. Physiol.* 273:L921–29

132. Fine A, Anderson NL, Rothstein TL, Williams MC, Gochuico BR. 1997. Fas expression in pulmonary alveolar type II cells. *Am. J. Physiol* 273:L64–71

133. Kitamura Y, Hashimoto S, Mizuta N, Kobayashi A, Kooguchi K, et al. 2001. Fas/FasL-dependent apoptosis of alveolar cells after lipopolysaccharide-induced lung injury in mice. *Am. J. Respir. Crit. Care Med.* 163:762–69

134. Matute-Bello G, Liles WC, Frevert CW, Nakamura M, Ballman K, et al. 2001. Recombinant human Fas ligand induces alveolar epithelial cell apoptosis and lung injury in rabbits. *Am. J. Physiol. Lung Cell Mol. Physiol.* 281:L328–35

135. Hotchkiss RS, Swanson PE, Freeman BD, Tinsley KW, Cobb JP, et al. 1999. Apoptotic cell death in patients with sepsis, shock, and multiple organ dysfunction. *Crit. Care Med.* 27:1230–51

136. Hotchkiss RS, Schmieg RE Jr, Swanson PE, Freeman BD, Tinsley KW, et al. 2000. Rapid onset of intestinal epithelial and lymphocyte apoptotic cell death in patients with trauma and shock. *Crit. Care Med.* 28:3207–17

137. Imai Y, de Parodo J, Kajikawa O, Perrot M, Fischer S, et al. 2003. Injurious mechanical ventilation and end-organ epithelial cell apoptosis and organ dysfunction in an experimental model of acute respiratory distress syndrome. *JAMA* 289:2104–12

138. Carney D, DiRocco J, Nieman G. 2005. Dynamic alveolar mechanics and ventilator-induced lung injury. *Crit. Care Med.* 33:S122–28

139. Ye SQ, Simon BA, Maloney JP, Zambelli-Weiner A, Gao L, et al. 2005. Pre-B-cell colony-enhancing factor as a potential novel biomarker in acute lung injury. *Am. J. Respir. Crit. Care Med.* 171:361–70

140. Marshall RP, Webb S, Bellingan GJ, Montgomery HE, Chaudhari B, et al. 2002. Angiotensin converting enzyme insertion/deletion polymorphism is associated with susceptibility and outcome in acute respiratory distress syndrome. *Am. J. Respir. Crit. Care Med.* 166:646–50

141. Nesslein LL, Melton KR, Ikegami M, Na CL, Wert SE, et al. 2005. Partial sp-b deficiency perturbs lung function and causes airspace abnormalities. *Am. J. Physiol. Lung Cell Mol. Physiol.* 288(6):L1154–61

142. McVerry BJ, Peng X, Hassoun PM, Sammani S, Simon BA, Garcia JG. 2004. Sphingosine 1-phosphate reduces vascular leak in murine and canine models of acute lung injury. *Am. J. Respir. Crit. Care Med.* 170:987–93

Annu. Rev. Physiol. 2006. 68:619–47
doi: 10.1146/annurev.physiol.68.040204.100431
Copyright © 2006 by Annual Reviews. All rights reserved
First published online as a Review in Advance on October 14, 2005

An Introduction to TRP Channels

I. Scott Ramsey, Markus Delling, and David E. Clapham

Howard Hughes Medical Institute, Cardiovascular Department, Children's Hospital Boston, Harvard Medical School, Boston, Massachusetts 02115;
email: sramsey@enders.tch.harvard.edu, mdelling@enders.tch.harvard.edu, dclapham@enders.tch.harvard.edu

Key Words   transient receptor potential, calcium permeable, voltage dependent, ligand gated, axon guidance, pharmacology, splice variants

■ **Abstract**   The aim of this review is to provide a basic framework for understanding the function of mammalian transient receptor potential (TRP) channels, particularly as they have been elucidated in heterologous expression systems. Mammalian TRP channel proteins form six-transmembrane (6-TM) cation-permeable channels that may be grouped into six subfamilies on the basis of amino acid sequence homology (TRPC, TRPV, TRPM, TRPA, TRPP, and TRPML). Selected functional properties of TRP channels from each subfamily are summarized in this review. Although a single defining characteristic of TRP channel function has not yet emerged, TRP channels may be generally described as calcium-permeable cation channels with polymodal activation properties. By integrating multiple concomitant stimuli and coupling their activity to downstream cellular signal amplification via calcium permeation and membrane depolarization, TRP channels appear well adapted to function in cellular sensation. Our review of recent literature implicating TRP channels in neuronal growth cone steering suggests that TRPs may function more widely in cellular guidance and chemotaxis. The TRP channel gene family and its nomenclature, the encoded proteins and alternatively spliced variants, and the rapidly expanding pharmacology of TRP channels are summarized in online supplemental material.

OVERVIEW OF MAMMALIAN TRP CHANNELS

TRP channels mediate the transmembrane flux of cations down their electrochemical gradients, thereby raising intracellular Ca^{2+} and Na^+ concentrations ($[Ca^{2+}]_i$ and $[Na^+]_i$, respectively) and depolarizing the cell. Changes in transmembrane voltage (V_m) underlie neuronal action potential propagation and muscle contraction (1). Voltage also plays a crucial role in nonexcitable cells both by directing the driving force for calcium entry through plasma membrane channels and by controlling the gating of voltage-dependent Ca^{2+}, K^+, and Cl^- channels. Calcium entry through plasma membrane channels is recognized as a cellular signaling event per se: Effector proteins sensitive to elevated $[Ca^{2+}]_i$ control a plethora of cellular events from transcriptional regulation to migration and proliferation (2).

As they are widely expressed in mammalian tissues, TRPs are well positioned to regulate $[Na^+]_i$, $[Ca^{2+}]_i$, and V_m in both excitable and nonexcitable cells.

The commonly accepted nomenclature for TRP channels (3, 4) subdivides 28 channel subunit genes into 7 subfamilies: TRPA, TRPC, TRPM, TRPML, TRPN, TRPP, and TRPV. Each subfamily contains at least one vertebrate representative; mammals bear representatives from all subfamilies except TRPN (5, 6). Bioinformatic searches of data from genome sequencing projects suggest that the mammalian TRP gene family tree is now essentially complete (Table S1; follow the Supplemental Material link from the Annual Reviews home page at http://www.annualreviews.org). Other than overall sequence homology, basic channel architecture, and cation selectivity, no particular feature defining the TRP family has yet emerged.

Many TRP channels are activated by multiple stimuli in expression systems. Sensitivity to polymodal activation suggests that the physiologically relevant stimulus for any given TRP will be governed by the specifics of cellular context (i.e., phosphorylation status, lipid environment, interacting proteins, and concentrations of relevant ligands). Furthermore, cooperativity intrinsic to TRP channels may result in allosteric coupling of distinct activation stimuli, blurring the definition of activator versus modulator. Nonetheless, the established modes of activation for expressed TRP channels may be divided into three general categories:

1. *Receptor activation.* G protein–coupled receptors (GPCRs) and receptor tyrosine kinases that activate phospholipases C (PLCs) can modulate TRP channel activity in at least three ways: (*a*) hydrolysis of phosphatidylinositol (4,5) bisphosphate (PIP$_2$), (*b*) production of diacylglycerol (DAG), or (*c*) production of inositol (1,4,5) trisphosphate (IP$_3$) and subsequent liberation of Ca^{2+} from intracellular stores (7). Evidence is strongest for the first two mechanisms, but many fundamental mechanistic questions remain to be answered.

2. *Ligand activation.* Ligands that activate TRP channels may be broadly classified as (*a*) exogenous small organic molecules, including synthetic compounds and natural products (capsaicin, icilin, 2-APB); (*b*) endogenous lipids or products of lipid metabolism (diacylglycerols, phosphoinositides, eicosanoids, anandamide); (*c*) purine nucleotides and their metabolites [adenosine diphosphoribose (ADP-ribose), βNAD^+]; or (*d*) inorganic ions, with Ca^{2+} and Mg^{2+} being the most likely to have physiological relevance. Although some TRP channels clearly function as chemosensors for exogenous ligands (i.e., capsaicin activation of TRPV1), relatively few endogenous chemical ligands with the capacity to activate TRP channels from the aqueous extracellular milieu are known (2-AG, anandamide). It is likely that endogenous molecules that function as TRP channel ligands remain to be discovered. Table S2 (follow the Supplemental Material link from the Annual Reviews home page at http://www.annualreviews.org) lists many compounds that exhibit agonist or antagonist activity toward TRP channels.

3. *Direct activation.* Changes in ambient temperature are strongly coupled ($Q_{10} > 10$) to the opening of TRPV1–TRPV3 and TRPM8 by poorly understood mechanisms. Other putative direct activators include mechanical stimuli, conformational coupling to IP_3 receptors, and channel phosphorylation. Heating and cell swelling may also act indirectly to activate TRP channels through second messengers or other unidentified mechanisms (8).

For TRP channels that are constitutively active in expression systems, acute posttranslational modification (i.e., phosphorylation) may be functionally equivalent to channel activation or deactivation. Although protein kinases A, C, and G (PKA, PKC, and PKG, respectively) have been demonstrated to modulate TRP channel activity, there is a paucity of data demonstrating direct phosphorylation and its effect on channel activity. Other cellular signaling mechanisms, such as regulation by Ca^{2+}/calmodulin (Ca^{2+}/CaM), have also been demonstrated to modulate TRP channel activity.

Indirect regulation also impinges on TRP channel function. For some TRPs, the number of functional channels in a given cell is regulated by plasma membrane insertion of vesicles carrying TRP proteins. As for classically defined GPCRs, the apparent potency of TRP channel agonists will be sensitive to both channel number and the degree of cellular signal amplification that is effected subsequent to agonist binding (9). Thus, agents that alter the number of spare channels may effectively sensitize or desensitize TRP channels to agonist stimulation. Likewise, signaling cascades that modulate cellular amplification downstream of TRP channels may alter the physiological meaning of channel activation.

Because most TRPs appear to be responsive to multiple distinct stimuli, they may be thought of as signal integrators. Channels are also high-gain signal amplifiers in that they function to enzymatically couple a single gating event (channel opening) to the flux of millions of ions per second. In the case of Ca^{2+}, each ion that permeates a TRP channel possesses some intrinsic signaling potential. If local Ca^{2+}-sensing effector proteins bind Ca^{2+} flowing through the channel, extremely high signal gain may be achieved. In most cells, low-density expression of such channels is warranted. The combined properties of signal integrator and amplifier suggest that TRPs are adapted to function as cellular coincidence detectors. As such, TRP channels are inherently suited to serving broadly defined roles as cellular sensors (5). Roles for TRP channels in growth cone guidance suggest that TRPs may also be utilized to control cellular chemotaxis.

STORE OPERATION?

Emptying of intracellular Ca^{2+} stores (primarily in the endoplasmic reticulum) activates influx of extracellular Ca^{2+} ($[Ca^{2+}]_o$) by an unknown mechanism to replenish the depleted stores (10). The experimental hallmark of the store-operated Ca^{2+} entry (SOCE) process is an $[Ca^{2+}]_o$-dependent plateau of elevated $[Ca^{2+}]_i$ subsequent to the receptor-activated $[Ca^{2+}]_i$ transient. One specific type of SOCE, described primarily in immune cells, is the calcium release–activated current (I_{CRAC}); it is distinguished by extremely high selectivity for Ca^{2+} over Na^+ among other unique biophysical properties (11). Initial reports of *Drosophila* TRP activation by store depletion (since refuted) suggested that mammalian TRP channels would be good candidates to mediate SOCE (12). Studies summarized elsewhere (6, 12, 13) have demonstrated that strategies designed to decrease expression or eliminate specific TRP proteins (antisense oligonucleotides, RNA interference, or gene knockouts) can attenuate native SOCE, leading investigators to conclude that TRP channels directly mediate SOCE. Similar results were reported for the cyclic nucleotide–gated channel CNGA2 (14), demonstrating that TRP channel knockdown is not unique in its ability to abrogate SOCE. Many studies of mammalian TRPs have been designed around the SOCE hypothesis, perhaps at the expense of investigating alternative mechanisms for SOC regulation or other molecular candidates as mediators of SOCE. These include other transporter or channel proteins that can modulate Ca^{2+} influx by directly permeating Ca^{2+}, controlling V_m, or affecting ionic gradients that are energetically coupled to Ca^{2+} homeostasis (12).

Although activation of some TRP channels certainly does accompany PLC activation, a requirement for Ca^{2+} store depletion has not been demonstrated in experiments in which both V_m and $[Ca^{2+}]_i$ were controlled. Store depletion protocols typically depolarize intact cells and elevate $[Ca^{2+}]_i$, leading to pleiotropic effects on intracellular signaling cascades. Experiments utilizing protocols to deplete Ca^{2+} stores (i.e., SERCA pump inhibitors such as thapsigargin) without controlling $[Ca^{2+}]_i$ and V_m should therefore be interpreted carefully. Despite abundant evidence that TRP channels are regulated by factors that either control or are influenced by Ca^{2+} store depletion, no mammalian TRP or other genetically defined channel has yet fulfilled the criteria set forth for unambiguous definition as a store-operated channel (12). If TRP proteins do form store-operated channels, other molecular components must endow these channel complexes with the activation and permeation properties of store-operated calcium currents, which are clearly distinct from the currents generated by expressed TRP channels. Even TRPV6, the TRP channel with biophysical properties most closely resembling I_{CRAC}, manifests biophysical features that are distinct from those of I_{CRAC} (15). Although TRPs may participate in SOCE, it is presently not accurate to refer to TRP channels as store-operated channels per se.

MAMMALIAN TRP CHANNEL STRUCTURE

Six-Transmembrane Channel Structure

The predicted structure of TRP proteins is nucleated by a channel-forming structure composed of six putative transmembrane (TM) domains with a pore domain (P) wedged between the fifth (S5) and sixth (S6) TMs. Hydropathy analyses predict a seventh N-terminal hydrophobic domain (h1) in many TRPC, TRPV, TRPM, TRPP, and TRPML family members, but h1 does not appear to form an authentic

TM domain in TRPC3, suggesting that other TRPs bearing an h1 domain also conform to the usual 6-TM structure (16). TRPP and TRPML proteins may possess an extended extracellular loop between S1 and S2, but experimental evidence for this structural model is lacking. Further testing of the proposed 6-TM architecture of TRP channels is required to refine our understanding of TRP channel structure.

Despite undetectable primary sequence homology, the basic 6-TM architecture characterizes a superfamily of ligand- and voltage-gated channels found in organisms from Archaea to plants and animals (17). Like other 6-TM channels, TRPs probably form a tetrameric quaternary structure (18), where each subunit contributes to a shared selectivity filter and ion-conducting pore similar to that seen in K^+ channels (19). For multisubunit ion channel complexes, heteromeric assemblies may manifest unique properties as compared to homomers. Although most TRPs exhibit little preference for Ca^{2+} over Na^+ ($P_{Ca}/P_{Na} = $ 1–10), Na^+-selective (TRPM3α1, TRPM4, TRPM5; $P_{Ca}/P_{Na} < 0.05$) and Ca^{2+}-selective (TRPM3α2, TRPV5, TRPV6; $P_{Ca}/P_{Na} > 100$) TRP channels are represented in the family. Details of TRP channel permeation properties are reviewed in great detail elsewhere in this volume (19a).

Ion flux through 6-TM channels is controlled by allosteric interactions among the constituent subunits to control one or more gates (1), but the location and structure of these gates in TRP channels is not well resolved. TRP channels lack a conserved series of arginine residues in S4 that form the sensor for TM electrical potential in archetypal 6-TM voltage-gated cation channels (6). But TRPs do exhibit voltage-dependent behavior, characterized by a shallow Boltzmann slope in channel conductance-voltage (G-V) relations (20). Voltage-dependent current relaxation following depolarizing voltage steps (i.e., tail current protocols) is now recognized as a general property of several TRP channels (20). TRP channel voltage sensitivity is reciprocally modulated by temperature or ligand binding in most instances. Although voltage-dependent gating in TRPs is likely to be subserved by a molecular mechanism distinct from that utilized by voltage-gated cation channels, the identity of critical voltage-sensitive residues in TRP proteins is presently unknown. Studies of TRP channel structure should be useful in elucidating whether coupling of voltage sensing to channel gating is similar to that deduced from the recently solved $K_v1.2$ structure (19).

Amino acid sequences flanking the pore in TRP channels are some of the most strongly conserved across the TRP channel family (6). S5, S6, and the proximal C-terminal tail (TRP domain) are strikingly similar even in disparate TRP channels (Figure 1). A mutation in the conserved region of TM6 functions as a dominant-negative in TRP channels (21). It therefore seems likely that amino acid sequences in TRP channels may constitute a core channel structure similar to that seen in $K_v1.2$ (19). But TRP channels are perhaps more aptly characterized by their divergent function than sequence conservation, so structural determinants that transduce gating information from the environment are probably more varied than the channel core itself. It will be intriguing to learn from future studies of TRP structure how channel gating and ion permeation are physically coupled in this family of proteins that can respond to myriad cellular stimuli.

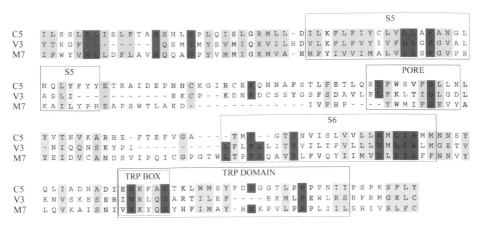

Figure 1   Alignment of selected mammalian TRP channel protein sequences. Amino acid (aa) sequences corresponding to S5 through the TRP domain of human TRPC5 (C5; aa 487–672), TRPV3 (V3; aa 565–721), and TRPM7 (M7; aa 970–1143) were aligned using the Clustal V algorithm (DNAStar, Lasergene). Identical residues are shaded red, strong conservation is indicated by orange shading, and green shading represents weak conservation. Note that regions of highest homology are located in S5, S6, and the TRP domain, whereas sequences flanking the pore region are notably divergent. Accession numbers for sequences used in the alignment were Q9UL62, Q8NET8, and Q96QT4 for human TRPC5, TRPV3, and TRPM7, respectively.

The term TRP domain refers to a homologous block of ~25 intracellular residues immediately C-terminal to S6 that is loosely conserved in all TRP mammalian subfamilies except TRPA and TRPP (5, 6). The TRP domain encompasses a highly conserved 6-amino acid TRP box (EWKFAR in TRPC channels); a proline-rich sequence within the TRP domain has been previously referred to as TRP box 2 (6). Versions of the TRP box, TRP domain, and proline-rich region are also found in TRPV and TRPM channels (Figure 1) (6). Recently, residues within the TRP domain were demonstrated to be required for PIP_2 binding and regulation of channel gating in TRPM8 and TRPV5 channels (22). PIP_2 is a nearly ubiquitous and physiologically relevant regulator of transporter and channel function (23), so it is perhaps not surprising that TRPs functionally interact with this polar lipid. It is presently unclear why PIP_2 appears to stimulate some TRP channels (TRPV5, TRPM5, TRPM7, TRPM8) (22, 24) while inhibiting others (TRPV1) (25, 26). TRP channels were recently shown to contain partial pleckstrin homology (PH) domain sequences that require intermolecular association with cognate sequences in other proteins to form a functional PH holodomain (27). The intriguing split-PH domain hypothesis may help to explain heterogeneity in lipid binding and responsiveness among TRP channels.

PIP_2 and calmodulin (CaM) interact so commonly with ion channels that they may be considered ion channel components (23, 28). Zhu, Birnbaumer, and colleagues (29, 30) demonstrated that TRPC proteins contain multiple putative CaM-binding sites in their N and C termini and that GST fusion proteins of all TRPC channels interact with CaM in vitro. Some CaM-binding sites overlap with an IP_3 receptor (IP_3R)–binding site (CRIB domain) (29). Ca^{2+}-dependent competition between CaM and IP_3R peptides for channel binding was demonstrated in vitro (29), suggesting that Ca^{2+}/CaM and IP_3Rs may dynamically regulate TRPC through competitive interactions. However, the IP_3R-dependent "conformational coupling" hypothesis for TRPC activation has not been tested directly, and other possible mechanisms have not been ruled out. The functional relevance of TRPC-IP_3R interactions is challenged by results in avian DT40 cells lacking all three IP_3R isoforms in which receptor-activated TRPC3 function is apparently normal (31).

In *Drosophila* photoreceptors, TRP channels are tethered into signaling complexes via PDZ interactions by the scaffolding protein INAD (32). Mammalian TRP channels may be similarly integrated into macromolecular assemblies by PLC, Homer, or other proteins (32). A C-terminal PDZ-binding motif (VTTRL) present in TRPC4 and TRPC5 mediates interactions with NHERF/EBP-50 and PLCβ1 (33). Although PDZ-binding motifs in the C terminus of TRP channels have been implicated in the control of plasma membrane localization (34–36), there are few data so far that directly implicate PDZ proteins in the control of TRP channel activity.

FUNCTIONAL PROPERTIES OF MAMMALIAN TRP CHANNELS

Functional properties common to many TRPs may endow them with the capacity to function as cellular sensors (5). The classical example was elucidated in the *Drosophila* visual system, in which the electrical response to light is attenuated in *trp* mutant flies (37). Predominant molecular themes in TRP channel biology that have emerged from work on fly TRP channels include control of channel activity by lipid metabolism (38) and physical linking of signal transduction pathways by scaffolding proteins such as INAD (39). As for many mammalian TRP channels, a precise understanding of the molecular mechanism responsible for TRP activation in the *Drosophila* eye remains elusive. We refer the reader to more detailed reviews of invertebrate TRP channels elsewhere in this volume (39a,b). We review additional details of the TRP gene family and its pharmacopoeia in online supplemental material (see Supplemental Text; follow the Supplemental Material link from the Annual Reviews home page at http://www.annualreviews.org).

THE TRPC SUBFAMILY

The TRPC subfamily was established following the identification and cloning of TRPC1, the first recognized mammalian TRP channel (40). Of the mammalian TRP channels, members of the TRPC subfamily are most closely related to *Drosophila* TRP (30–40% identity). The TRPC subfamily is readily divided into three groups on the basis of sequence alignments and functional comparisons: TRPC1/4/5, TRPC3/6/7, and TRPC2 (5). Others, placing TRPC1 by itself, have suggested that four groups exist (6, 41).

Although it was the first mammalian TRP channel identified, TRPC1 is still enigmatic. Lack of consensus appears to be the underlying theme in functional studies of TRPC1. Homomeric TRPC1 has been interpreted to function as a store-operated, receptor-operated, DAG-activated, IP_3R-gated cation channel, or alternatively to be a nonfunctional channel subunit. Evidence that TRPC1 also forms a stretch-activated channel was recently presented (42), suggesting that further heterogeneity in TRP channel function may remain to be discovered. The debate over homomeric TRPC1 function is emblematic of the apparent discrepancies regarding many TRP channels; resolution of such inconsistencies is an important challenge for future experimental design. It is generally agreed that when expressed in heteromeric complexes with TRPC3, TRPC4, or TRPC5, TRPC1 functions as part of a $G_{q/11}$ receptor–operated cation channel (43).

TRPC2 is distinct from other TRPC subfamily members. Because the human gene encodes a nonfunctional truncated protein and is considered a pseudogene (44), only rodent clones of TRPC2 have been expressed and functionally investigated. TRPC2 may be activated by DAG (45). In the rodent vomeronasal organ, TRPC2 is localized to neuronal microvilli and appears to be required for neuronal excitability in pheromone signal transduction (46).

TRPC4 and TRPC5 are highly homologous (64% identity) and form a group that is most closely related to TRPC1. TRPC4 and TRPC5 function as nonselective cation channels that are activated by $G_{q/11}$ family GPCRs and receptor tyrosine kinases (47, 48). Activation of these channels appears to require PLC enzymatic activity, but neither IP_3 nor DAG alone is sufficient to activate TRPC4 or TRPC5 (48, 49). Intracellular Ca^{2+} (>10 nM) is required for receptor-mediated channel activation, and increasing $[Ca^{2+}]_o$ above 5 mM facilitates channel activity (43, 48). When expressed alone, TRPC4 and TRPC5 each generate currents with an unusual doubly rectifying I-V relation (43, 48). Outward TRPC5 current complexity may involve voltage-dependent block by Mg^{2+} binding to an intracellular site (50).

TRPC4 and TRPC5 are widely expressed but particularly abundant in brain, where their overlapping distributions suggest heteromeric function in vivo (43, 51). Differences in the subunit composition of heteromeric TRPC1, TRPC4, and TRPC5 channels confer unique biophysical properties: Compared to homomeric TRPC4 or TRPC5 channels, heteromeric TRPC1+C4 and TRPC1+C5 channels exhibit a simpler I-V, with a gently negative slope at negative potentials and smooth outward rectification (43). The single-channel conductance of TRPC1+C5 channels is significantly smaller than for TRPC5 homomers (43). Uniquely among TRP channels, $G_{q/11}$ receptor–mediated activation of TRPC4 and TRPC5 (and of heteromers containing TRPC1) is potentiated by micromolar concentrations of the trivalent lanthanide cations La^{3+} or Gd^{3+} (43, 48), presumably by interacting with an extracellular Ca^{2+}-binding site (52).

Our knowledge of the repertoire of TRPC regulation has recently been expanded in intriguing new directions. TRPC5 regulates morphology of hippocampal growth cones, where its potential partner TRPC1 is excluded (53). Homomeric TRPC5 channels, but not TRPC1+C5 heteromers, are rapidly delivered to the plasma membrane following growth factor receptor stimulation in a Rac-, PI-3 kinase–, and PI-5 kinase–dependent manner (54). TRPC proteins were recently shown to control growth cone guidance in both mammalian and amphibian model systems (see discussion of growth cone guidance below). Localization of TRPV1, TRPC3, and TRPC6 also appears to be regulated by exocytosis (55–57). PIP$_3$ may be sufficient to rapidly activate Ca^{2+} entry in HEK cells expressing TRPC6 (58), but the details of this mechanism have not been elucidated. Signaling through PI pathways may therefore represent an important mode of control over the activity of TRP channels and their plasma membrane residence.

The second major TRPC group includes TRPC3, TRPC6, and TRPC7. These channels are 65–78% identical and generate similar nonselective and double-rectifying cation currents when activated by G$_{q/11}$-coupled receptors or DAG as either homomeric or heteromeric channels (49, 59). Muscarinic receptors activate TRPC6 in a highly cooperative fashion (Hill slope > 3) and synergize with the DAG analogs, suggesting that DAG production alone is insufficient to explain receptor-operated channel activity (60). As for other TRPC proteins, the details of receptor-mediated channel activation remain fundamentally mysterious.

TRPC3 and TRPC6 are mono- and diglycosylated on asparagine residues in the extracellular S1-S2 and S3-S4 loops; glycosylation status correlates with basal channel activity (61). The nonreceptor tyrosine kinases Src and Fyn positively regulate TRPC3 (62) and TRPC6 (63) activity, respectively. TRPC3 channels can be directly phosphorylated by PKG (64). Like PKG, PKC activity negatively regulates TRPC3 (65). TRPC6 and TRPC7 channels exhibit complex Ca^{2+} regulation. [Ca^{2+}]$_o$ stimulates TRPC6 activity and inhibits TRPC7 activity; likewise, [Ca^{2+}]$_i$ stimulates TRPC6 activity and inhibits TRPC7 activity via Ca^{2+}/CaM (66).

THE TRPV SUBFAMILY

The TRPV channel subfamily has six members divided into two groups: TRPV1–V4 and TRPV5 and TRPV6. TRPV1–V4 comprise the so-called thermo-TRPs that are activated by heating in heterologous expression systems. Although thermosensitive TRPV channels do appear to function in thermosensation (67, 68), these channels are also expressed in tissues in which thermoregulatory homeostasis should preclude dramatic temperature swings. Temperature may therefore play a permissive rather than instructive role in dictating these channels' activity (69).

The founding member, TRPV1, variously known as the capsaicin receptor or vanilloid receptor 1, is the most thoroughly studied channel among TRPV channels. TRPV1 is activated promiscuously and polymodally: TRPV1 agonists include capsaicin and resiniferatoxin, heat, H$^+$, endocannabinoid lipids such as anandamide, eicosanoids, and 2-APB (Table S2; follow the Supplemental Material link from the Annual Reviews home page at http://www.annualreviews.org). Expressed channels generate weakly Ca^{2+}-selective and outwardly rectifying cation currents (70), although the shape of the steady-state I-V relation is sensitive to agonist- and voltage-dependent modulation (71, 72).

TRPV1 appears to be sensitized by PKA, PKC, and receptor-activated PLC activity (25, 73–75), although the relative importance of these mechanisms remains to be clarified. PIP_2 appears to bind TRPV1 directly, causing channel inhibition that is relieved by PLC-catalyzed PIP_2 hydrolysis (25, 76). C-terminal TRPV1 truncations that disrupt the PIP_2-binding site also impair thermal responsiveness (77).

TRPV1 is widely expressed, but its function has been most thoroughly studied in sensory neurons, in which it was first identified (70, 78). TRPV1-null mice lack vanilloid sensitivity in nociceptive, inflammatory, and thermoregulatory assays; a crucial role for TRPV1 in thermal hyperalgesia is evident (79, 80). However, TRPV1 and TRPV2 double-knockout mice apparently have normal in vivo thermal nociceptor responses (81). A TRPV1 splice variant was reported to function as an amiloride-insensitive salt taste receptor; vanilloids and H^+ lower the temperature dependence of neuronal responses, allowing salty taste transduction at ambient temperature (69). TRPV1 is also required for mechanically evoked purinergic signaling in the bladder urothelium (82). TRPV1 also plays a role in a variety of other physiological functions from satiety (74) to hearing modulation (83) to gastrointestinal motility (84).

TRPV2 is 50% identical to TRPV1 and similarly forms a weakly Ca^{2+}-selective cation channel that is activated by cell swelling, heat, and 2-APB (85–87). TRPV2 associates with recombinase gene activator protein and may be regulated downstream of PKA activation in mast cells (88). TRPV2 protein is also found in the CNS (89), myenteric plexus, and nodose ganglion (90). In CHO cells and vascular smooth muscle, TRPV2 appears to function as a stretch-activated channel (85). Ankyrin repeats in the osmosensitive *Caenorhabditis elegans* channel OCR-2 are required for function, and TRPV2 can substitute for OCR-2 loss, suggesting conservation of function in mammals (91). TRPV2 has been proposed to mediate high-threshold noxious heat sensation, but its physiological role as such remains unsubstantiated (81).

Expressed TRPV3 also mediates a weakly Ca^{2+}-selective cationic conductance in response to non-noxious heat, 2-APB, or camphor (86, 92, 93). Heat-activated TRPV3 currents display a strong outward rectification, a striking thermal hysteresis, and sensitization following repeated activation (93). Prolonged exposure to high concentrations of 2-APB or heat causes the channel to enter a constitutively active $NMDG^+$-permeable conductance mode termed I_2 (94). TRPV1 also appears to be permeable to large organic dyes (95), suggesting that some TRP channels may permeate organic cations and thus function as organic solute transporters in vivo. A shift in the TRPV3 G-V relation to more negative potentials and linearization of the steady-state I-V relation accompanies the entry into I2 (94), consistent with the ligand-modulated voltage-dependent current relaxation described in other TRP channels (20). TRPV3 interacts with TRPV1 (96) and TRPV2 (97), but the effect of these interactions on TRPV3 function is not known. TRPV3 is reported to be widely expressed in humans and primates (93, 96) but skin-specific in mice (98); it remains to be determined whether observed differences in expression patterns are species-specific. Mouse keratinocytes do exhibit 2-APB and heat-evoked TRPV3 currents (92, 94), and TRPV3 knockout mice exhibit altered thermal preference (67), suggesting that TRPV3 function in keratinocytes may be important for thermosensation.

TRPV4 is a constitutively active Ca^{2+}-permeable cation channel that displays a gently outwardly rectifying I-V relation in response to moderate heating, hypotonic challenge, or the phorbol ester 4α-PDD (8, 99). The effect of hypotonicity on TRPV4 is attributable to swelling-induced production of the eicosanoid $5',6'$-epoxyeicosatrienoic acid ($5',6'$-EET), which directly activates TRPV4 channels (8, 100). Mutations in the N terminus of S3 can discriminate between TRPV4 gating by heat or phorbol esters versus swelling (8), suggesting that multiple, distinct activation mechanisms operate in TRPV4. However, allosteric crosstalk between the various gating mechanisms is likely because heat and hypotonic stimuli synergistically activate TRPV4 (101). TRPV4 activation by heat (102), but not by swelling (101), requires intact ankyrin repeats in the N terminus. Unlike TRPV1–TRPV3, TRPV4 is apparently insensitive to activation by 2-APB (86). Pore residues conferring cation selectivity in TRPV4 also alter voltage-dependent block by ruthenium red (RuR) (103). TRPV4 knockout mice exhibit reduced pressure and osmotic sensitivity, altered thermal selection, and hearing loss (104–107).

TRPV5 and TRPV6 share 74% amino acid identity but are much less similar (22–24% identity) to other TRPVs and therefore constitute a separate group (5, 6). The functional differences between TRPV5 and TRPV6 and other TRP channels are striking: These channels exhibit strong inward rectification, are highly Ca^{2+}-selective ($P_{Ca}/P_{Na} > 100$), and exhibit robust constitutive activity when overexpressed (108, 109). Calcium selectivity in TRPV5 and TRPV6 may be achieved by a ring of negatively charged aspartate residues in the selectivity filter like that proposed for other Ca^{2+}-selective channels (110).

N-terminal ankyrin repeats in TRPV6 were proposed to form a molecular zipper that stabilizes channel assembly and allows for interactions with intracellular partners (111, 112), although assembly and trafficking of the highly homologous TRPV5 channel appears to be more dependent on C-terminal sequences (113). TRPV5 and TRPV6 form homo- and heterotetrameric channels in expression systems via interactions in both the N and C termini; these channels may coassemble in kidney (18). Intra- and extracellular [Ca^{2+}] (109, 114) and CaM (115) regulate TRPV6 activity. TRPV5 and TRPV6 expression in small intestine and kidney is important for vitamin D–stimulated calcium uptake via epithelial cells (116, 117). Despite enhanced vitamin D levels, TRPV5 knockout mice exhibit diminished renal Ca^{2+} reabsorption and hypercalciuria (117).

THE TRPM SUBFAMILY

The genetically and functionally diverse TRPM subfamily is comprised of eight putative members divided into three main groups: TRPM1/3, TRPM4/5, and TRPM6/7; TRPM2 and TRPM8 exhibit low sequence homology and therefore do not seem to warrant grouping (5, 118). TRPM2, TRPM6, and TRPM7 are unique among known ion channels in that they encode enzymatically active protein domains fused to their ion channel structures.

TRPM1 (melastatin) was identified as a prognostic marker for metastasis of localized melanoma (119), but its channel activity has not been described definitively. TRPM2 is characterized by a nudix hydrolase domain that is highly homologous to the ADP pyrophosphatase NUDT9 (120). The TRPM2 nudix domain mediates adenosine diphosphoribose (ADP-ribose, ADPR) binding to the channel and is required for channel activation by ADP-ribose (120). Inhibitors of poly (ADP-ribose) polymerase, a potential enzymatic source of ADPR, prevent TRPM2 activation but do not block the channel directly (121). TRPM2 is nonselective and displays a nearly linear (i.e., voltage-independent) I–V relation in response to intracellular perfusion of ADPR, cyclic ADP-ribose (cADPR), and nicotinamide adenine dinucleotide (120, 122). $[Ca^{2+}]_i$, arachidonic acid, and cADPR potentiate channel activation by ADPR (123, 124), but the physiological relevance of TRPM2 activation by ADPR and cADPR remains elusive. The channel is also responsive to the reactive oxygen species H_2O_2, which may also cause release of ADPR from mitochondria, and to TNF-α, suggesting that TRPM2 functions as a cellular redox sensor (120, 125–128).

TRPM3 is alternatively spliced into multiple functional variants (129, 130); novel TRPM3 variants that differ only in the pore region were recently described (130). Mouse TRPM3α1 is monovalent-selective, whereas TRPM3α2 is divalent-selective, suggesting that in vivo TRPM3 function may depend largely on the relative abundance of these variants (130). Interestingly, TRPM3α1 and TRPM3α2 both exhibit constitutively active, outwardly rectifying currents that are blocked by intracellular Mg^{2+} (130), a feature reminiscent of TRPM6 and TRPM7 function. Hypotonicity also reportedly increases calcium entry through the 1325 amino acid TRPM3 variant (TRPM3$_{1325}$) when expressed in HEK293 cells (129). Recently, D-*erythro*-sphingosine, a metabolite of cellular sphingolipids, but not sphingosine-1-phosphate or ceramide, was shown to activate TRPM3$_{1325}$ (131). TRPM3 has been suggested to play a role in microglial (131) and choroid plexus function (130), although it is unclear which splice variants function in these tissues.

TRPM4 and TRPM5 are unique among TRP channels in that they manifest monovalent-selective currents with limited Ca^{2+} permeability (132, 133). TRPM4a probably encodes a nonfunctional channel; TRPM4b therefore appears to be the functionally relevant splice variant. Both TRPM4b and TRPM5 are activated by $[Ca^{2+}]_i$, albeit with widely varying apparent potency (EC_{50} = 300 nM–30 μM) and cooperativity (n_H = 1–6) (24, 132–134); the source of variability among various reports is unknown. In a parallel comparison, $[Ca^{2+}]_i$ activated mouse

TRPM5 $\sim$30-fold more potently than for mouse TRPM4, and the $[Ca^{2+}]_i$ sensitivity for both channels was dramatically reduced in inside-out patch recordings (135). TRPM4 possesses several CaM-binding sequences that impart complex $[Ca^{2+}]_i$-dependent channel regulation (136). $[Ca^{2+}]_i$ desensitizes TRPM5 channels, but PIP_2 is sufficient to restore their activity partially (24). Mutation of residues in the TRP domain that are required for PIP_2 binding to TRPM8 also abolishes $[Ca^{2+}]_i$-activated TRPM5 currents, implicating PIP_2 as a required cofactor for activation of functionally disparate TRP channels (22).

Both TRPM4 and TRPM5 exhibit prominent voltage-dependent deactivation at negative membrane potentials, resulting in a strongly rectifying steady-state I–V relation (137). Decavanadate increases the potency for ATP block tenfold and decreases the voltage sensitivity of TRPM4b, but not TRPM5, by binding to a C-terminal domain homologous to SERCA pumps that inhibits voltage-dependent channel closing (138, 139). TRPM4 channels are ubiquitously expressed; either alone or in combination with TRPM5 (which has a more restricted expression pattern), TRPM4 appears likely to mediate so-called Ca^{2+}-activated monovalent or Ca^{2+}-activated nonselective cation currents in vivo (132, 133). TRPM4b knockdown alters T cell receptor–activated $[Ca^{2+}]_i$ oscillations in T cells, indicating that TRPM4 depolarization normally serves to counterbalance I_{CRAC} by decreasing the driving force for Ca^{2+} entry (140). TRPM4 or TRPM5 currents also contribute to myogenic vasoconstriction of cerebral arteries (141), and TRPM5 is required for sweet, bitter, and umami taste transduction (142).

TRPM6 and TRPM7 are unique channels that possess both ion channel and protein kinase activities. The enzymatically active C-terminal kinase domain bears little sequence identity to eukaryotic protein kinases (e.g., PKA) but, except for the inclusion of a zinc-finger domain, is structurally homologous to other catalytically active protein kinases (143, 144). The kinase domain autophosphorylates TRPM7 (145) and annexin 1 (146), but the physiological significance of channel kinase activity is unclear. TRPM7 was identified as an interactor with $PLC\beta_1$ and is ubiquitously expressed (147, 148), whereas TRPM6 is selectively expressed in kidney and intestine (149). Overexpressed TRPM6 and TRPM7 currents rectify at positive potentials but pass very little inward current in the presence of divalent cations; both channels are inhibited by $[Mg^{2+}]_i$ (0.3–1.0 mM), so their currents are revealed by dialyzing out or chelating $[Mg^{2+}]_i$ (26, 147, 149). In contrast to voltage-modulated TRP channels that exhibit prominent time-dependent tail current relaxations in the absence of extracellular divalent cations (20), TRPM6 and TRPM7 currents are nearly linear in divalent-free solutions, indicating that the apparent voltage dependence of steady-state TRPM6 and TRPM7 I-V currents may arise from ionic interactions in the pore rather than a voltage-dependent modulation of channel gating.

TRPM6 and TRPM7 channels share important biophysical features with endogenous magnesium-inhibitable currents and are likely to represent the molecular correlates of these currents (155–159). Although the physiological mechanism for activation of TRPM7 is unknown, expressed TRPM7 currents are both decreased by PLC-catalyzed PIP_2 hydrolysis (26) and regulated through PKA by receptors coupled to adenylyl cyclase (150). TRPM7 may be regulated by reactive oxygen species to mediate anoxia-induced cell death in brain (151) and has been reported to be required for cell viability (152). TRPM6 mutations in human result in hypomagnesemia with secondary hypocalcemia (153, 154), and the high Mg^{2+} permeability of these channels (147) suggests that TRPM7 may be important for general Mg^{2+} homeostasis, whereas TRPM6 functions specifically in intestine and kidney (149).

Although first identified in prostate, in which it is proposed to be an androgen-responsive channel (160), TRPM8 has been most convincingly described as a cold- and menthol-activated nonselective cation channel with prominent voltage-dependent gating properties (71, 161–163). TRPM8 is expressed in sensory neurons, in which it may function as a cold thermosensor (161, 162). Channel agonists (cold, menthol, icilin) shift the voltage dependence of TRPM8 to more negative (physiologically relevant) potentials, similar to the shift in voltage dependence seen for TRPV1 (20, 71, 163, 164). PIP_2 activation of TRPM8 is sensitive to mutagenesis in the TRP domain, and cold, menthol, and depolarizing voltages increase the potency for PIP_2 channel TRPM8 activation (22). In light of the reported PIP_2 dependence of TRPM7 (26) and TRPM5 (24), the data from Rohacs et al. (22) imply that the TRP domain may function generally as a PIP_2-sensitive gating apparatus in divergent TRP channels.

TRPA1

TRPA1 is distinguished from other TRP channels by the presence of $\sim$14 ankyrin repeats in its N terminus. Although TRPA1 was initially described as a cold-sensitive nonselective cation channel (165), there is general agreement that it probably functions as a ligand-gated channel in expression systems and sensory neurons. The active constituents of mustard oil (allyl isothionate) and garlic (allicin) robustly activate TRPA1 currents (166–168), and TRPA1 appears to be regulated by PLC-coupled receptors (168). Thus, despite its residence in a separate subfamily, TRPA1 exhibits many functional characteristics of other TRP channels. But TRPA1 also exhibits unique properties: It has been tentatively identified as the elusive hearing transduction channel (169, 170). Multiple ankyrin repeats in TRPA1 have been proposed to function as a mechanical spring directly linking TRPA1 gating to cytoskeletal proteins (169–174).

THE TRPP SUBFAMILY

Polycystic kidney disease (PKD) proteins or polycystins form the TRPP subfamily of Ca^{2+}-permeant ion channels. Polycystin-1 (PC1, PKD1), polycystin-REJ, polycystin-1L1, and polycystin-1L2 are predicted to form 11-TM proteins with a

6-TM TRP-like channel domain at the C-terminal end of the polypeptide but are not explicitly included in the TRP channel family. In heteromeric complexes with TRPP2 (previously termed PKD2), PKD1 is reported to form a Ca^{2+}-permeable nonselective cation channel with a linear I-V relation (175). However, the possibility that PKD1 forms a channel by itself has not been ruled out. TRPP2 has also been suggested to form an intracellular channel (176). Phosphofurin acidic cluster sorting proteins interact with TRPP2 to regulate subcellular targeting; such interactions may explain observed differences in TRPP2 function and localization (177).

PKD1 and TRPP2 both appear to be targeted to primary cilia of renal epithelial cells, where the channel complex may be gated by fluid flow (178). Mutations in either TRPP2 or PKD1 cause alterations in polarization and function of cyst-lining epithelial cells, leading to autosomal dominant PKD. PKD1 and TRPP2 knockout mice also manifest cardiac septal defects and cystic nephrons and pancreatic ducts, and die in utero (179). TRPP2 and PKD1 are expressed on the primary cilia of ovarian granulosa cells and may play a role in ovarian follicle maturation and differentiation (180).

TRPP3 appears to be a large-conductance Ca^{2+}-permeable channel (181). The mouse homolog is deleted in *Krd* mutant mice that exhibit hypoplastic and cystic kidneys, but the contribution of TRPP3 to the *Krd* phenotype is unknown (182). There are no reports of channel activity for TRPP5.

THE TRPML SUBFAMILY

The mucolipins (TRPML1–3) are 6-TM channels that are probably localized to intracellular vesicles and excluded from the plasma membrane (183). TRPML1 is reported to be a nonselective channel that is inhibited by lowering pH; it may play a role in endosomal acidification (184). Mutations in TRPML1 are associated with the neurodegenerative lysosomal storage disorder mucolipidosis type IV that apparently results from abnormal late endocytic sorting or transport (183). TRPML3 is expressed in hair cells and sterocilia and its function is implicated in hearing; the *varitint-waddler* mouse, which possesses a TRPML1 mutation, exhibits deafness and pigmentation defects (187). The function of TRPML2 is unknown.

TRP CHANNELS IN GROWTH CONE GUIDANCE

Axons are guided to their respective targets in the migrating nervous system by diffusible and cell-bound guidance cues (188–192). The neuronal growth cone is a highly organized palm-like structure at the tips of developing neurites that modulates the rate and direction of neurite extension in response to attractive and repellent cues (Figure 2). Active sampling of the local environment is achieved by processive extension and retraction of filopodial protrusions on the growth

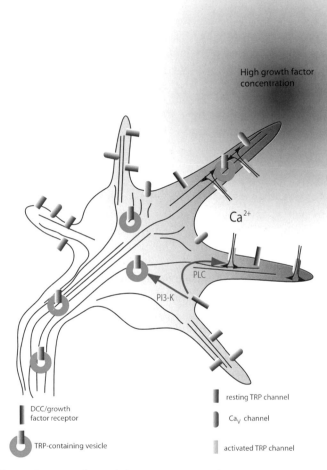

High growth factor
concentration

Ca^{2+}

PLC

PI3-K

DCC/growth
factor receptor

TRP-containing vesicle

resting TRP channel

Ca$_V$ channel

activated TRP channel

Figure 2   TRP channels in growth cone guidance. Growth cones turning in a gradient of an attractive growth factor (i.e., netrin-1; *red shading*) exhibit more elaborated filopodial structure in which growth factor concentrations are higher. Resting TRP channels (*blue rectangles*) are activated by receptors for growth factors or other guidance cues (*black rectangles*) that signal through PLC. Open TRP channels (*orange rectangles*) permeate Ca^{2+} directly and indirectly stimulate Ca^{2+} influx through voltage-dependent Ca^{2+} channels (Ca$_V$; *red ovals*) by membrane depolarization. Growth factor signaling through PI3-kinase (PI3-K) may also stimulate rapid insertion of TRP channels contained in vesicles (*green circles*) to augment TRP channel activity. Ca^{2+} influx presumably establishes a gradient of increased [Ca^{2+}]$_i$ (*blue shading*) across the growth cone that is instructive for turning.

cone. The molecular identity of several guidance cues and their corresponding surface receptors have been identified: netrin-1 and brain-derived neurotrophic factor (BDNF) are attractive, whereas myelin-associated glycoprotein and semaphorin 3A are repellent (191–194). [Ca^{2+}]$_i$ plays an essential role in growth cone guidance by regulating growth cone morphology (195–198), the cytoskeleton (199, 200), and trafficking of membrane precursor vesicles (201). Blocking plasma membrane Ca^{2+} channels inhibits growth cone motility and neurite extension (202) and can abolish growth cone steering in a gradient of netrin-1 (203). By contrast, activation of voltage-dependent Ca^{2+} channels (Ca_V) in the growth cone promotes attractive steering (194, 204–210). Ca_V are presumed to underlie Ca^{2+} spikes, which are characterized by high-amplitude, short-lived elevations in [Ca^{2+}]$_i$ (205). The molecular basis for Ca^{2+} waves has remained unclear until recently, but recent studies showing that TRP channels generate instructive [Ca^{2+}]$_i$ signals in growth cones strongly suggest that TRP channels underlie Ca^{2+} wave activity.

Seminal results from Greka et al. (53) suggested that TRPC5 controls neurite length and growth cone morphology by regulating Ca^{2+} influx. TRPC5 is enriched in the growth cones of cultured mouse hippocampal neurons, and overexpression of TRPC5 or a dominant-negative pore mutant of TRPC5 (DN-TRPC5) inhibits or enhances, respectively, the neurite outgrowth of the transfected neurons. A recent study found that Ca^{2+} influx through TRPC channels contributes to BDNF-induced elevation of [Ca^{2+}]$_i$ and cone attraction in cultured rat cerebellar granule cells (211). The BDNF receptor TrkB activates PLCγ, which induces TRPC-mediated Ca^{2+} influx and growth cone attraction toward BDNF. TRPC1, TRPC3, and TRPC6 are expressed in granular cells, but TRPC5 is absent; knockdown of endogenous TRPC3 using siRNA or by overexpression of DN-TRPC3 or DN-TRPC6 abolishes BDNF growth cone attraction in the cerebellar neurons (211). TRPC3 knockdown does not impair neurite outgrowth, indicating that Ca^{2+} influx through TRPC3 selectively controls growth cone guidance. In hippocampal neurons, TRPC5 expression levels affect neurite extension, suggesting that different TRP channels may subserve distinct regulatory functions in the growth cone by generating different spatiotemporal patterns of Ca^{2+} influx.

That TRP channels function as key players in growth cone guidance is supported by two recent studies of axon turning in *Xenopus* neurons (212, 213). XTRPC1, which is most closely related to the mammalian TRPC1, is essential for the turning of cultured *Xenopus* spinal neuron growth cones toward the attractive cues of netrin-1 and BDNF. Pharmacological block of TRP channels by SKF-96365 (2–15 μM) or La^{3+} (10–50 μM), expression of DN-XTRPC1, or knockdown of endogenous protein by morpholino injection renders the growth cone insensitive to netrin-1 attraction (212, 213). XTRPC1 protein was found to be enriched in the growth cones of the majority of cultured neurons; morpholinos substantially reduce XTRPC1 expression levels and also abolish the attractive guidance (212, 213). Interestingly, knockdown of XTRPC1 also affected the myelin-associated glycoprotein–induced repulsion of the growth cone, but semaphorin 3A repulsion was unaffected (213). The in vivo role of XTRPC1 in netrin-mediated growth cone guidance was further investigated in an established assay system for netrin-1-mediated growth cone guidance. Following injection of XTRPC1 morpholinos into the blastomere, the midline crossing of ventrally projecting commissural axons at stage 23–25 was reduced by 65% compared to controls. These results reinforce the notion that a TRP channel is essential for netrin-1-dependent axon guidance in vivo.

Netrin-1-mediated Ca^{2+} influx into *Xenopus* growth cones was reported to depend on Ca_v; with the exception of Cd^{2+}, Ca_v blockers (nimodipine, ryanodine, BHQ) do not completely block Ca^{2+} influx in growth cones (203, 204). However, Wang & Poo (213) found that with Ca_v blocked, netrin-1-activated nonselective cation current and Ca^{2+} influx remain; both are blocked by SKF-96365. These authors hypothesize that in amphibian spinal neurons, TRP channel activity suffices to elevate $[Ca^{2+}]_i$ but may also depolarize neurons enough to activate Ca_v, thereby serving as an initial trigger to amplify Ca_v-dependent Ca^{2+} influx (213). The data raise the intriguing possibility that TRP channels function to trigger Ca^{2+} entry in other neurons or tissues in which so far only Ca_v function has been implicated.

Control of TRP channels in specialized structures like growth cones may be mediated by factors distinct from outright receptor-mediated channel activation. TRPC5 is associated with Stathmin-2 in vesicular transport structures along the neurite and in mammalian growth cones (53), and TRPC5 is rapidly inserted into the plasma membrane after growth factor stimulation in both neurons and expression systems (54). Vesicles are thought to be preferentially inserted into the growth cones of actively extending neurites (201). Furthermore, synaptic vesicles move within individual growth cone filopodia and fuse with the plasma membrane after focal depolarization with high $[K^+]$ solution (214). Regulation of vesicle transport and fusion therefore represents an intriguing mechanism by which TRP channels can participate in steering growth cones or other chemotactic cells. Because heteromeric TRP channels are trafficked differently from monomers (54), it will be important to investigate the expression pattern and subcellular localization of individual TRP channels that are suspected to participate in cellular guidance.

**The *Annual Review of Physiology* is online at
http://physiol.annualreviews.org**

LITERATURE CITED

1. Hille B. 2001. *Ion Channels of Excitable Membranes*. Sunderland, MA: Sinauer

2. Berridge MJ, Bootman MD, Roderick HL. 2003. Calcium signalling: dynamics, homeostasis and remodelling. *Nat. Rev. Mol. Cell Biol.* 4:517–29

3. Montell C, Birnbaumer L, Flockerzi V, Bindels RJ, Bruford EA, et al. 2002. A unified nomenclature for the superfamily of TRP cation channels. *Mol. Cell* 9:229–31

4. Clapham DE, Montell C, Schultz G, Julius

D. 2003. International Union of Pharmacology. XLIII. Compendium of voltage-gated ion channels: transient receptor potential channels. *Pharmacol. Rev.* 55:591–96

5. Clapham DE. 2003. TRP channels as cellular sensors. *Nature* 426:517–24

6. Montell C. 2005. The TRP superfamily of cation channels. *Sci. STKE* 2005:re3

7. Clapham DE. 1995. Calcium signaling. *Cell* 80:259–68

8. Vriens J, Watanabe H, Janssens A, Droogmans G, Voets T, Nilius B. 2004. Cell swelling, heat, and chemical agonists use distinct pathways for the activation of the cation channel TRPV4. *Proc. Natl. Acad. Sci. USA* 101:396–401

9. Ross RA. 2003. Anandamide and vanilloid TRPV1 receptors. *Br. J. Pharmacol.* 140:790–801

10. Putney JW Jr. 1977. Muscarinic, alpha-adrenergic and peptide receptors regulate the same calcium influx sites in the parotid gland. *J. Physiol.* 268:139–49

11. Prakriya M, Lewis RS. 2003. CRAC channels: activation, permeation, and the search for a molecular identity. *Cell Calcium* 33:311–21

12. Nilius B. 2004. Store-operated Ca^{2+} entry channels: still elusive! *Sci. STKE* 2004:pe36

13. Clapham DE. 2002. Sorting out MIC, TRP, and CRAC ion channels. *J. Gen. Physiol.* 120:217–20

14. Zhang J, Xia SL, Block ER, Patel JM. 2002. NO upregulation of a cyclic nucleotide-gated channel contributes to calcium elevation in endothelial cells. *Am. J. Physiol. Cell Physiol.* 283:C1080–89

15. Voets T, Prenen J, Fleig A, Vennekens R, Watanabe H, et al. 2001. CaT1 and the calcium release-activated calcium channel manifest distinct pore properties. *J. Biol. Chem.* 276:47767–70

16. Vannier B, Zhu X, Brown D, Birnbaumer L. 1998. The membrane topology of human transient receptor potential 3 as inferred from glycosylation-scanning mutagenesis and epitope immunocytochemistry. *J. Biol. Chem.* 273:8675–79

17. Yu FH, Catterall WA. 2004. The VGL-chanome: a protein superfamily specialized for electrical signaling and ionic homeostasis. *Sci. STKE* 2004:re15

18. Hoenderop JG, Voets T, Hoefs S, Weidema F, Prenen J, et al. 2003. Homo- and heterotetrameric architecture of the epithelial Ca^{2+} channels TRPV5 and TRPV6. *EMBO J.* 22:776–85

19. Long SB, Campbell EB, Mackinnon R. 2005. Crystal structure of a mammalian voltage-dependent shaker family K^+ channel. *Science* 309:897–903

20. Nilius B, Talavera K, Owsianik G, Prenen J, Droogmans G, Voets T. 2005. Gating of TRP channels: a voltage connection? *J. Physiol.* 567:35–44

21. Kuzhikandathil EV, Wang H, Szabo T, Morozova N, Blumberg PM, Oxford GS. 2001. Functional analysis of capsaicin receptor (vanilloid receptor subtype 1) multimerization and agonist responsiveness using a dominant negative mutation. *J. Neurosci.* 21:8697–706

22. Rohacs T, Lopes CM, Michailidis I, Logothetis DE. 2005. PI(4,5)P_2 regulates the activation and desensitization of TRPM8 channels through the TRP domain. *Nat. Neurosci.* 8:626–34

23. Hilgemann DW, Feng S, Nasuhoglu C. 2001. The complex and intriguing lives of PIP_2 with ion channels and transporters. *Sci. STKE* 2001:RE19

24. Liu D, Liman ER. 2003. Intracellular Ca^{2+} and the phospholipid PIP_2 regulate the taste transduction ion channel TRPM5. *Proc. Natl. Acad. Sci. USA* 100:15160–65

25. Chuang HH, Prescott ED, Kong H, Shields S, Jordt SE, et al. 2001. Bradykinin and nerve growth factor release the capsaicin receptor from PtdIns(4,5)P_2-mediated inhibition. *Nature* 411:957–62

26. Runnels LW, Yue L, Clapham DE. 2002.

The TRPM7 channel is inactivated by PIP$_2$ hydrolysis. *Nat. Cell Biol.* 4:329–36

27. van Rossum DB, Patterson RL, Sharma S, Barrow RK, Kornberg M, et al. 2005. Phospholipase Cγ1 controls surface expression of TRPC3 through an inter-molecular PH domain. *Nature* 434:99–104

28. Saimi Y, Kung C. 2002. Calmodulin as an ion channel subunit. *Annu. Rev. Physiol.* 64:289–11

29. Tang J, Lin Y, Zhang Z, Tikunova S, Birnbaumer L, Zhu MX. 2001. Identification of common binding sites for calmodulin and inositol 1,4,5-trisphosphate receptors on the carboxyl termini of trp channels. *J. Biol. Chem.* 276:21303–10

30. Zhu MX. 2005. Multiple roles of calmodulin and other Ca^{2+}-binding proteins in the functional regulation of TRP channels. *Pflügers Arch.* 451:105–15

31. Venkatachalam K, Ma HT, Ford DL, Gill DL. 2001. Expression of functional receptor-coupled TRPC3 channels in DT40 triple receptor InsP3 knockout cells. *J. Biol. Chem.* 276:33980–85

32. Harteneck C. 2003. Proteins modulating TRP channel function. *Cell Calcium* 33:303–10

33. Tang Y, Tang J, Chen Z, Trost C, Flockerzi V, et al. 2000. Association of mammalian trp4 and phospholipase C isozymes with a PDZ domain-containing protein, NHERF. *J. Biol. Chem.* 275:37559–64

34. Palmada M, Poppendieck S, Embark HM, van de Graaf SF, Boehmer C, et al. 2005. Requirement of PDZ domains for the stimulation of the epithelial Ca^{2+} channel TRPV5 by the NHE regulating factor NHERF2 and the serum and glucocorticoid inducible kinase SGK1. *Cell Physiol. Biochem.* 15:175–82

35. Mery L, Strauss B, Dufour JF, Krause KH, Hoth M. 2002. The PDZ-interacting domain of TRPC4 controls its localization and surface expression in HEK293 cells. *J. Cell Sci.* 115:3497–508

36. Song X, Zhao Y, Narcisse L, Duffy H,

Kress Y, et al. 2005. Canonical transient receptor potential channel 4 (TRPC4) co-localizes with the scaffolding protein ZO-1 in human fetal astrocytes in culture. *Glia* 49:418–29

37. Montell C. 2003. The venerable inveterate invertebrate TRP channels. *Cell Calcium* 33:409–17

38. Hardie RC, Minke B. 1993. Novel Ca^{2+} channels underlying transduction in Drosophila photoreceptors: implications for phosphoinositide-mediated Ca^{2+} mobilization. *Trends Neurosci.* 16:371–76

39. Shieh BH, Zhu MY. 1996. Regulation of the TRP Ca^{2+} channel by INAD in Drosophila photoreceptors. *Neuron* 16:991–98

40. Wes PD, Chevesich J, Jeromin A, Rosenberg C, Stetten G, Montell C. 1995. TRPC1, a human homolog of a Drosophila store-operated channel. *Proc. Natl. Acad. Sci. USA* 92:9652–56

41. Plant TD, Schaefer M. 2005. Receptor-operated cation channels formed by TRPC4 and TRPC5. *Naunyn Schmiedebergs Arch. Pharmacol.* 371:266–76

42. Maroto R, Raso A, Wood TG, Kurosky A, Martinac B, Hamill OP. 2005. TRPC1 forms the stretch-activated cation channel in vertebrate cells. *Nat. Cell Biol.* 7:179–85

43. Strubing C, Krapivinsky G, Krapivinsky L, Clapham DE. 2001. TRPC1 and TRPC5 form a novel cation channel in mammalian brain. *Neuron* 29:645–55

44. Vannier B, Peyton M, Boulay G, Brown D, Qin N, et al. 1999. Mouse trp2, the homologue of the human trpc2 pseudogene, encodes mTrp2, a store depletion-activated capacitative Ca^{2+} entry channel. *Proc. Natl. Acad. Sci. USA* 96:2060–64

45. Lucas P, Ukhanov K, Leinders-Zufall T, Zufall F. 2003. A diacylglycerol-gated cation channel in vomeronasal neuron dendrites is impaired in TRPC2 mutant mice: mechanism of pheromone transduction. *Neuron* 40:551–61

46. Liman ER, Corey DP, Dulac C. 1999.

TRP2: a candidate transduction channel for mammalian pheromone sensory signaling. *Proc. Natl. Acad. Sci. USA* 96:5791–96

47. Okada T, Shimizu S, Wakamori M, Maeda A, Kurosaki T, et al. 1998. Molecular cloning and functional characterization of a novel receptor-activated TRP Ca^{2+} channel from mouse brain. *J. Biol. Chem.* 273:10279–87

48. Schaefer M, Plant TD, Obukhov AG, Hofmann T, Gudermann T, Schultz G. 2000. Receptor-mediated regulation of the nonselective cation channels TRPC4 and TRPC5. *J. Biol. Chem.* 275:17517–26

49. Hofmann T, Obukhov AG, Schaefer M, Harteneck C, Gudermann T, Schultz G. 1999. Direct activation of human TRPC6 and TRPC3 channels by diacylglycerol. *Nature* 397:259–63

50. Obukhov AG, Nowycky MC. 2005. A cytosolic residue mediates Mg^{2+} block and regulates inward current amplitude of a transient receptor potential channel. *J. Neurosci.* 25:1234–39

51. Strubing C, Krapivinsky G, Krapivinsky L, Clapham DE. 2003. Formation of novel TRPC channels by complex subunit interactions in embryonic brain. *J. Biol. Chem.* 278:39014–19

52. Jung S, Muhle A, Schaefer M, Strotmann R, Schultz G, Plant TD. 2003. Lanthanides potentiate TRPC5 currents by an action at extracellular sites close to the pore mouth. *J. Biol. Chem.* 278:3562–71

53. Greka A, Navarro B, Oancea E, Duggan A, Clapham DE. 2003. TRPC5 is a regulator of hippocampal neurite length and growth cone morphology. *Nat. Neurosci.* 6:837–45

54. Bezzerides VJ, Ramsey IS, Kotecha S, Greka A, Clapham DE. 2004. Rapid vesicular translocation and insertion of TRP channels. *Nat. Cell Biol.* 6:709–20

55. Singh BB, Lockwich TP, Bandyopadhyay BC, Liu X, Bollimuntha S, et al. 2004. VAMP2-dependent exocytosis regulates plasma membrane insertion of TRPC3 channels and contributes to agonist-stimulated Ca^{2+} influx. *Mol. Cell* 15:635–46

56. Cayouette S, Lussier MP, Mathieu EL, Bousquet SM, Boulay G. 2004. Exocytotic insertion of TRPC6 channel into the plasma membrane upon Gq protein-coupled receptor activation. *J. Biol. Chem.* 279:7241–46

57. Van Buren JJ, Bhat S, Rotello R, Pauza ME, Premkumar LS. 2005. Sensitization and translocation of TRPV1 by insulin and IGF-I. *Mol. Pain* 1:17

58. Tseng PH, Lin HP, Hu H, Wang C, Zhu MX, Chen CS. 2004. The canonical transient receptor potential 6 channel as a putative phosphatidylinositol 3,4,5-trisphosphate-sensitive calcium entry system. *Biochemistry* 43:11701–8

59. Okada T, Inoue R, Yamazaki K, Maeda A, Kurosaki T, et al. 1999. Molecular and functional characterization of a novel mouse transient receptor potential protein homologue TRP7. Ca^{2+}-permeable cation channel that is constitutively activated and enhanced by stimulation of G protein-coupled receptor. *J. Biol. Chem.* 274:27359–70

60. Estacion M, Li S, Sinkins WG, Gosling M, Bahra P, et al. 2004. Activation of human TRPC6 channels by receptor stimulation. *J. Biol. Chem.* 279:22047–56

61. Dietrich A, Mederos y Schnitzler M, Emmel J, Kalwa H, Hofmann T, Gudermann T. 2003. N-linked protein glycosylation is a major determinant for basal TRPC3 and TRPC6 channel activity. *J. Biol. Chem.* 278:47842–52

62. Vazquez G, Wedel BJ, Kawasaki BT, Bird GS, Putney JW Jr. 2004. Obligatory role of Src kinase in the signaling mechanism for TRPC3 cation channels. *J. Biol. Chem.* 279:40521–28

63. Hisatsune C, Kuroda Y, Nakamura K, Inoue T, Nakamura T, et al. 2004. Regulation of TRPC6 channel activity by tyrosine phosphorylation. *J. Biol. Chem.* 279:18887–94

64. Kwan HY, Huang Y, Yao X. 2004. Regulation of canonical transient receptor potential isoform 3 (TRPC3) channel by protein kinase G. *Proc. Natl. Acad. Sci. USA* 101:2625–30

65. Venkatachalam K, Zheng F, Gill DL. 2003. Regulation of canonical transient receptor potential (TRPC) channel function by diacylglycerol and protein kinase C. *J. Biol. Chem.* 278:29031–40

66. Shi J, Mori E, Mori Y, Mori M, Li J, et al. 2004. Multiple regulation by calcium of murine homologues of transient receptor potential proteins TRPC6 and TRPC7 expressed in HEK293 cells. *J. Physiol.* 561:415–32

67. Moqrich A, Hwang SW, Earley TJ, Petrus MJ, Murray AN, et al. 2005. Impaired thermosensation in mice lacking TRPV3, a heat and camphor sensor in the skin. *Science* 307:1468–72

68. Todaka H, Taniguchi J, Satoh J, Mizuno A, Suzuki M. 2004. Warm temperature-sensitive transient receptor potential vanilloid 4 (TRPV4) plays an essential role in thermal hyperalgesia. *J. Biol. Chem.* 279:35133–38

69. Lyall V, Heck GL, Vinnikova AK, Ghosh S, Phan TH, et al. 2004. The mammalian amiloride-insensitive non-specific salt taste receptor is a vanilloid receptor-1 variant. *J. Physiol.* 558:147–59

70. Caterina MJ, Schumacher MA, Tominaga M, Rosen TA, Levine JD, Julius D. 1997. The capsaicin receptor: a heat-activated ion channel in the pain pathway. *Nature* 389:816–24

71. Brauchi S, Orio P, Latorre R. 2004. Clues to understanding cold sensation: thermodynamics and electrophysiological analysis of the cold receptor TRPM8. *Proc. Natl. Acad. Sci. USA* 101:15494–99

72. Nilius B, Voets T. 2004. Diversity of TRP channel activation. *Novartis Found. Symp.* 258:140–49

73. Mohapatra DP, Nau C. 2003. Desensitization of capsaicin-activated currents in the vanilloid receptor TRPV1 is decreased by the cyclic AMP-dependent protein kinase pathway. *J. Biol. Chem.* 278:50080–90

74. Ahern GP. 2003. Activation of TRPV1 by the satiety factor oleoylethanolamide. *J. Biol. Chem.* 278:30429–34

75. Premkumar LS, Ahern GP. 2000. Induction of vanilloid receptor channel activity by protein kinase C. *Nature* 408:985–90

76. Prescott ED, Julius D. 2003. A modular PIP2 binding site as a determinant of capsaicin receptor sensitivity. *Science* 300:1284–88

77. Vlachova V, Teisinger J, Susankova K, Lyfenko A, Ettrich R, Vyklicky L. 2003. Functional role of C-terminal cytoplasmic tail of rat vanilloid receptor 1. *J. Neurosci.* 23:1340–50

78. Tominaga M, Caterina MJ, Malmberg AB, Rosen TA, Gilbert H, et al. 1998. The cloned capsaicin receptor integrates multiple pain-producing stimuli. *Neuron* 21:531–43

79. Caterina MJ, Leffler A, Malmberg AB, Martin WJ, Trafton J, et al. 2000. Impaired nociception and pain sensation in mice lacking the capsaicin receptor. *Science* 288:306–13

80. Davis JB, Gray J, Gunthorpe MJ, Hatcher JP, Davey PT, et al. 2000. Vanilloid receptor-1 is essential for inflammatory thermal hyperalgesia. *Nature* 405:183–87

81. Woodbury CJ, Zwick M, Wang S, Lawson JJ, Caterina MJ, et al. 2004. Nociceptors lacking TRPV1 and TRPV2 have normal heat responses. *J. Neurosci.* 24:6410–15

82. Birder LA, Nakamura Y, Kiss S, Nealen ML, Barrick S, et al. 2002. Altered urinary bladder function in mice lacking the vanilloid receptor TRPV1. *Nat. Neurosci.* 5:856–60

83. Zheng J, Dai C, Steyger PS, Kim Y, Vass Z, et al. 2003. Vanilloid receptors in hearing: altered cochlear sensitivity by vanilloids and expression of TRPV1 in the organ of corti. *J. Neurophysiol.* 90:444–55

84. Geppetti P, Trevisani M. 2004. Activation and sensitisation of the vanilloid receptor:

role in gastrointestinal inflammation and function. *Br. J. Pharmacol.* 141:1313–20

85. Muraki K, Iwata Y, Katanosaka Y, Ito T, Ohya S, et al. 2003. TRPV2 is a component of osmotically sensitive cation channels in murine aortic myocytes. *Circ. Res.* 93:829–38

86. Hu HZ, Gu Q, Wang C, Colton CK, Tang J, et al. 2004. 2-aminoethoxydiphenyl borate is a common activator of TRPV1, TRPV2, and TRPV3. *J. Biol. Chem.* 279: 35741–48

87. Caterina MJ, Rosen TA, Tominaga M, Brake AJ, Julius D. 1999. A capsaicin-receptor homologue with a high threshold for noxious heat. *Nature* 398:436–41

88. Stokes AJ, Wakano C, Del Carmen KA, Koblan-Huberson M, Turner H. 2005. Formation of a physiological complex between TRPV2 and RGA protein promotes cell surface expression of TRPV2. *J. Cell Biochem.* 94:669–83

89. Wainwright A, Rutter AR, Seabrook GR, Reilly K, Oliver KR. 2004. Discrete expression of TRPV2 within the hypothalamo-neurohypophysial system: Implications for regulatory activity within the hypothalamic-pituitary-adrenal axis. *J. Comp. Neurol.* 474:24–42

90. Kashiba H, Uchida Y, Takeda D, Nishigori A, Ueda Y, et al. 2004. TRPV2-immunoreactive intrinsic neurons in the rat intestine. *Neurosci. Lett.* 366:193–96

91. Sokolchik I, Tanabe T, Baldi PF, Sze JY. 2005. Polymodal sensory function of the *Caenorhabditis elegans* OCR-2 channel arises from distinct intrinsic determinants within the protein and is selectively conserved in mammalian TRPV proteins. *J. Neurosci.* 25:1015–23

92. Chung MK, Lee H, Mizuno A, Suzuki M, Caterina MJ. 2004. 2-aminoethoxydiphenyl borate activates and sensitizes the heat-gated ion channel TRPV3. *J. Neurosci.* 24:5177–82

93. Xu H, Ramsey IS, Kotecha SA, Moran MM, Chong JA, et al. 2002. TRPV3 is a calcium-permeable temperature-sensitive cation channel. *Nature* 418:181–86

94. Chung MK, Guler AD, Caterina MJ. 2005. Biphasic currents evoked by chemical or thermal activation of the heat-gated ion channel, TRPV3. *J. Biol. Chem.* 280:15928–41

95. Hellwig N, Plant TD, Janson W, Schafer M, Schultz G, Schaefer M. 2004. TRPV1 acts as proton channel to induce acidification in nociceptive neurons. *J. Biol. Chem.* 279:34553–61

96. Smith GD, Gunthorpe MJ, Kelsell RE, Hayes PD, Reilly P, et al. 2002. TRPV3 is a temperature-sensitive vanilloid receptor-like protein. *Nature* 418:186–90

97. Hellwig N, Albrecht N, Harteneck C, Schultz G, Schaefer M. 2005. Homo- and heteromeric assembly of TRPV channel subunits. *J. Cell Sci.* 118:917–28

98. Peier AM, Reeve AJ, Andersson DA, Moqrich A, Earley TJ, et al. 2002. A heat-sensitive TRP channel expressed in keratinocytes. *Science* 296:2046–49

99. Watanabe H, Davis JB, Smart D, Jerman JC, Smith GD, et al. 2002. Activation of TRPV4 channels (hVRL-2/mTRP12) by phorbol derivatives. *J. Biol. Chem.* 277:13569–77

100. Watanabe H, Vriens J, Prenen J, Droogmans G, Voets T, Nilius B. 2003. Anandamide and arachidonic acid use epoxyeicosatrienoic acids to activate TRPV4 channels. *Nature* 424:434–38

101. Liedtke W, Choe Y, Marti-Renom MA, Bell AM, Denis CS, et al. 2000. Vanilloid receptor-related osmotically activated channel (VR-OAC), a candidate vertebrate osmoreceptor. *Cell* 103:525–35

102. Watanabe H, Vriens J, Suh SH, Benham CD, Droogmans G, Nilius B. 2002. Heat-evoked activation of TRPV4 channels in a HEK293 cell expression system and in native mouse aorta endothelial cells. *J. Biol. Chem.* 277:47044–51

103. Voets T, Prenen J, Vriens J, Watanabe H, Janssens A, et al. 2002. Molecular determinants of permeation through the

cation channel TRPV4. *J. Biol. Chem.* 277:33704–10

104. Suzuki M, Mizuno A, Kodaira K, Imai M. 2003. Impaired pressure sensation in mice lacking TRPV4. *J. Biol. Chem.* 278: 22664–68

105. Tabuchi K, Suzuki M, Mizuno A, Hara A. 2005. Hearing impairment in TRPV4 knockout mice. *Neurosci. Lett.* 382:304–8

106. Liedtke W, Friedman JM. 2003. Abnormal osmotic regulation in trpv4$^{-/-}$ mice. *Proc. Natl. Acad. Sci. USA* 100:13698–703

107. Lee H, Iida T, Mizuno A, Suzuki M, Caterina MJ. 2005. Altered thermal selection behavior in mice lacking transient receptor potential vanilloid 4. *J. Neurosci.* 25:1304–10

108. Vennekens R, Hoenderop JG, Prenen J, Stuiver M, Willems PH, et al. 2000. Permeation and gating properties of the novel epithelial Ca^{2+} channel. *J. Biol. Chem.* 275:3963–69

109. Yue L, Peng JB, Hediger MA, Clapham DE. 2001. CaT1 manifests the pore properties of the calcium-release-activated calcium channel. *Nature* 410:705–9

110. Voets T, Janssens A, Droogmans G, Nilius B. 2004. Outer pore architecture of a Ca^{2+}-selective TRP channel. *J. Biol. Chem.* 279:15223–30

111. Niemeyer BA. 2005. Structure-function analysis of TRPV channels. *Naunyn Schmiedebergs Arch. Pharmacol.* 371: 285–94

112. Erler I, Hirnet D, Wissenbach U, Flockerzi V, Niemeyer BA. 2004. Ca^{2+}-selective transient receptor potential V channel architecture and function require a specific ankyrin repeat. *J. Biol. Chem.* 279: 34456–63

113. Chang Q, Gyftogianni E, van de Graaf SF, Hoefs S, Weidema FA, et al. 2004. Molecular determinants in TRPV5 channel assembly. *J. Biol. Chem.* 279:54304–11

114. Bodding M, Flockerzi V. 2004. Ca^{2+} dependence of the Ca^{2+}-selective TRPV6 channel. *J. Biol. Chem.* 279:36546–52

115. Lambers TT, Weidema AF, Nilius B, Hoenderop JG, Bindels RJ. 2004. Regulation of the mouse epithelial Ca^{2+} channel TRPV6 by the Ca^{2+}-sensor calmodulin. *J. Biol. Chem.* 279:28855–61

116. van Abel M, Hoenderop JG, Bindels RJ. 2005. The epithelial calcium channels TRPV5 and TRPV6: regulation and implications for disease. *Naunyn Schmiedebergs Arch. Pharmacol.* 371:295–306

117. Hoenderop JG, Nilius B, Bindels RJ. 2005. Calcium absorption across epithelia. *Physiol. Rev.* 85:373–422

118. Fleig A, Penner R. 2004. The TRPM ion channel subfamily: molecular, biophysical and functional features. *Trends Pharmacol. Sci.* 25:633–39

119. Duncan LM, Deeds J, Hunter J, Shao J, Holmgren LM, et al. 1998. Downregulation of the novel gene melastatin correlates with potential for melanoma metastasis. *Cancer Res.* 58:1515–20

120. Perraud AL, Fleig A, Dunn CA, Bagley LA, Launay P, et al. 2001. ADP-ribose gating of the calcium-permeable LTRPC2 channel revealed by Nudix motif homology. *Nature* 411:595–99

121. Fonfria E, Marshall IC, Benham CD, Boyfield I, Brown JD, et al. 2004. TRPM2 channel opening in response to oxidative stress is dependent on activation of poly(ADP-ribose) polymerase. *Br. J. Pharmacol.* 143:186–92

122. Sano Y, Inamura K, Miyake A, Mochizuki S, Yokoi H, et al. 2001. Immunocyte Ca^{2+} influx system mediated by LTRPC2. *Science* 293:1327–30

123. McHugh D, Flemming R, Xu SZ, Perraud AL, Beech DJ. 2003. Critical intracellular Ca^{2+} dependence of transient receptor potential melastatin 2 (TRPM2) cation channel activation. *J. Biol. Chem.* 278:11002–6

124. Kolisek M, Beck A, Fleig A, Penner R. 2005. Cyclic ADP-ribose and hydrogen peroxide synergize with ADP-ribose in the activation of TRPM2 channels. *Mol. Cell* 18:61–69

125. Perraud AL, Takanishi CL, Shen B, Kang S, Smith MK, et al. 2005. Accumulation of free ADP-ribose from mitochondria mediates oxidative stress-induced gating of TRPM2 cation channels. *J. Biol. Chem.* 280:6138–48

126. Hara Y, Wakamori M, Ishii M, Maeno E, Nishida M, et al. 2002. LTRPC2 Ca^{2+}-permeable channel activated by changes in redox status confers susceptibility to cell death. *Mol. Cell* 9:163–73

127. Wehage E, Eisfeld J, Heiner I, Jungling E, Zitt C, Luckhoff A. 2002. Activation of the cation channel long transient receptor potential channel 2 (LTRPC2) by hydrogen peroxide. A splice variant reveals a mode of activation independent of ADP-ribose. *J. Biol. Chem.* 277:23150–56

128. Heiner I, Eisfeld J, Halaszovich CR, Wehage E, Jungling E, et al. 2003. Expression profile of the transient receptor potential (TRP) family in neutrophil granulocytes: evidence for currents through long TRP channel 2 induced by ADP-ribose and NAD. *Biochem. J.* 371:1045–53

129. Grimm C, Kraft R, Sauerbruch S, Schultz G, Harteneck C. 2003. Molecular and functional characterization of the melastatin-related cation channel TRPM3. *J. Biol. Chem.* 278:21493–501

130. Oberwinkler J, Lis A, Giehl KM, Flockerzi V, Philipp SE. 2005. Alternative splicing switches the divalent cation selectivity of TRPM3 channels. *J. Biol. Chem.* 280:22540–48

131. Grimm C, Kraft R, Schultz G, Harteneck C. 2005. Activation of the melastatin-related cation channel TRPM3 by D-erythro-sphingosine. *Mol. Pharmacol.* 67:798–805

132. Hofmann T, Chubanov V, Gudermann T, Montell C. 2003. TRPM5 is a voltage-modulated and Ca^{2+}-activated monovalent selective cation channel. *Curr. Biol.* 13:1153–58

133. Launay P, Fleig A, Perraud AL, Scharenberg AM, Penner R, Kinet JP. 2002. TRPM4 is a Ca^{2+}-activated nonselective cation channel mediating cell membrane depolarization. *Cell* 109:397–407

134. Prawitt D, Monteilh-Zoller MK, Brixel L, Spangenberg C, Zabel B, et al. 2003. TRPM5 is a transient Ca^{2+}-activated cation channel responding to rapid changes in $[Ca^{2+}]i$. *Proc. Natl. Acad. Sci. USA* 100:15166–71

135. Ullrich ND, Voets T, Prenen J, Vennekens R, Talavera K, et al. 2005. Comparison of functional properties of the Ca^{2+}-activated cation channels TRPM4 and TRPM5 from mice. *Cell Calcium* 37:267–78

136. Nilius B, Prenen J, Tang J, Wang C, Owsianik G, et al. 2005. Regulation of the Ca^{2+} sensitivity of the nonselective cation channel TRPM4. *J. Biol. Chem.* 280:6423–33

137. Nilius B, Prenen J, Droogmans G, Voets T, Vennekens R, et al. 2003. Voltage dependence of the Ca^{2+}-activated cation channel TRPM4. *J. Biol. Chem.* 278:30813–20

138. Nilius B, Prenen J, Janssens A, Voets T, Droogmans G. 2004. Decavanadate modulates gating of TRPM4 cation channels. *J. Physiol.* 560:753–65

139. Nilius B, Prenen J, Voets T, Droogmans G. 2004. Intracellular nucleotides and polyamines inhibit the Ca^{2+}-activated cation channel TRPM4b. *Pflügers Arch.* 448:70–75

140. Launay P, Cheng H, Srivatsan S, Penner R, Fleig A, Kinet JP. 2004. TRPM4 regulates calcium oscillations after T cell activation. *Science* 306:1374–77

141. Earley S, Waldron BJ, Brayden JE. 2004. Critical role for transient receptor potential channel TRPM4 in myogenic constriction of cerebral arteries. *Circ. Res.* 95:922–29

142. Zhang Y, Hoon MA, Chandrashekar J, Mueller KL, Cook B, et al. 2003. Coding of sweet, bitter, and umami tastes: different receptor cells sharing similar signaling pathways. *Cell* 112:293–301

143. Yamaguchi H, Matsushita M, Nairn AC,

Kuriyan J. 2001. Crystal structure of the atypical protein kinase domain of a TRP channel with phosphotransferase activity. *Mol. Cell.* 7:1047–57

144. Ryazanova LV, Dorovkov MV, Ansari A, Ryazanov AG. 2004. Characterization of the protein kinase activity of TRPM7/ChaK1, a protein kinase fused to the transient receptor potential ion channel. *J. Biol. Chem.* 279:3708–16

145. Matsushita M, Kozak JA, Shimizu Y, McLachlin DT, Yamaguchi H, et al. 2005. Channel function is dissociated from the intrinsic kinase activity and autophosphorylation of TRPM7/ChaK1. *J. Biol. Chem.* 280:20793–803

146. Dorovkov MV, Ryazanov AG. 2004. Phosphorylation of annexin I by TRPM7 channel-kinase. *J. Biol. Chem.* 279:50643–46

147. Nadler MJ, Hermosura MC, Inabe K, Perraud AL, Zhu Q, et al. 2001. LTRPC7 is a Mg·ATP-regulated divalent cation channel required for cell viability. *Nature* 411:590–95

148. Runnels LW, Yue L, Clapham DE. 2001. TRP-PLIK, a bifunctional protein with kinase and ion channel activities. *Science* 291:1043–47

149. Voets T, Nilius B, Hoefs S, van der Kemp AW, Droogmans G, et al. 2004. TRPM6 forms the Mg^{2+} influx channel involved in intestinal and renal Mg^{2+} absorption. *J. Biol. Chem.* 279:19–25

150. Takezawa R, Schmitz C, Demeuse P, Scharenberg AM, Penner R, Fleig A. 2004. Receptor-mediated regulation of the TRPM7 channel through its endogenous protein kinase domain. *Proc. Natl. Acad. Sci. USA* 101:6009–14

151. Aarts M, Iihara K, Wei WL, Xiong ZG, Arundine M, et al. 2003. A key role for TRPM7 channels in anoxic neuronal death. *Cell* 115:863–77

152. Schmitz C, Perraud AL, Johnson CO, Inabe K, Smith MK, et al. 2003. Regulation of vertebrate cellular Mg^{2+} homeostasis by TRPM7. *Cell* 114:191–200

153. Walder RY, Landau D, Meyer P, Shalev H, Tsolia M, et al. 2002. Mutation of TRPM6 causes familial hypomagnesemia with secondary hypocalcemia. *Nat. Genet.* 31:171–74

154. Schlingmann KP, Weber S, Peters M, Niemann Nejsum L, Vitzthum H, et al. 2002. Hypomagnesemia with secondary hypocalcemia is caused by mutations in TRPM6, a new member of the TRPM gene family. *Nat. Genet.* 31:166–70

155. Gwanyanya A, Amuzescu B, Zakharov SI, Macianskiene R, Sipido KR, et al. 2004. Magnesium-inhibited, TRPM6/7-like channel in cardiac myocytes: permeation of divalent cations and pH-mediated regulation. *J. Physiol.* 559:761–76

156. Kozak JA, Cahalan MD. 2003. MIC channels are inhibited by internal divalent cations but not ATP. *Biophys. J.* 84:922–27

157. Kozak JA, Kerschbaum HH, Cahalan MD. 2002. Distinct properties of CRAC and MIC channels in RBL cells. *J. Gen. Physiol.* 120:221–35

158. Prakriya M, Lewis RS. 2002. Separation and characterization of currents through store-operated CRAC channels and Mg^{2+}-inhibited cation (MIC) channels. *J. Gen. Physiol.* 119:487–507

159. Hermosura MC, Monteilh-Zoller MK, Scharenberg AM, Penner R, Fleig A. 2002. Dissociation of the store-operated calcium current I(CRAC) and the Mg-nucleotide-regulated metal ion current MagNuM. *J. Physiol.* 539:445–58

160. Tsavaler L, Shapero MH, Morkowski S, Laus R. 2001. Trp-p8, a novel prostate-specific gene, is up-regulated in prostate cancer and other malignancies and shares high homology with transient receptor potential calcium channel proteins. *Cancer Res.* 61:3760–69

161. Peier AM, Moqrich A, Hergarden AC, Reeve AJ, Andersson DA, et al. 2002. A TRP channel that senses cold stimuli and menthol. *Cell* 108:705–15

162. McKemy DD, Neuhausser WM, Julius D. 2002. Identification of a cold receptor reveals a general role for TRP channels in thermosensation. *Nature* 416:52–58

163. Voets T, Droogmans G, Wissenbach U, Janssens A, Flockerzi V, Nilius B. 2004. The principle of temperature-dependent gating in cold- and heat-sensitive TRP channels. *Nature* 430:748–54

164. Weil A, Moore SE, Waite NJ, Randall A, Gunthorpe MJ. 2005. Conservation of functional and pharmacological properties in the distantly related temperature sensors TRPV1 and TRPM8. *Mol. Pharmacol.* 68:518–27

165. Story GM, Peier AM, Reeve AJ, Eid SR, Mosbacher J, et al. 2003. ANKTM1, a TRP-like channel expressed in nociceptive neurons, is activated by cold temperatures. *Cell* 112:819–29

166. Macpherson LJ, Geierstanger BH, Viswanath V, Bandell M, Eid SR, et al. 2005. The pungency of garlic: activation of TRPA1 and TRPV1 in response to allicin. *Curr. Biol.* 15:929–34

167. Jordt SE, Bautista DM, Chuang HH, McKemy DD, Zygmunt PM, et al. 2004. Mustard oils and cannabinoids excite sensory nerve fibres through the TRP channel ANKTM1. *Nature* 427:260–65

168. Bandell M, Story GM, Hwang SW, Viswanath V, Eid SR, et al. 2004. Noxious cold ion channel TRPA1 is activated by pungent compounds and bradykinin. *Neuron* 41:849–57

169. Nagata K, Duggan A, Kumar G, Garcia-Anoveros J. 2005. Nociceptor and hair cell transducer properties of TRPA1, a channel for pain and hearing. *J. Neurosci.* 25:4052–61

170. Corey DP, Garcia-Anoveros J, Holt JR, Kwan KY, Lin SY, et al. 2004. TRPA1 is a candidate for the mechanosensitive transduction channel of vertebrate hair cells. *Nature* 432:723–30

171. Zhou XL, Loukin SH, Coria R, Kung C, Saimi Y. 2005. Heterologously expressed fungal transient receptor potential channels retain mechanosensitivity in vitro and osmotic response in vivo. *Eur. Biophys. J.* 34:413–22

172. O'Neil RG, Heller S. 2005. The mechanosensitive nature of TRPV channels. *Pflügers Arch.* 451:191–203

173. Sotomayor M, Corey DP, Schulten K. 2005. In search of the hair-cell gating spring elastic properties of ankyrin and cadherin repeats. *Structure* 13:669–82

174. Lin SY, Corey DP. 2005. TRP channels in mechanosensation. *Curr. Opin. Neurobiol.* 15:350–57

175. Hanaoka K, Qian F, Boletta A, Bhunia AK, Piontek K, et al. 2000. Co-assembly of polycystin-1 and -2 produces unique cation-permeable currents. *Nature* 408:990–94

176. Koulen P, Cai Y, Geng L, Maeda Y, Nishimura S, et al. 2002. Polycystin-2 is an intracellular calcium release channel. *Nat. Cell Biol.* 4:191–97

177. Kottgen M, Benzing T, Simmen T, Tauber R, Buchholz B, et al. 2005. Trafficking of TRPP2 by PACS proteins represents a novel mechanism of ion channel regulation. *EMBO J.* 24:705–16

178. Nauli SM, Alenghat FJ, Luo Y, Williams E, Vassilev P, et al. 2003. Polycystins 1 and 2 mediate mechanosensation in the primary cilium of kidney cells. *Nat. Genet.* 33:129–37

179. Wu G, D'Agati V, Cai Y, Markowitz G, Park JH, et al. 1998. Somatic inactivation of Pkd2 results in polycystic kidney disease. *Cell* 93:177–88

180. Teilmann SC, Byskov AG, Pedersen PA, Wheatley DN, Pazour GJ, Christensen ST. 2005. Localization of transient receptor potential ion channels in primary and motile cilia of the female murine reproductive organs. *Mol. Reprod. Dev.* 71:444–52

181. Chen XZ, Vassilev PM, Basora N, Peng JB, Nomura H, et al. 1999. Polycystin-L is a calcium-regulated cation channel permeable to calcium ions. *Nature* 401:383–86

182. Nomura H, Turco AE, Pei Y, Kalaydjieva L, Schiavello T, et al. 1998. Identification of PKDL, a novel polycystic kidney disease 2-like gene whose murine homologue is deleted in mice with kidney and retinal defects. *J. Biol. Chem.* 273:25967–73

183. Bach G. 2005. Mucolipin 1: endocytosis and cation channel—a review. *Pflügers Arch.* 451:313–17

184. Raychowdhury MK, Gonzalez-Perrett S, Montalbetti N, Timpanaro GA, Chasan B, et al. 2004. Molecular pathophysiology of mucolipidosis type IV: pH dysregulation of the mucolipin-1 cation channel. *Hum. Mol. Genet.* 13:617–27

185. Deleted in proof

186. Deleted in proof

187. Di Palma F, Belyantseva IA, Kim HJ, Vogt TF, Kachar B, Noben-Trauth K. 2002. Mutations in Mcoln3 associated with deafness and pigmentation defects in varitint-waddler (Va) mice. *Proc. Natl. Acad. Sci. USA* 99:14994–99

188. Song H, Poo M. 2001. The cell biology of neuronal navigation. *Nat. Cell Biol.* 3:E81–88

189. Tessier-Lavigne M, Goodman CS. 1996. The molecular biology of axon guidance. *Science* 274:1123–33

190. Yu TW, Bargmann CI. 2001. Dynamic regulation of axon guidance. *Nat. Neurosci.* 4(Suppl.):1169–76

191. Huber AB, Kolodkin AL, Ginty DD, Cloutier JF. 2003. Signaling at the growth cone: ligand-receptor complexes and the control of axon growth and guidance. *Annu. Rev. Neurosci.* 26:509–63

192. Dickson BJ. 2002. Molecular mechanisms of axon guidance. *Science* 298:1959–64

193. Song H, Ming G, He Z, Lehmann M, McKerracher L, et al. 1998. Conversion of neuronal growth cone responses from repulsion to attraction by cyclic nucleotides. *Science* 281:1515–18

194. Ming G, Henley J, Tessier-Lavigne M, Song H, Poo M. 2001. Electrical activity modulates growth cone guidance by diffusible factors. *Neuron* 29:441–52

195. Gomez TM, Spitzer NC. 2000. Regulation of growth cone behavior by calcium: new dynamics to earlier perspectives. *J. Neurobiol.* 44:174–83

196. Song HJ, Poo MM. 1999. Signal transduction underlying growth cone guidance by diffusible factors. *Curr. Opin. Neurobiol.* 9:355–63

197. Gomez TM, Robles E, Poo M, Spitzer NC. 2001. Filopodial calcium transients promote substrate-dependent growth cone turning. *Science* 291:1983–87

198. Gomez TM, Spitzer NC. 1999. In vivo regulation of axon extension and pathfinding by growth-cone calcium transients. *Nature* 397:350–55

199. Henley J, Poo MM. 2004. Guiding neuronal growth cones using Ca^{2+} signals. *Trends Cell Biol.* 14:320–30

200. Dent EW, Gertler FB. 2003. Cytoskeletal dynamics and transport in growth cone motility and axon guidance. *Neuron* 40:209–27

201. Craig AM, Wyborski RJ, Banker G. 1995. Preferential addition of newly synthesized membrane protein at axonal growth cones. *Nature* 375:592–94

202. Mattson MP, Kater SB. 1987. Calcium regulation of neurite elongation and growth cone motility. *J. Neurosci.* 7:4034–43

203. Hong K, Nishiyama M, Henley J, Tessier-Lavigne M, Poo M. 2000. Calcium signalling in the guidance of nerve growth by netrin-1. *Nature* 403:93–98

204. Nishiyama M, Hoshino A, Tsai L, Henley JR, Goshima Y, et al. 2003. Cyclic AMP/GMP-dependent modulation of Ca^{2+} channels sets the polarity of nerve growth-cone turning. *Nature* 423:990–95

205. Spitzer NC, Lautermilch NJ, Smith RD, Gomez TM. 2000. Coding of neuronal differentiation by calcium transients. *Bioessays* 22:811–17

206. Klingauf J, Neher E. 1997. Modeling buffered Ca^{2+} diffusion near the membrane: implications for secretion in neuroendocrine cells. *Biophys. J.* 72:674–90

207. Monck JR, Robinson IM, Escobar AL, Vergara JL, Fernandez JM. 1994. Pulsed laser imaging of rapid Ca^{2+} gradients in excitable cells. *Biophys. J.* 67:505–14

208. Silver RA, Lamb AG, Bolsover SR. 1990. Calcium hotspots caused by L-channel clustering promote morphological changes in neuronal growth cones. *Nature* 343:751–54

209. Lipscombe D, Madison DV, Poenie M, Reuter H, Tsien RY, Tsien RW. 1988. Spatial distribution of calcium channels and cytosolic calcium transients in growth cones and cell bodies of sympathetic neurons. *Proc. Natl. Acad. Sci. USA.* 85: 2398–402

210. Davenport RW, Kater SB. 1992. Local increases in intracellular calcium elicit local filopodial responses in Helisoma neuronal growth cones. *Neuron* 9:405–16

211. Li Y, Jia YC, Cui K, Li N, Zheng ZY, et al. 2005. Essential role of TRPC channels in the guidance of nerve growth cones by brain-derived neurotrophic factor. *Nature* 434:894–98

212. Shim S, Goh EL, Ge S, Sailor K, Yuan JP, et al. 2005. XTRPC1-dependent chemotropic guidance of neuronal growth cones. *Nat. Neurosci.* 8:730–35

213. Wang GX, Poo MM. 2005. Requirement of TRPC channels in netrin-1-induced chemotropic turning of nerve growth cones. *Nature* 434:898–904

214. Sabo SL, McAllister AK. 2003. Mobility and cycling of synaptic protein-containing vesicles in axonal growth cone filopodia. *Nat. Neurosci.* 6:1264–69

Annu. Rev. Physiol. 2006. 68:649–84
doi: 10.1146/annurev.physiol.68.040204.100939
Copyright © 2006 by Annual Reviews. All rights reserved
First published online as a Review in Advance on September 7, 2005

INSIGHTS ON TRP CHANNELS FROM IN VIVO STUDIES IN *DROSOPHILA*

Baruch Minke and Moshe Parnas

Department of Physiology and the Kühne Minerva Center for Studies of Visual Transduction, The Hebrew University-Hadassah Medical School, Jerusalem 91120, Israel; email: minke@md.huji.ac.il, shiko@md.huji.ac.il

Key Words Ca^{2+} signaling, inositol lipid signaling, INAD, phospholipase C (PLC), phototransduction

■ **Abstract**   Transient receptor potential (TRP) channels mediate responses in a large variety of signaling mechanisms. Most studies on mammalian TRP channels rely on heterologous expression, but their relevance to in vivo tissues is not entirely clear. In contrast, *Drosophila* TRP and TRP-like (TRPL) channels allow direct analyses of in vivo function. In *Drosophila* photoreceptors, activation of TRP and TRPL is mediated via the phosphoinositide cascade, with both Ca^{2+} and diacylglycerol (DAG) essential for generating the light response. In tissue culture cells, TRPL channels are constitutively active, and lipid second messengers greatly facilitate this activity. Inhibition of phospholipase C (PLC) completely blocks lipid activation of TRPL, suggesting that lipid activation is mediated via PLC. In vivo studies in mutant *Drosophila* also reveal an acute requirement for lipid-producing enzyme, which may regulate PLC activity. Thus, PLC and its downstream second messengers, Ca^{2+} and DAG, constitute critical mediators of TRP/TRPL gating in vivo.

INTRODUCTION

Channel proteins are of prime importance for the survival and function of virtually every cell. Ca^{2+}-permeable channels are particularly important because Ca^{2+} is not only a charge carrier but also one of the most important second messengers. There are three main classes of Ca^{2+}-permeable channels: voltage gated, ligand gated, and TRP. TRP channels constitute a large and diverse family of proteins that are expressed in many tissues and cell types. The name TRP is derived from a spontaneously occurring *Drosophila* mutant lacking TRP that responded to a continuous light with a transient receptor potential [hence, TRP (2)]. The *Drosophila* TRP (3) was later used to isolate the first mammalian TRP homologs (4, 5). The TRP superfamily is conserved throughout evolution from nematodes to humans (1). TRP channels mediate responses to light, nerve growth factors, pheromones, olfaction, taste, mechanical changes, temperature, pH, osmolarity, vasorelaxation of blood vessels, and metabolic stress. Furthermore, mutations in members of the TRP family are responsible for several diseases (for reviews, see 6–11).

TRP channels are classified into seven related subfamilies designated TRPC (canonical or classical), TRPM (melastatin), TRPN (NOMPC), TRPV (vanilloid receptor), TRPA (ANKTM1), TRPP (polycystin), and TRPML (mucolipin) (for reviews, see 6–8, 10, 12). All these TRP subfamilies are represented in the *Drosophila* genome (13), and three of them were initially discovered in *Drosophila*. The pioneering research and basic concepts on the properties and function of the channels in two of them (i.e., TRP and NOMPC) were established in *Drosophila*. The three subfamilies are TRPC, which includes the founding member of the TRP superfamily (TRP); TRPA, which includes the temperature detectors painless (14), pyrexia (15), and ANKTM1 (12, 16); and the TRPN subfamily, in which the no mechanical potential C (NOMPC) is the first member (17). NOMPC currently includes only one vertebrate homolog that was found in Zebra fish (18).

A great deal is known today about members of the mammalian TRP channels. However, the exact physiological function and gating mechanisms of most channels are still elusive. Furthermore, a discrepancy exists in the mechanism of channel activation and channel properties between the native and the heterologously expressed channels. Therefore, the *Drosophila* TRPC channels, which are robustly expressed in the photoreceptor cells, still constitute one of the most useful model systems for studying TRP channels because the physiological function is well defined and understood. It has been well established that the function of the *Drosophila* TRP channels is to generate the light-induced current (LIC) that produces the photoreceptor potential and the sensation of light. In most mammalian systems that use TRP channels, the primary physiological role of the channels is not entirely clear. Even in the few systems in which the physiological roles of the channels have been determined [e.g., pain mechanism, taste pheromone and temperature detection (19–26)], the gating mechanism of the channels is still unclear. The robust properties of the response to light, which dictate the use of literally a million copies of TRP channels and a similar number of their activating molecules in a single cell, have turned out to be very useful for studying the mechanism underlying TRP channel activation. This may prove useful in directing future studies on mammalian TRPs.

The present review compares the properties of *Drosophila* light-sensitive channels in the native photoreceptor cells to those deduced by studies on the same channels expressed in heterologous systems. Because the physiological function of the native photoreceptor channels is known, this comparison may help in evaluating the vast literature on heterologously expressed mammalian TRP channels, which has been the major way to study these channels. A property common to both *Drosophila* and mammalian TRPs is also covered in this review. This property is manifested by signal-induced translocation of specific TRP channels, which has important implications for cellular functions.

DROSOPHILA PHOTOTRANSDUCTION USES LARGE QUANTITIES OF HIGHLY ORGANIZED SIGNALING MOLECULES

In the history of biological research, specialized cells and tissues have proved to be crucial for the discovery of fundamental principles and mechanisms. Illuminating examples are the insightful and pioneering studies in which the squid giant axon and the torpedo electroplax were used (27). *Drosophila* photoreceptor cells are formed as highly specialized cells designed to maximize the sensitivity to light; they can detect single photons (28) and respond at high speed, producing the photoreceptor potential in response to intense lights within 10 msec (29). These remarkable capabilities dictate concentrating large quantities of signaling molecules, including the TRP channels, in a small volume. Indeed, *Drosophila* photoreceptors are highly polarized cells composed of two well-defined compartments: a signaling compartment (termed rhabdomere) and the cell body. The rhabdomere contains $\sim$30,000 tightly packed actin-rich microvilli that harbor the signaling proteins required to generate the photoreceptor potential upon illumination. Each microvillus contains $\sim$1000 rhodopsin molecules and $\sim$100 TRP molecules leading to a total of $\sim$30,000,000 and 3,000,000 rhodopsin and TRP channel subunits, respectively, in a photoreceptor cell (30) (Figure 1).

The signaling membrane is composed of the microvilli and the nearby extension of smooth endoplasmic reticulum termed submicrovillar cisternae (SMC, Figure 1). It has been well established that flies in general and *Drosophila* photoreceptors in particular use the phosphoinositide cascade for vision (31, 32) (Figure 2). Upon absorption of light, the major rhodopsin [encoded by the *ninaE (neither inactivation nor after-potential E)* gene (33, 34)] is converted into the active metarhodopsin state, which activates a heterotrimeric G protein (dGq). This leads to activation of phospholipase C [PLCβ4, encoded by the *norpA (no-receptor-potential A)* gene (31)] and subsequent opening of two classes of light-sensitive channels, TRP and TRP-like [encoded by the *trp* and *trpl* genes (3, 35)] (for reviews see 7, 30, 36–38). Genetic elimination of these channels completely abolished the robust response to light, indicating that these proteins make up the light-sensitive channels (39, 40). A third channel subunit, designated TRPγ, was also found in the eye, but its function is still unclear (41). Deactivation of the channels and light adaptation are regulated in part by the eye-specific protein kinase C [PKC, encoded by the *inaC* gene (42)].

PLC catalyzes the hydrolysis of phosphatidylinositol-4,5-bisphosphate (PIP_2) into the soluble second messenger inositol-1,4,5-trisphosphate ($InsP_3$) and the membrane-bound diacylglycerol (DAG). DAG is recycled back to PIP_2 by the phosphatidyl inositol (PI) cycle in the SMC (43). DAG is converted to phosphatidic acid (PA) via DAG kinase [encoded by the *rdgA (retinal degeneration A)* gene (44)] and into CDP-DAG via CD synthetase [encoded by the *cds* gene (45)]. After conversion, PI is presumed to be transported back to the microvillar

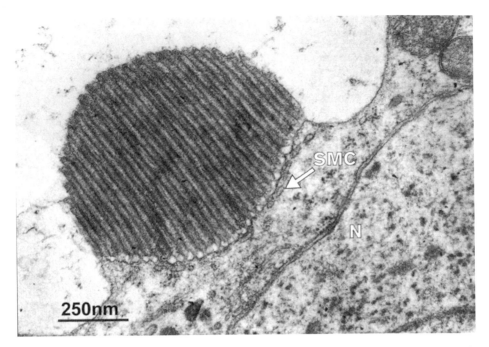

Figure 1   The *Drosophila* rhabdomere and submicrovillar cisternae (SMC) are shown in this electron micrograph. The electron-dense region shows the tightly packed microvilli. Each microvillus, which contains ~1000 rhodopsin and ~100 TRP molecules, is connected by a narrow neck to the cell body. A system of minute sac-like cisternae is located near the base of the microvilli (*arrow*). These represent the putative inositol 1,4,5-trisphosphate (InsP₃)-sensitive Ca²⁺ stores that might play a crucial role in phototransduction. A part of the nucleus (N) with a double membrane is also shown. (Modified from Reference 38 with permission from *Current Opinion in Neurobiology.*)

membrane by the PI transfer protein [PITP, encoded by the *rdgB* gene (46)]. Both RDGA and RDGB have been immunolocalized to the SMC (47, 48) (Figure 2).

The SMC at the base of the rhabdomere has been proposed by analogy to other insects to represent Ca²⁺ stores endowed with InsP₃ receptor (InsP₃R), which is an internal Ca²⁺ channel that opens and releases Ca²⁺ upon binding of InsP₃. The roles of InsP₃ and the InsP₃R in *Drosophila* phototransduction are still unclear (9).

The spatial localization of the signaling molecules in the microvilli is not random but highly organized by a signaling web composed of the multi-PDZ scaffold protein INAD (inactivation-no-after-potential D protein), which is anchored to the microvillar membrane via the TRP channel (49, 50). Genetic elimination of INAD resulted in mislocalization and degradation of the signaling molecules and strong suppression of the response to light (51).

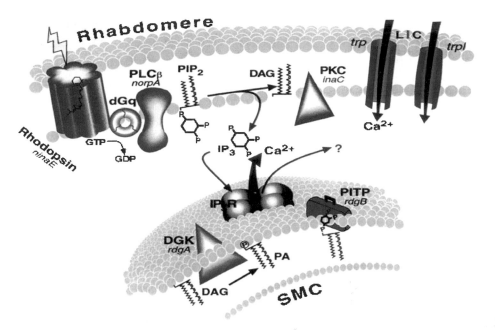

Figure 2   The phosphoinositide cascade of vision. Cloned genes (for all mutants that are available) are shown in italics, next to their corresponding proteins. Upon absorption of light, rhodopsin (*ninaE* gene) is converted to the active metarhodopsin state, which activates a heterotrimeric G protein (dGq). This leads to activation of phospholipase C (PLCβ, *norpA* gene) and subsequent opening of two classes of light-sensitive channels encoded at least in part by *trp* and *trpl* genes. PLC catalyzes the hydrolysis of phosphatidylinositol 4,5-bisphosphate (PIP$_2$) into the soluble second messenger InsP$_3$ and the membrane-bound diacylglycerol (DAG). DAG is recycled to PIP$_2$ by the phosphatidyl inositol (PI) cycle shown in an extension of the smooth endoplasmic reticulum called submicrovillar cisternae (SMC, shown on the *bottom*). DAG is converted to phosphatidic acid (PA) via DAG kinase (DGK, *rdgA* gene). After conversion to PI, PI is presumed to be transported back to the microvillar membrane by the PI transfer protein (PITP, encoded by the *rdgB* gene). The InsP$_3$ receptor (InsP$_3$R) is an internal Ca^{2+} channel that opens and releases Ca^{2+} upon binding of InsP$_3$. (Modified from Reference 7 with permission from *Physiological Reviews*.)

Organization in Signaling Complexes

The power of the molecular genetics of *Drosophila*, which has been applied in an unbiased manner by forward genetics to screen for visual mutants, led to the discovery of novel proteins. The scaffold protein INAD was discovered in a screen for mutants with abnormal prolonged depolarizing after-potential (PDA, 52). In this screen, conducted by Pak (53), a mutant designated *inaD* was isolated. The *inaD* gene was cloned and sequenced by Shieh & Zhu (54). Pioneering studies by Huber and colleagues (55) have shown that INAD binds not only TRP but also

PLC (NORPA) and eye-specific PKC (INAC). Investigators further found that the INAD scaffold protein consists of five ~90 amino acid protein interaction motifs called PDZ (PSD95, DLG, ZO1) domains (51). These domains are recognized as protein modules that bind to a diversity of signaling, cell adhesion, and cytoskeletal proteins (56–58) by specific binding to target sequences typically, though not always, in the final three residues of the carboxy-terminal. TRPL appears not to be a member of the complex because, unlike INAC, NORPA, or TRP, it remains strictly localized to the microvilli in the *inaD¹* null mutant (51) and it translocates without the signaling complex outside the rhabdomere upon illumination (59). However, coimmunoprecipitation studies by Montell and colleagues (60) have shown that TRP and TRPL interact directly when heterologously expressed in 293T cells as well as in *Drosophila* head extracts. Montell and colleagues have also shown that, in addition to PLC, PKC, and TRP, other signaling molecules such as TRPL, calmodulin (CaM), rhodopsin (61), and NINAC (62) bind to the INAD signaling complex. Recently, Minke and colleagues (62a) have found that the sole member of the ERM (ezrin/radixin/moesin) family in *Drosophila*, Dmoesin (63), binds to the TRP and TRPL channels in the dark. ERM proteins constitute a bridge between the actin cytoskeleton and the plasma membrane by binding to membrane proteins (64) [including mammalian TRP channels (65)]. Therefore, the association of proteins, which coimmunoprecipitate with the channels but do not bind directly to the PDZ domains of INAD (66), might be indirect via the actin cytoskeleton and Dmoesin in a dynamic process.

Additional evidence that TRPL functions as a channel in a native system without the need for INAD came from studies on neuropeptide-stimulated fluid transport in *Drosophila* Malpighian (renal) tubule (67). Interestingly, all photoreceptor channel proteins TRP, TRPL, and TRPγ are expressed in the Malpighian tubule. However, mutant analysis showed that only the TRPL channels are functional in neuropeptide-stimulated fluid transport and Ca^{2+} signaling in this system (67). Strikingly, INAD is not expressed in the Malpighian tubule, indicating that functional TRPL channels do not need INAD for function both in heterologous expression systems (68) and in a native system of the Malpighian tubule (67). This study further suggests that TRP does not form a functional channel without INAD.

Biochemical studies in *Calliphora* have revealed that both INAD and TRP are targets for phosphorylation by the nearby eye PKC (INAC, 69). Investigators have established an important aspect of the supramolecular complex formation by showing that TRP plays a major role in localizing the entire INAD multimolecular complex. INAD is correctly localized to the rhabdomeres in *inaC* mutants (where eye PKC is missing) and in *norpA* mutants (where PLC is missing), but is severely mislocalized in null *trp* mutants (70), thus indicating that TRP (but not PLC or PKC) is essential for localization of the signaling complex to the rhabdomere. The study of the above mutants was also used to show that TRP and INAD do not depend on each other to be targeted to the rhabdomeres, and thus that INAD-TRP interaction is not required for targeting but rather for anchoring the signaling complex (70). Additional experiments show that INAD has other functions in addition to anchoring the signaling complex (51). One important function is to preassemble the proteins of the signaling complex (49). Another important function, at least in the case of PLC, is to prevent degradation of the unbound signaling protein by a still unknown mechanism (49). The prevention of PLC degradation is important for keeping an accurate stochiometry among the signaling proteins (71). An accurate stochiometry between $G\alpha q$ and PLC is particularly important because PLC functions as a GTPase activating protein. Accordingly, the inactivation of $G\alpha q$ by its target, PLC, has proven essential for the high temporal and intensity resolution of the response to light (71).

The organization of the *Drosophila* photoreceptor's signaling proteins in a specific cellular compartment in the form of a multimolecular signaling complex seems to be a common mechanism. Recent studies in mammalian cells showed that the adaptor protein, termed Homer, facilitates a physical association between TRPC1 and the InsP$_3$R that is required for the TRPC1 channel to respond to signals. The TRPC1-Homer-InsP$_3$R complex is dynamic, and its disassembly parallels TRPC1 activation (72). In other studies, the mammalian TRPC3 has been found in a specific microdomain called caveoli. TRPC3 constitutes a part of a multimolecular signaling complex containing Ezrin and key Ca^{2+} signaling proteins, including PLCβ1 and $G\alpha q/11$, which are involved in Ca^{2+}-mediated regulation of TRPC3 channel activity and cytoskeletal reorganization (73). Another study reveals that PLCγ1 binds to and regulates the TRPC3 channel, although this interaction requires a partial pleckstrin homology (PH) domain of PLCγ1 and TRPC3 (74). Another study has shown that the ERM adaptor Na/H exchanger regulatory factor (NHERF, or EBP50) via its first PDZ domain associates with PLCβ, TRPC4, and TRPC5, and regulates channel activity and subcellular localization (65, 75, 76). Another scaffold protein, termed spinophillin, regulates Ca^{2+} signaling (77), but interaction with TRPC channels was not demonstrated for this protein. These data strongly suggest that Dmoesin-TRP and TRP-INAD interactions found in *Drosophila* are evolutionarily conserved mechanisms with important functional roles.

FUNCTIONAL TRPL CHANNELS WITH NATIVE PROPERTIES ARE READILY EXPRESSED IN HETEROLOGOUS SYSTEMS

TRP Channels with Native Properties Failed to be Expressed in Heterologous Systems

Both the *Drosophila* TRP and TRPL channels have been expressed in several heterologous systems. However, there is a marked difference in the outcome of these heterologous expressions between the TRP and TRPL channels. Although there is wide agreement that the expression of TRPL channels resulted in a functional TRPL channel in the host cells (68), this result does not occur in the TRP channel. The few reports on TRP expression seem to indicate that the expressed channels lead either to activation of endogenous host cell channels (see discussion in Reference 78) or to novel currents grossly different from those obtained in the native system; single-channel activity has not been published (60, 78, 79). These results strongly suggest that the measured currents did not arise from activation of *Drosophila* TRP channels.

Heterologous Expression of TRPL Channels

TRPL channels were expressed in several expression systems (60, 78, 80–84). The main functional characteristics of the expressed channels are, in general, similar to those of the native *Drosophila* TRPL channels. However, although the native TRPL channels are closed in the dark and open upon illumination, the expressed channels are constitutively active. Owing to the strong outward rectification of the TRPL-mediated current (see Figure 3C), it is not always obvious that these channels are already in their active state at physiological resting potential (i.e., $\sim$–50 mV). However, the active state of the channels is readily revealed when the holding voltage is stepped to positive values (i.e., $+150$ mV), unlike the closed state of the native channel in the dark (see Figure 3C, *right*). The open probability of the expressed channels can be modulated (enhanced or suppressed) pharmacologically (see Figure 3B), and this fact may give the impression that these modulations mimic the opening and closing of the channels under physiological condition. However, one has to bear in mind that these are already active channels and these modulations may have limited value in understanding the physiological gating mechanism.

Hardie and colleagues (81) showed that the native TRPL channel and the heterologously expressed TRPL channel in the S2 expression system have the same single-channel conductance, ion selectivity, inhibition by Mg^{2+}, power spectra of channel noise, and current-voltage relationship. Both the expressed and the native channels seem to require PLC for activation. Channel activation of the native channels by light requires PLC, as evidenced by a virtually complete and reversible block of channel activation by light in a temperature-sensitive PLC mutant (53, 85). Similarly, the constitutive activity of the expressed TRPL channels was greatly enhanced by activation of endogenous PLC or by external application of exogenous PLCs (68). Accordingly, coexpression of the TRPL channel along with either the muscarinic or the histamine (H1) receptors led to an increased channel activity upon application of carbachol (81), histamine, thrombin, and U46619 (80, 86–88). Consistent with this finding, activation of PLC via the G protein by application of GTPγS also activated the TRPL channels, and inhibition of the G protein pathway by GDPβS suppressed TRPL activity (87).

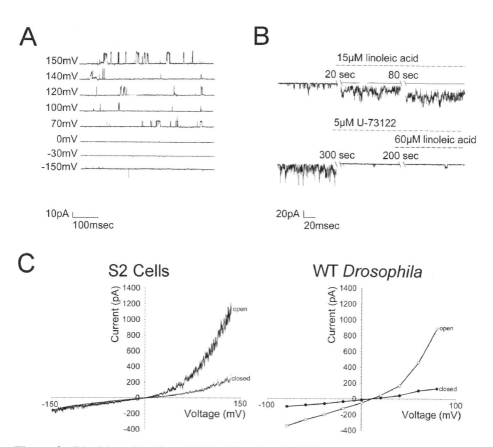

Figure 3   Linoleic acid affects TRPL channels via PLC. Recordings were made 24 h after induction of TRPL expression in S2 cells stably expressing *trpl*. (*A*) Traces of single-channel activity in cell-attached patch clamp recordings at various voltages as indicated at the left of each trace. (*B*) Single-channel recordings using cell-attached patch in physiological solution containing 0.5 mM EGTA (no Ca^{2+} added) (*top*). Application of 15 µM linoleic acid induced robust activation of single channels (*bottom*). The experiment shown in the top trace was repeated, except that application of the PLC inhibitor U-73122 (5 µM) resulted in a complete block of the excitatory effect of linoleic acid. (*C*) Current-voltage relationship measured during whole-cell recordings from S2 cell expressing the TRPL channel (*left*). The upper curve (*open*) was recorded during spontaneous activity of the TRPL channels, and the lower curve (*closed*) was recorded after an increase in cellular Ca^{2+} inactivated the open channels. Current-voltage relationship measured during whole-cell recordings from wild-type (WT) *Drosophila* photoreceptor (*right*). The lower curve (*closed*) was recorded in the dark, and the upper curve (*open*) was recorded after application of metabolic inhibitors.

Both Native and Heterologously Expressed TRPL Channels Reveal Antagonistic Effects of Ca^{2+} and Variable Dependence on the Intracellular Ca^{2+} Stores

THE ROLE OF Ca^{2+} IN ACTIVATION OF TRP AND TRPL CHANNELS IN THE NATIVE SYSTEM   Ca^{2+} is known both to facilitate (positive feedback effect) and to inactivate (negative feedback effect) the *Drosophila* response to light (89–91). Interestingly, there is a difference in the effect of Ca^{2+} between the TRP and the TRPL channels: Facilitation of the LIC is mediated via the TRP channels but not via the TRPL channels (90). Moreover, the Ca^{2+} influx via the TRP channels during the light response inactivates the TRPL channels. Therefore, the LIC in response to intense light is mediated almost exclusively via the TRP channels, and the TRPL channels are inactivated, making the LIC of normal *Drosophila* and that of *trpl* null mutant very similar (39, but see 92). The complex interaction of Ca^{2+} with the TRP and TRPL channels should critically depend on the spatial geometry of the signaling compartment, which is composed of tightly packed ∼60-nm-wide and ∼1-μm-long microvilli (Figure 1). The addition of only a few Ca^{2+} ions to a microvillus dramatically raises its Ca^{2+} concentration to the μM range. The local Ca^{2+} concentration in the microvilli depends on a delicate balance between a large influx via the channels and a very efficient and highly localized extrusion via the Na$^+$-Ca^{2+} exchanger (93, 94). This highly specialized structure should be taken into account when the role of Ca^{2+} is compared in *Drosophila* photoreceptors and in other cells and tissues that use TRP channels. This specialized structure is especially relevant when comparing the effect of Ca^{2+} on TRP channels in the native tissue and in cultured cells (95).

The role of Ca^{2+} in the generation of the LIC has been a controversial issue (96, 97). There is strong evidence that the presence of Ca^{2+} is necessary for the generation of the LIC. Accordingly, during a critical developmental time window, at the last pupa stages, *Drosophila* photoreceptors are incapable of responding to light. Strikingly, application of Ca^{2+} via the whole-cell recording pipette to blind pupa photoreceptors produces a transient response to light similar to that of the *trp* mutant (98). Consistent with this finding, prolonged (∼30 min) Ca^{2+} deprivation in the isolated ommatidia preparation (99, 100) and much shorter deprivation (<3 min) in intact flies (101) initially eliminates the steady state response to continuous light of wild-type flies, making the LIC similar to that of the *trp* mutant. The *trp* phenotype in Ca^{2+} buffered conditions is followed by total, but reversible, abolition of the LIC (99, 102). Investigators have suggested that these effects arise from illumination-dependent depletion of PIP$_2$ in the rhabdomere via a still unknown mechanism affecting PLC (102). Additional evidence showing that Ca^{2+} is necessary for the generation of the LIC came from experiments in which internal dialysis of the cell interior with a 10-mM-fast Ca^{2+} buffer, BAPTA, combined with zero external Ca^{2+} (1 mM EGTA) abolished the response to intense lights in a reversible manner (Figure 4). Dialysis of the cell interior with the slower Ca^{2+} buffer, EGTA, under the same conditions did not abolish the LIC (101), indicating that fast chelating of Ca^{2+} is required to abolish the LIC. This finding suggests that the Ca^{2+}-mediated effect on excitation is very fast. Because the Ca^{2+} affinity is similar for both BAPTA and EGTA, the above result further suggests that the source of Ca^{2+} for light excitation in the above experiments is from the cell interior (101). Experiments in which caged Ca^{2+} was photoreleased failed to excite the photoreceptors when applied to blind mutants without rhodopsin (*ninaE*) or PLC (*norpA*). However, in wild-type flies, photorelease of caged Ca^{2+} strongly facilitated the LIC when applied at the rising phase of the response to light (103).

Taken together, the present data strongly suggest that Ca^{2+} is necessary but not sufficient for excitation and that a combined action of Ca^{2+} and an additional factor, operating upstream to PLC, are required for excitation.

THE EFFECTS OF Ca^{2+} ON HETEROLOGOUSLY EXPRESSED TRPL CHANNELS   Heterologous expression of TRPL channels reported by several groups revealed opposite effects of Ca^{2+}: The TRPL response to histamine in Sf9 cells coexpressing the histamine receptor and TRPL was examined by Schultz and colleagues (104). In the range of 1–10 μM Ca^{2+} there was a marked inhibition in the response of the TRPL channels to histamine, which increased with Ca^{2+} concentration. However, the kinetics of TRPL activation by histamine showed the opposite effect: The response to histamine at low Ca^{2+} concentrations had a relatively long latency, whereas at high Ca^{2+} concentrations the activation by histamine had a short latency. These results suggest that Ca^{2+} has a dual effect on the TRPL channel, reminiscent of the effects Ca^{2+} has on the LIC in *Drosophila*. In *Drosophila* photoreceptors, Ca^{2+} suppressed the amplitude but sped up the kinetics of the LIC in a mechanism known as light adaptation. At zero external Ca^{2+} the kinetics of the LIC slows down because Ca^{2+} is also required for excitation.

Schilling and colleagues (68) have shown that after inhibition of the ER Ca^{2+} pump by thapsigargin, a high Ca^{2+} concentration (10 mM) has only an inhibitory effect on TRPL channels expressed at high levels in Sf9 cells. However, at low expression levels of TRPL channels, there is an initial facilitation of channel activity by a very high Ca^{2+} concentration (50 mM) followed by inhibition, suggesting a dual role for Ca^{2+} on the activity of TRPL channels.

S2 cells co-expressing the muscarinic receptor and the TRPL channels behave differently. In these experiments Hardie & Raghu (105) found that application of carbachol, which releases Ca^{2+} from internal stores, facilitates TRPL channel activity. However, internal perfusion with BAPTA (10 mM) reduced but did not abolish the response to carbachol, leading to the conclusion that Ca^{2+} is not the primary activator of the channels.

Flockerzi and colleagues (106) found that heterologously expressed TRPL channels in CHO cells do not show any channel activity at low cellular Ca^{2+} levels below 50 nM. However, the expressed TRPL channels showed a marked increase in channel activation by application of Ca^{2+} up to 3 μM. Interestingly, no inhibitory effect of Ca^{2+} was observed even at higher concentrations. In another

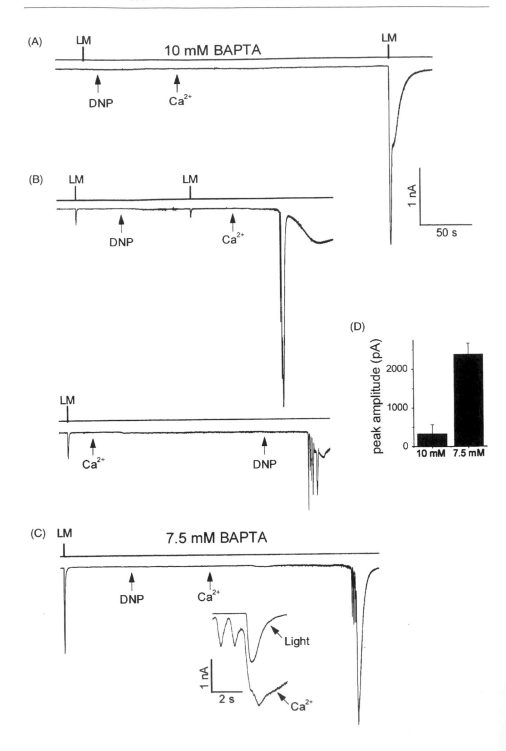

study, researchers reported that similar results were obtained when TRPL channels were expressed in COS cells (80). In addition, they showed that 10 mM BAPTA blocked the constitutive opening of the TRPL channels in COS cells. An additional study found, in the same experimental system, that a rise in cellular Ca^{2+} is required for activation of TRPL channels (84).

The apparently conflicting reports on the effects of Ca^{2+} on heterologously expressed TRPL channels may arise from overexpression of the channels in different expression systems (see below).

ACTIVATION OF THE TRPL CHANNELS BY MANIPULATION OF Ca^{2+} STORES   Investigators have clearly demonstrated that depletion of the $InsP_3$-sensitive Ca^{2+} stores by thapsigargin or by application of $InsP_3$ in cells deprived of Ca^{2+} by Ca^{2+} buffers activates several types of mammalian TRPC channels (107–109). In contrast, no activation of the light-sensitive channels during store depletion by thapsigargin was observed in *Drosophila* (96, 97).

There is wide agreement that modulation of the Ca^{2+} stores by thapsigargin affects the TRPL channels in a large variety of expression systems. TRPL channels expressed in Sf9, S2, CHO, and 293T cells increased their activity in response

---

Figure 4 Ca^{2+} is required for light- and DNP (2,4-dinitrophenol)-induced activation of the TRP and TRPL channels. (*A*) Whole-cell recordings in which the recording pipette contained 10 mM BAPTA but no ATP or NAD. Also, no Ca^{2+} but 1 mM EGTA was added to the external solution. Application of orange light $\sim$1 min after the beginning of the recordings elicited no response. Replacing the external solution by Ringer's solution containing 10 mM Ca^{2+} (*second arrow*; after application of DNP to the external medium, *first arrow*) had no effect in the dark in this particular cell, although the subsequent application of the orange light induced a large LIC. (*B, top*) The paradigm of (*A*) was repeated in another retina. The LIC was not abolished but became very small and relatively slow. Application of DNP (*arrow*) had no significant effect in the dark. Subsequent replacement of the bath solution with Ringer's solution containing 10 mM Ca^{2+} (*arrow*) elicited after a delay a robust, initially oscillating inward current with kinetics similar to that of the LIC under similar conditions (see *C*). (*B, bottom*) The paradigm of (*B, top*) was repeated in another cell, except that application of 10 mM Ca^{2+} (*first arrow*) was not preceded by application of DNP. In this case application of 10 mM Ca^{2+} did not induce any fast inward current in the dark; however, when DNP was applied (*second arrow*) a fast oscillating inward current was induced in the dark. (*C*) The paradigm of (*B, top*) was repeated, except that 7.5 mM BAPTA was included in the pipette. The inset shows the LIC and the oscillating response to application of Ca^{2+} in a fast time scale. (*D*) A histogram comparing the averaged peak amplitude of the LIC recorded from cells when 7.5 mM BAPTA (7.5 mM) or 10 mM BAPTA (10 mM) was added to the solution of the recording pipette. (From Reference 101, with permission from *Cell Calcium*.)

to store depletion by thapsigargin (60, 84, 110, 111). The only exception is the *Xenopus* oocytes expression system (87). The detailed studies of TRPL channel activation by thapsigargin gave different results in the different expression systems: Yagodin and colleagues (111) measured the level of TRPL activation after a 10 min incubation with 1 μM thapsigargin in the presence of 1 mM EGTA in the S2 expression system. This protocol resulted in opening of TRPL channels, suggesting that Ca^{2+} store depletion but not Ca^{2+} per se is required for channel activation. In contrast to this study, TRPL channels expressed in Sf9 cells responded to thapsigargin only when the extracellular medium contained Ca^{2+} (110). Application of thapsigargin in a divalent free medium did not lead to channel activation, whereas application of Ca^{2+} to the external medium resulted in activation of the channels. Researchers therefore concluded that the TRPL channels are not store-operated channels but rather channels that are activated by Ca^{2+} release from internal stores following receptor activation.

In CHO cells, store depletion by thapsigargin resulted in a minor activation of the TRPL channels but blocked the effect of exogenous application of $InsP_3$, which strongly activated the channels (84). Researchers concluded that the release of Ca^{2+} from internal stores is required for TRPL activation. $InsP_3$ was also shown to activate the TRPL channels in several expression systems. *Xenopus* oocytes expressing TRPL channels revealed an elevation in TRPL channel activity after application of $InsP_3$. Researchers suggested that TRPL opening resulted from direct binding of $InsP_3$ to the channels (88). In contrast, 10 mM BAPTA applied to CHO cells expressing TRPL channels completely blocked the effect of $InsP_3$ on TRPL openings. These experiments are consistent with the previous experiments, suggesting that release of Ca^{2+} from internal stores is required to open the TRPL channels. In contrast, Sf9 and S2 expression systems show no (or negligible) response of the TRPL channels to $InsP_3$ (110, 111).

Taken together, the results of the above studies clearly show that heterologously expressed TRPL channels respond differently to similar experimental protocols when expressed in different expression systems or when expressed at different levels in the same expression system. Importantly, the response of the TRPL channels in the native photoreceptor cells with similar experimental protocols is markedly different. Toward reconciliation of the conflicting results described above, we suggest that the TRPL channels expressed in heterologous systems are already in their activated state, and thus the various pharmacological manipulations only modulate the activity of already activated channels. Because the TRPL channels are expressed without additional exogenous proteins (except for the receptor), the channels obviously require participation of endogenous signaling proteins for their activation and regulation (e.g., Gq protein, PLC, CaM, protein kinases, and phosphatases). Therefore, the pharmacological modulations of the channels' activity most likely affected both the channels and their activator/modulator proteins of the host cells. For example, PLCβ1 was facilitated by Ca^{2+} in vitro (112). In contrast, the *Drosophila* PLC (most homologous to PLCβ4) has a bell-shaped response to Ca^{2+} in which low Ca^{2+} facilitates and high Ca^{2+} inhibits the PLC (113). Accordingly, high levels of Ca^{2+} are expected to facilitate TRPL channels in a specific expression system containing PLCβ1 but to inhibit TRPL channels in an expression system containing PLCβ4. Similarly, binding of CaM either activates or inhibits TRPL activity; therefore, the levels of CaM in the host cell can determine the facilitation/inhibition levels of channel activity by Ca^{2+}.

CALMODULIN INTERACTIONS WITH TRPL

The TRPL channel was first discovered by Kelly and colleagues (35) as a *Drosophila* gene that encodes a CaM-binding protein. These authors were also the first to show that the amino acid sequences of TRP and TRPL are typical for channel proteins (35). Two putative CaM-binding sites, designated CBS1 and CBS2, have been identified on the C-terminal of TRPL (35). In vitro experiments on CBS1 revealed that it binds to CaM in a Ca^{2+}-dependent manner at Ca^{2+} concentrations of 500 nM, which are significantly higher than the resting level (114). At this concentration, all four binding domains of CaM bind Ca^{2+}. Although CBS2 was initially reported to be Ca^{2+} independent, it was later found in one study to bind CaM in a Ca^{2+}-dependent manner (IC_{50} of 0.1 μM Ca^{2+} and 3.3 μM Ca^{2+} for CBS1 and CBS2, respectively) (114). The CaM IC_{50} for binding to CBS1 and CBS2 is very low (up to 200 nM), while CBS1 has a higher affinity to CaM than CBS2 (114). Because CaM concentrations in *Drosophila* photoreceptors are very high, reaching 500 μM (115), the only factor affecting CaM-binding to CBS1 and CBS2 is the cellular Ca^{2+} concentration (106).

Binding of CaM to CBS1 is regulated by phosphorylation. Accordingly, CBS1 has two phosphorylation sites: a PKC site and a PKA site. Phosphorylation at the PKC site showed no effect on the CaM-binding capability, whereas phosphorylation at the PKA site greatly reduced the CaM ability to bind to the channel. The phosphorylation by PKC, though, affects the phosphorylation by PKA and reduces it by 70% (114).

Zuker and colleagues (40) produced a *Drosophila* mutant lacking the CBS1 domain and an additional mutant lacking the CBS2 domain of TRPL. When placed on a *trpl;trp* null background, both mutants showed slow light responses relative to mutants on a similar double-null background that express the normal TRPL channel. The mutant lacking CBS1 showed slower responses than the mutant lacking CBS2. These results suggest that CaM-binding to TRPL exerts a negative feedback effect that speeds up the light response arising from activation of the TRPL channels (40).

The effect of CaM on the TRPL channels was tested in several expression systems. In contrast to the in vivo system, in heterologous expression studies both positive and negative effects of CaM on TRPL activity have been reported. Accordingly, TRPL channels expressed in *Xenopus* oocytes inactivated when CaM was blocked by CaM inhibitors. Injection of 1 μM CaM increased TRPL channel activity, while a higher concentration of 3 μM blocked TRPL activity altogether (87). Expression of a TRPL channel lacking CBS1 blocked the activating effect of CaM, whereas expression of a TRPL channel lacking CBS2 had no effect on CaM modulations. CaM inhibitors also affected the TRPL channel lacking CBS2, but had no effect on TRPL lacking CBS1 (88). In contrast to the above findings, TRPL channels expressed in Sf9 cells show a dramatic inhibitory effect of Ca^{2+} on channel activity. This effect is not mediated via CaM, however, as addition of CaM or CaM inhibitors (such as calmidazolium) had no effect on the TRPL activity (104).

In conclusion, CaM shows different effects on the TRPL channel in different expression systems. Because Ca^{2+} and CaM are tightly linked, it is difficult to know whether CaM is involved in TRPL channel activation or only in its inhibition. Furthermore, it is unclear if these two processes act solely on the TRPL channel or affect other stages of the signal transduction cascade.

Mammalian TRPs also have CaM-binding domains (116). Accordingly, TRPV1 channel expressed in *Xenopus* oocytes shows inhibition of the channel activity by Ca^{2+} with an IC_{50} of 60 μM. This effect is mediated by CaM, as application of Ca^{2+}-CaM, but not of Ca^{2+} or CaM separately, inhibits the channel. In this system, no activating effects of either Ca^{2+}-CaM or Ca^{2+} were observed. An additional study on TRPV6 revealed a Ca^{2+}-dependent binding of CaM at a unique site of TRPV6, which facilitated channel inactivation. Interestingly, facilitation of channel inactivation was counteracted by PKC-mediated phosphorylation of the CaM-binding site (117). In contrast to TRPL and TRPV6, TRPV1 binds CaM in a Ca^{2+}-independent manner at a site located at the N-terminal of the channel (116). In agreement with the inhibitory effects of Ca^{2+}-CaM, Zhu and colleagues (118, 118a) have demonstrated that TRPC3 and TRPC4 are inhibited by Ca^{2+}-CaM but not by Ca^{2+} or CaM alone. Interestingly, $InsP_3R$ binds to TRPC3 and TRPC4 at the same site as CaM (designated CIRB, $CaM/InsP_3$ receptor-binding) in a competitive manner. CIRB is conserved in all TRPC channels, including the *Drosophila* TRP channel. However, the *Drosophila* $InsP_3R$ has several modifications in the CIRB-binding domain that prevent TRP binding, suggesting that $InsP_3R$ does not bind to TRP or TRPL in *Drosophila* phototransduction (118).

ACTIVATION OF THE TRP AND TRPL CHANNELS IN THE PHOTORECEPTOR CELL IN THE DARK

To interpret the experimental results obtained from heterologously expressed TRPL channels, it is useful to compare these results to those obtained from studies on the TRPL channels of the native photoreceptor cells. Unfortunately, the signaling membranes of the photoreceptor cells (the rhabdomeres) where the light-sensitive channels reside are not accessible to the patch pipette under conditions that preserve the response to light. Therefore, the characteristics of the channels under physiological conditions have been derived indirectly from studies using whole-cell patch clamp recordings from isolated ommatidia, where the patch pipette was attached to the cell body and did not reveal single-channel activity. The use of shot noise analysis to derive single-channel properties during the light response under physiological conditions is problematic because current fluctuations during light are dominated by the relatively slow fluctuations of the summed responses to single photons (quantum pumps, 119, 120). To overcome these difficulties, Hardie & Minke (120) derived the TRP and TRPL channel properties indirectly from cells that undergo uncontrolled spontaneous activation of the TRP and TRPL channels in the dark during whole-cell recordings (see Figure 3*C* and Reference 120).

It is important to realize that in the isolated ommatidia preparation, which has been the major preparation for studying the properties of the TRP and TRPL channels in the native system, the channels are readily and spontaneously activated in the dark. This dark activation is strongly facilitated by prior illumination. The spontaneous channel opening is irreversible and accompanied by a marked shift of the positive reversal potential to ~0 voltage. The bump noise and the response to light are also lost during this process (120). Detailed studies on the mechanism underlying spontaneous activation of the channels revealed that it arises from metabolic exhaustion. Unlike the in vivo intact preparation in which the metabolites for ATP production in the photoreceptor cells come from the pigment (glia) cells (121), in the isolated ommatidia preparation used for whole-cell recordings the photoreceptor cells are stripped from their pigment cells. The lack of pigment cells limits ATP production, especially under illumination, which consumes large quantities of ATP. Application of ATP and its precursor, NAD, into the cells via the patch pipette is essential to prevent spontaneous activation of the channels, while intense light is still followed by spontaneous channel activation even in the presence of ATP and NAD in the pipette.

A study aiming to elucidate the mechanism underlying the spontaneous activation of the channels in the dark was carried out by Minke and colleagues (122). In this study they found that spontaneous openings of the channels can be obtained not only in the isolated ommatidia preparation but also in vivo in the intact fly by inducing anoxia with N_2. In the intact preparation, in contrast to the isolated ommatidia, channel opening in the dark by anoxia is completely reversible upon removal of anoxia. This study showed that anoxia rapidly and reversibly depolarizes the photoreceptors and induces Ca^{2+} influx into these cells in the dark. It further showed that openings of the light-sensitive channels, which mediate these effects, could also be obtained by mitochondrial uncouplers or by depletion of ATP in photoreceptor cells, whereas the effects of illumination and all forms of metabolic stress were additive. Effects similar to those found in wild-type flies were also found in mutants with strong defects in rhodopsin, Gq protein, or PLC. Genetic elimination of both TRP and TRPL channels prevented the effects of anoxia, mitochondrial uncouplers, and depletion of ATP, thus demonstrating that the TRP and TRPL channels are sensitive targets of metabolic stress. These results suggest that a constitutive ATP-dependent process is required to keep these channels closed in the dark, a requirement that would make them sensitive to metabolic stress (122).

In a follow-up study, Minke and colleagues (101) further showed that ATP depletion or inhibition of PKC activated the TRP channels, whereas photorelease of caged ATP or application of phorbol ester antagonized channel openings in the dark. Furthermore, Mg^{2+}-dependent stable phosphorylation events by ATPγS or protein phosphatase inhibition by calyculin A abolished activation of the TRP and TRPL channels either by light or by metabolic inhibition. Although a high reduction of cellular Ca^{2+} with 10 mM BAPTA abolished channel activation by light or metabolic inhibition, subsequent application of Ca^{2+} to the extracellular medium (in the presence of 10 mM BAPTA in the pipette) combined with ATP depletion induced a robust *dark* current that was reminiscent of light responses (Figure 4). These results suggest that the combined action of Ca^{2+} and phosphorylation-dephosphorylation reactions activates the TRP and TRPL channels in the dark (101).

A more specific mechanism for activation of the TRP and TRPL channels by metabolic inhibition in the dark has been proposed by Hardie and colleagues (123). According to this mechanism, activation of the channels by metabolic inhibition is primarily because DAG kinase failed to phosphorylate DAG owing to ATP depletion. Accumulation of DAG owing to intrinsic spontaneous activity (leak) of PLC, either during metabolic inhibition or in a mutant lacking DAG kinase (the *rdgA* mutant), activates the TRP and TRPL channels, which remain constitutively open and do not respond to light (124). The constitutive activity of the Ca^{2+}-permeable TRP and TRPL channels presumably leads to a toxic increase in cellular Ca^{2+} followed by photoreceptor degeneration in the *rdgA* mutant. This model suggests that DAG or its metabolites, polyunsaturated fatty acids (PUFAs), are second messengers of excitation (125). In support of this model, genetic elimination of the TRP channels partially rescued the degeneration and the LIC of the *rdgA* mutant (126). In addition, the LIC of PLC mutants with minimal PLC activity and with a very small residual response to light was partially rescued in double mutant combinations that also eliminated the DAG kinase (123). Although DAG surrogates activate heterologously expressed TRPL channels (110), so far activating the TRP and TRPL channels in the photoreceptor cells by exogenous application of DAG or its surrogates has proved impossible. In contrast, application of PUFAs to both isolated ommatidia of *Drosophila* and heterologously expressed TRPL channels in S2 cells did activate the TRP/TRPL or TRPL channels, respectively (125). The kinetics of channel activation by PUFAs in *Drosophila* is, however, slower by 2–3 orders of magnitude relative to activation by light and has similar kinetics and magnitude to activation by metabolic inhibition.

In summary, activation of the TRP and TRPL channels by metabolic inhibition in the dark has been a useful tool to decouple the activity of the TRP and TRPL channels from the phototransduction machinery and to study their properties in isolation in the native system. Furthermore, these studies show that a combined action of cellular Ca^{2+} and DAG is required for channel activation. It is not clear, however, if activation of the channels by metabolic inhibition is equivalent to the mechanism underlying physiological activation of the channels by light.

PLC HAS AN ESSENTIAL ROLE IN ACTIVATION OF THE TRP AND TRPL CHANNELS IN PHOTORECEPTOR CELLS

PLC has long been recognized as an essential enzyme for the generation of the response to light because the response to supersaturating light intensity is virtually abolished in the nearly null PLC mutant *norpA* (29, 53). Activation of PLC results in three molecular events: breakdown of PIP_2, production of DAG, and production of $InsP_3$. A role for DAG in light excitation has been demonstrated by Hardie and colleagues (126).

The Role of $InsP_3$

In *Drosophila*, the participation of $InsP_3$ in light excitation has been seriously challenged by genetic elimination of the apparently single $InsP_3R$, because $InsP_3R$ elimination had no effect on light excitation (127, 128). Because the expected role of $InsP_3$ is to release Ca^{2+} from internal stores via binding to the $InsP_3R$, it is still possible that light-induced production of $InsP_3$ leads, in a still unclear way, to a small, local Ca^{2+} elevation in the $InsP_3R$-defective mutant. Such a small localized increase in Ca^{2+} combined with production of DAG can initiate the positive feedback effect of Ca^{2+} on excitation by opening a few TRP channels located at the base of the microvilli, leading to a normal light response. The critical future test for the role of $InsP_3$ in light excitation is to examine the effects of manipulating the endogenous $InsP_3$ levels upon its production by light. It is interesting to note that the $InsP_3R$ antagonist, 2-APB, reversibly blocks the LIC upstream of channel activation (129).

It should be noted that in several invertebrate species, a clear role for $InsP_3$ in excitation via Ca^{2+} release from internal stores has been demonstrated (130–133). $InsP_3$ has been shown to participate in excitation of mammalian TRPC channels where its binding to $InsP_3R$ induces TRP channel openings via direct coupling of TRPC3 to the $InsP_3R$ (134).

The Role of PIP_2 Depletion

Although PIP_2 has been implicated in regulation of various channels (135), including mammalian TRPV, TRPM (136, 137), and the heterologously expressed TRPL channel (110), there is strong evidence that depletion of PIP_2 cannot account for activation of the TRP and TRPL channels in *Drosophila* in vivo. Although activity of TRPL channels expressed in Sf9 cells was suppressed by application of PIP_2, suggesting that breakdown of PIP_2 may open the channels (110), experiments in the fly point to the opposite conclusion. In *Drosophila*, PIP_2 levels were measured in vivo in transgenic flies by the activity of targeted expression of Kir2.1 channels, acting as biosensors for PIP_2 concentration (102). In these transgenic flies, depletion of PIP_2 by light in conditions that block its regeneration inhibits activation of the TRP and TRPL channels. These experiments showed that in the *trp* mutant or in wild-type flies under Ca^{2+} deprivation, continuous light strongly reduced PIP_2 levels in the rhabdomere and closed the TRP or TRPL channels in correlation with PIP_2 depletion (102). Additional Ca^{2+}-deprivation experiments in vivo in intact *Drosophila* strongly suggest that breakdown of PIP_2 does not activate the TRP channels as well (see figure 7 in Reference 122).

In summary, several experiments have indicated that application of exogenous Ca^{2+} in the dark or in blind mutants failed to activate the TRP and TRPL channels (96, 103). However, application of Ca^{2+} to Ca^{2+}-deprived cells when combined with metabolic inhibition excites the photoreceptors in the dark with magnitude and speed of the light response under similar conditions (Figure 4). These results together with the evidence obtained from studies of the *rdgA* mutant strongly suggest that a combined action of Ca^{2+} and DAG resulting from light activation of PLC mediates excitation.

PLC HAS AN EXCITATORY ROLE IN ACTIVATION OF HETEROLOGOUSLY EXPRESSED TRPL CHANNELS

TRPL channels expressed in various cell lines are spontaneously active. However, when the TRPL channel was coexpressed with G protein–coupled receptors, application of agonists greatly enhanced channel activity (81). These observations strongly suggest that TRPL is a promiscuous channel that uses the endogenous G protein and PLC of the host cell for its activation. A fact of significant importance was reported by Schilling and colleagues (110), who showed that the PLC inhibitor U-73122, but not its inactive analog, blocked the spontaneous TRPL activity in Sf9 cells as well as the enhanced channel activity by carbachol. M. Parnas and B. Minke (unpublished observations) have verified the striking blocking effect of U-73122 on TRPL channel activity in S2 cells (Figure 3). These results suggest that the spontaneous TRPL channel activity is mediated by spontaneous (leak) activity of PLC in the host cells. A support for the important role of PLC in activation of TRPL came from the experiments in which application of exogenous $PLC\beta2$ enhanced TRPL channel activity (110).

Importantly, Schilling and colleagues (110) demonstrated that the large facilitation of TRPL channel activity by PUFAs, initially reported by Hardie and colleagues (125) in S2 cells, was also inhibited by the PLC inhibitor U-73122 in Sf9 cells expressing TRPL. This striking result was also verified by M. Parnas and B. Minke (unpublished observations) in S2 cells (Figure 3*B*). The conclusion from these experiments is that activation of the TRPL channels by PUFAs is mediated via PLC and does not result from direct activation of the channels. A strong support for this notion came from recent experiments of Broadie and colleagues (138) in a *Drosophila* mutant with defects in an integral plasma membrane protein with a sequence of a putative DAG lipase designated rolling blackout (*rbo*). Photoreceptors are enriched with the RBO DAG lipase protein. Temperature-sensitive (ts) *rbo* mutants show reversible elimination of phototransduction in a light-dependent manner within minutes, demonstrating acute requirement for the protein. Conditional ts *rbo* mutants show activity-dependent depletion of DAG and concomitant accumulation of PIP and PIP_2 within minutes of induction at the restrictive temperature, strongly suggesting rapid downregulation of PLC activity. A strong indication that inhibition of phototransduction by the *rbo* mutation is not at the level of the TRP and TRPL channel activation came from induction of anoxia in the ts *rbo* mutants, which readily opened the channels when phototransduction was blocked. Additional support for lipid regulation of PLC activity came from a recent study by Mikoshiba and colleagues (139), who demonstrated a novel crosstalk between DAG and $InsP_3$-mediated Ca^{2+} signaling through PLC activation by a DAG analog in tissue culture cells. Taken together, the above results suggest that PUFAs, the products of DAG lipase, regulate PLC activity presumably via its partial PH domain (74) and do not activate the channels directly.

The difficulty in applying this conclusion to fly photoreceptors arises from the finding of Hardie and colleagues (123) that PUFAs activate the TRP and TRPL channels in the native system of a nearly null PLC mutant of *Drosophila* regardless of application time. In addition, they showed in this study that unlike the time-independent PUFAs' action, metabolic inhibitors induced openings of the channels in the PLC mutant only temporarily, during ∼20 min of residual PLC activity following illumination. A reconciliation of the apparently conflicting data on PUFAs' actions in vivo and in the tissue culture cells may come from many reports showing that PUFAs have multiple actions, including activation of PLC and inhibition of protein kinases, either directly (140) or by acting as metabolic inhibitors (141). Therefore, application of PUFAs to the PLC mutant is capable of inducing a residual PLC activation combined with metabolic inhibition, leading to channel activation via DAG accumulation. Accordingly, in the minimal system of the heterologously expressed TRPL channels, the specific targets of PUFAs' metabolic inhibition are most likely missing, and only the effect on PLC activation is realized, whereas in the native system of the fly both effects of PUFAs are realized.

In summary, although all studies on *Drosophila* phototransduction indicate that both DAG and Ca^{2+} are required for excitation, there are different views regarding the specific mechanism that triggers channel opening. A summary of these views is given in Table 1.

TABLE 1   The triggering mechanisms of TRP and TRPL channels

	Supporting evidence	Difficulties
Excitation by lipids	Inhibition of DAG kinase enhances channel activity in vivo (123, 124, 126)	Application of DAG has no effect on channel excitation in vivo (S. Frechter and B. Minke, unpublished data)
	The LIC of a PLC mutant with a very small residual response to light was partially rescued in the double mutant PLC/DAG kinase (123, 124)	Removal of a putative DAG lipase does not block channel opening (138)
	PUFAs open the channels both in vivo and in heterologous expression systems (124, 125)	The PUFAs' effect in heterologous expression systems is blocked upon inhibition of PLC (110)
		PUFAs' activation kinetics in vivo is slower by 2–3 orders of magnitude relative to the activation by light (123, 125)
Ca^{2+} is necessary but not sufficient for excitation	10 mM BAPTA but not 10 mM EGTA blocks channel activation by light and metabolic inhibition. In Ca^{2+} deprived cells application of Ca^{2+} combined with the metabolic inhibition mimics the response to light in the dark (kinetics and amplitude) (101)	Photorelease of caged Ca^{2+} in blind mutants does not open the channels (103)
	Embryonic *Drosophila* photoreceptors are incapable of responding to light unless first provided with Ca^{2+} via the whole-cell pipette (98)	Genetic elimination of the single known $InsP_3R$ in *Drosophila* has no effect on excitation (128)
	Application of the $InsP_3R$ antagonist 2APB reversibly blocks the light response (129)	2APB is known to have nonspecific effects (129)

LIGHT-REGULATED TRANSLOCATION OF THE TRPL CHANNEL

To achieve a very high gain, which is required to produce the single photon responses to dim lights, the photoreceptor cell maximizes the current produced by a single photon using both the TRP channels and a relatively high level of TRPL channels. To prevent saturation of the light response upon an increase in ambient light, the Ca^{2+} flow via the TRP channel shuts off the TRPL channels and also induces its translocation out of the signaling membranes (59, 91, 99). This fascinating and novel mechanism to fine-tune visual responses is also used by other cellular signaling mechanisms such as fine-tuning of growth cone pathfinding via the action of a growth factor on TRPC5 activation (142).

The Amount of TRPL but not of TRP, PLC, PKC, or INAD Present in the Rhabdomeres is Regulated by Light

A relatively high level of TRPL, observed in photoreceptor membranes of dark-raised flies, rapidly decreased when the flies were transferred to light (Figure 5). Furthermore, the amount of rhabdomeral TRPL of flies that were kept in light for 16 h was close to the detection limit of the Western blot analysis but increased to a high level within 1 h after the transfer of the flies to darkness (Figure 5). The amount of rhabdomeral TRP and also that of other components of the INAD signaling complex, PLCβ, ePKC, and INAD, was unaffected by light (Figures 5 and 6). This finding indicates that the changes in the rhabdomeral TRPL level is specific to this ion channel subunit and are not, for example, a consequence of a massive light-dependent photoreceptor membrane shedding. Radioimmune assays and Western blot analysis revealed that the reduction of TRPL levels in the rhabdomers results from translocation of TRPL rather than from its degradation and de novo synthesis (59).

The Mechanism that May Underlie TRPL Translocation

Direct visualization of intracellular movements of TRPL in photoreceptors upon illumination was obtained by immunolabeling of cross sections through *Calliphora* and *Drosophila* eyes. This study revealed that unlike TRPL, TRP and INAD are confined to the rhabdomeres, independent of whether the flies are kept in darkness or in light prior to sectioning (Figure 5). Antibodies directed against TRPL specifically labeled the rhabdomere area of cross-sectioned eyes obtained from dark-raised flies (Figure 6). In the eye sections of light-raised flies the TRPL-specific immunofluorescence was distributed over the cell body of the photoreceptor cell and was not detected in the rhabdomeres (Figure 6) in line with the Western blot analysis (59).

Because the translocation of TRPL depends on illumination, the question arises whether the response to light through activation of the TRP and TRPL channels is the trigger for TRPL movement to the cell body. Using immunocytochemistry, TRPL translocation from the rhabdomere to the cell body was tested in the nearly null PLC mutant, *norpA*^{P24} (31), the null mutant of TRP, *trp*^{P343} (40), and the null INAD mutant, *inaD*^{1}. Young *inaD* flies contain rhabdomeral TRP, which is most likely detached from the signaling complex and therefore degrades with age and is largely missing in older *inaD* flies (51). Light-induced translocation of TRPL to the cell body was observed in young *inaD* flies but did not occur in the *trp* null mutant or in old *inaD* flies. In *norpA*^{P24} mutants, translocation of TRPL tagged with green fluorescent protein (TRPL-eGFP) was blocked. These findings reveal that when TRP is missing, as in the *trp* mutant and in old *inaD* flies, TRPL internalization

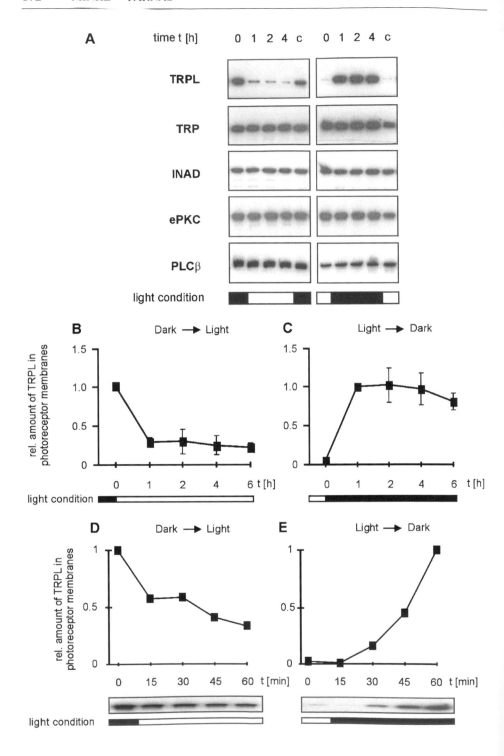

is not observed (59). Therefore, the presence of TRP seems to be required for TRPL internalization. If the lack of PLC blocks TRPL translocation as found in vivo, then the block of TRPL translocation, by the absence of either TRP or PLC, suggests that light-induced Ca^{2+} influx is the trigger for TRPL translocation. This is because lack of PLC completely blocks Ca^{2+} influx via the light-activated channels, whereas genetic elimination of TRP eliminates the main route of Ca^{2+} entry into the photoreceptor cell (100, 143).

Translocation of TRPL Affects the Characteristics of the Light-Induced Current

The LIC of wild-type flies is composed of two independent components arising from activation of the TRP and TRPL channels (91). Patch clamp whole-cell recordings in isolated ommatidia of *Drosophila* were used to examined two properties of the LIC that discriminate between the contribution of the TRP and TRPL channels to the LIC: the block by La^{3+} and the reversal potential, which is $\sim$15 mV and $\sim$0 mV for the TRP and TRPL channels, respectively (91).

Application of La^{3+} in micromolar concentration is known specifically to block the TRP but not the TRPL channels (100, 144, 145). In wild-type *Drosophila*, application of La^{3+} specifically blocks the TRP channels leaving a residual response, which is mediated by the TRPL channels and is indistinguishable from the response measured in *trp* mutant photoreceptors. In wild-type flies kept in light (light-raised), application of La^{3+} largely reduced the peak amplitude of the LIC in response to intense light and modified the waveform of the response to prolonged light displaying the typical *trp* phenotype. In dark-raised wild-type flies, application of La^{3+} under identical experimental conditions had a much smaller effect on the peak amplitude of the LIC in response to similar light intensity. The weak *trp* phenotype in dark-raised flies indicates a reduced effect for La^{3+}.

Figure 5 The amount of rhabdomeral TRPL but not TRP undergoes rapid light-induced changes. (*A*) Western blot analysis of phototransduction proteins present in the photoreceptor membrane of flies kept under different light conditions. Flies (*Calliphora*, mutant *chalky*) were raised in darkness or light for 16 h (t = 0 h) and were then switched from darkness to light (*left panels*) or vice versa (*right panels*). Photoreceptor membranes were isolated from dissected fly eyes at t = 0 and 1, 2, and 4 h (t = 1–4 h) after the change of light condition. A control group of flies (*C*) was kept under the original light condition during the experiment and was dissected at t = 5 h. (*B–E*) Time courses of the light-dependent increase and decrease of the TRPL level in rhabdomeral photoreceptor membranes. Flies were treated as described in (*A*). The time courses shown in (*B*) and (*C*) comprise 6 h after the change of light condition. Time courses revealing the changes occurring within the first hour after light change are shown in (*D*) and (*E*). (From Reference 59 with permission from *Neuron*.)

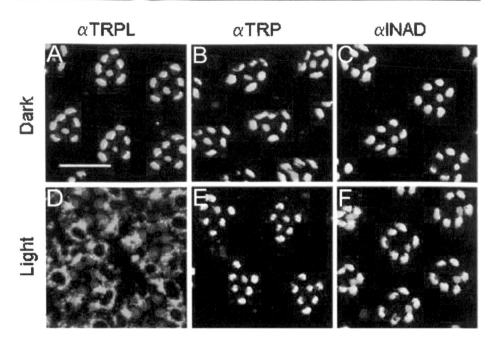

Figure 6   Light-dependent translocation of TRPL molecules in the photoreceptor cells of *Drosophila* compound eyes. Cross sections through wild-type *Drosophila* (*A–F*) eyes of light- and dark-raised flies were double labeled with rhodamin-coupled wheat germ agglutinin, which specifically labels rhabdomeral photoreceptor membranes (*red fluorescence*) and antibodies against TRPL, TRP, and INAD, as indicated. The overlay of both markers appears yellow. Scale bar in (*A*), 10 μm. (From Reference 59 with permission from *Neuron.*)

Quantitative analysis at dim lights shows that La^{3+} suppressed the LIC, suggesting a relative contribution of TRPL to the LIC of ~9% and 38%, in light- and dark-raised flies, respectively. Roughly similar conclusions were derived from measurements of the change in the reversal potential of light- and dark-raised flies, together suggesting that a significantly larger number of functional TRPL channels is present in dark-raised flies relative to light-raised flies (59).

Translocation of TRPL Induces Long-Term Adaptation

Measurements of the light sensitivity of dark- and light-raised flies in vivo revealed that wild-type flies kept in darkness are very sensitive to dim background lights and respond within a relatively wide dynamic range, having a relatively low sensitivity to small changes in stimulus intensity. Wild-type flies kept in light are different: (*a*) They are less sensitive to dim background lights, and (*b*) they have a smaller dynamic range, but (*c*) their photoreceptors are more sensitive to small changes in light intensity within their dynamic range. The fact that *trpl* mutants, when kept in either light or darkness, behave similarly to light-raised wild-type flies strongly suggests that translocation of TRPL underlies the fine-tuning of long-term adaptation (59).

Translocation of Mammalian TRP Channels

Recent studies on the mammalian TRPC5 channel revealed that growth factor stimulation initiated rapid (~2 min) incorporation into the plasma membrane of HEK-293 cells of expressed TRPC5 from vesicles held in reserve just under the plasma membrane. This incorporation was specific to TRPC5 and was not observed for TRPC1. Channel incorporation was measured by evanescent field of total internal reflection fluorescence microscopy and biotin surface labeling. Strikingly, TRPC5 incorporation into the plasma membrane resulted in a dramatic increase in the typical current produced by these channels, indicating insertion of functional channels. Epidermal growth factor–induced incorporation of functional TRPC5 channels requires phosphatidylinositol 3-kinase (PI3K), the Rho GTPase, Rac1, and phosphatidyl inositol 4-phosphate 5-kinase alpha (PIP5Kα). The role of ERM proteins, which are known to be regulated by Rho GTPase, was not investigated in the study. It is hypothesized that Rac1 mediates the rapid insertion of TRPC5 into the plasma membrane through stimulation of PIP5Kα and increased channel availability. This process also occurs in hippocampal neurons in primary culture, where it affects the process of growth cone extension, suggesting that growth factor–induced extension, which is known to be regulated by Ca^{2+} influx, is mediated by regulated insertion of the Ca^{2+}-permeable TRPC5 channels (142).

Internalization of TRPC3, together with its signaling complex Gq/11, PLCβ1, and caveolin-1, was reported to occur following treatment with the protein phosphatase inhibitor caliculin A, which also induced reorganization of the actin cytoskeleton (73). This channel internalization led to a marked attenuation in divalent cation (Sr^{2+}) entry induced by application of OAG, reminiscent of the reduction of TRPL-dependent currents in *Drosophila* upon translocation of TRPL. An additional study by Ambudkar and colleagues (146) revealed that carbachol-stimulation of epithelial cells resulted in translocation of TRPC3 to the plasma membrane in a mechanism that was inhibited by cleavage of the SNARE protein VAMP2 (vesicle-associated membrane protein 2). These data suggest that VAMP2-dependent exocytosis regulates plasma membrane insertion of TRPC3 channels and contributes to agonist-stimulation of Ca^{2+} influx (146). Another TRPC channel reported to be regulated by exocytosis is TRPC6. Stimulation of the M3 muscarinic receptor resulted in translocation of TRPC6 to the plasma membrane in a timescale that coincided with activation by Ca^{2+} influx (147).

Translocation of another mammalian TRP channel (TRPV2) was reported for Balb/c 3T3 cells. These cells expressed the TRPV2 channel, which is regulated by the insulin-like growth factor (IGF-1). Interestingly, stimulation of the cells by IGF-1 induced translocation of the channels from internal vesicles to the plasma membrane (148). No functional consequence of this translocation was tested, and similar experiments in heterologously expressed TRPV1 did not reveal any translocation (142).

In summary, translocation of TRP channels of both *Drosophila* and mammalian cells emerges as a novel and important regulatory mechanism with wide implications to a variety of signaling mechanisms.

CONCLUDING REMARKS

Channel proteins are capable of carrying relatively large currents, and, therefore, a relatively small number of channels is sufficient to produce large potential differences. Therefore, channel proteins are expressed in relatively small quantities in the native tissue. This rule applies to most TRP channels and makes in vivo studies of TRP channels difficult. In addition, activation and regulation of TRP channels seem to require multiple protein-protein interactions that are difficult to study in heterologous systems in which critical components are missing or exist in an abnormal stoichiometry. The *Drosophila* photoreceptor cell, which uses TRP channels and their activating proteins in huge quantities, combined with the great power of the *Drosophila* molecular genetics, is a unique signaling system. This system allows studies of TRP activation and gating mechanism in the native system of the eye.

Although a comparison of the single-channel properties of the native and expressed TRPL channels revealed many similarities, a profound difference between them is the constitutive activity of heterologously expressed TRPL channels, which is not observed in the native system in the dark. To account for this difference, investigators have suggested that the TRPγ subunit, which is missing in the heterologous system, is required to keep the channels closed in the photoreceptor cell (41). A different explanation based on substantial evidence (e.g., Figure 3) is that the endogenous PLC of the host cell (and not the TRPL channel itself) is constitutively active and therefore activates the expressed TRPL channel. A similar explanation may hold for mammalian TRP channels that also show constitutive activity in heterologous systems. In the photoreceptor cells, there seem to be powerful mechanisms that are missing in the heterologous system, preventing spontaneous activation of PLC or triggering slow inactivation of its products and the consequent generation of light response in the dark.

Unlike the promiscuous TRPL channel that can use endogenous or exogenous (110) PLCs for activation, the *Drosophila* TRP channel seems to be a tightly regulated channel that requires the native PLC (NORPA) and perhaps other TRP-interacting proteins of the INAD signaling complex to function in heterologous systems. Future experiments should resolve this possibility. Mammalian TRP channels may include both promiscuous and tightly regulated TRP channels, whereas only the promiscuous channels show constitutive activity when expressed in tissue culture cells.

ACKNOWLEDGMENTS

We thank Drs. Johannes Oberwinkler, Shmuel Muallem, Kendal Broadie, Barbara Niemeyer, Veit Flockerzi, Shaya Lev, and Ben Katz for critical reading of the manuscript and Johannes Oberwinkler for stimulating discussions. The experimental part of this review was supported by grants from the National Institutes of Health (EY 03529), the Israel Science Foundation (ISF), the Minerva Foundation, the U.S.-Israel Binational Science Foundation (BSF), and the German Israeli Foundation (GIF).

**The *Annual Review of Physiology* is online at
http://physiol.annualreviews.org**

LITERATURE CITED

1. Harteneck C, Plant TD, Schultz G. 2000. From worm to man: three subfamilies of TRP channels. *Trends Neurosci.* 23:159–66

2. Minke B, Wu C, Pak WL. 1975. Induction of photoreceptor voltage noise in the dark in *Drosophila* mutant. *Nature* 258:84–87

3. Montell C, Rubin GM. 1989. Molecular characterization of the *Drosophila trp* locus: a putative integral membrane protein required for phototransduction. *Neuron* 2:1313–23

4. Wes PD, Chevesich J, Jeromin A, Rosenberg C, Stetten G, Montell C. 1995. TRPC1, a human homolog of a *Drosophila* store-operated channel. *Proc. Natl. Acad. Sci. USA* 92:9652–56

5. Zhu X, Chu PB, Peyton M, Birnbaumer L. 1995. Molecular cloning of a widely expressed human homologue for the *Drosophila trp* gene. *FEBS Lett.* 373:193–98

6. Clapham DE. 2003. TRP channels as cellular sensors. *Nature* 426:517–24

7. Minke B, Cook B. 2002. TRP channel proteins and signal transduction. *Physiol. Rev.* 82:429–72

8. Corey DP. 2003. New TRP channels in hearing and mechanosensation. *Neuron* 39:585–88

9. Hardie RC. 2003. Regulation of TRP channels via lipid second messengers. *Annu. Rev. Physiol.* 65:735–59

10. Montell C. 2001. Physiology, phylogeny, and functions of the TRP superfamily of cation channel. *Sci. STKE* 90:Re1. http://stke.sciencemag.org/cgi/content/full/OC_sigtrans;2001/90/re1

11. Nilius B, Droogmans G. 2001. Ion channels and their functional role in vascular endothelium. *Physiol. Rev.* 81:1415–59

12. Patapoutian A, Peier AM, Story GM, Viswanath V. 2003. ThermoTRP channels and beyond: mechanisms of temperature sensation. *Nat. Rev. Neurosci.* 4:529–39

13. Montell C. 2005. The TRP superfamily of cation channels *Sci. STKE* 2005:re3

14. Tracey WD Jr, Wilson RI, Laurent G, Benzer S. 2003. painless, a *Drosophila* gene essential for nociception. *Cell* 113:261–73

15. Lee Y, Lee Y, Lee J, Bang S, Hyun S, et al. 2005. Pyrexia is a new thermal transient receptor potential channel endowing tolerance to high temperatures in *Drosophila melanogaster. Nat. Genet.* 37:305–10

16. Viswanath V, Story GM, Peier AM, Petrus MJ, Lee VM, et al. 2003. Opposite thermosensor in fruitfly and mouse. *Nature* 423:822–23

17. Walker RG, Willingham AT, Zuker CS. 2000. A *Drosophila* mechanosensory transduction channel. *Science* 287:2229–34

18. Sidi S, Friedrich RW, Nicolson T. 2003. NompC TRP channel required for vertebrate sensory hair cell mechanotransduction. *Science* 301.96–99

19. Caterina MJ, Schumacher MA, Tominaga M, Rosen TA, Levine JD, Julius D. 1997. The capsaicin receptor: a heat-activated ion channel in the pain pathway. *Nature* 389:816–24

20. Caterina MJ, Leffler A, Malmberg AB, Martin WJ, Trafton J, et al. 2000. Impaired nociception and pain sensation in mice lacking the capsaicin receptor. *Science* 288:306–13

21. Caterina MJ, Montell C. 2005. Take a TRP to beat the heat. *Genes Dev.* 19:415–18

22. Dulac C, Torello AT. 2003. Molecular detection of pheromone signals in mammals: from genes to behaviour. *Nat. Rev. Neurosci.* 4:551–62

23. Hoon MA, Adler E, Lindemeier J, Battey JF, Ryba NJ, Zuker CS. 1999. Putative mammalian taste receptors: a class of taste-specific GPCRs with distinct topographic selectivity. *Cell* 96:541–51

24. Peier AM, Moqrich A, Hergarden AC, Reeve AJ, Andersson DA, et al. 2002. A TRP channel that senses cold stimuli and menthol. *Cell* 108:705–15

25. Story GM, Peier AM, Reeve AJ, Eid SR, Mosbacher J, et al. 2003. ANKTM1, a TRP-like channel expressed in nociceptive neurons, is activated by cold temperatures. *Cell* 112:819–29

26. Zufall F, Munger SD. 2001. From odor and pheromone transduction to the organization of the sense of smell. *Trends Neurosci.* 24:191–93

27. Stryer L, Berg J, Tymoczko J. 2002. *Biochemistry*. New York: W.H. Freeman

28. Wu CF, Pak WL. 1975. Quantal basis of photoreceptor spectral sensitivity of *Drosophila melanogaster*. *J. Gen. Physiol.* 66:149–68

29. Minke B, Selinger Z. 1992. The inositol-lipid pathway is necessary for light excitation in fly photoreceptors. In *Sensory Transduction*, ed. D Corey, SD Roper, 47:202–17. New York: Rockefeller Univ. Press

30. Hardie RC, Raghu P. 2001. Visual transduction in *Drosophila*. *Nature* 413:186–93

31. Bloomquist BT, Shortridge RD, Schneuwly S, Perdew M, Montell C, et al. 1988. Isolation of a putative phospholipase C gene of *Drosophila*, *norpA*, and its role in phototransduction. *Cell* 54:723–33

32. Devary O, Heichal O, Blumenfeld A, Cassel D, Suss E, et al. 1987. Coupling of photoexcited rhodopsin to inositol phospholipid hydrolysis in fly photoreceptors. *Proc. Natl. Acad. Sci. USA* 84:6939–43

33. O'Tousa JE, Baehr W, Martin RL, Hirsh J, Pak WL, Applebury ML. 1985. The *Drosophila ninaE* gene encodes an opsin. *Cell* 40:839–50

34. Zuker CS, Cowman AF, Rubin GM. 1985. Isolation and structure of a rhodopsin gene from *D. melanogaster*. *Cell* 40:851–58

35. Phillips AM, Bull A, Kelly LE. 1992. Identification of a *Drosophila* gene encoding a calmodulin-binding protein with homology to the *trp* phototransduction gene. *Neuron* 8:631–42

36. Montell C. 1999. Visual transduction in *Drosophila*. *Annu. Rev. Cell Dev. Biol.* 15:231–68

37. Ranganathan R, Malicki DM, Zuker CS. 1995. Signal transduction in *Drosophila* photoreceptors. *Annu. Rev. Neurosci.* 18:283–317

38. Minke B, Selinger Z. 1996. The roles of *trp* and calcium in regulating photoreceptor function in *Drosophila*. *Curr. Opin. Neurobiol.* 6:459–66

39. Niemeyer BA, Suzuki E, Scott K, Jalink

K, Zuker CS. 1996. The *Drosophila* light-activated conductance is composed of the two channels TRP and TRPL. *Cell* 85:651–59

40. Scott K, Sun Y, Beckingham K, Zuker CS. 1997. Calmodulin regulation of *Drosophila* light-activated channels and receptor function mediates termination of the light response in vivo. *Cell* 91: 375–83

41. Xu XZ, Chien F, Butler A, Salkoff L, Montell C. 2000. TRPγ, a *Drosophila* TRP-related subunit, forms a regulated cation channel with TRPL. *Neuron* 26: 647–57

42. Smith DP, Ranganathan R, Hardy RW, Marx J, Tsuchida T, Zuker CS. 1991. Photoreceptor deactivation and retinal degeneration mediated by a photoreceptor-specific protein kinase C. *Science* 254:1478–84

43. Berridge MJ. 1993. Inositol trisphosphate and calcium signalling. *Nature* 361:315–25

44. Masai I, Okazaki A, Hosoya T, Hotta Y. 1993. *Drosophila* retinal degeneration A gene encodes an eye-specific diacylglycerol kinase with cysteine-rich zinc-finger motifs and ankyrin repeats. *Proc. Natl. Acad. Sci. USA* 90:11157–61

45. Wu L, Niemeyer B, Colley N, Socolich M, Zuker CS. 1995. Regulation of PLC-mediated signalling in vivo by CDP-diacylglycerol synthase. *Nature* 373:216–22

46. Vihtelic TS, Hyde DR, O'Tousa JE. 1991. Isolation and characterization of the *Drosophila* retinal degeneration B (*rdgB*) gene. *Genetics* 127:761–68

47. Masai I, Suzuki E, Yoon CS, Kohyama A, Hotta Y. 1997. Immunolocalization of *Drosophila* eye-specific diacylgylcerol kinase, *rdgA*, which is essential for the maintenance of the photoreceptor. *J. Neurobiol.* 32:695–706

48. Vihtelic TS, Goebl M, Milligan S, O'Tousa JE, Hyde DR. 1993. Localization of *Drosophila* retinal degener-

ation B, a membrane-associated phosphatidylinositol transfer protein. *J. Cell Biol.* 122:1013–22

49. Tsunoda S, Sun Y, Suzuki E, Zuker C. 2001. Independent anchoring and assembly mechanisms of INAD signaling complexes in *Drosophila* photoreceptors. *J. Neurosci.* 21:150–58

50. Xu XZ, Choudhury A, Li X, Montell C. 1998. Coordination of an array of signaling proteins through homo- and heteromeric interactions between PDZ domains and target proteins. *J. Cell Biol.* 142:545–55

51. Tsunoda S, Sierralta J, Sun Y, Bodner R, Suzuki E, et al. 1997. A multivalent PDZ-domain protein assembles signalling complexes in a G-protein-coupled cascade. *Nature* 388:243–49

52. Minke B. 1986. Photopigment-dependent adaptation in invertebrates—implications for vertebrates. In *The Molecular Mechanism of Photoreception*, ed. H Stieve, 34:241–65. Berlin: Springer-Verlag

53. Pak WL. 1995. *Drosophila* in vision research. The Friedenwald Lecture. *Invest. Ophthalmol. Vis. Sci.* 36:2340–57

54. Shieh BH, Zhu MY. 1996. Regulation of the TRP Ca^{2+} channel by INAD in *Drosophila* photoreceptors. *Neuron* 16: 991–98

55. Huber A, Sander P, Gobert A, Bahner M, Hermann R, Paulsen R. 1996. The transient receptor potential protein (Trp), a putative store-operated Ca^{2+} channel essential for phosphoinositide-mediated photoreception, forms a signaling complex with NorpA, InaC and InaD. *EMBO J.* 15:7036–45

56. Arnold DB, Clapham DE. 1999. Molecular determinants for subcellular localization of PSD-95 with an interacting K^+ channel. *Neuron* 23:149–57

57. Dimitratos SD, Woods DF, Stathakis DG, Bryant PJ. 1999. Signaling pathways are focused at specialized regions of the plasma membrane by scaffolding

proteins of the MAGUK family. *Bioessays* 21:912–21

58. Schillace RV, Scott JD. 1999. Organization of kinases, phosphatases, and receptor signaling complexes. *J. Clin. Invest.* 103:761–65

59. Bahner M, Frechter S, Da Silva N, Minke B, Paulsen R, Huber A. 2002. Light-regulated subcellular translocation of *Drosophila* TRPL channels induces long-term adaptation and modifies the light-induced current. *Neuron* 34:83–93

60. Xu XZS, Li HS, Guggino WB, Montell C. 1997. Coassembly of TRP and TRPL produces a distinct store-operated conductance. *Cell* 89:1155–64

61. Chevesich J, Kreuz AJ, Montell C. 1997. Requirement for the PDZ domain protein, INAD, for localization of the TRP store-operated channel to a signaling complex. *Neuron* 18:95–105

62. Wes PD, Xu XZ, Li HS, Chien F, Doberstein SK, Montell C. 1999. Termination of phototransduction requires binding of the NINAC myosin III and the PDZ protein INAD. *Nat. Neurosci.* 2:447–53

62a. Chorna-Ornan I, Tzarfaty V, Ankri-Eliahoo G, Joel-Almagor T, Meyer NE, et al. 2005. Light-regulated interaction of Dmoesin with TRP and TRPL channels is required for maintenance of photoreceptors. *J. Cell Biol.* 171:143–52

63. Polesello C, Delon I, Valenti P, Ferrer P, Payre F. 2002. Dmoesin controls actin-based cell shape and polarity during *Drosophila melanogaster* oogenesis. *Nat. Cell Biol.* 4:782–89

64. Bretscher A, Edwards K, Fehon RG. 2002. ERM proteins and merlin: integrators at the cell cortex. *Nat. Rev. Mol. Cell Biol.* 3:586–99

65. Mery L, Strauss B, Dufour JF, Krause KH, Hoth M. 2002. The PDZ-interacting domain of TRPC4 controls its localization and surface expression in HEK293 cells. *J. Cell Sci.* 115:3497–508

66. Montell C. 1998. TRP trapped in fly signaling web. *Curr. Opin. Neurobiol.* 8:389–97

67. Macpherson MR, Pollock VP, Kean L, Southall TD, Giannakou ME, et al. 2005. Transient receptor potential-like channels are essential for calcium signaling and fluid transport in a *Drosophila* epithelium. *Genetics* 169:1541–52

68. Estacion M, Sinkins WG, Schilling WP. 1999. Stimulation of *Drosophila* TrpL by capacitative Ca^{2+} entry. *Biochem. J.* 341(Pt. 1):41–49

69. Huber A, Sander P, Bahner M, Paulsen R. 1998. The TRP Ca^{2+} channel assembled in a signaling complex by the PDZ domain protein INAD is phosphorylated through the interaction with protein kinase C (ePKC). *FEBS Lett.* 425:317–22

70. Li HS, Montell C. 2000. TRP and the PDZ protein, INAD, form the core complex required for retention of the signalplex in *Drosophila* photoreceptor cells. *J. Cell Biol.* 150:1411–22

71. Cook B, Bar-Yaacov M, Cohen-Ben-Ami H, Goldstein RE, Paroush Z, et al. 2000. Phospholipase C and termination of G-protein-mediated signalling in vivo. *Nat. Cell Biol.* 2:296–301

72. Yuan JP, Kiselyov K, Shin DM, Chen J, Shcheynikov N, et al. 2003. Homer binds TRPC family channels and is required for gating of TRPC1 by IP_3 receptors. *Cell* 114:777–89

73. Lockwich T, Singh BB, Liu X, Ambudkar IS. 2001. Stabilization of cortical actin induces internalization of transient receptor potential 3 (Trp3)-associated caveolar Ca^{2+} signaling complex and loss of Ca^{2+} influx without disruption of Trp3-inositol trisphosphate receptor association. *J. Biol. Chem.* 276:42401–8

74. van Rossum DB, Patterson RL, Sharma S, Barrow RK, Kornberg M, et al. 2005. Phospholipase $C\gamma 1$ controls surface expression of TRPC3 through an intermolecular PH domain. *Nature* 434:99–104

75. Obukhov AG, Nowycky MC. 2004. TRPC5 activation kinetics are modulated by the scaffolding protein ezrin/radixin/moesin-binding phosphoprotein-50 (EBP50). *J. Cell Physiol.* 201: 227–35

76. Tang Y, Tang J, Chen Z, Trost C, Flockerzi V, et al. 2000. Association of mammalian Trp4 and phospholipase C isozymes with a PDZ domain-containing protein, NHERF. *J. Biol. Chem.* 275:37559–64

77. Wang X, Zeng W, Soyombo AA, Tang W, Ross EM, et al. 2005. Spinophilin regulates Ca^{2+} signalling by binding the N-terminal domain of RGS2 and the third intracellular loop of G-protein-coupled receptors. *Nat. Cell Biol.* 7:405–11

78. Gillo B, Chorna I, Cohen H, Cook B, Manistersky I, et al. 1996. Coexpression of *Drosophila* TRP and TRP-like proteins in *Xenopus* oocytes reconstitutes capacitative Ca^{2+} entry. *Proc. Natl. Acad. Sci. USA.* 93:14146–51

79. Vaca L, Sinkins WG, Hu Y, Kunze DL, Schilling WP. 1994. Activation of recombinant *trp* by thapsigargin in Sf9 insect cells. *Am. J. Physiol.* 267:C1501–5

80. Hambrecht J, Zimmer S, Flockerzi V, Cavalie A. 2000. Single-channel currents through transient-receptor-potential-like (TRPL) channels. *Pflugers Arch.* 440:418–26

81. Hardie RC, Reuss H, Lansdell SJ, Millar NS. 1997. Functional equivalence of native light-sensitive channels in the *Drosophila trp³⁰¹* mutant and TRPL cation channels expressed in a stably transfected *Drosophila* cell line. *Cell Calcium* 21:431–40

82. Hu Y, Vaca L, Zhu X, Birnbaumer L, Kunze DL, Schilling WP. 1994. Appearance of a novel Ca^{2+} influx pathway in Sf9 insect cells following expression of the transient receptor potential-like (*trpl*) protein of *Drosophila. Biochem. Biophys. Res. Commun.* 201:1050–56

83. Kunze DL, Sinkins WG, Vaca L, Schilling WP. 1997. Properties of single *Drosophila* Trpl channels expressed in Sf9 insect cells. *Am. J. Physiol.* 272: C27–34

84. Zimmer S, Trost C, Wissenbach U, Philipp S, Freichel M, et al. 2000. Modulation of recombinant transient-receptor-potential-like (TRPL) channels by cytosolic Ca^{2+}. *Pflugers Arch.* 440:409–17

85. Minke B, Selinger Z. 1991. Inositol lipid pathway in fly photoreceptors: excitation, calcium mobilization and retinal degeneration. In *Progress in Retinal Research*, ed. NA Osborne, GJ Chader, pp. 99–124. Oxford: Pergamon

86. Hu Y, Schilling WP. 1995. Receptor-mediated activation of recombinant Trpl expressed in Sf9 insect cells. *Biochem. J.* 305:605–11

87. Lan L, Bawden MJ, Auld AM, Barritt GJ. 1996. Expression of *Drosophila* trpl cRNA in Xenopus laevis oocytes leads to the appearance of a Ca^{2+} channel activated by Ca^{2+} and calmodulin, and by guanosine 5′[gamma-thio]triphosphate. *Biochem. J.* 316(Pt. 3):793–803

88. Lan L, Brereton H, Barritt GJ. 1998. The role of calmodulin-binding sites in the regulation of the *Drosophila* TRPL cation channel expressed in *Xenopus laevis* oocytes by Ca^{2+}, inositol 1,4,5-trisphosphate and GTP-binding proteins. *Biochem. J.* 330(Pt. 3):1149–58

89. Hardie RC. 1991. Whole-cell recordings of the light induced current in dissociated *Drosophila* photoreceptors: evidence for feedback by calcium permeating the light-sensitive channels. *Proc. R. Soc. London Ser. B* 245:203–10

90. Hardie RC, Minke B. 1994. Calcium-dependent inactivation of light-sensitive channels in *Drosophila* photoreceptors. *J. Gen. Physiol.* 103:409–27

91. Reuss H, Mojet MH, Chyb S, Hardie RC. 1997. In vivo analysis of the *Drosophila* light-sensitive channels, TRP and TRPL. *Neuron* 19:1249–59

92. Leung HT, Geng C, Pak WL. 2000. Phenotypes of *trpl* mutants and interactions between the transient receptor potential (TRP) and TRP-like channels in *Drosophila*. *J. Neurosci.* 20:6797–803

93. Oberwinkler J, Stavenga DG. 2000. Calcium imaging demonstrates colocalization of calcium influx and extrusion in fly photoreceptors. *Proc. Natl. Acad. Sci. USA* 97:8578–83

94. Wang T, Xu H, Oberwinkler J, Gu Y, Hardie RC, Montell C. 2005. Light activation, adaptation, and cell survival functions of the Na^+/Ca^{2+} exchanger CalX. *Neuron* 45:367–78

95. Oberwinkler J. 2002. Calcium homeostasis in fly photoreceptor cells. *Adv. Exp. Med. Biol.* 514:539–83

96. Hardie RC. 1996. Excitation of *Drosophila* photoreceptors by BAPTA and ionomycin: evidence for capacitative Ca^{2+} entry? *Cell Calcium* 20:315–27

97. Ranganathan R, Bacskai BJ, Tsien RY, Zuker CS. 1994. Cytosolic calcium transients: spatial localization and role in *Drosophila* photoreceptor cell function. *Neuron* 13:837–48

98. Hardie RC, Peretz A, Pollock JA, Minke B. 1993. Ca^{2+} limits the development of the light response in *Drosophila* photoreceptors. *Proc. R. Soc. London Ser. B Biol. Sci.* 252:223–29

99. Cook B, Minke B. 1999. TRP and calcium stores in *Drosophila* phototransduction. *Cell Calcium* 25:161–71

100. Hardie RC, Minke B. 1992. The *trp* gene is essential for a light-activated Ca^{2+} channel in *Drosophila* photoreceptors. *Neuron* 8:643–51

101. Agam K, Frechter S, Minke B. 2004. Activation of the *Drosophila* TRP and TRPL channels requires both Ca^{2+} and protein dephosphorylation. *Cell Calcium* 35:87–105

102. Hardie RC, Raghu P, Moore S, Juusola M, Baines RA, Sweeney ST. 2001. Calcium influx via TRP channels is required to maintain PIP_2 levels in *Drosophila* photoreceptors. *Neuron* 30:149–59

103. Hardie RC. 1995. Photolysis of caged Ca^{2+} facilitates and inactivates but does not directly excite light-sensitive channels in *Drosophila* photoreceptors. *J. Neurosci.* 15:889–902

104. Obukhov AG, Schultz G, Luckhoff A. 1998. Regulation of heterologously expressed transient receptor potential-like channels by calcium ions. *Neuroscience* 85:487–95

105. Hardie RC, Raghu P. 1998. Activation of heterologously expressed *Drosophila* TRPL channels: Ca^{2+} is not required and $InsP_3$ is not sufficient. *Cell Calcium* 24:153–63

106. Trost C, Marquart A, Zimmer S, Philipp S, Cavalie A, Flockerzi V. 1999. Ca^{2+}-dependent interaction of the *trpl* cation channel and calmodulin. *FEBS Lett.* 451:257–63

107. Putney JW Jr. 1999. TRP, inositol 1,4,5-trisphosphate receptors, and capacitative calcium entry. *Proc. Natl. Acad. Sci. USA* 96:14669–71

108. Venkatachalam K, van Rossum DB, Patterson RL, Ma HT, Gill DL. 2002. The cellular and molecular basis of store-operated calcium entry. *Nat. Cell Biol.* 4:E263–72

109. Vennekens R, Voets T, Bindels RJ, Droogmans G, Nilius B. 2002. Current understanding of mammalian TRP homologues. *Cell Calcium* 31:253–64

110. Estacion M, Sinkins WG, Schilling WP. 2001. Regulation of *Drosophila* transient receptor potential-like (TrpL) channels by phospholipase C-dependent mechanisms. *J. Physiol.* 530:1–19

111. Yagodin S, Hardie RC, Lansdell SJ, Millar NS, Mason WT, Sattelle DB. 1998. Thapsigargin and receptor-mediated activation of *Drosophila* TRPL channels stably expressed in a *Drosophila* S2 cell line. *Cell Calcium* 23:219–28

112. Biddlecome GH, Berstein G, Ross EM. 1996. Regulation of phospholipase

C-β1 by Gq and m1 muscarinic cholinergic receptor. Steady-state balance of receptor-mediated activation and GTPase-activating protein-promoted deactivation. *J. Biol. Chem.* 271:7999–8007

113. Running Deer JL, Hurley JB, Yarfitz SL. 1995. G protein control of *Drosophila* photoreceptor phospholipase C. *J. Biol. Chem.* 270:12623–28

114. Warr CG, Kelly LE. 1996. Identification and characterization of two distinct calmodulin-binding sites in the Trpl ion-channel protein of *Drosophila melanogaster*. *Biochem. J.* 314:497–503

115. Porter JA, Yu M, Doberstein SK, Pollard TD, Montell C. 1993. Dependence of calmodulin localization in the retina on the NINAC unconventional myosin. *Science* 262:1038–42

116. Rosenbaum T, Gordon-Shaag A, Munari M, Gordon SE. 2004. Ca^{2+}/calmodulin modulates TRPV1 activation by capsaicin. *J. Gen. Physiol.* 123:53–62

117. Niemeyer BA, Bergs C, Wissenbach U, Flockerzi V, Trost C. 2001. Competitive regulation of CaT-like-mediated Ca^{2+} entry by protein kinase C and calmodulin. *Proc. Natl. Acad. Sci. USA* 98:3600–5

118. Zhang Z, Tang Y, Zhu MX. 2001. Increased inwardly rectifying potassium currents in HEK-293 cells expressing murine transient receptor potential 4. *Biochem. J.* 354:717–25

118a. Zhang Z, Tang J, Tikunova S, Johnson JD, Chen Z, et al. 2001. Activation of Trp3 by inositol 1,4,5-trisphosphate receptors through displacement of inhibitory calmodulin from a common binding domain. *Proc. Natl. Acad. Sci. USA* 98:3168–73

119. Barash S, Minke B. 1994. Is the receptor potential of fly photoreceptors a summation of single-photon responses? *Comments Theor. Biol.* 3:229–63

120. Hardie RC, Minke B. 1994. Spontaneous activation of light-sensitive channels in *Drosophila* photoreceptors. *J. Gen. Physiol.* 103:389–407

121. Tsacopoulos M, Magistretti PJ. 1996. Metabolic coupling between glia and neurons. *J. Neurosci.* 16:877–85

122. Agam K, von Campenhausen M, Levy S, Ben-Ami HC, Cook B, et al. 2000. Metabolic stress reversibly activates the *Drosophila* light-sensitive channels TRP and TRPL in vivo. *J. Neurosci.* 20:5748–55

123. Hardie RC, Martin F, Chyb S, Raghu P. 2003. Rescue of light responses in the *Drosophila* "null" phospholipase C mutant, norpA^{P24}, by diacylglycerol kinase mutant, rdgA, and by metabolic inhibition. *J. Biol. Chem.* 278:18851–58

124. Hardie RC. 2003. TRP channels in *Drosophila* photoreceptors: the lipid connection. *Cell Calcium* 33:385–93

125. Chyb S, Raghu P, Hardie RC. 1999. Polyunsaturated fatty acids activate the *Drosophila* light-sensitive channels TRP and TRPL. *Nature* 397:255–59

126. Raghu P, Usher K, Jonas S, Chyb S, Polyanovsky A, Hardie RC. 2000. Constitutive activity of the light-sensitive channels TRP and TRPL in the *Drosophila* diacylglycerol kinase mutant, rdgA. *Neuron* 26:169–79

127. Acharya JK, Jalink K, Hardy RW, Hartenstein V, Zuker CS. 1997. InsP$_3$ receptor is essential for growth and differentiation but not for vision in *Drosophila*. *Neuron* 18:881–87

128. Raghu P, Colley NJ, Webel R, James T, Hasan G, et al. 2000. Normal phototransduction in *Drosophila* photoreceptors lacking an InsP$_3$ receptor gene. *Mol. Cell Neurosci.* 15:429–45

129. Chorna-Ornan I, Joel-Almagor T, Cohen-Ben-Ami H, Frechter S, Gillo B, et al. 2001. A common mechanism underlies vertebrate calcium signaling and *Drosophila* phototransduction. *J. Neurosci.* 21:2622–29

130. Brown JE, Rubin LJ, Ghalayini AJ, Tarver AP, Irvine RF, et al. 1984.

myo-Inositol polyphosphate may be a messenger for visual excitation in *Limulus* photoreceptors. *Nature* 311:160–63

131. del Pilar Gomez M, Nasi E. 1998. Membrane current induced by protein kinase C activators in rhabdomeric photoreceptors: implications for visual excitation. *Proc. Natl. Acad. Sci. USA* 93:11196–201

132. Fein A, Payne R, Corson DW, Berridge MJ, Irvine RF. 1984. Photoreceptor excitation and adaptation by inositol 1,4,5-trisphosphate. *Nature* 311:157–60

133. Walz B, Liebherr H, Ukhanov K. 2003. Ca^{2+}-dependent and Ca^{2+} release-dependent excitation in leech photoreceptors: evidence from a novel "inside-out" cell model. *Cell Calcium* 34:35–47

134. Kiselyov K, Xu X, Mozhayeva G, Kuo T, Pessah I, et al. 1998. Functional interaction between InsP$_3$ receptors and store-operated Htrp3 channels. *Nature* 396:478–82

135. Gomez MD, Nasi E. 2005. A direct signaling role for phosphatidylinositol 4,5-bisphosphate (PIP$_2$) in the visual excitation process of microvillar receptors. *J. Biol. Chem.* 280:16784–89

136. Chuang HH, Prescott ED, Kong H, Shields S, Jordt SE, et al. 2001. Bradykinin and nerve growth factor release the capsaicin receptor from PtdIns(4,5)P2-mediated inhibition. *Nature* 411:957–62

137. Runnels LW, Yue L, Clapham DE. 2002. The TRPM7 channel is inactivated by PIP$_2$ hydrolysis. *Nat. Cell Biol.* 4:329–36

138. Huang FD, Matthies HJ, Speese SD, Smith MA, Broadie K. 2004. Rolling blackout, a newly identified PIP$_2$-DAG pathway lipase required for *Drosophila* phototransduction. *Nat. Neurosci.* 7:1070–78

139. Hisatsune C, Nakamura K, Kuroda Y, Nakamura T, Mikoshiba K. 2005. Amplification of Ca^{2+} signaling by diacylglycerol-mediated inositol 1,4,5-trisphosphate production. *J. Biol. Chem.* 280:11723–30

140. Mirnikjoo B, Brown SE, Kim HF, Marangell LB, Sweatt JD, Weeber EJ. 2001. Protein kinase inhibition by ω-3 fatty acids. *J. Biol. Chem.* 276:10888–96

141. Arslan P, Corps AN, Hesketh TR, Metcalfe JC, Pozzan T. 1984. cis-Unsaturated fatty acids uncouple mitochondria and stimulate glycolysis in intact lymphocytes. *Biochem. J.* 217:419–25

142. Bezzerides VJ, Ramsey IS, Kotecha S, Greka A, Clapham DE. 2004. Rapid vesicular translocation and insertion of TRP channels. *Nat. Cell Biol.* 6:709–20

143. Peretz A, Suss-Toby E, Rom-Glas A, Arnon A, Payne R, Minke B. 1994. The light response of *Drosophila* photoreceptors is accompanied by an increase in cellular calcium: effects of specific mutations. *Neuron* 12:1257–67

144. Hochstrate P. 1989. Lanthanum mimicks the trp photoreceptor mutant of *Drosophila* in the blowfly *Calliphora*. *J. Comp. Physiol. A* 166:179–87

145. Suss-Toby E, Selinger Z, Minke B. 1991. Lanthanum reduces the excitation efficiency in fly photoreceptors. *J. Gen. Physiol.* 98:849–68

146. Singh BB, Lockwich TP, Bandyopadhyay BC, Liu X, Bollimuntha S, et al. 2004. VAMP2-dependent exocytosis regulates plasma membrane insertion of TRPC3 channels and contributes to agonist-stimulated Ca^{2+} influx. *Mol. Cell* 15:635–46

147. Cayouette S, Lussier MP, Mathieu EL, Bousquet SM, Boulay G. 2004. Exocytotic insertion of TRPC6 channel into the plasma membrane upon Gq protein-coupled receptor activation. *J. Biol. Chem.* 279:7241–46

148. Kanzaki M, Zhang YQ, Mashima H, Li L, Shibata H, Kojima I. 1999. Translocation of a calcium-permeable cation channel induced by insulin-like growth factor-I. *Nat. Cell Biol.* 1:165–70

Annu. Rev. Physiol. 2006. 68:685–717
doi: 10.1146/annurev.physiol.68.040204.101406
First published online as a Review in Advance on September 1, 2005

PERMEATION AND SELECTIVITY OF TRP CHANNELS

Grzegorz Owsianik, Karel Talavera, Thomas Voets, and Bernd Nilius

Laboratorium voor Fysiologie, Katholieke Universiteit Leuven, B-3000 Leuven, Belgium; email: grzegorz.owsianik@med.kuleuven.be, karel.talavera@med.kuleuven.be, thomas.voets@med.kuleuven.be, bernd.nilius@med.kuleuven.be

Key Words   transient receptor potential, cation channels, pores, selectivity filter, calcium permeation

■ **Abstract**   Ion channels are pore-forming transmembrane proteins that allow ions to permeate biological membranes. Pore structure plays a crucial role in determining the ion permeation and selectivity properties of particular channels. In the past few decades, efforts have been undertaken to identify key elements of the pore regions of different classes of ion channels. In this review, we summarize current knowledge about permeation and selectivity of channel proteins from the transient receptor potential (TRP) superfamily. Whereas all TRP channels are permeable for cations, only two TRP channels are impermeable for Ca^{2+} (TRPM4, TRPM5), and two others are highly Ca^{2+} permeable (TRPV5, TRPV6). Despite the great advances in the TRP channel field during the past decade, only a limited number of reports have dealt with functional characterization of pore properties, biophysical aspects of cation permeation, or description of pore structures of TRP channels. This review gives an overview of available experimental and theoretical data and discusses the functional impact of pore-structure modifications on TRP channel properties.

INTRODUCTION

Transient receptor potential (TRP) channels constitute a large and functionally versatile superfamily of cation channel proteins that are expressed in many cell types from yeast to mammals (for recent reviews, see References 1, 2). Contrasting with more classical ion-channel types, most TRP proteins were discovered after their encoding genes were cloned and subsequently characterized after their overexpression in heterologous cellular systems. This has posed two essential problems: establishing their function as genuine ion channels and determining their physiological roles.

The physiological implications of any given ion channel rely on two fundamental features: the regulation of its opening and closing (gating) and its ability to form an efficient pathway for certain ion species to cross cell membranes

0066-4278/06/0315-0685$20.00

(selectivity). Therefore, priorities in characterizing any channel protein should be to determine its permeation and selectivity properties and to provide a link between these properties and the pore structure. Besides satisfying the intellectual ambition of understanding how these remarkable proteins function, these studies have two major practical applications in the TRP field. First, experimental manipulations of the structure of a protein leading to the alteration of conduction properties provide the ultimate proof that this protein is an actual ion channel. It is well known that overexpression of exogenous or endogenous proteins in certain cellular systems can lead to the activation and/or upregulation of endogenous channels. Thus, successful measurements of channel activity in cells expressing a putative channel protein, even when this activity is not present in the control cells, does not necessarily imply that the expressed protein is the pore-forming channel subunit (3). Therefore, mutations of residues of a candidate channel protein that alter ion selectivity, single-channel conductance, or pore block provide conclusive evidence that the tested protein is indeed a pore-forming subunit of an ion channel. Second, conduction properties of ion channels are usually less sensitive to the expression system of choice than are the gating, modulation, and pharmacological properties. Therefore, conduction features provide reliable hallmarks that can be used to relate newly identified channel proteins with their native counterparts.

SHORT OVERVIEW OF THE TRP SUPERFAMILY

The TRP superfamily contains a growing number of proteins in vertebrates and invertebrates unified by their homology to the product of the *Drosophila trp* gene, which is involved in light perception in the fly eye (4). On the basis of structural homology, the superfamily can be subdivided into seven main subfamilies: TRPC (Canonical), TRPV (Vanilloid), TRPM (Melastatin), TRPP (Polycystin), TRPML (Mucolipin), TRPA (Ankyrin) and TRPN (no mechanoreceptor potential C, NOMPC) (2, 5, 6) (Table 1). All TRP channels comprise six putative transmembrane spans (6 TM) and cytosolic amino (N) and carboxy (C) termini. The length of these cytosolic tails varies greatly between TRP channel subfamilies, as do their different structural and functional domains (Figure 1) (for detailed reviews see References 1, 2). Like other 6 TM pore-forming proteins, functional TRP channels are either homo- or heteromultimers of four TRP subunits (7, 8). It is generally believed that the cation-permeable pore region is formed by a short hydrophobic stretch between TM5 and TM6.

In comparison with extensively studied families of ion channel proteins, including voltage-gated K^+, Na^+, and Ca^{2+} channels, chloride channels (cystic fibrosis transmembrane conductance regulator, CFTR, and chloride channels, CLC, of the CLC family), Na^+-selective channels of the epithelial sodium channel (ENaC)/acid-sensing ion channel (ASIC)/degenerin superfamily, and cyclic-nucleotide-gated (CNG) channels (9), insight into the pore structure and permeation properties of most TRP superfamily members is still extremely limited. Despite numerous functional studies on members of the TRPC, TRPV, and TRPM subfamilies (summarized in References 2, 5, 6, 10, 11), which have established the

TABLE 1   Main properties of TRP superfamily channels in mammals

Subfamily	Channel	ENSMBL[c]	Selectivity P_{Ca}/P_{Na}	Conductance (pS)	References
TRPC[a]	TRPC1	ENSG00000144935	Nonselective	16	(116)
	TRPC2	ENSMUSG00000058020	2.7	42	(73)
	TRPC3	ENSG00000138741	1.6	60–66	(117–120)
	TRPC4	ENSG00000100991	1.1	30–42	(80, 82, 121)
	TRPC5	ENSG00000072315	9	47–66	(81, 84, 85, 121, 122)
	TRPC6	ENSG00000137672	5	28–37	(120, 123)
	TRPC7	ENSG00000069018	0.5–5.4	25–50	(123, 124)
TRPV[b]	TRPV1	ENSG00000043316	~10	35–80	(125)
	TRPV2	ENSG00000154039	1–3	ND[d]	(29)
	TRPV3	ENSG00000167723	12	172	(31)
	TRPV4	ENSG00000111199	6	90	(126)
	TRPV5	ENSG00000127412	>100	75	(20)
	TRPV6	ENSG00000165125	>100	40–70	(127)
TRPM	TRPM1	ENSG00000134160	ND	ND	
	TRPM2	ENSG00000142185	0.5–1.6	52–80	(103)
	TRPM3	ENSG00000083067	0.1–10 Depends on splice variant	65(Ca^{2+})–130	(107, 110)
	TRPM4	ENSG00000130529	Nonpermeable	25	(100)
	TRPM5	ENSG00000070985	Nonpermeable	16–25	(89)
	TRPM6	ENSG00000119121	P_{Mg}/P_{Na} ~6	ND	(58)
	TRPM7	ENSG00000092439	3	40–105	(128)
	TRPM8	ENSG000000144481	1–3	83	(99)
TRPP	TRPP2	ENSG00000118762	1–5	40–177	(129–131)
	TRPP3	ENSG00000107593	4	137	(132)
	TRPP5	ENSG00000078795	1–5	300	(133)
TRPML	TRPML1	ENSG00000090674	~1	46–83	(134)
	TRPML2	ENSG00000153898	ND	ND	
	TRPML3	ENSG00000055732	ND	ND	
TRPA	TRPA1	ENSG00000104321	0.8–1.4	ND	(97)

[a]TRPC2 is a pseudogene in humans; therefore data refer to the mouse homologue.

[b]For all TRPV conductance from divalent-free solution.

[c]ENSMBL numbers are for human channels, except for mouse TRPC2; http://www.ensembl.org/.

[d]ND, not determined.

TRP channels as a family of genuine cation channels, the analysis of permeation and pore properties is far from complete. Little is known about the pore properties of the newly identified TRP subfamilies, TRPA, TRPML, TRPP, and TRPN. In this review, we recapitulate recent functional and theoretical data concerning the structure and permeation properties of TRP channel pores.

A TOOLKIT FOR CHARACTERIZING ION-CHANNEL PERMEATION AND SELECTIVITY

The conduction properties of ion channels have been studied with the aid of a handful of theories that assume particular models of ion-ion and ion-channel interactions (9). In this section, rather than provide a full description of these theories,

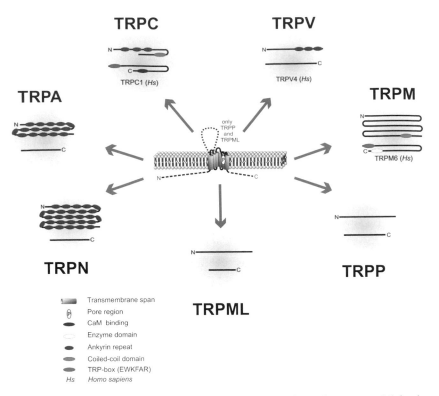

Figure 1   Schematic representation of the general TRP channel structure. Main domain characteristics for all subfamilies are annotated. CaM, calmodulin.

we give a succinct guide for their practical applications in the TRP field and highlight the fundamental ideas behind the mechanisms of selectivity of and permeation through ion channels. In general, narrow pores are able to discriminate between different cations on the basis of their passive sieve properties (selectivity by rejection) and their cation affinity (selectivity by binding) (Figure 2A).

Calculation of Permeability: The Goldman-Hodgkin-Katz Theory

The Goldman-Hodgkin-Katz (GHK) theory is the simplest and most widely used of all available theories. It lumps all ion-pore interactions into a single parameter, the ionic permeability, and assumes that ions do not interact with each other. The current carried by an ion species C is given by

$$I_C = P_C z_C^2 \frac{VF^2}{RT} \frac{[C]_i - [C]_e \exp(-z_C FV/RT)}{1 - \exp(-z_C FV/RT)},$$ 1.

where P_C is the permeability of the cation C, subscripts i and e denote intra- and extracellular concentrations, respectively. z_C is the valence, F is the Faraday constant, R is the gas constant, V is the membrane potential, and T is absolute temperature. At the reversal potential, the sum of the currents carried by all permeant ions equals zero. Using this definition, the relative permeability of monovalent cations can be calculated from the shift in reversal potential after complete substitution of extracellular Na$^+$ by the specific cation X$^+$:

$$P_X/P_{Na} = \exp(\Delta V_{rev} F/RT), \qquad\qquad 2.$$

where ΔV_{rev} is the measured shift in reversal potential and F, R, and T are the same as in Equation 1. Alternatively, mixtures of monovalent cations can be used to determine relative permeability:

$$\frac{P_X}{P_{Na}} = \frac{[Na^+]_e - [Na^+]_i \exp(V_{rev}F/RT)}{[X^+]_i \exp(V_{rev}F/RT) - [X^+]_e}. \qquad\qquad 3.$$

Before calculating the relative permeability, all reversal potentials should be corrected for liquid junction potentials (12):

$$V_{rev,corrected} = V_{rev,measured} - V_{LJ}, \qquad\qquad 4.$$

where V_{LJ}, the liquid junction potential, which can be calculated according to Reference 13, and $V_{rev,measured}$ (V_{rev}) are the measured reversal potentials (see also Reference 13a).

The analysis of TRP channel properties often involves the estimation of the permeability for divalent cations such as Ca^{2+} or Mg^{2+}. If Ca^{2+} is the only permeant cation in the bath solution and the internal solution comprises only a monovalent cation X$^+$, the relative permeability of Ca^{2+} can be calculated by

$$\frac{P_{Ca}}{P_X} = \frac{[X^+]_i}{4[Ca^{2+}]_e} \exp(V_{rev}F/RT)(\exp(V_{rev}F/RT) + 1). \qquad\qquad 5.$$

Such simple experimental conditions are mostly not applicable, as more than one permeant monovalent cation is present in the solutions, e.g., Cs$^+$ is often used to reduce background permeability through K$^+$ channels. In these cases, P_{Cs}/P_{Na} must be determined in the absence of permeant divalent cations using Equations 2 or 3. The permeability of the Ca^{2+} (or any other divalent cation) relative to Na$^+$ can then be calculated from the absolute reversal potential according to

$$\frac{P_{Ca}}{P_{Na}} = (1 + \exp(V_{rev}F/RT))$$

$$\times \frac{([Na^+]_i + \alpha[Cs^+]_i) \exp(V_{rev}F/RT) - [Na^+]_e - \alpha[Cs^+]_e}{4[Ca^{2+}]_e}, \qquad\qquad 6.$$

where α is P_{Cs}/P_{Na} (14–18).

Alternatively, the shift in the reversal potential (rather than the absolute reversal potential) upon changing the extracellular Ca^{2+} concentration in a solution that is otherwise devoid of permeant monovalent cations can be used to calculate P_{Ca}/P_{Na}. This is especially useful when currents become unstable in the absence of divalent cations. In this case, we measure the shift in reversal potential (ΔV_{rev}) when changing the extracellular Ca^{2+} concentration from $[Ca^{2+}]_{e,1}$ to $[Ca^{2+}]_{e,2}$ and calculate the relative permeability according to

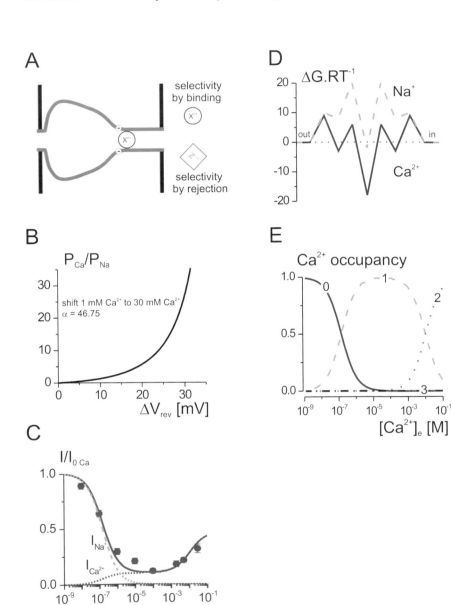

$$\frac{P_{Ca}}{P_{Na}} = \frac{\alpha\beta(\beta - 1)\left([Ca^{2+}]_{e,2} - [Ca^{2+}]_{e,1} - \beta[Ca^{2+}]_{e,1}\right)}{\left([Ca^{2+}]_{e,2} - \beta^2[Ca^{2+}]_{e,1}\right)^2}, \qquad 7.$$

where

$$\alpha = \frac{[Na^+]_i + [Cs^+]_i P_{Cs}/P_{Na}}{4}.$$

$$\beta = \exp(F\Delta V_{rev}/RT)$$

This method requires the measurement of P_{Cs}/P_{Na} unless Na^+ is used as the only intracellular cation. Figure 2*B* illustrates the relation between ΔV_{rev} and P_{Ca}/P_{Na} when $[Ca^{2+}]_e$ is increased from 1 to 30 mM.

Electrostatics of Ion Selectivity: Eisenman's Theory

From his works on ion-selective glasses, Eisenman showed that simple electro-statics provide a useful concept when analyzing ion-selectivity properties (19). He considered the transition from a hydrated cation in solution to a dehydrated cation interacting with a fixed negative site. The free energy for this transition is approximated by

$$\Delta G = \frac{z_{site}z_C q_e^2}{4\pi\varepsilon\varepsilon_0\left(r_{site} + r_{cation}\right)} - \frac{z_C^2 q_e^2}{8\pi\varepsilon_0 r_{cation}}\left(\frac{1}{\varepsilon} - \frac{1}{\varepsilon_0}\right), \qquad 8.$$

where q_e is the elementary charge, ε is the dielectric permittivity (polarizability) of water, ε_0 is the dielectric permittivity of the vacuum, $z_{site}q_e$ and $z_C q_e$ are the

---

Figure 2 Modeling of permeation and selectivity of the TRPV5 channel. (*A*) Schematic representation of the narrow pore (*gray*), which can discriminate between different cations by their binding properties (selectivity by binding) or their passive sieve properties (selectivity by rejection). (*B*) The relation between ΔV_{rev} and P_{Ca}/P_{Na} when $[Ca^{2+}]_e$ is increased from 1 to 30 mM (see text for details). (*C*) Mean normalized current values measured at -80 mV during linear voltage ramps in various extracellular Ca^{2+} concentrations ($[Ca^{2+}]_e$). Currents were normalized to the current value for the same cell in divalent-free solution containing 5 mM ethylene-glycol-bis(2-aminoethylether)-N,N,N',N'-tetraacetic acid (EGTA). The continuous blue line represents the current densities as predicted by a model with one high-affinity binding site flanked by a low-affinity binding site at each side, using the energy profiles for Ca^{2+} and Na^+ depicted in panel *D*. The green-dashed and red lines represent the fractions of the current carried by Ca^{2+} and Na^+, respectively. (*D*) Energy profiles of the TRPV5 pore along the path of the pore for Ca^{2+} and Na^+. (*E*) The predicted occupation of the TRPV5 pore by Ca^{2+} as a function of the Ca^{2+} concentration, i.e., the chance to find 1, 2, 3, or no Ca^{2+} ions bound within the pore. (Panels *C–E* are adapted from Reference 22 with copyright permission from *The Physiological Society*.)

TABLE 2   The Eisenman sequences for equilibrium ion exchange (19)

Weak-field-strength site	
I	$Cs^+ > Rb^+ > K^+ > Na^+ > Li^+$
II	$Rb^+ > Cs^+ > K^+ > Na^+ > Li^+$
III	$Rb^+ > K^+ > Cs^+ > Na^+ > Li^+$
IV	$K^+ > Rb^+ > Cs^+ > Na^+ > Li^+$
V	$K^+ > Rb^+ > Na^+ > Cs^+ > Li^+$
VI	$K^+ > Na^+ > Rb^+ > Cs^+ > Li^+$
VII	$Na^+ > K^+ > Rb^+ > Cs^+ > Li^+$
VIII	$Na^+ > K^+ > Rb^+ > Li^+ > Cs^+$
IX	$Na^+ > K^+ > Li^+ > Rb^+ > Cs^+$
X	$Na^+ > Li^+ > K^+ > Rb^+ > Cs^+$
XI	$Li^+ > Na^+ > K^+ > Rb^+ > Cs^+$
Strong-field-strength site	

effective charges of the binding site and the cation, respectively, and r_{site} and r_{cation} are the radii of the site and the cation, respectively. The first term of this equation represents the electrostatic energy of interaction (U) between a cation and the negative site (Coulomb's law), whereas the second term gives the dehydration energy of a cation (Born energy). Both terms are inversely proportional to the ionic radius. Analogously, permeation through an ion channel involves the transition of ions from solution to binding sites in a pore structure, and selective permeability of a specific cation depends on how energetically favorable its interaction with the pore is relative to other ion species. Using Equation 8, Eisenman was able to explain why only 11 cation selectivity sequences are common in biology and chemistry (Table 2). If r_{site} is large, U is small and ΔG will be dominated by the dehydration energy. In this case, easily dehydrated cations such as Cs^+ will be favored for permeation, as opposed to Li^+, which has a much higher dehydration energy. Such a site is called the weak-field-strength site and reflects Eisenman selectivity sequence I: $Cs^+ > Rb^+ > K^+ > Na^+ > Li^+$. Conversely, if r_{site} is very small, U is large compared with the dehydration energy. In this case, the smaller cations will be favored because they can come closer to the binding site. This reflects Eisenman sequence XI, or the strong-field-strength site, e.g., $Li^+ > Na^+ > K^+ > Rb^+ > Cs^+$. This approach has been successfully applied to classify pore properties of TRP channels (3, 14, 15, 20–23) and represents an important tool for comparing selectivity properties of different channels and for relating newly identified proteins to native counterparts.

A Hydrodynamic Approach: Estimations of Pore Diameter

It is sometimes instructive to consider pores as molecular sieves by neglecting the energy changes accompanying permeation. In such an approach, the diameter of the selectivity filter will cut off permeation of large particles. Thus, sizing permeable cations and measuring their permeability can be used to estimate the pore diameter (see Reference 23 for an example of this application on TRPV6). The relative permeability of a sized permeating cation relative to Na^+ can be plotted against the Stoke's diameter of the cation using the excluded volume models for the permeating pore, assuming that frictional effects of the ion in the pore are negligible:

$$\frac{P_X}{P_{Na}} = k \left(1 - \frac{a}{d} \right)^2 ,$$ 9.

where a is the diameter of the permeating cation, k is a constant factor, and d is the pore diameter (24). Fitting this equation to the permeation data yields an estimation of the pore diameter [for Stoke's cation diameters, see the classics (9, 25)]. If one takes the friction of the permeating cation with the pore wall into account, the following equation can be employed:

$$\frac{P_X}{P_{Na}} = \frac{k}{a} \left(1 - \frac{a}{d} \right)^2 .$$ 10.

Flux is Equal to Hopping over Barriers and Selectivity is Equal to Dwelling in a Deep Well

Eyring rate theory can be used to model ion movement through a channel as a series of hops between wells in a one-dimensional profile of Gibbs' free energy (9, 26). These wells are separated by barriers, whose heights determine the rate of ion translocation. The energy necessary to overcome these barriers (i.e., for ions to move in a certain direction) is provided by the thermal energy and the externally applied electric field (membrane potential). Eyring's model provided the first theory that permitted modeling of the basic features of ion permeation through biological channel pores, including single-file movement through a narrow pore, multi-ionic occupation, and ion-ion interactions (9). Within this theory, selective permeability for a certain ion species occurs when its energy profile across the pore exhibits deeper wells than that of other ion species. An application of Eyring models to TRPV channels is illustrated below.

PERMEATION AND SELECTIVITY PROPERTIES OF TRP CHANNELS

TRPV Channels

Within the TRP superfamily, permeation and selectivity properties are described in greatest detail for members of the TRPV subfamily. In vertebrates, six TRPV channels have been identified (Table 1). TRPV1 mediates nociception and contributes to the detection and integration of diverse chemical and thermal stimuli (27, 28). TRPV2 and TRPV3 open upon heating, activating in the warm and noxious heat range, respectively (29–32). TRPV4 plays a role in osmosensing, nociception, and warm sensing (for detailed reviews, see 33–35). TRPV5 and TRPV6 are highly Ca^{2+}-selective channels that play a role in Ca^{2+} reabsorption in the kidney and intestine (8, 14, 21, 22, 36–40).

TRPV1/2/3/4 are Ca^{2+}-permeable channels with a rather low discrimination between divalent and monovalent cations (P_{Ca}/P_{Na} between 1 and 10) (see Table 1) (3, 16, 41, 42). The relative permeability for monovalent cations for TRPV4 corresponds to Eisenman sequence IV, indicating a relatively weak-field-strength binding site. Notably, TRPV1 is also highly permeable to protons, which may underlie the intracellular acidification of nociceptive neurons after channel activation (43). In contrast, TRPV5 and TRPV6 are highly Ca^{2+}-selective channels, with P_{Ca}/P_{Na} values exceeding 100. Such high Ca^{2+} selectivity is unique within the TRP superfamily. TRPV5 and TRPV6 display a selectivity for monovalent cations corresponding to Eisenman sequence X or XI ($Na^+ \sim Li^+ > K^+ > Cs^+$). For divalent cations, a $Ca^{2+} > Ba^{2+} > Sr^{2+} > Mn^{2+}$ selectivity sequence has been reported (14, 15, 20–22, 44).

Eyring's Model of Permeation Applied to TRPVs

TRPV5 and TRPV6 share many permeation properties with the classical voltage-gated Ca^{2+} channels (9, 45–48). A characteristic feature of highly Ca^{2+}-selective channels is the anomalous mole fraction effect. At submicromolar Ca^{2+} concentrations, these channels are highly permeable to monovalent cations (Figure 2C). With increasing Ca^{2+} concentrations, the channels first become blocked and then start to conduct Ca^{2+}. The high ion transfer rate of the Ca^{2+} channel pore in the case of L-type voltage-gated Ca^{2+} channels has been explained by two models: the repulsion model, proposed by Almers & McCleskey (49) and Hess and colleagues (50, 51), and the step model, introduced by Dang & McCleskey (52). We demonstrated that such models can also be applied to the TRPV5/6 pore (22).

In the step model, we envision a channel pore in which two low-affinity binding sites flank a central high-affinity binding site (Figure 2D). In the absence of extracellular Ca^{2+}, the channel pore is available for monovalent permeation through the channel. However, in the presence of Ca^{2+} the binding sites within the pore will preferentially bind Ca^{2+} as a result of their higher binding affinity for Ca^{2+} compared with Na^+. Block of monovalent currents in nanomolar $[Ca^{2+}]_e$ occurs through the binding of a single Ca^{2+} ion in the pore, which functions as a plug occluding the channel. The Ca^{2+} flux at higher $[Ca^{2+}]_e$ with multiple occupancy of the channel pore, although the probability of finding the TRPV5 pore in the triple Ca^{2+}-occupied state is very low (Figure 2E). The drive for ion permeation results from the steps in binding affinity provided by the low-affinity

TABLE 3   Comparison of the step model for ion permeation for TRPV5 and L-type Ca^{2+} channels. Adapted from Reference 22 with copyright permission from *The Physiological Society*

	Outer barriers		Outer well		Inner barriers		Central well	
	Na^+	Ca^{2+}	Na^+	Ca^{2+}	Na^+	Ca^{2+}	Na^+	Ca^{2+}
TRPV5	10	9	8	−3	20	6	−2	−18
L-type Ca^{2+} channel	10	10	−2	−4	9.2	−2	−2	−15.5

sites, as if the flanking sites provide stair steps for the ion to exit the channel pore (52).

The clearest differences between the parameters for L-type Ca^{2+} channels and TRPV5 are the height of the inner barriers for Na^+ and the slightly higher Ca^{2+} affinity of the central well in the case of TRPV5 (Table 3). These differences account for the significantly higher current densities measured in micromolar Ca^{2+} concentrations in the case of TRPV5. This analysis underscores the similarities in permeation properties of TRPV5 and L-type voltage-gated Ca^{2+} channel properties, which can be described by mechanisms that reconcile channel specificity and high Ca^{2+} fluxes in a multiple-occupied single-file pore through steps in binding energy.

Models based on Eyring's rate theory do not imply an exact knowledge of underlying molecular structures. While allowing researchers to deduce general principles of ion permeation, this approach makes it difficult to correlate the experimental observations with the actual pore structure involved in the permeation process. To address this point, direct examination of the pore by mutagenesis and structural analysis is required.

Approaching the Pore of TRPVs

Although TRPVs were not the first mammalian TRP channels to be characterized, the structure-function analysis of their pores is the most advanced among all TRP subfamilies. As shown in Figure 3, the pore regions of the six mammalian TRPVs show significant sequence homology with the selectivity filter of the bacterial potassium ion channel (KcsA) [signature sequence TXXTXGYGD (53, 54)]. Neutralization of D^{546} of TRPV1 and corresponding D^{682} of TRPV4 strongly decreases the affinity of the channels to the voltage-dependent pore blocker ruthenium red and reduces the permeability for Ca^{2+} and Mg^{2+} (16, 55). An additional mutation at residue D^{672} of TRPV4 further reduces the selectivity for divalent cations and changes the monovalent permeability sequence from Eisenman IV to I (Figure 4) (16). Substitution of a negatively charged amino acid for M^{680} abolished Ca^{2+} and Mg^{2+} permeability in TRPV4 (16). Thus, these experiments clearly demonstrate

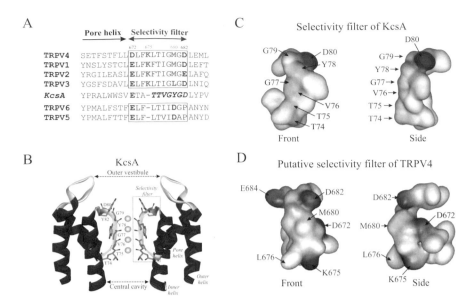

Figure 3   Structural homology between putative pore regions of human TRPV channels and the pore region of KcsA. (*A*) Multiple alignments of TRPVs and KcsA channels. Residues of the KcsA selectivity filter are noted in bold italic. Negatively charged residues next to or in the selectivity filter are indicated in red. The conserved lysine and the central methionine (leucine in TRPV3) residues in TRPV1/2/3/4 are in blue and green (violet for TRPV3 leucine), respectively. (*B*) Schematic ribbon model of the KcsA pore region (53). Only two KcsA subunits are visualized. Important residues of the selectivity filter (TTVGYG) are shown as ball-and-stick models. Large blue balls represent K$^+$ or water. (*C–D*) Distribution of electrostatic potentials on the surface of selectivity filters of (*C*) KcsA and (*D*) TRPV4.[1] The negative and positive charges are in red and blue, respectively. Localization of the main residues is annotated.

that the GM(L)GD sequence motif in the putative TRPV1/2/3/4 pore region determines permeation properties of the channels and is a functional counterpart of the GYGD sequence motif in the selectivity filter of K$^+$ channels. Surprisingly, mutation of a single basic residue in TRPV4, Lys^{675}, which is also present in TRPV1/2/3 selectivity filters but not in KcsA or TRPV5/6 (Figure 3), had no effect on the permeation properties of the channel (16).

---

[1]The modeling of TRP pore structures was done using the Swiss Model server (SIB Biozentrum Basel, http://swissmodel.expasy.org) (113–115). To model the TRPV4 (Q8NDY7), TRPV6 (Q9H1D0), TRPC1 (P48995), TRPC5 (Q9UL62), and TRPM4 (Q8TD43) pores (accession numbers enclosed in parentheses), we used the KscA pore as a template (53). All models were viewed using DS ViewerPro 5.0 software (Accelrys Inc., San Diego, California).

Single-Pore Aspartates Determine Permeation Properties of TRPV5/6 Channels

The first insights into the pore region of TRPV5 came from the work of Nilius and colleagues (15). They showed that an aspartate-to-alanine mutation at position D^{542} of TRPV5, which corresponds to D^{541} of TRPV6 (Figure 3A), abolishes Ca^{2+} permeation, Ca^{2+}-dependent current decay, and block by extracellular Mg^{2+}, whereas permeation of monovalent cations basically remains intact (Figure 5). Introduction of other uncharged or negatively charged amino acids at position D^{542} also abolished Ca^{2+} selectivity, whereas introduction of a basic amino acid (D^{542K}) resulted in a nonfunctional channel. These functional data established that D^{542} is a crucial determinant of the conductive properties of TRPV5 and a key element of the selectivity filter. Other negatively charged residues in the pore region of TRPV5 have less impact on pore properties. Mutations of E^{535} and D^{550} had only minor effects on Ca^{2+} permeation properties (15). It is interesting to note that E^{522}, located at the N terminus of the pore helix, functions as a putative pH sensor, regulating pH-dependent permeation properties of TRPV5/6 (21, 56). The glutamate-to-glutamine mutation, E522Q, decreases extracellular proton-dependent processes such as reduction of channel block by extracellular Mg^{2+} under more alkaline conditions and increases the efficiency of channel block by Ca^{2+} at low pH (56).

A characteristic feature of TRPV5/6 channels is the open pore block by intracellular Mg^{2+} (57, 58). In the absence of extracellular divalent cations, the monovalent currents display characteristic voltage-dependent gating and almost absolute inward rectification (Figure 6). These two features are contingent on the voltage-dependent block/unblock of the channel by intracellular Mg^{2+}. Mg^{2+} blocks the channel by binding to a single aspartate residue D^{541} within the TRPV6 pore (corresponding to D^{542} of TRPV5) (see Figure 3A), where it interacts with permeant cations. The block is relieved at positive potentials, indicating that under these conditions Mg^{2+} is able to permeate the selectivity filter of the channel. Neutralization of D^{541} abolishes the Mg^{2+} sensitivity of the channel, yielding voltage-independent, moderately inwardly rectifying monovalent currents in the presence of intracellular Mg^{2+}. These findings are crucial to understanding the pore properties of TRPV5/6.

Taken together, these data indicate that selectivity and permeation properties in TRPV5 and TRPV6 are mainly determined by a ring of four aspartate residues in the channel pore, similar to the ring of four negative residues (aspartates and/or glutamates) in the pore of voltage-gated Ca^{2+} channels (59–61). Sequence similarity is weak between the pore regions of voltage-gated Ca^{2+} channels and TRPV5/6 (Figure 7), suggesting that the highly similar functional properties are due mainly to the occurrence of a ring of negative charges in the selectivity filter. Nevertheless, as discussed below, the pore architecture of TRPV5/6 channels has the same general features as those suggested in a recent structural model of the voltage-dependent L-type Ca^{2+} channel (62).

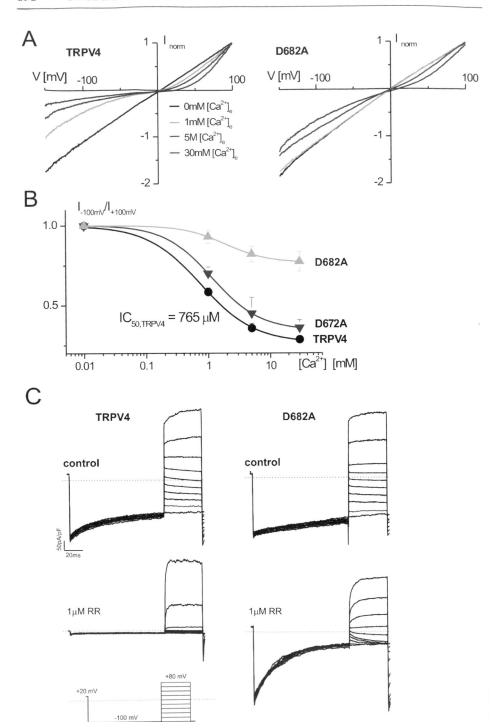

Pore Architecture of TRPV Channels

TRPV5 and TRPV6 can generate a pleiotropic set of functional homo- and heterotetrameric channels with different Ca^{2+} transport kinetics (8). More details of the pore structure were obtained using the substituted cysteine accessibility method (23, 63). Cysteines introduced in a region preceding $D^{542/541}$ displayed a cyclic pattern of reactivity, indicating that these residues form a pore helix, similar to that in the KcsA crystal structure. The pore helix is followed by the selectivity filter, which has a diameter of approximately 5.4 Å at its narrowest point as determined by measurements of permeability to cations of increasing size (23). This value is between those reported for the L-type (6.2 Å; see Reference 64) and T-type (5.1 Å; see Reference 65) Ca^{2+} channels. TRPV1 seems to have a much larger pore diameter (>6.8 Å), as it allows the flux of large monovalent cations such as N-methyl-d-glucamine ($NMDG^+$) and tetraethylammonium (TEA^+) (43). The TRPV6 pore diameter was significantly increased when D^{541} was replaced by amino acids with a shorter side chain, indicating that this aspartate lines the narrowest part of the selectivity filter and contributes to the sieving properties of the pore. On the basis of these experimental data and homology to KcsA, the first TRP channel pore structures have been presented (23, 63). As shown in Figure 8, the external vestibule in TRPV5/6 encompasses two main structural domains: a pore helix of approximately 15 amino acids followed by a nonhelical loop. In contrast to KcsA, the selectivity filter of TRPV6 appears to be lined by amino acid side chains rather than by backbone carbonyls.

Canonical Problem of TRPC Pores

Members of the TRPC subfamily share the highest homology with the first characterized TRP channels from *Drosophila* and mainly function as store- and/or PLC-dependent Ca^{2+}-permeable cation channels (2, 66–68). There are seven TRPC subfamily members in vertebrates (Table 1). Functional TRPC channels are either homo- or heterotetramers with other TRPCs (for a recent review, see Reference 69).

Figure 4 D^{682} in the putative selectivity filter of TRPV4 determines pore properties of the channel. (*A*) Ca^{2+}-dependent rectification of wild-type (*left*) and D^{682A}-mutant (*right*) TRPV4. Current-voltage relations in standard extracellular solution containing the indicated Ca^{2+} concentrations. Note the appearance of inward and outward rectification in the presence of Ca^{2+}. (*B*) Ca^{2+} dependence of rectification properties for wild-type and mutant TRPV4. The average ratios of the currents measured at -100 and $+100$ mV are shown. (*C*) Neutralization of D^{682} reduces TRPV4 block by ruthenium red (RR). Currents through wild-type TRPV4 (*left*) and the D^{682A} mutant (*right*) in the absence (control) and presence of 1-μM RR in the extracellular solution in response to the voltage protocol shown below. The extracellular solution contained 150-mM Na^+ and 5-mM Ca^{2+} (adapted from Reference 16 with copyright permission from *The American Society for Biochemistry and Molecular Biology*).

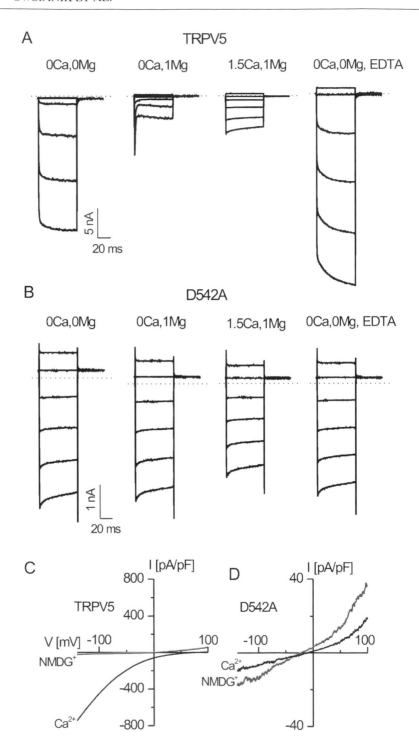

They are involved in various processes such as pheromone sensing (TRPC2), vasoregulation (TRPC3/4/5), signaling in the central nervous system (TRPC1/3/4), and functioning of smooth muscle cells (TRPC3/6/7) (70–76).

Few studies have been aimed at the description of the pore properties and the identification of the pore region of TRPC channels. Moreover, interpretation of the functional data obtained for members of the TRPC subfamily is often difficult and complex owing to a noisy background caused by endogenous cation-selective, Ca^{2+}-permeable channels regulated by store-depletion and/or products of PLC-dependent pathways (77–79). This has led to conflicting results, e.g., TRPC4 and TRPC5 have been described as either Ca^{2+} selective (80–82) or nonselective between mono- and divalent cations (83, 84).

A theoretical prediction of pore elements also seems to be ineffective. Unlike TRPV channels, the region between TM5 and TM6 of TRPC members does not share significant sequence homology to the pore segment of bacterial K^+ channels (Figure 9). This loss of homology to the bacterial archetype pore may signify that TRPC are phylogenetically younger than TRPV channels. Thus, it is possible that classical TRPCs were not the starting point for the evolution of other subfamilies of TRP channels.

TRPC1/5 Conduction: Regulation from Outside of the Pore

Greatest insight into the localization of the pore and the regulation of permeability of TRPC channels comes from studies of La^{3+} potentiation of TRPC5 (85) and functional identification of TRPC1 as one of the store-operated Ca^{2+} channels (86). To determine the molecular determinants of La^{3+} potentiation, Jung et al. (85) performed a systematic mutagenesis of all negatively charged residues localized in extracellular loops of TRPC5. Whereas neutralization of aspartates and glutamates in the loops between TM1-TM2 and TM3-TM4 had no obvious effects, neutralization of three of the five glutamates in the loop between TM5 and TM6

Figure 5   Phenotypes of the currents through wild-type and D^{542A}-mutant TRPV5 channels. Currents during voltage steps ranged from $+60$ mV to -140 mV (decrement, -40 mV; holding potential, $+20$ mV) in nominally divalent cation-free solution as well as solution plus either 1-mM Mg^{2+}, 1.5-mM Ca^{2+}, and 1-mM Mg^{2+}, or 0.1-mM ethylenediaminetetraacetic acid (EDTA). Data were obtained from HEK293 cells transfected with (*A*) wild-type TRPV5 or (*B*) D^{542A}. (*C* and *D*) Currents in response to a voltage ramp protocal from -150 to $+100$ mV (V_H, $+20$ mV; duration, 400 ms). In the absence of extracellular monovalent cations (*red line*, all substituted by *N*-methyl-D-glucamine$^+$), administration of 30-mM Ca^{2+} (*black line*) results in a large inward current carried by Ca^{2+} in the absence of another permeable cation. In the D^{542A} mutant, the background current in a monovalent cation-free solution was inhibited by the administration of 30-mM Ca^{2+} (adapted from Reference 15 with copyright permission from *The American Society for Biochemistry and Molecular Biology*).

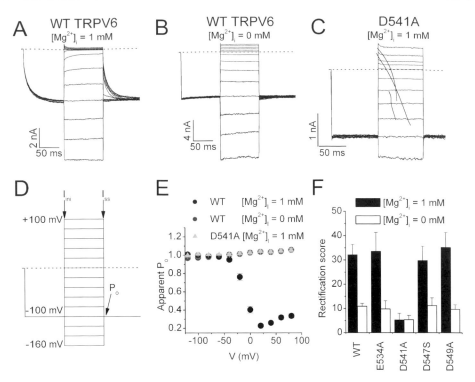

Figure 6 Asp[541] within the pore region of TRPV6 is required for Mg^{2+} binding and block of the channel. Currents measured in response to the voltage protocol shown in panel *D* and obtained with an intracellular solution containing (*A*) 1-mM free intracellular Mg^{2+} or (*B*) 10-mM EDTA and no Mg^{2+} through wild-type TRPV6. (*C*) Current trace obtained with 1-mM intracellular Mg^{2+} for D[541A] pore mutant using the protocol shown in panel *D*, except that test pulses ranged from -120 to 80 mV. (*D*) Voltage protocol used in panels *A–C* and subsequent figures. The time point at which the apparent open probability (P_o) was measured is indicated. (*E*) Voltage dependence of the apparent open probability (P_o) in the absence and presence of 1-mM intracellular Mg^{2+} for wild-type and D[541A]-mutant TRPV6. Apparent P_o was determined as the inward current measured immediately after the 100-ms test pulses normalized to the maximal inward current. (*F*) Comparison of the rectification scores for wild-type TRPV6 and four pore mutants in the presence and absence of 1-mM intracellular Mg^{2+} (reproduced from Reference 58 with copyright permission of *The Rockefeller University Press*).

(E[543], E[595], and E[598]) resulted in a loss of La^{2+} potentiation. Additionally, the E[595]/E[598] double mutant displayed altered single-channel properties (Figure 10*A*). In contrast, mutation of either E[559] or E[570] localized in the central part of this loop did not have any effect on the channel properties.

Expression of a TRPC1 mutant in which all seven negatively charged residues in the region between TM5 and TM6 are neutralized (D to N and E to Q)

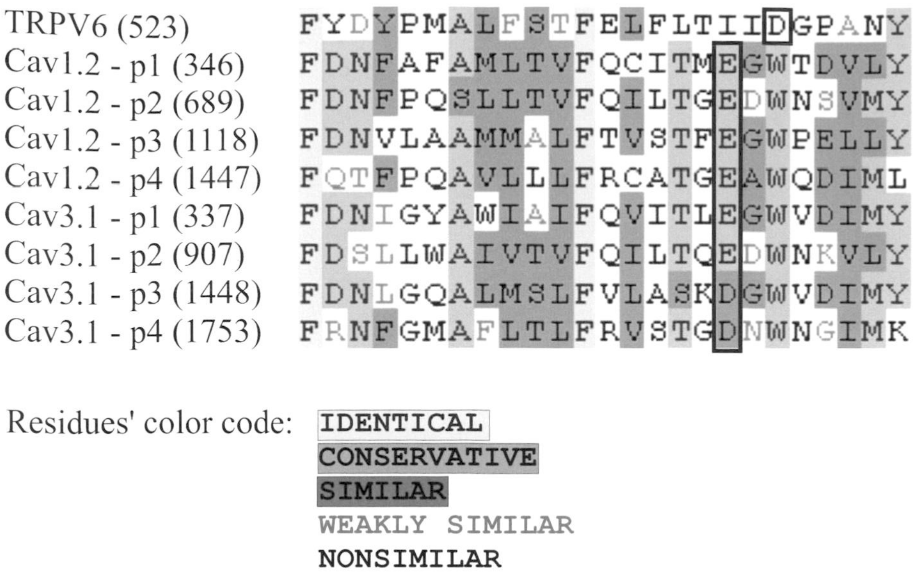

```
TRPV6 (523)         FYDYPMALFSTFELFLTIIDGPANY
Cav1.2 - p1 (346)   FDNFAFAMLTVFQCITMEGWTDVLY
Cav1.2 - p2 (689)   FDNFPQSLLTVFQILTGEDWNSVMY
Cav1.2 - p3 (1118)  FDNVLAAMMALFTVSTFEGWPELLY
Cav1.2 - p4 (1447)  FQTFPQAVLLLFRCATGEAWQDIML
Cav3.1 - p1 (337)   FDNIGYAWIAIFQVITLEGWVDIMY
Cav3.1 - p2 (907)   FDSLLWAIVTVFQILTQEDWNKVLY
Cav3.1 - p3 (1448)  FDNLGQALMSLFVLASKDGWVDIMY
Cav3.1 - p4 (1753)  FRNFGMAFLTLFRVSTGDNWNGIMK
```

Residues' color code: IDENTICAL
                       CONSERVATIVE
                       SIMILAR
                       WEAKLY SIMILAR
                       NONSIMILAR

Figure 7   The multiple alignment of TRPV6 with pore regions (p1–4) of human voltage-gated L-type (Cav1.2; accession number Q13936) and T-type (Cav3.1; accession number O43497) calcium channels. Localization of crucial aspartate and glutamate residues that determine permeation properties of the channels are indicated by red rectangles.

(Figure 10*A*) resulted in decreased Ca^{2+} but normal Na^+ currents through TRPC1, and induced changes in the reversal potential (86). These data suggest that the pore-forming region of TRPC1 is also localized between TM5 and TM6. Similar effects were induced by point mutations E^{576K} and D^{581K}, but not D^{581N} or E^{615K}. Analogous to the glutamate residues identified in TRPC5, the negatively charged residues that determine pore properties of TRPC1 seem to be located in the distal parts of the putative pore entrance.

A study of biogenesis and topology of TRPC1 (87) revealed that TRPC1 can potentially form eight hydrophobic α-helices, but only six of them span the membrane. One of the nonmembrane spanning α-helices is localized in the region between TM5 and TM6 (the TM nomenclature here is different from that proposed in Reference 87). Although this α-helix is long enough to span the membrane (23 amino acids; see Figure 10*A*), it only partially enters the membrane, suggesting that it might form a pore helix similar to what has been observed for KcsA and TRPV5/6. A striking feature of the topological model proposed by Dohke et al. (87) is the paucity of negatively charged residues in the pore region. In accordance with this, Obukhov & Nowycky (88) found that a crucial residue for current block by intracellular Mg^{2+} in TRPC5 homotetrameric channels, D^{633}, is situated intracellularly between the end of TM6 and the "TRP box" (Figures 10*B,C*) (88). Substitution of D^{633} by either noncharged or positively charged residues resulted

A

TRPV6 pore structure

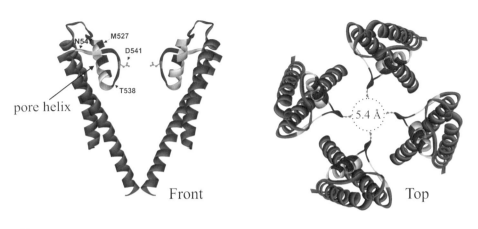

pore helix

Front   Top

B

Selectivity filter of TRPV6

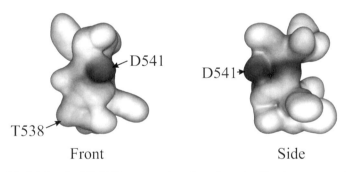

Front   Side

Figure 8   Model for the TRPV6 pore region, based on the KcsA structure. (*A*) Views of the structure are shown, looking sideways at two opposite subunits (*left*) or looking down from the external solution to the complete homotetrameric channel. At the narrowest point, formed by the acidic side chain of Asp[541] (*orange*), the pore has a diameter of 5.4 Å. Blue residues correspond to the residues in TM5 and TM6 (TM1 and TM2 in KcsA). Amino acids that were subjected to the substituted cysteine accessibility analysis (residues P[526] to N[547]) are colored in green, yellow, red, or gray: Residues in red reacted rapidly to Ag[+] (reaction rate $>5.10^6$ M^{-1}s^{-1}), residues in yellow reacted with Ag[+] (reaction rate $<5.10^6$ M^{-1}s^{-1}), and residues in green did not show significant reactivity to Ag[+]. Residues where cysteine substitution resulted in nonfunctional channels are colored in gray (adapted from Reference 23 with copyright permission from *The American Society for Biochemistry and Molecular Biology*). (*B*) Distribution of electrostatic potentials on the surface of the selectivity filters of TRPV6. The negative and positive charges are shown in red and blue, respectively.

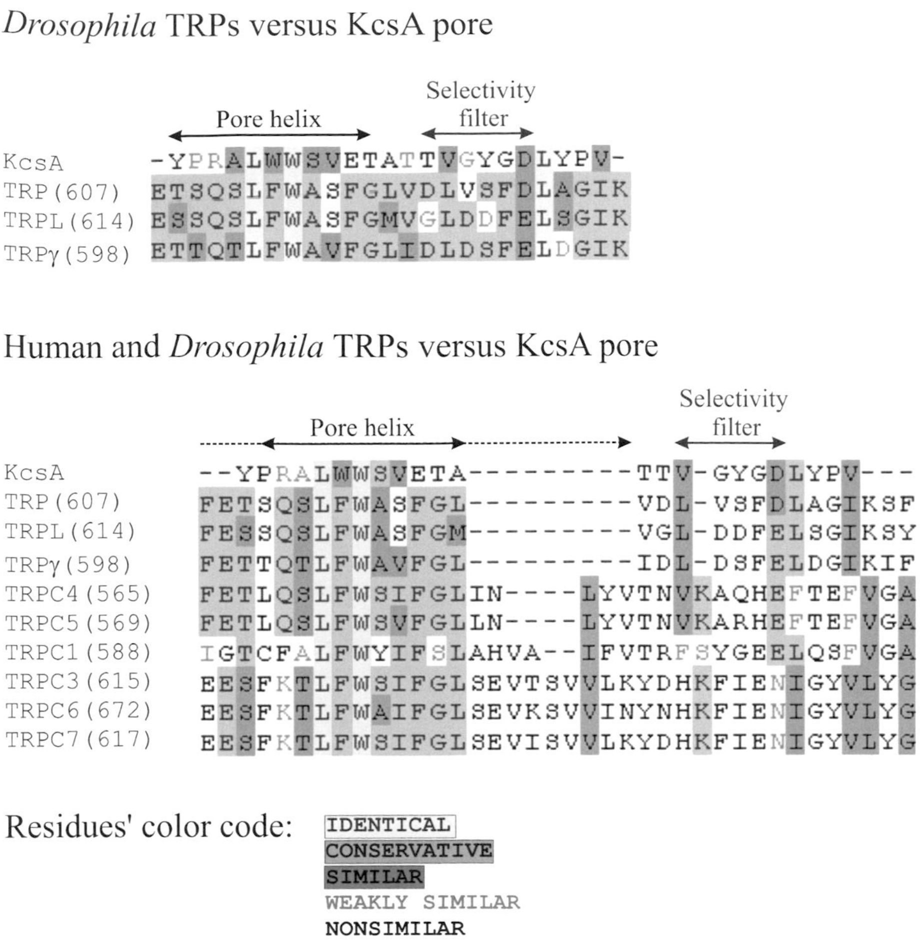

Drosophila TRPs versus KcsA pore

```
                             Selectivity
              Pore helix      filter
         ←――――――――――→    ←――――――→
KcsA     -YPRALWWSVETATTVGYGDLYPV-
TRP(607) ETSQSLFWASFGLVDLVSFDLAGIK
TRPL(614) ESSQSLFWASFGMVGLDDFELSGIK
TRPγ(598) ETTQTLFWAVFGLIDLDSFELDGIK
```

Human and *Drosophila* TRPs versus KcsA pore

```
                                            Selectivity
                   Pore helix                 filter
         ·········←――――――――――――――――――――→   ←――――――→
KcsA        --YPRALWWSVETA---------TTV-GYGDLYPV---
TRP(607)    FETSQSLFWASFGL---------VDL-VSFDLAGIKSF
TRPL(614)   FESSQSLFWASFGM---------VGL-DDFELSGIKSY
TRPγ(598)   FETTQTLFWAVFGL---------IDL-DSFELDGIKIF
TRPC4(565)  FETLQSLFWSIFGLIN----LYVTNVKAQHEFTEFVGA
TRPC5(569)  FETLQSLFWSVFGLLN----LYVTNVKARHEFTEFVGA
TRPC1(588)  IGTCFALFWYIFSLAHVA--IFVTRFSYGEELQSFVGA
TRPC3(615)  EESFKTLFWSIFGLSEVTSVVLKYDHKFIENIGYVLYG
TRPC6(672)  EESFKTLFWAIFGLSEVKSVVINYNHKFIENIGYVLYG
TRPC7(617)  EESFKTLFWSIFGLSEVISVVLKYDHKFIENIGYVLYG
```

Residues' color code: IDENTICAL
                         CONSERVATIVE
                         SIMILAR
                         WEAKLY SIMILAR
                         NONSIMILAR

Figure 9 Multiple alignments of KcsA and channels of the TRPC subfamily. Localizations of the putative pore helix and selectivity filter are annotated.

in markedly reduced inward currents and decreased Mg^{2+} block. In conclusion, available data indicate that permeation properties of TRPC1/5 are regulated by negatively charged residues that appear to be located close to but outside of the actual pore region.

Mysterious Pores of TRPMs

The TRPM subfamily comprises eight mammalian members, which play a role in processes as diverse as taste detection (89–91), Mg^{2+} homeostasis (92–95), cell proliferation (93, 96), and cold sensing (97–99) (Table 1). Similar to other TRP channels, the putative pore-forming region of TRPM channels may be located in the loop between TM5 and TM6. TRPM4 and TRPM5 are the only Ca^{2+}-impermeable

A

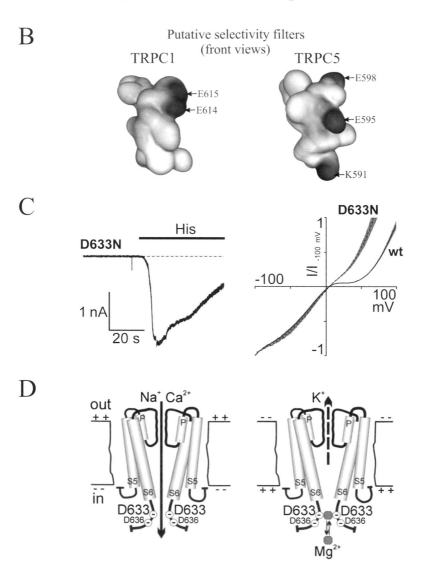

TRPC1
$\overset{576}{\downarrow}$ $\overset{581}{\downarrow}$ Pore helix $\overset{615}{\downarrow}$

TQLY**D**KGYTSK**E**QK**D**CVGIFC**E**QQSN**D**TFHS**FIGTCFALFWYIFSLAHVAIFV**TRF***SYGEE***LQSFVG

TRPC5
$\overset{543}{\downarrow}$ Pore helix $\overset{595}{\downarrow}$ $\overset{598}{\downarrow}$

YY**E**TRAI**DE**PNNCKGIRC**E**KQNNAF**STLFETLQSLFWSVFGLLNLYV**TNV***KARHE***FTEF

B

Putative selectivity filters
(front views)

TRPC1

TRPC5

C

D633N His

D633N

1 nA

20 s

1
I/I$_{-100\ mV}$

-100

100 mV

-1

wt

D

out
++

Na⁺ Ca²⁺

P P

S5 S6 S6 S5

D633 D633
D636 D636

in
--

++
--

K⁺

P P

S5 S6 S6 S5

D633 D633
D636 D636

Mg²⁺

--
++
--
++

TRP channels identified to date (89, 100–102). TRPM3, TRPM2, and TRPM8 are Ca^{2+}-permeable cation channels with low Ca^{2+} selectivity (98, 99, 103–107), and TRPM6 and TRPM7 are relatively highly permeable for divalent cations (3, 93, 94, 96). The functional properties of TRPM1 have not been reliably characterized.

Sequence comparison of the TM5-TM6 loop indicates that this region is well conserved among all members of the TRPM family (Figure 11A). This region consists of a hydrophobic stretch, which may correspond to the pore helix, followed by an invariant aspartate (site *D*), which may be located in the selectivity filter (108). These highly homologous regions of TRPMs also share limited homology to KcsaA and TRPVs channels. Three other sites contain partially conserved, negatively charged residues between the putative pore helix and the fully conserved aspartate. Taking the invariant aspartate as a reference, researchers found that these sites are localized at positions -7 (site *A*), -3 (site *B*), and -2 (site *C*) (Figure 11A). It is likely that this cluster of negative charges contributes to the pore properties of TRPM channels.

First Insights in the TRPM Pore Structure

The structure-function relation in TRPM channels has been experimentally addressed for the first time in two recent studies. Nilius et al. (109) exploited the theoretical predictions of the putative pore region of TRPM channels (Figure 11B) as a starting point for the identification of residues in TRPM4 that are responsible for the permeability of the pore and its blockade by intracellular spermine. Substitution of residues E^{981} to A^{986} with the selectivity filter of TRPV6 yielded a functional channel that combines the gating hallmarks of TRPM4 (activation by

---

Figure 10 Negatively charged residues located outside of the pore mouth determine permeation properties of TRPC1/5 channels. (*A*) Putative pore regions of TRPC1 and TRPC5 channels. All negatively charged residues are shown in red, and numbers annotate those residues that are involved in modifications of channel properties. Residues of the pore helix are indicated in green. The region that corresponds to the selectivity filter is in bold italic. (*B*) Distribution of electrostatic potentials on the surface of putative selectivity filters of TRPC1 (*left*) and TRPC5 (*right*). The negative and positive charges are shown in red and blue, respectively. Localization of the main residues is annotated. (*C*) Whole-cell D^{633N} currents at holding potential -60 mV (*left*). Averaged I–V values of wild-type TRPC5 and D633N recorded in Na^{+}, Ca^{2+}, and Mg^{2+}-free ($NMDG^{+}$-containing) extracellular solution (*right*). (*D*) A model of the TRPC5-cation conduction pathway at negative (*left*) and positive (*right*) potentials. For clarity, only two subunits are shown. D^{633} and D^{636} are designated as open circles, whereas Mg^{2+} is shown as an octagon. The bold arrows indicate the direction of cation fluxes. The thin arrows indicate that Mg^{2+} is a fast blocker. P, P-helix; S5, TM5; S6, TM6; double plus signs, positively charged; double minus signs, negatively charged. (Panels *C* and *D* are adapted from Reference 88, © 2005 by *The Society of Neurosciences*.)

A

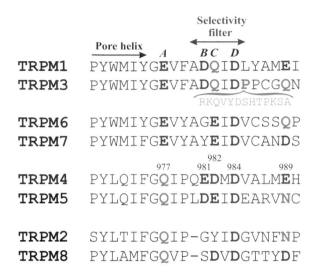

TRPM1   PYWMIYGEVFAD**Q**IDLYAM**E**I
TRPM3   PYWMIYGEVFAD**Q**IDPPCG**Q**N
                        RKQVYDSHTPKSA

TRPM6   PYWMIYGEVYAGE**I**DVCSS**Q**P
TRPM7   PYWMIFGEVYAYE**I**DVCAND**S**

TRPM4   PYLQIFG**Q**IPQ**E**DM**D**VALM**E**H
TRPM5   PYLQIFG**Q**IPL**DEI**DEARV**N**C

TRPM2   SYLTIFG**Q**IP-GYIDGVNF**N**P
TRPM8   PYLAMFG**Q**VP-SDVDGTTYDF

B

Putative selectivity filter of TRPM4

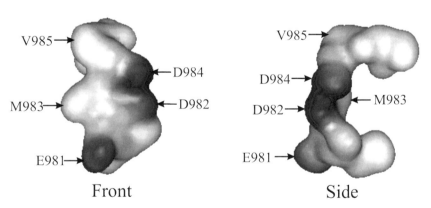

Front   Side

Figure 11   Putative pore regions of human TRPM channels. (*A*) Multiple alignments of TRPM-channel pore regions containing important pore elements. Localizations of the putative pore helix and selectivity filter are annotated. Capital italic letters *A–D* indicate the predominantly negatively charged residue sites. Aspartate/glutamate and asparagine/glutamine residues within these sites are shown in red and blue, respectively. For TRPM3, an alternative splice variant with an additional 12 amino acids in the putative pore region is shown in green. (*B*) Distribution of electrostatic potentials on the surface of the putative selectivity filter of TRPM4. The negative and positive charges are indicated in red and blue, respectively. Localization of the main residues is annotated.

Ca^{2+}, voltage dependence) with TRPV6-like sensitivity to block by extracellular Ca^{2+} and Mg^{2+}. This experiment provides the strongest indication of the location of TRPM4's selectivity filter. Substitution of E^{981} by alanine strongly reduced the channel's affinity to block by spermine, whereas mutations to the other negative charges in the region were ineffective (109). This indicates that E^{981} is located in the inner part of the pore where it is accessible to intracellular spermine. Surprisingly, mutations to the adjacent aspartates D^{982} and D^{984} strongly affected the rundown and voltage dependence of the channel. Substitution of Gln^{977} by a glutamate (site A in Figure 11A), the corresponding residue in divalent cation-permeable TRPM channels, altered the monovalent-cation permeability sequence and resulted in a pore with moderate Ca^{2+} permeability. Taken together, these results indicate that the putative pore region contains critical determinants of the pore properties of TRPM4.

In another recent study, several splice variants of TRPM3, termed TRPM3α $1-5$(110), were described. These splice variants differ in the length of the putative pore region, as one splice site is located in the TM5-TM6 loop. TRPM3α1, a variant that encompasses an optional stretch of 12 amino acids following the invariant aspartate (site D in Figure 11A), represents a cation channel with low permeability for divalent cations, whereas TRPM3α2, which lacks this stretch of amino acids, displays a more than 10-fold higher permeability to Ca^{2+} and Mg^{2+} and is blocked by extracellular monovalent cations. To our knowledge, this is the first example of an ion channel whose selectivity is regulated by alternative splicing.

CONCLUSIONS AND FUTURE DIRECTIONS

In this review we provide an overview of the relatively limited knowledge of the pore properties and pore structure of TRP channels. The emerging picture is that the architectural blueprint of TRP channel pores is similar to that of other 6-TM ion channels. Functional TRP channels consist of four identical or homologous subunits. Additionally, detailed studies of members of the TRPV superfamily have yielded convincing evidence that the loop between TM5 and TM6 forms a pore helix and selectivity filter, similar to what was observed in crystal structures of bacterial K^+ channels. It is probable, yet unproven, that the other TRP subfamilies have similarly structured pores. Given the limited sequence conservation in the putative pore region and the highly divergent pore properties in this large superfamily, the possibility of completely different pore architectures should not be dismissed.The study of TRP channel pores will be greatly facilitated when TRP channel structures become known. However, the production of sufficient TRP channel protein for crystallization purposes may be a serious obstacle, given the absence of TRP channel genes from bacterial genomes.

Studying the pore of an ion channel is aimed at understanding the molecular principles of ion conduction. Better knowledge of the pore properties is also highly instrumental for the molecular identification of TRP channels in their native cellular environment. It is probable that every mammalian cell expresses TRP genes and contains functional (Ca^{2+}-permeable) cation channels in its plasma membrane, but few studies have convincingly related the properties of an endogenous channel to a specific TRP channel gene. For example, Ca^{2+}-activated, Ca^{2+}-impermeable, nonselective, monovalent cation channels with a single-channel conductance around 25 pS have been described in several cell types, including astrocytes (111), vascular endothelial cells (18), and hamster vomeronasal sensory neurons (112). Comparison of these features with the pore properties of heterologously expressed TRP channels immediately points to TRPM4. We foresee that a full description of TRP channel pore properties, in conjunction with the increasing use of molecular genetics tools (siRNA, knockout mice, etc.), will lead to a better understanding of the functional expression of TRP channels in different cells and tissues and, eventually, of their participation in physiological processes.

ACKNOWLEDGMENTS

Grzegorz Owsianik and Karel Talavera share first authorship. B. Nilius is the corresponding author. This work was supported by the Human Frontiers Science Programme (research grant reference RGP 32/2004), the Belgian Federal Government, the Flemish Government, and the Onderzoeksraad KU Leuven (Grant Nos. GOA 2004/07, F.W.O. G.0214.99, F.W.O. G. 0136.00, and F.W.O. G.0172.03, Interuniversity Poles of Attraction Program, Prime Ministers Office IUAP).

**The *Annual Review of Physiology* is online at
http://physiol.annualreviews.org**

LITERATURE CITED

1. Vriens J, Owsianik G, Voets T, Droogmans G, Nilius B. 2004. Invertebrate TRP proteins as functional models for mammalian channels. *Pflügers Arch.* 449:213–26

2. Clapham DE. 2003. TRP channels as cellular sensors. *Nature* 426:517–24

3. Voets T, Nilius B. 2003. The pore of TRP channels: trivial or neglected? *Cell Calcium* 33:299–302

4. Montell C, Rubin GM. 1989. Molecular characterization of the Drosophila trp locus: a putative integral membrane protein required for phototransduction. *Neuron* 2: 1313–23

5. Vennekens R, Voets T, Bindels RJM, Droogmans G, Nilius B. 2002. Current

understanding of mammalian TRP homologues. *Cell Calcium* 31:253–64

6. Montell C. 2005. The TRP superfamily of cation channels. *Sci. STKE* 272:re3

7. Kuzhikandathil EV, Wang H, Szabo T, Morozova N, Blumberg PM, Oxford GS. 2001. Functional analysis of capsaicin receptor (vanilloid receptor subtype 1) multimerization and agonist responsiveness using a dominant negative mutation. *J. Neurosci.* 21:8697–706

8. Hoenderop JGJ, Voets T, Hoefs S, Weidema F, Prenen J, et al. 2003. Homo- and heterotetrameric architecture of the epithelial Ca^{2+} channels TRPV5 and TRPV6. *EMBO J.* 22:776–85

9. Hille B. 2001. *Ionic Channels of Excitable*

Membranes. Sunderland, MA: Sinauer Assoc. 3rd ed.

10. Montell C, Birnbaumer L, Flockerzi V. 2002. The TRP channels, a remarkable functional family. *Cell* 108:595–98

11. Clapham DE, Runnels LW, Strubing C. 2001. The trp ion channel family. *Nat. Rev. Neurosci.* 2:387–96

12. Neher E. 1992. Correction for liquid junction potentials in patch clamp experiments. *Methods Enzymol.* 207:123–31

13. Barry PH. 1994. JPCalc, a software package for calculating liquid junction potential corrections in patch-clamp, intracellular, epithelial and bilayer measurements and for correcting junction potential measurements. *J. Neurosci. Methods* 51:107–16

13a. Figl T, Lewis TM, Barry PH. 2003. Liquid junction potential corrections. *AxoBits* 39: 6–10

14. Vennekens R, Hoenderop JGJ, Prenen J, Stuiver M, Willems PHGM, et al. 2000. Permeation and gating properties of the novel epithelial Ca^{2+} channel. *J. Biol. Chem.* 275:3963–69

15. Nilius B, Vennekens R, Prenen J, Hoenderop JGJ, Droogmans G, Bindels RJM. 2001. The single pore residue Asp542 determines Ca^{2+} permeation and Mg^{2+} block of the epithelial Ca^{2+} channel. *J. Biol. Chem.* 276:1020–25

16. Voets T, Prenen J, Vriens J, Watanabe H, Janssens A, et al. 2002. Molecular determinants of permeation through the cation channel TRPV4. *J. Biol. Chem.* 277: 33704–10

17. Lewis CA. 1979. Ion-concentration dependence of the reversal potential and the single channel conductance of ion channels at the frog neuromuscular junction. *J. Physiol.* 286:417–45

18. Kamouchi M, Mamin A, Droogmans G, Nilius B. 1999. Nonselective cation channels in endothelial cells derived from human umbilical vein. *J. Membr. Biol.* 169: 29–38

19. Eisenman G. 1962. Cation selective glass electrodes and their mode of operation. *Biophys. J.* 2:259–323

20. Nilius B, Vennekens R, Prenen J, Hoenderop JGJ, Bindels RJM, Droogmans G. 2000. Whole-cell and single channel monovalent cation currents through the novel rabbit epithelial Ca^{2+} channel ECaC. *J. Physiol.* 527:239–48

21. Vennekens R, Prenen J, Hoenderop JGJ, Bindels RJM, Droogmans G, Nilius B. 2001. Modulation of the epithelial Ca^{2+} channel ECaC by extracellular pH. *Pflügers Arch.* 442:237–42

22. Vennekens R, Prenen J, Hoenderop JGJ, Bindels RJM, Droogmans G, Nilius B. 2001. Pore properties and ionic block of the rabbit epithelial calcium channel expressed in HEK 293 cells. *J. Physiol.* 530: 183–91

23. Voets T, Janssens A, Droogmans G, Nilius B. 2004. Outer pore architecture of a Ca2+-selective TRP channel. *J. Biol. Chem.* 279:15223–30

24. Bormann J, Hamill OP, Sakmann B. 1987. Mechanism of anion permeation through channels gated by glycine and gamma-aminobutyric acid in mouse cultured spinal neurones. *J. Physiol.* 385:243–86

25. Robinson RA, Stokes RH, eds. 1959. *Electrolyte Solution.* London: Butterworth. 571pp.

26. Woodbury JW. 1971. Eyring rate theory model of the current-voltage relationships of ion channels in excitable membranes. In *Chemical Dynamics:Papers in Honor of Henry Eyring*, ed. JO Hirschfelder, D Henderson, H Eyring, pp. 601–17. New York: Wiley

27. Caterina MJ, Leffler A, Malmberg AB, Martin WJ, Trafton J, et al. 2000. Impaired nociception and pain sensation in mice lacking the capsaicin receptor. *Science* 288:306–13

28. Jordt SE, Julius D. 2002. Molecular basis for species-specific sensitivity to "hot" chili peppers. *Cell* 108:421–30

29. Kanzaki M, Zhang YQ, Mashima HLL, Shibata H, Kojima I. 1999. Translocation

of a calcium-permeable cation channel induced by insulin-like growth factor-I. *Nat. Cell Biol.* 1:165–70

30. Smith GD, Gunthorpe J, Kelsell RE, Hayes PD, Reilly P, et al. 2002. TRPV3 is a temperature-sensitive vanilloid receptor-like protein. *Nature* 418:186–90

31. Xu HX, Ramsey IS, Kotecha SA, Moran MM, Chong JHA, et al. 2002. TRPV3 is a calcium-permeable temperature-sensitive cation channel. *Nature* 418:181–86

32. Peier AM, Reeve AJ, Andersson DA, Moqrich A, Earley TJ, et al. 2002. A heat-sensitive TRP channel expressed in keratinocytes. *Science* 296:2046–49

33. Liedtke W, Choe Y, Marti-Renom MA, Bell AM, Denis CS, et al. 2000. Vanilloid receptor-related osmotically activated channel (VR-OAC), a candidate vertebrate osmoreceptor. *Cell* 103:525–35

34. Liedtke W, Friedman JM. 2003. Abnormal osmotic regulation in trpv4-/- mice. *Proc. Natl. Acad. Sci. USA* 100:13698–703

35. Nilius B, Vriens J, Prenen J, Droogmans G, Voets T. 2004. TRPV4 calcium entry channel: a paradigm for gating diversity. *Am. J. Physiol. Cell Physiol.* 286:C195–205

36. den Dekker E, Hoenderop JGJ, Nilius B, Bindels RJM. 2003. The epithelial calcium channels, TRPV5 and TRPV6: from identification towards regulation. *Cell Calcium* 33:497–507

37. Hoenderop JGJ, Nilius B, Bindels RJM. 2002. ECaC: the gatekeeper of transepithelial Ca2+ transport. *Biochim. Biophys. Acta* 1600:6–11

38. Hoenderop JGJ, Nilius B, Bindels RJM. 2002. Molecular mechanism of active Ca^{2+} reabsorption in the distal nephron. *Annu. Rev. Physiol.* 64:529–49

39. Hoenderop JGJ, Nilius B, Bindels RJM. 2003. Epithelial calcium channels: from identification to function and regulation. *Pflügers Arch.* 446:304–8

40. Nijenhuis T, Hoenderop JGJ, Nilius B, Bindels RJM. 2003. (Patho)physiological implications of the novel epithelial Ca2+ channels TRPV5 and TRPV6. *Pflügers Arch.* 446:401–9

41. Gunthorpe MJ, Benham CD, Randall A, Davis JB. 2002. The diversity in the vanilloid (TRPV) receptor family of ion channels. *Trends Pharmacol. Sci.* 23:183–91

42. Benham CD, Davis JB, Randall AD. 2002. Vanilloid and TRP channels: a family of lipid-gated cation channels. *Neuropharmacology* 42:873–88

43. Hellwig N, Plant TD, Janson W, Schafer M, Schultz G, Schaefer M. 2004. TRPV1 acts as proton channel to induce acidification in nociceptive neurons. *J. Biol. Chem.* 279:34553–61

44. Hoenderop JGJ, Vennekens R, Müller D, Prenen J, Droogmans G, et al. 2001. Function and expression of the epithelial Ca^{2+} channel family: comparison of the epithelial Ca^{2+} channel 1 and 2. *J. Physiol.* 537:747–61

45. Varadi G, Strobeck M, Koch S, Caglioti L, Zucchi C, Palyi G. 1999. Molecular elements of ion permeation and selectivity within calcium channels. *Crit. Rev. Biochem. Mol. Biol.* 34:181–214

46. Tsien RW, Hess P, McCleskey EW, Rosenberg RL. 1987. Calcium channels: mechanisms of selectivity, permeation, and block. *Annu. Rev. Biophys. Biophys. Chem.* 16:265–90

47. Sather WA, McCleskey EW. 2003. Permeation and selectivity in calcium channels. *Annu. Rev. Physiol.* 65:133–59

48. McDonald TF, Pelzer S, Trautwein W, Pelzer DJ. 1994. Regulation and modulation of calcium channels in cardiac, skeletal, and smooth muscle cells. *Physiol. Rev.* 74:365–507

49. Almers W, McCleskey EW. 1984. Nonselective conductance in calcium channels of frog muscle: calcium selectivity in a single-file pore. *J. Physiol.* 353:585–608

50. Hess P, Tsien RW. 1984. Mechanism of ion permeation through calcium channels. *Nature* 309:453–56

51. Hess P, Lansman JB, Tsien RW. 1986. Calcium channel selectivity for divalent and monovalent cations. Voltage and concentration dependence of single channel current in ventricular heart cells. *J. Gen. Physiol.* 88:293–319

52. Dang TX, McCleskey EW. 1998. Ion channel selectivity through stepwise changes in binding affinity. *J. Gen. Physiol.* 111:185–93

53. Doyle DA, Morais Cabral J, Pfuetzner RA, Kuo A, Gulbis JM, et al. 1998. The structure of the potassium channel: molecular basis of K^+ conduction and selectivity. *Science* 280:69–77

54. Zhou Y, Morais-Cabral JH, Kaufman A, MacKinnon R. 2001. Chemistry of ion coordination and hydration revealed by a K^+ channel-Fab complex at 2.0 Å resolution. *Nature* 414:43–48

55. Garcia-Martinez C, Morenilla-Palao C, Planells-Cases R, Merino JM, Ferrer-Montiel A. 2000. Identification of an aspartic residue in the P-loop of the vanilloid receptor that modulates pore properties. *J. Biol. Chem.* 275:32552–58

56. Yeh BI, Sun TJ, Lee JZ, Chen HH, Huang CL. 2003. Mechanism and molecular determinant for regulation of rabbit transient receptor potential type 5 (TRPV5) channel by extracellular pH. *J. Biol. Chem.* 278:51044–52

57. Voets T, Prenen J, Fleig A, Vennekens R, Watanabe H, et al. 2001. CaT1 and the calcium release-activated calcium channel manifest distinct pore properties. *J. Biol. Chem.* 276:47767–70

58. Voets T, Janssens A, Prenen J, Droogmans G, Nilius B. 2003. Mg^{2+}-dependent gating and strong inward rectification of the cation channel TRPV6. *J. Gen. Physiol.* 121:245–60

59. Heinemann SH, Terlau H, Stuhmer W, Imoto K, Numa S. 1992. Calcium channel characteristics conferred on the sodium channel by single mutations. *Nature* 356:441–43

60. Talavera K, Staes M, Janssens A, Klugbauer N, Droogmans G, et al. 2001. Aspartate residues of the Glu-Glu-Asp-Asp (EEDD) pore locus control selectivity and permeation of the T-type Ca(2+) channel alpha(1G). *J. Biol. Chem.* 276:45628–35

61. Yang J, Ellinor PT, Sather WA, Zhang JF, Tsien RW. 1993. Molecular determinants of Ca2+ selectivity and ion permeation in L-type Ca2+ channels. *Nature* 366:158–61

62. Lipkind GM, Fozzard HA. 2001. Modeling of the outer vestibule and selectivity filter of the L-type Ca2+ channel. *Biochemistry* 40:6786–94

63. Dodier Y, Banderali U, Klein H, Topalak O, Dafi O, et al. 2004. Outer pore topology of the ECaC-TRPV5 channel by cysteine scan mutagenesis. *J. Biol. Chem.* 279:6853–62

64. McCleskey EW, Almers W. 1985. The Ca channel in skeletal muscle is a large pore. *Proc. Natl. Acad. Sci. USA* 82:7149–53

65. Cataldi M, Perez-Reyes E, Tsien RW. 2002. Differences in apparent pore sizes of low and high voltage-activated Ca2+ channels. *J. Biol. Chem.* 277:45969–76

66. Harteneck C, Plant TD, Schultz G. 2000. From worm to man: three subfamilies of TRP channels. *Trends Neurosci.* 23:159–66

67. Nilius B. 2003. From TRPs to SOCs, CCEs, and CRACs: consensus and controversies. *Cell Calcium* 33:293–98

68. Clapham DE, Montell C, Schultz G, Julius D. 2003. International Union of Pharmacology. XLIII. Compendium of voltage-gated ion channels: transient receptor potential channels. *Pharmacol. Rev.* 55:591–96

69. Vazquez G, Wedel BJ, Aziz O, Trebak M, Putney JW Jr. 2004. The mammalian TRPC cation channels. *Biochim. Biophys. Acta* 1742:21–36

70. Strubing C, Krapivinsky G, Krapivinsky L, Clapham DE. 2001. TRPC1 and TRPC5 form a novel cation channel in mammalian brain. *Neuron* 29:645–55

71. Sakura H, Ashcroft FM. 1997. Identification of four trp1 gene variants murine pancreatic beta-cells. *Diabetologia* 40:528–32

72. Kim SJ, Kim YS, Yuan JP, Petralia RS, Worley PF, Linden DJ. 2003. Activation of the TRPC1 cation channel by metabotropic glutamate receptor mGluR1. *Nature* 426:285–91

73. Lucas P, Ukhanov K, Leinders-Zufall T, Zufall F. 2003. A diacylglycerol-gated cation channel in vomeronasal neuron dendrites is impaired in TRPC2 mutant mice: mechanism of pheromone transduction. *Neuron* 40:551–61

74. Stowers L, Holy TE, Meister M, Dulac C, Koentges G. 2002. Loss of sex discrimination and male-male aggression in mice deficient for TRP2. *Science* 295:1493–500

75. Freichel M, Suh SH, Pfeifer A, Schweig U, Trost C, et al. 2001. Lack of an endothelial store-operated Ca^{2+} current impairs agonist-dependent vasorelaxation in TRP4-/- mice. *Nat. Cell Biol.* 3:121–27

76. Tiruppathi C, Freichel M, Vogel SM, Paria BC, Mehta D, et al. 2002. Impairment of store-operated Ca^{2+} entry in TRPC4(-/-) mice interferes with increase in lung microvascular permeability. *Circ. Res.* 91:70–76

77. Venkatachalam K, van Rossum DB, Patterson RL, Ma HT, Gill DL. 2002. The cellular and molecular basis of store-operated calcium entry. *Nat. Cell Biol.* 4: E263–72

78. Clapham DE. 1995. Calcium signaling. *Cell* 80:259–68

79. Parekh AB, Penner R. 1997. Store depletion and calcium influx. *Physiol. Rev.* 77: 901–30

80. Philipp S, Cavalie A, Freichel M, Wissenbach U, Zimmer S, et al. 1996. A mammalian capacitative calcium entry channel homologous to Drosophila TRP and TRPL. *EMBO J.* 15:6166–71

81. Philipp S, Hambrecht J, Braslavski L, Schroth G, Freichel M, et al. 1998. A novel capacitative calcium entry channel expressed in excitable cells. *EMBO J.* 17: 4274–82

82. Warnat J, Philipp S, Zimmer S, Flockerzi V, Cavalie A. 1999. Phenotype of a recombinant store-operated channel: highly selective permeation of Ca^{2+}. *J. Physiol.* 518:631–38

83. Plant TD, Schaefer M. 2003. TRPC4 and TRPC5: receptor-operated Ca2+-permeable nonselective cation channels. *Cell Calcium* 33:441–50

84. Okada T, Shimizu S, Wakamori M, Maeda A, Kurosaki T, et al. 1998. Molecular cloning and functional characterization of a novel receptor-activated TRP Ca^{2+} channel from mouse brain. *J. Biol. Chem.* 273:10279–87

85. Jung S, Muhle A, Schaefer M, Strotmann R, Schultz G, Plant TD. 2003. Lanthanides potentiate TRPC5 currents by an action at extracellular sites close to the pore mouth. *J. Biol. Chem.* 278:3562–71

86. Liu X, Singh BB, Ambudkar IS. 2003. TRPC1 is required for functional store-operated Ca2+ channels. Role of acidic amino acid residues in the S5-S6 region. *J. Biol. Chem.* 278:11337–43

87. Dohke Y, Oh YS, Ambudkar IS, Turner RJ. 2004. Biogenesis and topology of the transient receptor potential Ca2+ channel TRPC1. *J. Biol. Chem.* 279:12242–48

88. Obukhov AG, Nowycky MC. 2005. A cytosolic residue mediates Mg2+ block and regulates inward current amplitude of a transient receptor potential channel. *J. Neurosci.* 25:1234–39

89. Hofmann T, Chubanov V, Gudermann T, Montell C. 2003. TRPM5 is a voltage-modulated and Ca(2+)-activated monovalent selective cation channel. *Curr. Biol.* 13:1153–58

90. Perez CA, Huang L, Rong M, Kozak JA, Preuss AK, et al. 2002. A transient receptor potential channel expressed in taste receptor cells. *Nat. Neurosci.* 5:1169–76

91. Zhang Y, Hoon MA, Chandrashekar J,

Mueller KL, Cook B, et al. 2003. Coding of sweet, bitter, and umami tastes: different receptor cells sharing similar signaling pathways. *Cell* 112:293–301

92. Schlingmann KP, Weber S, Peters M, Niemann LN, Vitzthum H, et al. 2002. Hypomagnesemia with secondary hypocalcemia is caused by mutations in TRPM6, a new member of the TRPM gene family. *Nat. Genet.* 31:166–70

93. Nadler MJJ, Hermosura MC, Inabe K, Perraud AL, Zhu QQ, et al. 2001. LTRPC7 is a Mg.ATP-regulated divalent cation channel required for cell viability. *Nature* 411:590–95

94. Voets T, Nilius B, Hoefs S, van der Kemp AW, Droogmans G, et al. 2004. TRPM6 forms the Mg2+ influx channel involved in intestinal and renal Mg2+ absorption. *J. Biol. Chem.* 279:19–25

95. Walder RY, Landau D, Meyer P, Shalev H, Tsolia M, et al. 2002. Mutation of TRPM6 causes familial hypomagnesemia with secondary hypocalcemia. *Nat. Genet.* 31:171–74

96. Monteilh-Zoller MK, Hermosura MC, Nadler MJ, Scharenberg AM, Penner R, Fleig A. 2003. TRPM7 provides an ion channel mechanism for cellular entry of trace metal ions. *J. Gen. Physiol.* 121:49–60

97. Story GM, Peier AM, Reeve AJ, Eid SR, Mosbacher J, et al. 2003. ANKTM1, a TRP-like channel expressed in nociceptive neurons, is activated by cold temperatures. *Cell* 112:819–29

98. Peier AM, Moqrich A, Hergarden AC, Reeve AJ, Andersson DA, et al. 2002. A TRP channel that senses cold stimuli and menthol. *Cell* 108:705–15

99. McKemy DD, Neuhausser WM, Julius D. 2002. Identification of a cold receptor reveals a general role for TRP channels in thermosensation. *Nature* 416:52–58

100. Launay P, Fleig A, Perraud AL, Scharenberg AM, Penner R, Kinet JP. 2002. TRPM4 Is a Ca^{2+}-activated nonselective cation channel-mediating cell membrane depolarization. *Cell* 109:397–407

101. Prawitt D, Monteilh-Zoller MK, Brixel L, Spangenberg C, Zabel B, et al. 2003. TRPM5 is a transient Ca2+-activated cation channel responding to rapid changes in [Ca2+]i. *Proc. Natl. Acad. Sci. USA* 100:15166–71

102. Nilius B, Prenen J, Droogmans G, Voets T, Vennekens R, et al. 2003. Voltage dependence of the Ca2+-activated cation channel TRPM4. *J. Biol. Chem.* 278:30813–20

103. Perraud AL, Fleig A, Dunn CA, Bagley LA, Launay P, et al. 2001. ADP-ribose gating of the calcium-permeable LTRPC2 channel revealed by Nudix motif homology. *Nature* 411:595–99

104. Sano Y, Inamura K, Miyake A, Mochizuki S, Yokoi H, et al. 2001. Immunocyte Ca^{2+} influx system mediated by LTRPC2. *Science* 293:1327–30

105. Hara Y, Wakamori M, Ishii M, Maeno E, Nishida M, et al. 2002. LTRPC2 Ca^{2+}-permeable channel activated by changes in redox status confers susceptibility to cell death. *Mol. Cell* 9:163–73

106. Lee N, Chen J, Sun L, Wu SJ, Gray KR, et al. 2003. Expression and characterization of human transient receptor potential melastatin 3 (hTRPM3). *J. Biol. Chem.* 278:20890–97

107. Grimm C, Kraft R, Sauerbruch S, Schultz G, Harteneck C. 2003. Molecular and functional characterization of the melastatin-related cation channel TRPM3. *J. Biol. Chem.* 278:21493–501

108. Perraud AL, Schmitz C, Scharenberg AM. 2003. TRPM2 Ca2+ permeable cation channels: from gene to biological function. *Cell Calcium* 33:519–31

109. Nilius B, Prenen J, Janssens A, Owsianik G, Wang C, et al. 2005. The electivity filter of the cation channel TRPM4. *J. Biol. Chem.* 280:22899–906

110. Oberwinkler J, Lis A, Giehl KM, Flockerzi V, Philipp SE. 2005. Alternative

splicing switches the divalent cation selectivity of TRPM3 channels. *J. Biol. Chem.* 280:22540–48

111. Chen M, Simard JM. 2001. Cell swelling and a nonselective cation channel regulated by internal Ca2+ and ATP in native reactive astrocytes from adult rat brain. *J. Neurosci.* 21:6512–21

112. Liman ER. 2003. Regulation by voltage and adenine nucleotides of a Ca2+-activated cation channel from hamster vomeronasal sensory neurons. *J. Physiol.* 548:777–87

113. Schwede T, Kopp J, Guex N, Peitsch MC. 2003. SWISS-MODEL: an automated protein homology-modeling server. *Nucleic Acids Res.* 31:3381–85

114. Guex N, Peitsch MC. 1997. SWISS-MODEL and the Swiss-PdbViewer: an environment for comparative protein modeling. *Electrophoresis* 18:2714–23

115. Peitsch MC. 1995. Protein modeling by e-mail. *Bio/Technology* 13:658–60

116. Zitt C, Zobel A, Obukhov AG, Harteneck C, Kalkbrenner F, et al. 1996. Cloning and functional expression of a human Ca^{2+}-permeable cation channel activated by calcium store depletion. *Neuron* 16:1189–96

117. Kamouchi M, Philipp S, Flockerzi V, Wissenbach U, Mamin A, et al. 1999. Properties of heterologously expressed hTRP3 channels in bovine pulmonary artery endothelial cells. *J. Physiol.* 518:345–58

118. Kiselyov K, Xu X, Mozhayeva G, Kuo T, Pessah I, et al. 1998. Functional interaction between InsP$_3$ receptors and store-operated Htrp3 channels. *Nature* 396:478–82

119. Groschner K, Hingel S, Lintschinger B, Balzer M, Romanin C, et al. 1998. Trp proteins form store-operated cation channels in human vascular endothelial cells. *FEBS Lett.* 437:101–6

120. Hofmann T, Obukhov AG, Schaefer M, Harteneck C, Gudermann T, Schultz G. 1999. Direct activation of human TRPC6 and TRPC3 channels by diacylglycerol. *Nature* 397:259–63

121. Schaefer M, Plant TD, Obukhov AG, Hofmann T, Gudermann T, Schultz G. 2000. Receptor-mediated regulation of the nonselective cation channels TRPC4 and TRPC5. *J. Biol. Chem.* 275:17517–26

122. Yamada H, Wakamori M, Hara Y, Takahashi Y, Konishi K, et al. 2000. Spontaneous single-channel activity of neuronal TRP5 channel recombinantly expressed in HEK293 cells. *Neurosci. Lett.* 285:111–14

123. Shi J, Mori E, Mori Y, Mori M, Li J, et al. 2004. Multiple regulation by calcium of murine homologues of transient receptor potential proteins TRPC6 and TRPC7 expressed in HEK293 cells. *J. Physiol.* 561:415–32

124. Okada T, Inoue R, Yamazaki K, Maeda A, Kurosaki T, et al. 1999. Molecular and functional characterization of a novel mouse transient receptor potential protein homologue TRP7. Ca^{2+}-permeable cation channel that is constitutively activated and enhanced by stimulation of G protein-coupled receptor. *J. Biol. Chem.* 274:27359–70

125. Caterina MJ, Schumacher MA, Tominaga M, Rosen TA, Levine JD, Julius D. 1997. The capsaicin receptor: a heat-activated ion channel in the pain pathway. *Nature* 389:816–24

126. Strotmann R, Schultz G, Plant TD. 2003. Ca^{2+}-dependent potentiation of the nonselective cation channel TRPV4 is mediated by a carboxy terminal calmodulin binding site. *J. Biol. Chem.* 278:26541–49

127. Yue L, Peng JB, Hediger MA, Clapham DE. 2001. CaT1 manifests the pore properties of the calcium-release-activated calcium channel. *Nature* 410:705–9

128. Runnels LW, Yue L, Clapham DE. 2001. TRP-PLIK, a bifunctional protein with kinase and ion channel activities. *Science* 291:1043–47

129. Luo Y, Vassilev PM, Li X, Kawanabe Y, Zhou J. 2003. Native polycystin 2 functions as a plasma membrane Ca2+-permeable cation channel in renal epithelia. *Mol. Cell Biol.* 23:2600–7

130. Gonzalez-Perret S, Kim K, Ibarra C, Damiano AE, Zotta E, et al. 2000. Polycystin-2, the protein mutated in autosomal dominant polycystic kidney disease (ADPKD), is a Ca^{2+}-permeable nonselective cation channel. *Proc. Natl. Acad. Sci. USA* 98:1182–87

131. Koulen P, Cai Y, Geng L, Maeda Y, Nishimura S, et al. 2002. Polycystin-2 is an intracellular calcium release channel. *Nat. Cell Biol.* 4:191–97

132. Chen XZ, Vassilev PM, Basora N, Peng JB, Nomura H, et al. 1999. Polycystin-L is a calcium-regulated cation channel permeable to calcium ions. *Nature* 401:383–86

133. Volk T, Schwoerer AP, Thiessen S, Schultz JH, Ehmke H. 2003. A polycystin-2-like large conductance cation channel in rat left ventricular myocytes. *Cardiovasc. Res.* 58:76–88

134. LaPlante JM, Falardeau J, Sun M, Kanazirska M, Brown EM, et al. 2002. Identification and characterization of the single channel function of human mucolipin-1 implicated in mucolipidosis type IV, a disorder affecting the lysosomal pathway. *FEBS Lett.* 532:183–87

Annu. Rev. Physiol. 2006. 68:719–36
doi: 10.1146/annurev.physiol.68.040204.100715
First published online as a Review in Advance on September 16, 2005

TRP Channels in *C. elegans*

Amanda H. Kahn-Kirby[1] and Cornelia I. Bargmann[2]

[1]*Department of Physiology, University of California, San Francisco, California
94143-2240; email: amandak@phy.ucsf.edu*
[2]*The Rockefeller University, New York, New York 10021; email: cori@rockefeller.edu*

Key Words   sensory transduction, cation channels, Mg^{2+} homeostasis, cilia,
store-operated channel

■ **Abstract**   The TRP (transient receptor potential) superfamily of cation channels
is present in all eukaryotes, from yeast to mammals. Many TRP channels have been
studied in the nematode *Caenorhabditis elegans*, revealing novel biological functions,
regulatory modes, and mechanisms of localization. *C. elegans* TRPV channels function
in olfaction, mechanosensation, osmosensation, and activity-dependent gene regulation. Their activity is regulated by G protein signaling and polyunsaturated fatty acids.
C. elegans TRPPs related to human polycystic kidney disease genes are expressed
in male-specific neurons. The KLP-6 kinesin directs TRPP channels to cilia, where
they may interact with F0/F1 ATPases. A sperm-specific TRPC channel, TRP-3, is required for fertilization. Upon sperm activation, TRP-3 translocates from an intracellular
compartment to the plasma membrane to allow store-operated Ca^{2+} entry. The TRPM
channels GON-2 and GTL-2 regulate Mg^{2+} homeostasis and Mg^{2+} uptake by intestinal
cells; GON-2 is also required for gonad development. The TRPML CUP-5 promotes
normal lysosome biogenesis and prevents apoptosis. Dynamic, precise expression of
TRP proteins generates a remarkable range of cellular functions.

OVERVIEW: *C. ELEGANS* AS A MODEL SYSTEM FOR UNDERSTANDING TRP FUNCTION

An astonishing variety of biological functions are associated with the conserved
TRP (transient receptor potential) channel superfamily, a class of channels defined by sequence similarity to the *Drosophila* phototransduction channel TRP
(1, 2). TRPs assemble into homo- and heterotetramers to form cation-selective
ion channels that can be regulated by thermal stimuli, mechanical stimuli, lipids,
lipid derivatives, voltage, pH, phosphorylation, and intracellular Ca^{2+} stores (3).
Some TRPs serve as integrators of multiple regulatory pathways, and others are
activated by one predominant pathway. TRP proteins are linked to many sensory
modalities: vertebrate heat sensation (4–8), cold sensation (9, 10), osmosensation
(11, 12), pheromone sensation (13) and hearing (14); insect phototransduction (1,
15, 16), mechanosensation (17), thermosensation (18) and hearing (19); and nematode olfaction, mechanosensation, and osmosensation (20–22). Although TRP

functions have been studied most extensively in sensory neurons, vertebrate TRP channels also regulate cardiovascular (23, 24), renal (25), and lysosomal functions (26).

The TRP superfamily can be divided into seven families of channels based on sequence similarity. In humans, six TRP families encode a total of 28 channel subunits: TRPC (classical/short; seven members including a pseudogene), TRPV (vanilloid; six members), TRPM (melastatin/long; eight members), TRPML (mucolipin; three members), TRPP (polycystin; three members), and TRPA (one member) (27). Nonmammalian vertebrates also have a TRPN family (one member) (28). All TRP members have six predicted transmembrane domains; several families have a variable number of ankyrin motifs, suggested to participate in protein-protein interactions (Figure 1). Outside of these core regions, members of individual TRP families may share other motifs, such as coiled-coil domains. *Caenorhabditis elegans* has members of all seven known TRP families as well as novel TRP genes (Figure 1), and mutants are available for many of these loci. Because of the simple anatomy of *C. elegans*, the functions of these channels can be studied at single-cell resolution. Several *C. elegans* TRP channels have been studied in nonneural tissues, which may provide insight into analogous cellular functions of mammalian relatives. *C. elegans* studies of TRP channels have focused less on the molecular gating and biophysical properties of the channels and more on their integration into cellular pathways and neural circuits. As such, they are a useful complement to biophysical and pharmacological studies of mammalian counterparts.

C. ELEGANS TRPV CHANNELS FUNCTION IN OLFACTION AND NOCICEPTION

The TRPV gene *osm-9* was identified contemporaneously with its mammalian homolog TRPV1 (VR1), defining the first typical family beyond TRPC channels (4, 20). *osm-9* mutants have abnormal olfactory responses to all odors sensed by a class of ciliated neurons referred to as AWA neurons (20). The OSM-9 protein is localized to AWA sensory cilia (Figure 2), consistent with a role in olfactory signal transduction. In addition, *osm-9* mutants have a near-complete defect in the functions of ciliated sensory neurons called ASH neurons that act as polymodal nociceptors. ASH neurons mediate behavioral avoidance of high osmolarity, mechanical stimuli, noxious odors, heavy metals, bitter substances, and acid pH (20, 29–31). The OSM-9 protein is localized to ASH sensory cilia and is required for primary ASH sensory signal transduction—a nociceptive function analogous to the function of mammalian TRPV1 (32).

In addition to *osm-9*, the *C. elegans* genome encodes four other TRPV genes: *ocr-1*, *ocr-2*, *ocr-3*, and *ocr-4* (21). Each *ocr* gene is expressed in a subset of the cells that express *osm-9*, suggesting that *ocr* genes usually function together with *osm-9*. This prediction is valid for the *ocr-2* gene, which is expressed with *osm-9*

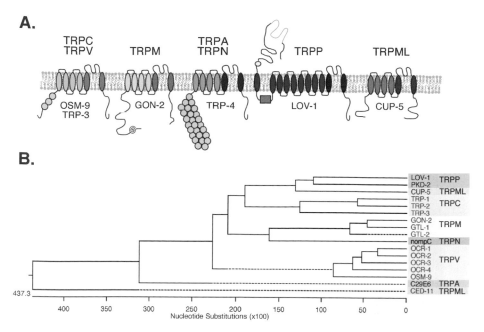

A.

B.

Figure 1 TRP structure and select *C. elegans* TRP channels. (*A*) Schematic domain structures for selected *C. elegans* TRP proteins. The cytoplasmic N' region of several TRP families contains a variable number of ankyrin repeats (*green circles*). The TRPP protein LOV-1 has a large extracellular domain containing serine/threonine-rich (*green*) and GPS (*yellow*) regions as well as a total of 11 predicted transmembrane domains. Channel regions have six transmembrane domains, with S5 and S6 gate domains flanking a pore-loop selectivity filter. The transmembrane domains and pore loop have the strongest conservation among TRP family channels. The cytoplasmic C' region varies among families and may contain lipid-binding motifs (*orange box*), coiled-coil domains (*red coil*), or other functional structures. (*B*) Alignment of *C. elegans* TRP channels. Conserved transmembrane regions were identified with SMART analysis and refined with NCBI CDD/reverse psi-BLAST. ClustalW was used to align transmembrane domains, and the results are presented as a phylogram. *C. elegans* has at least six candidate TRP genes from novel families that are entirely uncharacterized; these are omitted from the figure for clarity and are not discussed in the text.

in the ASH and AWA sensory neurons (Figure 2). Animals mutant for *ocr-2* have defects in nociception and olfaction that are similar to, though slightly less severe than, the defects in *osm-9* mutants. The similar mutant phenotypes of *osm-9* and *ocr-2* suggest that these genes may form heteromeric channels in ASH and AWA, although there is no direct biochemical evidence for this association. Increasing evidence suggests that many TRP channels may be heteromeric, including channels combining *Drosophila* TRP and TRPL as well as channels combining mammalian TRPC1 and TRPC5 (33, 34).

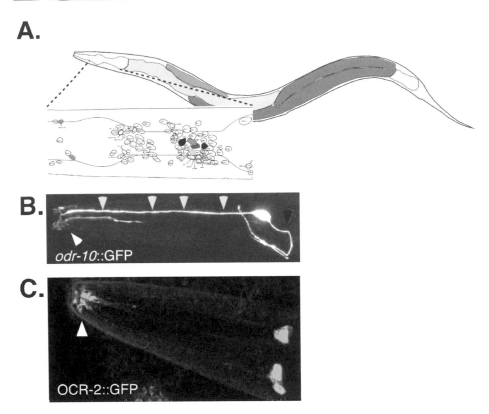

Figure 2 *C. elegans* TRPV proteins are expressed in sensory cilia. (*A*) Schematic diagram of a hermaphrodite *C. elegans*, highlighting a subset of anterior neurons with TRPV channel expression that are mentioned in the text (AWA, *green*; ASH, *orange*; ADF, *purple*; ASE, *pink*; AWC, *yellow*; OLQ, *blue*). (*B*) Confocal image of an AWA olfactory neuron expressing GFP driven by the *odr-10* promoter. The axon (*red arrow*), dendrite (*yellow arrows*), and sensory cilia (*white arrow*) are visible. (*C*) Confocal image of ASH and AWA sensory neurons expressing an OCR-2::GFP fusion protein. Note prominent expression of the OCR-2 protein in cilia (*arrowhead*) and cell bodies; the axons and dendrites have little OCR-2 protein.

In AWA and ASH neurons, both OSM-9 and OCR-2 proteins are enriched in sensory cilia (21). OSM-9 and OCR-2 mutually depend on each other for localization to cilia rather than to the cell body. Some neurons normally express the *osm-9* gene in the absence of any *ocr* gene, and in these neurons, OSM-9 protein is found in the cell body. However, ectopic expression of OCR-2 in one such cell class, the AWC chemosensory neurons, is sufficient to drive OSM-9 to the cilia. These findings suggest a physical interaction between the OSM-9 and OCR-2 subunits that mediates their localization.

G PROTEIN–COUPLED LIPID SIGNALING PATHWAYS REGULATE TRPV CHANNEL SIGNALING

OSM-9 and OCR-2 have not been amenable to electrophysiological analysis in heterologous cells, and as a result, the molecular regulation of OSM-9/OCR-2 is only partly understood. In AWA olfactory neurons, *osm-9* and *ocr-2* act downstream of G protein–coupled odorant receptors, probably as the olfactory transduction channel (20, 35). A similar role downstream of G protein–coupled receptors is likely in some forms of ASH nociception, particularly the avoidance of noxious odors (36). The role of OSM-9/OCR-2 in detecting physical stimuli such as high osmolarity and nose touch may mean that these channels directly sense force. The cytoplasmic OSM-10 protein is required only for osmosensation, suggesting that a specialized sensory apparatus helps OSM-9/OCR-2 sense osmotic stimuli (37).

Genetic analysis indicates that sensory G proteins may activate OSM-9 and OCR-2 by mobilizing specific polyunsaturated fatty acids (PUFAs). *C. elegans* mutants in the omega-3 lipid desaturase enzyme *fat-3* are deficient in long-chain PUFAs. Like TRPV mutants, *fat-3* mutants show pronounced defects in ASH nociceptive behaviors and AWA olfactory behaviors as well as primary defects in ASH sensory transduction measured by Ca^{2+} imaging (38). PUFAs stimulate rapid, TRPV-dependent Ca^{2+} transients in the ASH neurons and induce TRPV-dependent avoidance behaviors indicative of ASH activation. A battery of PUFA biosynthetic mutants, as well as acute rescue of *fat-3* mutants with dietary lipid supplementation, have implicated the omega-3 and omega-6 PUFAs arachidonic acid and eicosapentaenoic acid in OSM-9 TRPV signaling. The exact enzyme that mobilizes these PUFAs downstream of G proteins is unknown. The physiological mechanisms underlying the documented human health benefits of dietary omega-3 fatty acids are mysterious. By analogy with *C. elegans*, TRP channels that act in inflammation and cardiovascular regulation may represent molecular targets for dietary PUFAs in humans.

TRPV CHANNELS REGULATE TRANSCRIPTION AND MODULATE COMPLEX BEHAVIORS

In addition to their roles in sensory transduction, *osm-9*, *ocr-2*, and *ocr-1* regulate the transcription of sensory genes. *osm-9* and *ocr-2* mutants have reduced expression of the G protein–coupled receptor ODR-10 (which recognizes the odorant diacetyl) in AWA olfactory neurons (21). *ocr-1*, which is expressed in AWA but has no detectable role in AWA olfactory signaling, also affects the level of *odr-10* expression. *osm-9* and *ocr-2* also stimulate expression of the serotonin biosynthetic gene *tph-1* (encoding tryptophan hydroxylase) in ADF neurons, a pair of ciliated chemosensory neurons (39). *osm-9* and the *ocr* genes are likely to act in activity-dependent gene expression pathways that link sensory stimulation to patterns of gene expression. The Ca^{2+}/calmodulin-dependent kinase CaMKII functions downstream of OSM-9 and OCR-2 in the signaling pathway from sensory transduction to gene expression in ADF neurons (39). In one straightforward model, Ca^{2+} entry through OSM-9/OCR-2 channels could activate CaMKII to initiate transcriptional changes.

The ability of OCR-2 to regulate gene expression in ADF neurons can be separated from some of its other sensory functions. A point mutation in an N-terminal helical region of OCR-2 eliminates its ability to stimulate *tph-1* expression but does not diminish AWA olfactory function or cilia localization of the OCR-2 protein (40). Conversely, inserting the N-terminal helical region of OCR-2 into the related OCR-4 protein makes OCR-4 competent to stimulate *tph-1* expression.

In addition to their primary sensory roles, OSM-9 TRPV channels can affect sensory adaptation after prolonged exposure to an odor or taste. *osm-9* is expressed in many *C. elegans* ciliated neurons whose sensory transduction is mediated by cGMP and cGMP-gated channels (20). In two of these cGMP signaling neurons, the AWC olfactory neurons and the ASE gustatory neurons, *osm-9* is not required for primary sensory signaling but is required for sensory adaptation (41, 42). Neurons that use TRPV channels in adaptation express only *osm-9*, whereas neurons in which TRPV channels are primary transduction channels express both *osm-9* and at least one *ocr* gene. This distinction may be related to the preferential ciliary localization of OSM-9/OCR-2 complexes, as described above.

Another modulatory function for *osm-9* and *ocr-2* is their ability to regulate aggregation, or social behavior (43). Some natural isolates of *C. elegans* form aggregates of dozens of animals when they forage on bacteria, the "social" phenotype (44). Mutations in *osm-9* or *ocr-2* suppress aggregation, at least partly because of TRPV function in the ASH nociceptive neurons (43). Aggregation requires at least three different classes of sensory neurons, including TRPV-dependent nociceptive neurons, oxygen-sensing neurons that signal using a soluble guanylate cyclase (45), and a third neuronal class (46). TRPV-dependent nociception, oxygen sensation, and signals from food are integrated to produce context-dependent aggregation behavior.

PURSUING THE MAMMALIAN ANALOGY: ORTHOLOGY BETWEEN TRPVS?

Sequence analysis suggests that the common ancestor of mammals and invertebrates had one or two TRPV genes; there are no clear orthologies between individual mammalian and nematode TRPVs. Nonetheless, experiments using heterologous expression of mammalian channels in *C. elegans* neurons have revealed functional analogies between different mammalian TRPVs and *C. elegans* TRPVs.

The first experiment of this type involved the expression of rat TRPV1 in ASH nociceptive neurons (21). TRPV1 has a role in pain sensation, responding to irritants such as capsaicin. Rat TRPV1 expressed in ASH functions as a capsaicin-gated channel and can cause *C. elegans*, which is normally oblivious to capsaicin, to avoid the irritant. The behavioral response to this artificial activation of ASH is strikingly similar to the avoidance of repellents normally sensed by ASH. When expressed in ASH, rat TRPV1 functions independently of native ASH signal transduction pathways and cannot substitute for the normal functions of either *osm-9* or *ocr-2* (21).

By contrast, expression of the rat osmosensory channel TRPV4 in ASH nociceptive neurons can rescue the osmosensitivity and mechanosensitivity of *osm-9* mutants, although the channel cannot rescue their G protein–mediated odorant responses (47). TRPV4 requires endogenous ASH signaling molecules to perform this function, but it changes the threshold for osmosensation to match the mammalian threshold rather than that of *C. elegans*. Thus, TRPV4 functions as a partial *osm-9* ortholog, while retaining a distinct sensory signature. These results place TRPV4, and by implication OSM-9, very close to the primary event in osmosensation.

Finally, expression of mouse or human TRPV2 in ADF neurons can partially rescue the defect in *tph-1* gene expression that is observed in *ocr-2* mutants (40). As does endogenous *ocr-2*, TRPV2 requires *osm-9* for full function in ADF, again suggesting that TRPV2 can be integrated into endogenous *C. elegans* signaling pathways. The endogenous function of mammalian TRPV2 is not understood; potential analogies with OCR-2 may be explored further.

TRPP CHANNELS ARE REQUIRED FOR MALE MATING BEHAVIOR

Mutations in the two genes PKD1 and PKD2 account for 95% of the occurrences of human autosomal dominant polycystic kidney disease, one of the most common inherited genetic disorders (48). PKD1 and PKD2 encode the polycystins, large multidomain proteins that define the TRPP family of channels. Polycystic kidney disease is associated with fluid-filled cysts in the kidneys and other tissues. The mammalian polycystin-1 and polycystin-2 proteins are present in the cilia of renal cells (49, 50), where they have been proposed to act in intracellular traffic, fluid accumulation, or ion transport, or as generators or sensors of force. *C. elegans* has homologs of each of these genes, which are called *lov-1* (PKD1) and *pkd-2* (PKD2) (Figure 3). These TRPP proteins underlie the behavior that male worms exhibit when they encounter a hermaphrodite.

Male *C. elegans* use an elaborate sensory apparatus in their tail to execute a stereotyped search for the hermaphrodite vulva. This search is followed by spicule insertion and sperm release (51). *lov-1* and *pkd-2* males have defective responses to contact with a hermaphrodite, whereas other TRP mutants such as *osm-9* have normal male mating behavior (22, 52). *lov-1* and *pkd-2* are expressed in the cilia of the male-specific CEM, HOB, and ray neurons, which may have mechanosensory functions (22, 52). *lov-1; pkd-2* double mutants show the same behavioral defects as do the single mutants, consistent with the possibility that each has essential

A.

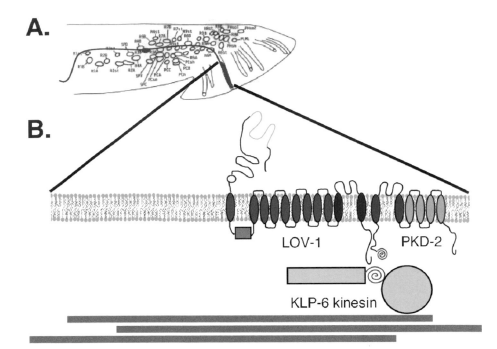

B.

Figure 3   Polycystin localization in the cilia of male-specific neurons: (*A*) Schematic diagram of male tail, highlighting one of the ray neurons that expresses *lov-1* and *pkd-2* TRPP proteins. The cilia extend into the fan-shaped male tail used for mating. (*B*) Enlarged view of ray neuron sensory cilia and proposed LOV-1/PKD-2 interactions. The KLP-6 kinesin is required for TRPP cilia localization, perhaps by transporting a LOV-1/PKD-2 complex to cilia along microtubules (*gray lines*). Red coils, coiled-coil domains; orange box, PLAT domain.

functions in the same signaling complex. Like their mammalian orthologs, PKD-2 and LOV-1 proteins are enriched in sensory cilia and require intact cilia for their function (53, 54). Cilia morphology appears normal in *lov-1* and *pkd-2* mutants, suggesting that they have an acute sensory role rather than a function in ciliogenesis.

The relationship between human kidney function and *C. elegans* mating is most easily explained by suggesting a special relationship between TRPP channels and force-sensing cilia. In this scenario, TRPP channels sense both fluid flow in the kidneys and mechanical stimuli during mating. A similar role is suggested by the role of TRPP channels in early vertebrate development. Ciliated cells in Hensen's node of vertebrates establish the left-right asymmetry of the developing embryo (55, 56). Some of these nodal cilia are motile, despite a 9 + 0 arrangement of microtubules that is typical for nonmotile cilia (a morphology shared by *C. elegans* cilia). Mouse polycystin-2 mutants have defects in left-right asymmetry, and polycystin-2 is expressed in nodal cilia, consistent with a role in left-right patterning (57). The polycystin complex is thus a candidate to generate or sense motility in nodal cilia.

TRPP CILIA LOCALIZATION AND A POSSIBLE RELATIONSHIP WITH THE F0/F1 ATPASE

The mechanisms by which membrane proteins such as LOV-1 and PKD-2 are localized to cilia are only partly understood. Targeted transport vesicles may carry G protein–coupled receptors and TRPV channels from the Golgi to the base of the cilia; an AP-1 adaptor complex appears to be essential for cilia-directed transport (58). Within cilia, proteins are transported by the intraflagellar transport (IFT) protein complex, with kinesins that move to the cilia tip and a dynein that moves back to the base of the cilia (59, 60). An uncharacterized transition occurs between the dendrite and the base of the cilia to allow membrane proteins access to the cilia. A genetic screen for mutants with *pkd-2*-like mating defects uncovered one potential player in this process, the kinesin KLP-6, which affects PKD-2 localization to cilia (61) (Figure 3). KLP-6 is related to the axonal synaptic vesicle transport kinesin UNC-104/Kif1A. In *klp-6* mutants, PKD-2 often accumulates at the base of the cilia rather than the cilia proper, and it is also more prevalent in the dendrites. Cilia morphology is normal in *klp-6* mutants, implicating *klp-6* in the function rather than development of the cilia. *lov-1*, *pkd-2*, and *klp-6* are all expressed in a subset of ciliated neurons, most prominently in the male mating neurons. These results raise the possibility that various motor proteins may have selective transport properties in different ciliated cells.

A priority in the further understanding of TRPPs is the identification of additional signaling components in the TRPP complex. Within the LOV-1 (polycystin-1) protein is a cytoplasmic loop called the PLAT (polycystin/lipoxygenase/a-toxin) domain. A yeast two-hybrid screen with the LOV-1 PLAT domain yielded an F1 ATP synthase subunit, ATP-2 (62). Human polycystin-1 can also bind ATP-2. The F0/F1 ATPase is an essential component of the mitochondrial respiratory chain, and because mitochondria are absent from cilia, this interaction looks odd, perhaps spurious. However, unlike other mitochondrial enzymes, both ATP-2 and the transmembrane F0 subunit ASG-2 can be detected in cilia, and surface expression of the F0/F1 ATPase has been reported in mammalian cells as well (63, 64). Reducing ATPase function in male sensory neurons with RNAi attenuates male mating, leading Hu & Barr (62) to suggest that ATPase function in cilia may promote LOV-1/PKD-2 function. The F0/F1 ATPase is best known for its coupling of a mitochondrial pH gradient to ATP production in respiration, and the presence of this ATPase in cilia may be indicative of a high ATPase requirement in this compartment (62). Alternatively, cilia may use the F0/F1 ATPase in a distinct capacity, such as its capacity to act as an ATP- and pH-regulated molecular motor (65, 66).

Genetic and biochemical studies should identify additional components of the TRPP signaling complex. For example, microarray analysis has identified four genes that are coexpressed with *lov-1* and *pkd-2* in male-specific neurons (67); Portman & Emmons (67) propose that these novel secreted proteins are components of an extracellular force-sensing matrix surrounding sensory cilia.

TRPC AND TRPM CHANNELS: A FERTILE FIELD

The TRPC protein encoded by *trp-3/spe-41* is found exclusively in sperm and is required for a late step in fertilization (68) (Figure 4). Both male and hermaphrodite *C. elegans* produce sterile sperm in *trp-3* mutants. Unlike TRPP *lov-1* and *pkd-2* mutants, male *trp-3* mutants execute normal behavioral mating and transfer sperm to hermaphrodites during mating. Moreover, *trp-3* mutants have motile, morphologically normal sperm. These sperm are even capable of competing with other sperm for a position in the spermatheca, a small compartment near the oocytes where hermaphrodites store sperm prior to fertilization. These experiments suggest that *trp-3* sperm have problems at a step between contact with the oocyte and fertilization.

Ca^{2+} imaging of normal *C. elegans* sperm reveals increased Ca^{2+} influx if internal Ca^{2+} stores are depleted with drugs such as thapsigargin (68). This influx is diagnostic of store-operated Ca^{2+} channels, an important class of homeostatic channels found in many cell types. In *trp-3* mutant sperm, this influx is lost, suggesting that TRP-3 functions as a store-operated channel in *C. elegans* sperm. When heterologously expressed in HEK293 cells, TRP-3 promotes Ca^{2+} influx in response to store depletion and Gq pathway activation. Studies of mammalian TRPC subunits have provided conflicting evidence about the role of Ca^{2+} stores in regulating this family (69), so it is gratifying to see that a native TRPC protein functions as a store-operated channel in its endogenous context.

The subcellular localization of TRP-3 is developmentally regulated, providing an additional layer of channel regulation (68). The TRP-3 protein is found in vesicular compartments of immature spermatids and translocates to the plasma membrane in mature sperm during sperm activation. Mammalian TRPC5 and *Drosophila* TRPL (TRP-like) dynamically regulate their subcellular localization upon stimulation, and similar possibilities have been suggested for other TRP channels as well (34, 70). Regulated surface expression may be an exciting common property of TRPC channels. It is intriguing that several TRPC subunits are expressed in human sperm, hinting that *C. elegans* and humans may share ancient cellular mechanisms of fertilization (71).

C. elegans has two other TRPC proteins encoded by *trp-1* and *trp-2*. Mutant phenotypes have not been described for these two genes. *trp-1* is expressed in many motor neurons and interneurons as well as vulval and intestinal muscles (20).

In a different fertility-related function, the TRPM family member *gon-2* is required during mid-larval stages for proper development of gonadal tissues (72). In *gon-2* mutants, germ cells fail to proliferate and mature, a defect that could be either intrinsic to the germ cells or associated with other tissues that regulate the germ line (73).

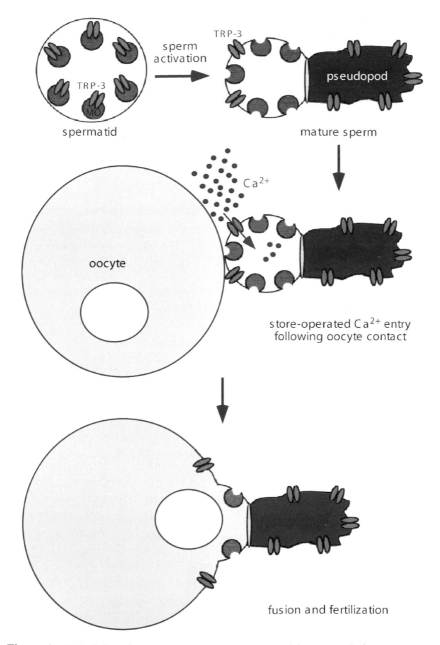

Figure 4 TRP-3 functions as a store-operated channel in sperm. In immature spermatids, TRP-3 (*green*) is sequestered inside cytoplasmic membranous organelles (*orange*). Following sperm activation, TRP-3 is found on the plasma membrane of sperm, including the pseudopod region. Upon contact with an oocyte, Ca^{2+} enters sperm through TRP-3 channels, followed by fusion with the oocyte and fertilization. Figure modified from Reference 68.

Mutations in *gem-4* suppress *gon-2* reduction-of-function alleles, restoring the mutants to fertility. *gem-4* encodes a widely expressed member of the copine family of Ca^{2+}-dependent phospholipid-binding proteins (74). *gem-4* fails to suppress the strongest *gon-2* mutations, suggesting that it acts by modulating *gon-2* activity. Because of its lipid-binding, Ca^{2+}-binding, and Mg^{2+}-binding motifs, GEM-4 has been suggested to regulate membrane trafficking of GON-2 (74).

Both GON-2 and the TRPM channel GTL-1 have roles in Mg^{2+} uptake by intestinal cells and in Mg^{2+} homeostasis (74a). These channels are localized to the apical surface of intestinal epithelial cells, in which they take up ions from dietary sources. GTL-1 appears to form a constitutively active channel for Ca^{2+} and Mg^{2+}, whereas GON-2 forms an outwardly rectifying channel for Ca^{2+} and Mg^{2+} that is strongly inhibited by intracellular Mg^{2+}. Animals with mutations in both genes exhibit arrested development in low Mg^{2+} but can be rescued if grown in high external Mg^{2+}. Mutations in the human TRPM6 gene result in familial hypomagnesemia with secondary hypocalcemia owing to poor Mg^{2+} uptake in the intestine. Thus, for TRPM channels, as for TRPML channels (see below), the physiological functions are strikingly comparable in nematodes and humans.

cup-5: A LYSOSOMAL TRPML WITH LINKS TO APOPTOSIS

Another family of TRP channels, the TRPMLs or mucolipins, is implicated in the rare human familial disorder mucolipidosis type IV. Human patients exhibit early-onset mental retardation and ophthalmic defects, including retinal degeneration, owing to lysosomal sorting and lysosomal storage defects. TRPML channels may be the most primitive of all of the TRPs, as yeast express a mechanosensitive, Ca^{2+}- and pH-regulated TRP channel in the lysosome-like vacuole (75–77).

C. elegans has one TRPML gene, *cup-5*, whose reduction-of-function mutants have an endocytosis defect in coelomocytes, scavenger cells that filter soluble proteins from the *C. elegans* body cavity (78). *cup-5* is expressed in many cell types and localizes to internal vesicles that are most likely to be lysosomes and late endosomes. *cup-5* mutants have abnormally large internal vacuoles and an inappropriate accumulation of proteins that should have been degraded in lysosomes. Animals bearing null mutants in *cup-5* have a maternal-effect lethal phenotype, with excessive apoptosis and many cells with large, malformed lysosomes and vacuoles (79). On the basis of these cell-biological criteria, the *C. elegans* phenotype closely matches the pathology of human mucolipidosis. Indeed, mammalian TRPML1 expressed from a heat-shock promoter rescues the lethality of *cup-5* mutants, consistent with an orthologous function (79). Similarly, coelomocyte expression of mammalian TRPML1 or TRPML3 rescues the *cup-5* endocytosis defects (80).

Detailed analysis of *cup-5* mutants, using subcellular markers and electron microscopy, indicates that their primary cellular defect is in lysosome biogenesis and that the accumulated organelles in *cup-5* mutants have mixed properties of late endosomes and lysosomes (79, 80). Thus, human and *C. elegans* mucolipins may share a function in lysosome biogenesis. Human TRPML1 expressed in liposomes forms a cation channel that is inhibited at low pH (81); perhaps a change in CUP-5/TRPML1 activity accompanies or defines the maturation of lysosomes.

cup-5 mutants have been isolated in a genetic screen for mutations that stimulate apoptosis (79). Animals bearing null mutants in *cup-5* have high levels of apoptosis even in the presence of a death-preventing *bcl2 (egl-9)* mutation. Hersh et al. (79) suggest that apoptosis is secondary to the *cup-5* defect in lysosome and vacuole formation. A worm TRPM subunit, *ced-11*, has been identified as an apoptosis mutant with abnormal cell corpses (G. Stanfield & H.R. Horvitz, personal communication), but the mechanism for this phenotype has not been described.

C. ELEGANS TRPS AND OPEN QUESTIONS

The study of *C. elegans* has already shed light on numerous aspects of TRP channel function and localization. The diverse *C. elegans* TRP channels offer avenues for illuminating additional questions. Because it is relatively easy to examine subcellular localization of *C. elegans* proteins in live animals, this should be a particularly valuable system for studying mechanisms of surface expression, trafficking to sensory cilia, and regulated translocation of TRP channels. An open area to explore is the relationship between channel trafficking and cellular function. For example, when channels such as TRP-3 in spermatids are contained in intracellular compartments, are they sequestered or are they actively producing cationic currents?

C. elegans has been an outstanding model for studying mechanosensation mediated by the Deg/EnaC channel family (reviewed in Reference 82); in the future, it may be a useful model with which to study possible mechanosensory functions of TRPs. One avenue to explore is the proposed mechanosensory and osmosensory function of OSM-9/OCR-2 channels in ASH nociception. Other TRP channels may also have mechanosensitive functions; for example, the TRPV gene *osm-9* is expressed, together with the uncharacterized TRPV gene *ocr-4*, in OLQ, PVD, and FLP mechanosensory neurons. Another candidate mechanosensor is the *C. elegans* TRPN protein Y71A12B.4/*trp-4*, the ortholog of *Drosophila* and zebrafish mechanoreceptive channels of the nompC family (17). *trp-4* is expressed in CEP and ADE neurons, which are thought to detect the light mechanosensory stimulus provided by a bacterial lawn (83). Mutants in *ocr-4* and *trp-4* have not yet been described.

Several *C. elegans* TRP genes are completely uncharacterized. These include one TRPA family member, C29E6.2, which shares 88% identity with the candidate *Drosophila* thermosensory channel ANKTM1, as well as several uncharacterized TRPC, TRPM, and TRPV family members. Six additional TRP genes in the *C. elegans* genome are apparently unrelated to the existing seven TRP families, and these could open up entirely new areas of TRP biology. The continuing analysis of

C. elegans TRP channels should raise and answer new questions while providing a physiological and cellular context for TRPs.

ACKNOWLEDGMENTS

We thank Miriam Goodman, Yun Zhang, Greg Lee, Andy Chang, and Massimo Hilliard for their critical comments on the manuscript; Gillian Stanfield and H. Robert Horvitz for sharing unpublished results; and Kouichi Iwasaki for sharing results prior to publication. C.I.B. is an Investigator of the Howard Hughes Medical Institute.

**The *Annual Review of Physiology* is online at
http://physiol.annualreviews.org**

LITERATURE CITED

1. Montell C, Rubin GM. 1989. Molecular characterization of the Drosophila *trp* locus: a putative integral membrane protein required for phototransduction. *Neuron* 2:1313–23

2. Hardie RC, Minke B. 1992. The *trp* gene is essential for a light-activated Ca^{2+} channel in Drosophila photoreceptors. *Neuron* 8:643–51

3. Clapham DE. 2003. TRP channels as cellular sensors. *Nature* 426:517–24

4. Caterina MJ, Schumacher MA, Tominaga M, Rosen TA, Levine JD, Julius D. 1997. The capsaicin receptor: a heat-activated ion channel in the pain pathway. *Nature* 389:816–24

5. Caterina MJ, Rosen TA, Tominaga M, Brake AJ, Julius D. 1999. A capsaicin-receptor homologue with a high threshold for noxious heat. *Nature* 398:436–41

6. Peier AM, Reeve AJ, Andersson DA, Moqrich A, Earley TJ, et al. 2002. A heat-sensitive TRP channel expressed in keratinocytes. *Science* 296:2046–49

7. Smith GD, Gunthorpe MJ, Kelsell RE, Hayes PD, Reilly P, et al. 2002. TRPV3 is a temperature-sensitive vanilloid receptor-like protein. *Nature* 418:186–90

8. Xu H, Ramsey IS, Kotecha SA, Moran MM, Chong JA, et al. 2002. TRPV3 is a calcium-permeable temperature-sensitive cation channel. *Nature* 418:181–86

9. McKemy DD, Neuhausser WM, Julius D. 2002. Identification of a cold receptor reveals a general role for TRP channels in thermosensation. *Nature* 416:52–58

10. Peier AM, Moqrich A, Hergarden AC, Reeve AJ, Andersson DA, et al. 2002. A TRP channel that senses cold stimuli and menthol. *Cell* 108:705–15

11. Liedtke W, Choe Y, Marti-Renom MA, Bell AM, Denis CS, et al. 2000. Vanilloid receptor-related osmotically activated channel (VR-OAC), a candidate vertebrate osmoreceptor. *Cell* 103:525–35

12. Strotmann R, Harteneck C, Nunnenmacher K, Schultz G, Plant TD. 2000. OTRPC4, a nonselective cation channel that confers sensitivity to extracellular osmolarity. *Nat. Cell Biol.* 2:695–702

13. Liman ER, Corey DP, Dulac C. 1999. TRP2: a candidate transduction channel for mammalian pheromone sensory signaling. *Proc. Natl. Acad. Sci. USA* 96:5791–96

14. Corey DP, Garcia-Anoveros J, Holt JR, Kwan KY, Lin SY, et al. 2004. TRPA1 is a candidate for the mechanosensitive transduction channel of vertebrate hair cells. *Nature* 432:723–30

15. Niemeyer BA, Suzuki E, Scott K, Jalink K, Zuker CS. 1996. The Drosophila light-activated conductance is composed of the two channels TRP and TRPL. *Cell* 85:651–59

16. Xu XZS, Chien F, Butler A, Salkoff L, Montell C. 2000. TRP gamma, a Drosophila TRP-related subunit, forms a regulated cation channel with TRPL. *Neuron* 26:647–57

17. Walker RG, Willingham AT, Zuker CS. 2000. A Drosophila mechanosensory transduction channel. *Science* 287:2229–34

18. Tracey WD Jr, Wilson RI, Laurent G, Benzer S. 2003. *painless*, a Drosophila gene essential for nociception. *Cell* 113:261–73

19. Kim J, Chung YD, Park DY, Choi S, Shin DW, et al. 2003. A TRPV family ion channel required for hearing in Drosophila. *Nature* 424:81–84

20. Colbert HA, Smith TL, Bargmann CI. 1997. OSM-9, a novel protein with structural similarity to channels, is required for olfaction, mechanosensation, and olfactory adaptation in *Caenorhabditis elegans*. *J. Neurosci.* 17:8259–69

21. Tobin DM, Madsen DM, Kahn-Kirby A, Peckol EL, Moulder G, et al. 2002. Combinatorial expression of TRPV channel proteins defines their sensory functions and subcellular localization in *C. elegans* neurons. *Neuron* 35:307–18

22. Barr MM, Sternberg PW. 1999. A polycystic kidney-disease gene homologue required for male mating behaviour in *C. elegans*. *Nature* 401:386–89

23. Zygmunt PM, Petersson J, Andersson DA, Chuang H-H, Sorgard M, et al. 1999. Vanilloid receptors on sensory nerves mediate the vasodilator action of anandamide. *Nature* 400:452–57

24. Hassock SR, Zhu MX, Trost C, Flockerzi V, Authi KS. 2002. Expression and role of TRPC proteins in human platelets: evidence that TRPC6 forms the store-independent calcium entry channel. *Blood* 100:2801–11

25. Mochizuki T, Wu G, Hayashi T, Xeno-phontos SL, Veldhuisen B, et al. 1996. PKD2, a gene for polycystic kidney disease that encodes an integral membrane protein. *Science* 272:1339–42

26. Sun M, Goldin E, Stahl S, Falardeau JL, Kennedy JC, et al. 2000. Mucolipidosis type IV is caused by mutations in a gene encoding a novel transient receptor potential channel. *Hum. Mol. Genet.* 9:2471–78

27. Montell C, Birnbaumer L, Flockerzi V, Bindels RJ, Bruford EA, et al. 2002. A unified nomenclature for the superfamily of TRP cation channels. *Mol. Cell* 9:229–31

28. Sidi S, Friedrich RW, Nicolson T. 2003. NompC TRP channel required for vertebrate sensory hair cell mechanotransduction. *Science* 301:96–99

29. Hilliard MA, Bargmann CI, Bazzicalupo P. 2002. *C. elegans* responds to chemical repellents by integrating sensory inputs from the head and the tail. *Curr. Biol.* 12:730–34

30. Sambongi Y, Nagae T, Liu Y, Yoshimizu T, Takeda K, et al. 1999. Sensing of cadmium and copper ions by externally exposed ADL, ASE, and ASH neurons elicits avoidance response in *Caenorhabditis elegans*. *NeuroReport* 10:753–57

31. Kaplan JM, Horvitz HR. 1993. A dual mechanosensory and chemosensory neuron in *Caenorhabditis elegans*. *Proc. Natl. Acad. Sci. USA* 90:2227–31

32. Hilliard MA, Bergamasco C, Arbucci S, Plasterk RH, Bazzicalupo P. 2004. Worms taste bitter: ASH neurons, QUI-1, GPA-3 and ODR-3 mediate quinine avoidance in *Caenorhabditis elegans*. *EMBO J.* 23:1101–11

33. Strubing C, Krapivinsky G, Krapivinsky L, Clapham DE. 2001. TRPC1 and TRPC5 form a novel cation channel in mammalian brain. *Neuron* 29:645–55

34. Bahner M, Frechter S, Da Silva N, Minke B, Paulsen R, Huber A. 2002. Light-regulated subcellular translocation of Drosophila TRPL channels induces long-term

adaptation and modifies the light-induced current. *Neuron* 34:83–93

35. Roayaie K, Crump JG, Sagasti A, Bargmann CI. 1998. The Gα protein ODR-3 mediates olfactory and nociceptive function and controls cilium morphogenesis in *C. elegans* olfactory neurons. *Neuron* 20: 55–67

36. Fukuto HS, Ferkey DM, Apicella AJ, Lans H, Sharmeen T, et al. 2004. G protein-coupled receptor kinase function is essential for chemosensation in *C. elegans*. *Neuron* 42:581–93

37. Hart AC, Kass J, Shapiro JE, Kaplan JM. 1999. Distinct signaling pathways mediate touch and osmosensory responses in a polymodal sensory neuron. *J. Neurosci.* 19: 1952–58

38. Kahn-Kirby AH, Dantzker JL, Apicella AJ, Schafer WR, Browse J, et al. 2004. Specific polyunsaturated fatty acids drive TRPV-dependent sensory signaling in vivo. *Cell* 119:889–900

39. Zhang S, Sokolchik I, Blanco G, Sze JY. 2004. *Caenorhabditis elegans* TRPV ion channel regulates 5HT biosynthesis in chemosensory neurons. *Development* 131: 1629–38

40. Sokolchik I, Tanabe T, Baldi PF, Sze JY. 2005. Polymodal sensory function of the Caenorhabditis elegans OCR-2 channel arises from distinct intrinsic determinants within the protein and is selectively conserved in mammalian TRPV proteins. *J. Neurosci.* 25:1015–23

41. Colbert HA, Bargmann CI. 1995. Odorant-specific adaptation pathways generate olfactory plasticity in *C. elegans*. *Neuron* 14: 803–12

42. Jansen G, Weinkove D, Plasterk RH. 2002. The G-protein gamma subunit *gpc-1* of the nematode *C. elegans* is involved in taste adaptation. *EMBO J.* 21:986–94

43. de Bono M, Tobin DM, Davis MW, Avery L, Bargmann CI. 2002. Social feeding in *Caenorhabditis elegans* is induced by neurons that detect aversive stimuli. *Nature* 419:899–903

44. de Bono M, Bargmann CI. 1998. Natural variation in a neuropeptide Y receptor homolog modifies social behavior and food response in *C. elegans*. *Cell* 94:679–89

45. Gray JM, Karow DS, Lu H, Chang AJ, Chang JS, et al. 2004. Oxygen sensation and social feeding mediated by a *C. elegans* guanylate cyclase homologue. *Nature* 430:317–22

46. Coates JC, de Bono M. 2002. Antagonistic pathways in neurons exposed to body fluid regulate social feeding in *Caenorhabditis elegans*. *Nature* 419:925–29

47. Liedtke W, Tobin DM, Bargmann CI, Friedman JM. 2003. Mammalian TRPV4 (VR-OAC) directs behavioral responses to osmotic and mechanical stimuli in *Caenorhabditis elegans*. *Proc. Natl. Acad. Sci. USA* 100(Suppl. 2):14531–36

48. Igarashi P, Somlo S. 2002. Genetics and pathogenesis of polycystic kidney disease. *J. Am. Soc. Nephrol.* 13:2384–98

49. Newby LJ, Streets AJ, Zhao Y, Harris PC, Ward CJ, Ong AC. 2002. Identification, characterization, and localization of a novel kidney polycystin-1-polycystin-2 complex. *J. Biol. Chem.* 277:20763–73

50. Yoder BK, Hou X, Guay-Woodford LM. 2002. The polycystic kidney disease proteins, polycystin-1, polycystin-2, polaris, and cystin, are co-localized in renal cilia. *J. Am. Soc. Nephrol.* 13:2508–16

51. Liu KS, Sternberg PW. 1995. Sensory regulation of male mating behavior in *Caenorhabditis elegans*. *Neuron* 14:79–89

52. Barr MM, DeModena J, Braun D, Nguyen CQ, Hall DH, Sternberg PW. 2001. The *Caenorhabditis elegans* autosomal dominant polycystic kidney disease gene homologs *lov-1* and *pkd-2* act in the same pathway. *Curr. Biol.* 11:1341–46

53. Qin H, Rosenbaum JL, Barr MM. 2001. An autosomal recessive polycystic kidney disease gene homolog is involved in intraflagellar transport in *C. elegans* ciliated sensory neurons. *Curr. Biol.* 11:457–61

54. Haycraft CJ, Swoboda P, Taulman PD,

Thomas JH, Yoder BK. 2001. The *C. elegans* homolog of the murine cystic kidney disease gene *Tg737* functions in a ciliogenic pathway and is disrupted in *osm-5* mutant worms. *Development* 128:1493–505

55. Nonaka S, Tanaka Y, Okada Y, Takeda S, Harada A, et al. 1998. Randomization of left-right asymmetry due to loss of nodal cilia generating leftward flow of extraembryonic fluid in mice lacking KIF3B motor protein. *Cell* 95:829–37

56. McGrath J, Somlo S, Makova S, Tian X, Brueckner M. 2003. Two populations of node monocilia initiate left-right asymmetry in the mouse. *Cell* 114:61–73

57. Pennekamp P, Karcher C, Fischer A, Schweickert A, Skryabin B, et al. 2002. The ion channel polycystin-2 is required for left-right axis determination in mice. *Curr. Biol.* 12:938–43

58. Dwyer ND, Adler CE, Crump JG, L'Etoile ND, Bargmann CI. 2001. Polarized dendritic transport and the AP-1 mu1 clathrin adaptor UNC-101 localize odorant receptors to olfactory cilia. *Neuron* 31:277–87

59. Vale RD. 2003. The molecular motor toolbox for intracellular transport. *Cell* 112:467–80

60. Orozco JT, Wedaman KP, Signor D, Brown H, Rose L, Scholey JM. 1999. Movement of motor and cargo along cilia. *Nature* 398:674

61. Peden EM, Barr MM. 2005. The KLP-6 kinesin is required for male mating behaviors and polycystin localization in *Caenorhabditis elegans*. *Curr. Biol.* 15:394–404

62. Hu J, Barr MM. 2005. ATP-2 interacts with the PLAT domain of LOV-1 and is involved in *Caenorhabditis elegans* polycystin signaling. *Mol. Biol. Cell* 16:458–69

63. Moser TL, Kenan DJ, Ashley TA, Roy JA, Goodman MD, et al. 2001. Endothelial cell surface F1-F0 ATP synthase is active in ATP synthesis and is inhibited by angiostatin. *Proc. Natl. Acad. Sci. USA* 98:6656–61

64. Martinez LO, Jacquet S, Esteve JP, Rolland C, Cabezon E, et al. 2003. Ectopic beta-chain of ATP synthase is an apolipoprotein A-I receptor in hepatic HDL endocytosis. *Nature* 421:75–79

65. Stock D, Leslie AG, Walker JE. 1999. Molecular architecture of the rotary motor in ATP synthase. *Science* 286:1700–5

66. Rondelez Y, Tresset G, Nakashima T, Kato-Yamada Y, Fujita H, et al. 2005. Highly coupled ATP synthesis by F1-ATPase single molecules. *Nature* 433:773–77

67. Portman DS, Emmons SW. 2004. Identification of *C. elegans* sensory ray genes using whole-genome expression profiling. *Dev. Biol.* 270:499–512

68. Xu XZS, Sternberg PW. 2003. A *C. elegans* sperm TRP protein required for sperm-egg interactions during fertilization. *Cell* 114:285–97

69. Nilius B. 2004. Store-operated Ca^{2+} entry channels: still elusive! *Science STKE* 2004:pe36

70. Bezzerides VJ, Ramsey IS, Kotecha S, Greka A, Clapham DE. 2004. Rapid vesicular translocation and insertion of TRP channels. *Nat. Cell Biol.* 6:709–20

71. Castellano LE, Trevino CL, Rodriguez D, Serrano CJ, Pacheco J, et al. 2003. Transient receptor potential (TRPC) channels in human sperm: expression, cellular localization and involvement in the regulation of flagellar motility. *FEBS Lett.* 541:69–74

72. Sun AY, Lambie EJ. 1997. *gon-2*, a gene required for gonadogenesis in *Caenorhabditis elegans*. *Genetics* 147:1077–89

73. West RJ, Sun AY, Church DL, Lambie EJ. 2001. The *C. elegans gon-2* gene encodes a putative TRP cation channel protein required for mitotic cell cycle progression. *Gene* 266:103–10

74. Church DL, Lambie EJ. 2003. The promotion of gonadal cell divisions by the *Caenorhabditis elegans* TRPM cation channel GON-2 is antagonized by GEM-4 copine. *Genetics* 165:563–74

74a. Teramoto T, Lambie EJ, Iwasaki K. 2005.

Differential regulation of TRPM channels governs electrolyte homeostasis in the *C. elegans* intestine. *Cell Metab.* 1:343–54

75. Palmer CP, Zhou XL, Lin J, Loukin SH, Kung C, Saimi Y. 2001. A TRP homolog in *Saccharomyces cerevisiae* forms an intracellular Ca^{2+}-permeable channel in the yeast vacuolar membrane. *Proc. Natl. Acad. Sci. USA* 98:7801–5

76. Zhou XL, Batiza AF, Loukin SH, Palmer CP, Kung C, Saimi Y. 2003. The transient receptor potential channel on the yeast vacuole is mechanosensitive. *Proc. Natl. Acad. Sci. USA* 100:7105–10

77. Denis V, Cyert MS. 2002. Internal Ca^{2+} release in yeast is triggered by hypertonic shock and mediated by a TRP channel homologue. *J. Cell Biol.* 156:29–34

78. Fares H, Greenwald I. 2001. Regulation of endocytosis by CUP-5, the *Caenorhabditis elegans* mucolipin-1 homolog. *Nat. Genet.* 28:64–68

79. Hersh BM, Hartwieg E, Horvitz HR. 2002. The *Caenorhabditis elegans* mucolipin-like gene *cup-5* is essential for viability and regulates lysosomes in multiple cell types. *Proc. Natl. Acad. Sci. USA* 99:4355–60

80. Treusch S, Knuth S, Slaugenhaupt SA, Goldin E, Grant BD, Fares H. 2004. *Caenorhabditis elegans* functional orthologue of human protein h-mucolipin-1 is required for lysosome biogenesis. *Proc. Natl. Acad. Sci. USA* 101:4483–88

81. Raychowdhury MK, Gonzalez-Perrett S, Montalbetti N, Timpanaro GA, Chasan B, et al. 2004. Molecular pathophysiology of mucolipidosis type IV: pH dysregulation of the mucolipin-1 cation channel. *Hum. Mol. Genet.* 13:617–27

82. Goodman MB, Schwarz EM. 2003. Transducing touch in *Caenorhabditis elegans*. *Annu. Rev. Physiol.* 65:429–52

83. Sawin ER, Ranganathan R, Horvitz HR. 2000. *C. elegans* locomotory rate is modulated by the environment through a dopaminergic pathway and by experience through a serotonergic pathway. *Neuron* 26:619–31

SUBJECT INDEX

A

Actin
superfast muscle
physiology and, 197
Activation
cardiac regeneration and,
30–33
CNG and HCN channels,
375–95
interstitial cells of Cajal as
pacemakers in GI tract
and, 307
signaling in smooth muscle
of the gut and, 363
superfast muscle
physiology and, 196–206
TRP channels in
C. elegans, 719
TRP channels in
Drosophila and, 658–59,
661–70
Acute respiratory distress
syndrome (ARDS)
biophysical lung injury and
biotrauma, 585–610
Adamussium colbecki
oxidative stress in marine
environments and, 267
Adaptation
superfast muscle
physiology and, 193
TRP channels in
Drosophila and, 674–75
Adaptive immunity
myocardial inflammatory
responses and, 74–86
Adenosine
airway surface liquid
volume and phasic shear
stress, 543–57
Adenosine triphosphate

(ATP)
airway surface liquid
volume and phasic shear
stress, 552
superfast muscle
physiology and, 200,
205–6, 208, 210–11,
213–15, 218
TRP channels in
Drosophila and, 665–66
ADF neurons
TRP channels in *C. elegans*
and, 724
ADO-CFTR pathway
airway surface liquid
volume and phasic shear
stress, 543–57
ADP ribose
signaling in smooth muscle
of the gut and, 350–51
β_2-adrenergic receptor
NHERF adaptor proteins
and, 493–94
Afferent gating
brainstem circuits
regulating gastric function
and, 287–93
Ag^+
TRP channel permeation
and selectivity, 704
AGC kinases
Sgk1 kinase and epithelial
transport, 462–63
Age groups
cardiac regeneration and,
29
mechanical force and
airways, 563–76
Agonist gating
brainstem circuits
regulating gastric function

and, 287–93
Airway surface liquid (ASL)
volume
phasic shear stress and
adenosine, 552–54
altered ion transport,
545–46
ATP, 552
Ca^{2+}-activated Cl^-
channel, 552
CFTR, 552–54
conclusions, 556–57
cystic fibrosis airway
disease, 545–46,
555–56
ENaC, 552–54
generating phasic shear
on airway surfaces,
550–51
introduction, 543–44
ion transport regulation,
548–52
lung defense, 554–55
modulating ion
transport, 552–54
Na^+, 552
normal tidal breathing,
546–47
phasic motion
rebalances ASL height,
548
salt transport, 544–45
transduction by airway
epithelia, 548–52
water transport, 544–45
Akt proteins
insulin resistance
syndromes and, 142
Aldosterone
Sgk1 kinase and epithelial
transport, 468, 472–76

CUMULATIVE INDEXES

CONTRIBUTING AUTHORS, VOLUMES 64–68

CHAPTER TITLES, VOLUMES 64–68

Cardiovascular Physiology

Cell Physiology

Comparative Physiology

Endocrinology

Gastrointestinal Physiology

Neurophysiology

Renal and Electrolyte Physiology

Respiratory Physiology